Methods in Enzymology

Volume 137
IMMOBILIZED ENZYMES AND CELLS
Part D

METHODS IN ENZYMOLOGY

EDITORS-IN-CHIEF

Sidney P. Colowick   Nathan O. Kaplan

Methods in Enzymology

Volume 137

Immobilized Enzymes and Cells

Part D

EDITED BY

Klaus Mosbach

PURE AND APPLIED BIOCHEMISTRY
CHEMICAL CENTER
UNIVERSITY OF LUND
LUND, SWEDEN

ACADEMIC PRESS, INC.
Harcourt Brace Jovanovich, Publishers
San Diego   New York   Berkeley   Boston
London   Sydney   Tokyo   Toronto

ACADEMIC PRESS, INC.
1250 Sixth Avenue
San Diego, California 92101

United Kingdom Edition published by
ACADEMIC PRESS INC. (LONDON) LTD.
24-28 Oval Road, London NW1 7DX

LIBRARY OF CONGRESS CATALOG CARD NUMBER: 54-9110

ISBN   0-12-182037-8   (alk. paper)

PRINTED IN THE UNITED STATES OF AMERICA
88 89 90 91   9 8 7 6 5 4 3 2 1

Table of Contents

Section I. Analytical Applications with Emphasis on Biosensors

A. Bioselective Electrodes

B. Hybrid, Cell, and Tissue Electrodes

v

F. Enzyme Reactors

G. Continuous Monitoring

H. Optical Methods

I. Miscellaneous New Techniques

Section II. Medical Applications

Section III. Novel Techniques for and Aspects of Immobilized Enzymes and Cells

Contributors to Volume 137

Article numbers are in parentheses following the names of contributors.
Affiliations listed are current.

MASUO AIZAWA (9), *Department of Bioengineering, Faculty of Engineering, Tokyo Institute of Technology, Ookayama, Meguro-ku, Tokyo 152, Japan*

HANS ARWIN (33), *Laboratory of Applied Physics, Department of Physics and Measurement Technology, Linköping Institute of Technology, S-581 83 Linköping, Sweden*

MAHMOOD R. AZARI (63), *Behring Diagnostics, Division of American Hoechst Corporation, La Jolla, California 92037*

HOWARD BERNSTEIN (46), *Department of Applied Biological Sciences, Massachusetts Institute of Technology, Cambridge, Massachusetts 02139*

STAFFAN BIRNBAUM (30, 57), *Pure and Applied Biochemistry, Chemical Center, University of Lund, S-221 00 Lund, Sweden*

J. D. BRYERS (65), *Center for Biochemical Engineering, Duke University, Durham, North Carolina 27706*

LEIF BÜLOW (30, 57), *Pure and Applied Biochemistry, Chemical Center, University of Lund, S-221 00 Lund, Sweden*

STEVE CARAS (21), *Chemistry Division, Naval Research Laboratory, Department of the Navy, Washington, D.C. 20375*

GIACOMO CARREA (14), *Istituto di Chimica degli Ormoni, Consiglio Nazionale delle Ricerche (CNR), 20131 Milan, Italy*

WAYNE L. CHANDLER (43), *Department of Laboratory Medicine, Coagulation Division, University of Washington, School of Medicine, Seattle, Washington 98195*

T. M. S. CHANG (40), *Artificial Cells and Organs Research Centre, Faculty of Medicine, McGill University, Montreal, Quebec, Canada H3G 1Y6*

LELAND C. CLARK, JR. (6), *Children's Hospital Research Foundation, Children's Hospital Medical Center, Cincinnati, Ohio 45229*

NEIL CLELAND (26), *Department of Biochemistry and Biotechnology, The Royal Institute of Technology, S-100 44 Stockholm, Sweden*

DIDIER COMBES (53), *Département de Génie Biochimique et Alimentaire, U.A. 544 Centre National de la Recherche Scientifique (CNRS), Institut National des Sciences Appliquees, F-31077 Toulouse Cedex, France*

CHARLES L. COONEY (46), *Department of Applied Biological Sciences, Massachusetts Institute of Technology, Cambridge, Massachusetts 02139*

BENGT DANIELSSON (1, 16, 20, 27, 30), *Pure and Applied Biochemistry, Chemical Center, University of Lund, S-221 00 Lund, Sweden*

GEORG DECRISTOFORO (17), *Research and Development Department, Biochemie Gesellschaft m.b.H., A-6250 Kundl, Austria*

PETER EDMAN (44), *Pharmacia AB, S-751 82 Uppsala, Sweden*

SVEN-OLOF ENFORS (26), *Department of Biochemistry and Biotechnology, The Royal Institute of Technology, S-100 44 Stockholm, Sweden*

HANSRUEDI FELIX (58), *AGRO Research, Sandoz Ltd., CH-4002 Basel, Switzerland*

CECILIA FÖRBERG (56), *Department of Biochemistry and Biotechnology, The Royal Institute of Technology, S-100 44 Stockholm, Sweden*

CHRISTIAN FREIBURGHAUS (41), *Gambro AB, S-220 10 Lund, Sweden*

MOTOHISA FURUSAWA (19), *Department of Chemistry, Faculty of Engineering, Yamanashi University, Kofu 400, Japan*

SEVERINO GHINI (14), *Istituto di Scienze Chimiche, Facoltà di Farmacia, Università di Bologna, 40127 Bologna, Italy*

STEFANO GIROTTI (14), *Istituto di Scienze Chimiche, Facoltà di Farmacia, Università di Bologna, 40127 Bologna, Italy*

MICHAEL J. GOLDFINCH (31), *Plum Tree Cottage, Farley, Salisbury SP5 1AH, England*

GEORGE G. GUILBAULT (2), *Department of Chemistry, University of New Orleans, New Orleans, Louisiana 70148*

LENA HÄGGSTRÖM (56), *Department of Biochemistry and Biotechnology, The Royal Institute of Technology, S-100 44 Stockholm, Sweden*

BÄRBEL HAHN-HÄGERDAL (59), *Department of Applied Microbiology, Chemical Center, University of Lund, S-221 00 Lund, Sweden*

HAKAN HAKANSON (28), *Research Department, Gambro AB, S-220 10 Lund, Sweden*

G. HAMER (65), *Institute of Aquatic Sciences, Swiss Federal Institutes of Technology—Zürich, CH-8600 Dübendorf, Switzerland*

TIINA HEINONEN (15), *Wallac Biochemical Laboratory, Wallac Oy, SF-20101 Turku, Finland*

MOTOHIKO HIKUMA (10), *Central Research Laboratories, Ajinomoto Co., Inc., Kawasaki-ku, Kawasaki 210, Japan*

TSUNETOSHI HINO (48), *Research Institute for Production and Development, Kyoto 606, Japan*

MAT H. HO (24), *Department of Chemistry, University of Alabama at Birmingham, Birmingham, Alabama 35294*

YOSHIHITO IKARIYAMA (9), *Research Institute, National Rehabilitation Center for the Disabled, Namiki, Tokotozawa, Saitama 359, Japan*

JIŘÍ JANATA (21), *Materials Science and Engineering, The University of Utah, Salt Lake City, Utah 84112*

ULF JÖNSSON (34), *Pharmacia AB, S-751 82 Uppsala, Sweden*

JUNICHI KAMBAYASHI (47), *Osaka University Medical School, Sumiyoshi, Osaka 558, Japan*

ISAO KARUBE (11, 22, 62), *Research Laboratory of Resources Utilization, Tokyo Institute of Technology, Midori-ku, Yokohama 227, Japan*

RYUZO KAWAMORI (29), *The First Department of Medicine, Osaka University Medical School, Fukushima-ku, Osaka 553, Japan*

NOBUTOSHI KIBA (19), *Department of Chemistry, Faculty of Engineering, Yamanashi University, Kofu 400, Japan*

DAVID J. KING (63), *Celltech Limited, Slough, Berkshire SL1 4EN, England*

OUTI KOLHINEN (15), *Department of Biochemistry, University of Turku, SF-20500 Turku 50, Finland*

KALEVI KURKIJÄRVI (15), *Wallac Biochemical Laboratory, Wallac Oy, SF-20101 Turku, Finland*

ROBERT LANGER (46), *Department of Applied Biological Sciences, Massachusetts Institute of Technology, Cambridge, Massachusetts 02139*

G.-X. LI (64), *Institute of Microbiology, Academia Sinica, Beijing, People's Republic of China*

P. LINKO (64), *Laboratory of Biotechnology and Food Engineering, Helsinki University of Technology, SF-02150 Espoo, Finland*

SUSAN LINKO (64), *Laboratory of Biotechnology and Food Engineering, Helsinki University of Technology, SF-02150 Espoo, Finland*

YU-YEN LINKO (64), *Laboratory of Biotechnology and Food Engineering, Helsinki University of Technology, SF-02150 Espoo, Finland*

Mou Chung Liu (47), *Kitahama Clinic, Osaka University Medical School, Sumiyoshi, Osaka 558, Japan*

Timo Lövgren (15), *Wallac Biochemical Laboratory, Wallac Oy, SF-20101 Turku, Finland*

Christopher R. Lowe (31), *Institute of Biotechnology, University of Cambridge, Cambridge CB2 3EF, England*

Arne Lundin (15), *Research Centre and Department of Medicine, Karolinska Institute, Huddinge University Hospital, S-141 86 Huddinge, Sweden*

Ingemar Lundström (20, 33, 35), *Laboratory of Applied Physics, Department of Physics and Measurement Technology, Linköping Institute of Technology, S-581 83 Linköping, Sweden*

A. V. Maksimenko (49), *Institute of Experimental Cardiology, Cardiology Research Center of the USSR, Academy of Medical Sciences, Moscow 121552, USSR*

Magnus Malmqvist (34), *Pharmacia AB, S-751 82 Uppsala, Sweden*

Carl Fredrik Mandenius (27, 35), *Pure and Applied Biochemistry, Chemical Center, University of Lund, S-221 00 Lund, Sweden*

Sohrab Mansouri (32), *Medical Instruments Systems Research Division, Lilly Research Laboratories, A Division of Eli Lilly and Company, Lilly Corporate Center, Indianapolis, Indiana 46285*

Karel Martinek (55), *Institute of Organic Chemistry and Biochemistry, Czechoslovak Academy of Sciences, 166 10 Prague 6, Czechoslovakia*

Bo Mattiasson (60, 61), *Department of Biotechnology, Chemical Center, University of Lund, S-221 00 Lund, Sweden*

A. V. Mazaev (49), *Institute of Clinical Cardiology, Cardiology Research Center of the USSR, Academy of Medical Sciences, Moscow 121552, USSR*

Marian L. Miller (6), *Department of Environmental Health, University of Cincin-nati, College of Medicine, Cincinnati, Ohio 45267*

Pierre Monsan (53), *BioEurope, F-31400 Toulouse, France*

Takesada Mori (47), *Osaka University Medical School, Sumiyoshi, Osaka 558, Japan*

Toyosaka Moriizumi (22), *Department of Electrical and Electronic Engineering, Tokyo Institute of Technology, Meguro-ku, Tokyo 152, Japan*

Klaus Mosbach (1, 16, 20, 30, 35, 39, 52, 57), *Pure and Applied Biochemistry, Chemical Center, University of Lund, S-221 00 Lund, Sweden*

Takashi Murachi (23, 48), *Department of Clinical Science and Laboratory Medicine, Faculty of Medicine, Kyoto University Hospital, Sakyo-ku, Kyoto 606, Japan*

Krishna Narasimhan (8), *Department of Pharmacology, University of Pittsburgh, School of Medicine, Pittsburgh, Pennsylvania 15261*

Inga Marie Nilsson (41), *Department for Coagulation Disorders, University of Lund, General Hospital, S-214 01 Malmö, Sweden*

Linda K. Noyes (6), *Children's Hospital Research Foundation, Children's Hospital Medical Center, Cincinnati, Ohio 45229*

Ulf Nylén (44), *Biochemistry Department, Excorim KB, S-220 10 Lund, Sweden*

Sten Ohlson (41), *Research and Development, Perstorp Biolytica AB, S-223 70 Lund, Sweden*

Takeshi Ohshiro (47), *Hanwa General Hospital, Osaka University Medical School, Sumiyoshi, Osaka 558, Japan*

Göran Olofsson (34), *Laboratory of Applied Physics, National Defence Research Institute, S-901 82 Umeå, Sweden*

T. S. Parker (42), *Research Lipid Laboratories, The Rogosin Institute, Medical Re-*

search and Health Care, New York, New York 10021

MARK J. POZNANSKY (50), Department of Physiology, Faculty of Medicine, University of Alberta, Edmonton, Alberta, Canada T6G 2H7

RAIMO RAUNIO (15), Department of Biochemistry, University of Turku, SF-20500 Turku 50, Finland

GARRY A. RECHNITZ (12), Biosensor Research Laboratory, Departments of Chemistry and Biochemistry, University of Delaware, Newark, Delaware 19716

REINHARD RENNEBERG (3), Central Institute of Molecular Biology, Academy of Sciences of the DDR, DDR-1115 Berlin-Buch, German Democratic Republic

ALDO RODA (14), Istituto di Scienze Chimiche, Facoltà di Farmacia, Università di Bologna, 40127 Bologna, Italy

J. L. ROMETTE (4), Laboratoire de Technologie Enzymatique, Université de Technologie de Compiègne, F-60206 Compiègne Cedex, France

INGER RÖNNBERG (34), Pharmacia AB, S-751 82 Uppsala, Sweden

IKUO SATOH (18), Department of Industrial and Engineering Chemistry, Ikutoku Technical University, Atsugi-shi, Kanagawa-ken 243-02, Japan

FRIEDER W. SCHELLER (3, 13), Central Institute of Molecular Biology, Academy of Sciences of the DDR, DDR-1115 Berlin-Buch, German Democratic Republic

GOTTFRIED SCHMER (43), Department of Laboratory Medicine, Coagulation Division, University of Washington, School of Medicine, Seattle, Washington 98195

FLORIAN SCHUBERT (3, 13), Central Institute of Molecular Biology, Academy of Sciences of the DDR, DDR-1115 Berlin-Buch, German Democratic Republic

JEROME S. SCHULTZ (32), Center for Biotechnology and Bioengineering, University of Pittsburgh, Pittsburgh, Pennsylvania 15260

ZE'EV SHAKED (54), CODON Inc., South San Francisco, California 94080

MOTOAKI SHICHIRI (29), Department of Metabolic Medicine, Kumamoto University Medical School, Kumamoto 860, Japan

YASUHIKO SHIMIZU (48), Department of Experimental Surgery, Research Center for Medical Polymers and Biomaterials, Kyoto University, Sakyo-ku, Kyoto 606, Japan

INGVAR SJÖHOLM (44), Department of Drugs, National Board of Health and Welfare, S-751 25 Uppsala, Sweden

ROBERT B. SPOKANE (6), Children's Hospital Research Foundation, Children's Hospital Medical Center, Cincinnati, Ohio 45229

J. F. STUDEBAKER (42), Engineering/Scientific Regional Support, IBM Corporation, Princeton, New Jersey 08540

RANJAN SUDAN (6), Children's Hospital Research Foundation, Children's Hospital Medical Center, Cincinnati, Ohio 45229

ANTHONY M. SUN (51), Islet Research, Connaught Research Institute, Willowdale, Ontario, Canada M2R 3T4, and Department of Physiology, University of Toronto, Toronto, Ontario, Canada M5S 1A8

P. V. SUNDARAM (25), National Bureau of Standards, Gaithersburg, Maryland 20899, and V. H. S. Medical Centre, Adyar, Madras 600113, India

SHUICHI SUZUKI (62), Department of Environmental Engineering, The Saitama Institute of Technology, Okabe, Oosatogun, Saitama Prefecture 369102, Japan

MASAYOSHI TABATA (23), College of Medical Technology, Kyoto University Hospital, Sakyo-ku, Kyoto 606, Japan

TAKASHI TERAMATSU (48), Department of Thoracic Surgery, Chest Disease Research Institute, Kyoto University, Kyoto 606, Japan

DAVID S. TERMAN (45), 25371 Outlook Drive, Carmel, California 93923

D. THOMAS (4), Laboratoire de Technologie Enzymatique, Université de Technologie de Compiègne, F-60206 Compiègne Cedex, France

V. P. TORCHILIN (49, 55), *Institute of Experimental Cardiology, Cardiology Research Center of the USSR, Academy of Medical Sciences, Moscow 121552, USSR*

ANTHONY P. F. TURNER (7), *The Biotechnology Centre, Cranfield Institute of Technology, Cranfield, Bedford MK43 0AL, England*

PEKKA TURUNEN (15), *Wallac Biochemical Laboratory, Wallac Oy, SF-20101 Turku, Finland*

S. D. VARFOLOMEEV (38), *A. N. Belozersky Laboratory of Molecular Biology and Bioorganic Chemistry, Department of Biokinetics, Moscow State University, Moscow 117234, USSR*

BERT WALTER (36), *Ames Division, Miles Laboratories, Inc., Elkhart, Indiana 46515*

SATOSHI WATANABE (48), *Department of Experimental Surgery, Research Center for Medical Polymers and Biomaterials, Kyoto University, Sakyo-ku, Kyoto 606, Japan*

JAMES C. WEAVER (37), *Harvard University–Massachusetts Institute of Technology Division of Health, Sciences and Technology, Massachusetts Institute of Technology, Cambridge, Massachusetts 02139*

STEFAN WELIN (35), *Laboratory of Applied Physics, Linköping Institute of Technology, S-581 83 Linköping, Sweden*

LEMUEL B. WINGARD, JR. (8), *Department of Pharmacology, University of Pittsburgh, School of Medicine, Pittsburgh, Pennsylvania 15261*

FREDRIK WINQUIST (20), *Laboratory of Applied Physics, Department of Physics and Measurement Technology, Linköping Institute of Technology, S-581 83 Linköping, Sweden*

ALAN WISEMAN (63), *Department of Biochemistry, University of Surrey, Guildford, Surrey GU2 5XH, England*

SIDNEY WOLFE (54), *Cetus Corporation, Emeryville, California 94608*

YOSHIMITSU YAMASAKI (29), *The First Department of Medicine, Osaka University Medical School, Fukushima-ku, Osaka 553, Japan*

VICTOR C. YANG (46), *College of Pharmacy, University of Michigan, Ann Arbor, Michigan 48109*

TAKEO YASUDA (10), *Central Research Laboratories, Ajinomoto Co., Inc., Kawasaki-ku, Kawasaki 210, Japan*

KENTARO YODA (5), *Katata Research Institute, Toyobo Co., Ltd., Shiga 520-02, Japan*

LI-CHAN ZHONG (64), *Institute of Microbiology, Academia Sinica, Beijing, People's Republic of China*

Preface

Volumes 135 through 137 of *Methods in Enzymology,* Immobilized Enzymes and Cells, Parts B through D, include the following sections: (1) Immobilization Techniques for Enzymes; (2) Immobilization Techniques for Cells/Organelles; (3) Application of Immobilized Enzymes/Cells to Fundamental Studies; (4) Multistep Enzyme Systems and Coenzymes; (5) Immobilized Enzymes/Cells in Organic Synthesis; (6) Enzyme Engineering (Enzyme Technology); (7) Analytical Applications with Emphasis on Biosensors; (8) Medical Applications; and (9) Novel Techniques for and Aspects of Immobilized Enzymes and Cells. The first three sections appear in Volume 135, the next three in Volume 136, and the last three in Volume 137.

Immobilization techniques for enzymes, Section (1), has already been treated in Volume XLIV of this series. Immobilization techniques for cells/organelles, Section (2), an area which seems to have great potential, especially for the application of immobilized yeast and plant and animal cells, is covered for the first time in these volumes. Sections (3) and (4) have been dealt with previously. Section (5), the use of immobilized enzymes/cells in organic synthesis, has probably not been covered before. It is my firm opinion that in the not too distant future we will see a number of processes employed which are based, in part, on the examples given in this section. Section (6) on industrial uses updates the material presented in Volume XLIV. The examples given are, to the best of my knowledge, in operational use today or, at least, on a pilot plant level. Section (7), analytical applications with emphasis on biosensors, is the subject of a great deal of research at present, and it may very well be that in the not too distant future we will witness a breakthrough, i.e., many applications of a number of such devices. The medical area, covered in Section (8), seems promising, but certainly more research is required to fully exploit any underlying potential. Finally, in Section (9), I have collected a number of contributions that did not seem to fit in any of the other sections, but do address important and novel developments.

I would like to note that although major emphasis in these volumes has been placed on immobilization in its strictest sense, preferentially, covalent attachment of enzymes or entrapment of cells, one should not view immobilized systems in too limited a manner. In fact, bioreactors confined by ultrafilter membranes or hollow fiber systems belong in this category, and the various systems appear to overlap. Immobilization techniques as applied to affinity chromatography or immunoassays such as ELISA are not included to any extent in these volumes since they have been adequately covered in other volumes of this series (e.g., Volumes XXXIV and 104 on affinity techniques).

An area that was originally scheduled for inclusion is synzymes or artificial enzymes. These include attempts to create catalysts mimicking enzymes by coupling of functional groups to, for instance, cyclodextrin [e.g., D'Souza et al. (*Biochem. Biophys. Res. Commun.* **129,** 727–732, 1985) and Breslow et al. (*J. Am. Chem. Soc.* **108,** 1969, 1986)], to crown ethers [Cram et al. (*J. Am. Chem. Soc.* **107,** 3645, 1985)], or to solid matrices [Nilsson and Mosbach (*J. Solid-Phase Biochem.* **4,** 271, 1979) and Leonhardt and Mosbach (*Reactive Polymers,* in press)].

Related to these studies are attempts to create cavities in polymers with substrate-binding properties [notably by Wulff et al. (e.g., *Reactive Polymers* **3,** 261, 1985; and previous publications by these authors) and Arshady and Mosbach (*Makromol. Chem.* **182,** 687, 1981)]. This exciting area is presently in a rapid state of development, and the methodology involved should soon be made available in a more comprehensive context.

Mention should be made of the developments in the utilization of recombinant DNA technology for the immobilization (and affinity purification) of biomolecules. I refer to the reported fusion of "affinity tails" as polyarginine (Smith et al., *Gene* **32,** 321, 1984), of polycysteine [Bülow and Mosbach, Proceedings of the VIII International Conference on Enzyme Engineering, *Annals of the New York Academy of Sciences,* in press (presented 1985)], or of protein A (Nilsson et al., *EMBO J.* **4,** 1075, 1985) to enzymes facilitating their purification and immobilization. These preparations can be obtained by fusion of the respective groups as "tail" to the NH_2 or COOH termini of the enzyme or by site-directed mutagenesis leading to substitution on the enzyme structure. DNA technology can also be usefully employed to create new multienzyme complexes, fusing enzymes acting in sequence to one another (Bülow et al., *Bio/Technology* **3,** 821, 1985) as an alternative to their co-immobilization on supports; similarly, attachment of "tails" allowing reversible coenzyme binding may be accomplished. The same technology has also been used recently in attempts to prepare esterase mimics from the ground up (Bülow and Mosbach, *FEBS Lett.* **210,** 147, 1987).

Since this is such a rapidly moving area, I advise the reader, apart from the usual standard books in this area, to read the proceedings of the Enzyme Engineering Conferences 1–8 (Wiley, first conference; Plenum Press, second–sixth conferences; and Annals of the New York Academy of Sciences, seventh and eighth conferences); Biochemical Engineering, Volumes I–III and subsequent volumes; Annals of the New York Academy of Sciences, 1983; the patent book "Enzyme Technology, Recent Advances" (S. Torrey, ed.), Noyes Data Corporation, Park Ridge, New

Jersey, 1983; and *Biotechnology Review* no. 2. In addition, in the following journals many articles relating to immobilized enzyme and cell research can be found: *Biotechnology and Bioengineering* (John Wiley & Sons); *Trends in Biotechnology* (Elsevier, The Netherlands); *Bio/Technology* (Nature Publishing Co., U.S.); *Applied Biochemistry and Biotechnology* (The Humana Press, Inc., U.S.); *Applied Biochemistry with Special Emphasis on Biotechnology*; *Biotechnology Letters* (Science and Technology Letters, England); *Applied Microbiology and Biotechnology* (Springer-Verlag, Germany); *Enzyme and Microbial Technology* (Butterworth Scientific Limited, England); *Biosensors* (Elsevier Applied Science Publishing Ltd., England).

In studies with immobilized systems, sometimes useful, not immediately obvious "by-products" may be obtained. I refer to the finding that immobilized *Escherichia coli* cells, when kept in media without selection pressure, show improved plasmid stability (de Taxis du Poët, P., Dhulster, P., Barbotin, J.-N., and Thomas, D., *J. Bact.* **165,** 871, 1986). An additional example would be the improved regeneration of plants using immobilized protoplasts discussed in Section (2).

I would like to express the hope that these volumes present an overview of the various areas in which immobilized enzymes and cells are used, act as a stimulus for further research, and provide methodological "know-how." The proper choice of support and/or immobilization technique for a particular application may not always be easily accomplished, but I hope that guidance to do so is found in these volumes.

Putting these volumes together has been a time-consuming and, at times, frustrating undertaking. Without the coeditors, Drs. Lars Andersson, Peter Brodelius, Bengt Danielsson, Stina Gestrelius, and Mats-Olle Månsson, the volumes would not have materialized. Because of the number of coeditors, some heterogeneity in the editing has resulted. Contributors to the various sections are from substantially different disciplines, and again this has contributed to the heterogeneity that can be found. Part of the editing of the three volumes was carried out in Zürich, where I held a chair in biotechnology at the Swiss Federal Institute of Technology. Without the enormous efforts and skills of the staff of Academic Press, these volumes would never have reached production. I also owe much gratitude to my secretaries, notably Ingrid Nilsson, for their highly qualified help. Finally, I would like to thank the contributors for their efforts.

These volumes are dedicated to the memory of the late Professors N. O. Kaplan and S. P. Colowick, with whom I had highly fruitful discussions, especially at the beginning of this undertaking.

KLAUS MOSBACH

METHODS IN ENZYMOLOGY

EDITED BY

Sidney P. Colowick and Nathan O. Kaplan

VANDERBILT UNIVERSITY
SCHOOL OF MEDICINE
NASHVILLE, TENNESSEE

DEPARTMENT OF CHEMISTRY
UNIVERSITY OF CALIFORNIA
AT SAN DIEGO
LA JOLLA, CALIFORNIA

METHODS IN ENZYMOLOGY

EDITORS-IN-CHIEF

Sidney P. Colowick and Nathan O. Kaplan

xxiii

VOLUME XXXII. Biomembranes (Part B)
Edited by SIDNEY FLEISCHER AND LESTER PACKER

VOLUME XXXIII. Cumulative Subject Index Volumes I–XXX
Edited by MARTHA G. DENNIS AND EDWARD A. DENNIS

VOLUME XXXIV. Affinity Techniques (Enzyme Purification: Part B)
Edited by WILLIAM B. JAKOBY AND MEIR WILCHEK

VOLUME XXXV. Lipids (Part B)
Edited by JOHN M. LOWENSTEIN

VOLUME XXXVI. Hormone Action (Part A: Steroid Hormones)
Edited by BERT W. O'MALLEY AND JOEL G. HARDMAN

VOLUME XXXVII. Hormone Action (Part B: Peptide Hormones)
Edited by BERT W. O'MALLEY AND JOEL G. HARDMAN

VOLUME XXXVIII. Hormone Action (Part C: Cyclic Nucleotides)
Edited by JOEL G. HARDMAN AND BERT W. O'MALLEY

VOLUME XXXIX. Hormone Action (Part D: Isolated Cells, Tissues, and Organ Systems)
Edited by JOEL G. HARDMAN AND BERT W. O'MALLEY

VOLUME XL. Hormone Action (Part E: Nuclear Structure and Function)
Edited by BERT W. O'MALLEY AND JOEL G. HARDMAN

VOLUME XLI. Carbohydrate Metabolism (Part B)
Edited by W. A. WOOD

VOLUME XLII. Carbohydrate Metabolism (Part C)
Edited by W. A. WOOD

VOLUME XLIII. Antibiotics
Edited by JOHN H. HASH

VOLUME XLIV. Immobilized Enzymes
Edited by KLAUS MOSBACH

VOLUME XLV. Proteolytic Enzymes (Part B)
Edited by LASZLO LORAND

VOLUME LX. Nucleic Acids and Protein Synthesis (Part H)
Edited by KIVIE MOLDAVE AND LAWRENCE GROSSMAN

VOLUME 61. Enzyme Structure (Part H)
Edited by C. H. W. HIRS AND SERGE N. TIMASHEFF

VOLUME 62. Vitamins and Coenzymes (Part D)
Edited by DONALD B. MCCORMICK AND LEMUEL D. WRIGHT

VOLUME 63. Enzyme Kinetics and Mechanism (Part A: Initial Rate and Inhibitor Methods)
Edited by DANIEL L. PURICH

VOLUME 64. Enzyme Kinetics and Mechanism (Part B: Isotopic Probes and Complex Enzyme Systems)
Edited by DANIEL L. PURICH

VOLUME 65. Nucleic Acids (Part I)
Edited by LAWRENCE GROSSMAN AND KIVIE MOLDAVE

VOLUME 66. Vitamins and Coenzymes (Part E)
Edited by DONALD B. MCCORMICK AND LEMUEL D. WRIGHT

VOLUME 67. Vitamins and Coenzymes (Part F)
Edited by DONALD B. MCCORMICK AND LEMUEL D. WRIGHT

VOLUME 68. Recombinant DNA
Edited by RAY WU

VOLUME 69. Photosynthesis and Nitrogen Fixation (Part C)
Edited by ANTHONY SAN PIETRO

VOLUME 70. Immunochemical Techniques (Part A)
Edited by HELEN VAN VUNAKIS AND JOHN J. LANGONE

VOLUME 71. Lipids (Part C)
Edited by JOHN M. LOWENSTEIN

VOLUME 72. Lipids (Part D)
Edited by JOHN M. LOWENSTEIN

Section I

Analytical Applications with Emphasis on Biosensors

Coeditors

Bengt Danielsson* and Klaus Mosbach

*Pure and Applied Biochemistry
Chemical Center
University of Lund
Lund, Sweden

[1] Introduction

By B. Danielsson and K. Mosbach

Analytical applications of immobilized enzymes and cells are probably more widespread than any other use of immobilized systems. In particular, biosensors, in which the biocatalyst is held in close contact with the transducer, are receiving increasing attention, as evidenced by the large number of contributors to this section. Table I lists the various assays presented in this section to allow a better overall view. Not all of the contributions are strictly methodological; this is probably due to the fact that a methodological "treatment" in the traditional sense of the devices discussed and their applications is difficult. We hope, however, that sufficient methodology is provided in each chapter to allow the reader to repeat the experiments.

Before discussing some of the contributions in this section, we would like to consider the definition of a biosensor. As mentioned above, close proximity, usually accomplished by immobilization of an enzyme on or around an electrode, constitutes a biosensor. Another example is the bioluminescence monitor, in which the light produced by the immobilized system is directly detected by a photosensor. In the enzyme thermistor system the temperature sensor is placed directly in the heat flow carried by the flow stream from a small enzyme reactor. In some other cases, however, it is less obvious that the device in question can be regarded as a biosensor. When the product of the biocatalytic activity of a small enzyme reactor is measured with a remote, although on-line, transducer, for example, in the flow cell of a spectrophotometer, this is not a biosensor in the strict sense. However, as in most situations, there may be overlap and we do not intend to make an issue out of a semantic question. Therefore, some examples of this type have been included. What we particularly want to stress in this section is the usefulness of immobilized enzyme/cell systems. Table I has been compiled in order to give easier access to the various procedures proposed for the determination of a specific analyte and serves as starting point in the search for a particular method. Useful information may also be found elsewhere in this volume.

Bioselective Electrodes. In this volume Guilbault [2] gives a survey of many different enzyme electrode configurations. An enzyme electrode can be more or less complex depending on the number of enzymes utilized. Scheller *et al.* [3] and Yoda [5] describe systems in which several different enzymes are used to give the desired signal. Scheller *et al.* [3]

TABLE I

Compilation of Assays for Specific Analytes Described in Section I[a]

Analyte	Chapter no.	Immobilized biocatalyst	Detection principle; remarks
Adenosine	[12]	Mouse small intestine; tissue	NH_3 probe
Albumin	[31]	Dye; affinity	Optoelectronic
Alanine aminotransferase	[36]	Pyruvate oxidase + POD; dry reagent	Colorimetric
AMP	[12]	Rabbit muscle; tissue	NH_3 probe
α-Amylase activity	[3]	Glucan 1,4-α-glucosidase + GOD; membrane	H_2O_2 or O_2 probe
	[5]	α-Glucosidase + GOD; membrane	H_2O_2 probe
Ascorbic acid	[13]	Microsomes; membrane	O_2 probe
	[24]	L-Ascorbate oxidase; reactor	Electrochemical
L-Asparagine	[20]	L-Asparaginase; reactor	NH_3 sensor (IrMOS capacitor)
Aspartate aminotransferase activity	[36]	Oxaloacetate decarboxylase + pyruvate oxidase + POD; dry reagent	Colorimetric
Bilirubin	[3]	GOD + POD; membrane	H_2O_2 electrode
Biotin, dethiobiotin	[9]	Avidin, catalase; membrane	O_2 probe; bioaffinity
BOD	[10]	Trichosporon cutaneum	O_2 probe
Cephalosporins	[17]	β-Lactamase; reactor	Photometric
Cholesterol	[18]	Cholesterol oxidase + catalase; reactor	Enthalpimetric
Cholesterol (esters)	[36]	Cholesterol oxidase + cholesterol esterase; dry reagent	Colorimetric
Cholic acid	[14]	Hydroxysteroid dehydrogenase +luciferase; nylon tubing	Luminometer
Choline	[15]	Luciferase; reactor	Luminometer
Creatinine	[5]	Creatininase + creatinase + sarcosine oxidase; membrane	H_2O_2 electrode
	[11]	Creatinine deiminase + Nitrosomonas sp. + Nitrobacter sp.; membrane	O_2 probe
	[20]	Creatinine deiminase; reactor	NH_3 sensor (IrMOS capacitor)

	[36]	Creatinine deiminase + pH indicator; dry reagent	Colorimetric
Cysteine	[12]	Cucumber leaves	NH₃ probe
Cytochrome P-450 substrates	[13]	Microsomes; membrane	Electrochemical determination of quinoneimine
Dextrin	[3]	Glucan 1,4-α-glucosidase + GOD; membrane	H₂O₂ probe
Ethanol	[4]	Alcohol oxidase; membrane	O₂ electrode
	[15]	Luciferase; reactor	Luminometer
	[24]	Alcohol dehydrogenase; reactor	Photometric
	[16]	Alcohol oxidase; reactor	Enthalpimetric
FMN/NADH	[14]	NAD(P)H dehydrogenase (FMN) + luciferase; nylon tubing	Luminometer
Gentamycin	[36]	IgG, immunochemical; dry reagent	Fluorimetric
Glucose	[3]	GOD; membrane	O₂ or H₂O₂ probe
	[4]	GOD; membrane	O₂ probe
	[6]	GOD; membrane	H₂O₂ probe; long-term, in vivo
	[7]	GOD + mediator	Amperometric carbon probe
	[32]	Concanavalin A + fluorescein-labeled dextran	Optical fiber, fluorimetric (affinity sensor)
	[15]	Luciferase; reactor	Luminometer
	[19]	GOD; reactor	Enthalpimetric
	[21]	GOD; membrane	ISFET
	[22]	GOD; membrane	Micro-O₂ sensor
	[23]	GOD; reactor	Colorimetric or chemiluminescence
	[25]	Aldehyde dehydrogenase + GOD; nylon tubing	Colorimetric
	[24]	GOD + POD; reactor	F⁻ probe
	[26]	GOD; membrane	Externally buffered O₂ probe
	[28]	GOD; reactor	O₂ probe, bedside monitoring
	[29]	GOD; membrane	H₂O₂ electrode, needle-type probe
	[36]	GOD + POD; dry reagent	Colorimetric

(continued)

TABLE I (*continued*)

Analyte	Chapter no.	Immobilized biocatalyst	Detection principle; remarks
Glucose (*cont.*)	[27]	GOD + catalase; reactor	Enthalpimetric, continuous monitoring
	[16]	GOD + catalase; reactor	Enthalpimetric, continuous monitoring
Glucose 6-phosphate	[3]	Potato slice + GOD; membrane	H_2O_2 electrode
L-Glutamate	[10]	*Escherichia coli*; lyophilized	CO_2 probe
	[12]	Yellow squash; tissue	CO_2 probe
	[15]	Luciferase; reactor	Luminometer
Glutamine	[12]	Porcine kidney; tissue	NH_3 probe
Glycerol	[16]	*Gluconobacter oxydans*; reactor	Enthalpimetric
Guanine	[12]	Rabbit liver; tissue	NH_3 probe
IgG	[34]	Protein A; surface layer	Ellipsometric
Heavy metals	[16]	Urease; reactor	Enthalpimetric
Hydrogen peroxide	[15]	Luciferase; reactor	Luminometer
D-Lactate	[4]	*E. coli* K12; membrane	O_2 probe
L-Lactate	[4]	L-Lactate 2-monooxygenase; membrane	O_2 probe
	[16]	L-Lactate 2-monooxygenase; membrane	Enthalpimetric
	[16]	L-Lactate oxidase + catalase + L-lactate dehydrogenase; reactor	Enthalpimetric; highly sensitive recycling
Lactose	[4]	Lactase + GOD; membrane	O_2 probe
L-Lysine	[4]	L-Lysine oxidase; membrane	O_2 probe
Maltose	[3]	Glucan 1,4-α-glucosidase + GOD + POD; membrane	H_2O_2 probe
Methanol	[7]	Methanol dehydrogenase + mediator	Amperometric carbon probe
NAD/NADH	[13]	Microsomes; membrane	O_2 probe (reduced pyridine nucleotide)

Oxalate	Oxalate oxidase; reactor	[16]	Enthalpimetric
Penicillin	β-Lactamase + bromocresol green	[31]	Optoelectronic
	β-Lactamase; reactor	[17]	Enthalpimetric
Phenol	Lactase; membrane	[4]	O_2 probe
Phosphatidylcholine	Phospholipase D + choline oxidase + catalase; reactor	[18]	Enthalpimetric
Proinsulin	IgG + proinsulin catalase; reactor	[30]	Enthalpimetric (TELISA)
Pyruvate	Corn kernels	[12]	CO_2 probe
	L-Lactate dehydrogenase + lactate oxidase + catalase; reactor	[16]	Enthalpimetric; highly sensitive recycling
Sucrose	β-Fructofuranosidase (invertase) + GOD; membrane	[3]	H_2O_2 probe
	β-Fructofuranosidase (invertase) + aldose 1-epimerase (mutarotase) + GOD; membrane	[4]	O_2 probe
	β-Fructofuranosidase (invertase); reactor	[27]	Enthalpimetric, continuous monitoring
Theophylline	β-Fructofuranosidase (invertase); reactor	[16]	Enthalpimetric
Thyroxine (T4)	IgG + GOD/POD label; dry reagent	[36]	Colorimetric
	IgG + GOD/POD label; dry reagent	[36]	Colorimetric
Triglycerides	Lipoprotein lipase; reactor	[18]	Enthalpimetric
Urea	Urease + nitrifying bacteria; membrane	[11]	O_2 probe
	Urease + bromthymol blue; membrane	[31]	Optoelectronic
	Urease; membrane	[22]	ISFET
	Urease; membrane	[16]	Enthalpimetric
	Urease; reactor	[20]	NH_3 sensor (IrMOS capacitor)
Uric acid	Urate oxidase (uricase); reactor	[23]	Colorimetric or chemiluminescence
	Urate oxidase (uricase) + aldehyde dehydrogenase; nylon tubing	[25]	Colorimetric

[a] Abbreviations used: POD, peroxidase; GOD, glucose oxidase; IrMOS, iridium metal oxide semiconductor.

use enzymes in antiinterference layers to remove disturbing components from a sample, in recycling systems to increase the sensitivity, and in competitive enzymatic routes to increase the applicability of, for example, a glucose sensor. Romette and Thomas [4] as well as Hikuma and Yasuda [10] describe systems which are used for industrial monitoring of various compounds. Clark *et al.* [6], Shichiri *et al.* [29], and Håkansson [28] describe systems intended for bedside monitoring of blood glucose, especially insulin administration for diabetes. Today, blood glucose determination for this purpose and for decentralized medical diagnosis is the major target of most of the efforts in this field (medical application of enzyme electrodes).

Much effort is currently being directed toward the development of mediator-modified electrodes and electron-transporting mechanisms to make better use of cofactor-requiring enzymes and to facilitate direct contact between the enzyme and the electrode. Turner [7], Wingard and Narasimhan [8], and Varfolomeev [38] present examples applicable to implantability. Unfortunately, at present these systems display rather low stability and a short lifetime, but they can be successfully employed in disposable electrodes (Turner [7]).

In many cases it is possible to use immobilized cells or organelles (Schubert and Scheller [13]) as biocatalysts. Then there is no need for isolation of enzymes which can be difficult or even practically impossible. Moreover, coupled reactions and cofactor requirements are solved naturally. Hikuma and Yasuda [10] describe a commercial instrument for biochemical oxygen demand (BOD) measurements based on immobilized cells. Rechnitz [12] takes this a step further in utilizing tissue slices in combination with electrodes.

Affinity Sensors. Bioaffinity can be used to increase the selectivity of a sensor in which antibody–antigen or lectin–carbohydrate binding is used to separate the analyte from a complex mixture. Detection can be accomplished via enzyme-amplified reactions. Several chapters in this section illustrate this technique (Ikariyama and Aizawa [9], Lowe and Goldfinch [31], and Schultz and Mansouri [32]). Enzyme immunoassays, the most widely used of these techniques, have been omitted as they are treated comprehensively in other volumes of this series (Vols. 70, 73, and 92 and to some extent also Vols. 74 and 84). We have only included one example, the TELISA (thermometric enzyme-linked immunosorbent assay) procedure as described by Birnbaum *et al.* [30], in which an enzyme thermistor unit is used as the detector. This is an example of semicontinuous monitoring of biotechnological processes producing, e.g., hormones or antibodies, an analytical field of increasing importance. For other examples based on affinity binding see Mattiasson [60]. Using chemi- or bioluminescence, highly sensitive methods can be designed as demonstrated by Roda *et al.* [14], Kurkijärvi *et al.* [15], and Murachi and Tabata [23].

Thermistor Probes. The temperature sensor, probably the most general transducer in biosensors, is utilized in enzyme thermistors. Danielsson and Mosbach [16] give a general description of enzyme thermistors; Decristoforo [17], Satoh [18], and Kiba and Furusawa [19] give more specialized views. Continuous monitoring with enzyme thermistors is treated by Mandenius and Danielsson [27]. The flow–injection analysis (FIA) procedure described by Decristoforo [17] shows that enzyme thermistors can solve routine, analytical problems in a very satisfactory manner. Satoh [18] discusses how even lipids, which are difficult analytes to handle, can be analyzed with accuracy.

Solid-State Sensors. The use of solid-state sensors in chemistry, e.g., ISFETs (ion-sensitive field effect transistors) and ChemFETs, is of interest in industrial research laboratories. ChemFETs for the assay of Na^+ and K^+ electrolytes, pH, and pO_2 in physiological fluids are in use already. Attempts to combine semiconductor sensors with biocatalysts are described by Winquist *et al.* [20], Caras and Janata [21], and Karube and Moriizumi [22]. Special advantages of these devices are small size, low signal impedance, possibility for multisensor applications, direct integration with electronics, and potential implantability. The gas sensors described by Winquist *et al.* [20] differ somewhat from the other devices since the enzyme is immobilized in a reactor and not placed directly on the semiconductor structure. These sensors are highly sensitive, and the gas-phase operation prevents fouling of the sensor.

Enzyme Reactors. Enzyme reactors can be combined with various transducers such as amperometric electrodes, electrochemical detectors, conductivity detectors, and thermal detectors. They are highly flexible and can be loaded extensively with enzyme, which leads to high operational stability, reliability, and speed. They can easily be adapted to suitable flow systems. Analytical uses of enzyme reactors are described here by Murachi and Tabata [23], Ho [24], and Sundaram [25]. Enzyme reactors also constitute fundamental parts of other systems treated in this section, such as enzyme thermistors.

Continuous Monitoring. Many of the analytical systems described here are suitable for continuous monitoring of analyte concentrations in various fluids. There is an increasing demand for this type of monitoring in medicine, especially for glucose and for substrates and products involved in fermentations and other biotechnological processes. Special problems encountered in sampling are treated by Enfors and Cleland [26] and Mandenius and Danielsson [27], who also give examples from monitoring in fermentation. Håkansson [28], Clark *et al.* [6], and Shichiri *et al.*

[29] give examples from blood glucose monitoring; articles [6] and [29] discuss implanted amperometric sensors.

Other Techniques. Many other analytical techniques employ immobilized enzymes and cells. Some of these are rather new and not widely used. The combination of mass spectrometry with immobilized enzymes or cells is presented by Weaver [37]. Dry reagent chemistry (see Walter [36]) is becoming increasingly important, especially in clinical chemistry. Optical techniques suitable for studies on surface interactions are discussed by Arwin and Lundström [33], Jönsson *et al.* [34], and Mandenius *et al.* [35] and include ellipsometry, reflectometry, and related techniques. The use of optical fibers as an emerging analytical technique is presented by Schultz and Mansouri [32].

In selecting the proper biosensor or analytical concept based on immobilized enzymes or cells for a specific task, several choices have to be made. These include the choice between enzyme and whole cells, selection of transducer, immobilization technique, the attachment or adaptation of the biocatalyst to the transducer, and the use of a membrane or reactor configuration.

Enzymes or Whole Cells

Normally enzymes are to be preferred to cells since they give higher specificity, higher sensitivity, quicker response, and usually better operational stability. Some enzymes may, however, be difficult to isolate with retained activity or the detecting system could involve several enzymatic steps or enzymes which require cofactors. In such cases, immobilized cells (or maybe organelles or even tissue preparations) could be a useful alternative. They could also be very suitable for preliminary tests because of the simple preparation technique required.

Immobilization Techniques

It is difficult to provide simple recommendations for the selection of proper immobilization techniques as this depends on the type of carrier or support material used and the application, e.g., in a reactor or a membrane configuration. A number of immobilization techniques are comprehensively treated in Vol. 135 of this series, but valuable information can be gained from the contributions in this section as well. Immobilization inside nylon tubing is treated by Sundaram [25] and Roda *et al.* [14]; immobilization in membranes by Yoda [5] and Romette and Thomas [4]. Ho [24] describes the preparation of whisker-walled glass tube reactors and single-bead string reactors, while Lowe and Goldfinch [31] describe enzyme–dye membranes. Surface immobilization techniques are discussed by Jönsson *et al.* [34] and Mandenius *et al.* [35]. The most widely used combination of support–immobilization techniques to date consists of alkylaminosilanized controlled-pore glass and glutaraldehyde binding.

Membrane or Reactor; Batch or Flow Technique

A "bioprobe" which can be simply dipped in the sample solution is readily obtained by binding or entrapping the biocatalyst to or within a membrane, which is then closely adapted to a suitable transducer. Such a "batch method" is acceptable for preliminary tests or for measurements on a limited number of samples. For more premanent use a flow system is recommended since it is easier to handle and allows better control of measurement conditions. It is possible to use a probe with a biospecific membrane in a continuous flow system, but it may be advantageous to apply the biocatalyst immobilized in a reactor and mount the transducer in a flow cell. One obvious advantage is that a larger amount of the biocatalyst can be utilized compared to the membrane configuration which provides for higher operational stability.

Selection of Transducer

Again it is difficult to give definite recommendations as much depends on tradition and availability. Oxidases suitable for analytical work are commercially available, and they are usually highly specific, active, and stable. Oxygen and hydrogen peroxide sensors are popular transducers. Temperature transducers, such as thermistors, are the most general type of transducer and can be used with most enzymes. They often need only one enzyme to work adequately, and they can be used with immobilized cells as well.

Bioluminescence techniques can be used in numerous ways and have very high sensitivity. Highly sensitive methods can also be designed for electrode or thermistor devices using enzyme amplification by substrate or coenzyme recycling. Semiconductor sensors have the advantage of being small and easily adaptable to associated electronics. Surface techniques are useful for the detection of macromolecules in affinity systems, and they can be designed to allow for continuous measurements. With these systems the interaction between antigen–antibody or receptor–cell is visualized directly.

In most laboratories photometric detection of the products formed in an immobilized enzyme reactor is straightforward and can be used with a large number of enzymes, namely dehydrogenases and oxidases (combined with peroxidase). Assays based on soluble enzymes are usually designed for photometric detection, and many useful enzymes are readily

TABLE II
COMMERCIALLY AND SEMICOMMERCIALLY AVAILABLE BIOSENSORS

Manufacturer	Instrument, model	Analyte	Concept
Ajinomoto Co.; sold by Nisshin Electric Co., Japan	BOD sensor	BOD	Microbial electrode
Analytical Instruments Co.; sold by Toyo Jozo, Japan	AD-300, AS-200, M-100	Glucose, alcohol, glycerol, lactate	O_2 electrode + enzyme membrane
Daiichi Kagaku, Japan	GA-1120, GM-1320	Glucose	GOD + H_2O_2 electrode
Denki Kagaku, Japan	—	Alcohol, acetic acid	—
Fuji Electric Co., Japan	Gluco-20A, UA-500 A	Glucose, α-amylase, uric acid	Enzyme membrane + H_2O_2 electrode
Gambro, Sweden	Blood glucose analyzer	Glucose	GOD reactor + O_2 electrode
Genetics International, United Kingdom	Exactech	Glucose	GOD layer + amperometric probe
Lithuanian Academy of Sciences, Soviet Union	Enzalyst	Glucose	GOD
Midwest Research Institute, United States	IEM, CAM	Pesticides	Cholinesterase + electrochemical detector
Mitsubishi Chemical Industries, Japan	GL-101	Glucose	GOD reactor + H_2O_2 detector
Omron Toyobo Co., Japan	Diagluca, HER-100	Glucose, lactate	Enzyme membrane + H_2O_2 electrode

Oriental Electric Co., Japan	KV 101, ST-1	Fish, freshness	O_2 electrode + enzyme membrane, O_2 electrode + soluble enzyme
Radelkis, Hungary	OP-CL-71105	Glucose	GOD
Seres, France	Enzymat	Glucose, lysine, alcohol	Enzyme membrane + O_2 electrode
Thorn EMI Simtec, Ltd., United Kingdom	NAIAD	Nerve gas	Cholinesterase inhibition
Toa Electric Co., Japan	—	Glucose	H_2O_2 electrode + enzyme membrane
Universal Sensors, United States	—	Glucose, urea, lactate, amino acids, ethanol	Enzyme membrane + O_2 electrode
University of Lund, Sweden	Enzyme thermistor	Various sugars, alcohol. lactate, penicillin, urea, oxalate, triglycerides, etc.	Enzyme reactor + thermistor
Wolverine Medical Inc., United States	—	Lactate	—
Yellow Springs Instruments Co., United States	23A and 27	Glucose, lactate, alcohol, sucrose, lactose, fructose	Enzyme membrane + H_2O_2 electrode
Zentrum für Wissenschaftlichen Gerätebau, German Democratic Republic	GKM	Glucose, uric acid, etc.	Enzyme membrane + H_2O_2 electrode

available. It is also very easy to adapt a spectrophotometer for continuous flow measurements.

The number of commercially available biosensors is to the best of our knowledge somewhat limited and are listed in Table II. However, we are confident that many such devices will be developed. In addition in many laboratories researchers are building biosensors for their own use. Finally, in the long-range planning toward the development of "biochips," it may very well be that some of the concepts and immobilization techniques, such as surface attachment, discussed in this section might be utilized or at least serve as a starting point.

It is not possible to cover within this section all aspects and new developments in the rapidly advancing area of biosensors. For further insights the reader is referred to journals which regularly cover various aspects of biosensors. These include *Analytical Biochemistry, Analytical Chemistry, Analytica Chimica Acta, Biotechnology and Bioengineering, Biotechnology Letters, Bio/Technology*, and the specialized journal *Biosensors*. Some books on this topic include P. W. Carr and L. B. Bowers, *Immobilized Enzymes in Analytical and Clinical Chemistry: Fundamentals and Application*, Wiley, New York, 1980; G. G. Guilbault, *Analytical Uses of Immobilized Enzymes*, Dekker, New York, 1984; and A. P. F. Turner, I. Karube, and G. S. Wilson (eds.), *Biosensors: Fundamentals and Application*, Oxford Univ. Press, 1987. The reader is also referred to a recent survey entitled "Biosensors: Major opportunities in the 1990s" by R. F. Taylor, *Spectrum, Biotechnology Products and Technologies*, Arthur D. Little Decision Resources, D.S., 1987.

[2] Enzyme Electrode Probes

By GEORGE G. GUILBAULT

Introduction

The enzyme electrode is a combination of an ion-selective electrode base sensor with an immobilized (insolubilized) enzyme, which provides a highly selective and sensitive method for the determination of a given substrate. Clark and Lyons[1] first introduced the concept of the "soluble" enzyme electrode, but the first working electrode was reported by Updike and Hicks,[2] who used glucose oxidase immobilized in a gel over a polarographic oxygen electrode to measure the concentration of glucose in bio-

---

[1] L. Clark and C. Lyons, *Ann. N.Y. Acad. Sci.* **102**, 29 (1962).
[2] S. J. Updike and G. P. Hicks, *Nature (London)* **214**, 986 (1971).

METHODS IN ENZYMOLOGY, VOL. 137

logical solutions and tissues. These were both voltammetric or amperometric probes, i.e., the current produced on application of a constant applied voltage was measured. The first potentiometric (no applied voltage, the voltage produced is monitored) enzyme electrode was described by Guilbault and Montalvo for urea in 1969.[3] Since then over 100 different electrodes have appeared in the literature; a summary of these can be found in Ref. 4 and Table I.

In enzyme electrodes, the enzyme is usually immobilized, thus reducing the amount of material required to perform a routine analysis and eliminating the need for frequent assay of the enzyme preparation in order to obtain reproducible results. Furthermore, the stability of the enzyme is often improved when it is incorporated in a suitable gel matrix. For example, an electrode for the determination of glucose prepared by covering a platinum electrode with chemically bound glucose oxidase has been used for over 300 days.[5] Of the two methods used to immobilize an enzyme—(1) the chemical modification of the molecules by the introduction of insolubilizing groups and (2) the physical entrapment of the enzyme in an inert matrix, such as starch or polyacrylamide—the technique of chemical immobilization is the best for making electrode probes.

Experimental

Apparatus

The enzyme electrode, once constructed, is used like any other ion-selective electrode. A typical apparatus is shown in Fig. 1. Potentiometric probes, e.g., for urea, amino acids, and penicillin, are plugged directly into a digital voltmeter (e.g., Orion, Corning, Sargent). The millivoltages read for each concentration tested are then plotted against concentration in a linear log plot. To use amperometric electrodes (which are based on the use of a Pt or O_2 electrode), e.g., for alcohol or glucose a polarographic apparatus, like the Princeton Applied Research Model 170 or 174, can be used. A potential of either ±0.85 V is applied to the electrode, and the current (μA) is recorded. A plot of current (μA) versus concentration (M) is a linear plot.

To eliminate the necessity of using a polarograph to monitor the amperometric electrode, Universal Sensors, Inc. (P.O. Box 736, New Orleans, LA 70148) markets an electrode adapter, which is a device that will simultaneously apply a potential to a amperometric glucose or alcohol

[3] G. Guilbault and J. Montalvo, *J. Am. Chem. Soc.* **91**, 2164 (1969).
[4] G. Guilbault, "Handbook of Immobilized Enzymes." Dekker, New York, 1984.
[5] G. G. Guilbault and G. Lubrano, *Anal. Chim. Acta* **64**, 439 (1973).

TABLE I
TYPICAL ELECTRODES AND THEIR CHARACTERISTICS

Type	Enzyme	Sensor	Immobilization[a]	Stability	Response time	Amount of enzyme (U)	Range (mol/liter)[b]
Urea	(EC 3.5.1.5)	Cation	Physical	3 weeks	0.5–1 min	25	10^{-2}–5×10^{-5}
		Cation	Physical	2 weeks	1–2 min	75	10^{-2}–10^{-4}
		Cation	Chemical	>4 months	1–2 min	10	10^{-2}–10^{-4}
		pH	Physical	3 weeks	5–10 min	100	5×10^{-3}–5×10^{-5}
		Gas (NH_3)	Chemical	4 months	2–4 min	10	5×10^{-2}–5×10^{-5}
		Gas (NH_3)	Chemical	20 days	1–4 min	0.5	10^{-2}–10^{-4}
		Gas (CO_2)	Physical	3 weeks	1–2 min	25	10^{-2}–10^{-4}
Glucose	Glucose oxidase (EC 1.1.3.4)	pH	Soluble	1 week	5–10 min	100	10^{-1}–10^{-3}
		Pt (H_2O_2)	Physical	6 months	12 sec kinetic[c]	10	2×10^{-2}–10^{-4}
		Pt (H_2O_2)	Chemical	>14 months	1 min steady state[c]	10	2×10^{-2}–10^{-4}
		Pt (H_2O_2)	Soluble	—	1–2 min	10	10^{-2}–10^{-4}
		Pt (quinone)	Soluble	<1 week[d]	3–10 min	10	2×10^{-2}–10^{-3}
		Pt (O_2)	Chemical	>4 months	1 min	10	10^{-1}–10^{-5}
		I^-	Chemical	>1 month	2–8 min	10	10^{-3}–10^{-4}
		Gas (O_2)	Physical	3 weeks	2–5 min	20	10^{-2}–10^{-4}
	Glucose oxidase (EC 1.1.3.4) and catalase (EC 1.11.1.6)	Gas (O_2)	Chemical	>3 weeks	2–5 min	10	2×10^{-2}–10^{-4}
L-Amino acids (general)[e]	L-Amino-acid oxidase (EC 1.4.3.2)	Pt (H_2O_2)	Chemical	4–6 months	12 sec kinetic[c]	10	10^{-3}–10^{-5}
		Gas (O_2)	Chemical	—	2 min	10	10^{-2}–10^{-4}
		Pt (O_2)	Chemical	>4 months	1 min	10	10^{-2}–10^{-4}
		Cation	Physical	2 weeks	1–2 min	10	10^{-2}–10^{-4}
		NH_4^+	Chemical	>1 month	1–3 min	10	10^{-2}–10^{-4}
		I^-	Chemical	>1 month	1–3 min	10	10^{-3}–10^{-4}

Substance	Enzyme	Sensor	Immobilization	Stability	Response time		Range
L-Tyrosine	L-Tyrosine decarboxylase (EC 4.1.1.25)	Gas (CO_2)	Physical	3 weeks	1–2 min	25	10^{-1}–10^{-4}
L-Glutamine	Glutaminase (EC 3.5.1.2)	Cation	Soluble	2 days[d]	1 min	50	10^{-1}–10^{-4}
L-Glutamic acid	Glutamate dehydrogenase (NAP(P)⁺) (EC 1.4.1.3)	Cation	Soluble	2 days[d]	1 min	50	10^{-1}–10^{-4}
L-Asparagine	Asparaginase (EC 3.5.1.1)	Cation	Physical	1 month	1 min	50	10^{-2}–5×10^{-5}
D-Amino acids (general)[f]	D-Amino-acid oxidase (EC 1.4.3.3)	Cation	Physical	1 month	1 min	50	10^{-2}–5×10^{-5}
Lactic acid	L-Lactate dehydrogenase (EC 1.1.1.27)	Pt [Fe(CN)$_6^{4-}$]	Soluble	<1 week	3–10 min	2	2×10^{-3}–10^{-4}
Succinic acid	Succinate dehydrogenase (EC 1.3.99.1)	Pt (O_2)	Physical	1 week	1 min	10	10^{-2}–10^{-4}
Acetic, formic acids	Alcohol oxidase (EC 1.1.3.13)	Pt (O_2)	Chemical	>4 months	30 sec	10	10^{-1}–10^{-4}
Alcohols	Alcohol oxidase (EC 1.1.3.13)	Pt (H_2O_2)	Soluble	1 week	12 sec kinetic[c]	10	0.5–100 mg%
		Pt (H_2O_2)	Soluble	1 day[d]	1 min	1	0.5–50 mg%
Penicillin	β-Lactamase penicillinase, (EC 3.5.2.6)	pH	Physical	1–2 weeks	0.5–2 min	400	10^{-2}–10^{-4}
			Soluble	3 weeks	2 min	1000	10^{-2}–10^{-4}
Uric acid	Urate oxidase (uricase, EC 1.7.3.3)	Pt (O_2)	Chemical	4 months	30 sec	10	10^{-2}–10^{-4}
Amygdalin	β-Glucosidase (EC 3.2.1.21)	CN⁻	Physical	3 days[g]	10–20 min	100	10^{-2}–10^{-5}

(continued)

TABLE I (continued)

Type	Enzyme	Sensor	Immobilization[a]	Stability	Response time	Amount of enzyme (U)	Range (mol/liter)[b]
Cholesterol	Cholesterol oxidase (EC 1.1.3.6)	Pt (H_2O_2)	Soluble	—	2 min		10^{-2}–10^{-4}
Phosphate	Alkaline phosphatase (EC 3.1.3.1) and glucose oxidase (EC 1.1.3.4)	Pt (O_2)	Chemical	4 months	1 min	10 each	10^{-2}–10^{-4}
Nitrate	Nitrate reductase (NADH) (EC 1.6.6.1) and nitrite reductase (NAD(P)H) (EC 1.6.6.4)	NH_4^+	Soluble	—	2–3 min	10	10^{-2}–10^{-4}
Nitrite	Nitite reductase (NAD(P)H) (EC 1.6.6.4)	NH_3(gas)	Chemical	3–4 months	2–3 min	10	5×10^{-2}–10^{-4}
Sulfate	Arylsulfatase (EC 3.1.6.1)	Pt	Chemical	1 month	1 min	10	10^{-1}–10^{-4}

[a] "Physical" refers to polyacrylamide gel entrapment in all cases; "chemical" is attachment chemically to glutaraldehyde with albumin, to polyacrylic acid, or to acrylamide, followed by physical entrapment.

[b] Analytically useful range, either linear or with reasonable change if curvature is observed.

[c] "Kinetic," rate of change in current measured after 12 sec; "steady state," current reaches a maximum in 1 min.

[d] Preparation lacks stability as evidenced by constant decrease in signal each day.

[e] Electrode responds to L-cysteine, L-leucine, L-tyrosine, L-tryptophan, L-phenylalanine, and L-methionine.

[f] Electrode responds to D-phenylalanine, D-alanine, D-valine, D-methionine, D-leucine, D-norleucine, and D-isoleucine.

[g] Time required for signal to return to baseline before reuse.

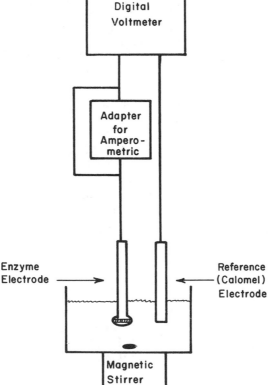

FIG. 1. Equipment for measurement with an enzyme electrode.

probe, take the resulting current generated, and convert it to a voltage which can then be read directly on any voltmeter (Orion, Corning, etc.), provided the scale is ±1.7 V. This simplifies the operational use of all enzyme electrodes by requiring only a voltameter plus an adaptor (the latter sells for about $125).

A reference electrode, generally a calomel electrode, is used with the enzyme electrode. Alternatively, the reference electrode can be combined as an integral part of the enzyme electrode as is the case with the NH_3, CO_2, or O_2 electrode base sensors (Universal Sensors) used in the urea, amino acid, glucose, or alcohol probes. Finally, the electrode must be kept in a solution which is stirred at a *constant* rate, since it has been shown (see Ref. 6) that a change in stirring rate will change the potential of the electrode measured.

[6] G. G. Guilbault, "Handbook of Enzymatic Analysis." Dekker, New York, 1974.

Materials Required

To construct an enzyme electrode one must first select the appropriate enzyme. From information in standard reference books on enzymology (Refs. 4 and 6) find an enzyme system which is suitable for your determination, i.e., one which is based on the primary function of the enzyme, the main substrate–enzyme reaction. For example, to construct a glucose electrode, use glucose oxidase; for a urea electrode, use urease. In some cases, you may use an enzyme which acts on the compound of interest as a secondary substrate, e.g., alcohol oxidase for an acetic acid electrode.

If the enzyme can be purchased from a commercial supplier (Sigma, Boehringer Mannheim, Worthington, Calbiochem, Miles), pay special attention to the purity of the enzyme in the specifications in order to avoid cross-reactivity with other substrates. In some cases, impure enzymes such as jack bean urease and glucose oxidase for the food industry (General Mills) can be used directly in a pseudo-"immobilized" form, e.g., as a liquid covered with a dialysis membrane. To be useful in an electrode, the activity of the enzyme should be at least 10 units/mg of protein. Excellent tables giving sources of enzymes can be found in Refs. 4 and 6.

If the enzyme required for your electrode is not available commercially, contact any large biochemical supply house and inquire whether it will isolate and purify the enzyme needed. The New England Enzyme Center of Tufts University (Boston, MA) also specializes in such a service. You may want to prepare the enzyme yourself after ascertaining the isolation and purification methods given in the literature. In most cases, the techniques can be routinely performed.

After immobilizing the enzyme, by one of the techniques described below, assemble the electrode using the ion-selective electrode chosen. The better the enzyme is immobilized, the more stable it is, and, hence, it can be used longer and for more assays. The entrapment of soluble enzyme is the simplest form of immobilization and could be tried first. Physical entrapment or chemical attachment can also be used. The method recommended by the author is glutaraldehyde attachment which is simple, fast, and often gives highly desirable results.

Preparation of Electrode

The electrodes are prepared as described in Fig. 2, using configuration A for physically entrapped enzymes and configuration B for chemically bound or soluble entrapped enzymes. Choose the base sensor electrode, according to the enzyme reaction to be studied. It must respond to either one of the products or reactants of the enzyme system. The base elec-

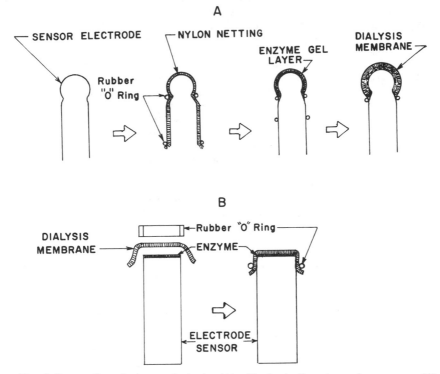

FIG. 2. Preparation of enzyme electrodes (A) with physically entrapped enzymes or (B) with chemically bound or soluble enzymes.

trode and the types of enzyme electrodes that can be constructed from it are given in Tables I and II.

Dialysis Membrane Electrode. To prepare a dialysis membrane electrode, take the base electrode sensor, chosen from Table I or II, and turn it upside down (see Fig. 2B). Place 10–15 units of the soluble enzyme, in the form of a thick paste of freeze-dried powder, on the surface of the base sensor. (Note: for best results, the base sensor should be flat, not round.) Cover with a piece of dialysis membrane (cellophane 20–25 μm thick, Will Scientific, Arthur H. Thomas, Sigma, etc.) about twice the diameter of the electrode sensor. Place a rubber O ring having a diameter that fits the electrode snugly around the cellophane membrane (Fig. 2B), and gently push the O ring into the electrode body so that the enzyme forms a uniform layer on top of the electrode surface. Place the electrode in a buffer solution overnight to allow penetration of buffer into the enzyme layer and to remove entrapped air. Store the electrode in buffer (optimum for the enzyme system) in a refrigerator between use.

TABLE II
POSSIBLE ELECTRODE SENSORS USEFUL IN CONSTRUCTION OF ENZYME ELECTRODES

Sensor	Useful for
Potentiometric	
NH_3	Urea, amino acids, glutamine, glutamic acid, nitrate, nitrite, creatinine, lyase, and deaminase enzymes
CO_2	Urea, amino acids, decarboxylative enzyme systems
pH	Penicillin, RNA, DNA, glucose, enzyme reactions giving pH changes
I^-	Glucose, amino acids, cholesterol, alcohols
CN^-	Amygdalin
Voltammetric (amperometric)	
O_2	Glucose, amino acids, organic acids, alcohols, uric acid, cholesterol, phosphate, all O_2-consuming enzymes
Pt or C	All redox enzymes, sulfate, uric acid, cholesterol, alcohols, glucose, amino acids, organic acids, NADH/NADPH systems

Physically Entrapped Electrode. To prepare a physically entrapped electrode, place the electrode sensor upside down (Fig. 2A) and cover with a thin nylon net (about 50 μm thick—a sheer nylon stocking is satisfactory) which is secured with a rubber O ring in the manner shown. This serves as a support for the enzyme gel solution. Prepare the gel as follows: dissolve N,N'-methylenebisacrylamide (Eastman Chemical Co.) (1.15 g) in phosphate buffer (0.1 M, pH 6.8, 40 ml) by heating to 60°. Cool to 35°, and 6.06 g of acrylamide monomer (Eastman Chemical Co.), and filter into a 50-ml volumetric flask containing 5.5 mg of riboflavin (Eastman Chemical Co.) and 5.5 mg of potassium persulfate. This solution is stable for months if stored in the dark.

Prepare the enzyme gel solution by mixing 100 mg of enzyme (purity at least 10 units/mg) with 1.0 ml of gel solution. Gently pour the enzyme gel solution onto the nylon net in a thin film, making sure all of the pores of the net are saturated. One milliliter of this solution should be enough for several electrodes.

Place the electrode in a water-jacketed cell at 0.5° and remove oxygen, which inhibits polymerization, by purging with N_2 before and during the polymerization. Complete the polymerization by irradiating with a 150-W Westinghouse projector spotlight for 1 hr. The enzyme layer should be dry and hard. Place a piece of dialysis membrane over the outside of the nylon net for further protection and secure with a second O ring. Store the electrode in buffer in a refrigerator overnight before use and between usage.

Chemically Attached Enzymes. The enzyme is chemically attached to a solid support, and the immobilized enzyme is placed onto the base sensor (Fig. 2B) and covered with a dialysis membrane. Four methods are described below, that can be utilized. The glutaraldehyde (procedure 1) and collagen membrane (procedure 2) methods are the most easily carried out and are highly recommended. If denaturation of the enzyme occurs in immobilization using either of these two methods (e.g., the enzyme electrode has very low activity), follow procedures 3 or 4 below.

PROCEDURE 1. GLUTARALDEHYDE ATTACHED ENZYMES. Turn the electrode base sensor upside down, as shown in Fig. 2B. Place a piece of pig intestine membrane (Universal Sensors) on the sensor and secure with an O ring. Add 50 μl of a solution containing phosphate buffer (0.1 M, pH 6.8, 2.7 ml), bovine serum albumin (Sigma) (17.5%, 1.5 ml), and 50 mg of enzyme (at least 10 units/mg). Completely wet the membrane with the solution. Add 10 μl of a 10% solution of glutaraldehyde while rapidly mixing with a stirring rod. The enzyme layer should harden immediately. Let the membrane set overnight in a refrigerator, upside down, to ensure the layer is completely formed, then cover with a dialysis membrane, and store in the buffer solution used with the enzyme system. The general reaction for this preparation is given by Eq. (1).

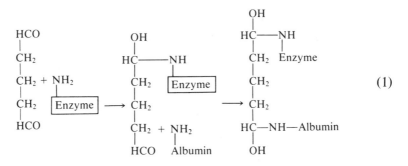

$$(1)$$

PROCEDURE 2. COLLAGEN MEMBRANE ELECTRODES. The procedure described below is the same as reported by Coulet and co-workers, using a highly polymerized collagen, prepared under industrial conditions, as the binding site of various enzymes. Because of its protein nature, collagen has free amino groups available for covalent coupling with enzymes. Collagen membranes also possess a high polyol content (33%) which enhances film hydrophilicity, making it very supple and mechanically strong.

Collagen membranes can be obtained from Centre Technique du Cuir (Lyon, France) or Rutgers University (New Brunswick, NJ) (diameter 2.5 cm and 0.1 mm thick in the dry state, 0.3 to 0.5 mm thick when swollen). Immerse these collagen membranes in 60 ml of 100% methanol containing 0.2 M HCl, for 3 days at room temperature. Rinse five of these membranes thoroughly (the others can be kept in solution) with distilled water. Then place them in 100 ml of 1% hydrazine and keep immersed for 12 hr at room temperature. Wash with distilled water at 0°, then immerse in a mixture of 0.5 M KNO$_2$ and 0.3 M HCl (in an ice bath at 0°) for 15 min. Wash the membranes with buffer solution (phosphate buffer, pH 7.5, 50 mM), then place these membranes in a solution containing at least 50 units of enzyme in 2 ml of phosphate buffer (pH 7.5, 50 mM). Store in the refrigerator overnight at 4°. Wash the membranes with buffer and store in the same buffer until used. The general scheme for immobilization of enzymes on collagen is shown in Eqs. (2)–(5).

Activation: Acyl Azide Formation

$$\equiv\!\!-COOH \xrightarrow[\text{23°, 3 days}]{\text{CH}_3\text{OH/0.2 } M \text{ HCl}} \equiv\!\!-COOCH_3 \tag{2}$$

Wash in doubly distilled water, 23°

$$\equiv\!\!-COOCH_3 \xrightarrow[\text{23°, 12 hr}]{\text{1\% NH}_2-\text{NH}_2} \equiv\!\!-CONH-NH_2 \tag{3}$$

Wash in doubly distilled water, 0°

$$\equiv\!\!-CONH-NH_2 \xrightarrow[\text{0°, 15 min}]{\text{0.5 } M \text{ NaNO}_2/\text{0.3 } M \text{ HCl}} \equiv\!\!-CON_3 \tag{4}$$

Wash in phosphate buffer, pH 7.5, 50 mM

Coupling of the enzymes

$$\equiv\!\!-CON_3 \xrightarrow[\text{4°, 12 hr}]{\text{H}_2\text{N}-\text{Enzyme}} \equiv\!\!-CO-NH-\text{Enzymes} \tag{5}$$

Wash in phosphate buffer, pH 7.5, 50 mM;
store enzymatically active film in the same
buffer at 4°

The collagen enzyme layer is mounted on the base electrode to be used (e.g., a CO$_2$ or NH$_3$ gas membrane electrode or a Pt electrode), and the enzyme layer is secured in place with a rubber O ring of appropriate diameter. The electrode is placed in 2 ml of buffer containing the sample to be assayed (e.g., glucose), and the substrate concentration is measured from an appropriate calibration curve. After each analysis, the collagen membrane is washed several times with 50 mM phosphate buffer, pH 7.5.

PROCEDURE 3. PREPARATION OF COVALENTLY BOUND GLUCOSE OXIDASE VIA ACYL AZIDE DERIVATIVES OF POLYACRYLAMIDE. This glucose oxidase preparation is a modification of one given by Inman and Dintzis (see Ref. 6).

a. **Preparation of polyacrylamide beads.** Dissolve acrylamide (21 g), N,N'-methylenebisacrylamide (0.55 g), tris(hydroxymethyl)aminomethane (Sigma) (13.6 g), N,N,N',N'-tetramethylethylenediamine (Eastman Chemical Co.) (86 μl), and HCl (1 M, 18 ml) in 300 ml of distilled water. Then add ammonium persulfate as a catalyst. [Free radical polymerization is brought about by stirring magnetically under a nitrogen atmosphere and irradiating with a 150-W projector spotlight (Westinghouse).] After polymerization is complete, remove the stirring bar and break the polyacrylamide into large pieces. Prepare small spherical beads of the polymer by blending at high speed with 300 ml water for 5 min. Store the resulting bead suspension in a refrigerator.

b. **Preparation of hydrazide derivative.** Place 300 ml of the polyacrylamide bead suspension in a siliconized three-necked round-bottomed flask. Immerse the flask and a beaker containing anhydrous hydrazine (Matheson, Coleman and Bell) (30 ml) in a 50 ° constant-temperature oil bath. After about 45 min add the hydrazine to the gel. Stopper the flask and stir the mixture magnetically for 12 hr at 50°. Centrifuge the gel and discard the supernatant. Wash the gel with 0.1 M NaCl by stirring magnetically for several minutes, centrifuging, and discarding the supernatant. Repeat the washing procedure until the supernatant is essentially free of hydrazine as indicated by a pale violet color after 5 min when tested by mixing 5 ml with a few drops of a 3% solution of sodium 2,4,6-trinitrobenzene sulfonate (Eastman Chemical Co.) and 1 ml of saturated sodium tetraborate (Mallinckrodt). Store the hydrazide derivative under refrigeration.

c. **Coupling of glucose oxidase.** Place the hydrazide derivative (100 mg) in a plastic centrifuge tube and wash with 0.3 M HCl. Suspend in HCl (0.3 M, 15 ml), cool to 0°, and add sodium nitrite (1 M, 1 ml), also at 0°. After stirring magnetically in an ice bath for 2 min, rapidly wash the azide derivative formed with phosphate buffer (0.1 M, pH 6.8) at 0° by centrifugation and decantation until the pH of a supernatant is close to 6.8. Then suspend the azide gel in phosphate buffer (0.1 M, pH 6.8, 10 ml) containing 100 mg glucose oxidase (Sigma, Type II, from *Aspergillus niger*). Stir the mixture magnetically for 60 min at 0°, after which time add glycine (10 ml, 0.5 M) to couple with unreacted azide groups. After stirring for an additional 60 min at 0°, wash the enzyme gel several times with phosphate buffer (0.1 M, pH 6.8) and store in a refrigerator. The general reaction scheme for this preparation is given by Eq. (6)–(9).

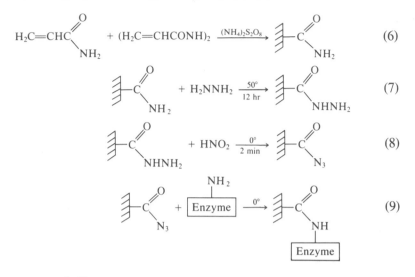

$$\text{(6)}$$

$$\text{(7)}$$

$$\text{(8)}$$

$$\text{(9)}$$

PROCEDURE 4. PREPARATION VIA DIAZONIUM DERIVATIVES OF POLY-ACRYLIC ACID

a. Polymerization of acrylic acid. Dissolve approximately 50 ml of reagent-grade acrylic acid (Aldrich Chemical Co.) in 20 ml hexane and place in a round-bottomed flask. Add a few milligrams of ammonium persulfate as a free radical initiator and keep the system in a dry nitrogen atmosphere. Heat the flask with a heating mantle until rapid polymerization is observed. Then quickly remove the mantle and allow the flask to cool to room temperature.

b. Preparation of copolymer. Break the polymer into small particles and neutralize with sodium hydroxide. Evaporate the sodium salt to dryness in a rotary evaporator and grind to a fine powder. Suspend the powder (~3.6g) in 6 ml hexane and cool to approximately 4°. Convert the acid to the acyl chloride by the addition of $SOCl_2$ (2.8 ml) while stirring in an ice bath for 1 hr (remove generated gases by suction). Wash the acyl chloride polymer with ether and dry under vacuum. Then add nitroaniline (0.5 g) and ether (6 ml) and allow the mixture to be stirred overnight. Filter the product formed, wash with ether, and air-dry.

c. Coupling of glucose oxidase. Dissolve the p-nitroaniline derivative (150 mg) in 10 ml of distilled water and adjust the solution to pH 5 with dilute acetic acid. Add ethylenediamine slowly with stirring until a fine white precipitate is observed. Then wash the precipitate 3 times with distilled water and suspend in 5 ml of distilled water. Reduce the polymer by the addition of TiCl₃ and wash several times with distilled water by centrifugation and decantation. Convert the reduced derivative to the diazonium salt by addition of nitrous acid (0.5 M, 10 ml) at approximately 4° while stirring for 2 min. Flush the diazonium salt intermediate with cold distilled water and rapidly wash several times with phosphate buffer (0.1 M, pH 6.8) by centrifugation and decantation. Then mix with a phosphate buffer solution (0.1 M, pH 6.8) containing 100 mg glucose oxidase (Sigma, Type II) at approximately 4° for 1 hr. Wash the resulting gel several times with buffer and store in a refrigerator. The general reaction scheme for this preparation is given by the Eqs. (10)–(15).

How to Use the Enzyme Electrode

Store the enzyme electrode in buffer solution (type, pH, and optimum concentration dictated by the enzyme system used) in a refrigerator when not in use. With soluble enzyme the electrode should be useful for about 50 assays over 7 days, with physically entrapped enzyme about 100–200 assays over 14–21 days, and with chemically bound enzyme 200–1000 assays over 6–14 months. The dialysis membrane is added (1) to help hold

$$H_2C{=}CHC\overset{O}{\underset{OH}{\diagup}} \xrightarrow[\text{(NH}_4)_2S_2O_8]{80°-110°C} {\Big\}}{-}C\overset{O}{\underset{OH}{\diagup}} \tag{10}$$

$$ {\Big\}}{-}C\overset{O}{\underset{OH}{\diagup}} \xrightarrow{\text{SOCl}_2} {\Big\}}{-}C\overset{O}{\underset{Cl}{\diagup}} \tag{11}$$

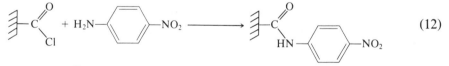

$$ {\Big\}}{-}C\overset{O}{\underset{Cl}{\diagup}} + H_2N{-}\langle\!\!\!\bigcirc\!\!\!\rangle{-}NO_2 \longrightarrow {\Big\}}{-}C\overset{O}{\diagdown}_{HN{-}\langle\!\!\!\bigcirc\!\!\!\rangle{-}NO_2} \tag{12}$$

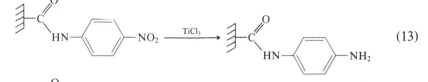

$$ {\Big\}}{-}\overset{O}{\overset{\|}{C}}{\diagdown}_{HN{-}\langle\!\!\!\bigcirc\!\!\!\rangle{-}NO_2} \xrightarrow{\text{TiCl}_3} {\Big\}}{-}\overset{O}{\overset{\|}{C}}{\diagdown}_{HN{-}\langle\!\!\!\bigcirc\!\!\!\rangle{-}NH_2} \tag{13}$$

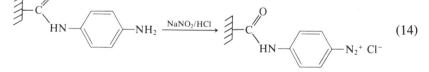

$$ {\Big\}}{-}\overset{O}{\overset{\|}{C}}{\diagdown}_{HN{-}\langle\!\!\!\bigcirc\!\!\!\rangle{-}NH_2} \xrightarrow{\text{NaNO}_2/\text{HCl}} {\Big\}}{-}\overset{O}{\overset{\|}{C}}{\diagdown}_{HN{-}\langle\!\!\!\bigcirc\!\!\!\rangle{-}N_2^+ \; Cl^-} \tag{14}$$

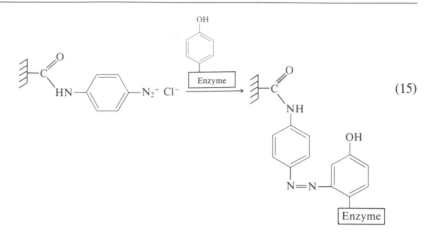

$$(15)$$

the enzyme on the electrode and (2) to keep bacteria out of the enzyme layer.

To use the electrode, couple the enzyme electrode to a digital voltmeter, through the Universal Sensors adaptor if an amperometric probe (e.g., glucose). (See Fig. 1.) Plug the calomel electrode into the digital voltmeter, if the enzyme sensor does not have its own reference electrode built in. Prepare two standard solutions of the substance to be assayed, at concentrations of 10^{-2} and 10^{-4} M in the buffer solution used. First place the enzyme and reference electrodes in the buffer solution to read the background potential, then insert the electrodes in the standard solutions and read the potential. Prepare a calibration plot of potential versus log concentration of substance to be assayed. Then insert the electrode pair in the solution(s) to be assayed, and read the potential. Calculate the concentration of the unknown solution(s) from the calibration curve. When the response of the enzyme electrode has deteriorated remove the enzyme membrane, and replace it with a new membrane.

Commercial Availability of Probes

Immobilized enzymes, together with electrochemical sensors, are used in several instruments available commercially. Owens-Illinois (Kimble) has designed a urea instrument using immobilized urease and an NH_3 electrode probe as well as a glucose instrument using insolubilized glucose oxidase and a Pt electrode. Patent rights to this system have been purchased by Technicon, who markets the instrument in Europe. Yellow Springs Instrument Co. (Ohio) markets a glucose instrument with an immobilized glucose oxidase pad placed on a Pt electrode, and the company has instruments available for triglycerides, lipase, cholesterol, and amylase. Fuji Electric (Tokyo) has a glucose instrument, similar in design to the Yellow Springs Instrument.

Self-contained electrode probes are available from two companies: Tacussel (Lyon, France), which sells a glucose electrode based on the collagen immobilization of Coulet, and Universal Sensors (P.O. Box 736, New Orleans, LA 70148), which offers probes for urea, glucose, creatinine, amino acids, alcohols, and others on request, based on pig intestine immobilization. (See also Table II in [1].)

[3] Coupled Enzyme Reactions in Enzyme Electrodes Using Sequence, Amplification, Competition, and Antiinterference Principles

By FRIEDER W. SCHELLER, REINHARD RENNEBERG, and FLORIAN SCHUBERT

The number of substances which can be directly determined by mono-enzyme electrodes is limited because in many enzyme-catalyzed reactions the cosubstrates involved and the products formed are electrochemically inactive. Therefore readily measurable substances have to be formed in coupled subsequent (sequential) or parallel (competitive) enzyme reactions. Furthermore, the sensitivity of cosubstrate-dependent electrodes is increased when the substrate is shuttled in a cyclic enzyme sequence to give multiple products at the expense of the cosubstrate excess. Interfering substrates, which particularly pose problems when multienzyme systems are used, may be advantageously eliminated in enzymatic antiinterference layers.

Beside the pH and temperature dependence of the functional parameters, in the development of enzyme electrodes the coupling of substrate transport and enzyme-catalyzed reactions has to be characterized at different enzyme loadings by measuring the electrode signal, the effectiveness factor and turnover number, and the degree of substrate conversion.

In this chapter the development of a family of sensors is described based on a well-characterized glucose oxidase membrane and different coupling principles. The sensors are suited for the determination of saccharides, cofactors, and peroxidase substrates as well as for activities of amylases.

Experimental

Preparation of Gelatin-Entrapped Enzyme Layers

All enzyme layers for electrode covering are prepared according to the following procedure: 50 mg of acid photogelatin (VEB Gelatinewerk, Calbe, GDR) is suspended in 0.5 ml doubly distilled water and allowed to swell for 60 min at room temperature. Then 0.5 ml of 50 mM phosphate buffer, pH 6.5, is added; the solution thoroughly mixed and heated in a water bath at 40° for another 60 min. The powder or solution of the respective enzyme (at least 25 U) is added, and the mixture is gently stirred and cast on a plane poly(vinyl chloride) support. The liquid is spread on a total area 5 × 10 cm using a glass rod. The enzyme layer is allowed to dry at room temperature over a period of 6 hr and is then removed from the support. The layer has a thickness of 20–30 μm. It is placed between two sheets of filter paper and stored at 4°.

Preparation of Glutaraldehyde Cross-Linked Enzyme Layers

To a mixture of 500 U disaccharidase (e.g., β-fructofuranosidase, invertase, EC 3.2.1.26) dissolved in 400 μl 20 mM phosphate buffer, pH 6.8, is added 30 μl of 30% bovine albumin solution, followed by 60 μl of 25% glutaraldehyde solution. The solution is stirred for 1 min. This enzyme mixture is homogenously spread on 5 × 10 cm stretched silk. After drying, a piece 2 × 1 cm is cut off and attached by two O rings to the rack under the lid of the measuring cell.

Apparatus

To prepare the enzyme electrode 4 × 4 mm of the enzyme layer is sandwiched between two cellulose membranes 15–30 μm thick (Nephrophan, VEB CK, Bitterfeld, GDR). This layered membrane is held by an O ring on the tip of the electrode holder of the oxygen electrode SMZ (VEB Metra, Radebeul, GDR). One-half milliliter of 0.1 M KCl solution is inserted into the holder to assure electrolytic contact between the Pt indicator electrode (diameter 0.5 mm) and the Ag/AgCl counter electrode. The electrode body is then placed in the holder and the indicator electrode pressed against the enzyme membrane. The complete enzyme electrode is introduced in a horizontal position into the cell of the GLUKOMETER GKM 01 (Zentrum für Wissenschaftlichen Gerätebau der Akademie der Wissenschaften, Berlin, GDR). Two milliliters of the appropriate air-saturated buffer solution is filled into the cell which is equipped with a magnetic stirring bar and thermostated at $25 \pm 0.1°$. The sample volume varies between 10 and 100 μl.

The indicator electrode is poised at a potential of $+600$ mV versus the Ag/AgCl electrode in the H_2O_2 indicator setup or at -600 mV for an oxygen consumption measurement. The dialysis membrane system is also used for oxygen measurement. At -600 mV potential almost no interfering electrode reactions have been found. Either the current–time curves are recorded (stationary measuring principle) or the numerical value of the maximum of the first derivative of the current–time curve is displayed digitally (kinetic measuring principle).

Characterization of the Enzyme Membrane

Recovery of Enzyme Activity

Determination of the intrinsic activity of the immobilized enzyme is complicated since diffusion processes and distribution equilibria may influence the measured (apparent) value. These effects were eliminated by measuring the remaining glucose oxidase (GOD, EC 1.1.3.4) activity after resolubilization of a definite part of the enzyme membrane by heating in 50 mM phosphate buffer, pH 5.5, to $40°$. The initial rate of H_2O_2 formation in the measuring cell after addition of a portion of the resolubilized enzyme membrane containing about 1 U of GOD to the air-saturated 5 mM glucose solution in 50 mM phosphate buffer, pH 5.5, was measured at $37°$. In these experiments the platinum indicator electrode set at $+600$ mV was covered only by two Nephrophan membranes. Its sensitivity was calibrated using H_2O_2 standard solutions. About 90% of the original GOD activity was recovered after the immobilization and resolubilization steps.[1] This very high activity recovered may be ascribed to the mild conditions of the immobilization procedure and the "native milieu" of the gelatin matrix.

Characterization of Reaction–Transport Coupling

In homogeneous solutions the initial rate of substrate conversion rises linearly with increasing concentration of the enzyme. However, because of diffusion the overall rate with immobilized enzymes is constant at high enzyme loadings. Thus only part of the enzyme is used in accelerating the

[1] F. Scheller, D. Pfeiffer, I. Seyer, D. Kirstein, T. Schulmeister, and J. Nentwig, *Bioelectrochem. Bioenerg.* **11**, 155 (1983).

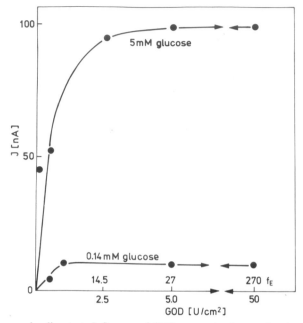

FIG. 1. Enzyme loading test: Influence of GOD concentration on the signal of the GOD sensor at 0.14 and 5 mM glucose. Buffer: 66 mM phosphate buffer, pH 7.0. Polarization voltage: +600 mV.

substrate conversion. The effectiveness factor η reflects the influence of transport processes. It is defined by the ratio of the measured (apparent) activity and the reaction rate in the presence of the same amount of enzyme in homogeneous solution. Using the GOD–gelatin membrane, different parameters characterizing the reaction–transport coupling have been studied.

Enzyme Loading Test. The steady-state current at both 0.14 and 5 mM glucose increases linearly from 0.046 U GOD/cm^2 to about 1 U/cm^2 (Fig. 1). At higher GOD loadings the electrode current reaches a saturation value. Using the following parameters for the GOD–gelatin membrane, the enzyme loading factor f_E was calculated according to Carr and Bowers[2]:

$$f_E = k_2 c L^2 / K_m / D$$

where k_2 is the rate constant, c the enzyme concentration, L the membrane thickness, K_m the Michaelis constant, and D the diffusion coefficient.

[2] P. Carr and L. D. Bowers, *Chem. Anal. (N.Y.)* **56**, 218 (1980).

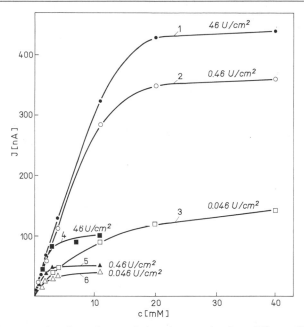

FIG. 2. Concentration dependence of the glucose signal at different GOD loadings. Conditions are as in Fig. 1. Plots 1–3, oxygen-saturated solution; plots 4–6, air-saturated solution.

The parameters are L = 30 μm, k_2 = 320 sec^{-1},[3] and K_M (glucose) = 10^{-2} M. In accordance with theoretical considerations,[4] the transient from the linear region to the saturation proceeds at about f_E = 10 (Fig. 1).

Dependence of Current on Substrate Concentration. When plotting the values of anodic steady-state currents obtained for different substrate concentrations, Michaelis–Menten-type curves are obtained at different enzyme loadings (Fig. 2). The glucose concentration leading to half-maximal current was estimated to be between 1.4 and 1.8 mM when air-saturated. The linear region extends up to 2 mM glucose in the measuring solution, the slope of the linear part being 1.5 μA/mM/cm^2. With oxygen saturation of the solution, both linearity and sensitivity are considerably extended.

The linear dependence of the reciprocal of the current plotted against the reciprocal of the bulk glucose concentration (Fig. 3) at low enzyme loading gives evidence that the current is controlled by the enzyme reac-

[3] W. Kühn, D. Kirstein, and P. Mohr, *Acta Biol. Med. Germ.* **39,** 1121 (1980).
[4] D. A. Gough and J. K. Leypoldt, *Appl. Biochem. Bioeng.* **3,** 181 (1981).

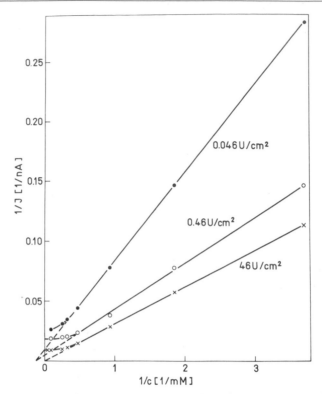

FIG. 3. Electrochemical Lineweaver–Burk plot of steady-state currents at different enzyme loadings. Conditions are as in Fig. 1.

tion. The small curvature at extremely high glucose concentrations is due to limitation by consumption of the cosubstrate, oxygen. The apparent K_m value of glucose (when air-saturated) determined by these "electrochemical Lineweaver–Burk plots" is 7.5 mM. The corresponding value of soluble GOD was determined to be 3.8 mM. At high enzyme loading the electrochemical Lineweaver–Burk plot is curved, and extrapolation passes through the origin. It should be noted that the Lineweaver–Burk plot of the *kinetic* signal (di/dt) yields a considerably higher K_m value, i.e., 22 mM, as expected for this measuring system.[5]

Effectiveness Factor. In order to characterize the apparent activity of the GOD membrane, the initial rate of H_2O_2 accumulation or the rate of

[5] G. G. Guilbault and G. J. Lubrano, *Anal. Chim. Acta* **64,** 439 (1973).

oxygen or glucose consumption in the bulk solution is measured.[1] For this purpose a double measuring cell containing the glucose-indicating enzyme electrode and a nonenzymatic H_2O_2- or O_2-indicating Pt electrode is used. As it is less influenced by H_2O_2 destruction, the O_2 consumption mode seems to be more accurate than measuring of H_2O_2 accumulation.[6] In these experiments a closed measuring cell completely filled with air-saturated 66 mM phosphate buffer, pH 7.0, is used. After obtaining a constant baseline of the oxygen electrode, 5 mM glucose is introduced into the stirred solution. The initial rate of oxygen consumption by glucose conversion catalyzed by the part of the GOD membrane contacting the glucose solution (0.14 cm^2) is measured.

The initial rate of oxygen consumption (and also of glucose conversion) in the solution at saturating glucose concentrations is 14 nmol/min at 46 U GOD/cm^2. This value is below 1% of the biocatalytic activity immobilized in the membrane. This result underlines the limitation set by the internal diffusion. On the other hand, the activity of the GOD membrane containing 0.046 U/cm^2 amounts to about 70% of the original value, which is very near kinetic control.

The relatively low apparent GOD activity is mainly caused by the shielding effect of the external dialysis membrane. Its diffusional resistance for glucose is almost 10 times higher than that for the cosubstrate, oxygen. Nevertheless, the apparent activity of this GOD sandwich membrane with 0.1 U/cm^2 at 25° compares well with GOD membranes prepared by covalent enzyme fixation. The membrane prepared by Tsuchida and Yoda,[7] who used GOD bound in pores of an asymmetric cellulose acetate layer, has an apparent activity of 0.34 U/cm^2 at 37°. The GOD-collagen membrane developed by Thevenot et al.[8] contains an apparent activity of 0.06 U/cm^2 at 30°.

On the other hand, the low apparent activity of the GOD membrane causes only very small glucose consumption during measurement. Within 10 min about 2% of the bulk glucose is converted by the GOD membrane. This amount can be neglected during the kinetic measuring time of only 10 sec. Using an enzyme concentration of 46 U/cm^2 in the measurement of a 0.14 mM glucose sample, which corresponds (at 1:40 dilution) to the normal value of blood glucose, each GOD molecule converts one glucose molecule during a measuring time of 10 sec.

[6] P. R. Coulet, R. Sternberg, and D. Thevenot, *Biochim. Biophys. Acta* **612,** 317 (1980).
[7] T. Tsuchida and K. Yoda, *Enzyme Microb. Technol.* **3,** 326 (1981).
[8] D. R. Thevenot, R. Sternberg, P. R. Coulet, J. Laurent, and D. C. Gautheron, *Anal. Chem.* **51,** 96 (1979).

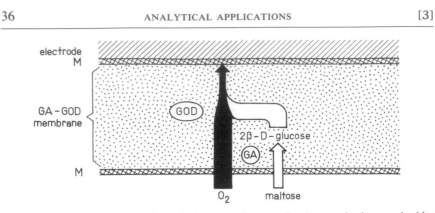

FIG. 4. Enzyme sequence electrode for determination of maltose and other saccharides using a glucoamylase–GOD membrane.

Linear Enzyme Sequences

Glucoamylase–GOD Electrode

One milligram glucoamylase (GA, glucan 1,4-α-glucosidase, EC 3.2.1.3) (50 U) is coimmobilized with 0.5 mg GOD (23 U) per cm^2 in gelatin according to the procedure described in the section "Experimental." GA splits off β-D-glucose units from saccharides; therefore the GOD can use the glucose formed directly without spontaneous or enzymatic mutarotation (Fig. 4). The respective enzyme electrode responds to glucose, maltose, and dextrins, and it indicates the sum of all these substrates. The sensitivity for maltose in comparison to glucose is about 50%. This decreased signal might be caused by the approximately doubled diffusional resistance of the membrane to maltose owing to its larger molecular cross section. The linear range extends up to 2 mM maltose.

The endohydrolase, α-amylase (EC 3.2.1.1), splits starch into maltose via dextrin fragments. These products are indicated by the glucoamylase–GOD electrode. For determining the α-amylase activity in serum, 2 ml of 5% soluble starch in 50 mM phosphate buffer, pH 6.0, is pipetted into the measuring cell and equilibrated at 37°. When a constant baseline is reached, the reaction is started by addition of 100 μl of serum. After a rapid initial current increase corresponding to the endogenous glucose, the current continues to rise, albeit slower. The slope of this part of the current–time curve (1 min after sample addition) depends linearly on the activity of α-amylase.[9] The activity of glucoamylase can be directly deter-

[9] D. Pfeiffer, F. Scheller, M. Jänchen, K. Bertermann, and H. Weise, *Anal. Lett.* **13,** 1179 (1980).

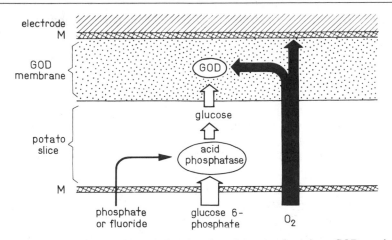

FIG. 5. Enzyme sequence electrode for phosphate determination using a GOD membrane and a slice of potato tissue.

mined with the GOD electrode by measuring the β-D-glucose split off from soluble starch in the measuring solution.[9]

Potato Slice–GOD Hybrid Electrode

For the potato slice–GOD hybrid sensor a slice of about 0.1 mm thickness is cut with a razor blade from potato (*Solanum tuberosum*) tissue and placed directly on the GOD membrane (Fig. 5). Based on the sequential action of potato acid phosphatase (EC 3.1.3.2) and GOD, this sensor responds to glucose 6-phosphate. Since the response is controlled by the acid phosphatase reaction,[10] the sensor is applicable to the determination of the phosphatase inhibitor, inorganic phosphate. To obtain a linear phosphate concentration dependence, the phosphate-containing sample is added to the measuring solution (0.1 M citric acid–sodium citrate buffer, pH 6.0) prior to injection of glucose 6-phosphate (0.37 mM). The reciprocal of the kinetic electrode signal is then proportional to inhibitor concentration between 0.025 and 1.5 mM.

GOD–Peroxidase Electrode

A piece of an enzyme layer containing coimmobilized horseradish peroxidase (HRP, EC 1.11.1.7) (32 U/cm²) and GOD (46 U/cm²) is sandwiched between two dialysis membranes. The sensor is applied to the

[10] F. Schubert, L. Kirstein, R. Renneberg, and F. Scheller, *Anal. Chem.* **56,** 1677 (1984).

determination of the substrate for HRP, bilirubin. The reactions proceed as follows:

$$Glucose + O_2 \xrightarrow{GOD} gluconolactone + H_2O_2$$

$$2\ H_2O_2 + bilirubin \xrightarrow{HRP} 2\ H_2O + product$$

In contrast to the glucoamylase–GOD and potato slice–GOD electrodes, GOD is not used for glucose indication in this case but for generation of H_2O_2.[11] At 10 mM, glucose in the background solution (0.1 M phosphate buffer, pH 7.4) causes the complete conversion of dissolved oxygen in the enzyme layer to H_2O_2. Therefore the anodic current at +600 mV is not influenced by addition of a sample, e.g., serum. When the anodic current reaches a constant level, 50 μl of the bilirubin-containing sample solution is added to the measuring cell. Part of the hydrogen peroxide formed in the GOD reaction inside the membrane is consumed by the HRP-catalyzed conversion of bilirubin, and the anodic current is decreased depending on the bilirubin concentration. The slope of the calibration curve is 0.8 μA/mM/cm^2 between 0.005 and 0.05 mM bilirubin.

Invertase–GOD Reactor Electrode

For successive measurements of glucose and disaccharides, the GOD electrode described may be combined with enzymatic disaccharide hydrolysis catalyzed by an immobilized disaccharidase in the measuring cell. For sucrose determination, invertase is used. The enzyme is fixed on silk which is then installed on a rack mounted under the lid of the measuring cell (Fig. 6). Before the immobilized invertase is immersed in the measuring solution, the endogenous glucose concentration of the sample is determined. Addition of the sample causes a current increase of the GOD-covered electrode which is completed after 30 sec. At this time the immobilized invertase is inserted in the stirred solution, catalyzing the splitting of sucrose into D-fructose and α-D-glucose which spontaneously mutarotates to form β-D-glucose. The rate of β-D-glucose formation depends linearly on the sucrose concentration in the measuring cell up to 10 mM. Glucose concentrations up to 0.5 mM do not interfere.[12]

This principle of successive determination can be extended to several substances in one and the same sample by applying other immobilized disaccharidases. The procedure combines the advantages of enzyme

[11] R. Renneberg, D. Pfeiffer, F. Scheller, and M. Jänchen, *Anal. Chim. Acta* **134,** 359 (1982).
[12] F. Scheller and C. Karsten, *Anal. Chim. Acta* **155,** 29 (1983).

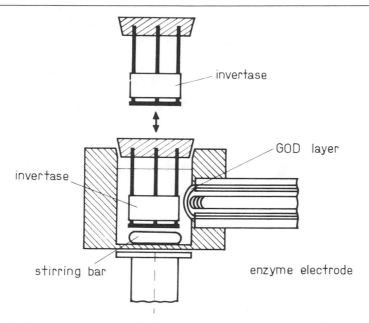

Fig. 6. Enzyme reactor electrode for measurement of sucrose and glucose using an immobilized invertase reactor and a GOD electrode.

membrane electrodes with those of sequential reactors, i.e., short response time for glucose measurement, simple setup of the stirred measuring cell, and subsequent interference-free sucrose determination.

Substrate Amplification

If very low substrate concentrations are to be determined, the sensitivity of enzyme electrodes can be enhanced by using enzymatic substrate amplification. A dehydrogenase and an oxidase, both acting on the same substrate, may be used for this purpose. Operational conditions have to be adjusted in such a way that the dehydrogenase catalyzes the regeneration of the oxidase substrate.

The substrate is continuously shuttled between the two enzymes. As a consequence, the electrochemically active product is formed in much higher amount than that of substrate diffusing into the membrane. Therefore the current exceeds that of a diffusion-controlled process. Substrate amplification is exemplified by the GOD–glucose dehydrogenase (GDH, EC 1.1.1.47) system. The gluconolactone formed in the GOD-catalyzed

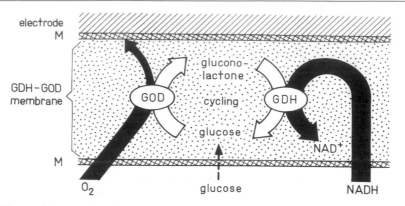

FIG. 7. Substrate amplification principle demonstrated by a GOD–GDH electrode for measurement of glucose.

reaction is reconverted to glucose in the GDH-catalyzed reaction with NADH (Fig. 7).

Twenty units GOD and 10 U GDH/cm² of the gelatin layer are coimmobilized. The glucose measurements are performed in 0.1 M phosphate buffer, pH 7.0, containing variable amounts of NADH (Fig. 8). The electrode is polarized to −600 mV. The glucose sample is added after attaining a constant baseline. An amplification factor of 8 is obtained in the solution containing 10 mM cofactor. The limit of detection is 0.8 μM glucose.

Amplification makes use of a part of the GOD excess which is unused in the absence of NADH. The overall reaction represents the oxidation of NADH by oxygen, with glucose acting as the mediator between the two enzymes.

Enzyme Competition

The problem of measuring cofactor concentrations with electrochemical sensors may be solved if a cofactor-dependent enzyme is coupled with an oxidase, with both competing for the same substrate. The competition of GOD and GDH for glucose was used to develop a bienzyme electrode for determination of the oxidized cofactor, NAD⁺ (Fig. 9). The reaction between glucose and NAD⁺ is characterized by the equilibrium constant, $K = 3 \times 10^7\ M$.[13] The reaction proceeds in both directions depending on the concentrations of the partners. The above-described GOD–GDH

[13] C. C. Liu and A. K. Chen, *Process Biochem.* **Sept./Oct.,** p. 12 (1982).

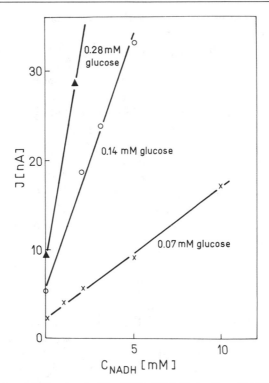

Fig. 8. Dependence of the signal of the GOD–GDH electrode on NADH concentration at different glucose concentrations. Conditions are as in Fig. 1.

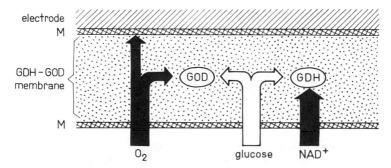

Fig. 9. Enzyme competition electrode for measurement of NAD$^+$ using a GOD–GDH membrane.

membrane is used in combination with electrochemical measurement of oxygen consumption at -600 mV.

Glucose, 1 mM in 0.1 M phosphate buffer, pH 7.0, is used as the background solution. Glucose conversion by GOD causes a decrease in the current. The addition of the NAD^+-containing sample switches on GDH, which competes with GOD for the common substrate, glucose. The degree of the NAD^+-dependent reaction is reflected by a current increase based on the diminished oxygen consumption in the GOD reaction. The calibration curve for NAD^+ is linear up to a final concentration of 1 mM NAD^+; however, the sensitivity is only about 3% as compared with that of glucose, i.e., 0.045 μA/mM/cm^2. Furthermore the functional stability is limited to 3 days of operation.

An enzyme electrode based on the competition between GOD and hexokinase has also been developed, measuring ATP in the concentration range 0.05–1 mM.[9]

Antiinterference Layer

Interference caused by different substances (substrates, inhibitors, activators) is a serious problem for the practical application of enzyme sensors. For coupled enzyme reactions the substrate of each particular enzyme may interfere if present in the sample. These interfering substrates can be eliminated by covering the enzyme sensor with an antiinterference layer. This layer contains immobilized enzymes which convert the disturbing substances to noninterfering products.

In all enzyme electrodes using GOD, endogenous glucose from biological or food samples will interfere. This problem is overcome by converting the glucose by GOD and catalase (EC 1.11.1.6) to the electrode-inactive products, gluconolactone and water. The antiinterference layer and the GOD-based enzyme membrane are two spatially separated linear enzyme sequences (Fig. 10). For the purpose of disaccharide measurement in the presence of glucose, the bienzyme layer (e.g., glucoamylase–GOD or invertase–GOD) is covered with an antiinterference layer [containing GOD (46 U/cm^2) and catalase (320 U/cm^2)], in which β-D-glucose (up to 2 mM glucose in the measuring solution) is completely eliminated.[14] Sucrose, maltose, starch, and α-amylase[15] are directly measurable in the presence of up to 2 mM glucose in juice of sugar beets or in samples of

[14] F. Scheller and R. Renneberg, *Anal. Chim. Acta* **152,** 265 (1983).
[15] R. Renneberg, F. Scheller, K. Riedel, E. Litschko, and M. Richter, *Anal. Lett.* **16,** 877 (1983).

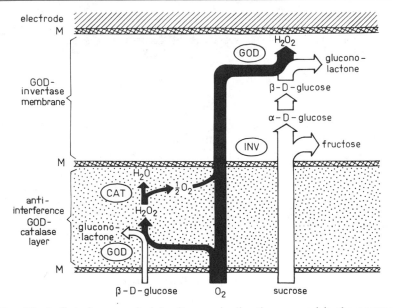

FIG. 10. Antiinterference principle. Sucrose is directly measured in the presence of glucose.

instant cocoa. The linear measuring range for sucrose extends up to 15 mM with a slope of 0.15 μA/mM/cm^2.

Concluding Remarks

Coupled enzyme reactions obviously simulate the metabolic situation in organelles and microorganisms[16] which are frequently employed in biospecific electrodes.[17] However, the application of mixtures of isolated enzymes offers several advantages. Because of the absence of contaminant enzyme activities, the selectivity of enzyme sensors is higher than that of organelle or microbial electrodes. Owing to the higher specific enzyme activity applicable in multienzyme electrodes, these are also superior with regard to sensitivity and response time.

[16] P. A. Srere, B. Mattiasson, and K. Mosbach, *Proc. Natl. Acad. Sci. U.S.A.* **70,** 2534 (1973).

[17] M. Hikuma and T. Yasuda [10], I. Karube [11], and F. Schubert and F. W. Scheller [13], this volume.

[4] Industrial Applications of Enzyme Electrodes Including Instrumentation

By J. L. Romette and D. Thomas

Perhaps the most interesting application of immobilized enzymes has been their use as the active element of an electrochemical probe or sensor. The analytical principle of the enzyme electrode and its potential use as a tool for biotechnological analysis have been discussed.[1]

Enzyme electrodes, which generally consist of an electrochemical sensor with an immobilized enzyme in close contact, have several advantages over other methods of analysis: (1) Immobilization generally stabilizes the enzyme, enables repeated measurements, and offers the possibility of continuous analysis which is a major advantage in any process control. (2) Classical analytical procedures require the precipitation or dialytic separation of proteins. In contrast, electrochemical monitoring with enzyme electrodes can be carried out on whole fermentation juice or biological media, thus eliminating preparation of the sample. The use of gas-sensing membrane electrodes has led to the development of better enzyme electrodes[2,3] which are free of interference from ions, proteins, and electrochemically active compounds.

The use of an enzyme as a functional element of an electrochemical device was first reported by Clark and Lyons.[4] However, the earliest electrode which incorporated an enzyme was described by Updike and Hicks[5] and was followed by extensive work of Guilbault and co-workers.[6] A review of the literature shows numerous references to enzyme electrode production, but only a few enzyme electrodes are commercially available.

In fact, a major problem is the application of theoretical procedures to practical measurements. For this important reason, experiments have been focused on the use of oxidase enzymes which are very specific and which do not need any cofactor added to the sample media. However, in this case another difficulty exists, namely, the oxygen content of the

[1] M. M. Fishman, *Anal. Chem.* **52,** 185 (1980).
[2] C. L. Di Paolantonia, M. A. Arnold, and G. A. Rechnitz, *Anal. Chim. Acta* **128,** 121 (1981).
[3] R. R. Walters, P. A. Johnson, and R. P. Buck, *Anal. Chem.* **52,** 1684 (1980).
[4] L. C. Clark and C. Lyons, *Ann. N.Y. Acad. Sci.* **102,** 29 (1962).
[5] S. J. Updike and J. P. Hicks, *Nature (London)* **214,** 986 (1967).
[6] G. G. Guilbault, this series, Vol. 44, p. 579.

sample. Measurements in samples exhibiting high variability in oxygen content present an important problem to solve. Different solutions exist, all of which are based on the stabilization of the sample pO_2 before or during the measurement. For instance, commercially available glucose sensor equipment stabilizes the pO_2 of the sample by bubbling air through a vibrating semipermeable membrane.

Theoretical studies[7] have shown that when an enzyme is immobilized inside a matrix, the rates of diffusion and reaction are quite different than those in solution, mainly as an effect of levels of all parameters which modify enzyme activity. So it is not evident that, even under conditions of aeration, the reaction rate becomes independent of the amount of O_2 available inside the support.

An example of such behavior is given for a glucose oxidase (GOD, EC 1.1.3.4) membrane in the case of a glucose sensor application. This type of enzyme is not classically Michaelien but exhibits a ping-pong mechanism. So there is not one Michaelis constant (K_m) but two, one for the main substrate, the other for the oxygen. The reaction rate is defined in Eq. (1):

$$\beta\text{-D-Glucose} + O_2 \xrightarrow{\text{glucose oxidase}} \text{gluconic acid} + H_2O_2$$

$$v = V_{max} \frac{1}{1 + K_S/S + K_A/A} \tag{1}$$

where S and A denote glucose and oxygen concentrations at the enzyme level, K_S and K_A the Michaelis constants for each substrate, and V_{max} the maximal velocity.

We have tried to solve the O_2 problem by using an enzyme membrane in which oxygen solubility is far higher than that in aqueous solution.[8] In this way, the oxygen used during measurement is the oxygen dissolved in the membrane, not the oxygen present in the sample.

Apparatus

Figure 1 illustrates one version of the laboratory prototype used during the development of this technology. An Apple II Plus computer system was applied to the automation of the enzyme electrode. The system comprises an Apple II microcomputer (Apple Computer, Inc., Cupertino, CA) with 48K of main memory. The internal peripheral bus of the computer enables the addition of a variety of peripheral units. We have added

[7] D. Thomas and G. Broun, this series, Vol. 44, p. 901.
[8] J. C. Quenesson and D. Thomas, French Patent 7,715,616 (1977).

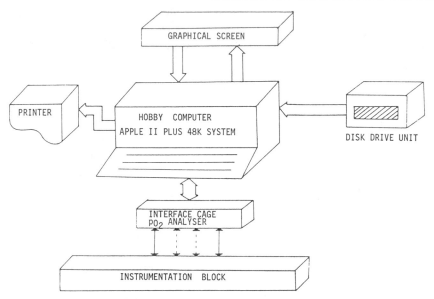

FIG. 1. Schematic diagram of the laboratory prototype.

a disk controller card and parallel interface cards (Apple Computer, Inc.) for the printer and the interface card cage. The interface card cage (U.T.C. Electronic Department, Compiègne, France) is mounted to the computer through a ribbon cable and contains circuit boards for signal conditioning and buffering, an analog to a digital card based on an Analo-

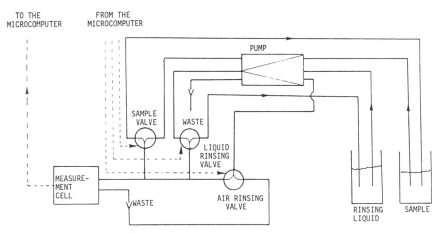

FIG. 2. Flow block system of the laboratory prototype.

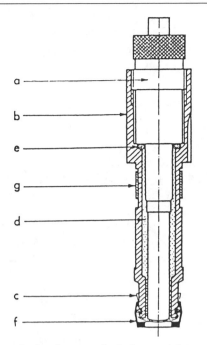

FIG. 3. pO_2 sensor. a, Body of sensor; b, jacket; e, joint; g, joint; d, electrolyte; c, selective film; f, joint of the measurement cell.

gic/Digital converter Datel EK 12 B (Datel Intersil, Mansfield, OH), a pO_2 analyzer, and four logic outputs for valve driving.

The flow block system (Fig. 2), connected to the pO_2 analyzer, is composed of a measurement cell (Fig. 5) connected to the pump through electromagnetic valves which control the circuit. The pO_2 sensor (Figs. 3 and 4) is a Radiometer electrode model E.5046 (Radiometer, Copenhagen,

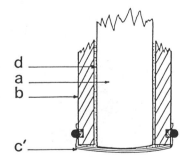

FIG. 4. Detail of the enzyme electrode tip. a, Body of sensor; b, jacket; d, electrolyte; c, active-selective bilayer.

Denmark). The selective gas membrane is of polypropylene, 6 μm thick, produced by Bollore Manufactory (Bollore Inc., Paris, France). The temperature is controlled by a water bath, Haake model FE (Haake Company, Karlsruhe, FRG) set at 25°.

Electrode Preparation

The sensor in the electrode is a Clark electrode used for amperometric measurement of pO_2. The selective hydrophobic film is a polypropylene membrane on which the active layer containing the enzyme is spread (Fig. 4). The inert protein solution is prepared from a 250 bloom gelatin from bone (Rousselot Company, Paris, France), solubilized as a 5% w/w solution in 50 mM phosphate buffer, pH 7.5, at 50°.

After solubilization, the required amount of enzyme (Table I) is added to the gelatin solution after the temperature has decreased to 30°. One milliliter of the gelatin–enzyme solution is poured on 35 cm^2 of the poly-

TABLE I
AMOUNT OF ENZYME INCORPORATED INTO THE ACTIVE LAYER OF THE ELECTRODE

Parameter	Enzyme	Amount incorporated[a]	Source of enzyme
Glucose	GOD	10 IU	Grade II from *Aspergillus niger* (Sigma, St. Louis, MO)
Sucrose	β-Fructofuranosidase (invertase)	50 IU	Grade VII from bakers yeast (Sigma, St. Louis, MO)
	Aldose 1-epimerase (mutarotase)	150 IU	Porcine kidney (Sigma, St. Louis, MO)
	GOD	50 IU	Grade II from *A. niger* (Sigma, St. Louis, MO)
Lactose	Lactase	7 IU	*Saccharomyces fragilis* (Sigma, St. Louis, MO)
	GOD	10 IU	Grade II from *A. niger* (Sigma, St. Louis, MO)
Ethanol	Ethanol oxidase	6 IU	*Hansenula polymorpha* (LTE, Compiègne, France)
L-Lysine	L-Lysine oxidase	8 IU	*Trichoderma viride* (Yamasa Shoyu Choshi, Chiba, Japan)
L-Lactate	Lactate oxidase	12 IU	*Mycobacterium smegmatis* (Sigma, St. Louis, MO)
D-Lactate	*Escherichia coli* K12 strain 3300	50 mg/ml	*E. coli* (LTE, Compiégne, France)
Phenol	Laccase	0.9 AU	*Botrytis cinerea* (LTE, Compiégne, France)

[a] Per 35 cm^2 of active film.

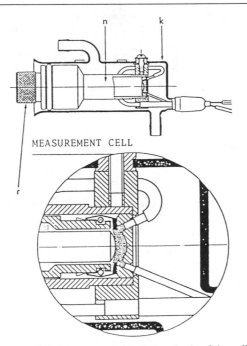

Fig. 5. Measurement cell. k, Thermostated jacket; n, body of the cell; r, screw for fixing the enzyme electrode inside the measurement cell.

propylene film fixed on a glass plate. The solution spread on the polypropylene is dried at 25° for 1 hr under air flow and then immersed in 2.5% v/v glutaraldehyde in 50 mM phosphate buffer, pH 6.8, for 3 min in order to insolubilize the matrix. After rinsing, the selective active composite membrane is attached to the jacket of the pO_2 sensor with an O ring (Fig. 4). Then the enzyme electrode is introduced in the measurement cell (Fig. 5).

Electrode Measurements

The electrode is exposed sequentially to air, to a sample containing the substrate, and finally to a buffer. In order to feed the enzyme support with oxygen quickly, a rinsing step with air is performed. The concentration of oxygen in the active layer is roughly 20 times higher than in water for the same partial pressure as shown in the tabulation below.[8] During the measurement step, the oxygen inside the enzyme support is consumed. Rinsing with buffer removes the remaining substrate from the membrane by diffusion.

Medium	Oxygen content (mM)
Substrate solution	0.26
Albumin support	0.26
Gelatin support	6.5

Computerized Electrode[9]

Under software control (Fig. 6), the air valve is opened until a plateau is obtained for the electrode signal. Then the air valve is shut and the sample valve opened. Within seconds after the admission of sample to the mesurement cell, the electrode signal is sampled. The liquid rinsing step is started at a given time after the beginning of the measurement. After a given time, air rinsing is resumed, and so on. During the O_2 consumption step, the electrode signal is sampled by the microcomputer (30 measurement data points during 9 sec).

The dotted line electrode signal $z(t)$ behaves as arc AB in Fig. 6, and is well approximated by a third-order polynomial:

$$z(t) = \alpha_0 + \alpha_1 t + \alpha_2 t^2 + \alpha_3 t^3 \tag{2}$$

The coefficients α_0, α_1, α_2, and α_3 are obtained by a least squares method that involves minimizing the function

$$J(\alpha_0, \alpha_1, \alpha_2, \alpha_3) = \frac{1}{2} \sum_{k=1}^{M} (z_k - \alpha_0 - \alpha_1 k - \alpha_2 k^2 - \alpha_3 k^3)^2 \tag{3}$$

where z_k is the measured value at time $t = k$ ($k = 1, 2, \ldots, M$). The necessary conditions

$$\partial J/\partial \alpha_i = 0, \quad i = 0, 1, 2, 3$$

can be written as $A\alpha = b$, where $\alpha = (\alpha_0 \alpha_1 \alpha_2 \alpha_3)^t$ and the coefficients a_{ij} of A and b_i of b are given by

$$a_{ij} = \sum_{k=1}^{M} k^{i+j} \quad \text{and} \quad b_i = \sum_{k=1}^{M} z_k k^i \quad (0 < i, j < 3)$$

This fourth-order symmetric linear system is easily solved by a direct method. Once the coefficients α_i are obtained, the inflection point and the slope at this point are determined by $z''(t) = 0$, where

$$t = -\alpha_2/(3\alpha_3) \quad \text{and} \quad z'(t) = \alpha_1 - \alpha_2^2/(3\alpha_3)$$

[9] J. P. Kernevez, L. Konate, and J. L. Romette, *Biotechnol. Bioeng.* **25**, 845 (1983).

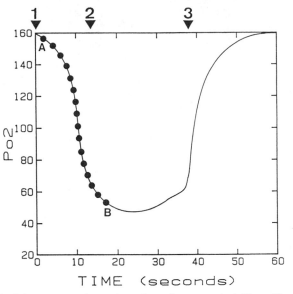

FIG. 6. Signal of the enzyme electrode under software control. From Kernevez *et al.*[9]

It is important to remark that, as numerical and experimental studies have shown, this slope depends only on the sample (substrate) concentration.

Calibration of the Enzyme Electrode

Given $(m + 1)$ samples with known substrate concentrations y_0, y_1, . . . , y_m, we determine the corresponding slopes x_0, x_1, . . . , x_m by the above procedure and obtain a calibration curve as shown in Fig. 7. This curve $y = f(x)$ is well approximated by cubic spline functions, defined as follows. In interval (x_{i-1}, x_1), $i = 1, 2, . . . , m$, $f(x)$ is given by the third degree polynomial

$$f(x) = ty_i + (1 - t)y_{i-1} + h_i t(1 - t) [(k_{i-1} - d_i)(1 - t) - (k_i - d_i)t]$$

where

$$t = (x - x_{i-1})/h_i, \qquad h_i = x_i - x_{i-1}, \qquad d_i = (y_i - y_{i-1})/h_i$$

and k_0, k_1, . . . , k_m satisfies the linear tridiagonal system of equations:

$$h_{i+1}k_{i-1} + 2(h_i + h_{i+1})k_i + h_i(k_{i+1}) = 3(h_i d_{i+1} + h_{i+1}d_i)$$
$$(i = 1, 2, . . . , m - 1)$$

and the two additional conditions:

$$2k_0 + k_1 = 3d_1 \qquad \text{and} \qquad k_{m-1} + 2k_m = 3d_m$$

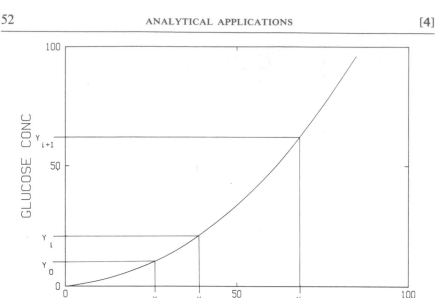

FIG. 7. Theoretical calibration curve, e.g., glucose calibration.

Results

Glucose Determination[10]

Figure 8 gives the calibration curve obtained for glucose.

$$\beta\text{-D-Glucose} + O_2 \xrightarrow{\text{GOD}} \text{gluconic acid} + H_2O_2$$

Lactose and Sucrose Determinations[10]

Figure 9 gives the calibration curve obtained for lactose.

$$\text{Lactose} \xrightarrow{\text{lactase}} \text{galactose} + \beta\text{-D-glucose}$$

$$\beta\text{-D-Glucose} + O_2 \xrightarrow{\text{GOD}} \text{gluconic acid} + H_2O_2$$

Figure 10 gives the calibration curve obtained for sucrose.

$$\text{Sucrose} \xrightarrow{\text{invertase}} \alpha\text{-D-glucose} + \text{fructose}$$

$$\alpha\text{-D-Glucose} \xrightarrow{\text{mutarotase}} \beta\text{-D-glucose}$$

$$\beta\text{-D-Glucose} + O_2 \xrightarrow{\text{GOD}} \text{gluconic acid} + H_2O_2$$

[10] J. L. Romette, Ph.D. thesis. University of Compiègne, Compiègne, France (1980).

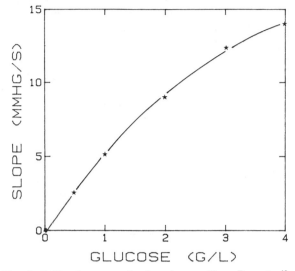

FIG. 8. Calibration curve for β-D-glucose. From Romette.[10]

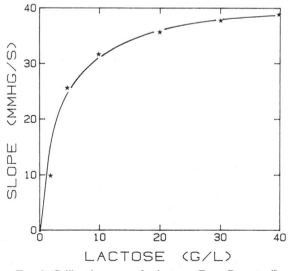

FIG. 9. Calibration curve for lactose. From Romette.[10]

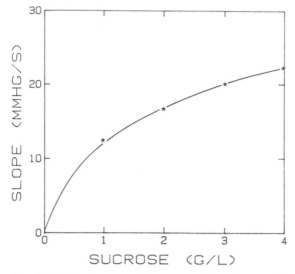

FIG. 10. Calibration curve for sucrose. From Romette.[10]

Ethanol Determination[11]

The calibration curve for ethanol is given in Fig. 11. The enzyme which reacts is the methanol/ethanol oxidase (alcohol oxidase, EC 1.1.3.13):

$$\text{Ethanol} + O_2 \xrightarrow{\text{ethanol oxidase}} \text{acetaldehyde} + H_2O_2$$

L-Lysine Determination[12]

Figure 12 gives the calibration curve for L-lysine. The enzyme reaction is

$$\text{L-Lysine} + O_2 + H_2O \xrightarrow{\text{lysine oxidase}} 2\ \text{oxo-}\varepsilon\text{-aminocaproate} + NH_3 + H_2O_2$$

L-Lactate Determination[13]

For the determination of L-lactate, the active element of the sensor is composed of bacteria, *Escherichia coli* K12 (Strain 3300), immobilized

[11] H. Belghith and J. L. Romette, "Determination of Ethanol by Oxidase Enzyme Electrode." Paper presented at the International Symposium on Analytical Methods and Problems in Biotechnology, Delft, The Netherlands, April 17–19, 1984.

[12] J. L. Romette, J. S. Yang, H. Kusakabe, and D. Thomas, *Biotechnol. Bioeng.* **25,** 2556 (1983).

[13] C. Burstein, J. L. Romette, E. Adamowicz, K. Boucherit, and C. Rabouille, French Patent 8,500,728 (1985).

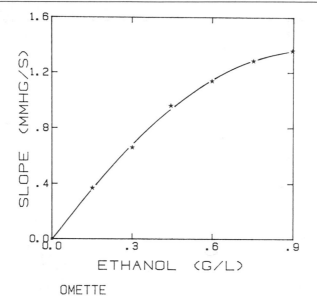

OMETTE

FIG. 11. Calibration curve for ethanol. From Belghith and Romette.[11]

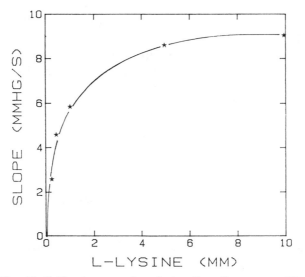

FIG. 12. Calibration curve for L-lysine. From Romette *et al.*[12]

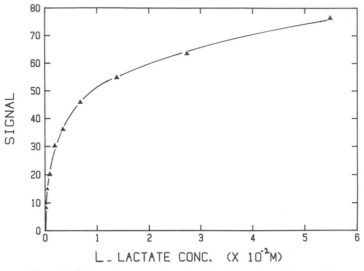

FIG. 13. Calibration curve for L-lactate. From Burstein *et al.*[13]

inside the gelatin support. In this case whole cells were used, and the respiratory chain was involved in the O_2 consumption as described.

$$\text{L-Lactate} \xrightarrow{\text{L-lactate ''oxidase''}} \text{flavoprotein}$$
$$\rightarrow \text{coenzyme Q} \rightarrow \text{cytochrome } b \rightarrow \text{cytochrome } o \rightarrow O_2$$

The specificity of the system was obtained by growing the *E. coli* on a medium inducing the flavoprotein specific for L-lactate. Figure 13 gives the calibration curve for L-lactate. Other molecules can be detected such as L-alanine, ornithine, succinate, NADH, NADPH, choline, phenol, glutamate, D-lactate, L-malate, pyruvate, and fumarate.

Analytical Characteristics[14]

Oxygen Dependence

The gelatin support acts as an O_2 supplier for the enzymatic reaction. So the signal of the electrode is correlated only to the main substrate concentration. An example of such behavior is given in Fig. 14. The test is performed with a glucose sensor. The same solution of glucose (2 g/liter) is introduced inside the measurement cell, but with three different pO_2

[14] J. L. Romette, B. Fromet, and D. Thomas, *Clin. Chim. Acta* **95,** 249 (1979).

ASSAY N		1	2	3	4	5	MV
0	mmHG	25.6	24.4	26	24.8	24.9	25.1
160	mmHG	24.8	22.4	25	23	24.6	24
760	mmHG	26	25.6	25.2	23	24.9	24.9

FIG. 14. Variability of the sensor response under different conditions of pO_2, e.g., glucose measurement. From Romette et al.[14]

conditions (nitrogen bubbling, air bubbling, and pure oxygen bubbling). The variability of the response under these conditions is less than 4%.

Stability

An important characteristic of such sensors is their stability, which must be studied under conditions of both storage and operation. When stored at 5° in buffer, the enzyme membrane remains stable for at least 1 year. The operational stability was examined by repeatedly measuring the electrode response (measurement every 2 min). The results of the test realized are given in Fig. 15.

The enzyme seems to be inactivated by the reaction. One molecule of enzyme can transform only a limited number of molecules of substrate.

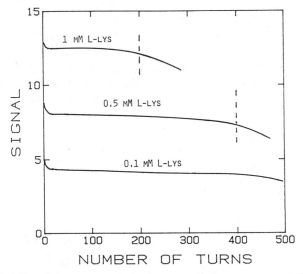

FIG. 15. Stability of the enzyme electrode, e.g., L-lysine measurement. From Romette et al.[12]

With 1 mM lysine, 200 measurements can be done; with 0.5 mM lysine, 400 measurements can be performed. This type of inactivation is very similar to that observed for the glucose oxidase enzyme reaction.[15] It seems to be a common feature of FAD oxidase enzymes. When used in dynamic state response, 3000 measurements were performed successfully for L-lysine and 5000 for β-D-glucose.

Selectivity

One of the main advantages of this technology is the selectivity of the method, given by the use of enzymes. The selectivity spectrum differs from one enzyme to the other. An example is given in Fig. 16 for the L-lysine electrode. For possible application of this electrode, interfering compounds have been reduced to three amino acids, L-arginine, L-phenylalanine, and L-ornithine. The selectivity of the electrode, at the electrochemical level, is absolute for the gases because of the hydrophobic selective gas membrane, in contrast to other electrochemical sensors such as those for H_2O_2.

pH Dependence and Ionic Strength Dependence[16]

The protein support of the enzyme has different characteristics of ion exchange as a function of the buffer pH value. The behavior of the electrode, when the pH of the medium is changing, can differ according to whether the substrate of the enzyme included inside the support is charged. It should be mentioned that the pH dependence of the enzyme in this case is more pronounced when the enzyme is immobilized than when the enzyme is free in solution. That consideration is to be linked to the diffusion limitation which occurs inside the enzyme matrix.

When using a sensor the pH and ionic strength must be controlled and stabilized. The pH optima found for various analytes are listed below.

Parameter	pH optimum
Glucose	6.5–7.5
Sucrose	6–8
Lactose	4.5
Ethanol	7.5
L-Lysine	8–9
L-Lactate	6
D-Lactate	7.7
Phenol	4

[15] C. Boudillon, T. Vaughan, and D. Thomas, *Enzyme Microb. Technol.* **4,** 175 (1982).
[16] A. Friboulet and D. Thomas, *Biophys. Chem.* **16,** 139 (1982).

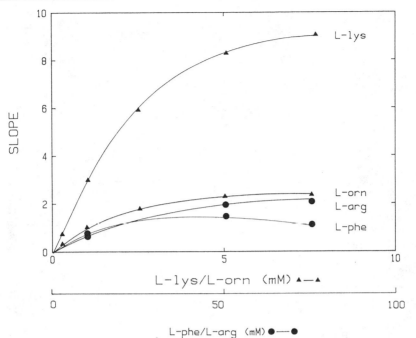

FIG. 16. Selectivity of the enzyme electrode, e.g., L-lysine measurement.

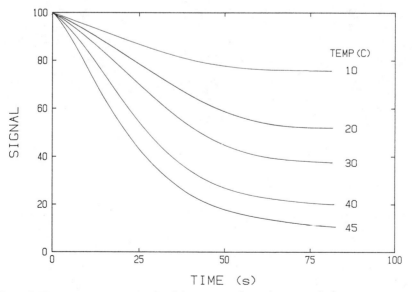

FIG. 17. Temperature dependence of the enzyme electrode, e.g., L-lysine measurement.

Fig. 18. Photograph of a commercial analyzer.

Temperature Dependence

The copolymerization process stabilizes the enzyme against thermal denaturation much more than when the enzyme is physically entrapped or when it is linked at the surface of the support. The co-cross-linking procedure limits the unfolding phenomenon which occurs during the thermal inactivation of an enzyme. For instance, in the case of lysine determination, the thermostability of the active support allows the use of the electrode at 50° for 2000 measurements. Thus it is possible to test the analytical system in the control of a fermentation process involving thermophilic bacteria. Figure 17 shows the performance of the L-lysine electrode with increasing the temperature of the medium. These curves can be repeated 2000 times at least.

Industrial and Commercial Development of the Enzyme Electrode

These experimental results provide an opportunity for industrial and biomedical application. An autoanalyzer has now been developed to satisfy the significant demands coming from biotechnological industries. The initial prototype has been modified in order to produce a more integrated apparatus. Figure 18 shows a commercially available analyzer.

Conclusion

It now seems possible to anticipate further development of these analytical techniques because of their facile application in clinical analysis as well as in biotechnology. The cost of measurement is appreciably reduced, which is an important consideration, e.g., in the health industry. The system also has significant potential. At present only a few enzymes are commercially available for analytical application, but in the near future, advances in microbiology and genetics should permit the production of more enzymes which may be utilized in new sensors.

[5] Multienzyme Membrane Electrodes

By KENTARO YODA

Since the concept of the enzyme electrode was reported by Clark and Lyons[1] and Updike and Hicks,[2] a number of enzyme electrodes have been developed and presently some are commercially available. Enzyme electrodes present specific, sensitive, and simple analytical methods for the determination of organic and inorganic compounds in complex samples

[1] L. C. Clark and C. Lyons, *Ann. N.Y. Acad. Sci.* **102**, 39 (1962).
[2] S. J. Updike and G. P. Hicks, *Nature (London)* **214**, 986 (1967).

like body fluids. Multienzyme membrane electrodes, in which more than two different types of enzymes are utilized, have been developed for the assay of α-amylase,[3] lactose,[4] maltose,[5] sucrose,[6] creatinine,[7] and the like.

In this chapter, α-amylase sensors and creatinine sensors will be introduced. It is most important in enzyme electrodes to combine a permselective immobilized enzyme membrane with a special electrode. The combination acts selectively toward substances which interfere with the electrode reaction. The permselectivity toward products or reagents of the reaction, such as hydrogen peroxide, is the most important function of the membrane in an enzyme electrode. If a polarographic probe consisting of a silver cathode and a platinum anode is used as an electrochemical sensor for hydrogen peroxide, the membrane covering the probe needs to be permeable to hydrogen peroxide but must exclude other electrically oxidative substances such as uric acid and ascorbic acid. In addition, the membrane needs to have mechanical strength. An asymmetrically structured cellulose acetate membrane has these properties.[8] Moreover, the porous layer of the asymmetric membrane, with its large surface area, is well suited as a support for the immobilization of a large amount of enzyme per unit area.

Materials

α-Glucosidase (EC 3.2.1.20), glucose oxidase (EC 1.1.3.4), creatinine amidohydrolase (creatininase, EC 3.5.2.10), and creatine amidinohydrolase (creatinase, EC 3.5.3.3) were obtained from the Biochemical Operations Division, Toyobo Co., Ltd., Osaka, Japan. Sarcosine oxidase (EC 1.5.3.1) was purchased from Seishin Pharmaceutical Co., Ltd., Chiba, Japan. Purified human salivary α-amylase (EC 3.2.1.1) (Type IX-A) and crystaline bovine serum albumin were purchased from Sigma Chemical Co., St. Louis, MO. Acetyl cellulose (Type 394-30) was purchased from Tennessee Eastman, Kingsport, TN. Porous polycarbonate membranes (pore size 0.05 μm) were purchased from Nucleopore Corp., Pleasanton, CA. Control sera validates were purchased from Warner-Lambert Co.,

---

[3] K. Yoda and T. Tsuchida, *Proc. Int. Meet. Chem. Sensors,* p. 648 (1983).

[4] M. Cordonier, F. Lawny, D. Chapot, and D. Thomas, *FEBS Lett.* **59,** 263 (1975).

[5] F. Cheng and G. Christian, *Analyst (London)* **102,** 124 (1977).

[6] I. Satoh, I. Karube, and S. Suzuki, *Biotechnol. Bioeng.* **18,** 269 (1976).

[7] T. Tsuchida and K. Yoda, *Clin. Chem.* **29,** 51 (1983).

[8] T. Tsuchida and K. Yoda, *Enzyme Microb. Technol.* **3,** 326 (1981).

Morris Plains, NJ. Amylase Test Wako was purchased from Wako Pure Chemicals Industries, Ltd., Osaka, Japan.

Preparation of Asymmetric Cellulose Acetate Membranes

A polymer solution consisting of 39.6 g of acetyl cellulose, 0.4 g of poly(vinyl acetate), 600 ml of acetone, and 400 ml of cyclohexanone is cast in a thickness of 150 μm on a glass plate. This glass plate is placed in n-hexane for a few hours, then dried in air. The 11-μm thick asymmetric membrane thus obtained is peeled from the glass plate in distilled water and then wound around a glass rod. The membrane is unwound and spread onto a polyester film support. Before coupling with enzymes, the membrane is treated with 3-aminopropyltriethoxysilane. A mixture of 100 μl of 3-aminopropyltriethoxysilane, 30 μl of acetic acid, and 200 μl of distilled water is spread over a 10 $\times$ 40 cm area of the porous side of the asymmetric membrane. After drying in air, the membrane is immersed in 0.1 M sodium hydroxide for 10 min at room temperature and then throughly rinsed with distilled water.

Preparation of Immobilized Multienzyme Membranes

The 0.1 M phosphate buffer, pH 7.0 (150 μl), containing 15 mg (945 U) of α-glucosidase, 5 mg (265 U) of glucose oxidase, and 5 mg of bovine serum albumin is mixed quickly with 30 μl of 40 g/liter glutaraldehyde. Without delay, the mixed solution is spread over a 5 $\times$ 5 cm area of the porous side of the asymmetric membrane. On standing at 4° for 10 min in air, the enzymes cross-link with the membrane. The coimmobilized enzyme membrane is finally rinsed with 0.1 M potassium phosphate buffer. The enzyme membrane thus obtained is covered with a porous polycarbonate membrane and dried at 4° in air.

The 0.1 M phosphate buffer mixture, pH 7.0 (150 μl), containing 5 mg (1000 U) of creatinine amidohydrolase, 15 mg (105 U) of creatine amidinohydrolase, 10 mg (38 U) of sarcosine oxidase, and 5 mg of bovine serum albumin is mixed quickly with 50 μl of 40 g/liter glutaraldehyde. The mixed solution is spread over a 50 $\times$ 63 mm area of the porous side of the asymmetric membrane. After letting it stand at 4° for 1 hr in air, the membrane is treated with a 50 mM potassium phosphate buffer, containing only glycine (1 M), at 4° for 16 hr. The coimmobilized trienzyme membrane is treated with a mixture of glycerol and 50 mM potassium phosphate buffer (50/950 by volume). The immobilized enzyme membrane thus obtained is covered with porous polycarbonate membrane and dried at 4° in air.

The immobilized bienzyme membrane (creatine amidinohydrolase and sarcosine oxidase) is prepared in a similar manner.

α-Amylase Sensor[3]

The α-amylase activity in standard solutions is determined as follows. A YSI Model 2510 hydrogen peroxide electrode probe (Yellow Springs Instrument Co., Yellow Springs, OH) is covered with the coimmobilized α-glucosidase/glucose oxidase membrane and immersed in a cell filled with 10 ml of potassium phosphate buffer containing 50 g/liter of soluble starch as a substrate (Fig. 1). The buffer solution (0.1 M) contains 1 mM of disodium ehtylenediaminetetraacetate, 5 mM of sodium azide, and 8.76 g of sodium chloride per liter. The pH is adjusted at 7.0. Into the same cell is injected 0.5 ml of an α-amylase standard solution. Human salivary α-amylase is used as a standard. The current generated in the polarized electrode is proportional to the hydrogen peroxide formed as a result of

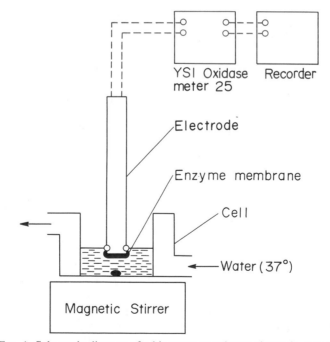

FIG. 1. Schematic diagram of a bienzyme membrane electrode system.

the following sequential reactions, that is, the current increase is proportional to the α-amylase activity.

$$Starch \xrightarrow{\alpha\text{-amylase}} maltose + maltotriose + oligosaccharide$$

$$Maltose + maltotriose + oligosaccharide \xrightarrow{\alpha\text{-glucosidase}} glucose$$

$$Glucose + O_2 + H_2O \xrightarrow{glucose\ oxidase} gluconic\ acid + H_2O_2$$

$$2\ H_2O_2 \longrightarrow 4\ H^+ + 2\ O_2 + 4\ e^- \quad (anode)$$

$$4\ H^+ + O_2 + 4\ e^- \longrightarrow 2\ H_2O \quad (cathode)$$

Figure 2 shows response curve for the α-amylase assay. The bienzyme electrode responds linearly up to 2500 U/dl (Caraway Units[9]) of human salivary α-amylase activity.

α-Amylase activity in human serum is determined with an immobilized bienzyme membrane mounted on an electrode of YSI Glucose Analyzer 23 A. In this case, the immobilized bienzyme membrane and phosphate buffer containing 1 g of soluble starch per liter are used instead of a glucose oxidase membrane and a phosphate buffer based on the manual for glucose measurement. Exactly 30 sec after the injection of 25 μl of human serum, the current increase is recorded for 30 sec. The total assay time is 100 sec, the response time with linear rise of the current being 30 sec, the recording time of the additional increase being 30 sec, and the recovery time on rinsing with buffer solution being 40 sec.

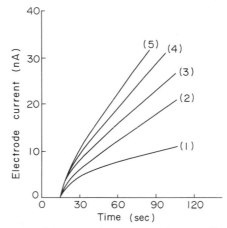

Fig. 2. Response curves of the α-amylase sensor to human salivary α-amylase. α-Amylase (U/dl): (1) 625, (2) 1250, (3) 1875, (4) 2500, (5) 3125.

TABLE I
WITHIN-DAY PRECISION WITH THREE
HUMAN SERA[a]

α-Amylase parameter	Serum 1	Serum 2	Serum 3
$\bar{x}$ (U/dl)	93.2	226.8	269.3
SD (U/dl)	6.8	10.3	11.9
CV (%)	7.3	4.5	4.4

[a] $n = 10$ each.

No interference on the measurement of α-amylase activity is observed in the presence of substances which are commonly used as preservatives to prevent glycolysis and blood coagulation, and nor in the presence of oxidative materials which interfere with electrode reactions. Endogenous saccharides such as glucose do not interfere with the assay. Within-day precisions in the analysis of α-amylase activity in three human sera are shown in Table I. Repeated assays of control human sera supplemented by human salivary α-amylase over 17 days show a between-day coefficient of variation (CV) of 5.0% for a mean α-amylase activity of 367 U/dl (Caraway Units). Control sera are used to assess the analytical recovery. Human salivary α-amylase is added to the validate resulting in serial five final activities from 39 U/dl to 371 U/dl. The analytical recoveries range from 91 to 113% (average, 103.4%) for the present method.

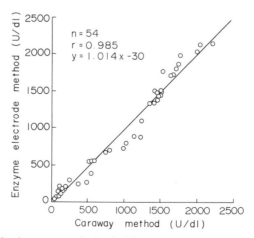

FIG. 3. Correlation between results by the bienzyme membrane electrode method and by the Caraway method.[9]

The correlation of the present method with the Caraway method is shown in Fig. 3. Specimens are sera from 54 patients, and the standard is human salivary α-amylase in 0.876% sodium chloride aqueous solution containing 1 mM calcium chloride. The Wako Amylase Test is used for the determination of α-amylase activity by the Caraway method.[9]

The durability of the α-glucosidase/glucose oxidase membrane electrode is checked by repeated assays of α-amylase activity in control human sera supplemented by human salivary α-amylase. The apparent activity of the membrane electrode decreases very slowly, and 70% of the initial activity is still present after 1000 assays, in over 17 days. The α-amylase sensor is useful in the clinical laboratory for the determination of α-amylase activity with high sensitivity, as well as with excellent precision and durability, in body fluids such as serum, whole blood, and urine.

Creatinine Sensor[7]

Two types of enzyme electrodes were constructed, a combination of either the trienzyme membrane (creatinine amidohydrolase, creatine amidinohydrolase, and sarcosine oxidase) or the bienzyme membrane (creatine amidinohydrolase and sarcosine oxidase) and a polarographic electrode for sensing hydrogen peroxide. Creatinine and creatine in standard solution are determined as follows. An electrode probe is covered with either the trienzyme membrane or the bienzyme membrane and immersed in a cell containing 10 ml of potassium phosphate buffer. At 32°, 0.5 ml of the standard solution is injected into the cell by stirring. The current generated in the polarized electrode is proportional to the hydrogen peroxide formed as a result of the following enzyme-catalyzed reactions:

$$\text{Creatinine} + H_2O \xrightarrow{\text{creatinine amidohydrolase}} \text{creatine}$$

$$\text{Creatine} + H_2O \xrightarrow{\text{creatine amidinohydrolase}} \text{sarcosine} + \text{urea}$$

$$\text{Sarcosine} + O_2 + H_2O \xrightarrow{\text{sarcosine oxidase}} \text{formaldehyde} + \text{glycine} + H_2O_2$$

Figure 4 shows the response curves of the trienzyme membrane electrode for creatinine. In the end-point method, the addition of creatinine causes a rapid increase in current, which reaches a steady state within 2 min. This current is proportional to the creatinine concentration. In the rate mode, addition of a creatinine solution causes a rapid increase in the reaction rate, reaching the maximum within about 20 seconds, and this increase is proportional to the creatinine concentration. In both tech-

[9] W. T. Caraway, *Am. J. Clin. Pathol.* **32,** 97 (1959).

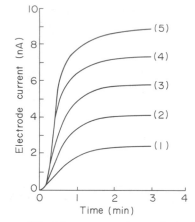

FIG. 4. Response curves of the trienzyme membrane electrode to creatinine in the end-point method. Creatinine (mg/liter): (1) 20, (2) 40, (3) 60, (4) 80, (5) 100.

niques, the calibration curves are excellent, being linear with concentration to 100 mg/liter of creatinine.

The trienzyme membrane electrode also responds to creatine as well as the bienzyme membrane electrode. Therefore, in the determination of creatinine concentration in complex samples which contain both creatinine and creatine, we use a dual electrode system in which a creatinine sensor and a creatine sensor are utilized. Real creatinine concentration in serum is obtained by subtracting the creatine value from the total creatinine value (creatinine plus creatine) by applying differential circuit with the dual electrode system.[7]

[6] Long-Term Implantation of Voltammetric Oxidase/Peroxide Glucose Sensors in the Rat Peritoneum

By LELAND C. CLARK, JR., LINDA K. NOYES, ROBERT B. SPOKANE, RANJAN SUDAN, and MARIAN L. MILLER

Introduction

Although enzymes had been used in analytical chemistry for many years, and polarographic and potentiometric electrodes were well known, it was not until 1962 that Clark and Lyons[1] showed that potentiometric,

---

[1] L. C. Clark, Jr. and C. Lyons, *Ann. N.Y. Acad. Sci.* **102**, 29 (1962).

polarographic, and conductometric electrodes could be made highly specific by interposing a thin layer of enzyme between the electrode and the analyte. Generally, the enzyme is selected so that its substrate is the analyte and either oxygen or the product is one to which the electrode is responsive. Typically, for example, glucose can be measured with glucose oxidase by measuring the peroxide generated, the oxygen consumed, the hydrogen ions generated, or the change in conductivity. In the glucose oxidase–glucose electrode the enzyme is held in juxtaposition to the peroxide-measuring anode by a glucose-permeable membrane. This membrane serves other functions as well, such as excluding catalase and other large molecules with catalytic activity. In our work reported here the membrane also serves to exclude marauding cells, proteolytic enzymes, and microorganisms.

The first and most widely used glucose sensor utilizing immobilized enzymes is the YSI–Clark glucose sensor. Here, the enzyme is separated from the platinum surface by a special peroxide-permeable, acetaminophen-impermeable, high-density cellulose acetate film. In other systems such as the Beckman glucose analyzer the enzyme is simply dissolved in solution with the analyte and a Clark oxygen electrode is used as the sensor. Additional systems have been reported which depend on enzyme transduction of the substrate with multiple enzymes.

Millions of glucose measurements are made annually by enzyme electrodes in whole blood, plasma, and other biological liquids. Other oxidase-based electrodes for measurement of alcohol, lactate, and oxalate have been described and are being used in clinical medicine and the food industry. Racine's lactate-sensing electrode[2] which depends on lactate dehydrogenase and a ferrocyanide/ferricyanide electron couple is also commercially available. Over the past two decades, thousands of publications on "enzyme electrodes," as the combination came to be called, have appeared.

Because of our long and extensive experience with the measurement of glucose with peroxide-sensitive anodes,[3,4] which function well in the presence of whole blood cells and are stable for relatively long periods at body temperature, we are conducting an investigation into the possibility of using such a glucose sensor as the sensing element in a microcomputer/ insulin pump artificial pancreas or β cell. We have found that platinum electrodes, in the form of bare wire, function for months or years when implanted in the brain of cats.[5] The glucose sensor because of its platinum

[2] P. Racine, R. Englehardt, J. C. Higeline, and W. Mindt, *Med. Instrum.* **9**, 11 (1975).
[3] L. C. Clark, Jr., this series, Vol. 46, p. 448.
[4] L. C. Clark, Jr., U.S. Patent 3,539,455 (1970).
[5] L. C. Clark, Jr., G. Misrahy, and R. P. Fox, *J. Appl. Physiol.* **13**, 85 (1958).

heart lends itself to other measurements, particularly those of blood circulation near its surface. For example, potentiometric[6-8] and polarographic[9] measurements can be used to assess blood flow by hydrogen washout curves.[10-12] In addition, as will be seen from the results presented here, a rough estimation of tissue pO_2 can be made, especially if the glucose level is known and is in the normal range. For the sensor described in this chapter the reaction is

$$\text{Glucose} + \text{oxygen} + H_2O \xrightarrow[\text{oxidase}]{\text{glucose}} \text{gluconic acid} + \text{hydrogen peroxide}$$

In order for an implanted glucose sensor based on this reaction to function quantitatively it is necessary that the geometry and microenvironment of the platinum anode be glucose limited and not affected, or at least not limited, by the partial pressure of oxygen. Lucisano *et al.*[13] have outlined a method to decrease glucose diffusion and to increase oxygen diffusion to a glucose sensor. Fisher and Abel[14] have decreased glucose diffusion to an oxidase sensor by mechanically creating a pinhole aperture. In addition, it is necessary that the sensor function for many months or possibly years after implantation. Further, its functional state should be capable of electrochemical characterization at anytime by noninvasive techniques. Another version of a potentially implantable enzyme electrode depends on measuring the difference in pO_2 between an oxidase-coated and a noncoated Clark-type electrode, the pO_2 difference being glucose dependent.[15-17]

For our implantation work, the rat peritoneum was selected because it seems to be the most hostile environment for the sensor. The peritoneal milieu has temperatures, oxygen and carbon dioxide tensions, pH, and

[6] G. A. Misrahy and L. C. Clark, Jr., *Proc. Int. Physiol. Congr., 20th, Brussels, Belgium, July 30–August 4* (1956).

[7] L. C. Clark, Jr. and G. A. Misrahy, *Proc. Int. Physiol. Congr., 20th, Brussels, Belgium, July 30–August 4* (1956).

[8] L. C. Clark, Jr. and L. M. Bargeron, Jr., *Science* **130,** 709 (1959).

[9] L. C. Clark, Jr., *in* "Biomedical Sciences Instrumentation" (W. E. Murray and P. F. Salisbury, eds.), Vol. 2, p. 165. Plenum, New York, 1964.

[10] W. Young, *Stroke* **11,** 552 (1980).

[11] P. J. Feustel, M. J. Stafford, J. S. Allen, and J. W. Severinghaus, *J. Appl. Physiol.* **56,** 150 (1984).

[12] L. C. Clark, Jr. and L. K. Noyes, *Proc. Symp. Biosensors,* p. 69 (1984).

[13] J. Y. Lucisano, J. C. Armour, and D. A. Gough, *Proc. Symp. Biosensors,* p. 78 (1984).

[14] U. Fisher and P. Abel, *Trans. Am. Soc. Artif. Intern. Organs* **28,** 245 (1982).

[15] S. J. Updike and G. P. Hicks, *Nature (London)* **214,** 986 (1967).

[16] L. C. Clark, Jr. and G. Sachs, *Ann. N.Y. Acad. Sci.* **148,** 133 (1968).

[17] S. P. Bessman and R. D. Schultz, *Trans. Am. Soc. Artif. Intern. Organs* **19,** 361 (1973).

ionic composition typical of tissue implant sites, as well as extracellular fluid compositions which have a wide variety of potentially interfering substances. It is expected that implanted glucose sensors will be encapsulated by avascular collagen generated by fibroblasts and fibrocytes.[18] Such collagen membranes are, however, expected to be glucose permeable. Hence the host response to the sensor may be put to use, instead of being resisted or overcome in developing the best sensor design.

Thevenot et al.[19] have published an excellent study of the stability of glucose oxidase in vitro. The glucose oxidase (β-D-glucose : oxygen 1-oxidoreductase, EC 1.1.3.4) used in our research is derived from Aspergillus niger. It may be a mixture of six closely related enzymes and may contain 10–16% carbohydrate.[20]

It is the purpose of this chapter to describe the methods developed and the salient results obtained so far and to indicate future work needed to perfect a long-term stable glucose sensor.

Methods

Principle

We have chosen to collect and present the polarographic data in the fashion shown in Fig. 1. In the upper right quadrant are electroreductive processes, typified by oxygen reduction. In the lower left quadrant are oxidation (electron removal from electrode reactant) processes, typified by the oxidation of hydrogen peroxide. We have selected voltage ranges below that of hydrogen, chlorine, or oxygen evolution, and for most research we now limit the scan range from +0.9 V to −0.6 V.

Instrumentation

Conventional cathodic and anodic two- and three-electrode polarograms are generated with a Princeton Applied Research (Model 174A) polarograph. Cyclic polarograms are generated with an IBM (Model EC/225) polarograph. A Hewlett-Packard XY recorder (Model 7040A), a Linseis XYT recorder (Model LY18100), or a Fisher Recordall (Series 5000) is used for pen and ink recording. Voltages are monitored with a Keithley electrometer (Model 602). An Apple IIe computer with a Techmar digital-to-analog converter (Model DA101) is used to generate voltages to the

[18] S. C. Woodward, Diabetes Care 5, 278 (1982).
[19] D. R. Thevenot, R. Sternberg, and P. Coulet, Diabetes Care 5, 203 (1982).
[20] S. Hayashi and S. Nakamura, Biochim. Biophys. Acta 657, 40 (1981).

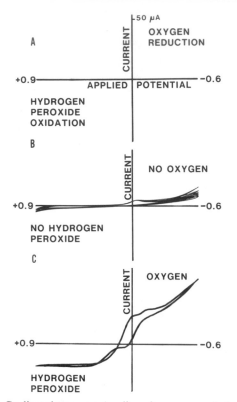

FIG. 1. Principle. Cyclic polarograms (cyclic voltammograms) showing the two quadrants involved in the oxygen oxidoreductase electrode. (A) In the upper right quadrant oxygen reduction (addition of electrons to oxygen) occurs; in the lower left, oxidation of enzyme-generated hydrogen peroxide occurs. (B) Typical tracing of a glucose sensor in a glucose-free buffer at a pO_2 near zero is shown. Electrogram (C) is that observed after saturating the buffer with air and adding 200 mg/100 ml glucose. The voltage is applied in a triangular waveform at a scan rate of 20 mV/sec. The negative limit of 0.6 V is below that of hydrogen gas formation, and the positive limit of 0.9 V is below that of oxygen or chlorine gas formation. In conventional dc polarography the pen direction is the reverse of that shown in the lower left quadrant.

IBM polarograph. An Appligration II software package (Dynamic Solutions Corp.) with some modification is used to generate step-type voltage waveforms. Figure 2 shows the general scheme used in electrochemical monitoring. A Yellow Springs glucose analyzer (Model 23A) and a Beckman analyzer (Model 6517) were used for analyzing blood and buffer samples for glucose.

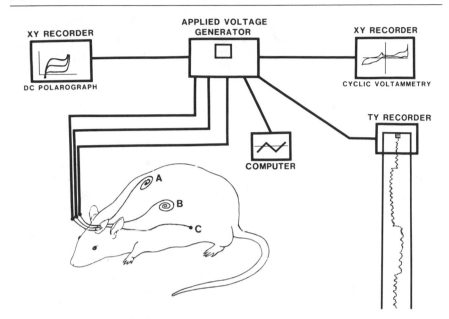

FIG. 2. Representation of the systems used for continuous *in vivo* monitoring in the unanesthetized rat. A and B are subcutaneous counter and reference electrodes; C is the glucose sensor which is in the peritoneal cavity. A TY recorder is used for continuous recording of current at a fixed applied potential and continuous recording of current and applied potential during cyclic polarography.

Reagents and Materials

Chemicals are the highest commercial grade available and are not purified further. Gomori buffer (0.1 M) with 50 mM NaCl at pH 7.4 is made with NaH_2PO_4, Na_2HPO_4, and NaCl (Fisher Scientific) dissolved in distilled or deionized water. Glutaraldehyde (25% G-6257), special glutaraldehyde (25% G-5882), and glucose oxidase (*Aspergillus niger*) are obtained from Sigma. The enzyme (Sigma G-2133 or G6125) is used directly from the bottle or is purified by dialysis for 1 day against stirred isotonic sodium chloride at room temperature. The dialysand is concentrated by pervaporation or by freeze-drying. Platinum and silver wire (28 gauge) are obtained from Englehard (Carteret, NJ). Medical grade Silastic tubing (Dow Corning, 0.020 in. × 0.037 in.) and Silastic adhesive is used. Cellulose membranes are cut from Spectrapor (3787-F45, 32 mm) dialysis tubing (Thomas Scientific, Philadelphia, PA). Gore-Tex (W. L. Gore) membrane is microporous tetrafluoroethylene (Teflon), having nominal pore sizes of 0.2 or 0.02 μm. Coria sausage casing with a flattened diameter of 3.0 cm is obtained from Tee-Pak, Inc. (Oakbrook, IL). Silux light-cured dental restorative material kits are obtained from 3M (St. Paul, MN).

Glucose Sensors: Fabrication

Glucose sensors are made from platinum wire, silicone elastomeric tubing, cellulose membrane, glucose oxidase, and silk thread, as diagramed in Fig. 5. The platinum electrode is made by forming a 1.5-mm bead on the end of a 10-cm wire by holding it vertically with a tip in a gas/oxygen flame. The wire is threaded through the Silastic tubing until the bead fits snugly against the end. The yellow enzyme concentrate is painted or smeared on the bead to give a thin uniform yellow coat which is allowed to air-dry. A piece of cellulose membrane is wet with saline, pulled tightly over the coated bead, and fastened with many turns of silk thread. The same procedure is used for fastening microporous fluorocarbon membranes. For chronic implantation the Silastic tubing on the non-sensor end of the electrode is covered with a Silastic cap sealed with Silastic adhesive. The glucose sensor (about 8 cm in length) is placed on a rack above a 25% solution of glutaraldehyde for at least 12 hr. The vapor insolubilizes the enzyme. Over 100 sensors were made for this research.

Glucose Sensors: Preimplant Characterization

Each sensor is tested before implantation by recording two successive scans at each glucose level of 0, 100, 200, 300, and 400 mg/100 ml for a total of 10 scans between 0 and +0.9 V. A 3-cm silver reference, a 5-cm platinum counter, and the working (sensor) electrode are mounted in stirred, air-saturated Gomori/chloride buffer at 37°. A scan rate of 10 mV/sec and a full-scale sensitivity of 10 to 20 μA is used. The sensors are tested in the same way after explantion. If they are to be stored overnight for retesting, sensors are kept in Gomori/chloride solution with 200 mg/100 ml glucose at 4°. Over 6,000 polarograms have been generated and evaluated for this research.

Glucose Sensor Implantation

Intraperitoneal Long-Term Stability Testing. Female Sprague–Dawley rats 50–70 grams in weight are anesthetized with intraperitoneal sodium pentobarbital (40 mg/kg), shaved over the abdominal area, and the peritoneal space exposed by a 1-cm longitudinal incision to one side of the midline. Xylocaine (1%) is infiltrated locally if necessary. The glucose sensor, coiled in a circle, is implanted, and secured loosely with 5-0 silk to the inside of the abdominal wall. The animal is given oxygen to breathe, the electroencephalogram (ECG) is monitored, and the body temperature is maintained with a rectal thermistor controlling a heating pad or an infrared light by means of a YSI proportioning circuit (Model 72). Antibiotics, 200,000 units of Bicillin (Wyeth, Philadelphia, PA), are given intramuscularly. X-rays of each rat are taken directly after sensor implantation and before explanation. Forty-three glucose sensors have been chronically implanted. After recovery, the rats are maintained in an air-conditioned room and fed *ad libitum* with Purina Rat Chow in individual suspended, automatically watered cages. For electrode evaluation, the rat is again anesthetized, the sensor dissected free and removed, and the rat sacrificed. The sensor is tested directly after removal and, if found to be active, is replaced in another rat.

Continuous Recording from Two- and Three-Electrode Implanted Rats. As illustrated in Fig. 2, glucose electrodes are implanted intraperitoneally with counter and reference electrodes placed subcutaneously on either side of the thorax. Female Sprague–Dawley rats (150 g) are anesthetized with sodium pentobarbital at a dose of 40 mg/kg. In some animals, the wires are anchored to the skull using small stainless steel screws and fast-setting cold cure (repair) dental acrylic. Later, the procedure was improved so that the skull is exposed, the periosteal tissue scraped away, the bony surface etched with 37% phosphoric acid gel, and Scotchbond dental bonding liquid applied. This is followed by anchoring the insulated electrode wires with light-polymerized dental restorative plastic. The restorative polymer is applied to the skull and to the wires in several coats with high intensity light curing between each application. Tight bonding to the skull, without the use of screws, is usually attained if care is taken to keep the field dry.

Since the restorative polymer does not adhere to Silastic, a small collar is cemented on the tubing where it passes through the plastic. Two or three centimeters of tubing extend vertically for attachment of the polarographic lead wires. A long electrically nonconductive path along a clean dry silicone (or Teflon) surface must be used to avoid spurious currents. The animals tolerate these procedures well and seem to be unaware of the fixation on the skull or the implanted electrodes. In some animals a chronic heparin-filled polyvinyl microcatheter in the jugular vein is mounted on the same unit on the skull. It is preferable to leave a loop of wire and/or catheter under the skin to give more leeway as the rats grow.

Applied Potentials. Most of the glucose sensor testing is conducted with scan rates of 10 mV/sec, with a range from 0 to +0.9 V, using a dc

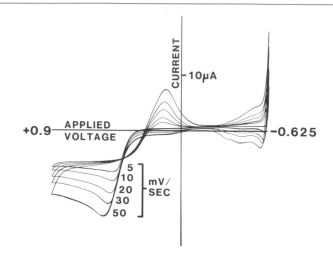

FIG. 3. The effect of scan rate on the cyclic voltammogram. The tracings were made with the three-electrode system, using a glucose sensor, a silver reference, and a platinum counterelectrode in glucose-containing Gomori/chloride buffer at 37° and pH 7.4. Electrochemical equilibrium with the sensor is obtained at rates below 1 mV/sec or at a fixed applied potential. A scan rate of 10 or 20 mV/sec is used as a compromise. Note the cathodic hydrogen formation (sharp upward curves) above −0.60 V.

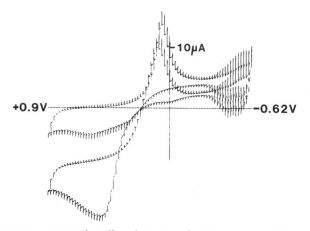

FIG. 4. Computer-generated cyclic polarogram of a glucose sensor. The applied voltage is a triangular ramp of about 80 steps from +0.90 V to −0.62 V and return. The scan rate is 20 mV/sec in 20-mV steps, generated by the Apple IIe. The top curve is for zero glucose, the bottom curve for 200 mg/100 ml. A decrease in current in the oxygen quadrant can be seen at the higher glucose concentration.

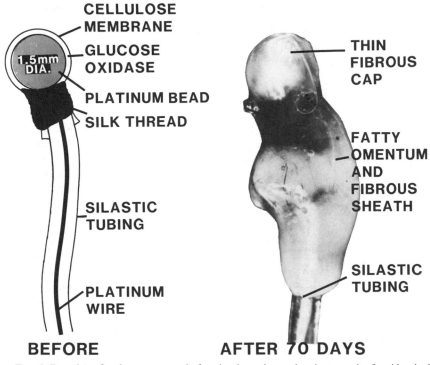

FIG. 5. Drawing of a glucose sensor before implantation and a photograph of an identical sensor explanted after 70 days in the peritoneal space of a rat. The anodic polarogram at 70 days showed a *higher* reactivity (current) toward the glucose standards than before implantation. The yellow color of the enzyme could be seen through the whitish fibrous cap. It appears that the fibrous sheath begins growing along the Silastic tubing and over the tip from the silk thread. After testing, the glucose-reactive sensor was reimplanted in a new rat.

linear ramp. Recently we have begun to use cyclic voltammetry more. The effect of scan rate is shown in Fig. 3. Triangular and sine wave applied voltage forms are being evaluated. In addition we have a programmable potential applicator, typical scans for which are shown in Fig. 4.

Histology. Samples of tissue immediately surrounding implanted glucose sensors are placed in phosphate-buffered glutaraldehyde–paraformaldehyde fixative. They are mounted and stained for light and/or electron microscopy. The thin translucent cup-shaped tissue seen on the end of the electrode tip after implantation as shown in Fig. 5 is readily dissected free and is studied separately. It is not adherent to the cellulosic membrane. Failed sensors, electrodes, and membranes, after removal, are tested for sterility by culture, microscopic examination, and other methods.

Results

When a glucose sensor having a glucose oxidase layer which has not been treated with glutaraldehyde is implanted, it loses activity in a few days. It can be seen after explantation that the yellow color of the oxidase layer is greatly diminished or absent. This same effect can be seen *in vitro* when a sensor is placed in isotonic saline or Ringer's solution. Apparently, osmotic forces either drive the enzyme through spaces where the membrane is fastened to the Silastic or cause the membrane to swell and to force enzyme from microscopic cracks in the cellulose. One of the most important aspects of the glutaraldehyde treatment is therefore to render it *insoluble* so as to abolish osmotic (oncotic) forces. For this reason all the sensors described here are treated with glutaraldehyde vapor. This cross-linking probably also serves to stabilize the enzyme against degradation by heat, proteolytic enzymes, and hydrolysis.[21]

When a glucose sensor is implanted in body fluid containing glucose and oxygen, it produces hydrogen peroxide and gluconate which diffuse from the enzyme layer, through the cellulose membrane, and back to the body fluid. The glucose sensor acts as a sink for oxygen and glucose, and the nearby circulation carries away the gluconate and decomposes the peroxide. An implanted glucose sensor, then, is identical to a functioning measuring polarographic glucose sensor except for the polarizing potential ($+0.9$ V) and the flow of current (10^{-6} A). Self-contained, unconnected glucose sensors implanted in the peritoneal or other spaces in the body should yield the same information as to tissue reaction and sensor longevity as polarized sensors where the current is flowing.

The thin fibrous cap which forms on the tip of the glucose sensor after a few weeks (Fig. 5) is readily cleanly dissected free, fixed, and examined. Polarograms (not shown) have been run on many such explanted electrodes before and after removing the membrane cap. The cap is glucose permeable. The first reaction to the membrane surface of the cellulose membrane on the sensor, which, bear in mind is continuously exuding peroxide and gluconate, is shown in Fig. 6. Peritoneal macrophages cover the surface within a few hours. Following this, fibroblasts, fibrocytes, and, finally, the beautiful collagen membrane shown in Fig. 7 is formed.

Figure 8 shows that the activity of a sensor explanted from the peritoneum of the rat on day 37 remains very active. This sensor was rinsed with saline and reimplanted. After an additional 62 days it was again

---

[21] M. Salmona, C. Saronio, and S. Garattini (eds.), "Insolubilized Enzymes." Raven, New York, 1974.

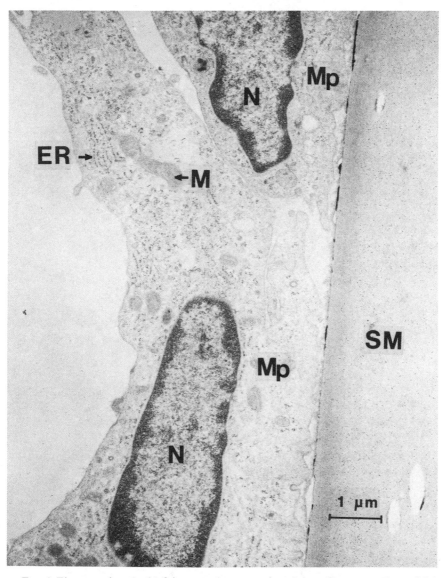

FIG. 6. Electron micrograph of tissue on the sensor tip at 2 days. Two macrophages (Mp) have attached to the exposed (peritoneal side) of the sensor membrane (SM). During the first 2 days these cells spread out to a monolayer. Inflammatory cells are not present. Nuclei (N), mitochondria (M), and rough endoplasmic reticulum (ER) of the macrophages are seen. Stain: Lead citrate, uranyl acetate. Magnification: ×22,500.

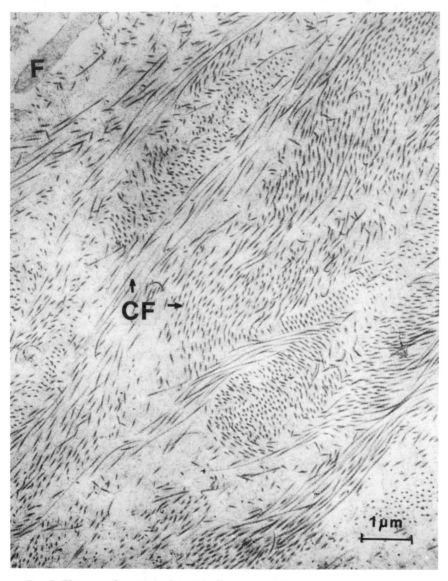

Fig. 7. Electron micrograph of the thin fibrous cap at 70 days. A few fibroblasts (F) and loosely packed collagen fibers (CF) are the main components of the cap, which formed on an electrode implanted 70 days prior. Stain: Lead citrate, uranyl acetate. Magnification: ×22,500. The electrode surface is in the lower right.

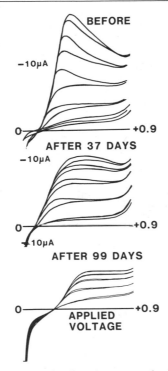

FIG. 8. *In vivo* stability of sensor. Anodic polarograms of a glucose sensor before implantation, after explanation from the first rat at 37 days, and after explanation from the second rat 62 days later. All curves were made with the sensor in air-saturated stirred Gomori/chloride buffer at 37° at pH 7.4. Two voltammograms are shown for additions of 0, 100, 200, 300, and 400 mg/100 ml glucose. 10 μA indicates the current calibration for the three sets of curves.

removed and tested, with the result shown in the lower curves. The activity had decreased but remained in the range of a usable glucose sensor. It is possible that the second rat reacted to the used sensor as a more foreign body than the first rat did to the new sensor. After 99 days the yellow color of the oxidase layer, while lighter, was still grossly visible. We have observed active explanted glucose sensors where the enzyme layer appeared colorless to the naked eye.

Inactive explanted sensors are often observed to have a whitish layer around the electrode near the thread, and *under* the cellulose membrane. We have not seen an active electrode with such a submembrane leucodeposit. In order to determine if this material had catalase-type activity we compared the peroxide response of a new glucose sensor with such a nonreactive glucose sensor. As can be seen in Fig. 9 the used sensor was

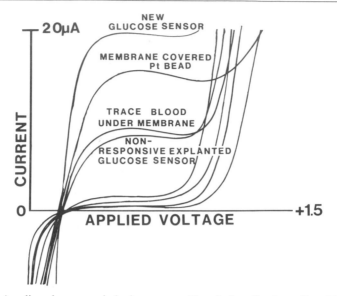

FIG. 9. Anodic polarograms in hydrogen peroxide solution. Catalase effect. The bottom four curves are for the four types of sensors (1–4) when tested in peroxide-free buffer. (1) The top curve is for a new glucose sensor in hydrogen peroxide. (2) The explanted glucose sensor, *which did not respond to glucose,* shows a lowered response to hydrogen peroxide. (3) A membrane-covered (enzyme-free) platinum bead is compared with (4), the same bead coated with a trace of fresh blood before the membrane is affixed. If catalase concentrate is used between the membrane and the bead instead of blood, the anode becomes completely unresponsive to hydrogen peroxide added to the buffer.

not as responsive to peroxide as the new sensor. Also compared in Fig. 9 is the response of a cellulose-covered anode with the same electrode after a trace of fresh human blood had been trapped between the membrane and the bead. The peroxide response is indeed decreased, indicating that at least one cause of the loss in activity of peritoneally implanted electrodes in the rat may be associated with encroachment of catalase activity into the oxidase layer. Such activity may be due to penetration of catalytic protein activity, blood cells, or microorganisms.

Figure 10 is included in this chapter to show that collagen membrane, in this case a commercial glutaraldehyde-treated regenerated collagen sausage casing, forms an excellent membrane for a glucose sensor, being both oxygen and glucose permeable.

The oxygen tension in the peritoneal cavity of an air-breathing rat averages about 46.5 torr; after oxygen breathing it increases to about 65.1 torr. This was determined by injecting perfluorotributylamine neat liquid intraperitoneally, waiting several hours, and withdrawing samples for

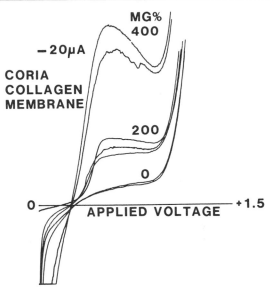

FIG. 10. Anodic polarograms of a Coria membrane-covered glucose sensor. Parameter: glucose concentration in mg/100 ml (mg%).

analysis on a Radiometer ABL 30 blood gas analyzer. In the range of peritoneal oxygen tensions in the rat there is clearly an effect (Fig. 11) on the glucose (anodic) current when a cellulose membrane is used. This is due to the low diffusion of oxygen through wet cellophane, compared with the high diffusion of glucose.

As part of a continuing search for membranes which have a high permeability to oxygen and a lower permeability to glucose we are investigating microporous fluorocarbon polymeric and silicone polymeric films of various types. In Fig. 12 is shown the response of a glucose sensor (as in Fig. 5) but having a Gore-Tex membrane. It shows a remarkably high current response to glucose standards.

The use of platinum anodes and platinizied platinum potentiometric electrodes to measure circulation by means of hydrogen washout curves is well known. Figure 13 demonstrates that the platinum anode even when encapsulated by a layer of cross-linked enzyme and cellulose film is still responsive to hydrogen. Hence blood circulation in the vicinity of an implanted glucose sensor can be quantitatively measured by inhalation of this inert (and highly insoluble) gas. In Fig. 14 is shown a cyclic polarogram of another anodically active substance, vitamin C, obtained with an actual glucose sensor, not a bare or coated platinum electrode.[12] This high anodic current shows that circulation can be evaluated following intrave-

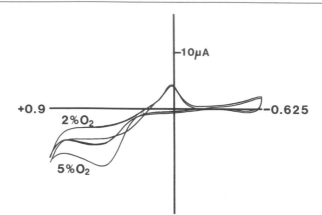

FIG. 11. Glucose currents at low oxygen tensions. This cyclic polarogram of a glucose sensor in Gomori/chloride buffer with 200 mg/100 ml glucose added was run at a scan rate of 20 mV/sec between the potentials indicated. The stirred solutions were bubbled first with 2% oxygen/98% nitrogen, then with 5% oxygen/95% nitrogen.

nous vitamin C administration, and blood levels of vitamin C are not expected to introduce significant error. The sensitivity of the platinum anode to phenols, such as acetaminophen, can be eliminated by the use of high density inner cellulose acetate membrane. Diabetics with an implanted glucose sensor should not use Tylenol.

We have chosen an implanted Gore-Tex-type glucose sensor to illustrate the nature of the recordings obtainable from a normal awake rat. The current obtained from a steady applied, as well as a cyclic triangular,

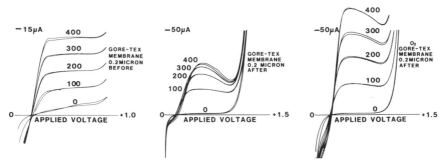

FIG. 12. Anodic polarograms of a microporous polytetrafluoroethylene membrane glucose sensor. The set of polarograms on a Gore-Tex-type sensor before implantation were obtained in air-saturated buffer. The sensor was implanted in the rat used for the continuous recording shown in Fig. 15. The sensor was removed on day 13. The polarograms after explantation were obtained in air-saturated and oxygen-saturated buffer. Scan rate: 10 mV/sec. Temperature: 37°. Parameter: glucose concentration in mg/100 ml.

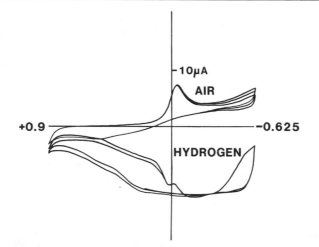

Fig. 13. Cyclic polagoram of the glucose sensor in air-saturated and hydrogen-bubbled buffer. The top polarogram is in air-saturated stirred Gomori/chloride buffer at pH 7.4 and 37°. The bottom curve is after bubbling to near saturation with hydrogen. Scan rate: 20 mV/sec.

potential are shown (Fig. 15). The wavelike variations in current indicates there is an active circulation near the electrode. The anodic current responds to glucose injection. The current on the cathodic side (upper right) of the cyclic polarogram is a reflection of the pO_2 in the sensor, while those on the lower (anodic) side are almost entirely a measure of glucose levels and fluctuations. A set of polarograms from this sensor before implantation and two after explantation are shown in Fig. 12.

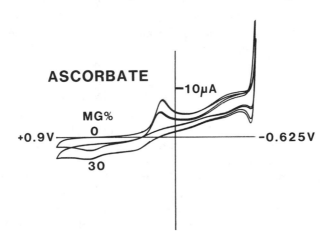

Fig. 14. Response of the glucose sensor to ascorbate (0 and 30 mg/100 ml).

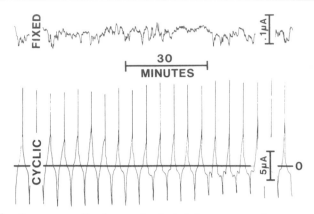

FIG. 15. Continuous recording from an implanted glucose sensor in a normal rat. The top tracing is that obtained during a 105-min portion of a 177.5-hr recording. The electrode is held at a fixed potential of +0.8 V. The bottom tracing was obtained during a 105-min portion of a 33.7-hr recording. The average current for the fixed potential readings was about 0.4 μA. The electrode was cycled between +0.9 and −0.6 V at a rate of 20 mV/sec. Altogether 223 hr of recording were obtained from this rat. The portions of the recording shown occurred near midnight.

Discussion

Commercially available three-electrode instruments are not suited in the long run for research in animals because, if the reference electrode is disconnected, the voltage applied to the electrode through the compensating circuit greatly increases. For example, dc potentials as high as 40 V (IBM) or 90 V (PAR) can result in destruction of the enzyme activity of the electrode or injury to the tissue in the vicinity of the electrodes. A future instrument is envisioned to be battery operated, voltage limited, implantable, and programmable from outside the body as are modern cardiac pacemakers.

Glutaraldehyde is notoriously unstable. For this reason, for cross-linking in the liquid phase, we have used fresh material which is stored at −80° in the dark. Glutaraldehyde vapor is proving to be very effective for insolubilizing enzymes after electrode fabrication. The vapor readily permeates cellulose membranes and protein layers and, by virtue of its being a vapor, tends to be purer than that in solution.

Most vectors decrease the activity of glucose oxidase-dependent glucose sensors. Such factors as body temperature, peroxide formation, glucose turnover per se, solubility of the enzyme, mixture with catalase, mixture with peroxide-destroying ions, presence of peroxidases, generation of superoxide anions, presence of proteolytic enzymes, generation of hydrogen ions, excessively high applied voltages, loss of prosthetic groups, presence of carbohydrases or proteases, invasion of macrophages, bacteria, or fungi, and exposure to light lead to decreased activity. This is not to mention the naturally occurring substances which may coat or "poison" the surface of the platinum electrode. A surprisingly large number of these antisensor factors are eliminated by use of a tightly sealed low porosity membrane. Glucose and gluconic acid may help to stabilize the sensor while the peroxide generated while in use may serve to sterilize it. It is possible that there are enzyme-stabilizing substances in body juices. We are accumulating data on injection of sodium ascorbate boluses. Fluctuation is the normal evidence that cathodic treatment often reactivates a glucose sensor which has partially lost activity, hence making continuous cycling desirable for this and a number of other reasons.[12]

When we used silk thread to fasten the regenerated cellulose membrane, a fibrous cap formed consisting, after a few weeks, mainly of collagen. In the immediate vicinity of this kind of sensor tip, there was very little microcirculation. When Gore-Tex was used, small capillaries could be seen developing, and the wavelike activity of the glucose current suggested an active circulation. This membrane will not keep proteolytic or catalase activity, or probably even macrophages, from entering the enzyme layer. It may be, however, that the resistance of cross-linked enzymes to proteolytic enzymes[22] will be sufficient to protect the sensors from gradual inactivation. It is not yet certain, then, whether it is better to use surface active materials[23] or inert materials[24]; more research is required.

At the beginning of this research the two main concerns were (1) that the *lifetime* of the enzyme would be too short to make a long-term implantable sensor feasible and (2) that the *oxygen tension dependency* of the sensor would be so high that quantitative *in vivo* glucose measurements could not be made. Once these two problems are solved, and our present findings encourage us to believe this possible, a systematic search for the most reliable and useful sites for implantation will be made, the electrochemical circuits can be optimized, and the glucose sensor engineered to control an internal or external insulin infusion system. Knowledge about stabilizing implantable enzymes will also be valuable in other forms of diagnostic devices, such as glucose and lactate catheters, in

[22] Y. Morikawa, I. Karube, S. Suzuki, Y. Nakano, and T. Taguchi, *Biotechnol. Bioeng.* **20,** 1143 (1978).

[23] L. L. Hench and J. Wilson, *Science* **226,** 630 (1984).

[24] M. L. Miller, R. E. Moore, and L. C. Clark, Jr., *Proc. Int. Symp. Perfluorochem. Blood Substitutes, 4th,* p. 81 (1979).

cancer chemotherapy, and in enzyme replacement therapy. Increased knowledge about stabilizing implantable enzyme systems may finally lead to an implantable artificial liver.

The lifetime of the explanted sensor can be estimated by measuring peroxide generated in the presence of standard glucose concentrations in the laboratory. Alternatively, the sensor's glucose responsivity can be estimated *in situ* by use of a subcutaneous reference electrode, a polarographic circuit, and injections of glucose solutions. Of course, an implanted glucose sensor can be calibrated *in vivo* while *in situ* by comparing the glucose current with blood glucose levels. Two- and three-electrode conventional dc polarography and triangular cyclic voltammetry are used. Continuous recordings for several hours or even for days are obtained at various times in normal awake rats, having implanted sensors and reference and/or counter electrodes, in order to analyze and understand the characteristics of the glucose current. If the oxidase is not insolubilized its osmotic activity causes the enzyme to be washed away from under the membrane and the enzyme layer to fill with water and expand, or even rupture, the membrane. When the enzyme is rendered insoluble by crosslinking or other means, loss of glucose sensor responsivity due to these hydraulic factors does not occur.

The lifespan of glucose sensors chronically implanted in the rat peritoneum varies from sensor to sensor but is remarkably long, extending at least up to 3 months. This surprising lifespan may be further extended as the nature of the inactivation process, such as intrusion of catalase activity, is better understood and prevented. By designing the sensor so as to regulate glucose and oxygen membrane permeability and electrochemical programming, the influence of variations in pO_2 in the tissue on the glucose current can be virtually eliminated. Lifespan can also be extended by careful control of polarizing voltages. By appropriate orchestration of biochemical, physiological, and electrochemical factors, oxidase-type glucose sensors suitable for controlling an insulin pump via a microcomputer seem entirely feasible.

Summary

Methods for designing, fabricating, testing *in vitro* and *in vivo,* and improving chronically implantable oxidase/peroxide-type polarographic glucose sensors are described. Voltammetric means to evaluate oxygen supply to the sensor and to measure the nearby microcirculation with hydrogen washout techniques using the implanted glucose sensor are outlined. Because some peritoneally implanted sensors have, perhaps surprisingly, remained functional for months, such devices may prove with further development to be useful as the sensing components in artificial pancreatic β cells for the control of diabetes.

Addendum[25-32]

We have now established that glucose sensors survive over 11 months of peritoneal implantation in the mouse with retention of full enzymatic activity. Bacterial infection of the sensor is not a problem because the glutaraldehyde vapor used for the oxidase insolubilization sterilizes the entire assembly.

Acknowledgments

The authors are grateful for the assistance of Eleanor Clark, Jackie Grupp, Estelle Riley, and Barbara Williams in preparing the manuscript. John Erickson performed some of the surgical implants. Ann Maloney assisted in the laboratory. Some of the original artwork is by Luis Alicea, the rest by Linda Noyes. We are indebted to Mary Gilchrist, for examinations of microbiological specimens from implanted sensors. We are grateful to John Kutt at IBM Instruments for his assistance. We are indebted to W. L. Gore (Elkton, MD) for samples of microporous membrane. This work is supported by Grant R01 AM31054 from the National Institutes of Health.

[25] L. C. Clark, Jr. and C. A. Duggan, *Diabetes Care* **5**, 174 (1980).

[26] L. C. Clark, Jr. and L. K. Noyes, in "Proceedings of the Symposium on Biosensors" (A. R. Potvin and M. R. Neuman, eds.), p. 69. Institute of Electrical and Electronics Engineers (IEEE), New York.

[27] L. C. Clark, Jr., L. K. Noyes, R. B. Spokane, R. Sudan, and M. L. Miller, *Ann. N.Y. Acad. Sci.* **501**, 534 (1986).

[28] L. C. Clark, Jr., in "Biosensors: Fundamentals and Applications" (A. P. F. Turner, I. Karube, and G. S. Wilson, eds.), p. 3. Oxford Univ. Press, New York and London, 1986.

[29] L. C. Clark, Jr., R. B. Spokane, R. Sudan, and M. Homan, *Trans. Am. Soc. Artif. Inter. Organs Abstr.* **16**, 67 (1987).

[30] L. C. Clark, Jr., R. B. Spokane, R. Sudan, and T. L. Stroup, *Trans. Am. Soc. Artif. Intern. Organs Abstr.* **16**, 68 (1987).

[31] L. C. Clark, Jr., R. B. Spokane, R. Sudan, and T. L. Stroup, *Trans. Am. Soc. Artif. Intern. Organs,* in press (1987).

[32] L. C. Clark, Jr., *Trans. Am. Soc. Artif. Intern. Organs,* in press (1987).

[7] Amperometric Biosensors Based on Mediator-Modified Electrodes

By ANTHONY P. F. TURNER

Introduction

The multifarious descriptions of novel biosensors appearing in the literature have been extensively reviewed.[1,2] Electrochemical detection has clearly proved the favored method to date for use in these devices. Most of the electrochemical sensors used may be assigned to two broad categories; potentiometric devices measure variously derived voltages, and amperometric systems record the current that flows when the voltage is held at a constant value. The former category includes pH electrodes, ion-selective electrodes, and ion-sensitive field effect transistors. Familiar amperometric instruments include the Clark oxygen electrode and detectors for high-performance liquid chromatography (HPLC). In addition, impedimetric techniques have found limited application, most notably for biomass estimation.

A further classification of amperometric biosensors may be made into direct and indirect systems. Indirect sensors exploit conventional detectors to measure the metabolic substrate or product of biological material. Electrodes most commonly used are the oxygen electrode or, for the detection of hydrogen peroxide, a platinum electrode held at 600–700 mV versus the saturated calomel electrode (SCE). The biological element may be intact microorganisms, plant cells, animal tisue, or isolated enzymes, and it is usually immobilized in the vicinity of the electrode. Direct amperometric biosensors are the subject of this chapter and involve attempts to achieve a more intimate relationship between biology and electrochemistry. The technique harnesses biological redox reactions by substituting modified electrodes for the natural electron donor or, more usually, acceptor. It is intended that the simplicity of this approach will lead to cheaper, more reliable, and highly sensitive sensors for clinical, industrial, and environmental applications.

The ideal situation for a direct amperometric biosensor would be where electron transfer occurred freely between the redox center of a catalytic protein and an amperometric circuit. Cytochrome c exhibits

[1] S. L. Brooks and A. P. F. Turner, *Meas. Control* **20,** 37 (1987).
[2] A. P. F. Turner, I. Karube, and G. S. Wilson, "Biosensors: Fundamentals and Applications." Oxford Univ. Press, London and New York, 1987.

METHODS IN ENZYMOLOGY, VOL. 137

reversible electrochemistry at modified gold electrodes; this reaction has been used to couple reductive[3] and oxidative[4] enzymes to electrodes in sensor configurations. Practical sensors based on direct electron transfer between a redox enzyme and an electrode, however, have not been demonstrated. An expedient is the use of mediators which shuttle electrons between the protein and the electrode. Compounds such as ferrocene and its derivatives (and more recently, tetrathiafulvalene[5] and tetracyanoquinodimethane[6] and their derivatives) have been shown to be well suited to this role, facilitating electrochemical coupling of a range of oxidases and non-NAD-linked dehydrogenases.[7] This approach to the construction of biosensors will be illustrated below by reference to glucose and alcohol sensors; the technique, however, may in principle be applied to any oxidoreductase, cell component, or cell that will undergo rapid exchange of electrons with an electrochemically active intermediate. The homogeneous coulometric systems also provide an alternative to conventional spectrophotometric determinations of enzyme kinetics.

Equipment

A major attraction of amperometric techniques is their relatively low cost. The components necessary to construct a simple electronic package to operate a well-defined biosensor can cost as little as a hundred dollars.

The basic requirement is for a means of holding a working electrode at a steady potential and measuring any current flow. This may be achieved in a two-electrode configuration by using a defined voltage source to hold the potential of a working electrode constant with respect to a reference electrode. A simple circuit designed for this purpose is outlined in Fig. 1. In practice, two electrode configurations are usually quite adequate, but more precise control of the potential at a working electrode may be achieved by using a potentiostat and a third electrode as a reference.[8] Suitable potentiostats may be purchased from many companies including

[3] H. A. O. Hill, N. J. Walton, and I. J. Higgins, *FEBS Lett.* **126,** 282 (1981).
[4] A. P. F. Turner, W. J. Aston, J. Bell, J. Colby, G. Davis, I. J. Higgins, and H. A. O. Hill, *Anal. Chim. Acta* **163,** 161 (1984).
[5] A. P. F. Turner, S. P. Hendry, and M. F. Cardosi, "Biosensors, Instrumentation and Processing," p. 125. Online Publications, Pinner, England, 1987.
[6] S. P. Hendry and A. P. F. Turner, *Horm. Metab. Res.,* in press (1987).
[7] M. F. Cardosi and A. P. F. Turner, *in* "Biosensors: Fundamentals and Applications" (A. P. F. Turner, I. Karabe, and G. S. Wilson, eds.), p. 257. Oxford Univ. Press, London and New York, 1987.
[8] A. J. Bard, and L. R. Faulkner, "Electrochemical Methods." Wiley, New York, 1980.

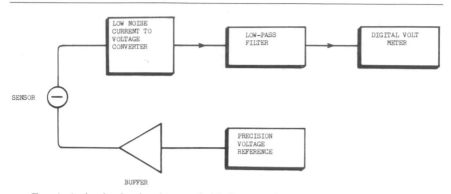

FIG. 1. A simple circuit scheme suitable for use with an amperometric biosensor.

Thompson Electrochem Limited (P.O. Box 6, Forest Hall, Newcastle upon Tyne, Tyne and Wear NE12 9BG, England), EG&G Instruments (Dancastle House, Dancastle Road, Bracknell, Berkshire RG1Z4PG, England), BAS (2701 Kent Avenue, Purdue Research Park, West Lafayette, Indiana 47906), and Linton Instrumentation (Hysol, Harlow, Essex CM18 6QZ, England).

Investigations and applications of biosensors are facilitated by interfacing to microprocessors. Software control of the voltage regime and sophisticated multichannel data analysis can considerably speed up characterization of sensors and allow checking of alogirthms when the sensors are in use. The author, in association with Artek (59, Langlands, Lavedon, Bucks., MK46 4EP), has designed a low cost programmable interface package which is commercially available. An outline of the interface design is shown in Fig. 2. The units may be used with IBM clones, such as the Amstrad 1512, or with the BBC range of microcomputers (Acron Computers, 645 Newmarket Road, Cambridge, CB5 8TD) and provide 12-bit software control of the voltage applied concurrently to four, eight, sixteen, or twenty-four biosensors. The amperometric response of each electrode may be continuously displayed, with statistical treatments of the data being carried out within programs written principally in BBC BASIC. The interfaces may be used in conjunction with a suitable potentiostat (e.g., the MP81, Bank Elektronik, 34 Goettingen, Werner-Von-Siemens-Strasse 3, West Germany) allowing the use of a working, auxiliary, and reference electrode in each cell. More sophisticated microprocessor-based systems, capable of a full range of electroanalytical techniques, are available from, for example, BAS (2701 Kent Avenue, West Lafayette, IN 47906) and EG&G Princeton Applied Research (P.O. Box 2565, Princeton, NJ 08540).

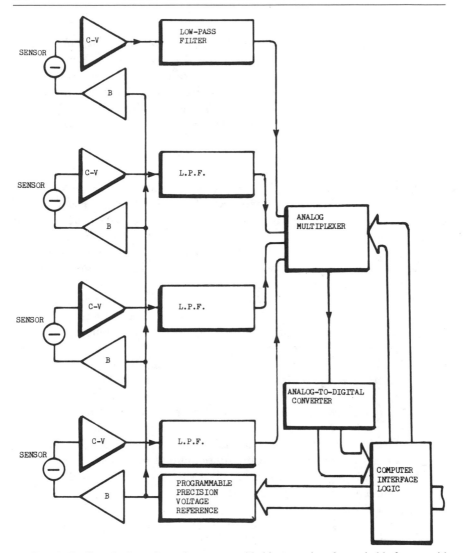

Fig. 2. Outline of a four-channel programmable biosensor interface suitable for use with a microcomputer.

Amperometric biosensors are temperature sensitive with their output typically varying by about 4%/°C. For accurate measurements it is therefore necessary to either control temperature, using, for example, a jacketed reaction vessel with a thermocirculator, or to measure the sample temperature and compensate the reading.

Enzymes

Mediated amperometric biosensors may be readily constructed with oxidoreductases that donate electrons to electrochemically active artificial electron acceptors. Mediated electron transfer from an enzyme to an electrode may be studied in rapid systems by using direct current (dc) cyclic voltammetry and the reaction kinetics established.[4,9,10] Two enzymes that are amenable to this approach are glucose oxidase and methanol dehydrogenase.

Glucose oxidase (β-D-glucose : oxygen 1-oxidoreductase, EC 1.1.3.4, from *Aspergillus niger*) obtained from three commercial UK sources (Boehringer Mannheim, Bell Lane, Lewes, East Sussex, BN7 1LG; Sigma Chemical Company, Fancy Road, Poole, Dorset, BH17 7NH; Sturge, Enzymes Denison Road, Selby, N. Yorkshire, Y08 8EF) has been used to construct enzyme electrodes. Glucox PS from the last source produced the best results. The enzyme may be further purified by gel filtration HPLC for electrochemical characterization, but this is not essential for the construction of electrodes.

Glucose oxidase is a flavoprotein of molecular weight approximately 186,000 which catalyzes the oxidation of β-D-glucose to D-δ-gluconolactone. The natural electron acceptor is oxygen, which is reduced to hydrogen peroxide. This reaction has a pH optimum of 5.6. However, glucose oxidase will use a variety of artificial electron acceptors with the reactions exhibiting elevated pH optima. The enzyme requires neither cofactors nor activators and is readily immobilized, making it particularly suitable for use in enzyme electrodes. It is described here as an example of the group of flavoprotein oxidases that will donate electrons to mediators.[11]

Methanol dehydrogenase [alcohol : (acceptor) oxidoreductase, EC 1.1.99.8] may be prepared from methylotrophic bacteria such as *Methylophilus methylotrophus*[12] or *Pseudomonas extorquens*.[13] The enzyme is commercially available only from Sigma. The most convenient method of purification of the enzyme from cell-free extracts of methanol-grown bacteria is by aqueous two-phase partition.[12] An aqueous two-phase system

[9] A. E. G. Cass, G. Davis, G. D. Francis, H. A. O. Hill, W. J. Aston, I. J. Higgins, E. V. Plotkin, L. D. L. Scott, and A. P. F. Turner, *Anal. Chem.* **56,** 667 (1984).

[10] G. Davis, in "Biosensors: Fundamentals and Applications" (A. P. F. Turner, I. Karube, and G. S. Wilson, eds.), p. 247. Oxford Univ. Press, London and New York, 1987.

[11] L. C. Clark, *Biotechnol. Bioeng. Symp.* **3,** 377 (1972).

[12] I. J. Higgins, W. J. Aston, D. J. Best, A. P. F. Turner, S. G. Jezequel, and H. A. O. Hill, in "Microbial Growth on C₁ Compounds" (R. L. Crawford and R. S. Hanson, eds.), p. 297. Am. Soc. Microbiol., Washington, D.C., 1984.

[13] G. Davis, H. A. O. Hill, W. J. Aston, I. J. Higgins, and A. P. F. Turner, *Enzyme Microb. Technol.* **5,** 383 (1983).

may be constructed in a vigorously stirred vessel containing the following: ruptured cell suspension (about 24 g of protein in 50 mM phosphate buffer, pH 7.0), 600 ml; polyethylene glycol solution (MW 1000, 50% v/v), 1400 ml; potassium phosphate solution (50% w/v, pH 7.0), 1050 ml; methanol (100 mM), 350 ml. The system is brought to equilibrium by stirring for 5 min, and the phases are separated by centrifugation (5 min at 2,500 g). The two phases should be carefully decanted into a separating funnel and the bottom, clear, pale yellow layer containing the enzyme removed. Diafiltration against 10 liters of phosphate buffer (50 mM, pH 7.0, containing 10 mM methanol) and concentration to about 200 ml may be achieved using a Peillicon Cassette System (Millipore, 11-15 Peterborough Road, Harrow, Middx., HA1 2YH) with a molecular weight cutoff of 10,000. About 600 mg of 95% pure enzyme (specific activity 5–12 units/mg) is produced by this procedure. The high nucleic acid content of this preparation may be reduced by protamine sulfate (1% w/v) precipitation and further purification achieved by gel filtration HPLC.

Methanol dehydrogenase is a dimeric quinoprotein of molecular weight around 120,000 which catalyzes the oxidation of methanol (via formaldehyde) to formate. The natural electron acceptor appears to be cytochrome c, but this couple is only retained in anaerobic preparations.[14] Aerobically purified enzyme requires ammonium ions (or primary amines) as an activator and exhibits an elevated pH optimum, from pH 7.0 in the coupled system to in excess of pH 9.0. Methanol dehydrogenase is described here as an example of a class of dehydrogenases that contain pyrroloquinoline quinone and use mediators as electron acceptors.[14] Methanol dehydrogenase is the best characterized quinoprotein, but so far has not proved amenable to covalent immobilization, is unstable in the absence of methanol, and requires an activator. Other quinoproteins (e.g., glucose dehydrogenase) do not suffer these disadvantages.[15]

Mediators

A mediator should (1) readily participate in redox reactions with both the biological component and the electrode, effecting rapid electron transfer; (2) be stable under the required assay conditions; (3) not participate in side reactions during the transfer of electrons e.g., reduction of oxygen; (4) have an appropriate redox potential away from that of other electrochemically active species that may be present in samples; (5) be unaffected by a wide range of pH; (6) preferably be nontoxic; (7) be amenable

[14] J. A. Duine and J. Frank, *Trends Biochem. Sci.* **Oct.,** 278 (1981).
[15] E. J. D'Costa, A. P. F. Turner, and I. J. Higgins, *Biosensors* **2,** 71 (1986).

to immobilization. These criteria are met to varying extents by ferrocene and some of its derivatives.[7] Of the commercially available ferrocenes (Strem Chemicals, 7 Mulliken Way, Dexter Industrial Park, P.O. Box 108, Newburyport, MA 01950), ferrocene and 1,1′-dimethylferrocene, with $E°$ values of 165 and 100 mV (versus SCE), respectively, are most suitable for incorporation into immobilized enzyme anodes. The soluble derivative, ferrocene monocarboxylic acid ($E° = 275$ mV versus SCE), is a useful alternative to hexacyanoferrate in homogeneous assays and other ferrocene derivatives have been synthesized for incorporation into commercial devices.[16] Other soluble mediators such as ferricyanide, $N,N,N′,N′$-tetramethyl-4-phenylenediamine (TMPD), and benzoquinone remain useful soluble mediators in certain assays.[7] More recent work has shown that tetrathiafulvalene (TTF)[5] and tetracyanoquinodimethane (TCNQ)[6] derivatives offer useful alternatives to ferrocene and its derivatives for incorporation in enzyme electrodes and immunoassays.

The potential of the working electrode should be held sufficiently positive of the $E°$ value of the mediator to ensure rapid regeneration of the oxidized species and to minimize the effect of any variation in poised potential. In many biological samples, however, a compromise is necessary in order to avoid oxidation of other components; a poised potential of between 200 and 220 mV (versus Ag/AgCl) is generally suitable for 1,1′-dimethylferrocene-modified electrodes.

Electrodes

A suitable reference electrode is an Ag/AgCl electrode which generates a reference voltage (-45 mV versus SCE) while allowing the passage of current. The electrode must be of sufficient size to cope with the maximum current produced by the biosensor; a reference electrode with as large an area as possible is preferable, but one of approximately the same area as the working electrode is often adequate. The Ag/AgCl electrode requires the presence of chloride ions in the sample and usually needs to be regenerated after continual use for 1 or 2 days in biological samples. A simple Ag/AgCl electrode can be produced or regenerated by immersing silver foil (0.13 mm; BDH Chemicals, Broom Road, Poole, Dorset, BH12 4NN) in either 0.1 M HCl or 0.1 M KCl and holding the potential at $+400$ mV (versus SCE) for approximately 30 min at a current density of about 0.4 mA/cm^2. Connections to the foil may be soldered, but should be insulated by, for example, coating in epoxy resin (Ciba-Geigy

[16] J. M. McCann, "Pharmaceuticals and Health Care." Online Publications, Pinner, England, 1987.

Plastics and Additives, Duxford, Cambridge, CB2 4QA). More elaborate electrodes may be produced by various coating, painting, etching, or printing techniques, but generally these involve sophisticated equipment designed for mass production rather than research.

Working electrodes may be constructed from either gold, platinum, or carbon. The last two materials are favored, with carbon being the most appropriate for enzyme immobilization. Platinum electrodes produce substantially lower background currents, however, and are therefore particularly useful when low limits of detection are required.

Platinum guaze (50–80 mesh) and foil (24.5 μm upward) are suitable electrode materials and may be obtained from, e.g., Johnson Matthey Chemicals (Orchard Road, Royston, Herts., SG8 58G), Englehard Industries (Saint Nicholas House, Nicholas Road, Sutton, Surrey, SM1 1EN), and BDH Chemicals. Gauze electrodes have a large surface area, but they are difficult to clean, requiring electrochemical cycling (-1.0 V to $+1.0$ V at 0.2 V/sec) in 0.5 M sulfuric acid for about 15 min prior to use. Platinum foil may be cleaned with cotton wool and aluminum oxide/water paste (particle size ~0.3 μm). The thinner foils, however, must be supported by gluing to a glass or plastic backing using epoxy resin.

Carbon may be purchased in a variety of forms suitable for the construction of biosensors: charcoal (BDH Chemicals); graphitized carbon felt (Le Carbonne, South Street, Portslade, Sussex, BN4 2LX); reticulated viterous carbon (Hitemp Materials, 3 Cedars Avenue, Mitcham, Surrey, CR4 1HN); carbon fiber (Courtaulds Carbon Fibres, P.O. Box 16, Coventry, CV6 5AE); "Ultracarbon" rod (Union Carbide, Fountain Precinct, Balm Green, Sheffield, S1 3AE); Papyex and Graphoil 0.5-mm carbon foil (Le Carbonne and Union Carbide, respectively). Carbon paste and carbon foil electrodes have proved the most useful in the laboratory.

Carbon paste electrodes, suitable for use with entrapped enzymes, can be conveniently housed inside the wide end of a Pasteur pipet or other forms of glass tube.[4,8] A 25-mm length should be cut from the end of a pipet and the sharp edges smoothed with emery paper. Electrical connection may be made to the paste using a platinum disk (5.0 mm) bonded to a wire with conductive epoxy resin (Johnson Matthey Chemicals). The disk is glued inside the glass tube approximately 2 mm from the end, using nonconductive epoxy resin (Ciba-Geigy Plastics and Additives), with the connecting wire protruding from the long end of the tube. When the resin is set the platinum should be cleaned with aluminum oxide paste and the cavity packed with a carbon paste of the following composition: charcoal, 2.5 g; 1,1'-dimethylferrocene (Strem Chemicals), 125 mg; liquid paraffin, 1.5 ml. The exposed face of the paste should be smoothed to leave a shallow hollow in which soluble enzyme may be placed.

Carbon foil electrodes may also be constructed using glass tubing as a base.[6] Figure 3A illustrates a general purpose electrode suitable for use with a covalently attached enzyme. A carbon foil disk (4 mm) cut with a cork borer (number 5) is bonded to the end of a glass tube using nonconductive epoxy resin (Ciba-Geigy Plastics and Additives). Electrical connection is made by gluing a wire to the electrode using a conductive epoxy resin (Johnson Matthey Chemicals). The electrical connection is strengthened and insulated by filling the area immediately behind the electrode with nonconductive resin. Heating in an oven for 1.5 hr at 60° helps to set the resin. The electrode is modified by applying 15 μl of 0.1 M 1,1'-dimethylferrocene (or other mediator, such as TCNQ or TTF) dissolved in toluene or acetone and air-drying.

Figure 3B illustrates an alternative electrode configuration incorporating both a working and a reference electrode. The electrodes may be cut out from foil and bonded onto a plastic backing. Connecting leads should be insulated with resin. Various more elaborate manufacturing techniques may be used to deposit electrodes on to nonconductive supports.[16]

The reference electrode should be placed in close proximity to the working electrode in order to minimize any error in the measured potential due to the electrical resistance of the solution. The ohmic potential drop in a cell is a function of the current and the solution resistance (iR_s).

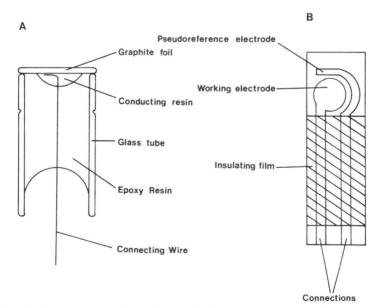

FIG. 3. Two designs for electrodes suitable for covalent attachment of enzymes.

The biosensors described are usually unaffected by the iR_s drop when operated at a sufficiently high potential (i.e., within the plateau region of the current versus voltage plot) in biological solutions. In experiments where iR_s may be high, a three-electrode cell arrangement with a potentiostat is preferable.[8] A reference electrode (e.g., SCE) can be brought into effective close contact with the working electrode using a saturated potassium chloride solution bridge ending in a Luggin capillary. The auxiliary electrode (normally platinum) should be housed behind a medium glass frit in order to reduce fouling and minimize the required reaction mixture volume. Glass frits may also prove useful in protecting reference electrodes when two electrode configurations are used in the presence of high concentrations of protein.

Immobilization

Bioelectrochemical assays may be performed using enzyme and mediator dissolved in an electrolyte. Suitable media for glucose oxidase are 100 mM phosphate/perchlorate buffer, pH 7.5, and 50 mM phosphate buffer, pH 7.5, containing 150 mM NaCl. Methanol dehydrogenase assays are optimal in 250 mM borate buffer, pH 10.5, containing 50 mM NH$_4$Cl. The reaction mixture is stirred over the electrodes, and the current/time integral (i.e., number of coulombs) determined on addition of either substrate or enzyme. This coulometric method is extremely sensitve,[12] but lacks the convenience and speed of immobilized configurations. Assays at high substrate concentration can be particularly tedious since all the substrate must be consumed. The linear range of the technique, however, is very wide.

When enzyme is retained at an electrode surface a steady-state current may be established, which is proportional to substrate concentration over a certain range. Major factors in determining the performance of an enzyme electrode are the enzyme loading (both amount and activity of enzyme retained per unit area) and the diffusion barrier offered by the immobilized layer or any associated membrane. The immobilization will modify the apparent enzyme kinetics. As a result, some enzyme electrodes exhibit a linear range extending into substrate concentrations that would saturate a conventional homogeneous enzyme assay.

Enzymes for which no satisfactory chemical immobilization procedure exists, such as methanol dehydrogenase, may be retained at a carbon paste electrode using either dialysis (MW cutoff 10,000) or polycarbonate (0.03 μm pore size) membrane.[4,13] A disk of membrane (12 mm) is held in place over the end of the glass tube with a neoprene O ring trapping 20–40 μg of enzyme. Care must be taken in the construction of the probe to ensure that the enzyme does not leak out during use. Methanol dehydrogenase probes must be operated in a high pH buffer containing ammonium ions and stored in the presence of methanol. The endogenous current due to methanol must be allowed to decay prior to use, but it should be noted that the probe is unstable in the total absence of methanol. These latter problems are specific to the use of methanol dehydrogenase and have not been encountered with other soluble enzyme probes. Methanol sensors prepared by this procedure respond very rapidly to alcohol concentrations up to 0.2 mM (Table I) and are unaffected by variation in oxygen concentration.

Glucose oxidase is strongly adsorbed to carbon foil electrodes placed in a 12.5 mg/ml solution of the enzyme and may be held in place by cross-linking with glutaraldehyde (2.5% v/v in 200 mM phosphate buffer, pH 7.0). A method for covalent attachment of enzyme to ferrocene-modified carbon electrodes has been described using water-soluble carbodiimide.[9] The electrode tip is immersed in 1 ml of 150 mM 1-cyclohexyl-3-(2-morpholinoethyl)carbodiimide metho-p-toluene sulfonate (Sigma Chemical Company) in 100 mM acetate buffer, pH 4.5, for 80 min at room temperature. After washing with distilled water, the electrode is placed in a stirred solution of 0.1 M carbonate buffer, pH 9.5, containing 12.5 mg/ml glucose oxidase, for 90 min. The electrodes may be conveniently stored at $-20°$ in 50 mM phosphate buffer, pH 7.5. Electrodes prepared by this procedure are relatively stable and will respond rapidly to a broad range of glucose concentrations (Table I). Their response is largely independent of pH over the range 6–9 and is insensitive to variation in oxygen concentration. The sensors will operate without further modification in water, plasma, heparinized blood, and some fermentation media.

Recent work has shown that the stability of mediated enzyme electrodes can be markedly improved by adopting a better immobilization technique[17] based on the method of Barbaric et al.[18] Glucose oxidase (100 mg) and 10 mg of sodium-*meta*-periodate are stirred together in 5 ml of 0.2 M sodium acetate buffer, pH 5.5 overnight at 4° in a darkened vial. The

[17] S. L. Brooks, R. E. Ashby, A. P. F. Turner, M. R. Cadder, and D. J. Clarke, *Biosensors* **3**, 45 (1987).

[18] S. Barbaric, B. Kozulie, I. Leustek, B. Pavlovic, V. Cesi, and P. Mildner, *Eur. Congr. Biotechnol. 3rd*, **1**, 307 (1984).

[19] A. Z. Preneta, "Studies on lactate oxidizing enzymes and their application to ferrolene-based enzyme electrodes for lactate" Ph.D. Thesis, Cranfield Institute of Technology, 1987.

[20] J. M. Dicks, W. J. Aston, G. Davis, and A. P. F. Turner, *Anal. Chim. Acta* **182**, 103 (1986).

[21] B. H. Schneider "Biosensor and Bioelectrocatalysis studies of enzymes immobilized on graphite electrode materials" Ph.D. Thesis, Canfield Institute of Technology, 1987.

TABLE I
PERFORMANCE OF SOME BIOSENSORS BASED ON FERROCENE-MODIFIED ELECTRODES

Substrate	Enzyme	Immobilization	Range (mM)	Linearty (mM)	Response time (sec to 95%)	Half-life in continuous use at 30° (hr)	Ref.
Glucose	Glucose oxidase (EC 1.1.3.4)	Covalent	<0.1 to 70	0 to 30	60 to 90	24 to >600[b]	9
Glucose	Glucose dehydrogenase (pyrroloquinoline-quinone) (EC 1.1.99.17)	Covalent	<0.25 to 4.0[a]	0 to 4	<30	36	15
Methanol	Alcohol dehydrogenase (acceptor) (EC 1.1.99.8)	Entrapped	1×10^{-3} to 0.2	0 to 0.1	<20	1.5	12
Carbon monoxide	Carbon monoxide dehydrogenase (EC 1.2.99.2)	Entrapped	$<2 \times 10^{-5}$ to $>7 \times 10^{-2}$	0 to 0.068	<10	~6	4
L-Lactate	L-Lactate oxidase	Covalent	1 to 4	0 to 1	75	~10	19
Amino acid	L-Amino acid oxidase (EC 1.4.3.2)	Entrapped	<1 to 15	0 to 4	120	18	20
Glycolate	(S)-2-Hydroxy-acid oxidase (EC 1.1.3.15)	Entrapped	<1 to 20	0 to 7	180	3	20
Galactose	Galactose oxidase (EC 1.1.3.9)	Entrapped	<1 to 40	0 to 20	180	~1.5	20
NADH	Dihydrolipoamide dehydrogenase (EC 1.8.1.4)	Covalent	0.1 to 3	0 to 1	60	ND[c]	21

[a] Up to 20 mM using glutaraldehyde immobilization.
[b] Unpublished results.
[c] ND, Not determined.

enzyme is desalted into 0.2 M acetate buffer using Sephradex G-25 (e.g., Pharmacia PD-10 prepacked columns). The eluent normally contains ca 10–12 mg glucose oxidase/ml and may be stored at 4° for up to 2 weeks. Carbon electrodes are prepared in the normal way and dipped in a solution of hexadecylamine (1 mg/ml in chloroform) for 15 min. The electrodes are air dried and modified with mediator as before. The modified electrodes are then placed in the periodate-modified glucose oxidase solution for 90 min resulting in a covalent immobilization (via a Schiff base) to the amine coated electrode. The stability can be further enhanced by cross-linking the enzyme using a solution of 14 mM adipic dihydrazide in 0.2 M acetate buffer, pH 5.5.

Sensor Development

The two biosensors detailed above are given as examples of a variety of possible configurations. Table I summarizes the performance of some immobilized enzyme electrodes.[4,6,8,11–13] There is considerable scope for improving the characteristics of these sensors, however, and this development work is being pursued in collaboration with commercial companies. (Recently, a pen-sized mediated glucose sensor has been developed by Genetics International and launched in the United States.[11b])

Some general points of relevance to the application of these biosensors in the laboratory, however, may be made. The amperometric response of the more stable configurations (e.g., glucose oxidase-based probes) tends to decay approximately exponentially with time. Since the electrodes described yield relatively large currents (10^{-5} A), a preconditioning step, involving overnight running in 10 mM glucose, may be used to produce more stable electrodes. This technique, however, is less applicable to miniaturized electrodes where the current density becomes a critical factor.

Electrochemical interferences may be compensated for by using a second cell containing a working electrode lacking active enzyme. Providing the correct circuitry is used, several working electrodes may be used with a single reference electrode and the necessary compensation carried out in the software. An additional means of reducing interference is by the use of selectively permeable or absorbing membranes. Some batches of polycarbonate membrane (e.g., 0.03 μm from Nucleopore, 7035 Commerce Circle, Pleasanton, California 94566), for example, restrict the access of ascorbate while allowing the relatively free passage of glucose. At present, this is not a sufficiently consistent property of the commercially available product. Various membranes may also be used to reduce fouling by proteins and microorganisms in practical situations. The substrate range of enzyme electrodes and their behavior in the presence of bio-chemical inhibitors generally reflects that of the native enzyme. In the case of oxidases, however, oxygen is no longer required for the reaction and the probes will function anaerobically.

Both the speed with which the response is read and the range of an enzyme electrode may be improved using a transient instead of steady-state measurements. A transient response may be elicited either by flow-ing samples past the electrode in short pulses or by intermittent applica-tion of the working potential. Such techniques may also improve the stability of sensors. Several sensors may be used concurrently to improve reproducibility, and cross-checking alogirthms may be included in the software. Well-characterized sensors may also be amenable to drift cor-rections carried out in the software. Most biosensors have a limited linear range but respond nonlinearly to higher substrate concentrations. The full range of response may be exploited by the use of nonlinear regression analysis or lookup tables within programs.

Acknowledgments

The author is a Senior Research Fellow of the British Diabetic Association. He would like to thank his present and past colleagues at Cranfield Institute of Technology for helpful comments and in particular Dr. G. Davis, Dr. D. J. Best, Dr. W. J. Aston, Dr. G. Ramsay, Dr. E. J. D'Costa, Mr. S. L. Brooks, and Prof. I. J. Higgins.

[8] Immobilized Flavin Coenzyme Electrodes for Analytical Applications

By LEMUEL B. WINGARD, JR. and KRISHNA NARASIMHAN

Approaches for Flavin Electrodes

Flavins are compounds that typically contain the isoalloxazine ring system (Fig. 1). The flavins serve as the coenzyme for well over 100 oxidoreductase enzymes and in addition are active redox compounds by themselves. The application of flavins in analytical chemistry can be based either on their coenzyme properties or on their nonenzymatic chemical reactivity. This chapter is limited to approaches based only on flavin coenzyme properties.

The objective in combining flavin coenzymes with electrodes is to obtain some type of electrical readout when the coenzyme, in conjunction with a suitable apoenzyme, encounters an appropriate substrate. Nor-

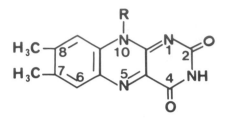

FIG. 1. Isoalloxazine ring system shown in oxidized form.

mally the electrical readout takes the form of a current (amperometric) or a potential (potentiometric) measurement. Both types of readout find use with enzyme electrodes; however, the amperometric approach is of primary interest in this chapter since it has been the method used in many of the flavin-modified electrode studies carried out so far.[1-4] A flavin–enzyme electrode by definition must contain a coenzyme, an apoenzyme, and a substrate. When the substrate undergoes oxidation, the electrons (and often protons) are transferred to the coenzyme to give reduced coenzyme. With flavoenzymes, the coenzyme usually remains attached to the apoenzyme during the course of the enzymatic reaction. Several approaches can be used to measure either the amount or the rate of formation of the reduced coenzyme. A small molecular weight mediator molecule can be selected to react with the reduced coenzyme to form a reduced mediator. The mediator then must diffuse to the electrode surface and be sufficiently electroactive to undergo oxidation at the electrode surface and thereby produce a current.[5] This method sometimes is practical; however, mediator diffusional transport resistances and the presence of other materials that might interact with the mediator can cause problems. A second approach is to apply an appropriate potential to the electrode surface with the expectation that the flavin coenzyme, that now is embedded within the apoenzyme matrix, will undergo oxidation and transfer the released electrons to the electrode surface. This approach is sound in principle, but, in practice, it has been very difficult to realize such direct electron transfer from oxidoreductase enzymes in which the coenzyme is tightly bound to the apoenzyme.[3]

    An alternative approach that may overcome some of the above problems is to attach the flavin coenzyme to the electrode surface, either by

[1] L. B. Wingard, Jr., and J. L. Gurecka, Jr., *J. Mol. Catal.* **9,** 209 (1980).
[2] K. Narasimhan and L. B. Wingard, Jr., *J. Mol. Catal.* **9,** 253 and 263 (1986).
[3] R. M. Ianniello, T. J. Lindsay, and A. M. Yacynych, *Anal. Chem.* **54,** 1098 (1982).
[4] L. Gorton and G. Johansson, *J. Electroanal. Chem.* **113,** 151 (1980).
[5] G. Nagy, L. H. Von Storp, and G. G. Guilbault, *Anal. Chim. Acta* **66,** 443 (1973).

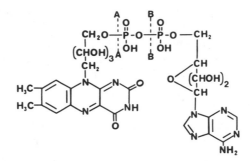

FIG. 2. Structural formulas for riboflavins, FMN, and FAD (see text for explanation).

adsorption or by covalent coupling.[6] This omits the need for a mediator; direct electron transfer from the coenzyme to the electrode surface should be relatively easy to accomplish. A key requirement of this approach is that the attached coenzyme must be capable of reassociation with the apoenzyme to give at least partial reconstitution of enzymatic activity. Several questions arise. Where on the coenzyme molecule can covalent attachment to the electrode surface be done and still meet the requirements of rapid electron transfer plus reconstitution of enzyme activity? Second, does the presence of the support cause steric interference for reconstitution of enzymatic activity, and can a spacer arm be incorporated to eliminate this steric interference? And third, how does the coenzyme undergo reoxidation and give up electrons to produce a current? This chapter deals mainly with the first question regarding the position on the flavin coenzyme molecule for making the attachment to the electrode surface so as to enable reconstitution of enzyme activity. The other two questions, for the most part, have not yet been addressed experimentally.

The structures of the three most relevant flavin compounds are shown in Fig. 2. The isoalloxazine ring system and the other portion up to A–A constitutes riboflavin with that to B–B being FMN. The entire structure in Fig. 2 is FAD. Riboflavin serves as a coenzyme for at least one natural enzyme.[7] Flavin mononucleotide (FMN) is a coenzyme for several enzymes, and flavin adenine dinucleotide (FAD) acts as the coenzyme for a large majority of the flavoenzymes.[8]

In regard to the position of attachment, X-ray crystallographic data

[6] L. B. Wingard, Jr., *in* "Enzyme Engineering: Future Directions" (L. B. Wingard, Jr., I. V. Berezin, and A. A. Klyosov, eds.), p. 339. Plenum, New York, 1980.

[7] S. Takemori, K. Suzuki, and M. Katagiri, *in* "Flavins and Flavoproteins" (T. P. Singer, ed.), p. 178. Elsevier, Amsterdam, 1976.

[8] M. Dixon, E. C. Webb, C. J. R. Thorne, and K. F. Tipton, "Enzymes," 3rd Ed., p. 480. Academic Press, New York, 1979.

suggest that the adenine portion of FAD is buried deep in the apoenzyme in *p*-hydroxybenzoate hydroxylase[9] (4-hydroxybenzoate 3-monooxygenase) and less so in glutathione reductase.[10] These are two of the very few flavoenzymes for which X-ray crystallographic data have been reported. None of the oxidase flavoenzymes have been subject to X-ray crystallographic studies, although preliminary work should be available soon on glycolate oxidase [(*S*)-2-hydroxy-acid oxidase].[11] We have attached FAD covalently to carbon electrodes through the FAD adenine amino or ribityl hydroxyl groups; however, no reconstitution of enzyme activity occurred on addition of the apoenzyme for glucose oxidase.[12] Position 8 of the isoalloxazine ring system also is a possible coupling site. There are several natural flavoenzymes where position 8 is coupled covalently to a histidine group on the apoenzyme.[13] Another study, aimed at demonstrating the solvent accessibility of position 8 of the isoalloxazine ring system, has shown this position to be very close to the protein surface in several flavoenzymes, including D-amino-acid oxidase, flavodoxin, and *p*-hydroxybenzoate hydroxylase.[14] This suggests that position 8 may not be needed for substrate or coenzyme binding to the apoenzyme and thus might be a suitable point for immobilization to an electrode surface. Furthermore, attachment at position 8 should maintain the π-orbital electron delocalization with the redox center of the isoalloxazine ring system. This may be an important factor for rapid electron transfer onto the electrode surface.

Immobilization at Position 8 of Isoalloxazine Ring System

Immobilization at position 8 constitutes a major focus of our work. The procedure is described below.[2] Studies to use other sites for coupling are in progress. Since most flavoenzymes use FAD as the coenzyme, the covalent attachment of FAD at position 8 has been the goal. However, FAD is relatively labile, thus limiting the options available for coupling at position 8; therefore, the overall scheme involves attaching riboflavin to

[9] R. K. Wierenga, K. H. Kalk, J. M. van der Laan, J. Drenth, J. Hofsteenge, W. J. Weijer, P. A. Jekel, J. J. Beintema, F. Muller, and W. J. H. van Berkel, *in* "Flavins and Flavoproteins" (V. Massey and C. H. Williams, eds.), p. 11. Elsevier/North-Holland, Amsterdam, 1982.

[10] G. E. Schulz, R. H. Schirmer, and E. F. Pai, *J. Mol. Biol.* **160,** 287 (1982).

[11] Y. Lindquist and C.-I. Bränden, *in* "Flavins and Flavoproteins" (R. C. Bray, P. C. Engel, and S. G. Mayhew, eds.), p. 277. de Gruyter, New York, 1984.

[12] O. Miyawaki and L. B. Wingard, Jr., *Biochim. Biophys. Acta* **838,** 60 (1985).

[13] D. E. Edmondson and T. P. Singer, *FEBS Lett.* **64,** 255 (1976).

[14] L. M. Schopfer, V. Massey, and A. Claiborne, *J. Biol. Chem.* **256,** 7329 (1981).

an aldehyde-derivatized electrode and then converting the immobilized riboflavin to immobilized FAD by chemical means.[2] Glassy carbon is selected as the electrode material because it is available with a very low porosity and also has been well characterized electrochemically. The low porosity is necessary so that adsorbed flavins can be eliminated completely from the flavin electrodes. In addition, the surface of glassy carbon can be derivatized with aldehyde groups. None of the procedures designed to couple intact FAD directly to aldehyde-derivatized glassy carbon have been successful,[2] so that riboflavin remains as the most useful starting material for obtaining immobilized FAD on the glassy carbon surface.

Tokai type GC-A glassy carbon rods, about 0.5 cm in diameter, are polished on the flat end, finally with 1 μm diamond powder. They are then cleaned ultrasonically, extracted overnight with methanol, and dried under vacuum. The rods can be activated by treating for 15 min in nitric acid at room temperature followed by 1 hr at 170° in sulfuric acid to generate surface carboxylic acid groups. After washing and another methanol extraction, the surface groups are converted to the acyl chlorides. This is done by placing the rods in refluxing anhydrous benzene that contains 10% (v/v) thionyl chloride. Reduction of the acyl chlorides using an excess of lithium tri-*tert*-butoxyaluminum hydride in dry diglyme at −78° gives the desired aldehyde groups on the glassy carbon surface. Care must be taken to exclude moisture since the hydride reacts violently with water. The rods, after removal from the hydride–diglyme mixture, are washed with water and dried under vacuum.

Riboflavin can be selectively activated at the 8-methyl position by bromination. Subsequent conversion of the 8-bromo derivative to the 8-lithium salt gives an intermediate that condenses readily with aldehydes to form a double bond linkage. In order to carry out this sequence, riboflavin is first acetylated to protect the ribityl hydroxyl groups from subsequent bromination. The acetylation is carried out using riboflavin (3.76 g) in dry pyridine (20 ml) and 4-dimethylaminopyridine (122 mg) as catalyst in a reaction flask protected from light and moisture and equipped for stirring. Acetic anhydride (15 ml, freshly distilled) is added dropwise over 0.5 hr. Then the reaction mix is stirred 30 hr at ambient temperature. Excess acetic anhydride is reacted by the addition of 5 g ice. The solvent can be removed at room temperature by evaporation under vacuum. Then water is added to precipitate riboflavin tetraacetate.

The recrystallized material (0.5 mmol) (from ethanol) is dissolved in dry dioxane that contains anhydrous potassium carbonate. To this mixture is added dropwise a solution of bromine (0.55 mmol) dissolved in carbon tetrachloride. Selective monobromination of position 8 has been verified by special NMR techniques.[2] The methyl group at position 7 is unchanged. The potassium carbonate acts to take up any HBr that is released and thereby greatly reduces the formation of dibromo- and tribromotetraacetate. 8-Bromoriboflavin tetraacetate (1 mmol) is dissolved in 10 ml anhydrous, deoxygenated tetrahydrofuran, and the solution is cooled to $-78°$. A solution (660 μ) of 1.55 M n-butyllithium in hexane is added to form the lithium complex at the 8-bromo position. After 10 min, the aldehyde-derivatized glassy carbon rods are added and held for 1 hr at $-78°$. By this treatment, coupling takes place to generate attached tetraacetylriboflavin. Any remaining lithium complex is destroyed by washing the rods with water. Adsorbed riboflavin tetraacetate can be removed from glassy carbon by extraction (12–24 hr) using water and methanol.[12]

The presence of attached flavin, either by adsorption or by covalent bonding, can be verified using cyclic or the much more sensitive differential pulse voltammetry. An example of some differential pulse voltammograms is shown in Fig. 3. The difference in peak potentials between the immobilized and solution tetraacetylriboflavin voltammograms can be attributed to electron rearrangement caused by covalent attachment of the isoalloxazine ring system to the glassy carbon electrode. Little response would be expected for the two control electrodes shown in Fig. 3. With the differential pulse technique much less than monolayer coverage of flavin on the glassy carbon electrode can be detected. A typical loading of covalently attached riboflavin tetraacetate is 3×10^{-11} mol/cm^2, which is slightly less than monolayer coverage. The cyclic and differential pulse voltammetry measurements are made using a three-electrode electrochemical cell. The flavin-modified glassy carbon serves as the working electrode, with a 1-cm^2 platinum screen as the auxiliary electrode and an Ag/AgCl or calomel reference electrode as the third component. For the cell electrolyte, 1 M KCl or 0.1 M Tris buffer are typical solutions. The value of the formal potential and the influence of scan rate on the peak current are two electrochemical measurements that can be used to characterize the immobilized flavin.

In order to utilize the covalently attached flavin as a coenzyme for a wide variety of enzymes, it is necessary to convert the attached riboflavin tetraacetate to attached FAD. The conversion to attached FAD can be carried out chemically, using methodology that has been checked out in solution. It may also be possible to carry out this conversion to attached FAD enzymatically[15]; although the presence of the glassy carbon may cause adsorption or modifications in the FAD synthetase complex. The chemical route begins by acid hydrolysis to remove the tetraacetate

[15] R. Spencer, J. Fisher, and C. Walsh, *Biochemistry* 15, 1043 (1976).

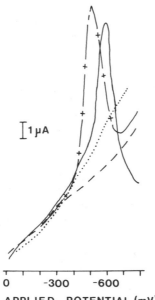

APPLIED   POTENTIAL (mV)

Fig. 3. Differential pulse voltammogram obtained with glassy carbon electrodes at 25° in 1 M KCl, pH 7.0, in the absence of oxygen and at a sweep rate of 5 mV/sec, pulse amplitude of 25 mV, and time between pulses of 0.5 sec. The potentials are measured with respect to a Ag/AgCl (1 M KCl) reference electrode. Glassy carbon electrode with attached tetraacetylriboflavin (———); untreated glassy carbon electrode with no flavin in solution (– – –) or with tetraacetylriboflavin in solution (— + —); untreated glassy carbon electrode that was exposed to the organolithium–riboflavin intermediate and then washed prior to electrochemical testing (····). Reproduced by permission from Ref. 2.

groups; this is done by holding the rods at 80° for 1 hr in 2 M HCl. The next step is phosphorylation at the ribityl 5′-position. Ice-cold phosphorus oxychloride and roughly an equal volume of methanol is mixed slowly over 0.5 hr and then stirred for 16 hr at room temperature. Then the riboflavin–glassy carbon rod electrodes are added, and the mixture is agitated gently for another 8 hr at room temperature. Phosphorylation likely occurs at multiple ribityl hydroxyl groups; but the non-5′ centers can be selectively hydrolyzed by washing the electrodes with an equal volume mixture of dioxane and water cooled at 0°. After further washing, overnight extraction, and drying under vacuum, the FMN-derivatized electrodes are ready for conversion to FAD electrodes.

The approach is to generate 5′-pyridinium phosphate-derivatized FMN on the electrodes and then condense that with a morpholidate salt that contains the phosphate–ribose–adenine group to be added. FMN

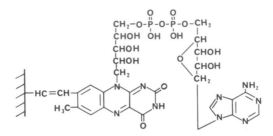

FIG. 4. Proposed structure of FAD–glassy carbon electrode.

electrodes are treated with lithium hydroxide followed by ion exchange of pyridinum for lithium. The resulting 5'-pyridinum phosphate FMN electrodes are placed in a solution of anhydrous 4-morpholino-*N*,*N*'-dicyclohexylcarboxamidinium adenosine 5'-phosphoromorpholidate (104 mg/electrode) in anhydrous pyridine. The electrodes are held in the above solution for 3 days at room temperature, protected from light and moisture. After 3 days, the electrodes are washed with water, neutralized quickly with ammonium hydroxide, extracted, and dried. The proposed FAD–electrode coupling is shown in Fig. 4.

Direct structural evidence for the conversion of attached riboflavin to attached FAD is extremely difficult to obtain because of the presence of the glassy carbon surface and the low loading. Although the riboflavin-to-FAD procedure proved efficient when carried out in solution, it would be helpful to have NMR, IR, or other spectroscopic data to verify the product on the electrode surface; so far such data have not been obtainable at less than monolayer loading. The convincing evidence that FAD is present on the electrode surface comes from the interaction of the electrode with the apoenzyme for glucose oxidase. The apoenzyme is prepared in 12% aqueous glycerol at pH 7.0.[16] If an electrode (3.0 cm²) is held overnight at room temperature in a solution containing 0.15 mg apoglucose oxidase, 0.002 units of reconstituted glucose oxidase activity can be measured after washing away unattached apoenzyme. A variety of control tess show that the activity comes from FAD, that the FAD must originate with the electrode, and that the activity could not result from any FAD from the apoenzyme.

The enzymatic activity of an FAD–apoenzyme electrode is measured by placing the electrode in a solution of 18% glucose buffered at pH 6.5 and held at 25°. Aliquots (1.5 ml) are taken at 2-min intervals and immediately added to cuvettes preloaded with 1 ml of 0.02% *o*-dianisidine in 0.1

[16] D. L. Morris, P. B. Ellis, R. J. Carrico, F. M. Yeager, H. R. Schroder, J. P. Albarella, R. C. Boguslaki, W. E. Hornby, and D. Rawson, *Anal. Chem.* **53**, 658 (1981).

M sodium phosphate buffer, pH 6.0, and 0.1 ml of 0.02% peroxidase in water. The absorbance is read at 460 nm, corrected for the removal of sampling volumes, and plotted versus time to obtain the rate of formation of H_2O_2. One unit of activity is defined as the rate of change of absorbance equivalent to the formation of 1 μmol H_2O_2/min.

A wide variety of studies remain to be carried out to obtain enhanced loading and rapid electron transfer as well as improved reconstituted glucose oxidase activity. In addition, other flavoapoenzymes need to be tested. Because of the large number of flavoenzymes, it may be possible to develop analytical flavin electrodes in which the substrate specificity is controlled by the choice of apoenzyme, and the current produced is a measure of the concentration of substrate. The preparation of FAD electrodes with FAD immobilized presumably at position 8 and with partial retention of coenzyme activity with the apoenzyme of glucose oxidase is one of several steps that need to be carried out initially for such flavin electrodes to find practical applications in analytical chemistry.

Acknowledgments

This work was supported by grants from the National Science Foundation and from the Army Research Office.

[9] Bioaffinity Sensors

By Yoshihito Ikariyama and Masuo Aizawa

Immobilization techniques employing antibodies and binding proteins for the determination of antigens and haptens have stimulated the new field of solid-phase immunoassay.[1,2] In many cases, the use of enzyme labels assures comparable sensitivity to radioisotope labels, yet has the advantages of convenience, safety, and economy. Such sensing devices as enzyme immunosensors[3,4] and immunoelectrodes[5,6] have been devel-

[1] E. Engvall and P. Perlman, *Immunochemistry* **8,** 871 (1971).

[2] B. van Weeman and A. H. W. M. Schuurs, *FEBS Lett.* **15,** 232 (1971).

[3] M. Aizawa, A. Morikawa, H. Matsuoka, and S. Suzuki, *J. Solid-Phase Biochem.* **1,** 319 (1976).

[4] M. Aizawa, A. Morioka, and S. Suzuki, *J. Membr. Sci.* **4,** 221 (1978).

[5] N. Yamamoto, Y. Nagasawa, Y. Shutto, M. Sawai, T. Sudo, and H. Tsubomura, *Chem. Lett.* **1978,** 245 (1978).

[6] N. Yamamoto, Y. Nagasawa, M. Sawai, T. Sudo, and H. Tsubomura, *J. Immunol. Methods* **22,** 309 (1978).

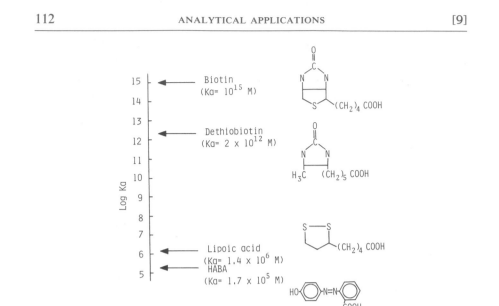

FIG. 1. Association constants (K_a) of avidin with biotin-related molecules (structures shown on the right-hand side).

oped for rapid determination of antigens. In the case of enzyme immunosensors trace amounts (10^{-9} g/ml) of antigenic protein can be determined with the use of an enzyme as an amplifiable label. The usefulness of these sensing techniques may, however, be demonstrated in the determination of antigens of large molecular size, since small molecular antigens (haptens) such as hormones and vitamins are susceptible to degradation during labeling with enzyme. It is the purpose of this chapter to introduce a new sensing method, bioaffinity sensing, and to demonstrate applicability to the sensitive determination of biotin by using membrane-immobilized ligand and enzyme-labeled avidin.

Avidin, which forms a very stable complex with vitamin H (biotin), is an egg white protein of MW 60,000. However, avidin also forms a complex of low affinity with analog compounds such as 2-(4-hydroxyphenylazo)benzoic acid (HABA) and lipoic acid.[7,8] It is a well-accepted fact that these complexes dissociate on exposure to biotin with the resulting formation of a more stable complex. Figure 1 shows the association constants of biotin and related compounds in the protein-binding reaction.

Figure 2 illustrates a generalized scheme of a bioaffinity sensor. The

[7] N. M. Green, *Biochem. J.* **89,** 585 (1963).
[8] N. M. Green, this series, Vol. 18, p. 418.

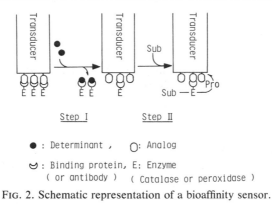

Step I   Step II

● : Determinant ,   ○: Analog

∪ : Binding protein, E: Enzyme
(or antibody)   (Catalase or peroxidase)

FIG. 2. Schematic representation of a bioaffinity sensor.

sensor is composed of a porous membrane, on which a molecular complex is formed from an analog molecule of the determinant and an enzyme-labeled binding protein, and a transducer such as a Clark-type oxygen electrode.

In this chapter we demonstrate that a metastable complex can be used as a receptor for a small molecular substance of biochemical importance.[9] A membrane-bound analog is complexed with enzyme-labeled avidin. The protein component of the metastable molecular complex is released from the membrane and then undergoes conversion to a very stable complex when it encounters a biotin molecule. The displacement is dependent on biotin concentration. Thus biotin is determined by measuring the remaining enzyme activity at the membrane matrix. The bioaffinity sensor produces high sensitivity by employing enzyme amplification techniques. The sensitivity of such a biosensor is dependent on the turnover number of the enzyme and the method employed to detect the product of the catalyzed reaction. With catalase as a label, an oxygen electrode can be used to quantitate directly the amount of catalase-labeled binding protein remaining on a membrane matrix by continuously measuring the rate of oxygen produced from hydrogen peroxide as substrate.

Results presented here give evidence that a molecular complex of low affinity is a good choice for sensing small biomolecules. The authors further discuss the effects of bioaffinity difference on sensitivity based on the results for thyroxine and insulin[10,11] and those for metabolites.[12]

[9] Y. Ikariyama, M. Furuki, and M. Aizawa, *Proc. Int. Meet. Chem. Sensors,* p. 693 (1983).

[10] Y. Ikariyama and M. Aizawa, *Proc. Sensor Symp., 2nd,* p. 97 (1982).

[11] Y. Ikariyama and M. Aizawa, *Proc. Sensor Symp., 3rd,* p. 17 (1983).

[12] J. S. Schultz and G. Sims, *Biotechnol. Bioeng. Symp.* **9,** 65 (1979).

Experimental Section

Materials

HABA [2-(4-hydroxyphenylazo)benzoic acid] and lipoic acid were products of Nakarai (Kyoto). Water-soluble carbodiimide [1-cyclohexyl-3-(2-morpholinoethyl)carbodiimide-p-toluene sulfonate] was obtained from Kokusan Kagaku (Tokyo). Cellulose triacetate was the product of Eastman Kodak (Rochester, NY). Avidin and ovalbumin (Grade III) were from Sigma (St. Louis, MO), and catalase from Tokyo Kasei (Tokyo). 4-Aminomethyl-1,8-octanediamine was supplied by Asahi Chemical (Tokyo). A phosphate buffer (0.1 M KH$_2$PO$_4$, 0.1 M Na$_2$HPO$_4$, pH 7.0), was used as a working buffer throughout. The ionic strength of this buffer was 0.1 M. All solutions were prepared with reagent grade chemicals and distilled, deionized water.

Preparation of Membrane-Bound Analog

4-Aminomethyl-1,8-octanediamine (2 ml) and 50% glutaraldehyde (400 μl) are added to cellulose triacetate (500 mg) dissolved in 10 ml of dichloromethane, and then cast on a glass plate. The aldehyde cross-links throughout the cellulose triacetate layer. After drying at room temperature for a few days the membrane becomes sticky and turns pink due to the formation of Schiff base. The membrane is cut into small pieces on the glass plate. The membrane pieces are easily peeled off the glass plate when immersed in water. The membranes are incubated for 1 hr in 1% glutaraldehyde, to introduce aldehyde groups to the membrane, and then transferred to a 1% ovalbumin solution in the phosphate buffer for 24 hr, to immobilize the protein at the membrane surface. When biotins are bound to the ε-amino group of the enzymes, biotinyl enzymes are inhibited by avidin.[13,14] Therefore, biotin analogs are immobilized to the membrane via the ε-amino group of a protein molecule to reduce steric hindrance.

Analog compounds such as HABA and lipoic acid are immobilized on the membrane by the carbodiimide method. Membranes (24–36) are incubated with 50 mg of HABA (or lipoic acid) in the presence of 90 mg of the carbodiimide. During the immobilization process the pH is controlled at 4.5. After a few hours' incubation the polymer membranes are reduced with sodium borohydride maintaining pH 9 using NaH$_2$PO$_4$. The noncovalently bound HABA (or lipoic acid) molecules are washed out with 0.1 M carbonate buffer (pH 10), and then stored in 1% ovalbumin solution (pH 7)

[13] Y. Kajiro and S. Ochoa, *J. Biol. Chem.* **236,** 3131 (1961).
[14] T. Hashimoto, H. Isano, N. Iritani, and S. Numa, *Eur. J. Biochem.* **24,** 128 (1971).

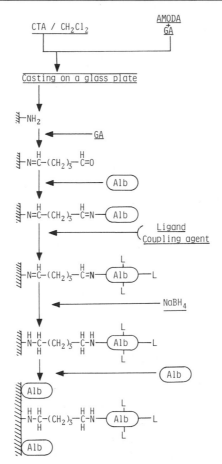

FIG. 3. Schematic representation of membrane preparation. CTA, Cellulose triacetate; AMODA, 4-aminomethyl-1,8-octanediamine; GA, glutaraldehyde; Alb, ovalbumin (or bovine serum albumin); coupling agent, carbodiimide (or glutaraldehyde).

to saturate nonspecific binding sites with albumin. UV spectroscopy shows that 3 μmol of HABA was covalently immobilized to the membrane through the ovalbumin. Figure 3 is a schematic representation of the membrane preparation.

Preparation of Catalase-Labeled Avidin

The method of glutaraldehyde (GA) conjugation used here is based on the techniques previously reported by Avrameas[15] for preparation of en-

[15] S. Avrameas, *Immunochemistry* **6**, 43 (1969).

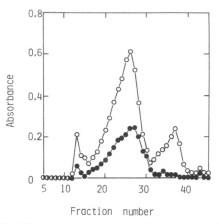

FIG. 4. Sepharose CL-6B column chromatogram of catalase-labeled avidin. Every fraction (2 ml) was monitored at 280 nm (○) and 406 nm (●).

zyme–protein conjugates. GA is a dialdehyde, and theoretically it can cross-link two proteins via the ε-amino groups of lysine by formation of a Schiff base.

Catalse (4 mg) in 1 ml of phosphate buffer is incubated with 50 μl of 2% GA for 10 min. Avidin (1.5 mg) previously dissolved in 1 ml of phosphate buffer is added to the catalase solution in a unimolar ratio and then incubated for another 20 min. After reduction with sodium borohydride of the Schiff bases between catalse and avidin, the mixture is packed in a dialysis tube to be concentrated with a hydroscopic gel [Sunwet IM-300 of Sankyo Kasei (Kyoto)]. The tube is surrounded by the gel for a few hours. The sample is then chromatographed on a Sepharose CL-6B column. Figure 4 shows a Sepharose CL-6B column chromatogram of the catalase-labeled avidin. Every fraction (2 ml) is monitored at 280 nm for the protein moieties of catalase and avidin and at 406 nm for the heme protein moiety (catalase). The first and second peaks in the chromatogram are ascribed to catalase-labeled avidin. The second peak (fractions 16 to 25) is transferred to a dialysis tube, and then concentrated to 4 ml with Sunwet IM-300. The catalase/avidin molar ratio of the conjugate is estimated for the second peak from the absorbances at 280 and 406 nm on the assumption that E_{280} of catalase and avidin and E_{406} of catalase are 14.6, 15.5, and 18.7, respectively.[16] The approximate catalase/avidin molar ratio was 1.2.

The binding capacity of catalase-labeled avidin was studied with HABA. This test is based on the use of HABA, which binds only avidin

[16] T. Yamakawa (ed.), "Data Book of Biochemistry," Vol. 1, pp. 94 and 100. Tokyo Kagaku Dojin, Tokyo, 1980.

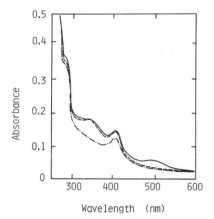

Wavelength (nm)

Fig. 5. Colorimetric assay of the binding capacity of the catalase-labeled avidin. Catalase-labeled avidin (—·—) was mixed with HABA. The complex (——) was then displaced
with biotin (– – –).

and can therefore be used as an indicator for estimating binding sites.
Addition of HABA to an excess of avidin under conditions where almost
all the HABA was bound gives a new absorption band at 500 nm (ε_{500}
increased from 600 to 34,500) and a change of color from yellow to red. At
the same time the 348-nm band of the free HABA anion ($\varepsilon = 20,700$)
almost disappears.[17] Figure 5 shows the appearance of the 500-nm absorption band when labeled avidin (0.2 mg) is mixed with free HABA (14.6 ×
10^{-9} mol) in 1 ml of phosphate buffer. The new absorption band at 500 nm
is compared with that of native avidin complexed with free HABA. The
binding capacity of the binding site of the avidin moiety is found to be
considerably decreased after conjugation. However, approximately 80%
of the labeled avidin seems to be able to bind HABA immobilized on the
membrane, since avidin has four binding sites in itself. The absorption
band disappears immediately when free biotin is added to the avidin–
HABA complex. The catalase activity in the conjugate is also determined
using a spectroscopic method.[18] The enzyme activity of catalase-labeled
avidin is approximately 40% of that of free catalase.

Assembling the Bioaffinity Sensor

An example of how to assemble a bioaffinity sensor shows the bioaffinity sensor composed of a Clark-type oxygen electrode (diameter of Pt
cathode 3 mm) and the HABA-immobilized membrane. The cathode is

[17] N. M. Green, *Biochem. J.* **94,** 23c (1965).
[18] R. F. Beers, Jr., and I. W. Sizer, *J. Biol. Chem.* **194,** 133 (1952).

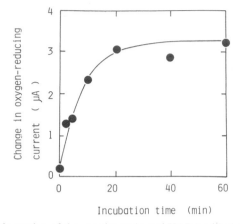

Incubation time (min)

FIG. 6. Complex formation of the membrane-bound HABA with the labeled avidin. The catalase-labeled avidin bound on the HABA-immobilized membrane was determined in the presence of hydrogen peroxide (3 mM).

wrapped with a Teflon membrane, and the HABA-bound membrane is tightly attached to the Teflon membrane. Then the electrode is immersed in phosphate buffer (1 ml) containing 0.66 mg of catalase-labeled avidin in the presence of 1% ovalbumin. The time course of complex formation is shown in Fig. 6. It seems that there are two adsorption processes during receptor preparation. The first and rapid process is a specific binding process, i.e., complexation between membrane-bound HABA and catalase-labeled avidin. This process is completed within 20 min. The second and slow process is a nonspecific adsorption of catalase-labeled avidin to the membrane surface. The second process is completed within 60 min. The time needed for receptor preparation is set at 60 min in later experiments.

Typical Response of Bioaffinity Sensors

The time course of molecular recognition with the proposed bioaffinity sensor was studied. The sensor is immersed in a biotin solution (10^{-5} g/ml) in the presence of 1% ovalbumin at 37°. Either HABA or lipoic acid is used as the analog molecule to be immobilized to the membrane. After the biosensing (see above), the sensor is transferred to the measuring medium. The change in sensor output on addition of hydrogen peroxide is recorded. Figure 7 illustrates the relation between the incubation time and the change in sensor output. The current change is caused by the molecular complex receptor remaining on the sensor. Biotin recognition by the

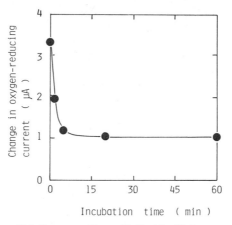

FIG. 7. Time course of biotin recognition with the bioaffinity sensor. Biotin recognition by the molecular complex receptor (Step I in Fig. 2) terminated within 10 min.

receptor (Step I of Fig. 2) is completed within 10 min after the start of biosensing.

The sensor is then immersed for 1 hr in a sample solution containing a given amount of biotin (or dethiobiotin) in the presence of 1% ovalbumin. After biosensing, the sensor is transferred to 0.1 M phosphate buffer (pH 7). The output of the oxygen electrode reaches a steady-state current within a few minutes with magnetic stirring. Then hydrogen peroxide (30 mM, 5 ml) is injected to the phosphate buffer to make the final concentration 3 mM, and the current versus time curve is recorded. In previous work[19] we have shown that the activity of several solid-phase matrix-adsorbed enzymes including catalase and glucose oxidase can readily and sensitively be determined using an amperometric-type oxygen electrode. In the case of the catalase label the change in current arises from the production of molecular oxygen according to the following catalase-catalyzed and electrode reactions:

Enzyme reaction: $2\ H_2O_2 \xrightarrow{\text{catalase}} 2\ H_2O + O_2$

Cathode reaction: $O_2 + 2\ H_2O + 4e^- \xrightarrow{\text{Pt}} 4\ OH^-$

Anode reaction: $Pb + 4\ OH^- \longrightarrow PbO_2^{2-} + 2\ H_2O + 2\ e^-$

The sensor output increases to reach another constant value within 1 min. Figure 8 shows a typical current versus time curve. The resulting

[19] M. Aizawa, A. Morioka, and S. Suzuki, *Anal. Chim. Acta* **115**, 61 (1980).

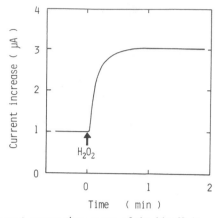

FIG. 8. Typical current versus time curve of the bioaffinity sensor. The sensor was applied to a sample solution whose biotin concentration was 20×10^{-9} g/ml, and the residual molecular complex was then determined. H_2O_2 was added at the point indicated.

change in current arises from the production of oxygen gas at the electrode surface. The change in sensor output becomes larger at higher biotin concentration. The net change in sensor output, as well as the rate of current change, correlates well with the biotin concentration. The rate portion of the curve, expressed in microamperes (μA) per minute, can be calculated by graphically extending the straight-line portion of the curve.

Results

Determination of Biotin with Membrane-Bound HABA Complexed with Catalase-Labeled Avidin

Figure 9 shows a calibration curve for biotin obtained with a bioaffinity sensor equipped with membrane-bound HABA complexed with catalase-labeled avidin. The change in sensor output decreases as the concentration of biotin is increased. The concentration of biotin ranged from 10^{-9} (90% response) to 10^{-7} g/ml (10% response). The midpoint concentration was 10^{-8} g/ml. In the case of $n = 10$, the coefficient of variation was $\pm 11\%$. The catalase-labeled avidin is not fully dissociated from the membrane. This indicates that there are two binding states of the catalase-labeled avidin, i.e., specifically adsorbed and nonspecifically adsorbed avidin. Most of the specifically bound labeled avidin may be dissociated, depending on the biotin concentration. It seems that the nonspecifically adsorbed labeled avidin is hardly dissociated even in the presence of a large amount of biotin.

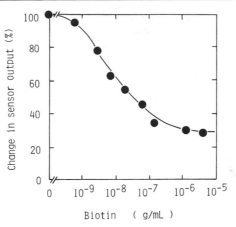

FIG. 9. Calibration curve for biotin. The receptor was membrane-bound HABA complexed with catalase-labeled avidin.

Determination of Biotin with Membrane-Bound Lipoate Complexed with Catalase-Labeled Avidin

Figure 10 shows a typical calibration curve for biotin determined by membrane-bound lipoate complexed with catalase-labeled avidin. Comparison with Fig. 9 suggests some improvement in sensitivity. However, the molecular complex receptor becomes a little bit dissociable; approximately 50% of the adsorbed labeled avidin still remains undissociated.

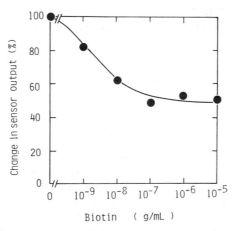

FIG. 10. Calibration curve for biotin. The receptor was membrane-bound lipoate complexed with catalase-labeled avidin.

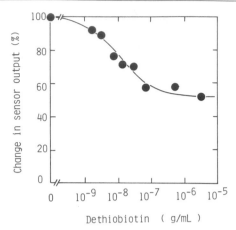

Fig. 11. Calibration curve for dethiobiotin. The receptor was membrane-bound HABA complexed with catalase-labeled avidin.

Biotin was thus determined in the concentration range 5×10^{-10} (90% response) to 5×10^{-8} g/ml (10% response).

Determination of Dethiobiotin with Membrane-Bound HABA Complexed with Catalase-Labeled Avidin

Figure 11 shows a calibration curve for dethiobiotin using membrane-bound HABA complexed with catalase-labeled avidin. Dethiobiotin is a metabolic precursor of biotin. The precursor also strongly binds avidin with an association constant of 2×10^{12} M. The concentration of dethiobiotin ranged from 10^{-9} (90% response) to 10^{-7} g/ml (10% response), respectively, whereas the midpoint dethiobiotin concentration was 10^{-8} g/ml. In the HABA–dethiobiotin combination, approximately one-half of the catalase-labeled avidin was displaced in the presence of an excess amount of biotin.

Concluding Remarks

In previous work[10] we demonstrated that thyroxine was determined in the concentration range 10^{-8} to 10^{-6} g/ml with an electrode-based bioaffinity sensor. This bioaffinity sensor was assembled from membrane-bound thyroxine complexed with catalase-labeled antibody and a Clark-type oxygen electrode. In general, the association constant in a hapten–antibody binding reaction ranges from 10^5 to 10^8 M. In a system where both membrane-bound T4 (modified T4) and free T4 and enzyme-labeled antibody are present, it is estimated that the equilibrium lies to the right (free T4–

TABLE I

DETERMINATIONS USING BIOAFFINITY SENSORS

Determinant	Membrane-bound molecule	Binding protein	Labeling enzyme	Determinable range (g/ml)	Detector
Biotin	HABA	Avidin	Catalase	$10^{-9} – 10^{-7}$	O_2 electrode
	Lipoate	Avidin	Catalase	$5 \times 10^{-10} – 5 \times 10^{-8}$	O_2 electrode
Dethiobiotin	HABA	Avidin	Catalase	$10^{-9} – 10^{-7}$	O_2 electrode
Thyroxine	Thyroxine (modified)	Antibody	Catalase	$10^{-8} – 10^{-6}$	O_2 electrode
Insulin (bovine)	Bovine insulin	Antibody	Peroxidase	$10^{-8} – 10^{-6}$	Photon counter
	Porcine insulin	Antiody	Peroxidase	$10^{-7} – 10^{-5}$	Photon counter

antibody formation) since free T4 shows higher affinity for the corresponding antibody than the modified T4 does.

In the work of Schultz and Sims, although their approach was different from ours, glucose was determined in the concentration range 10^{-3} to 10^{-2} M.[12] The affinity of glucose for the binding protein concanavalin A (Con A) is not as strong ($K_a = 10^3$ M) so that neither glucose–Con A formation nor FITC–dextran–Con A formation is favored when FITC–dextran is employed as a competitor of glucose. Table I shows the results of bioaffinity determinations of biotin-related molecules,[9] thyroxine,[10] and bovine insulin.[11]

From the present work, biotin and dethiobiotin can be determined with a sensitivity that cannot be easily attained with conventional chemical analysis.[8,20] This sensitive determination is achieved by the use of the metastable molecular complex which easily undergoes dissociation to form very stable molecular complexes with biotin (or dethiobiotin). In every combination, i.e., biotin–HABA, dethiobiotin–HABA, and biotin–lipoic acid, the affinity ratio of a determinant to an analog compound is greater than 10^7. The equilibrium seems to lie overwhelmingly to the right (complex formation between determinant and avidin). The high affinity ratio reflects the sensitivity of the proposed sensing system. Although a few points in the procedure still need to be improved, the approach looks very promising because no reagents need to be added to the sample and the receptor membrane can be used repeatedly.

[20] D. B. McCormick and J. A. Roth, *Anal. Biochem.* **34,** 226 (1970).

[10] Microbial Sensors for Estimation of Biochemical Oxygen Demand and Determination of Glutamate

By Motohiko Hikuma and Takeo Yasuda

In the fermentation industry determinations of various organic compounds are very important for control of the process. Recently many microbial sensors have been developed, and they provide useful tools for automatic determinations. Two different types of microbial sensors are described in this chapter.

Biochemical oxygen demand (BOD) is one of the most widely used and important tests in the measurement of organic pollution. The conventional BOD test requires a 5-day incubation period, and test results tend to depend on the skill of the operator. Therefore rapid and reproducible methods are desirable for the BOD test. A microbial sensor consisting of an immobilized living yeast membrane and an oxygen probe was used for the estimation of the BOD in wastewater from the fermentation process.[1-3] When the microbial sensor was inserted in a sample solution containing organic compounds, the compounds were assimilated by the yeast in the membrane. The respiration (oxygen uptake) of the yeast was raised, and the dissolved oxygen was, consequently, decreased around the membrane. The BOD of the sample solution could be estimated from the current decrease of the oxygen probe caused by the decrease in dissolved oxygen.

A large quantity of glutamic acid is produced by fermentation. Automatic determination of glutamic acid in the fermentation broth is required for control of the process. The conventional method based on an enzymatic reaction can be employed for the determination. However, this method has a major disadvantage: consumption of the expensive enzyme. A microbial sensor was constructed with immobilized microorganisms containing glutamate decarboxylase and a carbon dioxide probe.[4] When the microbial sensor was inserted in a sample solution, carbon dioxide was produced from glutamic acid by the immobilized microorganisms in the membrane. The concentration of glutamic acid could be determined from the potential increase of the carbon dioxide probe.

BOD Sensor

Preparation of the Microbial Membrane. *Trichosporon cutaneum* (AJ 4816) is cultured in 50 ml of medium (pH 6.0) containing 0.1% glucose, 0.3% malt extract (Difco Labs., Detroit, MI), 0.3% polypeptone, and 0.3% yeast extract (Takeda-yakuhin Co., Ohsaka, Japan) under aerobic conditions at 30° for 36 hr. Four vinyl spacers (O.D. 14 mm, I.D. 6 mm, thickness 50 μm) are attached to an acetylcellulose membrane (type HA, diameter 47 mm, pore size 0.45 μm, Millipore Co.). Three milliliters of culture broth is poured on the membrane as shown in Fig. 1, and the microorganisms are adsorbed on the surface of the acetylcellulose membrane using suction. The microbial membrane is cut along the spacer to be mounted to an oxygen probe.

[1] I. Karube, S. Mitsuda, T. Matsunaga, and S. Suzuki, *J. Ferment. Technol.* **55,** 243 (1977).
[2] I. Karube, T. Matsunaga, S. Mitsuda, and S. Suzuki, *Biotechnol. Bioeng.* **19,** 1535 (1977).
[3] M. Hikuma, H. Suzuki, T. Yasuda, I. Karube, and S. Suzuki, *Eur. J. Appl. Microbiol. Biotechnol.* **8,** 289 (1979).
[4] M. Hikuma, H. Obana, T. Yasuda, I. Karube, and S. Suzuki, *Anal. Chim. Acta* **116,** 61 (1980).

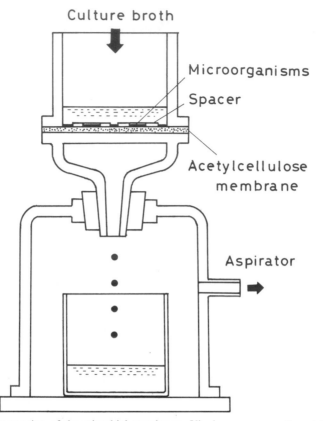

FIG. 1. Preparation of the microbial membrane. Vinyl spacers are adhered to an acetylcellulose membrane prior to the treatment. Culture broth of *T. cutaneum* is filtered
through the membrane, and the yeast cells are adsorbed on the membrane.

Assembly of the Microbial Electrode. The microbial sensor is illustrated in Fig. 2. The oxygen probe (Type 3021, Denkikagaku keiki Co.,
Tokyo, Japan) consists of a Teflon membrane (thickness 50 μm), a platinum cathode, an aluminum anode, and saturated potassium chloride electrolyte. The microbial membrane is placed on the Teflon membrane of the
oxygen probe and held in place with a screw cap as shown in Fig. 2.

Apparatus and Procedures. Figure 3 shows a schematic diagram of the
sensor system which consists of a jacketed flow cell (diameter 1.7 cm,
height 0.6 cm, volume 1.4 ml) containing a microbial sensor, a peristalic
pump (model MHRE-22, Watson Marlow Ltd., Falmouth, Cornwall,
UK), an automatic sampler (Model SC-160FA, Toyoh-kagaku Sangyo
Co., Tokyo, Japan), and a current recorder (Model ER-181, Yokogawa

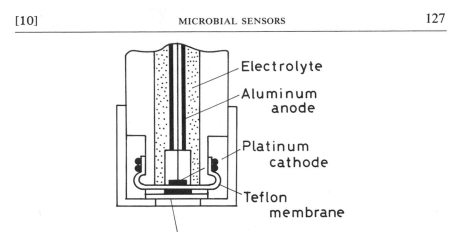

FIG. 2. Design of the microbial sensor for BOD estimation. The sensor consists of a galvanic-type oxygen probe and the membrane containing living yeast.

Electric Works Co., Japan). The temperature of the microbial sensor is maintained at 30 ± 0.2° by passing warm water through the jacket. A phosphate buffer solution (10 mM, pH 7.0) is transferred to the flow cell at a flow rate of 1.0 ml/min together with air. The flow rate of air is kept at 250 ml/min in order to obtain stable mass transfer rates of dissolved oxygen and substrates to the microbial membrane by agitation of the

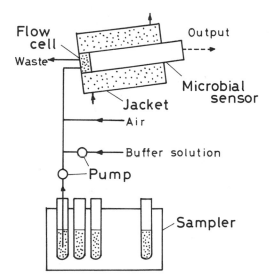

FIG. 3. Schematic diagram of the BOD sensor system. The current recorder is not illustrated.

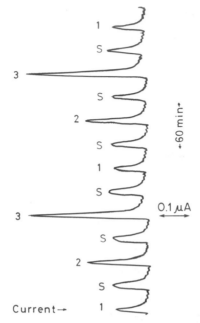

FIG. 4. Response curves of the BOD sensor for standard solutions [BOD: 50 mg/liter (1), 25 mg/liter (2), 12.5 mg/liter (3)] and wastewater from the fermentation factory (S).

liquid in the flow cell. When the output current of the microbial sensor reaches a steady-state value (approximately 1 day after the preparation of the microbial membrane), sample solutions are pumped into the flow cell at a flow rate of 0.2 ml/min for 7 min at 30-min intervals.

A solution containing glucose (150 mg/liter) and glutamic acid (150 mg/liter) is employed as a standard (5-day BOD: 220 mg/liter) for the microbial sensor. The 5-day BOD of wastewater is determined by the Japanese industrial standard method.[5]

Results and Discussion. Figure 4 shows response curves for the microbial sensor using immobilized living *T. cutaneum.* The base-line current is obtained with a buffer solution which corresponds to the endogenous respiration level of the immobilized yeast. When the standard solution containing glucose and glutamic acid is pumped into the flow cell of the microbial sensor, the organic compounds permeate through the porous membrane and are assimilated by the immobilized yeast. Consumption of oxygen by the yeast begins and causes a decrease in dis-

[5] Japanese Industrial Standard Committee, "Testing Methods for Industrial Waste Water," JIS K 01020, p. 33. Jpn. Ind. Stand. Comm., Tokyo, Japan, 1974.

solved oxygen around the membrane. As a result, the output current of the oxygen probe decreases markedly with time until a steady state is reached within 18 min (steady-state current is not shown in Fig. 4). The steady state indicates that oxygen consumed by the immobilized yeast and oxygen diffused from the sample solution to the membrane are in equilibrium. When a sufficient number of the living yeast cells are maintained in the membrane, the consumption of oxygen caused by assimilation of organic compounds depends on the diffusion rate of the organic compounds to the membrane. The diffusion rate corresponds to the concentration of the organic pollutants. Therefore, the steady-state current depends on the BOD of the sample solution. When only buffer solution is transferred to the flow cell, the output current of the microbial sensor returns to its initial level within about 20 min. The time required to obtain the steady-state current is too long for practical use. Therefore, a pulse method (a certain volume of a sample is pumped for a certain time) is employed for the estimation of BOD.

Figure 4 shows the response curves of the BOD sensor for standard solutions and wastewater. The current decreases are proportional to the BOD of standard solutions up to 60 mg/liter. Reproducibility is good enough for practical use (standard deviation 3%). The BOD of a sample solution can be estimated from the relationship between the current decrease and the BOD value (calibration curve), which is obtained with the standard solutions. The microbial sensor was applied to estimation of 5-day BOD of various wastewaters, and good comparative results were observed for the two methods [$r = 0.97$ for wastewaters from the fermentation process ($n = 18$), $r = 0.86$ for domestic sewage ($n = 20$), and $r = 0.82$ for organic wastewater from an electronic factory ($n = 37$)].[6] The yeast membrane can be used for more than 1400 assays a month.

Glutamate Sensor

Preparation of the Microbial Sensor. Escherichia coli (ATCC 8739) is cultured in 50 ml of medium containing 1.0% glucose, 1.0% casamino acid (Difco Labs.), 0.5% K_2HPO_4, 0.2% monosodium glutamate, and 0.2% yeast extract under aerobic conditions at 30° for 20 hr. The cells are harvested by centrifuging at 6000 g and washed twice with 0.2% potassium chloride solution. The cells are freeze-dried at $-30°$ and stored in a refrigerator.

A carbon dioxide probe (Model 5036, Radiometer Copenhagen, Denmark) consisting of a silicone rubber membrane, a combined glass elec-

[6] K. Harita, Y. Ohotani, M. Hikuma, and T. Yasuda, *Proc. Workshop Instrum. Control Water Wastewater Treat. Transp. Syst., 4th,* 529 (1985).

trode, and buffer solution is used for the sensor. About 3 mg of the freeze-dried cells are mixed with one drop of water and coated on both sides of a nylon mesh (60 mesh, diameter 7 mm). The nylon mesh containing the microorganisms is placed on the silicone rubber membrane of the probe and covered with a cellophane membrane (Type C, Technicon, Tarry-town, NY) to entrap the microorganisms between the two membranes.

Apparatus and Procedures. The apparatus for the gluatamate sensor is nearly identical with that for the BOD sensor shown in Fig. 3. A buffer solution containing 0.8% pyridine, 0.5% KH_2PO_4, and 0.5% NaCl (pH adjusted to 4.4 with HCl) is transferred to the flow cell (temperature kept at 30°) at a flow rate of 1.7 ml/min with a flow rate of nitrogen gas kept at 500 ml/min. This removes dissolved carbon dioxide from the sample solution and maintains a stable mass transfer rate of substrate to the microbial membrane as described previously. Sample or standard solutions are pumped at a flow rate of 0.7 ml/min for 3-min periods at 6-min intervals.

Results and Discussion. When a sample solution containing glutamic acid is pumped into the flow cell, glutamic acid permeates through the cellophane membrane and is metabolized by the microorganisms to produce carbon dioxide[7]:

$$HOOCCH_2CH_2CH(NH_2)COOH \xrightarrow{\text{glutamate decarboxylase}} HOOCCH_2CH_2CH_2NH_2 + CO_2$$

The enzyme reaction is carried out at pH 4.4 which is sufficiently below the pK value of carbon dioxide (6.34 at 25°) to allow the partial pressure of carbon dioxide around the membrane to increase. As a result, the potential of the carbon dioxide probe increases with time. The maximal potential observed for the sample injection period of 3 min is 96% of the steady-state value which was obtained with sample injection for more than 5 min. When only buffer solution begins to flow after sample injection, the potential of the sensor returns to its initial value. A linear relationship is observed between the maximal potentials of the carbon dioxide probe (from 50 to 100 mV) and the logarithm of the glutamic acid concentrations (from 100 to 800 mg/liter).

The microbial sensor scarcely responds to other amino acids except for glutamine. The response to glutamine can be decreased, if necessary, using acetone-treated *E. coli*. Under anaerobic conditions with nitrogen gas bubbling through the flow cell, the sensor does not respond to organic substances such as glucose (7800 mg/liter). The influence of inorganic salts on the response is negligible under usual conditions. The microbial sensor is used for the determination of glutamic acid in fermenta-

[7] E. F. Gale, *Adv. Enzymol.* **6**, 1 (1946).

tion broth. Good agreement is obtained between results from the microbial sensor and the conventional method[8] (correlation coefficient 0.99 for 45 experiments). The response of the microbial sensor is almost constant for more than 3 weeks and 1500 assays.

Conclusions

Various microbial sensors using immobilized microoganisms and electrochemical devices have been developed for fermentation processes.[9] The advantages of these sensors are as follows: (1) The sample solution can be measured directly without any pretreatment such as filtration and dialysis. (2) No reagent is required except for a buffer solution. (3) Usually, the microbial sensors are stable for more than 3 weeks. (4) Microbial sensors can be readily constructed according to the method proposed above.

[8] Technicon Industrial Systems, "Monosodium Glutamate in Fermentor Solutions," Industrial Method No. 210-72A. Technicon Ind. Syst., Tarrytown, New York, 1973.
[9] M. Hikuma, T. Yasuda, I. Karube, and S. Suzuki, *Ann. N.Y. Acad. Sci.* **369,** 307 (1981).

[11] Hybrid Biosensors for Clinical Analysis and Fermentation Control

By ISAO KARUBE

Introduction

The determinations of urea and creatinine in biological fluids are diagnostically important tests. External dialysis (artificial kidney) has become so widely practiced that precise, rapid, and simple methods for creatinine and urea determinations are very desirable. The determination of urea is also important in the fermentation industry.

Spectrophotometric methods have conventionally been used for measuring creatinine and urea.[1,2] However, these methods are time-consuming and complicated, and other substances present in a sample solution can interfere seriously. Electrochemical monitoring systems for urea and creatinine have been developed for clinical analysis.[3,4] These electrode

[1] M. Jaffé, *Z. Physiol. Chem.* **10,** 391 (1886).
[2] R. L. LeMar and D. Bootzin, *Anal. Chem.* **29,** 1233 (1957).
[3] T. Huvin and G. A. Rechnitz, *Anal. Chem.* **46,** 246 (1974).
[4] G. G. Guilbault and G. Nagy, *Anal. Chem.* **45,** 417 (1975).

systems consist of immobilized enzymes and ammonia gas-sensing electrodes or ammonium ion-sensing electrodes. However, ions or volatile compounds such as amines sometimes interfere with the determination of ammonia and ammonium ion.

A microbial sensor consisting of immobilized nitrifying bacteria and an oxygen electrode has been developed for the amperometric determination of ammonia.[5,6] This chapter describes new sensors for the amperometric determination of urea and creatinine.[7,8] They are based on amalgamation of enzyme reactions and bacterial metabolism.

Creatinine Sensor

Creatinine deiminase (EC 3.5.4.21) hydrolyzes creatinine to N-methylhydantoin and ammonium ion, and the ammonia produced is successively oxidized to nitrite and nitrate by nitrifying bacteria, which have already seen use in an ammonia sensor.[5,6] The bacteria have not been completely characterized but are known to be a mixed culture of *Nitrosomonas* sp. and *Nitrobacter* sp. The sequence of reactions is as follows:

$$\text{Creatinine} + \text{H}_2\text{O} \xrightarrow[\text{deiminase}]{\text{creatinine}} \text{NH}_4^+ + N\text{-methylhydantoin}$$

$$\text{NH}_4^+ \xrightarrow{\textit{Nitrosomonas} \text{ sp.}} \text{NH}_2^- \xrightarrow{\textit{Nitrobacter} \text{ sp.}} \text{NO}_3^-$$

The reacting bacteria consume oxygen, so that the oxygen decrease may be detected by an oxygen electrode. The hybrid creatinine sensor thus consists of a cellulose dialysis membrane, immobilized creatinine deiminase, immobilized nitrifying bacteria, and an oxygen electrode.

Experimental Procedures

Culture of Bacteria. The microorganisms used in this study are nitrifying bacteria. Activated sludges containing nitrifying bacteria were obtained from a fermentation factory (Ajinomoto Central Research Laboratory, Kawasaki, Japan). Nitrifying bacteria are cultured in 1 liter of a medium (pH 8.0) containing the following: $(\text{NH}_4)_2\text{SO}_4$, 6 g; K_2HPO_4, 0.5 g; $\text{MgSO}_4 \cdot 7\text{H}_2\text{O}$, 0.05 g; $\text{CaCl}_2 \cdot 2\text{H}_2\text{O}$, 4 mg; CaCO_3, 10 g. The medium is placed in a 2-liter suction flask, and the bacteria are cultured under aerobic conditions [one-third volume of gas per volume of culture medium per min (1/3 VVM) with forced aeration] for more than 2 weeks at 25°.

[5] M. Hikuma, T. Kubo, T. Yasuda, I. Karube, and S. Suzuki, *Anal. Chem.* **52**, 1020 (1980).
[6] I. Karube, T. Okada, and S. Suzuki, *Anal. Chem.* **53**, 1852 (1981).
[7] I. Kubo, I. Karube, and S. Suzuki, *Anal. Chim. Acta* **151**, 371 (1983).
[8] T. Okada, I. Karube, and S. Suzuki, *Eur. J. Appl. Microbiol. Bioetchnol.* **14**, 149 (1982).

Calcium and magnesium salts are sterilized separately to avoid precipitation. Fresh culture medium of the same composition described above is added as a pH indicator. When the color of the indicator in the broth changes from pink to yellow due to nitrite produced by bacteria, sterilized potassium carbonate solution (1 M) is added to the broth to keep the pH constant (pH 8.0).

Immobilization of the Bacteria. The bacteria (300 mg wet weight) are suspended in 5 ml of sterilized water, and the suspension is dripped onto a porous acetylcellulose membrane (Millipore Co., type HA, 0.45 μm pore size, 4.7 mm diameter, 150 μm thickness) with slight suction. The bacteria are retained on the acetylcellulose membrane, which is then rinsed with 6 ml of pH 8.5 buffer solution (10 mM borate with hydrochloric acid).

Immobilization of Creatininase. A triamine membrane is used to immobilize creatinine deiminase (2.3 U/mg, donated by Kyowa Hakko Kogyo Co., Tokyo, Japan). It is a triacetylcellulose membrane containing 1,8-diamino-4-aminomethyloctane. After being cut to the proper size, the membrane is immersed in a 1% glutaraldehyde solution for 1 hr, rinsed in distilled water, and immersed in a solution containing creatinine deiminase (1 mg/ml in 10 mM phosphate buffer, pH 7.0) for more than 15 hr at 4°.

Apparatus. A schematic diagram of the sensor is shown in Fig. 1. The oxygen electrode (DG-5, Ishikawa Seisakusho, Tokyo, Japan) consists of a Teflon membrane, a platinum cathode, a lead anode, and 30% sodium hydroxide as electrolyte. The acetylcellulose membrane retaining nitrifying bacteria is carefully attached to the Teflon membrane of the oxygen electrode so that the bacteria are between the two membranes. The immobilized enzyme membrane is attached over the acetylcellulose membrane.

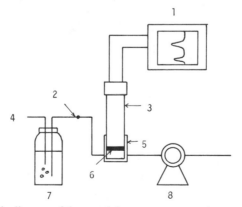

FIG. 1. Schematic diagram of the creatinine sensor system: 1, recorder; 2, injection port; 3, oxygen electrode; 4, air; 5, flow cell; 6, enzyme–bacteria membrane; 7, carrier solution; 8, peristaltic pump.

These membranes are covered with a dialysis membrane (seamless cellulose tubing, Visking Co., USA) and fastened with a rubber O ring. The analytical system consists of the sensor inserted in a flow cell, a carrier solution, a peristaltic pump (Model SJ-1211, Mitsumi Scientific Industry, Tokyo, Japan), an injection port, and a recorder (Model EPR 20 A, TOA Electronics, Tokyo, Japan).

Procedure. The temperature of the carrier solution and the flow cell is maintained at 30 ± 1°. The carrier solution, borate buffer (pH 8.5, containing 1 mg/liter chloramphenicol), is saturated with dissolved oxygen and transferred to the flow cell by the peristaltic pump. When the current reaches a steady state, 0.1 ml of sample solution is injected into the system, after which the current gradually decreases and reaches a minimum value. The flow cell is then washed with buffer solution, and the current gradually returns to the initial level. The maximum current decrease is used as a measure of the creatinine concentration.

Results and Discussion

Figure 2 shows typical response curves, obtained by the procedure described above. The initial steady current indicates an endogenous respiration level of the immobilized nitrifying bacteria. When a sample solution containing creatinine is applied to the sensor system, creatinine permeates through the dialysis membrane and is decomposed to ammonia and N-methylhydantoin. The ammonia is assimilated by the immobilized bacteria. At the same time, the bacteria consume dissolved oxygen from

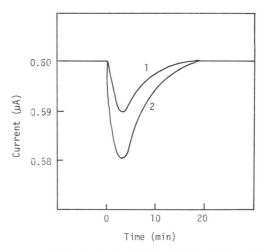

FIG. 2. Response to creatinine: 1, 5 mg/dl; 2, 10 mg/dl. Conditions: 30°, pH 8.5, flow rate 1 ml/min.

around the membrane, so that the current from the oxygen electrode decreases markedly and reaches a minimum value within 3 min. Carrier solution is continuously transferred to the system; therefore, the current gradually returns to the initial value within 15 min. When the bacteria and the enzyme are sufficiently active, the current difference between the initial and minimum values depends on the creatinine concentration.

There is a linear relationship between the current difference and the concentration of creatinine below 100 mg/dl, with a maximum current difference of approximately 0.23 μA at 50 mg/dl. The minimum detectable concentration of creatinine using this sensor is 5 mg/dl (0.44 mM).

The reproducibility of the sensor was examined with a 50 mg/dl solution. The determination was done about 20 times a day. The current difference is reproducible to within 6.7% (within-day) and 8.8% (day-to-day). The selectivity of the hybrid biosensor for creatinine was examined with solutions containing other organic compounds. The sensor does not respond (current = 0.00 μA) to urea (5 mg/liter), uric acid (10 mg/dl), citrate (50 mg/dl), pyruvate (5 mg/dl), glucose (100 mg/dl), arginine (2 mg/dl), glutamine (10 mg/dl), or EDTA (80 mg/dl). Therefore, the selectivity of the hybrid biosensor is satisfactory.

The reusability of the creatinine sensor was examined with a sample solution containing 50 mg/dl creatinine. The current output of the sensor eventually decreases, but it can be used for more than 3 weeks and 300 assays.

Urea Sensor

Urease (EC 3.5.1.5) hydrolyzes urea to ammonium ion and carbon dioxide. The improved ammonium ion sensor has been developed and applied to the determination of ammonium ion.[9] A urea sensor therefore consists of immobilized urease and an ammonium ion sensor.

The ammonium ion is utilized as the sole source of energy by nitrifying bacteria. The nitrifying bacteria require dissolved oxygen for nitrification, and the consumption of oxygen is determined by an oxygen electrode. Therefore the concentration of urea can be indirectly determined from the current decrease of the oxygen electrode.

Experimental Procedures

Culture and Immobilization of Bacteria. The cultivation of nitrifying bacteria and immobilization of the bacteria are performed as described above.

[9] T. Okada, I. Karube, and S. Suzuki, *Anal. Chim. Acta* **135,** 159 (1982).

Preparation of the Urease–Collagen Membrane. Collagen fibrils are well dispersed with a homogenizer (Ikemoto Scientific Co., Tokyo, Japan). The suspension (40 g) containing urease (0.3%) and collagen (0.7%) is cast on a Teflon plate and dried at 20° for 10 hr. The membrane is immersed in phosphate buffer solution (pH 7.0) containing 2.0% glutaraldehyde for 5 min, at 4°, washed with cold water, and dried again at room temperature. The membrane thickness is about 25 μm.

Assembly of the Microbial Electrode. The porous membrane retaining immobilized bacteria is cut into a circle (0.8 cm diameter) and soaked in buffer solution (100 mM KH_2PO_4–NaOH buffer, pH 8.0), because nitrifying bacteria are activated in alkaline conditions (pH 7–9). The bacterial membrane is fixed on the surface of the Teflon membrane of the oxygen electrode. The bacterial membrane is covered with a gas-permeable Teflon membrane (Millipore Co., Type FH, 0.5 μm pore size) and fastened with rubber O rings. An alkaline bed (glycine–NaCl–NaOH buffer, pH 10), a cation-exchange membrane (Selemion Type CMV, Asahi Glass Co., Tokyo, Japan), and a urease membrane are also fixed on the electrode. The microorganisms are immobilized between the two porous membranes (Teflon membrane and gas-permeable membrane).

Procedures. The system consists of a cell (100 ml) with a microbial electrode, a magnetic stirrer (1,000 rpm), and a recorder (Model CDR-11A, TOA Electronics). The microbial electrode is inserted into a sample solution (phosphate buffer, pH 7.0, 50 ml). The sample solution is saturated with dissolved oxygen and stirred magnetically during the measurement. The temperature of the cell is controlled at 30° ± 0.1° by a thermostatted bath. The current obtained from the electrode is directly displayed on the recorder through 2 kΩ resistance.

Results and Discussion

The initial steady-state level of the output current corresponds to the endogenous respiration activity of the immobilized bacteria in the phosphate buffer (pH 7.0). When the sensor is placed in a sample solution containing urea, the output current decreases due to oxygen consumption by immobilized nitrifying bacteria. This output current reaches a steady-state level within 7 min. The steady-state level of the output current depends on the concentration of urea. When the sensor is inserted in tap water, the output current of the sensor returns to its initial level within 2 min. Thus the total time required for an assay of urea is 9 min by the steady-state method.

Inhibition of nitrifying bacteria with nitrite formed by the microorganisms is not observed during the experiments (150 assays). The response of the sensor for urea does not change with the harvesting time of the nitrifying bacteria. The current decrease (current difference between initial and steady-state currents) was plotted against the concentration of urea. Figure 3 shows the calibration curve of the sensor. A linear relationship is observed between the current decrease and the concentration of urea in the range of 2–200 mM.

The reproducibility of the current decrease was examined using the same sample. The current decrease is reproducible within ±5% in 25 experiments when a sample solution containing 150 mM of urea is employed. The standard deviation is 7.5 mM (in 25 experiments, 150 mM urea). The microbial sensor may be applied to the determination of urea in human urine. Urine is diluted with the phosphate buffer and employed for experiments. The concentration of urea was determined by the biosensor and a conventional method.[2] A good agreement is obtained between results from the sensor and the conventional method (correlation coefficient 0.97).

The long-term stability of the sensor was examined with a sample solution containing 150 mM of urea. The output current of the sensor is almost constant for more than 10 days and 150 assays. Therefore, the microbial sensor can be used to assay urea for long periods of time. The sensor proposed gives an economical and reliable method for the assay of urea in biological fluids.

The selectivity of the sensor for urea was examined. The sensor does not respond to amines (diethylamine, propylamine, and butylamine) nor to nonvolatile compounds such as glucose, amino acids, and metal ions

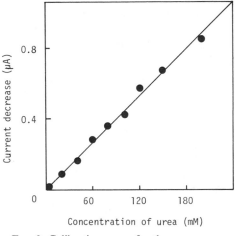

FIG. 3. Calibration curve for the urea sensor.

(sodium, potassium, calcium). Therefore, the selectivity of the sensor is satisfactory.

Conclusion

In conclusion, the use of hybrid biosensors appears to be quite promising for the amperometric determination of creatinine and urea.

Acknowledgment

The author is grateful to Prof. Shuichi Suzuki, The Saitama Institute of Technology, Dr. Tadashi Okada, and Mrs. Izumi Kubo, Tokyo Institute of Technology, for their help and encouragement during this study.

[12] Bioselective Membrane Electrodes Using Tissue Materials as Biocatalysts

By Garry A. Rechnitz

Introduction

The entire field of potentiometric membrane electrodes has been greatly stimulated by a renewed emphasis on biosensors.[1] The pattern of electrode development and the relationship of the several active subareas has been viewed[2] in terms of a "family tree"; in this context it can also be seen that the enzyme electrodes of the 1960s and early 1970s represent a subset of a more extended hierarchy of biocatalytic membrane electrodes which includes biosensors using mitochondria, bacterial cells, or tissue materials.[3] At the present time it is still too early to say whether the use of such natural materials will be of major practical value in analysis or of general interest to the wider scientific community. To date, most of the published work in this field has come from just a few laboratories.

Scientific Basis

When tissue-based potentiometric membrane electrodes were first introduced in 1978, considerable skepticism was expressed as to whether such biosensors could work at all or yield an analytical response. Could

[1] M. A. Arnold and M. E. Meyerhoff, *Anal. Chem.* **56**, 20R (1984).
[2] G. A. Rechnitz, *Anal. Chem.* **54**, 1194A (1982).
[3] G. A. Rechnitz, *Science* **214**, 287 (1981).

adequate biocatalytic activity, selectivity, and stability be obtained with animal or plant tissue sections as immobilized biocatalysts at potentiometric membrane electrodes? Although a 1980 comparison study[4] of potentiometric glutamine sensors using enzyme, mitochondria, bacterial cells, and tissue slices as biocatalysts put to rest some of these concerns, it must be admitted that the success of tissue-based sensors results partly from a combination of fortunate circumstances.

Such electrodes could not function, for example, if the time scale of events involving the biocatalytic layer were not compatible with the potentiometric sensing elements (to date, most of the reported electrodes have used gas-sensing probes with response times of the order of minutes). Since the relaxation time for translocation of substrates into intact cells is in the $10^1–10^3$ sec range, much slower than enzymatic reaction and turnover, the combination of tissue biocatalysts with potentiometric gas sensors turns out to be a happy one.

Another fortunate coincidence is the relative constancy of the rates of gas diffusion in both mammalian and nonmammalian tissues. Diffusion constants for CO_2 in rat cerebrum, frog skeletal muscle, dog diaphragm, cat urinary bladder, and rat skeletal muscle, for example, differ by less than 10% at physiological temperatures. This ensures a predictable response for tissue electrodes where the measured product is gaseous (most electrodes have used either CO_2 or NH_3 sensors). Moreover, the input/output characteristics of such biocatalysts display strict chemical stoichiometry. This was elegantly demonstrated in a recent study by LaNoue et al.[5] where NH_3 formation in liver mitochondria was exactly equal to the substrate influx/efflux difference under all conditions studied. Thus, analytical response proportionality can be expected for such sensors.

Frankly, none of these factors was even considered in our early (starting in 1975) empirical research on whole cell biosensors. The first experiments with cultured cells and animal tissue sections concentrated primarily on *finding* desired biocatalytic activity which could be coupled to potentiometric membrane electrodes. By good fortune we readily found both bacterial strains and mammalian organ tissue sections (from beef liver and porcine kidney) with high, stable biocatalytic activity for amino acids. In retrospect it is clear that other tissue materials, which might have been chosen for initial investigation, would have been much less satisfactory in terms of substrate selectivity and time stability. Indeed, such problems were encountered a few years later when the focus of our work shifted to plant tissue biocatalysts.

[4] M. A. Arnold and G. A. Rechnitz, *Anal. Chem.* **52,** 1170 (1980).
[5] K. F. LaNoue, A. C. Schoolwerth, and A. J. Pease, *J. Biol. Chem.* **258,** 1726 (1983).

The recent history of tissue-based potentiometric sensors has brought many improvements in construction, tissue selection, selectivity enhancement, and optimization of other experimental variables so that several electrode systems of practical analytical utility can now be found in the literature. Nevertheless, this research area is still very much in its infancy and much needs to be done in terms of both practical development and fundamental investigation. As is the case with any new research field, it is always tempting to expand the outer boundaries of possible investigation rather than to explore in-depth the mundane, but more fundamental, aspects of sensor operation. Thus, we find ourselves in a situation where the empirical use of novel and ever more exotic tissue materials has substantially outrun our biochemical and electrochemical understanding of the molecular mechanisms involved.

General Construction of Tissue Electrodes

All the tissue-based electrodes developed in our laboratory, whether from plant or animal sources, have utilized gas-sensing membrane sensors because of inherent advantages in simplicity of operation and excellent selectivity properties. Typically, a thin slice of tissue material is mechanically entrapped at the tip of the gas sensor, although fragile tissues may need to be stabilized with an immobilization procedure (*vide infra*). This arrangement is no more tedious than the preparation of conventional enzyme electrodes and, often, is much simpler. Table I gives an overview of the tissue electrodes to be discussed in more detail.

TABLE I
TISSUE-BASED BIOCATALYTIC MEMBRANE ELECTRODES

Substrate measured	Tissue employed	Sensing electrode	Comments
Glutamine	Porcine kidney	NH_3	First tissue electrode developed
Adenosine	Mouse small intestine	NH_3	Selectivity enhancement strategy
Adenosine mono-phosphate	Rabbit muscle	NH_3	Biocatalytic activity increase
Glutamate	Yellow squash	CO_2	First use of plant tissue
Guanine	Rabbit liver	NH_3	—
Pyruvate	Corn kernels	CO_2	Localization of activity
Cysteine	Cucumber leaves	NH_3	First leaf-based electrode

Animal Tissue-Based Membrane Electrodes

Porcine Kidney

Mammalian tissues are a rich source of enzymes; indeed, many commercially available enzymes are isolated from such tissues. Thus, early development of tissue-based electrodes focused on the use of liver and kidney slices as biocatalysts, and, because of the ready availability of fresh organs from food animals in markets and slaughterhouses, initial experiments on such tissues employed those from bovine and porcine sources. To the writer's best recollection the first working tissue electrode, developed in the summer of 1978, used bovine liver at an NH_3 gas sensor. However, this electrode was soon superseded, because it requires an auxiliary enzyme, by the simpler porcine kidney electrode with excellent response and selectivity properties for the amino acid glutamine. Formal publication followed[6] in early 1979 and a patent was subsequently granted.[7]

Porcine kidney is known to contain substantial levels of the enzyme glutaminase (EC 3.5.1.2) which catalyzes the reaction shown in Eq. (1).

$$\text{Glutamine} + H_2O \rightleftharpoons \text{glutamate} + NH_3 \tag{1}$$

Thus, the combination of porcine kidney tissue and an NH_3 gas sensor permits construction of a biocatalytic electrode for glutamine.

To prepare this electrode, a thin (0.3–0.5 mm thick) slice of cortex tissue from fresh porcine kidney is placed between a dialysis membrane and a monofilament nylon net (105 μm pore size). The resulting tissue "sandwich" is placed on the sensing tip of the NH_3 gas-sensing electrode with the dialysis membrane toward the electrode and the nylon mesh facing the sample solution. The screw cap of the electrode assembly serves to hold the membranes and tissue slice snugly in place if care is taken to match the circumference of the tissue slice with that of the electrode tip. The electrode can be conditioned by exposure, with rapid stirring, to 25–50 ml of working buffer (0.1 M phosphate buffer at pH 8.0 containing 0.02% sodium azide preservative) until base-line potentials are obtained. The integrity of the porcine kidney slice biocatalyst can be tested using the strategy outlined in Fig. 1. It can be seen from the results summarized at the bottom of Fig. 1 that the immobilized kidney slice maintains its integrity well under operating conditions with minimal cell

[6] G. A. Rechnitz, M. A. Arnold, and M. E. Meyerhoff, *Nature (London)* **278,** 466 (1979).
[7] G. A. Rechnitz, M. A. Arnold, and M. E. Meyerhoff, *U.S. Patent* 4,216,065 (1980).

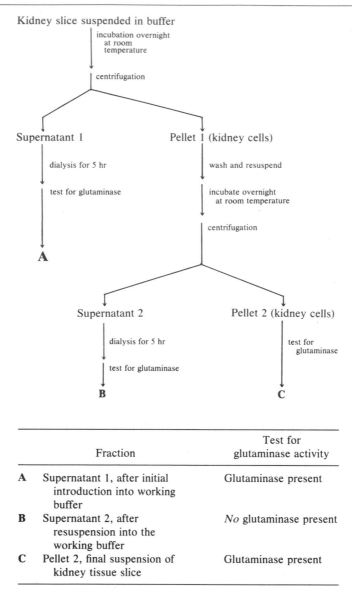

Kidney slice suspended in buffer

incubation overnight
at room
temperature

centrifugation

Supernatant 1   Pellet 1 (kidney cells)

dialysis for 5 hr   wash and resuspend

test for glutaminase   incubate overnight
                                 at room temperature

                                 centrifugation

A

Supernatant 2   Pellet 2 (kidney cells)

dialysis for 5 hr   test for
                                 glutaminase

test for glutaminase

B   **C**

	Fraction	Test for glutaminase activity
A	Supernatant 1, after initial introduction into working buffer	Glutaminase present
B	Supernatant 2, after resuspension into the working buffer	*No* glutaminase present
C	Pellet 2, final suspension of kidney tissue slice	Glutaminase present

FIG. 1. Schematic sequence of a study to determine the integrity of porcine kidney cells in a pH 7.8, 0.1 M phosphate, 0.02% sodium azide buffer at room temperature.

breakage or enzyme release. This suggests that this tissue biosensor should have favorable lifetime properties, and indeed useful operating lifetimes of 30–60 days have been routinely obtained.

Calibration curves can be constructed by plotting the steady-state potential versus the negative logarithm of standard glutamine concentrations. Best results are obtained when pH, temperature, stirring rate, and sample volume are carefully controlled. No special precautions are necessary when working with purely aqueous samples or synthetic laboratory standards, but biological or physiological samples may require additional treatment. For example, human control serum contains unacceptably high levels of background ammonia and must be pretreated with Dowex 50W-X8 cation-exchange resin, while cerebrospinal fluid needs to be diluted with working buffer, owing to viscosity problems, prior to measurement.

The choice of operating pH for tissue-based sensors often involves a compromise between the requirements of the electrode element and the biocatalyst. In the case of the porcine kidney sensor for glutamine, the pH optimum of the biocatalyst falls at 7.8 while the NH_3 gas sensor has its best response at a pH above 10. However, since the NH_3 sensor can be effectively used at lower pH values it is practical to operate the kidney electrode system at pH 7.8 to take advantage of the optimal biocatalytic activity.

Once the operating pH has been selected, the choice among possible buffer components to achieve that pH may involve additional biochemical considerations. In the case of the porcine kidney, it has been reported[8] that the glutaminase biocatalytic activity involves both a phosphate-dependent and a phosphate-independent pathway. Since the former is both activated and stabilized by phosphate, it is reasonable to employ a phosphate buffer at pH 7.8.

Sodium azide is added to the buffer system as an antimicrobial agent and tissue preservative. Without use of sodium azide, a visible film of bacterial contamination builds up on the tissue slice with time. Since these contaminating bacteria can introduce competing biocatalytic pathways, the proper response and selectivity of the electrode will be compromised and eventually lost. Fortunately, this problem is completely eliminated by the addition of even 0.02% sodium azide to the buffer. (Caution: Sodium azide solutions must be kept alkaline to avoid the formation of toxic gases.)

---

[8] M. Crompton, J. D. McGivan, and J. B. Chappell, *Biochem. J.* **132,** 27 (1973).

Under these operating conditions, the porcine kidney-based biosensor gives an excellent potentiometric response to glutamine over the 6.4×10^{-5} to 5.2×10^{-3} M range with a detection limit of less than 2×10^{-5} M. Response slopes of at least 50 mV per concentration decade are readily obtained with response times of 5–7 min to steady-state potentials.

The selectivity of the porcine kidney electrode for glutamine is remarkable. Tests of urea, creatinine, glycine, L-alanine, L-arginine, L-glutamic acid, L-histidine, L-valine, L-serine, L-aspartic acid, L-asparagine, D-alanine, and D-asparagine as possible interferents showed negligible effects even at 6.6×10^{-3} M concentration levels. It had been reported[9] that porcine kidney contains significant activity of the enzyme D-amino-acid oxidase (EC 1.4.3.3) which catalyzes the reaction shown in Eq. (2)

$$\text{D-Amino acid} + H_2O + O_2 \rightleftharpoons \text{a 2-oxo acid} + NH_3 + H_2O_2 \qquad (2)$$

and, thus, would result in interference from D-amino acids. However, such interference was not found in our investigation, probably because the oxidase enzyme requires flavin adenine dinucleotide (FAD) for its activity and the weakly bound[10] FAD is dialyzed off during electrode preconditioning.

Under proper storage conditions, the porcine kidney tissue material retains its selective biocatalytic activity for a surprisingly long time. We have routinely prepared electrodes with excellent response characteristics from stock tissue which had been stored at $-25°$ for 6 months.

A possible bioanalytical application of porcine kidney tissue electrodes involves the measurement of glutamine in human cerebrospinal fluid (CSF). Glutamine levels in CSF are elevated in such disease states as Reye's syndrome, and quantitative information about such levels aids in differential diagnosis. In early experiments with commercially available control samples of CSF, an unexpected interference from high glucose levels was found, apparently due to glycolysis. The interference manifested itself via nonlinear calibration curves and generally *low* results; it demonstrates once again that extra care must always be taken in going from synthetic laboratory standards to complex biological samples if unpleasant surprises are to be avoided. Fortunately, in this case the interference can easily be overcome by addition of millimolar iodoacetamide, a glycolysis inhibitor, which specifically inactivates the enzyme glyceraldehyde-3-phosphate dehydrogenase (EC 1.2.1.12). With these precautions, the porcine kidney tissue electrode yields excellent quantitative glutamine

[9] M. Dixon and K. Kleppe, *Biochim. Biophys. Acta* **96,** 368 (1965).
[10] G. G. Guilbault and E. Hrabankova, *Anal. Chim. Acta* **56,** 285 (1971).

measurements in human cerebrospinal fluid in good agreement with the results obtained by a colorimetric enzyme method.[11]

Mouse Small Intestine

In most cases, tissue materials will contain numerous enzymes and will be capable of sustaining multiple metabolic pathways. Thus, it becomes necessary to enhance the desired biocatalytic activity or to suppress competing pathways to achieve adequate substrate selectivity for analytical purposes. The development of an adenosine-selective biosensor using mouse small intestine tissue in conjunction with an ammonia gas-sensing membrane electrode illustrates a possible strategy for dealing with such a situation.

The enzyme adenosine deaminase (EC 3.5.4.4) is frequently isolated from intestinal mucosal cells,[12] but a tissue electrode constructed with such cells also exhibits response to some other adenosine-containing nucleotides. This interfering response could be the result either of pathways involving other deaminating enzymes or of the enzymatic formation of intermediates degraded by the primary enzyme. Since the choice between these alternatives is not obvious, the successful elimination of nucleotide interference requires experimental verification of the undesirable metabolic pathway or pathways and suppression by the use of inhibitors or stereospecific blocking agents to achieve an enhanced electrode selectivity for adenosine.

Mouse small intestine mucosal cells are obtained by cutting open a 1-cm segment of the intestine, rinsing with working buffer (0.1 M Tris–HCl, 0.2 M phosphate, and 0.02% sodium azide at pH 9), and scraping off the mucosal cells from the support tissue with a razor blade.[13] The cells are then immobilized by BSA–glutaraldehyde cross-linking[14] as follows. The cell suspension and 25 μl of 15% BSA solution are placed directly on the gas-permeable membrane of the ammonia electrode and stirred with 10 μl of 25% glutaraldehyde solution until the mixture becomes viscous. The membrane is allowed to dry for 20 min, then placed in water for 15 min, and finally in 0.1 M glycine solution for 15 min. The electrode assembly can then be stored in the pH 9 working buffer. When the mouse mucosal electrode is operated in Tris buffer, it shows excellent response to adenosine (54.5 mV per decade slope and 1.9×10^{-5} M detection limit) but

[11] A. M. Glasgow and K. Dhiensiri, *Clin. Chem.* **20,** 642 (1974).
[12] T. G. Brady and C. I. O'Donovan, *Comp. Biochem. Physiol.* **14,** 101 (1965).
[13] F. Dickens and H. Weil-Malherbe, *Biochem. J.* **35,** 7 (1941).
[14] M. Mascini and G. G. Guilbault, *Anal. Chem.* **49,** 795 (1977).

serious interferences by adenosine mono-, di-, and triphosphate nucleotides. Elimination of these interferences and enhancement of the desired selectivity for adenosine requires a multifaceted strategy based on a knowledge of biochemical and metabolic pathways.

A very simple adjustment involves the choice of operating pH. One can readily show that the pH optima for the interfering nucleotides do not coincide with that of adenosine. Thus, pH 9.0 is a good choice from the point of view of selectivity enhancement, but this parameter alone is not sufficient to adequately suppress the interferences.

It, therefore, becomes necessary to identify the principal interfering pathway in operation. Three possibilities may be considered, e.g., nonspecific deamination,

$$AXP + H_2O \rightarrow IXP + NH_3 \tag{3}$$

specific deaminating pathways,

$$AMP + H_2O \xrightarrow[\text{(EC 3.5.4.6)}]{\text{AMP deaminase}} IMP + NH_3 \tag{4}$$

or an alkaline phosphatase pathway,

$$AMP + H_2O \xrightarrow[\text{(EC 3.1.3.1)}]{\text{alkaline phosphatase}} \text{adenosine} + \text{phosphate} \tag{5}$$

$$\text{Adenosine} + H_2O \xrightarrow[\text{(EC 3.5.4.4)}]{\text{adenosine deaminase}} \text{inosine} + NH_3 \tag{6}$$

By working with various activators and inhibitors, it can be shown that the third alternative is operative in this case. Since it is known[15] that phosphate inhibits the activity of alkaline phosphatase, the simple addition of 0.2 M K_2HPO_4 to the buffer serves to suppress the interference. Under such operating conditions, adenosine can be selectively measured in the presence of adenosine mono-, di-, and triphosphate or such substances as adenine, guanosine, and guanine. Of course, a fair amount of biochemical detective work had to be carried out in order to achieve this goal.

Rabbit Muscle

There are situations in the preparation of biocatalytic potentiometric membrane electrodes where the specific activity of an enzyme per available surface area at the tip of the electrode is not great enough to give good response characteristics. In such cases, it may be possible to use intact tissue material, where the biocatalytic is already optimized, as

[15] H. N. Fernley and P. G. Walker, *Biochem. J.* **104,** 1011 (1967).

alternative biocatalysts.[16] Such a situation is encountered in the case of the rabbit muscle-based biosensor for adenosine 5'-monophosphate (AMP).

Both the enzyme-based and tissue-based AMP electrodes utilize the biocatalytic process shown in Eq. (7). The commercially available en-

$$AMP + H_2O \xrightarrow{\text{biocatalyst}} \text{inosine 5'-monophosphate} + NH_3 \qquad (7)$$

zyme, AMP deaminase (EC 3.5.4.6) can be used to construct electrodes, but the specific activity of the enzyme preparation is too low to be used directly and tedious concentration steps are required. By using a thin slice of rabbit muscle rich in the biocatalytic activity of interest, the concentration steps are eliminated and an AMP sensor with excellent response characteristics and a long useful lifetime can be prepared.

Quantitative determinations of the specific biocatalytic activities of rabbit muscle tissue and the enzyme preparation show that the tissue has approximately 50 times the specific activity of the enzyme suspension on an equal volume basis. This has two desirable consequences in favor of the tissue material. First, the tissue-based electrode has a much longer useful lifetime (>28 days) than the enzyme-based electrode (~4 days). It should also be noted that the rabbit muscle tissue can be stored for at least 7 months at −25° without loss of biocatalytic activity. Second, the high specific activity of the tissue allows thin slices to be used for electrode construction; this results in sensors with improved dynamic properties both in terms of response times and return-to-base-line (recovery) times.

Rabbit Liver

Currently available theoretical treatments are not yet able to predict fully the behavior of tissue-based biocatalytic membrane electrodes as a function of operating conditions. Thus, it is necessary to experimentally investigate and optimize the principal variables of selectivity, sensitivity, response time, electrode lifetime, and operating conditions. Such parameters can be viewed[17] in terms of two subsets of rate processes, e.g., those of substrate and product diffusion and those of the biocatalytic reaction or reactions, which were evaluated for the case of an electrode system in which slices of rabbit liver provide biocatalytic activity for guanine at an ammonia gas-sensing membrane electrode [Eq. (8)].

$$\text{Guanine} + H_2O \xrightarrow{\text{rabbit liver}} \text{xanthine} + NH_3 \qquad (8)$$

[16] M. A. Arnold, Ph.D. thesis. University of Delaware, Newark, Delaware, 1982.
[17] M. A. Arnold and G. A. Rechnitz, *Anal. Chem.* **54,** 777 (1982).

Since rabbit liver tissue is rather soft, the biocatalytic layer was constructed by sandwiching the tissue between two dialysis membranes; this arrangement also protects the gas-sensing electrode from contamination and leakage. The resulting guanine electrode displays two unusual features. First, the guanine substrate is of limited solubility so that the electrode cannot be operated at concentrations above $3 \times 10^{-4} M$; this is not a problem in practice since samples can always be diluted. Second, and more important, the electrode cannot be effectively operated at its optimum pH of 9.5. At that pH (borate buffer, 0.02% sodium azide) the biocatalytic activity drops to 20% of its initial value in just 7 hr and becomes negligible within a single day. On the other hand, at pH 8, where only 50% of the maximal activity is initially available, there is almost no loss in biocatalytic activity over a period of 2 weeks. Thus, in this case the most practical operating pH is not the optimum pH.

The attainment of proper guanine selectivity for the rabbit liver-based sensor also presents some interesting challenges. The two main types of interferences likely to be encountered involve either the generation of the measured product from another substrate or a localized alteration in the pH at the sensor element through interfering reactions. The latter can usually be suppressed through the use of a medium with high buffer capacity, but the former requires specific biochemical manipulation. In the case of the guanine electrode, we are fortunate that none of the following interfere, even at the millimolar level: inosine, adenine, GMP, IMP, creatinine, creatine, ornithine, asparagine, serine, urea, glutamine, glutamate, threonine, lysine, valine, glycine, and arginine.

Initial work in mixed phosphate–borate buffer revealed an interference from guanosine. This interference was traced to the action of the enzyme guanosine phosphorylase (EC 2.4.2.15) via the reaction of Eq. (9). Since this reaction requires phosphate, the elimination of phosphate from the buffer also eliminated the interference.

$$\text{Guanosine + phosphate} \rightarrow \text{D-ribose 1-phosphate + guanine} \qquad (9)$$

The rabbit liver-based guanine electrode also showed interference due to adenosine and adenosine-containing nucleotides, probably via deamination. Fortunately, a search of the literature revealed[18] that manganese(II) ions inhibit such deamination at pH 8, and, indeed, we were able to suppress these interferences completely by addition of $10^{-2} M$ MnCl$_2$ to the buffer medium. It might be expected that such inhibition would require extensive incubation of the tissue with manganese(II), but in fact the inhibition (and its reversal) takes place within minutes. The combination

[18] T. Aikawa, Y. Aikawa and T. G. Brady, *Int. J. Biochem.* **12**, 493 (1980).

of such biochemical "tuning" steps with appropriate selection of membrane materials, tissue thicknesses, and immobilization procedures represents the essential elements of the best strategy for electrode optimization available at the present time.

Plant Tissue-Based Membrane Electrodes

Yellow Squash

The first potentiometric plant tissue-based electrode to be developed[19] used yellow squash as biocatalyst. Yellow squash contains the enzyme glutamate decarboxylase (EC 4.1.1.15) which, in conjunction with the coenzyme pyridoxal-5'-phosphate (PLP), degrades glutamate according to Eq. (10). Thus, by coupling yellow squash tissue with a potentiometric

$$\text{L-Glutamic acid} \longrightarrow \gamma\text{-aminobutyric acid} + CO_2 \tag{10}$$

CO_2 electrode, it is possible to construct a biosensor for L-glutamic acid. In practice, it is desirable to use the mesocarp layer of the squash for best activity and to stabilize the tissue with BSA–glutaraldehyde immobilization. The mesocarp of the squash is located just between the outer hard pericarp skin and the pulpy endocarp portions of the interior.

To prepare the sensor, one cuts a 0.3-mm layer of tissue with a razor blade and dips the slice into a solution composed of 90 μl of 11.1% BSA and 10 μl of 8% glutaraldehyde which had previously been mixed on a glass plate for 3 min. The treated squash tissue is then immobilized on the gas-permeable membrane of the CO_2 electrode with 15 μl of the BSA-glutaraldehyde solution and allowed to stand for 25 min at room temperature.

At pH 5.5 and 35° the resulting sensor gives a glutamate response over the 10^{-2} to 10^{-4} M range with slopes of 48 mV/decade and steady-state response times of 10 min. The biocatalytic activity of the tissue layer remains unchanged for at least 7 days provided some precautions are taken. Specifically, it is advantageous to add 40% glycerol to the buffer since glycerol stabilizes the activity of plant enzymes. The PLP coenzyme (3×10^{-4} M) is necessary to maintain biocatalytic activity; since PLP is only weakly bound and easily lost from the tissue, it is best to add the PLP to the buffer. Chlorhexidine diacetate (0.002%) was found to be useful as a preservative without decreasing the biocatalytic activity.

The squash tissue sensor has surprisingly good selectivity for L-glutamic acid. Pyruvic acid also gives a modest response, but D-glutamic

[19] S. Kuriyama and G. A. Rechnitz, *Anal. Chim. Acta* **131**, 91 (1981).

acid, L-glutamine, N-acetyl-L-glutamic acid, D-glucose, urea, α-ketoglu-taric acid, L-isocitric acid, L-malic acid, L-aspartic acid, L-asparagine, L-alanine, L-valine, L-leucine, L-isoleucine, L-phenylalanine, L-tryptophan, L-methionine, glycine, L-serine, L-threonine, L-tyrosine, L-lysine, L-arginine, or L-histidine do not interfere, even at the $9.1 \times 10^{-3}\ M$ concentration level. This selectivity is comparable to conventional enzyme electrodes.

Corn Kernels

The inhomogeneity of biocatalytic activity in plant materials becomes strikingly apparent in attempts to develop a pyruvate-sensing electrode using corn kernel portions in conjunction with a CO_2 gas electrode. Corn kernels are known to contain the enzyme pyruvate decarboxylase (EC 4.1.1.1); in fact, the enzyme is often isolated from this source. Thus, either the commercially isolated enzyme or corn kernel tissue can be used in conjunction with a potentiometric CO_2 electrode to construct a biosensor for pyruvate via the biocatalytic step [Eq. (11)] and, moreover, can provide a direct comparison of the isolated enzyme and tissue cases.

$$\text{Pyruvate} + \text{H}_2\text{O} \rightleftharpoons \text{acetaldehyde} + \text{CO}_2 \tag{11}$$

The immobilization procedure is the same as already described. Corn kernel tissue segments can be most conveniently obtained by freezing the kernel and taking longitudinal slices (~0.3 mm thick) with a razor blade. In fresh corn, the biocatalytic activity is fairly evenly distributed within the endosperm, aleurone, and germ portions with no activity in the pericarp and testa (which can be peeled off and discarded). On freezing there is a separation of fluid from the tissue so that the fluid is highest in biocatalytic activity at the expense of the aleurone.

The activity of the native tissue or isolated enzyme can be enhanced through the use of thiamin pyrophosphate (TPP) and magnesium ions as activators. We found a buffer medium of 0.1 M phosphate, 40% glycerol, 5 mM TPP, 5 mM MgCl$_2$, and 0.002% chlorhexidine diacetate at pH 6.5 to be optimal. Chlorhexidine diacetate serves as a preservative, and the glycerol stabilizes biocatalytic activity without adversely affecting electrode response.

Under these operating conditions, the corn kernel tissue electrode yielded pyruvate responses over the 8×10^{-5} to $3.0 \times 10^{-3}\ M$ range with slopes of 45–50 mV per concentration decade. These response characteristics are substantially superior to those obtained with the comparable electrode prepared from isolated enzyme, although the enzyme electrode does have a shorter response time than the tissue electrode. Most important is the fact that the tissue electrode has a useful lifetime of approximately 7 days, while the enzyme electrode persists for less than 1 day under similar conditions; clearly, the tissue matrix serves to stabilize the biocatalytic activity. The corn tissue electrode has good selectivity with no interference from α-ketoglutarate, L-isocitrate, D-glucose, urea, L-glutamine, or L-glutamate at the 3.2×10^{-3} M concentration level. Full details have been published elsewhere.[20]

Cucumber Leaves

The latest development in tissue-based electrodes is the use of plant leaves as biocatalysts. Plant leaves offer a particularly attractive natural structural arrangement for possible use as biocatalysts at potentiometric gas-sensing electrodes. Many leaves have a multilayer structure consisting of a waxy coating (cuticle) at the outer surface, a layer of epidermal cells, followed by a third layer (spongy mesophyll) directly under the epidermis, with the same arrangement repeated in reverse on the other side of the leaf. The cuticle is hydrophobic in nature but permits the passage of gases; gas exchange takes place through small surface openings called stomata. The spongy mesophyll layer is the most active in metabolic processes involving gases. For the construction of biocatalytic membrane electrodes, the cuticle can be detached from either the upper or lower epidermal layer and the remaining leaf structure fixed at the surface of a gas-sensing potentiometric electrode with the exposed epidermal layer contacting the sample and the gas permeable waxy cuticle facing the internal elements of the sensor.

The principle has been demonstrated[21] with the use of cucumber leaves at an NH_3 electrode to construct a sensor for L-cysteine. Such leaves have biocatalytic activity involving the enzyme L-cysteine desulfhydrolase (cystathionine γ-lyase, EC 4.4.1.1) according to Eq. (12).

$$\text{L-Cysteine} + H_2O \rightarrow \text{pyruvate} + H_2S + NH_3 \qquad (12)$$

Thus, sensors could be constructed using either NH_3 or H_2S gas-sensing electrodes, but the NH_3 case is perferable on chemical grounds.

The technique is quite simple. Cucumber plants (*Cucumis saturis*) are grown from seed in Fertilite seed starter soil. Mature leaves are detached when needed and soaked in water for 45 min; this soaking softens the cuticle and permits ready removal to expose the biochemically active epidermis. This procedure is necessary because the substrate, L-cysteine,

[20] S. Kuriyama, M. A. Arnold, and G. A. Rechnitz, *Anal. Chim. Acta* **12**, 269 (1983).
[21] N. Smit and G. A. Rechnitz, *Biotechnol. Lett.* **6**, 209 (1984).

cannot readily diffuse through the waxy cuticle layer. Leaf disks are then cut to fit the gas-sensor tip and held there with a dialysis membrane. Such a biosensor gives a response to L-cysteine in pH 7.6 phosphate buffer between approximately 10^{-3} and 10^{-5} M with a slope of about 35 mV per decade. The relatively poor slope and fairly long response times of this sensor show that further development is needed, but the long useful lifetime (up to 4 weeks) and extremely low cost of the biocatalyst suggest that leaf materials could be attractive alternatives to immobilized enzymes or cells.

Future Prospects

The field of tissue-based potentiometric biosensors is still very much in its infancy. Many new types of possible biocatalytic materials need yet to be explored and superior combinations of biocatalyst and electrochemical sensor devised. It may well be that this resarch field will become simply a laboratory curiosity, or some fruitful practical utility could emerge. At this time it is impossible to predict which of the possible alternatives, e.g., enzymes, organelles, microorganisms, or tissue materials, will ultimately emerge as the biocatalyst of choice. There seem to be intriguing possibilities not only for the animal and plant tissue biocatalysts described above but also for such exotic materials as blossoms, seeds, insect materials,and receptor tissues.

Acknowledgment

The support of the National Science Foundation is greatly appreciated.

[13] Organelle Electrodes

By FLORIAN SCHUBERT and FRIEDER W. SCHELLER

Subcellular organelles (lysosomes, chloroplasts, peroxisomes, mitochondria, microsomes) carry out several essential cell functions ranging from protein synthesis and degradation to foreign compound detoxication. In spite of this versatility, so far only mitochondrial[1-3] and microsomal[4-7] fractions have been employed in biocatalytic electrodes.

[1] G. G. Guilbault, "Handbook of Enzymatic Methods of Analysis," p. 507. Dekker, New York, 1976.

In this chapter the use of liver microsomes in biosensors is discussed. These organelle vesicles with a mean diameter of 200 nm are obtained from the endoplasmic reticulum.[8] Of about 50 different enzyme activities,[9] liver microsomes contain a monooxygenase system consisting of cytochrome *P*-450 (unspecified monooxygenase, EC 1.14.14.1), NADPH–cytochrome *P*-450 reductase (NADPH–ferrihemoprotein reductase, EC 1.6.2.4), and phospholipid. This system, which has been extensively covered in several monographs,[10–12] catalyzes the hydroxylation of a large number of both foreign and endogenous compounds (e.g., drugs, fatty acids, and steroid hormones) in a mixed-function oxidation reaction [Eq. (1)]. The system is also responsible for the aerobic oxidation of NADPH in the absence of substrates that can be hydroxylated [Eq. (2)].[6,13] The very sensitive cytochrome *P*-450 enzyme exhibits its highest

$$SH + H^+ + NADPH + O_2 \longrightarrow SOH + NADP^+ + H_2O \qquad (1)$$

$$NADPH + O_2 + H^+ \longrightarrow NADP^+ + H_2O_2 \qquad (2)$$

in vitro stability in its natural microsomal environment so that the intact organelle is best suited for application.

It will be shown how these cytochrome *P*-450 and other microsomal enzyme activities can be used to determine several substrates with organelle electrodes. By combination with coimmobilized enzymes, hybrid organelle electrodes are assembled in which a desired substrate is coupled to the cytochrome *P*-450 reaction. For a particular application a nonenzymatic microsomal reaction can also be used in an organelle sensor.

Preparation and Operation of Organelle Electrodes

Microsomes

Liver microsomes from phenobarbital-treated[14] rats or rabbits are prepared by established methods[15,16] and immediately frozen in liquid nitro-

[2] M. Aizawa, M. Wada, S. Kato, and S. Suzuki, *Biotechnol. Bioeng.* **22,** 1769 (1980).
[3] M. A. Arnold and G. A. Rechnitz, *Anal. Chem.* **52,** 1170 (1980).
[4] F. Schubert, D. Kirstein, F. Scheller, and P. Mohr, *Anal. Lett.* **13,** 1167 (1980).
[5] F. Schubert, F. Scheller, and D. Kirstein, *Anal. Chim. Acta* **141,** 15 (1982).
[6] F. Schubert, F. Scheller, P. Mohr, and W. Scheler, *Anal. Lett.* **15,** 681 (1982).
[7] I. Karube, S. Sogabe, T. Matsunaga, and S. Suzuki, *Eur. J. Appl. Microbiol. Biotechnol.* **17,** 216 (1983).
[8] G. E. Palade and P. Siekevitz, *Biophys. Biochem. Cytol.* **2,** 171 (1956).
[9] H.-U. Schulze and H. Staudinger, *Naturwissenschaften* **62,** 331 (1975).
[10] S. Fleischer and L. Packer (eds.), this series, Vol. 52.
[11] R. Sato and T. Omura (eds.), "Cytochrome *P*-450." Academic Press, Tokyo, 1978.
[12] K. Ruckpaul and H. Rein (eds.), "Cytochrome *P*-450." Akademie Verlag, Berlin, 1984.

gen. They are stable for several months when stored at $-18°$. The cytochrome P-450 content is determined as described elsewhere.[17] Microsomes containing 3.1–3.6 nmol cytochrome P-450 per milligram protein are used. NADPH oxidase activity [Eq. (2)], determined spectrophotometrically in 0.2 M Tris–HCl, pH 7.4, at 20°, is generally between 20 and 25 nmol/min/mg protein.

Membranes

For entrapment of microsomes in gelatin membranes, a 5% gelatin (acid photogelatin, VEB Gelatinewerk, Calbe, GDR) solution in doubly distilled water at 37° is prepared. The solution is removed from the water bath and immediately mixed with the microsomal suspension in a ratio of 1 : 0.65. The mixture is cast on a plane polyester or poly(vinyl chloride) support to yield 2.5–2.7 mg/cm² microsomal protein. After drying at 4° for 18–20 hr, the membrane is removed and either used immediately or stored dry at 4°. Essentially no monooxygenase or NADPH oxidase activity is lost during storage for 1 year.

The thickness of the dry membrane is between 100 and 110 μm as determined with a micrometer screw. Membranes are subject to considerable swelling when in contact with aqueous solution.

Organelle Electrodes

Organelle electrodes are assembled by fixing a piece of membrane (4 × 4 mm) to the surface of an appropriate electrode using a dialysis membrane covering (cellulose, molecular weight cutoff ~5000, VEB CK, Bitterfeld, GDR) and an O ring.

Operation

Organelle electrodes are connected to a polarograph (Zentrum für Wissenschaftlichen Gerätebau der Akademie der Wissenschaften, Berlin, GDR) equipped with a recorder and are operated as those described by Scheller et al. in this volume[18] except that the measuring cell is thermostatted at 20°. Stationary current changes are measured.

---

[13] H. Kuthan and V. Ullrich, Eur. J. Biochem. **126,** 583 (1982).
[14] Phenobarbital induces cytochrome P-450, leading to a 3-fold increase in its concentration in liver microsomes [R. W. Estabrook, M. R. Franklin, B. Cohen, A. Shigamatzu, and A. G. Hildebrand, Metabolism **20,** 187 (1971)].
[15] Y. Imai and R. Sato, Eur. J. Biochem. **1,** 419 (1967).
[16] D. Baess, G.-R. Jänig, and K. Ruckpaul, Acta Biol. Med. Ger. **34,** 1745 (1975).
[17] T. Omura and R. Sato, J. Biol. Chem. **239,** 2370 (1964).
[18] F. W. Scheller, R. Renneberg, and F. Schubert, this volume [3].

Activity Yield and Recovery

Apparent monooxygenase activity of gelatin-immobilized microsomes is determined by measuring N-demethylation of a typical cytochrome *P*-450 substrate, e.g., aminopyrine, after shaking at room temperature in the following reaction mixture[4]:

> 10 ml 50 m*M* Tris–HCl, pH 7.4, air saturated
> 6 m*M* aminopyrine
> 5 m*M* semicarbazide–HCl
> A piece of membrane corresponding to 2.8 μ*M* cytochrome *P*-450
> 1.5 m*M* NADPH to start

The liberation of the reaction product, formaldehyde, is followed by the procedure of Nash.[19] In a typical experiment 1.27 nmol HCHO/min/mg protein or 3.43 nmol HCHO/min/cm^2 membrane surface were produced. This apparent activity amounts to 60% of that of free microsomes.[4]

Recovery is determined by comparison of the activity of native microsomes with that of a redissolved membrane. Gelatin membranes are solubilized by gently shaking a given piece of membrane in 0.2 *M* Tris–HCl, pH 7.4, at 37° for 5 min. The results for the recovery of NADPH oxidase are shown in Table I. The specific NADPH oxidase activity of the gelatin-immobilized microsomes is about 35 mU/cm^2. Good numerical agreement is obtained between the percentage of monooxygenase yield and NADPH oxidase recovery.

Microsomal Electrodes for Cytochrome *P*-450 Substrates and Reduced
  Pyridine Nucleotides

Aniline Sensor

An organelle sensor for the cytochrome *P*-450 monooxygenase substrate, aniline, is prepared by fixing immobilized microsomes to a rotating glassy carbon disk electrode (2.5 mm diameter, plastic-sealed) to which a potential of +0.25 V versus a saturated calomel electrode is applied. Measurements are performed in 0.1 *M* phosphate buffer, pH 7.0, containing 0.1 m*M* NADPH as cofactor for aniline hydroxylation. Electrochemical indication is based on the anodic oxidation of the hydroxylation product, *p*-aminophenol, to the corresponding quinoneimine.[20] The electrode current increases on addition of aniline, the increase being proportional in the concentration range between 0.05 and 0.5 m*M*. The response time of the electrode is 5 min, which is comparable to other [2,3,7] organelle sensors.

[19] T. Nash, *Biochem. J.* **55**, 416 (1953).
[20] L. A. Sternson and J. Hes, *Anal. Biochem.* **67**, 74 (1975).

TABLE I
RECOVERY OF NADPH OXIDASE IN IMMOBILIZED
LIVER MICROSOMES[a]

Preparation no.	Activity (nmol NADPH/min/mg/ protein)		Recovery (%)
	Microsomes	Dissolved membrane	
1	23.1	15.0	64.9
2	22.6	12.9	57.1
3	20.9	12.4	59.3

[a] Determined in 0.2 M Tris–HCl, pH 7.4, containing 0.2% NaN$_3$ and 146 μM NADPH.

According to Eq. (1) an oxygen probe should be applicable as an electrochemical sensor for cytochrome P-450 monooxygenase substrates. The major obstacle to this approach is the high NADPH oxidase activity [Eq. (2)] which is influenced to varying degrees by different substrates, thus making a correlation between substrate concentration and O$_2$ consumption impossible. The product-indicating sensor described above circumvents this problem but is only applicable to substrates which are hydroxylated to electrochemically distinguishable products.

Reduced Pyridine Nucleotide Sensors

For organelle electrodes based on the microsomal NADPH oxidase, immobilized microsomes are fixed to the polyethylene membrane of a commercial Clark-type oxygen electrode. For the present experiments a platinum/silver–silver chloride (0.1 M KCl) probe with a Pt diameter of 3.5 mm (Forschungsinstitut Meinsberg, Meinsberg, GDR) is used.

The parameters of NADPH determination with this sensor are given in Table II. The sensitivity of the electrode may be enhanced by inhibiting the microsomal contaminant, catalase, by addition of sodium azide to the measuring buffer. An increase by a factor of 1.75 approaches the theoretical value of 2 for a homogeneous system. However, routine inhibition of catalase is not advisable since H$_2$O$_2$ inactivates hemoproteins[21] and would thus lower the operational lifetime of the sensor.

The organelle electrode is also suited for determination of NADH. Whereas the free microsomal suspension is only 20–25% as effective in

[21] F. P. Guengerich, *Biochemistry* **17**, 3633 (1978).

TABLE II
PARAMETERS OF THE MICROSOMAL
NADPH ELECTRODE[a]

Parameter	Value
Sensitivity (nA/μM)	0.140; 0.244[b]
Linear range (mM)	0.05–1.0
Response time (min)	2–3
Coefficient of varia-tion (%, $N = 12$)	5
Useful lifetime[c] (days)	14

[a] Determined in 0.2 M Tris–HCl, pH 7.4.
[b] Measuring solution contains 0.2% NaN_3.
[c] Storage overnight in measuring buffer at 4°.

oxidizing NADH as NADPH, the organelle electrode is 1.6 times more sensitive to NADH compared to NADPH.[5]

Ascorbate Sensor

Ascorbate stimulates nonenzymatic lipid peroxidation in liver microsomes, a process which is accompanied by the disappearance of polyunsaturated fatty acids from microsomal phospholipids and the consumption of O_2.[22] Since gelatin-immobilized microsomes retain this function, ascorbate can be determined with the organelle electrode. The linear range is between 0.2 and 2.5 mM. The sensor can be made highly specific for ascorbate by heating the microsomes for 10 min to 90° before the membrane is prepared. However, the activity of the ascorbate sensor drops to 20% of the initial value after 4 days of operation. This low stability is probably caused by the depletion of polyunsaturated fatty acids.

Endogenous Enzyme Coupling in Organelle Electrodes

In the organelle electrodes described by Arnold and Rechnitz[3] and Karube et al.[7] only single enzyme activities, i.e., mitochondrial glutaminase and microsomal sulfite oxidase, respectively, have been utilized. In contrast, mitochondrial sensors for NADH[2] and succinate[1,2] make use of endogenously coupled enzyme reactions involving the mitochondrial electron transfer chain. Similarly, in microsomal sensors the

[22] E. D. Wills, *Biochem. J.* **113**, 315 (1969).

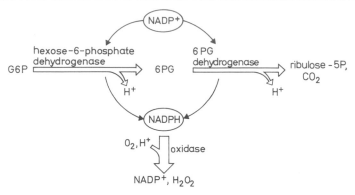

FIG. 1. Endogenous enzyme coupling in the microsomal electrode for determination of glucose 6-phosphate (G6P) and 6-phosphogluconate (6PG). From Schubert *et al.*[5]

coupling of the cytochrome *P*-450 and its reductase can be extended by employing the interaction with other microsomal enzyme activities.

The interrelationship between cytochrome *P*-450 and microsomal dehydrogenases[23,24] as depicted in Fig. 1 is utilized in an organelle electrode for determination of glucose 6-phosphate and 6-phosphogluconate. NADP$^+$ as a cofactor of the dehydrogenase must be added to the measuring buffer, the saturating concentrations being above 2 m*M*. Between 0.02 and 0.8 m*M* of either dehydrogenase substrate can be measured. The response of the sensor is nonlinear over this concentration range. The organelle electrode is almost twice as sensitive to glucose-6-phosphate as it is to both 6-phosphogluconate and NADPH.

Typical values (means ± S.E.M. of three experiments) with 2 m*M* NADP$^+$ are as follows:

Glucose 6-phosphate	(0.233 ± 0.011) nA/μM
6-Phosphogluconate	(0.124 ± 0.004) nA/μM
NADPH	(0.128 ± 0.008) nA/μM

This confirms the function of the enzyme sequence shown in Fig. 1. For each glucose 6-phosphate molecule consumed two molecules of NADP$^+$ are reduced.

If the organelle sensor is operated in a solution containing 10 m*M* glucose 6-phosphate it responds highly sensitively to NADP$^+$. This is due to recyclization of the cofactor between the dehydrogenases and the NADPH oxidase. Under these conditions the lower detection limit for

[23] A. P. Kulkarni and E. Hodgson, *Biochem. Pharmacol.* **31**, 1131 (1982).
[24] C. Bublitz, *Biochem. Biophys. Res. Commun.* **98**, 588 (1981).

TABLE III
PARAMETERS OF HYBRID ORGANELLE ELECTRODES[a]

Enzyme(s) coimmobilized with microsomes	Activity applied (U/cm^2)	Substrate	Linear range (mM)	Response time (min)	Coefficient of variation (%, N = 12)
Glucose-6-phosphate dehydrogenase (EC 1.1.1.49)	1.4	Glucose 6-phosphate	0.01–0.2	2–3	5
Glucose-6-phosphate dehydrogenase + hexokinase (EC 2.7.1.1)	1.4 5.6	ATP[b]	0.01–0.25	2–3	9
D-Isocitrate dehydrogenase [isocitrate dehydrogenase (NADP$^+$), EC 1.1.1.42]	0.3	Isocitrate[c]	0.01–0.27	3–4	4
L-Lactate dehydrogenase (EC 1.1.1.27)	6.9	L-Lactate	0.5–10.0	10	Not determined

[a] Determined in 0.2 M Tris–HCl, pH 7.4, containing 2 mM NADP$^+$ or NAD$^+$.
[b] Measuring solution contains 7 mM MgCl$_2$.
[c] Measuring solution contains 3 mM MgCl$_2$.

NADP$^+$ is 0.4 μM. A similar amplification by cofactor recycling has been described for NAD$^+$ with coimmobilized dehydrogenases.[25]

Hybrid Organelle Electrodes

The potentials of microsome-based sensors can be expanded by coimmobilization with isolated enzymes. This was shown with an aniline electrode involving cytochrome P-450–cosubstrate generation by coimmobilized glucose oxidase.[4] Some properties of hybrid organelle electrodes based on microsomal pyridine nucleotide oxidases and different coimmobilized enzymes are given in Table III. For hybrid membrane preparation, the enzymes are simply mixed with microsomes prior to gelatin addition.

In contrast to the organelle electrode for glucose 6-phosphate using an internal, microsomal, enzyme sequence, the hybrid sensor with coimmobilized glucose-6-phosphate dehydrogenase exhibits a linear current–concentration dependence. The half-lives of the hybrid electrodes are in the range of 14 days, indicating that stability is determined by the microsomal oxidase.

[25] B. Danielsson and K. Mosbach, this series, Vol. 44, p. 453.

Reaction or Diffusion Control in Organelle Electrodes?

For substrate measurement with biosensors, diffusion control of the sensor is desired. Although in the case of a microsomal sulfite oxidase electrode diffusional limitation has been described,[7] it can generally be expected that the enzyme activities of the organelle, even if inducible, are too low to achieve diffusion control, at least if organelle loading of the electrodes is kept low enough to assure reasonable response times.

To study the behavior of the microsomal NADPH oxidase electrode, the quantity of NADPH penetrating the organelle membrane was determined by anodic oxidation at a modified oxygen electrode. Modifications consisted of replacement of the polyethylene membrane by a dialysis membrane and application of an anodic potential of +0.6 V to the Pt tip. To assure complete destruction of the reaction product H_2O_2 of the NADPH oxidase, which would interfere with the electrochemical NADPH oxidation, additional catalase (EC 1.11.1.6) was coimmobilized with the microsomes. The catalase (10 U/cm^2) was again simply mixed with the microsomes prior to membrane preparation. The membrane thus obtained is impermeable to H_2O_2 up to a concentration of 9 mM. Comparison of the anodic NADPH oxidation current at the electrode before and after inhibition[26] of the microsomal NADPH oxidase showed that only 36% of the substrate diffusing into the membrane is biocatalytically oxidized therein. Consequently the enzymatic reaction limits the overall process at the NADPH sensor.

The low enzyme activity in organelle electrodes which would tend to decrease their lifetime is compensated by the increased stability of the enzymes in their natural organelle environment. Useful lifetimes of organelle (microsomal and mitochondrial) electrodes between 7 and 14 days are generally obtained.

Selectivity

Most organelle electrodes appear less selective than enzyme electrodes. This can be expected from the complexity of the microsomal sensors, but was also shown for those using mitochondria.[2,3] For enhancement of the selectivity, specific inhibitors of interfering pathways have been successfully used.[2] On the other hand, a promising approach would be to make use of the low selectivity by applying organelle electrodes for the detection of complex processes, e.g., those caused by mutagenic or toxic chemicals.

[26] Complete inhibition is obtained with 1 mM p-chloromercuribenzoate.

[14] Continuous-Flow Assays with Nylon Tube-Immobilized Bioluminescent Enzymes

By ALDO RODA, STEFANO GIROTTI, SEVERINO GHINI, and GIACOMO CARREA

Introduction

The assay of NAD(P)H and NAD(P)H-generating metabolites with bacterial bioluminescent enzymes is rapid, sensitive, and specific.[1,2] Most assays use soluble enzymes, but the stability and reusability of immobilized enzymes make them favorable low-cost tools for the determination of compounds in biological fluids. The use of Sepharose-immobilized bioluminescent enzymes for analytical purposes has been described by several investigators,[3-5] and attempts have also been made to automate these assays using Sepharose columns.[6,7] The continuous-flow determination of NAD(P)H and bile acids with nylon tube-immobilized bioluminescent enzymes is described in this chapter.

The determination of NAD(P)H[8] is based on an enzymatic system, consisting of an NAD(P)H : FMN oxidoreductase [NAD(P)H dehydrogenase] and a luciferase which emits light in the presence of FMN, NAD(P)H, a long-chain aldehyde and molecular oxygen, according to the following reactions:

$$NAD(P)H + FMN + H^+ \xrightleftharpoons{\text{oxidoreductase}} NAD(P)^+ + FMNH_2 \qquad (1)$$

$$FMNH_2 + RCHO + O_2 \xrightarrow{\text{luciferase}} FMN + RCOOH + H_2O + \text{light} \qquad (2)$$

[1] M. DeLuca and W. D. McElroy (eds.), "Bioluminescence and Chemiluminescence: Basic Chemistry and Analytical Applications." Academic Press, New York, 1981.

[2] L. J. Kricka and T. J. N. Carter (eds.), "Clinical and Biochemical Luminescence." Dekker, New York, 1982.

[3] J. Ford and M. DeLuca, *Anal. Biochem.* **110,** 43 (1980).

[4] A. Roda, L. J. Kricka, M. DeLuca, and A. F. Hofmann, *J. Lipid Res.* **23,** 1354 (1982).

[5] G. Wienhausen, L. J. Kricka, and M. DeLuca, this series, Vol. 136, [8].

[6] K. Kurkijärvi, R. Raunio, and T. Korpela, *Anal. Biochem.* **125,** 415 (1982).

[7] L. J. Kricka, G. Wienhausen, J. E. Hinkley, and M. DeLuca, *Anal. Biochem.* **129,** 392 (1983).

[8] S. Girotti, A. Roda, S. Ghini, B. Grigolo, G. Carrea, and R. Bovara, *Anal. Lett.* **17,** 1 (1984).

The determination of bile acids,[9] whose concentrations in serum are an important index of liver function,[10] is based on the coupling of reactions (1) and (2) with the following NAD(P)H-generating reaction:

$$\text{Hydroxy-bile acid} + \text{NAD(P)}^+ \underset{}{\overset{\text{HSDH}}{\rightleftharpoons}} \text{oxo-bile acid} + \text{NAD(P)H} + \text{H}^+ \qquad (3)$$

where HSDH is immobilized 3α-, 7α-, or 12α-hydroxysteroid dehydrogenase for the assay of 3α-, 7α-, or 12α-hydroxy-bile acids, respectively.

Experimental Methods

Materials

Luciferase from *Photobacterium fischeri* (EC 1.14.14.3, alkanal monooxygenase) (specific activity 12 mU/mg), NAD(P)H : FMN oxidoreductase from *Photobacterium fischeri* [EC 1.6.8.1, NAD(P)H dehydrogenase (FMN)] (specific activity 4 U/mg), NAD$^+$ (lithium salt), NADP$^+$, and FMN were purchased from Boehringer Mannheim (FRG). 3α-Hydroxysteroid dehydrogenase (EC 1.1.1.50) chromatographically purified (specific activity 2.5 U/mg), 7α-hydroxysteroid dehydrogenase (EC 1.1.1.159) (specific activity 6 U/mg), decanal, and dithiothreitol were obtained from Sigma Chemical Co. (St. Louis, MO). 12α-Hydroxysteroid dehydrogenase (specific activity 1.5 U/mg) was extracted from *Clostridium* group P as described by Macdonald *et al.*[11] Bile acids were purchased from Calbiochem-Behring (San Diego, CA) and were crystallized before use. Glutaraldehyde (25% aqueous solution) was obtained from Merck (Darmstadt, FRG). All solutions were made with apyrogenic reagent-grade water prepared with a Milli-Q System (Millipore). Nylon 6 tubes with 1 mm internal diameter were obtained from Snia Viscosa (Italy). All other reagents and compounds were of analytical grade.

Enzyme Assays

The activity of free enzymes is measured in a 3-ml cuvette by spectrophotometrically monitoring (340 nm) the formation or consumption of NAD(P)H. The conditions for the various assays are as follows: NAD(P)H : FMN oxidoreductase in 0.1 M potassium phosphate buffer, pH 7, containing 0.15 mM NADH, 0.2 mM FMN, and 1 mM dithiothreitol; 3α- and 7α-hydroxysteroid dehydrogenase in 0.1 M potassium phosphate

[9] A. Roda, S. Girotti, S. Ghini, B. Grigolo, G. Carrea, and R. Bovara, *Clin. Chem.* **30,** 206 (1984).

[10] D. Festi, A. M. Morselli Labate, A. Roda, F. Bazzoli, F. Fabroni, R. P. Rucci, F. Taroni, R. Aldini, E. Roda, and L. Barbara, *Hepatology* **3,** 707 (1983).

[11] I. A. Macdonald, J. F. Jellet, and D. E. Mahony, *J. Lipid. Res.* **20,** 234 (1979).

buffer, pH 9, containing 0.5 mM NAD$^+$ and 1.5 mM cholic acid; 12α-hydroxysteroid dehydrogenase in 0.1 M potassium phosphate buffer, pH 8, containing 0.5 mM NADP$^+$ and 1.5 mM cholic acid. The activity of immobilized enzymes is determined by spectrophotometrically monitoring (340 nm) the eluate from the nylon tubes. Flow rates of 20–100 ml/hr and tubes of 25–100 cm length are used. The composition of assay buffers is identical to that used with free enzymes.

Enzyme Immobilization

Nylon coils (1 cm diameter) are formed by heating tubes at 100° for 15 min. Nylon tubes (3–5 m) are then O-alkylated through triethyloxonium tetrafluoroborate[12] which is prepared as follows: 12 ml of 1-chloro-2,3-epoxypropane is slowly added to 150 ml of 15% (v/v) boron trifluoride in dry ether. The mixture is stirred under reflux for 1 hr, and then the precipitated triethyloxonium tetrafluoroborate is washed 3 times with 100-ml aliquots of dry ether and finally dissolved in dry dichloromethane (final volume 200 ml). Within 24 hr, nylon tubes (3–5 m) are filled by suction with the triethyloxonium tetrafluoroborate solution and incubated at 25° for 10 min.

The O-alkylated tubes are washed with dichloromethane, filled immediately with a solution of 1,6-diaminohexane in methanol (10%, w/v), and incubated for 1 hr at 30°C. After extensive washing with water the tubes are activated, within 48 hr, by perfusion with 5% (w/v) glutaraldehyde in 0.1 M borate buffer, pH 8.5, for 15 min at 20°. Thereafter, the tubes are washed with 0.1 M potassium phosphate buffer, pH 8, filled (1-m portions) with solutions of enzyme in 0.1 M potassium phosphate buffer, pH 8, 0.2 mM dithiothreitol, 0.5 mM NAD$^+$, and left overnight at 4°. After removal of the enzyme solutions, the tubes are washed thoroughly with 0.1 M potassium phosphate buffer, pH 7, to remove proteins which are not covalently linked. The proportion of enzyme immobilized is calculated by subtracting the unbound enzyme activity from the total added activity.

The immobilized enzymes are stored in 0.1 M potassium phosphate buffer, pH 7, 1% bovine serum albumin, 1 mM DTT, and 0.02% sodium azide, at 4°.

Continuous-Flow Assays

Apparatus. The manifold developed for bioluminescent continuous-flow assay is shown in Fig. 1. For the analysis of NAD(P)H or bile acids by means of coimmobilized enzymes, the flow system involves two

---

[12] W. E. Hornby and L. Goldstein, this series, Vol. 44, p. 118.

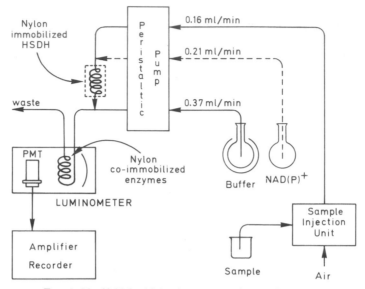

FIG. 1. Manifold for bioluminescent continuous-flow assay.

streams (Fig. 1, solid line): the first is the working bioluminescent solution and the second a continuous flow of air into which a known volume of sample is intermittently added. With hydroxysteroid dehydrogenases immobilized separately from bioluminescent enzymes, there is a third stream (Fig. 1, dashed line) supplying NAD(P)$^+$ to immobilized hydroxysteroid dehydrogenases (1-m coil) placed outside the luminometer. A multichannel peristaltic pump (Minipuls HP4, Gilson, Villiers-le-Bel, France) and calibrated tubes of different diameters are used to produce different flow rates. The bioluminescent reactor—a 0.5- to 1-m coil of nylon tube containing coimmobilized luciferase, NAD(P)H : FMN oxidoreductase, and, if necessary, hydroxysteroid dehydrogenase—is wound around a plexiglass support and positioned inside the luminometer in front of the photomultiplier window (PMT). Before reaching the reactor the stream passes through a stainless steel coil (0.8 mm i.d.) which mixes the stream and prevents a possible "optical fiber" light-diffusion effect[9]; a similar steel coil is also inserted after the reactor.

The luminometer we used is the Model 1250 (LKB, Wallac, Bromma, Sweden), which required only slight modifications of the original light-recording system. The sampling system, which is very simple and based on the use of commercial, calibrated pipets, is discontinuous (manual), but it can easily be automated with the employment of a standard sampler. It consists of a terminal made up of a micropipet tip mounted vertically and working like a funnel. The sample (5–100 μl), which is added into the funnel with a standard pipet, is aspirated uniformly without fragmentation.[8,9] Steady-state (stable background) operation is reached about 5 min after a preliminary washing with 0.1 M potassium phosphate buffer, pH 7, containing 0.5 mM dithiothreitol.

Solutions

NAD(P)H Assay. The nylon coil (0.5–1 m) contains coimmobilized luciferase and NAD(P)H : FMN oxidoreductase. The working bioluminescent solution is 0.1 M potassium phosphate buffer, pH 7, containing 10 μM FMN, 27 μM decanal, and 0.5 mM dithiothreitol. The solution is prepared 20–30 min before analysis. Decanal is previously dissolved in 2-propanol (0.05%, v/v) and remains stable for several weeks at 4°. The working bioluminescent solution shows no remarkable alteration after 8–10 hr at room temperature, in the dark. NAD(P)H standard solutions are 0.1–100 μM.

Bile Acid Assay with Coimmobilized Enzymes. The nylon coil (1 m) cointains coimmobilized luciferase, NAD(P)H : FMN oxidoreductase and 7α-hydroxysteroid dehydrogenase. The working bioluminescent solution is like that for NAD(P)H assay, plus 1 mM NAD$^+$. Bile acid standard solutions are 0.1–100 μM. Serum samples are filtered through a Millipore filter, 0.22 μm average pore size, and stored at −20°. Before the analysis they are diluted (5- to 10-fold) with 0.1 M potassium phosphate buffer, pH 7.

Bile Acid Assay with Separately Immobilized Enzymes. A nylon coil (0.5–1 m) placed into the luminometer contains luciferase and oxidoreductase and a second coil (1 m) placed outside the luminometer contains the specific hydroxysteroid dehydrogenase (Fig. 1). The working bioluminescent solution is the same as for NAD(P)H assay. NAD$^+$ solution (1 mM) in 10 mM potassium phosphate buffer, pH 9, is used for 3α- or 7α-hydroxy-bile acid assays whereas NADP$^+$ solution (1 mM) in 10 mM potassium phosphate buffer, pH 8, is used for 12α-hydroxy-bile acid assays (Fig. 1, dashed line). Serum samples are diluted, after filtration, with 10 mM potassium phosphate buffer, pH 9 (for 3α- or 7α-hydroxy-bile acid assays), or with 10 mM potassium phosphate buffer, pH 8 (for 12α-hydroxy-bile acid assays).

Properties of Immobilized Enzymes

The activity and stability of immobilized enzymes are shown in Table I. Activity recoveries, which varied from 1.5 to 15% depending on the

TABLE I
ACTIVITY AND STABILITY OF NYLON TUBE-IMMOBILIZED ENZYMES

Enzyme	Added enzyme[a] (U/m nylon tube)	Immobilized enzyme (U/m nylon tube)	Activity recovery (%)	Stability[b] (half-life, days)
NAD(P)H : FMN oxidoreductase[c]	4.1	0.06	1.5	20
7α-Hydroxysteroid dehydrogenase[d]	8.4	0.76	9.0	30
3α-Hydroxysteroid dehydrogenase	2.5	0.38	15.2	>30
12α-Hydroxysteroid dehydrogenase	2.2	0.12	5.4	>30

[a] The total amount of enzyme present in the coupling solution was covalently linked by the nylon tube.

[b] At room temperature in 0.1 M potassium phosphate buffer, pH 7, 1 mM dithiothreitol, and 0.02% sodium azide.

[c] NAD(P)H : FMN oxidoreductase was coimmobilized with 7 mU of luciferase.

[d] 7α-Hydroxysteroid dehydrogenase (7 U) was also coimmobilized with NAD-(P)H : FMN oxidoreductase (5 U) and luciferase (9 mU).

enzyme, were in the range of those found with other nylon tube-immobilized enzymes.[13] Variations of immobilization conditions including time of alkylation with triethyloxonium tetrafluoroborate (5–20 min), age of commercial glutaraldehyde (fresh or after 1 year storage at 4°), and use of a bisimidate[12] (dimethyl pimelimidate) instead of glutaraldehyde scarcely influenced the activity recovery of hydroxysteroid dehydrogenases. Instead, the activity recovery of NAD(P)H : FMN oxidoreductase was somewhat erratic and without a strict relationship with immobilization conditions.

The K_m values for substrates and coenzymes of immobilized enzymes were higher than those of the free ones, except for NAD(P)H : FMN oxidoreductase, where the values were similar (Table II). The increased K_m values should be due to diffusional limitations frequently present in immobilized-enzyme systems.[14–16]

Determination of the activity, stability, and K_m values of singly immobilized luciferase was not possible since, to our knowledge, no reliable method is available for the flow assay of the enzyme. On the other hand,

[13] D. L. Morris, J. Campbell, and W. E. Hornby, *Biochem. J.* **147,** 593 (1975).
[14] L. Goldstein, this series, Vol. 44, p. 397.
[15] M. A. Mazid and K. J. Laidler, *Biochim. Biophys. Acta* **614,** 225 (1980).
[16] G. Carrea, R. Bovara, and P. Cremonesi, *Anal. Biochem.* **136,** 328 (1984).

TABLE II
MICHAELIS CONSTANTS OF FREE AND NYLON TUBE-IMMOBILIZED ENZYMES

Enzyme	Substate or coenzyme	K_m of free enzyme[a] (M)	$K_{m, app}$ of immobilized enzyme[a] (M)
NAD(P)H : FMN oxidoreductase	FMN	4.2×10^{-5}	3.9×10^{-5}
	NADH	2.0×10^{-4}	4.0×10^{-4}
7α-Hydroxysteroid dehydrogenase	Cholic acid	3.2×10^{-4}	7.1×10^{-4}
	NAD+	1.0×10^{-3}	1.4×10^{-3}
3α-Hydroxysteroid dehydrogenase	Cholic acid	2.5×10^{-6}	1.8×10^{-4}
	NAD+	1.1×10^{-4}	3.7×10^{-4}
12α-Hydroxysteroid dehydrogenase	Cholic acid	1.1×10^{-4}	3.2×10^{-4}
	NADP+	2.5×10^{-5}	1.0×10^{-4}

[a] The Michaelis constants were obtained from Lineweaver–Burk plots.

the intensity of the light emitted by coimmobilized NAD(P)H : FMN oxidoreductase and luciferase (see the following section) is a function of the activity of both enzymes, and, therefore, even so, no quantitative information on the properties of immobilized luciferase can be obtained.

NAD(P)H Assay

The effect of several parameters such as FMN and decanal concentrations, flow rate, and sample volume on the performance of the luminescent reactor is shown in Table III. The best signal-to-noise ratio was obtained with 10 μM FMN and 27 μM decanal. The working bioluminescent solution prepared by previously dissolving decanal in 2-propanol was more stable and gave a better signal-to-noise ratio than those prepared using methanol or a suspension of decanal in water. The NAD(P)H assay was independent of total flow rate in the range 0.36–1.50 ml/min. Also sample volume and coil length did not influence the response provided the total volume resulting from sample plus working bioluminescent solution was smaller than the reactor volume.

The sensitivity of the NADH assay was very high since as little as 1 pmol of standard was detected (signal-to-noise ratio 3 : 1); the assay was linear from 1 to 1000 pmol. The NADPH assay was about 3 times less sensitive owing to the lower activity of NAD(P)H : FMN oxidoreductase toward NADPH. Up to 30 samples per hour were analyzed. Washing between samples was not strictly necessary since in its absence no appreciable carryover was observed. Representative traces obtained in the NADH assay are shown in Fig. 2. Two samples with low (2.5 pmol) and high (200 pmol) NADH levels were assayed to determine intra- and

TABLE III
EFFECT OF SOME PARAMETERS ON
CONTINUOUS-FLOW ASSAY OF NADH[a]

Parameter	Relative response	Signal-to-noise ratio
FMN (μM)		
3	84	143
10	100	150
25	86	73
50	78	55
100	62	45
Decanal (μM)		
7	56	131
27	70	150
60	85	114
110	100	66
270	96	44
Flow rate (ml/min)		
0.18	67	
0.36	95	
0.70	100	
1.50	100	
3.10	89	
Sample volume (μl)		
25	100	
50	100	
100	100	
150	100	
200	80	

[a] Unless otherwise stated the conditions were as follows: working bioluminescent solution: 0.1 M potassium phosphate buffer, pH 7, containing 10 μM FMN, 27 μM decanal, and 0.5 mM dithiothreitol; flow rate: 0.53 ml/min; sample: 50 μl of 1 μM NADH; nylon coil length: 100 cm.

interassay variations. The coefficients of variation were less than 9 or 5% at the low or high level, respectively.

The residual activity of a reactor used for 2 months, analyzing more than 50 samples per day, was about 20%.

Bile Acid Assay

The assay of primary bile acids carried out using 7α-hydroxysteroid dehydrogenase coimmobilized with luciferase and NAD(P)H : FMN ox-

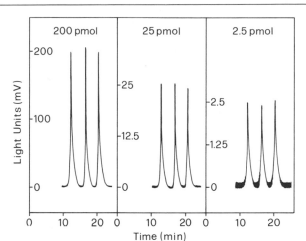

Fig. 2. Representative traces obtained with the bioluminescent NADH assay.

idoreductase had a sensitivity of 10 pmol for any 7α-hydroxy-bile acid. The assay was linear from 10 to 1000 pmol (Fig. 3). When separately immobilized 7α-hydroxysteroid dehydrogenase was used the sensitivity was 1 pmol of bile acid (Fig. 3). The increase in sensitivity was due to the fact that bile acid oxidation was carried out at pH 9 where the transformation of substrate was practically complete. High flow rates (>0.5 ml/min) in the hydroxysteroid dehydrogenase reactor decreased assay sensitivity by decreasing bile acid transformation. Potential enzyme contamination with aldehyde dehydrogenase,[3] which would increase noise values in the presence of an aldehyde, could not affect the assay based on separately immobilized enzymes since decanal was not present in the buffer feeding the immobilized hydroxysteroid dehydrogenase.

Serum samples ($n = 30$) analyzed by the bioluminescence method gave results in good agreement with those obtained by radioimmunoassay, enzyme immunoassay, and high-performance liquid chromatography.[9] Two serum samples with low (2.5 μM) and high (22 μM) concentration of bile acids were assayed to determine intra- and interassay variations. The coefficients of variation were lower than 8 or 10% at low or high concentration, respectively. Up to 20 samples per hour were analyzed with no carryover.

The determination of total bile acids by means of separately immobilized 3α-hydroxysteroid dehydrogenase gave similar results, whereas the assay of 12α-hydroxy-bile acids by means of NADP-dependent 12α-hydroxysteroid dehydrogenase was less sensitive owing to the lower activity of NAD(P)H : FMN oxidoreductase toward NADPH (Fig. 3).

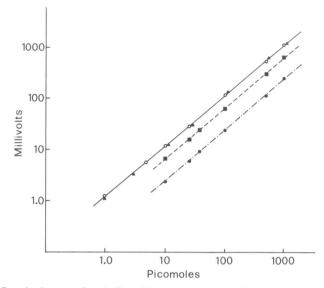

FIG. 3. Standard curves for cholic acid obtained with the bioluminescent assay. (●) 7α-Hydroxysteroid dehydrogenase coimmobilized with luciferase and NAD(P)H : FMN oxidoreductase. (○) 7α-Hydroxysteroid dehydrogenase, (▲) 3α-hydroxysteroid dehydrogenase, and (■) 12α-hydroxysteroid dehydrogenase immobilized separately from luciferase and NAD(P)H : FMN oxidoreductase.

Conclusions

Nylon tube-immobilized bioluminescent enzymes make it possible to specifically assay NAD(P)H and bile acids at picomole levels. The precision of the method, as well as its correlation with other methods such as radioimmunoassay, enzyme immunoassay, and high-performance liquid chromatography, is satisfactory.

The adopted continuous-flow system is simple, requires only minor modifications of a commercial detector, and allows analyzing of about 20–30 samples per hour. Unlike Sepharose columns,[6,7] nylon reactors present no problems with packing or disruption of the gel matrix, nor bacterial growth, which markedly enhances background light level.[7] This, together with its handiness, makes the nylon tube a very suitable enzyme support for continuous-flow analysis, in spite of the relatively low activity recovery of immobilized enzymes. Up to 500–700 bile acid samples were analyzed with use of only a few milligrams of the enzymes, and therefore this bioluminescent method appears highly competitive with other methods such as radioimmunoassay, enzyme immunoassay, high-performance liquid chromatography, and fluorometry where radioactive materials, separation steps, sample manipulation, or expensive equipment are needed.

Potentially, a variety of other NAD(P)H-generating metabolites could be analyzed using the bioluminescent reactor (coimmobilized NAD-(P)H : FMN oxidoreductase and luciferase) coupled with a proper immobilized dehydrogenase.

[15] Flow–Injection Analysis with Immobilized Chemiluminescent and Bioluminescent Columns

By KALEVI KURKIJÄRVI, PEKKA TURUNEN, TIINA HEINONEN, OUTI KOLHINEN, RAIMO RAUNIO, ARNE LUNDIN, and TIMO LÖVGREN

Introduction

The use of purified bioluminescent enzymes from marine bacteria [Eqs. (1) and (2)] and fireflies [Eq. (3)] are well documented in the sensitive measurement of NAD(P)H and ATP, respectively, as well as the use of luminol reaction [Eq. (4)] to measure peroxides[1–4]:

$$NAD(P)H + FMN \xrightarrow{\text{oxidoreductase}} NAD(P) + FMNH_2 \qquad (1)$$

$$FMNH_2 + RCHO + O_2 \xrightarrow{\text{luciferase}} FMN + RCOOH + H_2O + \text{light} \qquad (2)$$

$$ATP + \text{luciferin} + O_2 \xrightarrow{\text{luciferase}} AMP + \text{oxyluciferin} + PP_i + CO_2 + \text{light} \quad (3)$$

$$\text{Luminol} + 2 H_2O_2 \xrightarrow{\text{peroxidase}} \alpha\text{-aminophthalate} + H_2O + N_2 + \text{light} \qquad (4)$$

The above-mentioned analytes are in a key position since many biochemical reactions can be coupled to their conversion.

The use of immobilized enzymes in analytical chemistry has gained increasing interest during the last decade. The main reason for this is the improved stability of enzymes on immobilization. The immobilized enzymes are reusable, enabling multiple analyses with the same preparation.[5] Immobilized enzymes can be used in flow reactors through which

[1] A. Lundin, A. Rickardsson, and A. Thore, *Anal. Biochem.* **75,** 611 (1976).
[2] F. R. Leach, *J. Appl. Biochem.* **3,** 473 (1981).
[3] L. J. Kricka and G. H. G. Thorpe, *Analyst* **108,** 1274 (1983).
[4] K. Kurkijärvi, R. Raunio, J. Lavi, and T. Lövgren, in "Bioluminescence and Chemiluminescence: Instruments and Applications" (K. Van Dyke, ed.), Vol. II, p. 167. CRC Press, Boca Raton, Florida, 1985.
[5] M. DeLuca, in "Analytical Applications of Bioluminescence and Chemiluminescence" (L. J. Kricka, P. E. Stanley, G. H. G. Thorpe, and T. P. Whitehead, eds.), p. 111. Academic Press, London, 1984.

the sample is pumped continuously or as a front, making easy and low cost automation of the analytical system possible.[6] Additionally, if the analytical system requires several enzyme reactions all the enzymes can be immobilized on the same matrix. These coimmobilized enzymes have been found to be much more efficient in catalyzing the coupled reactions than the soluble ones.[7] However, in some cases better analytical results can be achieved with separately immobilized enzymes than with coimmobilized ones.[4]

There are many matrices and many different procedures used for enzyme immobilization which have been well reviewed.[8,9] Which of the methods is the best for a particular enzyme is not so easy to say as it is always a process of trial and error. High coupling efficiency, high remaining activity, good stability of the final product during storage and continuous use, and good mechanical stability of the support are the main criteria when the immobilization procedure is optimized. For chemi- and bioluminescent systems there is also a special demand for the matrix and the coupling reaction: the enzyme–matrix conjugate should not absorb any light at 400–600 nm.[10] In this chapter we describe the immobilization of peroxidase, choline oxidase, firefly luciferase, and the bacterial bioluminescence enzymes with different dehydrogenases, including the use of resulting enzyme preparations in flow–injection analysis of some metabolites.

Experimental Methods

Immobilization of Enzymes

Bacterial Bioluminescence Enzymes. One vial of NADH monitoring reagent (Wallac Oy, Turku, Finland) consisting of the bacterial luciferase (alkanal monooxygenase), NADH:FMN oxidoreductase from *Vibrio harveyi*, and the stabilizers is reconstituted in 1 ml of doubly distilled water and dialyzed against 2 liters of deaerated 0.1 M potassium phosphate, 20 μM FMN, pH 7.0, to remove dithiothreitol (DTT) (Cleland's reagent), since it has been found to interfere with the immobilization.[11] CNBr-activated Sepharose 4B (Pharmacia Fine Chemicals, Uppsala, Sweden; 0.35 g lyophilized powder or about 1 ml in the swollen state), treated

[6] W. R. Seitz, *CRC Crit. Rev. Anal. Chem.* **13**(1), 1 (1981).
[7] N. Siegbahn and K. Mosbach, *FEBS Lett.* **137**, 6 (1982).
[8] L. Goldstein and G. Manecke, *Appl. Biochem. Bioeng.* **1**, 23 (1976).
[9] K. Mosbach (ed.), this series, Vol. 44.
[10] K. Kurkijärvi, R. Raunio, and T. Korpela, *Anal. Biochem.* **125**, 415 (1982).
[11] E. Jablonski and M. DeLuca, *Proc. Natl. Acad. Sci. U.S.A.* **73**, 3848 (1976).

according to the supplier, is suspended in the dialyzate and shaken gently at 4° for 20 hr. When coimmobilizing glutamate dehydrogenase, alcohol dehydrogenase, or glucose dehydrogenase (all from Sigma Chemical Co., St. Louis, MO), the enzyme is added to the above-mentioned immobilization solution in small aliquots (50 μl each, 3.2, 16.0, and 6.0 IU, respectively). After incubation the gel is washed with 50 ml of cold 0.1 M potassium phosphate, 0.5 mM DTT, pH 7.0, followed by 50 ml of the same buffer containing additionally 1 M KCl, and finally with 50 ml of the first solution. The washed enzyme–gel conjugates are stored as a suspension in the phosphate–DTT buffer, pH 7.0 (1 ml buffer/1 ml gel) at 4° in the dark. The immobilized enzymes can also be stored deep freezed in the presence of glycerol.[12]

Firefly Luciferase. Inorganic porous glass (Sigma, mesh size 80–120 Å, mean pore diameter 330 Å, pore volume 1.15 cm^3/g, surface area 68 m^2/g) is first silanized and activated as follows[13]: 10 g of glass is shaken and suspended in 100 ml of 10% 3-aminopropyltriethoxysilane (Sigma)–water solution, pH 3.5 (adjusted with HCl), at 75° for 3 hr. After silanization the glass is washed with distilled water and dried at 100° overnight. One gram of dry alkyl-glass is activated by shaking in 5 ml of a 1% glutaraldehyde solution at 4° for 2 hr followed by washing with 200 ml of distilled water.

Thirty-eight milligrams of purified firefly luciferase (Photinus-luciferin 4-monooxygenase) (Wallac Oy) in 4 ml of 0.1 M potassium phosphate, pH 7.5, is shaken with 1 g of moist activated alkyl glass at 4° for 1.5 hr. The resulting immobilized luciferase preparation is first washed with 10 ml of 0.1 M potassium phosphate, 0.5 mM DTT, pH 7.0, followed by 10 ml of the same buffer containing 1 M KCl, and finally with 10 ml of distilled water. Before the first washing the bonds between glass aldehyde groups and enzyme amino groups are reduced by incubating the enzyme–glass conjugate in the immobilization supernatant for 2 min in the presence of 200 mg solid NaBH$_4$. The immobilized firefly luciferase is stored at 4° in the dark suspended in 0.1 M Tris–acetate, 2 mM EDTA, 10 mM DTT, pH 7.5.

Peroxidase and Choline Oxidase. Peroxidase. Two grams of silica gel (Sigma) is aminopropylated as described above for firefly luciferase. The dried aminopropyl-silica gel is suspended in 5 ml of acetone containing 200 mg cyanuric chloride (Merck, Darmstadt, FRG), and the suspension is mixed for 5 min at room temperature followed by addition of 10 ml of 1 M Na$_2$CO$_3$ and another 5 min mixing at room temperature. The activated

[12] G. K. Wienhausen, L. J. Kricka, J. E. Hinkley, and M. DeLuca, Appl. Biochem. Biotechnol. 7, 463 (1982).
[13] Y. Li, H. Jiay, and X. X. Z. Shuzeng, Appl. Biochem. Biotechnol. 7, 325 (1982).

silica gel is then washed with 25 ml of 50% acetone–water followed by 25 ml of distilled water, and finally the washed activated gel is immediately suspended in 4 ml of 0.5 M borate buffer, pH 9.0, containing 20 mg of horseradish peroxidase (HRP VI, 275 IU/mg, Sigma). The suspension is gently shaken at 4° overnight. The resulting enzyme–gel conjugate is washed with 25 ml of 0.5 M borate buffer, pH 9.0, followed by 25 ml of the same buffer with 1 M KCl, and finally with distilled water until no peroxidase activity is found in the washing solution. The immobilized enzyme is stored at 4° in the dark as a suspension in the borate buffer.

Choline oxidase. Two grams of silica gel is aminopropylated as described above. The dried aminopropyl-silica gel is suspended in 10 ml of a 10% glutaraldehyde–water solution, and the suspension is gently shaken at 4° for 2 hr followed by washing with 100 ml of distilled water. The moist gel is transferred to the immobilization buffer (0.1 M potassium phosphate, pH 7.4) containing 40 mg of choline oxidase (Sigma, 15 IU/mg). The reaction is allowed to proceed with gentle shaking at 4° overnight. Then the suspension is transferred on a glass sinter, and 10 mg of solid NaBH$_4$ is added to reduce the bonds between the enzyme and matrix. After 2 min incubation with mixing, the immobilized choline oxidase is

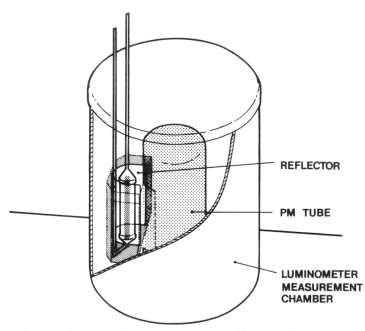

FIG. 1. Construction of the flow-through column inside the luminometer measurement chamber. The flow inlet is at the lower part of the column and the outlet at the upper part (see also text).

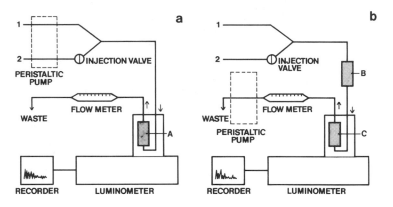

Fig. 2. Scheme of the flow–injection system (a) for bacterial and firefly bioluminescence packed-bed reactors and (b) for luminol reaction-based enzyme columns.

washed with 50 ml of 0.1 M potassium phosphate, pH 7.4, with 50 ml of the same buffer containing 0.5 M KCl, and with 50 ml of the first washing solution. The immobilized choline oxidase is stored at 4° in the dark suspended in the potassium phosphate buffer, pH 7.4.

Construction of the Enzyme Reactors

The columns consist of a glass tube (height 25 mm, outer and inner diameters 7 and 3 mm, respectively) fixed to two sintered plastic funnels. The gel-filled column is installed in the sample holder of the luminometer (LKB-Wallac 1250 Luminometer) modified with a tight holder for the lower column funnel and a channel for solution outlet. The light-emitting enzyme reactor is placed in the front of the photomultiplier (PM) tube, as close as possible (Fig. 1). The joints of the column are leakage proof, and the columns are able to be emptied and refilled.

In Fig. 2a is presented the manifold of the flow injection system. Column A contains the bacterial bioluminescent enzymes (and the coimmobilized glutamate dehydrogenase, alcohol dehydrogenase, glucose dehydrogenase) or firefly luciferase. In Fig. 2b, column C contains the immobilized peroxidase and column B the immobilized choline oxidase. In all measurements the total flow rate from 0.42 to 0.8 ml/min is used. The solutions pumped through the flow injection system with each analyte are as follows:

NADH   1: 10 μM FMN, 0.001% decanal, 0.1 M DTT, and
                    0.1 M KCl in 50 mM potassium phosphate
                    buffer, pH 7.0

	2: 50 mM potassium phosphate buffer, pH 7.0 (+10 mg/ml NAD for glutamate, ethanol, and glucose) plus sample injection
ATP	1: 4 mM EDTA, 10 mM Mg^{2+}, 80 μM luciferin in 50 μM Tris–acetate buffer, pH 8.0
	2: 50 mM Tris–acetate buffer, pH 8.0, plus sample injection
H_2O_2 and choline	1: 200 μM luminol in 50 mM borate buffer, pH 9
	2: 50 mM borate buffer, pH 9.0, plus sample injection

Results

Bacterial Bioluminescence

Table I shows the immobilization results of bacterial luciferase with the additional enzymes. When used according to March et al.,[14] activated agarose as a matrix results in 50% lower recovery of coupled activity[12] because of the incomplete washing of the unreacted CNBr after activation. As high recovery of light-emitting activity as with commercial CNBr-activated Sepharose can be achieved[4] by activating the agarose with the cyano-transfer method of Kohn and Wilchek.[15]

Figure 3 presents the standard curves for all four analytes. The linear ranges are several orders of magnitude with sensitivities at picomole levels. The precision of the flow–injection method is excellent ($CV < 5\%$). No significant sample dispersion occurs when injection volumes from 5 to 50 μl are used. At least 30 samples per hour can be analyzed without any significant carryover.

The immobilized enzymes are stable when stored in the presence of fresh DTT at 4° in the dark. No significant decrease in activities is found during 4 months of storage. The operational stability of the immobilized column(s) is good as well. From 600 to more than 1000 samples can be analyzed during several weeks with no analytically significant decrease in the responses.[16]

Firefly Luciferase

Approximately 80% (31 mg/g) of the used luciferase is bound to the alkylated glass retaining 12% of its original activity. By using less enzyme

[14] S. C. March, I. Parikh, and P. Cuatrecasas, Anal. Biochem. 60, 149 (1974).
[15] J. Kohn and M. Wilchek, Biochem. Biophys. Res. Commun. 107, 878 (1982).
[16] K. Kurkijärvi, T. Heinonen, T. Lövgren, J. Lavi, and R. Raunio, in "Analytical Applications of Bioluminescence and Chemiluminescence" (L. J. Kricka, P. E. Stanley, G. H. G. Thorpe, and T. P. Whitehead, eds.), p. 125. Academic Press, London, 1984.

TABLE I
IMMOBILIZATION OF BACTERIAL BIOLUMINESCENCE SYSTEM

	Protein			Activity		
Enzyme(s)	Used (mg)	Bound (mg)	Yield (%)	Used	Bound	Recovery (%)
terial luciferase + xidoreductase	12[a]	10	85	70 V[b]	50 ± 5 V	71 ± 7
tamate dehydrogenase	0.08	0.08	100	3.2 IU	1.12 ± 0.16 IU	35 ± 5
ohol dehydrogenase	0.04	0.04	100	16 IU	5.6 ± 0.8 IU	35 ± 5
cose dehydrogenase	0.03	0.03	100	6 IU	2.1 ± 0.3 IU	35 ± 5

[a] The protein content of the NADH monitoring reagent is 12 mg (biuret assay) including albumin.

[b] In the presence of 2 μM NADH, a constant light intensity of 70 V (calculated to the whole reagent) was found. Correspondingly, the immobilized preparation emitted light at a level of 50 ± 5 V.

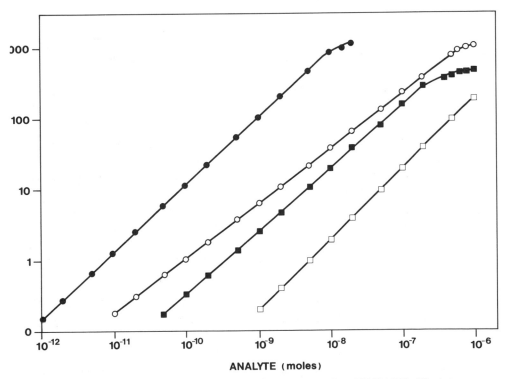

FIG. 3. Peak light intensity as function of analyte concentration. (●) NADH; (○) glutamate; (■) glucose; (□) ethanol. A total flow rate of 0.6 ml/min was used.

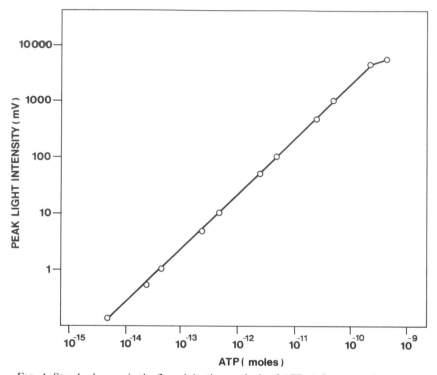

FiG. 4. Standard curve in the flow–injection analysis of ATP. A flow rate of 0.42 ml/min was used.

(a few milligrams) per immobilization the relative recovery of activity is better, but to get the highest sensitivity for ATP measurement a high amount of luciferase is needed. The luciferase column responded linearly to ATP from 5 fmol to 250 pmol (Fig. 4), showing good precision in the whole concentration range ($CV < 10\%$). The best immobilization results with firefly luciferase have been found by using polysaccharides as the matrix.[17,18] However, these are not very stable matrices in flow analysis with immobilized enzymes because of the poor mechanical stability of the support.[10]

The half-life of the immobilized luciferase, when stored at 4° suspended in Tris–acetate buffer, pH 7.5, containing 2 mM EDTA and 10 mM DTT, is about 10 days (Fig. 5). On lyophilization and deep freezing,

[17] N. N. Ugarova, L. Y. Brovko, and I. V. Berezin, *Anal. Lett.* **13,** 881 (1980).
[18] N. N. Ugarova, L. Y. Brovko, and N. V. Kost, *Enzyme Microbiol. Technol.* **4,** 224 (1982).

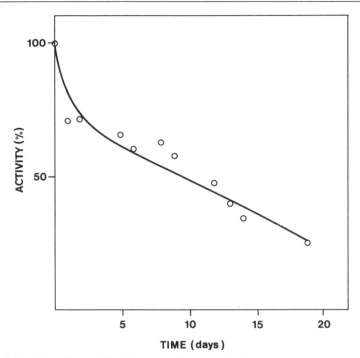

FIG. 5. Stability of immobilized firefly luciferase at 4°. The enzyme preparation was kept as a suspension in the dark.

the immobilized luciferase retains 37 and 90% of its original activity, respectively. Both preparations are stable during storage.

Peroxidase and Choline Oxidase

Table II presents the immobilization results for peroxidase and choline oxidase. The peroxidase gave better recovery of activity by immobiliza-

TABLE II
IMMOBILIZATION OF PEROXIDASE AND CHOLINE OXIDASE

Enzyme	Protein			Activity		
	Used (mg)	Bound (mg)	Yield (%)	Used (IU)	Bound (IU)	Recovery (%)
Peroxidase	20	18.6	93	5500	693	12.6
Choline oxidase	40	32.8	82	600	19600	3267

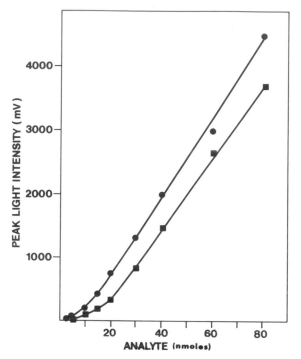

FIG. 6. The response of light as a function of H_2O_2 (■) and choline (●) concentration. A flow rate of 0.8 ml/min was used.

tion to carbonyldiimidazole-activated Dynosphere XP-4001,[19] but this matrix is unsuitable for packed-bed enzyme reactors. The immobilized choline oxidase is 30 times more active than the corresponding soluble enzyme for some unclear reasons. One possible explanation is that the soluble enzyme preparation contains some inhibitor; another one is the more favorable microenvironment of the immobilized enzyme for catalysis.

The sensitivity of the flow–injection system for choline and hydrogen peroxide is 2.5 and 5 nmol, respectively (Fig. 6). The sensitivity for choline is two times higher than for hydrogen peroxide because in the choline oxidase reaction where choline is converted to betaine 2 mol of H_2O_2 is formed per mole of choline. The unlinearity in the lower part of the standard curve is due to mass transfer limitations.[20] Despite the small unlinearity the precision of the assay is excellent, showing CV values less than 5% even at low concentrations.

[19] P. Turunen, T. Lövgren, and K. Kurkijärvi, in press.
[20] K. J. Laidler and P. S. Bunting, this series, Vol. 64, p. 227.

The immobilized peroxidase and choline oxidase are stable, and no significant decrease in activity occurs during 3 months of storage at 4°. In continuous use the half-life of the columns is approximately 20 hr, during which time more than 500 measurements can be performed. This relatively short half-life is mainly caused by the oxidative effect of hydrogen peroxide on the solid-phase enzyme conjugate.

Concluding Remarks

Flow analysis using immobilized chemi- and bioluminescent packed-bed enzyme reactors as detectors is rapid, sensitive, and precise. The reusability and stability of the immobilized enzyme columns make them potential and low-cost tools for automated determinations of different analytes in both research and routine analysis. A few interesting examples have been published.[4,10,16,21-24] Work is still needed, however, to improve these flow reactors especially the operational stability and design of the packed-bed reactor system.

[21] M. Tabata, C. Fukunaga, M. Oxyabu, and T. Murachi, *J. Appl. Biochem.* **6**, 251 (1984).
[22] L. J. Kricka, G. K. Wienhausen, J. E. Hinkley, and M. DeLuca, *Anal. Biochem.* **129**, 392 (1983).
[23] A. Roda, S. Girotti, S. Ghini, B. Grigolo, G. Carrea, and R. Bovara, *Clin. Chem.* **30**, 206 (1984).
[24] G. Carrea, R. Bovara, and P. Cremonesi, *Anal. Biochem.* **136**, 328 (1984).

[16] Enzyme Thermistors

By BENGT DANIELSSON and KLAUS MOSBACH

Introduction

About 10 years ago several different calorimetric devices were introduced which combined the general detection principle of calorimetry with the specificity of immobilized enzymes.[1-5] Additional advantages were reusability of the biocatalyst, possibility to work with continuous-flow

[1] K. Mosbach and B. Danielsson, *Biochim. Biophys. Acta* **364**, 140 (1974).
[2] C. L. Cooney, J. C. Weaver, S. R. Tannenbaum, D. V. Faller, A. Shields, and M. Jahnke, *Enzyme Eng.* **2**, 411 (1974).
[3] S. N. Pennington, *Enzyme Technol. Digest* **3**, 105 (1974).
[4] L. M. Canning, Jr. and P. W. Carr, *Anal. Lett.* **8**, 359 (1975).
[5] H.-L. Schmidt, G. Krisam, and G. Grenner, *Biochim. Biophys. Acta* **429**, 283 (1976).

systems, insensitivity to the optical properties of the sample, and simple procedures. Since most enzymatic reactions are associated with rather high enthalpy changes in the range of 20–100 kJ/mol, it is often possible to work with only one enzymatic step in contrast to other techniques where detection is based on, for instance, the change in concentration of colored reactants. In such cases it is usually necessary to couple the primary reaction with one or several subsequent (enzymatic) reactions in order to obtain measurable changes.

The appealing possibilities of bioanalytical calorimetry were recognized early in studies by conventional calorimetry.[6] The microcalorimeters normally used for biochemical studies were, however, rather sophisticated and expensive instruments with relatively slow response. They were consequently unsuitable for rapid routine analysis. In contrast, the flow enthalpimetric analyzers described below, based on the use of immobilized enzymes, permit rapid analyses with a relatively simple and inexpensive instrument.

Instrumentation

In our initial studies we used different types of simple plexiglass constructions containing the immobilized enzyme column. These devices were thermostatted in accurate water baths, and the temperature at the exit of the column was monitored with a small thermistor connected to a commercial Wheatstone bridge constructed for temperature measurements and osmometry. Later, we developed our own more sensitive instruments for temperature monitoring, and the water baths were replaced by a carefully temperature-controlled metal block which contained the enzyme column. The enzyme thermistor concept itself is patented in several major countries, for instance, U.S. Patent 4,021,307 (K. Mosbach).

These simple plexiglass devices are surprisingly useful and can be employed for determinations down to, in favorable cases, 0.01 mM. An example of such a simple device will therefore be described here in some detail (Fig. 1). The plastic column, which can hold up to 1 ml of the immobilized enzyme preparation, is mounted in a plexiglass holder, leaving an insulating airspace around the column. The heat exchanger consists of acid-proof steel tubing (i.d. 0.8–1 mm, about 50 cm long) which is coiled and placed in a water-filled cup. The whole device is placed in a water bath (Heto Type 02 PT623 UO, Birkeroed, Denmark) with a temperature stability of at least 0.01°. The cap surrounding the heat ex-

[6] C. Spink and I. Wadsö, *Methods Biochem. Anal.* **23,** 1 (1976).

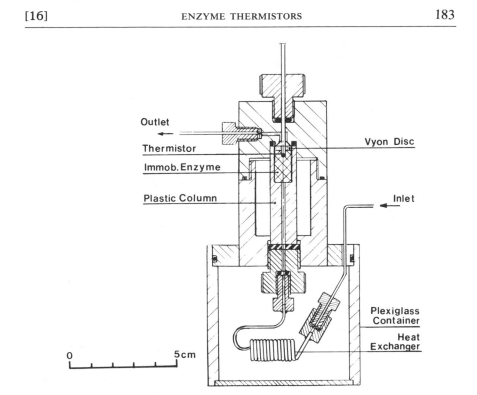

FIG. 1. A simple plexiglass version of the enzyme thermistor. The plexiglass container is filled with water, and the whole unit is placed in a water bath. The vyon disc consists of sintered polyethylene, pore size about 0.1 mm (Porvair Ltd., Kings Linn, England).

changer considerably reduce the temperature fluctuations and improved the baseline.

The temperature is measured at the top of the column with a thermistor (Veco Type 41A28, 10 kohm at 25°, 1.5 × 6 mm, Victory Engineering Corporation, Springfield, NJ or equivalent) epoxied at the tip of a 2-mm (o.d.) acid-proof steel tube. The temperature is measured as the unbalance signal of a sensitive Wheatstone bridge (Knauer Temperature Measuring Instrument; Knauer Wissenschaftlicher Geraetebau, W. Berlin, FRG). At the most sensitive setting the recorder output produces 100 mV at a temperature change of 0.01°. Placing the temperature probe at the very top of the column rather than in the effluent outside the column reduces the turbulence around the thermistor and gives a more stable temperature recording.

Solution is pumped through the system at a flow rate of the order of 1 ml/min with a peristaltic pump (LKB Varioperpex pump, Bromma, Sweden or a Gilson Minipulse, Villiers-le-Bel, France). The sample (0.1–1 ml) is introduced with a three-way valve or a chromatographic sample loop valve. The height of the resulting temperature peak is used as a measure and is found to be linear with substrate concentration over wide ranges, typically 0.01–100 mM, if not limited by amount of enzyme or deficiency in any of the reactants.

As an example, this type of instrument is adequate for the determination of urea in clinical samples.[7] The sensitivity is high enough to permit 10-fold dilution of the samples, which eliminates problems with nonspecific heat. The resolution is consequently about 0.1 mM, and up to 30 samples can be measured per hour.

For more sensitive determinations, we have developed a two-channel instrument in which the water bath is replaced by a carefully thermostatted metal block. A specially designed Wheatstone bridge permits temperature determinations with a resolution of sensitivity of 100 mV/0.001°. The calorimeter (Fig. 2) is placed in a container insulated by polyurethane foam. It consists of an outer aluminum cylinder which can be thermostatted at 25, 30, or 37° with a stability of at least ±0.01°. Inside is a second aluminum cylinder with cavities for two columns and a pocket for a reference thermistor. Before entering the column, the solution passes through a thin-walled acid-proof steel tube (i.d. 0.8 mm) two-thirds of which act as a coarse heat exchanger in contact with the outer cylinder, while the last third is in contact with the inner cylinder. This has a rather high heat capacity and is separated from the thermostatted jacket by an airspace. Consequently, the column is surrounded by a very constant temperature, and the temperature fluctuations of the solution become exceedingly small.

The columns are attached to the end of plastic tubes by which they are inserted into the calorimeter (see the enlarged part of Fig. 2). Columns can thereby be readily changed with a minimum disturbance of the temperature equilibrium. Inside the plastic tube is the effluent tubing and the leads to the thermistor which are fastened to a short piece of gold capillary with heat-conducting, electrically insulating epoxy resin. Presently, Veco Type A 395 thermistors (16 kohm at 25°, temperature coefficient 3.9%/°) are used. These are very small, dual-bead isotherm thermistors with 1% accuracy. This means that they are interchangeable, comparatively well matched, and follow the same temperature response curve (within 1%). An identical thermistor is mounted in the reference probe.

The Wheatstone bridge is built with precision resistors with low temperature coefficient (Econistor Type 8E16; 0.1%; temperature coefficient

[7] B. Danielsson, K. Gadd, B. Mattiasson, and K. Mosbach, *Anal. Lett.* **9,** 987 (1976).

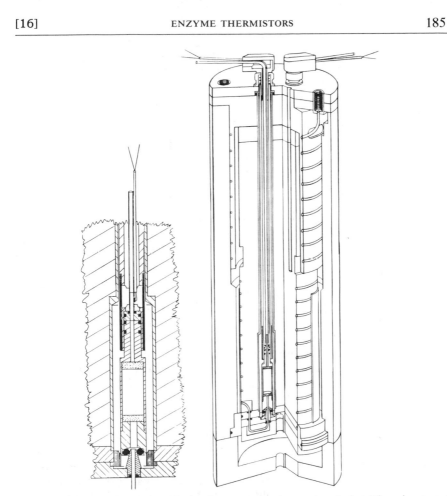

FIG. 2. Enzyme thermistor with aluminum constant temperature jacket. The enlargement at left shows the attachment of a column.

3 ppm; General Resistance, Bronx, NY) and is equipped with a chopper-stabilized operational amplifier (MP 221 from Analogic Corp., Wakefield, MA). This bridge maximally produces a 100-mV change in the recorder signal for a temperature change of 0.001°. The lowest practically useful temperature range is, however, limited mainly by temperature fluctuations caused by friction and turbulence in the column to typically 0.005–0.01°. The thermistor resistance is differentially monitored either versus a reference thermistor inserted in a pocket in the inner aluminum block or versus an identical thermistor probe with an inactive reference column. The latter arrangement is useful when nonspecific heat effects (e.g., due to solvation or dilution heats) are encountered. The sample is then equally

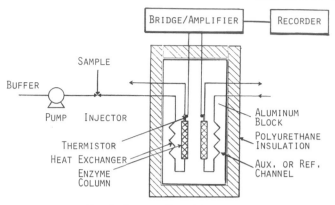

FIG. 3. Enzyme thermistor setup.

split between the enzyme column and the reference column.[8] Alternatively, the second channel can be reversed for another enzyme preparation, permitting a quick change of enzymatic analysis. Some instruments have even been equipped with a dual Wheatstone bridge enabling two different, independent analyses to be carried out simultaneously. In total, over 20 instruments of the newer type have been assembled at the workshop of our institute for use in industry and in research institutes.

Procedure

A typical instrumental setup is shown in Fig. 3. A peristaltic pump produces a continuous buffer stream at a flow rate of 0.5–4 ml/min through the injection valve (usually a Type 5020 valve from Rheodyne, Cotati, CA). Sample loops holding 0.1–0.5 ml of sample are normally used. In some cases we have used a septum injection valve for injection of sample volumes less than 25 μl. Consequently, it is possible to inject an undiluted sample. If larger sample volumes are used it is recommended that the samples be diluted at least 5- to 10-fold by the buffer used to avoid nonspecific heat from solvation or dilution effects. Such disturbances may also be eliminated by the split-flow technique[8] which involves a reference column as mentioned above.

The enzyme column contains maximally 1 ml of the immobilized enzyme preparation (at 7 mm i.d.). It is also possible to adapt a piece of nylon tubing (length ≤1 m) as enzyme support, which is particularly

[8] B. Mattiasson, B. Danielsson, and K. Mosbach, *Anal. Lett.* **9,** 867 (1976).

advantageous in analysis of crude samples containing particles.[9] It suffers, however, from low enzyme loading capacity. The carrier material normally used is CPG (controlled pore glass) which is available in many different pore and particle size ranges and with different ligands. CPG offers high enzyme coupling capacity, good mechanical, chemical, and microbial stability, as well as relatively simple coupling procedures (see, e.g., Vol. 44, this series). Other materials that have been used include Sepharose CL-6B (Pharmacia Fine Chemicals, Uppsala, Sweden) and Eupergit (Röhm Pharma, Weiterstadt, FRG). Sepharose can, however, only be used at flow rates below 0.5–0.7 ml/min or the column will rapidly become clogged. The standard conditions are to use propylamino-derivatized CPG with a pore size in the range 500–2000 Å and particle size around 80 mesh (0.18 mm) with a large excess of enzyme (often 100 units or more) attached to the ligand with glutardialdehyde (Vol. 44, this series).

Under these circumstances, good operational stability is achieved with unchanged performance over long series of samples (thousands) or extended periods of continuous monitoring. The column may be functional for several months. The limiting factor for column life appears to be mechanical obstruction. The column gradually clogs due to fines created by attrition and by particles and adsorbed material from the sample. It is therefore recommended to use pumps with negligible pulsation or to use a pulse damper, which in addition improves the stability of the temperature signal. Furthermore, the solutions used as well, as the samples, should be filtered through a 1–5 μm filter. Finally, the samples should be diluted as much as possible to prolong the life of the column.

Another factor that may have profound impact on the performance is bacterial growth. Thus, a bacteriostatic agent should be included in the buffer solution during longer experiments or the solutions should be filtered through a sterile filter before use. Bacteria present in the enzyme column or in the fluid lines could affect the results in several different ways: if substrate is consumed by the bacteria, an artificially low reading will be obtained; if consumption of components in the sample other than the enzyme substrate occurs, bacteria present in the column will produce additional heat and artificially high readings will be obtained.

Because of the high heat of vaporization of water, the apparatus should be handled to avoid spillage inside the calorimeter. For the same reason, gas bubbles should be prevented from entering the system. Some

[9] B. Mattiasson, B. Danielsson, F. Winquist, H. Nilsson, and K. Mosbach, *Appl. Environ. Microbiol.* **41**, 903 (1981).

pulse damper constructions are able to trap incoming gas bubbles as well. Solutions can rapidly be degassed in an ultrasonic bath under vacuum before use.

Normally the thermograms are evaluated by peak height determination. This is simple to perform both manually from the recorder diagram and automatically by an integrator or computer. The peak height as well as the area under the peak and the ascending slope of the peak have been found to be linearly related to the substrate concentration.[10] The last two are also easily retrieved by automatic digital methods, but under normal conditions it has been found that the peak height calculation gives the best precision.[11] It should be noted that the normally large excess of enzyme results in virtually total substrate conversion. The peak height and the area determination will therefore represent an end-point determination. The slope, on the other hand, more closely resembles a kinetic determination. This can be useful to reduce the influence of any undesirable enzymatic side reactions which are likely to proceed at a much lower rate than the main reaction. In an end-point determination, such reactions will be more notable.

Applications

Thermal biosensors have been used in a considerable number of applications over the years.[12,13] A number of special applications are separately treated in this volume ([17]–[19], [27], [30]). Some additional enzyme-based applications are briefly described in this chapter to illustrate the simplicity of the technique. Furthermore, some applications based on immobilized cells and the use of the enzyme thermistor unit as a detector in chromatographic monitoring and in studies of microbial cell metabolism are presented.

Enzyme Thermistor Applications Based on Immobilized Enzymes

Glucose. To 1 ml of sedimented glass beads (CPG from Corning Glass Works, Corning, NY; 40–80 mesh, mean pore diameter 550 Å), γ-aminopropyltriethoxysilanized and glutardialdehyde activated, is added 4 mg or

---

[10] B. Danielsson, B. Mattiasson, and K. Mosbach, *Appl. Biochem. Bioeng.* **3**, 97 (1981).

[11] G. Decristoforo and B. Danielsson, *Anal. Chem.* **56**, 263 (1984).

[12] B. Danielsson and K. Mosbach, *in* "Biosensors: Fundamentals and Applications" (A. P. F. Turner, I. Karube, and G. Wilson, eds.), p. 575. Oxford Univ. Press, Oxford, 1986.

[13] K. Mosbach and B. Danielsson, *Anal. Chem.* **53**, 83A (1981).

1000 U of glucose oxidase (EC 1.1.3.4 from *Aspergillus niger,* type I, from Boehringer Mannheim, FRG) and 130,000 U of catalase (EC 1.11.1.6 from horse liver, 65,000 U/mg, Boehringer Mannheim). After gentle mixing overnight at 4°, the preparation is washed on a glass filter and packed in a 1-ml column. This preparation is stable for several months. The assay is usually run in 0.1 or 0.2 M sodium phosphate, pH 7.0. The linear range is limited by oxygen availability to about 0.7 mM. The detectability is better 0.005 mM. As shown by Kiba and Furusawa [19], the linear range can be considerably extended by running the assay in the presence of benzoquinone. We have found the oxygen-based method useful for the determination of glucose in blood serum samples diluted 50-fold.[14] If the blood is collected in tubes containing sodium fluoride, the running buffer is supplemented with 0.25 M NaF to mask any dilution effects of the fluoride. This method has also been used in many biotechnological studies in continuous monitoring[15] as well as with an automated flow–injection technique.[16]

Urea. 600 IU of urease (EC 3.5.1.5) type III from jackbeans (Sigma Chemical Co., St. Louis, MO) is immobilized to CPG or to Eupergit C (250 mg) in 50 mM potassium phosphate, pH 7.0, and packed into a 1-ml column. The assay is run in 0.1 M phosphate buffer, pH 7.0, at a flow rate of 1 ml/min. The calibration curve is linear in the range 0.01–250 mM. Addition of 1 mM EDTA and 1 mM reduced glutathione has been found to increase the stability of the enzyme. The enzyme column could be used for several weeks. Serum samples are diluted 10-fold with the running buffer prior to analysis. The results correlate well with those obtained by conventional colorimetric techniques.[7]

L-*Lactate.* For L-lactate determination two different enzymes have been used with good results. Originally we used lactate 2-monooxygenase (EC 1.13.12.4) from *Mycobacterium smegmatis* (Boehringer Mannheim). About 50 units of the enzyme is applied to a CPG column, and the assay is run in 0.2 M sodium phosphate buffer, pH 7.0. This assay is rather sensitive; with 0.5- to 1-ml samples and a flow rate of 1 ml/min the calibration curve is linear in the range of 0.005–1 mM. Alternatively, lactate oxidase (EC 1.1.3.2) from *Pediococcus pseudomonas* (Calbiochem, San Diego, CA) can be used, preferably coimmobilized with catalase. Typically, we apply 25 U of lactate oxidase and 130,000 U of catalase to 1 ml of CPG (1350 Å). The assay is run in 0.1 M sodium phosphate, pH 7.0, with

---

[14] B. Danielsson, K. Gadd, B. Mattiasson, and K. Mosbach, *Clin. Chim. Acta* **81,** 163 (1977).

[15] B. Danielsson, B. Mattiasson, R. Karlsson, and F. Winquist, *Biotechnol. Bioeng.* **21,** 1749 (1979).

[16] C. F. Mandenius, L. Bülow, B. Danielsson, and K. Mosbach, *Appl. Microbiol. Biotechnol.* **21,** 135 (1985).

a linear response below 1 mM L-lactate and with a detectability of 0.005 mM or better. Both enzyme preparations are stable for several months.

Oxalate. For oxalate determination we have either purified oxalate oxidase (EC 1.2.3.4) from barley seedlings[17] or purchased the same enzyme from Boehringer Mannheim. To 1 ml of glutaraldehyde-activated propylamino-CPG 10 (800 Å, 80–120 mesh) is added 4 U of oxalate oxidase. The assay is run in 0.1 M sodium citrate buffer, pH 3.5, containing 0.8 mM 8-hydroxyquinoline and 2 mM EDTA. The enzyme preparation can be used up to 3–4 weeks with a useful operating range of 0.005–0.5 mM. The assay is suitable for urine samples diluted 10-fold or for beverage and food samples. The determination of oxalate in urine, preferably preceded by passing the sample through a C_{18} cartridge (Sep-Pak, Waters Associates, Milford, MA), is considerably simpler and quicker than most methods used to date, which usually involve tedious precipitation steps.[17]

Ethanol. Ethanol is measured with alcohol oxidase from *Candida boidinii* (EC 1.1.3.13), obtained from Boehringer Mannheim. To 1 g of CPG (1350 Å) we add 50 U of alcohol oxidase purified by gel filtration on Sephadex G-25 (Pharmacia) and 130,000 U of catalase. Coimmobilization with catalase has been found to have a prominent effect on the stability of the enzyme column which with catalase should be stable for several months with an operating range of 0.01–2 mM when run in 0.1 M sodium phosphate, pH 7.0. The assay is useful, for instance, with samples from beverages or with minute blood samples.[18]

Sucrose. The enzyme thermistor offers a very simple and highly specific assay for sucrose in the range of 0.1–100 mM, using the enzyme invertase (EC 3.2.1.26, β-fructofuranosidase). To 1 g of glutaraldehyde-activated propylamino-CPG (1350 Å) suspended in 0.1 M sodium phosphate, pH 7.0, is added 10,000 U of β-fructofuranosidase (invertase) purchased from Boehringer Mannheim (300 U/mg lyophilizate), yielding enzyme columns stable for several months. The assay is run in 0.1 M sodium citrate, pH 4.6. Advantages of this method include high specificity, high operational stability, large linear range, and the use of a single enzyme, which reduces problems with interferences and obviates the need for any differential measurement to account for endogeneous glucose as with most electrode-based methods. A continuously running variation was used in a recent study on monitoring and control of enzymatic sucrose hydrolysis using on-line biosensors.[16]

[17] F. Winquist, B. Danielsson, J.-Y. Malpote, L. Persson, and M.-B. Larsson, *Anal. Lett.* **18,** 573 (1985).
[18] G. G. Guilbault, B. Danielsson, C. F. Mandenius, and K. Mosbach, *Anal. Chem.* **55,** 1582 (1983).

Enzymatic and Chemical Amplification

One of the main advantages of enzyme calorimetry is that the primary enzymatic reaction often generates enough heat for determinations with adequate sensitivity. Thus, an enzymatic reaction with a molar enthalpy change of -80 kJ/mol (such as that for glucose oxidase) will produce a peak height of 0.01° or more for a 1-min pulse of a 1 mM substrate solution at 1 ml/min. A temperature resolution of 0.0001° will then give an accuracy of 1% which is obtainable even with the simple device described above. The sensitivity of the oxidase reaction can be approximately doubled by coimmobilization with catalase ($\Delta H = -100$ kJ/mol). Additional advantages are that the oxygen consumption is reduced by 50%, which extends the linear range correspondingly, and, furthermore, deleterious effects by the hydrogen peroxide are eliminated or reduced.[18] The reproducibility or operational stability of the system will also be improved since oxidase preparations often contain catalase as a contaminant.

Instead of coimmobilizing enzymes, bi- or polyfunctional enzymes obtained by gene fusion can be used directly.[19,20] Thus, in preliminary studies, the system β-galactosidase–galactokinase was used in the enzyme thermistor unit, giving rise to an increase in the sensitivity of lactose on simultaneous addition of ATP to the system.[21]

Hydrolytic enzymes, such as β-galactosidase and β-glucosidase, usually produce little heat. For higher sensitivity, the β-galactosidase or β-glucosidase preparation should be used in a precolumn to an enzyme thermistor device (ET) containing glucose oxidase/catalase. A drawback is that endogenous glucose must be accounted for by a separate measurement or by complete elimination prior to the determination. Another way of increasing the sensitivity in hydrolytic or other enzymatic reactions where protons are generated is to employ a buffer with high protonation enthalpy, such as Tris buffer (-47.5 kJ/mol as compared to -4.7 kJ/mol for phosphate buffer).[6,10]

Highly increased sensitivity can be obtained by substrate or coenzyme recycling. Thus, we have demonstrated a 1000-fold enhancement of the detectability for lactate or pyruvate by using coimmobilized lactate oxidase/catalase and lactate dehydrogenase (LDH).[22] L-Lactate is oxidized by lactate oxidase to pyruvate, which is reduced to lactate by the LDH, etc. The total enthalpy of the system is further increased by the catalase reaction. The total enthalpy change is approximately the same as for the

[19] L. Bülow, P. Ljungcrantz, and K. Mosbach, *Bio/Technology* **3**, 821 (1985).
[20] L. Bülow, *Eur. J. Biochem.* **163**, 443 (1987).
[21] Bengt Danielsson, Leif Bülow, and Klaus Mosbach, to be published.
[22] F. Scheller, N. Siegbahn, B. Danielsson, and K. Mosbach, *Anal. Chem.* **57**, 1740 (1985).

highly exothermic oxidation of NADH ($\Delta H = -225$ kJ/mol). With 0.5-ml samples and a flow rate of 1 ml/min, concentrations as low as 10 nM can be determined. The sensitivity can be further increased by reducing the flow rate thereby permitting a larger number of cycles to occur during the passage of a sample through the column. With a similar technique the detectability for NAD(H) was improved 80-fold by coenzyme recycling using coimmobilized LDH and glucose-6-phosphate dehydrogenase (GDH) (Fig. 4).

Another system recently studied, allowing the determination of ADP/ATP down to 10^{-8} M, involves the simultaneous participation of two coimmobilized enzyme couples, pyruvate kinase–hexokinase, for ADP/ATP recycling and lactate dehydrogenase–lactate oxidase for recycling of the formed pyruvate/lactate.[23]

Environmental Control Applications

The thermometric detection principle can be exploited in two principally different ways in environmental control analysis.[10] The inhibitory effect of a pollutant on an immobilized enzyme[24] or immobilized cell preparation[25] in the ET device can be detected. Alternatively, the heat of an enzymatic or cell metabolic route converting the pollutant to some product can be detected. In the latter case, the sensitivity will frequently be too low for measurements on samples from air or water recipients, but could be of interest for determinations in process streams involving potential pollutants. The inhibitory effect of a pollutant, on the other hand, can be established with very high sensitivity, and it could be argued that the concept of environmental control analysis by measuring the biological effects should be an ideal one. For highest sensitivity the amount of enzyme immobilized should be low. If the inhibition is irreversible, the reversible immobilization technique described by Mattiasson (this volume, [60] and [61]) offers a convenient way of replacing the enzyme when "consumed." This can, for instance, be accomplished by using an immobilized lectin in the ET column, which binds the enzyme, if it is a glucoprotein, in such a way that it can easily be washed out and replaced by fresh enzyme. In many cases, as in the example described below, the inhibition is reversible, and the inhibitor can be washed out after the measurement and the ET column restored.

For the determination of heavy metals (Hg^{2+}, Cu^{2+}, and Ag^+) we employ a 0.5-ml column of urease bound to propylamino-CPG. To 1 ml of

[23] D. Kirstein, B. Danielsson, F. Scheller, and K. Mosbach, to be published.
[24] B. Mattiasson, B. Danielsson, C. Hermansson, and K. Mosbach, *FEBS Lett.* **85,** 203 (1978).
[25] B. Mattiasson, P.-O. Larsson, and K. Mosbach, *Nature (London)* **268,** 519 (1977).

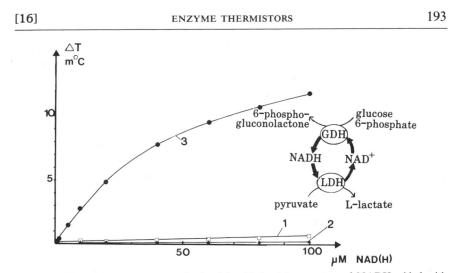

FIG. 4. Temperature response obtained for (1) 5 mM pyruvate and NADH added with only LDH working; (2) 5 mM glucose 6-phosphate and NAD$^+$ added with only GDH working; and (3) pyruvate, glucose 6-phosphate, and NADH added with both enzymes working. Reproduced from the Ph.D. thesis of N. Siegbahn, Univ. of Lund, Sweden, "Use of Immobilized Enzymes as Models for Organized Enzyme Systems" (1985), with permission of the author.

CPG, 38 mg of urease (Type III, 28 U/mg; Sigma Chemical Co.) is added. Coupling is carried out as described above. The response obtained for a 0.5-ml pulse of 0.5 M sodium phosphate buffer, pH 7.0, at 1 ml/min is noted, after which 0.5 ml of the sample is injected. Exactly 30 sec after the sample pulse (timing is important for reproducible results) a new 0.5-ml pulse of 0.5 M urea is introduced, and the temperature response is compared with the noninhibited peak. A calibration curve can be constructed with samples of known heavy metal ion concentration, and the concentration of an unknown sample can be estimated from such a curve. After a sample, the initial response can be established after washing with 0.1–0.3 M NaI plus 50 mM EDTA for 3 min. At the conditions given, a 50% lower response was recorded after the sample had been introduced, representing 50% inhibition with 0.04–0.05 mM Hg^{2+} or Ag$^+$ or 0.3 mM Cu^{2+}.[24] Longer sample pulses result in a considerably higher sensitivity. Thus, a 5-min pulse of 10^{-9} M HgCl$_2$ (0.2 ppb) resulted in 25% inhibition in one experiment. If more than one heavy metal ion species is present, a special selective washing procedure can be employed to give increased selectivity.[24]

Enzyme Activity Determinations Using the ET Unit

With a small modification of the flow system and with an empty column (preferably made of Teflon) or a piece of Teflon tubing as reaction chamber, the ET unit can be used for the determination of soluble enzyme activity. The sample solution and a solution containing an appropriate substrate to the enzyme in excess are each passed through a heat exchanger prior to mixing, and then the mixture is rapidly passed through one of the inner, short heat exchangers to eliminate mixing heats. Each flow channel normally includes a longer heat exchanger tube in contact with the outer aluminum block and a shorter heat exchanger in contact with the inner heat sink for final temperature equilibration. A valve has been installed by which the two outer heat exchangers can be connected to one of the inner ones to facilitate conversion between normal operation and enzyme activity determination. The temperature at the outlet of the reaction chamber is continuously monitored with one of the temperature probes as described in the Instrumentation section. Linear correlation over a wide range with a sensitivity of at least 0.01–0.1 U/ml has been observed between temperature response and enzyme activity for a variety of enzymes.[26]

This technique could be useful in clinical chemistry and should be of particular interest for monitoring enzyme purification processes. The absolute sensitivity is rather low, about the same as for A_{280} monitoring, but this drawback is balanced by the advantages of the technique, which is a direct, continuous-flow method that can be used on crude samples and with inexpensive substrates.

Calorimetric detection of the eluted enzyme activity has definite advantages for specific monitoring in chromatography, as was demonstrated in a study including gel filtration, ion-exchange chromatography, and affinity chromatography.[27] In this study it was shown that the ET unit could be used for direct on-line identification and localization of a specific enzyme in complex chromatograms with a sensitivity comparable to that of a UV monitor. It was especially valuable in affinity chromatography eluted with coenzymes which have a strong UV absorbance.

Alternatively, the enzyme present in the sample solution can first be enriched by affinity binding in a small affinity column inside the ET unit (for instance, binding to Concanavalin A (Con A)–Sepharose or to an adsorbent containing hydrophobic groups). Preferably this binding step should be reversible (compare Mattiasson, this volume, [60] and [61]). This arrangement resembles the conventional ET configuration, and the enzymatic activity is determined by introducing substrate in excess amounts. The adsorbent is then regenerated, and a new sample can be

[26] B. Danielsson and K. Mosbach, *FEBS Lett.* **101,** 47 (1979).
[27] B. Danielsson, L. Bülow, C. R. Lowe, I. Satoh, and K. Mosbach, *Anal. Biochem.* **117,** 84 (1981).

introduced, etc. The realization of this technique has been demonstrated by the determination of choline esterase (EC 3.1.1.8) activity in dilute solutions.[28] Choline esterase is dissolved in 50 mM KH$_2$PO$_4$/K$_2$HPO$_4$, pH 7.0, containing 1 mM CaCl$_2$, 1 mM MgCl$_2$, and 1 mM MnCl$_2$ and introduced into an ET column containing 0.7 ml of Con A–Sepharose. The enzyme is assayed by a pulse of 10 mM butyrylcholine for 1 min in the same buffer. Regeneration is accomplished by a pulse of 0.2 M glycine–HCl, pH 2.2, for 2 min. One unit of choline esterase produces a temperature peak of 0.01°.

A third alternative is to incubate the enzyme sample with substrate solution for an appropriate time, whereafter the remaining substrate concentration or the amount of product formed is assayed in the usual way with an ET. This approach could also give a high sensitivity if a long incubation time is used.

Applications Based on the Use of Immobilized Cells

Occasionally it could be advantageous to use immobilized cells instead of enzymes. This is the case when the pure enzyme is too unstable or difficult to isolate or when the desired reaction involves enzyme sequences. Furthermore, immobilized whole cells can be used in assays based on the metabolic effect of an agent. An obvious disadvantage is the lack of specificity of whole cells in comparison with enzymes.

Different aspects of utilizing *Gluconobacter oxydans* in calorimetric determinations have been studied by Satoh *et al.*[29] For the determination of glycerol, 1 g of a cell suspension containing 0.2 g/ml (dry weight) of *Gluconobacter oxydans* is mixed with 4 g of 1% alginate solution and extruded through a 1-mm orifice in 0.1 M CaCl$_2$ solution. The preparation is allowed to stand for 2 hr whereafter it is cut into 2-mm pieces and packed into a 1-ml ET column. The assay is run in 0.1 M sodium succinate plus 10 mM CaCl$_2$, pH 5.0, at a flow rate of 0.9 ml/min. For sample pulses the calibration curve is linear up to 2 mM glycerol with a sensitivity of 0.003°/mM, whereas continuous sample introduction results in a sensitivity of 0.008°/mM with linearity up to 1 mM. The response of this column is almost as rapid as that of an enzyme column and allows up to 15–20 samples per hour to be analyzed.

If, instead of an immobilized cell preparation, a suspension of *Gluconobacter oxydans* (0.3 g/ml) is mixed with glycerol and a solution (flow rate 0.9 ml/min of both streams) as described above under Enzyme Activ-

[28] B. Danielsson, B. Mattiasson, and K. Mosbach, *Prepr. Eur. Congr. Biotechnol. 1st*, p. 1/107 (1978).
[29] I. Satoh, B. Danielsson, and B. Mattiasson, unpublished results.

ity Determinations Using the ET Unit, a linear temperature response is again obtained with respect to glycerol concentration. Linearity is obtained up to 10 mM glycerol, but at a lower sensitivity ($0.0005°$/mM). If, on the other hand, the glycerol concentration is kept constant, the same arrangement permits determination of cell density of the bacteria suspension: at 50 mM glycerol linear temperature response is obtained as a function of the cell density in the range 100–500 mg/ml.

The effects of various agents on the metabolism of a microorganism can be studied by feeding a microorganism suspensions containing suitable substrates through an ET unit. The temperature level recorded represents the total metabolism of the microorganism, and changes in the metabolism due to the presence of, for instance, an inhibitor are directly detectable as changes in the temperature response. Thus, Hörnsten *et al.*,[30] in a study on the ampicillin susceptibility of *Escherichia coli,* could clearly observe changes in the metabolic behavior at cultivation under standard conditions even in the presence of ampicillin at a concentration 10 times lower than the MIC (minimal inhibitory concentration).

Similar studies can be undertaken on immobilized cell preparations placed in an ET column. For example, *Lactobacillus plantarum* cultivated in a medium deficient of the vitamin nicotinic acid has been immobilized by entrapment in gelatin beads (see Vol. 135, this series, for suitable entrapment procedures) and packed in a 1-ml ET column. Pumping nicotinic acid-deficient medium through the ET results in a constant temperature level which is considerably increased within 10 min even when very small amounts of nicotinic acid are included in the medium.[31] In principle, this technique can be used to measure, for instance, vitamin concentrations in food samples. The sensitivity is adequate, but the slow response is a considerable disadvantage, and the column can only be used for one positive test. Multiple samples can, however, be analyzed by a flow technique involving a suspension of the bacteria.

General Comments

As shown, the enzyme thermistor has found application in clinical analysis and fermentation control, including environmental control as a toxiguard, and will probably also be useful in general biochemical studies. Its relatively small size compared to that of conventional microcalorimeters and the possibility of using the device in continuous flow-through

[30] E. G. Hörnsten, B. Danielsson, H. Elwing, and I. Lundström, *Appl. Microbiol. Biotechnol.* **24,** 117 (1986).
[31] K. Jakobsen and B. Danielsson, unpublished results.

conditions and in measuring many discrete samples per time unit make it highly convenient as a general biosensor.

Other aspects worth mentioning, which could be studied conveniently with this sensor because of its measuring capabilities and relative simplicity, include the following: (a) Measurements in organic solvent water mixtures, miscible as well as immiscible (not much interference is expected from turbid solutions). In addition, as preliminary data indicate, amplified signals may be obtained due to increased heat capacity of organic solvents. (b) Systematic on-line screening of an enzyme for a great variety of commonly found and unusual substrates may easily be carried out. (There is no dependence, for instance, on chromophoric substrates.) Similarly, screening for competitive (and, less conveniently, for noncompetitive) inhibitors and cofactors, including metals, is easily carried out. Furthermore, artificial enzymes might be conveniently screened and assayed provided their specific activity is sufficiently high. (c) Enzyme kinetics can be conveniently studied.

In summary, the enzyme thermistor devices described should find wide applications.

[17] Flow–Injection Analysis for Automated Determination of β-Lactams Using Immobilized Enzyme Reactors with Thermistor or Ultraviolet Spectrophotometric Detection

By Georg Decristoforo

Immobilized enzymes have found widespread application in analytical chemistry within the last decade.[1] The use of immobilized enzyme reactors in flow–injection analysis (FIA) has several advantages compared with the conventional FIA technique, where usually a chemical reaction is performed by mixing reagents and substrate in a reaction coil and the reaction product is measured by a detector.[2–7] Most detectors for liquid

[1] P. W. Carr and L. D. Bowers, "Immobilized Enzymes in Analytical and Clinical Chemistry," Chs. 4 and 8. Wiley (Interscience), New York, 1980.
[2] J. Růžička and E. H. Hansen, *Anal. Chim. Acta* **99**, 37 (1978).
[3] J. Růžička and E. H. Hansen, *NBS Spec. Publ. (U.S.)* **519**, 501 (1979).
[4] E. H. Hansen, J. Růžička, and B. Rietz, *Anal. Chim. Acta* **89**, 241 (1977).
[5] J. W. B. Stewart and J. Růžička, *Anal. Chim. Acta* **82**, 137 (1976).
[6] J. Růžička, E. H. Hansen, A. K. Ghose, and H. A. Mottola, *Anal. Chem.* **51**, 199 (1979).
[7] H. Baadenhuijsen and H. E. H. Seuren-Jacobs, *Clin. Chem.* **25**, 443 (1979).

METHODS IN ENZYMOLOGY, VOL. 137

chromatography are suitable for FIA (e.g., ultraviolet, fluorimetric, refractometric, mass spectrometric, electrochemical, and enthalpimetric detectors).[8] Solid-bed enzyme reactors permit reagentless determination of various substrates by use of an excess of enzyme in relation to sample amount, yielding total substrate conversion.[9-17]

Due to the high specificity of enzyme reactions, universal and nonselective detection principles may be chosen. In this chapter the design of a FIA system is described, which is operated with different detectors in direct and subtractive modes. Examples for the automated determination of various β-lactams in fermentation broth and pure solutions by means of a thermistor detector and by subtractive UV spectrophotometric detection are given.[18,19]

Theoretical Considerations

In continuous-flow systems two general detection principles are viable; either measuring the change of nonspecific physicochemical property of the whole solution or monitoring a property exclusively specific to the solute to be detected. The combination of a detector with an enzyme reactor enhances the selectivity of measurement significantly, as is demonstrated by enthalpimetric detection of penicillins with an enzyme thermistor detector, where the enthalpy of an enzymatic reaction is directly related to the amount of converted substrate. In case of unspecific detection such as the described subtractive UV spectrophotometric determination of cephalosporins, the difference between the signals obtained for the whole solution and that derived from the solvent matrix serves as a sole measure of the amount of substrate to be determined.

Principle of Enzyme Thermistor–FIA

Appropriately prepared sample solutions are injected into the FIA system via an automatic sampler. The substrate contained in the sample

[8] J. Růžička and E. H. Hansen, "Flow Injection Analysis," Chs. 2, 4, 5, and 6. Wiley (Interscience), New York, 1981.
[9] B. Mattiasson, K. Mosbach, and Å. Svensson, *Biotechnol. Bioeng.* **19,** 1556 (1977).
[10] L. Ögren and G. Johansson, *Anal. Chim. Acta* **96,** 1 (1978).
[11] D. R. Senn, P. W. Carr, and L. N. Klatt, *Anal. Chem.* **48,** 954 (1976).
[12] L. J. Forrester, D. M. Yourtee, and H. D. Brown, *Anal. Lett.* **7,** 599 (1974).
[13] G. Johansson, K. Edström, and L. Ögren, *Anal. Chim. Acta* **85,** 55 (1976).
[14] B. Mattiasson, B. Danielsson, and K. Mosbach, *Anal. Lett.* **9,** 217 (1976).
[15] D. C. Williams, G. F. Huff, and W. R. Seitz, *Clin. Chem.* **22,** 372 (1976).
[16] L. P. Leon, M. Sansur, L. R. Snyder, and C. Horvath, *Clin. Chem.* **24,** 1556 (1977).
[17] B. Danielsson, K. Gadd, B. Mattiasson, and K. Mosbach, *Anal. Lett.* **9,** 987 (1976).
[18] G. Decristoforo and B. Danielsson, *Anal. Chem.* **56,** 263 (1984).
[19] G. Decristoforo and F. Knauseder, *Anal. Chim. Acta* **163,** 73 (1984).

solutions is selectively converted by a specific carrier-bound enzyme while flowing through a packed-bed enzyme reactor, exactly thermostatted to 30°. The heat generated by the enzymatic reaction is transported to the end of the reactor by the liquid stream and monitored there by a thermistor (NTC type) fixed to a gold capillary against a reference thermistor. The reaction enthalpy is measured in an adiabatic system causing a change of thermistor resistance which is in linear relationship to substrate concentration over a wide range. The resistance values are measured on a precision Wheatstone bridge, and the maximum achieved temperature resolution is approximately 10^{-5}°.

Principle of FIA Determination of Cephalosporins by Subtractive UV Spectrophotometric Detection

The cleavage of the β-lactam ring (see Scheme 1) of Δ^3-cephalosporins by nucleophilic agents such as water, hydroxide, amines, and enzymes has a different mechanism than the hydrolysis of penicillins, which leads to relatively stable penicilloic acid derivatives.[20,21] The special behavior of cephalosporins (I) is primarily due to their Δ^3-double bond; the enamine resonance possible thereby weakens the β-lactam amide bond for nucleophilic attack. If there is a substituent at the C-3 position of the molecule, which can act as a suitable leaving group (L), intermediate compounds with an *exo*-methylene moiety (II) are formed; these intermediates exhibit an absorption maximum at 230 nm. If no suitable C-3 substituents are present, cephalosporoic acid derivatives (λ_{max} 260 nm; III) are formed, most of which are unstable and undergo further degradation to various fragments absorbing at other wavelengths. In all these cases, degradation of the cephalosporin molecule is followed by destruction of the UV chromophore with, typically, an absorption maximum at 260 nm, i.e., the absorbance at 260 nm decreases significantly and sometimes new chromophores can be formed. An exception is deacetylcephalosporin C lactone, an α,β-unsaturated γ-lactone, where enamine resonance cannot develop; after hydrolytic cleavage of the β-lactam ring, the cephalosporoic acid structure (λ_{max} 265 nm; IV) is stabilized, and the absorbance at 260 nm is largely maintained.[22] Extensive NMR investigations[23] have shown that the cephalosporoic acid derivative is the cause for the maintenance of

[20] G. G. F. Newton, E. P. Abraham, and S. Kuwabara, *Proc. Intersci. Conf. Antimicrob. Agents Chemother, 7th, Chicago*, p. 449 (1967).

[21] M. A. Schwartz, *in* "Beta-Lactamases" (J. M. T. Hamilton-Miller and J. T. Smith, eds.), p. 51. Academic Press, London, 1979.

[22] J. M. T. Hamilton-Miller, G. G. F. Newton, and E. P. Abraham. *Biochem. J.* **116,** 371 (1970).

[23] J. M. T. Hamilton-Miller, E. Richards, and E. P. Abraham, *Biochem. J.* **116,** 385 (1970).

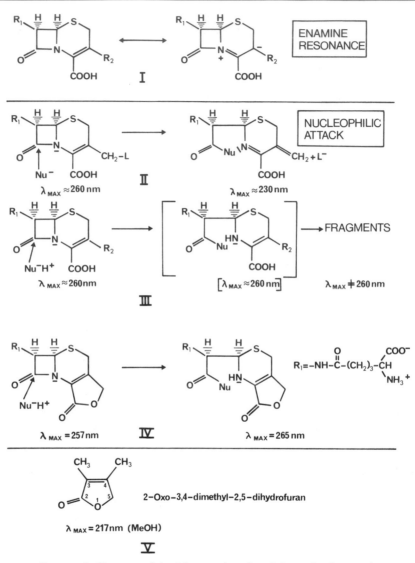

SCHEME 1. Cleavage of the β-lactam ring of cephalosporins (see text).

absorbance at 265 nm and not the α,β-unsaturated γ-lactone ring. Substituted α,β-unsaturated γ-lactones, e.g., 2-oxo-3,4-dimethyl-2,5-dihydrofuran (**V**), exhibit absorption maxima between 210 and 220 nm.[24]

These observations are utilized in the described FIA system for the

[24] D. E. Ames, R. E. Bowman, and T. F. Grey, *J. Chem. Soc.* p. 375 (1954).

determination of cephalosporins. The samples are injected into the flow system, where the cephalosporins are degraded selectively by immobilized β-lactamase in an enzyme reactor, so that the UV chromophore (λ_{max} 260 nm) is destroyed. Each sample is injected twice and passes through a dummy column of exactly the same geometric dimensions as the enzyme reactor, filled with inert glass carrier material during the first run. In this run, the total absorbance of the sample and sample matrix is measured at 260 nm. Before the second injection of the sample, the enzyme reactor is switched into the flow line; only the cephalosporins in the sample lose their chromophore. The difference of the A_{260} values between untreated and enzymatically degraded samples is proportional to the concentration of cephalosporins. Peak areas are measured, in order to overcome the differences in sample dispersion caused by varying kinetics.

Experimental Methods

Apparatus

Enzyme Thermistor (ET). Enthalpy is measured in an aluminum thermostat as described by Danielsson et al.,[25] and resistance changes are registrated with a precision Wheatstone bridge, both constructed and assembled at the University of Lund (Lund, Sweden) [16]. The thermistors are isotherm matched pairs with NTC characteristics (Type A 395, 16 $k\Omega$ at 25°) and were purchased from Veco (Springfield, NJ). Buffer solution is pumped through the flow system by a peristaltic pump Minipuls 2-HP 8 from Gilson (Villiers-le-Bel, France). Sampling is done by a cooled Perkin-Elmer ISS-100 sampler (Perkin-Elmer, Bodenseewerk, Überlingen, FRG), and cooling is performed by a Haake D3G constant-temperature circulator (Haake, Karlsruhe, FRG). A microcomputer interface based on an INTEL 8085 CPU for total ET–FIA control and synchronization of auto sampler and data system was designed and built by the Electronic Department of Biochemie Ges.m.b.H. (Kundl, Austria). A Servogor 210 pen recorder (BBC-Goerz, Vienna, Austria) is used for immediate visualization of the temperature signals. Digitizing is done by a Hewlett-Packard HP-18652 A/D converter (ADC), and the data are evaluated by a HP-3357 data system. For method comparison a HPLC apparatus (HP-1084-B) with variable-wavelength detector and automatic sampler is used.

Connecting tubing is of PTFE (0.5 mm i.d.) with Swagelok nylon joints. The total length of tubing from the injection valve to the detector is

[25] B. Danielsson, B. Mattiasson, and K. Mosbach, *Appl. Biochem. Bioeng.* **3**, 97 (1981).

approximately 80 cm. The reactor columns for the enzyme thermistor (with the dimensions 0.8 cm o.d., 0.7 cm i.d., 1.5 cm bed height) are of delrin (polyacetal).

Enzyme Photometer (EP). The instrumental design of the FIA system with UV photometer detector is similar to that of the ET–FIA apparatus described above because of their modular construction. Block switching diagrams for both flow systems are given in Fig. 1. Absorbance is measured with a fixed-wavelength detector at 254 nm (HP-1036 A, Hewlett-Packard, Böblingen, FRG). The buffer solution is pumped through the system with a HPLC pump (Altex 110, Altex/Beckman, Berkeley, CA) at a flow rate of 3 ml/min. Sampler, cooling thermostat, recorder, ADC, and data system are identical to those used for enzyme thermistor–FIA. The INTEL 8085 CPU-based microcomputer interface is also the same for both ET–FIA and EP–FIA, but different assembler language programs had to be used for runoff control. The reactor columns for EP–FIA are of stainless steel tubing (0.64 cm o.d., 0.47 cm i.d., 5.0 cm length) fixed at both ends with Swagelok reductions (1/4 to 1/16 inch) and sealed with 2-μm stainless steel frits. Connecting tubing was of stainless steel (0.2 mm i.d.).

Reagents and Chemicals

For enzyme immobilization controlled-pore glass (CPG) beads (Pierce Chemical Co., Rockford, IL) with a pore diameter of 50 nm and a particle size of 125–175 μm are used. Different enzymes are bound covalently to

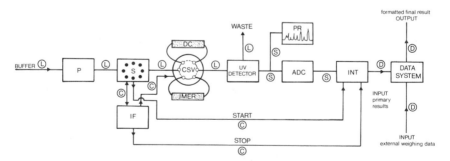

FIG. 1. Block switching diagram of the automated enzyme photometer flow–injection (EP–FIA) system. When the column-switching valve and the UV detector are replaced by the aluminum thermostat and Wheatstone bridge, the same diagram is valid for the enzyme thermistor flow system (ET–FIA). P, Pump; S, sampler; IF, interface; CSV, column-switching valve; DC, dummy column; IMER, immobilized enzyme reactor; PR, pen recorder; ADC, analog-digital converter; INT, integrator; L, liquid line; C, control line; S, signal line. For details, see Fig. 2 and text.

this glass: penicillinase P0389 (type I) and penicillinase P6018 (type II) with 70 IU/mg, both from *Bacillus cereus* (Sigma Chemical Co., St. Louis, MO), bactopenicillinase concentrate, 65 IU/ml, from *Bacillus cereus* (0346-62) (Difco Labs., Detroit, MI), and cephalosporinase from *Enterobacter cloacae* (P99) with 66 IU/mg (Miles Labs., Elkhart, IN). 3-(Triethoxysilyl)propylamine (Merck, Darmstadt, FRG) is used for CPG activation as received. Penicillin V (potassium salt, charge 07037/1) with a content of 96.1% (w/w), cephalosporin C (sodium salt, dihydrate, 97.5% pure), and deacetylcephalosporin C (sodium salt, dihydrate, 94.0% compared with the cephalosporin C salt) produced by Biochemie G.m.b.H. are used as working standards. Deacetoxycephalosporin C (sodium salt, dihydrate, 92.9% pure) is obtained from Proter (Milan, Italy). Other compounds used are ceftriaxone (disodium salt, Rocephin) and cefalexin (both Biochemie G.m.b.H.); cefazolin, cefamandole, cefachlor, and cefalotin (all sodium salts, Eli Lilly, Indianapolis, IN); cefoxitin (sodium salt, Merck, Sharp & Dohme, Rockway, NY); cefaloridine (Sigma Chemical Co.); and cephaloglycine (a gift from Sandoz Research Institute, Vienna, Austria). Methanol (p.a. quality; Riedel de Haen, Seelze, FRG) and tetrabutylammonium bromide (Fluka AG, Buchs, Switzerland) are used for HPLC. The stationary phase for HPLC is Nucleosil C_{18} reverse-phase material (5 μm; Macherey & Nagel, Düren, FRG). Other chemicals and reagents are p.a. grade (Merck).

Procedures

Enzyme Immobilization. The enzymes used are bound covalently to activated porous glass. The glass is washed in boiling 5% nitric acid, reacted with 3-(triethoxysilyl)propylamine to give the alkylamino glass, and treated with 2.5% glutardialdehyde in an aqueous system.[26] The penicillinases are bound to the pretreated glass by gentle shaking at ambient temperature for 8 hr, whereas the cephalosporinases are immobilized by shaking for 4 hr at 4°. Concentrations of 5–15 mg protein/g derivatized glass yield sufficient enzyme concentrations. Stirring of the carrier material in the enzyme solution should be avoided, because of the risk of breaking up the porous glass beads, which makes the reactor less resistant to clogging.

Sample Preparation. Biological samples for ET–FIA are pretreated with special care in order to prevent particle contamination, which results in instabilities of the thermosignal and leads to continuous clogging of the columns. Samples from the fermentation process containing analyte in

---

[26] H. H. Weetall, this series, Vol. 44, p. 134.

concentrations between 5 and 30 mg/ml are diluted with bidistilled water (1 : 7), centrifuged at 15,000 rpm for 5 min and finally pipetted directly into the sampler vials. Standard solutions are prepared with twice-distilled water in the concentration range 0.5–10 mg/ml.

Far fewer problems arise when preparing biological materials for EP–FIA determinations, because the sensitivity of the method permits high dilutions to be used. Interfering effects by the biological matrix and particle contaminations are therefore kept very low. Samples for EP–FIA from the fermentation process are diluted by a factor 100, centrifuged, and pipetted into glass sample vials. The concentrations of the standard solutions in water are in the range 100–500 μg/ml.

Flow–Injection Methods. The FIA conditions for both ET and EP systems are summarized in Fig. 2. The eluent consists of a 0.15 M phosphate buffer (pH 7) containing sodium azide (0.065 g/liter, 1 mM) which is pumped continuously at 3.0 ml/min in both systems. Sample volumes (100 μl) are injected into the buffer stream by an automatic sampler, which holds up to 100 samples and is cooled to 4°. The maximum back pressure for both flow systems with the packed reactors in place is 0.6–0.8 MPa (6–8 bar). A very versatile microcomputer-based interface for communication between sampler, ADC, and data system counts the sample injections, controls switching valves, starts and stops both sampler and ADC,

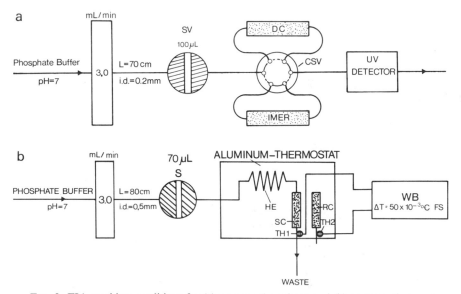

FIG. 2. FIA working conditions for (a) enzyme thermistor and (b) enzyme photometer.

and synchronizes the whole procedure. The interface timing diagrams for ET and EP applications are shown in Fig. 3. The enzyme reactions within the enzyme thermistor are carried out at $30 \pm 0.002°$, those in the enzyme photometer are done at ambient temperature.

In EP–FIA the switching between the enzyme reactor and the dummy reactor is performed with a pneumatic 6-port valve. Before the first sample is injected, the column-switching valve is automatically set to a position such that the buffer solution flows through the dummy column and such that the absorbance of the whole untreated sample is measured. Before the second injection of the sample, the enzyme reactor is switched into the flow stream, and the residual absorbance of the reacted sample is recorded. Before each further injection, the pneumatic valve is switched to the other position. A preset number of single injections (2 per sample) is used (40 injections, see below). When utilizing the ET–FIA device, no column switching is required, and the number of single injections is set to 20. After the last preset injection the sampler is stopped after a delay of 10 sec. Then a second delay time of 90 sec is running off in order to obtain the last peak and to regain the integration base-line. Then the ADC is stopped for result output, and a third delay is started in order to finish the

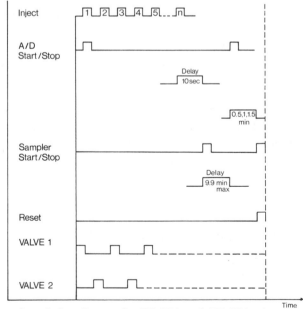

FIG. 3. Interface timing diagram for ET–FIA and EP–FIA, showing the cycles for injection, integrator start/stop, sampler start/stop, and, in case of EP–FIA, the column-switching valve cycles, which are omitted with the ET system.

data evaluation. A new start command by the interface activates the cycle again for the next sample set.

For the EP the detector sensitivity is set to 1.024 absorbance units full scale, the recorder scale is 10 mV, and the chart speed is 0.5 cm/min. For ET measurements the Wheatstone bridge is run in differential mode against a reference thermistor. Bridge sensitivity is adjusted to $5 \times 10^{-2\circ}$ full scale; the recorder scale is 100 mV, and the chart speed is set to 0.5 cm/min.

Results and Discussion

Data Evaluation

The signals generated by both FIA methods correspond approximately to a normal distribution and may therefore be processed like chromatographic peaks. ET peaks are evaluated by height because of peak asymmetry due to heat dispersion, whereas EP signals are processed by area rather than height because of different sample dispersion within the dummy and enzyme reactors.

The sample concentrations are calculated by means of a multiple internal-standard method. The data for 3 standards and 17 samples are collected in a group, a procedure which allows the handling of a large number of samples in a minimal overall analysis time. The results of the 20 runs (40 single injections for the EP) are then evaluated by the data system, analogously to a single chromatogram with 20 (40) peaks. Typical recorder outputs resulting from a penicillin V (potassium salt) run on the ET and from a cephalosporin C run on the EP are demonstrated in Fig. 4. The size of the sample groups and the number, concentrations, and reactions of the standard solutions in the sampler can be freely chosen during method development. The sample concentrations are calculated on the basis of a corrected calibration curve, which is derived from the standards contained in every particular group of samples. A statistical test is applied to all internal calibration functions automatically by application of curve fitting by the least-squares method. The correlation coefficient (r) is displayed in each final report as a measure of the quality of the results; the value of r must be greater than 0.999 for three internal standards, otherwise the results for the block of sample runs are discarded because of statistical unreliability.

Enzyme Thermistor in β-Lactam FIA

The determination of β-lactams in biological samples and pure aqueous solutions is performed very rapidly and specifically with an ET detec-

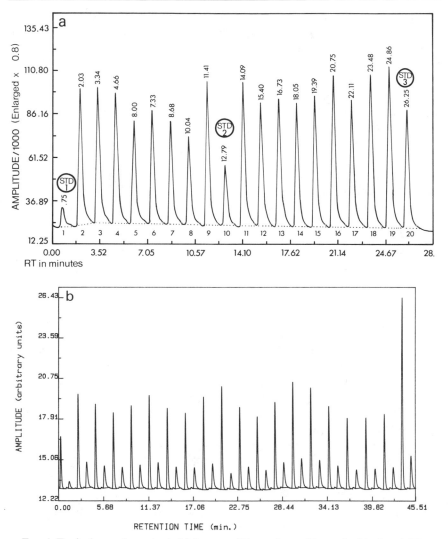

FIG. 4. Typical recorder outputs (a) for a run (17 samples and 3 standards) of penicillin V (potassium salt) samples with the ET–FIA system and (b) for a run of cephalosporin C (sodium salt) samples with the EP–FIA system.

tor in flow–injection analysis. The extent of substrate conversion within the enzyme reactor was investigated under the described ET–FIA conditions. For this purpose standard solutions of penicillin V (potassium salt) and cephalosporin C (sodium salt) with concentrations of 6 and 10 mg/ml are injected into the ET–FIA system, the effluent collected, and the amount of unreacted β-lactam determined by HPLC. Less than 0.1% of β-lactam could be detected; thus the rate of conversion was higher than 99.9% even for the most concentrated standards.

Response time for a single sample is approximately 80 sec. To examine the long-term stability of the enzyme reactors two hundred penicillin samples from fermentation process were analyzed after pretreatment every day for 1 week. No significant decrease in enzyme activity can be detected after 1000 sample injections. The temperature response for three penicillin V (potassium salt) standard solutions with different concentrations (0.961, 2.883, and 5.766 mg/ml) is taken as a measure of enzyme stability. Nevertheless, after throughput of more than 1000 samples a distinct reduction of column efficiency was realized due to clogging of the enzyme columns arising from particulate matter contamination. The decreased permeability of the reactor columns results in instabilities of the thermosignals.

Determination of Penicillin V (Potassium Salt). Penicillin V (3,3-dimethyl-7-oxo-6-[(phenoxyacetyl)amino]-4-thia-1-azabicyclo[3.2.0]heptane-2-carboxylic acid) (potassium salt) was determined according to the method described in "Procedures."

Statistical accuracy of the ET–FIA determinations was examined by method comparison with HPLC. The HPLC procedure is performed on a reverse-phase C_{18} column with methanol/50 mM ammonium carbonate solution in water (30/70) as eluent and has been described in literature earlier.[27] Sixty samples from the fermentation process containing penicillin V in a concentration range between 1 and 20 mg/ml are diluted according to the respective method and analyzed simultaneously by both ET–FIA and HPLC for the evaluation of method correlation. The correlation curve follows a linear equation $y = kx + d$ ($n = 60$, $k = 1.035$, $d = -0.02$). Excellent correspondence between the ET–FIA and the HPLC method is demonstrated by the coefficient of correlation r ($r = 0.9993$). The minimal amount of penicillin V which can be detected by ET–FIA is 10^{-6} g. The data pair t-test was applied to the results of 21 additional parallel determinations of penicillin V by both ET–FIA and HPLC. No statistical difference between the two methods can be detected at a significance level of 5% ($n = 21$, $r_{\alpha(2p=0.05)} = 1.23 < 2.09$).[28,29]

Precision of the method was determined by repeating the whole assay

[27] R. E. White, M. A. Carroll, J. E. Zarembo, and D. A. Bender, *J. Antibiot.* **28,** 205 (1975).
[28] R. Kaiser and G. Gottschalk, "Elemtare Tests zur Beurteilung von Messdaten." Bibliographisches Institut, Mannheim, Federal Republic of Germany, 1972.
[29] Ciba-Geigy Ltd. (ed.), "Wissenschaftliche Tabellen," 7th Ed., p. 32. Ciba-Geigy, Basel, Switzerland, 1971.

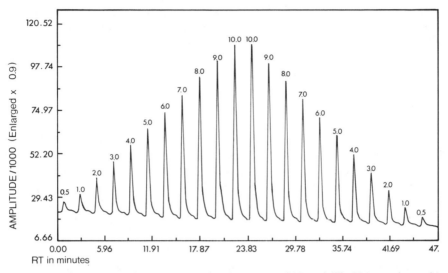

FIG. 5. Linear range of a standard calibration curve of 22 penicillin V (potassium salt) standard solutions with concentrations from 0.5 to 10.0 mg/ml determined with ET–FIA.

10 times under the same conditions for the same sample. The relative standard deviation (RSD) is found to be $\pm 0.8\%$ ($n = 10$, $\bar{x} = 11.76$ mg/ml, $s = \pm 0.098$, RSD $= \pm 0.8\%$). The linear range of the standard calibration curve is given in Fig. 5 ($n = 22$, $k = 10.8$, $d = -0.53$, $r = 0.9998$). The coefficient of quality (GK), used instead of the correlation coefficient r for characterization of the calibration curve fit, is $GK = \pm 1.27\%$.[30]

Some results of penicillin V determinations with ET–FIA yielded 1–5% higher values compared with HPLC due to simultaneous comonitoring of β-lactam by-products like p-hydroxypenicillin V. Since no separation is performed between the investigated substances and the interfering by-products before the calorimetric detection within the enzyme thermistor, carrier-bound enzymes with very high specifity toward the substrates to be analyzed are required. For that reason only highly purified enzymes are utilized for satisfactory performance of the ET–FIA system. To achieve selective determinations of various substrates it is necessary to select suitable enzymes in the course of method development.

Determination of Cephalosporin C. The cephalosporinase used for the cephalosporin C assay (from *Enterobacter cloacae* P99) does not exclusively affect the cephalosporin C (3-[(acetyloxy)methyl]-7-[(5-amino-

[30] J. Knecht and G. Stork, *Z. Anal. Chem.* **270,** 97 (1974).

5-carboxy-1-oxopentyl)-amino]-8-oxo-5-thia-1-azabicyclo[4.2.0]oct-2-ene-2-carboxylic acid, sodium salt) (CC) β-lactam ring; it also exhibits reactivity against deacetylcephalosporin C (DACC) and deacetoxycephalosporin C (DAOCC). Therefore only group-specific determinations of these cephalosporins in common could be performed with this enzyme. Experiments for enhancement of selectivity with a more CC-specific β-lactamase are reported under "Enzyme Photometer for Determination of Different Cephalosporins."[31]

Penicillin N, an intermediate metabolite in CC biosynthesis, and components from the fermentation medium also undergo nonspecific enzyme reactions with cephalosporinase (Miles). A precolumn filled with an immobilized β-lactamase of type I (from *Bacillus cereus*) was therefore inserted in the liquid stream before the thermostat block of the enzyme thermistor in order to eliminate these unwanted side effects. CC is decomposed by β-lactamase of type I only to an extent of 0.01% compared with penicillin V (100%) and penicillin N (38%).[32]

Comparison of CC determinations performed with both the ET method and a HPLC procedure yielded good correlation.[33] In Table I the results of CC determinations, analyzed simultaneously by means of ET–FIA and HPLC, are summarized. On an average the values for the ET–FIA results are around 1.5% higher than the results from HPLC for the sum of CC, DACC, and DAOCC. Relative standard deviation for ET–FIA results compared with HPLC was RSD = ±4%.

Enzyme Photometer for Determination of Different Cephalosporins

Determination of Cephalosporin C. Different cephalosporins can be determined with the described EP–FIA system rapidly and selectively. The extent of substrate conversion was tested as described above (see "Enzyme Thermistor in β-Lactam FIA") and found to be greater than 99.9%. In biological samples, which contained both DACC and CC, both substances were detected together.

The response time of the EP–FIA system taken from injection of the sample is only 10 sec, and a peak is complete after 30 sec. Nevertheless, the time required for one sample is approximately 2 min, because each sample is injected twice and the automatic sampler used is limited in cycle time by its rinsing and sample uptake processes. The long-term stability of

[31] E. P. Abraham and S. G. Waley, *in* "Beta-Lactamases" (J. M. T. Hamilton-Miller and J. T. Smith, eds.), p. 334. Academic Press, London, 1979.
[32] E. P. Abraham and S. G. Waley, *in* "Beta-Lactamases" (J. M. T. Hamilton-Miller and J. T. Smith, eds.), p. 324. Academic Press, London, 1979.
[33] R. E. White and J. E. Zarembo, *J. Antibiot.* **34**, 836 (1981).

TABLE I
CEPHALOSPORIN C DETERMINATIONS IN
FERMENTATION SAMPLES SIMULTANEOUSLY
PERFORMED BY ET–FIA AND HPLC

Sample	Concentration of CC, DACC, and DAOCC in biological samples (mg/ml)		
	HPLC	ET–FIA	Δ%
1	5.76	6.06	+5.2
2	5.68	6.24	+9.9
3	19.86	21.09	+6.2
4	14.28	13.29	−6.9
5	13.80	13.95	+1.1
6	19.56	19.47	−0.5
7	19.71	19.80	+0.5
8	20.72	20.79	+0.3
9	22.33	23.07	+3.3
10	19.29	19.50	+1.1
11	15.59	15.33	−1.7
12	16.56	16.32	−1.4
13	17.70	17.82	+0.7
14	16.88	17.28	+2.4

the enzyme reactor is excellent, even better compared with the ET device, due to the high sensitivity of the method and the high dilutions. Several thousand samples can be processed without any remarkable decrease in signal yield with one reactor (Fig. 6). The minimal detectable amount of CC is 2×10^{-7} g. The linear range of the standard calibration curve was measured at concentrations between 10 and 1000 μg/ml; the calibration graph became nonlinear above 800 μg/ml ($n = 16$, $k = 193.85$, $d = 2357$, $r = 0.998$). Precision of the method was determined by 15-fold injection of an identical sample. The relative standard deviation was ±0.7% ($n = 15$, $\bar{x} = 207.23$ μg/ml, SD = ±1.368 μg/ml).

Method Comparison between EP–FIA and HPLC. The accuracy of the described EP–FIA method was determined by comparison with a HPLC method: 44 samples from CC fermentations, containing CC in concentrations of 0.1–13 mg/g of culture broth, are analyzed simultaneously by the two methods. HPLC separations are done on an octadecylsilica reverse-phase column (Nucleosil C_{18}, 5 μm) with an eluent consisting of 10 mM tetrabutylammonium bromide in water/methanol (3/1). The chromatographic column is 10 cm long (0.64 cm o.d., 0.47 cm i.d.) and thermostatted at 40°; the flow rate is 2 ml/min and the back pressure

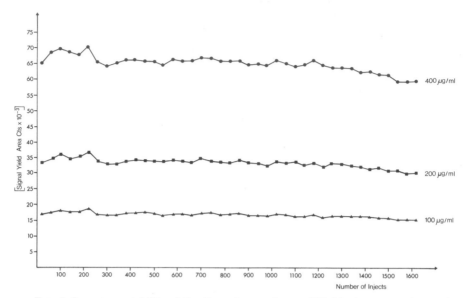

Fig. 6. Long-term stability of the *Enterobacter cloacae* P99 β-lactamase reactor used with the EP–FIA system, with varying cephalosporin C concentrations.

TABLE II
PARALLEL DETERMINATIONS OF CEPHALOSPORIN
C (PLUS DEACETYLCEPHALOSPORIN C) IN
FERMENTATION SAMPLES BY HPLC AND
EP–FIA

| Sample | Concentration (mg/g) (CC + DACC) | | |
	HPLC	EP–FIA	Δ%
1	3.41	3.32	−2.64
2	3.71	3.77	+1.62
3	4.24	4.40	+3.77
4	4.59	4.39	−4.36
5	5.24	5.75	+9.73
6	6.23	6.28	+0.8
7	6.65	6.61	−0.6
8	7.03	7.28	+3.56
9	7.63	7.78	+1.97
10	8.36	8.26	−1.20
11	9.63	9.13	−5.19
12	9.84	9.39	−4.57
13	10.64	10.67	+0.28
14	10.77	10.51	−2.41
15	10.96	11.25	+2.65

around 230 bar (23.0 MPa). The detector wavelength is 260 nm, and the injection volume is 30 μl. The correlation plot between the two methods is linear ($y = kx + d$; $n = 44$, $k = 0.9805$, $d = 0.192$), and the correlation coefficient ($r = 0.993$) shows good correspondence between the flow injection and HPLC methods.

A double-sided data pair t-test was also applied to a further 40 samples analyzed by both methods. There was no difference between the results of the two methods at a 5% significance level ($n = 40$, $t_{39,\alpha(2P=0.05)} = 1.33 <$ 2.023). The results from 15 parallel determinations of fermentation samples are collected in Table II.

Enhancement of Selectivity for Cephalosporin C in the Presence of DACC and DAOCC. The β-lactamase used, from *Enterobacter cloacae* P99, permitted only the determination of CC and DACC together. Attempts to enhance selectivity for CC were therefore made with a β-lactamase of type II from *Bacillus cereus,* which has been reported to be reactive toward CC (80% compared with benzylpenicillin as substrate) but not to affect DACC (0.02%) or DAOCC (0.4%).[31] Immobilization on the CPG carrier is done as described under "Experimental Methods." A series of CC sample solutions containing different amounts of DACC was

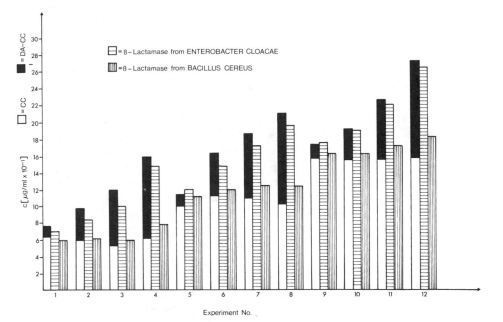

FIG. 7. Diagram showing the selective determination of cephalosporin C in the presence of deacetylcephalosporin C in standard solutions by means of cephalosporinase from *Bacillus cereus* 569/H.

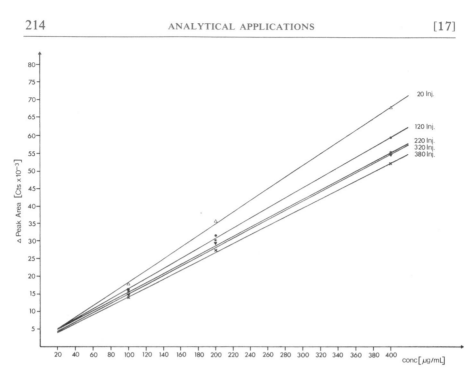

FIG. 8. Decrease in column efficiency of the *Bacillus cereus* 569/H β-lactamase reactor in relation to the number of samples processed.

investigated by EP–FIA. The selectivity for CC is significantly enhanced as illustrated in Fig. 7. Although the results for CC in the presence of fermentation by-products such as DACC and DAOCC were improved by the use of the type II β-lactamase, these enzyme reactors turned out to be far less stable than the original reactors with β-lactamase from *Enterobacter cloacae*. The loss in enzyme activity was approximately 20% after 380 subsequent injections on the *Bacillus cereus* 569/H β-lactamase reactor (see Fig. 8) but was scarcely detectable when the *Enterobacter* P99 β-lactamase reactor was used for the same number of samples.

Application of EP–FIA to Different Cephalosporin Derivatives. Twelve commercially available cephalosporins were investigated with the EP–FIA apparatus for their relative susceptibility to the *Enterobacter cloacae* enzyme reactor. Table III lists the chemical structures and the absorption maxima of the investigated cephalosporins in order of their reactivities.[34] Solutions of these compounds (~100 μg/ml) were analyzed at 254 nm, and the percentage differences in the absorbances between the

[34] E. E. Roets, J. H. Hoogmartens, and H. J. Vanderhaeghe, *J. Assoc. Off. Anal. Chem.* **64,** 166 (1981).

TABLE III
Chemical Structures of Cephalosporins Investigated by EP–FIA

$$R_1-\overset{\displaystyle O}{\overset{\|}{C}}-HN \quad \underset{8}{\overset{7}{}} \quad \text{(cephalosporin nucleus)} \quad R_2$$

COOH(Na)

Substance	R_1	R_2	R_3	λ_{max} (nm)
Cephalosporin C	$-CH_2CH_2CH_2CH\!\!\begin{smallmatrix}COO^-\\ \\NH_3^+\end{smallmatrix}$	$-CH_2O\overset{\displaystyle O}{\overset{\|}{C}}CH_3$	H	260
Cefalotin	$-CH_2$–(thienyl)	$-CH_2O\overset{\displaystyle O}{\overset{\|}{C}}CH_3$	H	260
Cefaloglycine	$-CH(NH_2)$–(phenyl)	$-CH_2O\overset{\displaystyle O}{\overset{\|}{C}}CH_3$	H	260
Cefaloridine	$-CH_2$–(thienyl)	$-CH_2N^+$–(pyridinium)	H	255
Cefaclor	$-CH(NH_2)$–(phenyl)	Cl	H	262
Cefalexin	$-CH(NH_2)$–(phenyl)	CH_3	H	260
DAOCC	$-CH_2CH_2CH_2CH\!\!\begin{smallmatrix}COO^-\\ \\NH_3^+\end{smallmatrix}$	CH_3	H	260
Cefazolin	$-CH_2N$–(tetrazolyl)	$-CH_2S$–(thiadiazolyl)$-CH_3$	H	271
DACC	$-CH_2CH_2CH_2CH\!\!\begin{smallmatrix}COO^-\\ \\NH_3^+\end{smallmatrix}$	CH_2OH	H	260
Cefamandole	$-CH(OH)$–(phenyl)	$-CH_2S$–(N-CH_3 tetrazolyl)	H	265
Ceftriaxone (Rocephin)	(aminothiazole methoxyimino group, H_3CO)	$-CH_2S$–(triazinonyl)	H	242, 272
Cefoxitin	$-CH_2$–(thienyl)	$-CH_2O\overset{\displaystyle}{C}NH_2$ ($\|O$)	OCH_3	262.5

TABLE IV
RELATIVE REACTIVITY OF DIFFERENT CEPHALOSPORINS AGAINST A β-LACTAMASE FROM
Enterobacter cloacae P99, MEASURED WITH THE EP–FIA SYSTEM AT 254 nm

Substance	$E_{untreated}$ peak area (arbitrary units)	$\Delta E_{after\ treatment}$ peak area (arbitrary units)	Reactivity (%)[a]
Cephalosporin C	23,500	18,200	100.0
Cefalotin	33,600	23,900	91.7
Cefaloglycine	25,000	17,450	90.1
Cefaloridine	52,900	33,900	82.7
Cefaclor	24,300	15,450	82.1
Cefalexin	27,800	17,300	80.3
DAOCC	19,200	11,900	80.0
Cefazolin	29,500	18,150	79.4
DACC	26,950	16,100	77.0
Cefamandole	26,300	10,300	50.6
Ceftriaxone (Rocephin)	26,300	9200	19.1
Cefoxitin	28,300	3100	14.2

[a] Compared with cephalosporin C (sodium salt, dihydrate) = 100%.

untreated and degraded cephalosporins were related to their reactivity with the immobilized enzyme. Cephalosporin C exhibited the highest relative reactivity and was taken as reference with 100%. The results are listed in Table IV. Although the nature of the leaving group in the 3 position is primarily affecting the enzymatic cleavage of cephalosporin, the substituents at the 7 position also play an important role in the mechanism of further degradation of the molecule.

As an example, cefoxitin differs from cefalotin only in the 7 position with a methoxy moiety instead of hydrogen and the carbamoyl instead of the acetyl function at the 3 position, but its reactivity against P99 β-lactamase is only 13% compared with cefalotin. The acetoxy moiety seems to be the most effective leaving group, followed by the pyridine molecule in cefaloridine. The heterocyclic substituents of the cefazolin, cefamandole, and ceftriaxone molecules do not show leaving group properties to the same extent. The influence of the leaving group in the 3 position is also clearly demonstrated in the series CC, DACC, and DAOCC. A special case is DACC lactone, which shows only a small difference in its UV absorbance. Although DACC lactone is affected by the P99 β-lactamase, the product is a relatively stable cephalosporoic acid derivative which exhibits nearly the same UV absorption as the original compound.

Conclusions

The modular construction and the use of a versatile microcomputer interface for total flow–injection control considerably extend the applicability of the described immobilized enzyme reactor–flow injection analysis (IMER–FIA) system beyond merely β-lactam analysis. Minor modifications of the instrumental setup and of the controlling microcomputer programs permit automated determinations of a wide spectrum of other different compounds such as sugars, peptides, proteins, esters, carboxylic acids, and urea, for which selective and sensitive enzymes are available. These determinations can be performed either in direct or in subtractive mode according to the detection system employed.

Acknowledgments

The author is grateful to Mr. Dietmar Ascher and Mr. Josef Lettenbichler for construction and programming of the microcomputer interface, to Mr. Johann Patka for writing the data evaluation program, and to Mrs. Irene Röck for excellent technical assistance. Parts of the text and figures have been presented elsewhere[18,19] and are used after modifications with permission of the American Chemical Society and Elsevier/North Holland.

[18] Biomedical Applications of the Enzyme Thermistor in Lipid Determinations

By Ikuo Satoh

Determination of lipids in human body fluids is of great importance in medical health care. Measurement of lipid content, e.g., cholesterol, triglycerides and phospholipids, provides important information in the diagnosis of diseases. Total cholesterol and triglyceride levels in serum are valuable indicators of abnormalities in lipid metabolism, arteriosclerosis, and hypertension. Phospholipids in serum are also important parameters of obstructive jaundice and cirrhosis. Furthermore, assay for phospholipids, especially the ratio of lecithin to sphingomyelin, gives an assessment of fetal lung maturity during pregnancy which is useful in predicting the risk of respiratory distress syndrome.

The determination of these lipids is generally performed by enzymatic spectrophotometric methods. These methods, however, involve multistep reactions to obtain a measurable optical change and therefore involve rather long operational times and the use of expensive enzymes. Calorimetric methods can eliminate the need for auxiliary enzymes and cofactors. Since most enzymatic reactions are associated with a considerable enthalpy change, the primary reaction itself usually produces sufficient heat for reliable measurements. The enzyme thermistor is a unique flow bioanalyzer, equipped with an immobilized enzyme reactor for highly specific recognition of substrates and a thermometric transducer. In this chapter, applications of the enzyme thermistor in biomedicine are presented, with special focus on the determination of serum lipids.

Apparatus

The enzyme thermistor device developed by Mosbach *et al.* has previously been described[1] (see also this volume [16]). In an earlier model of the enzyme thermistor used for cholesterol analysis, the assemblies are completely immersed in a water bath, which allows control of temperature with accuracy better than ±0.01°. Temperature registration is made by a Wheatstone bridge equipped with an amplifier (Knauer Temperature Measuring Instrument, Wissenschaftlicher Geraetebau, W. Berlin, FRG) connected to a potentiometric recorder. At the most sensitive measuring range a change in the output of the bridge of 100 mV corresponds to a temperature change of 0.02°. The water bath temperature is held at 27°. When the apparatus is placed in the water bath, it takes 1–2 hr to establish a stable base-line. During that time buffer is continuously pumped through the system. Sample introduction without stopping the pump is accomplished by means of a three-way valve at the inlet side of the pump. The thermistors used are matched Veco Type 41A28 thermistors with a resistance of 10 kΩ at 25° (Victory Engineering Corp., Springfield, NJ).

A more recent design of the apparatus is described elsewhere in this volume [16] and can be used for determination of triglycerides and phospholipids in serum. The new apparatus, which has an aluminum constant-temperature jacket instead of the water bath, is a twin system that can be used with one enzyme column and one reference column. The thermostat can be operated at any temperature (25, 30, or 37°) with temperature variation of ±0.002° when equilibrated. At maximum sensitivity, it produces a 100-mV change in the recorder signal for a temperature change of 0.001°.

Reagents and Standard Solutions

Phospholipase D

Phospholipase D (phosphatidylcholine phosphatidohydrolase, EC 3.1.4.4, from *Streptomyces chromofuscus,* 60.5 units/mg) was obtained

[1] B. Danielsson and K. Mosbach, this series, Vol. 44, p. 667.

from Toyo Jozo Co. (Ohito, Japan). Phospholipase D is used in the free state as a catalytic reagent for hydrolysis of phosphatidylcholine to generate phosphatidic acid and choline.

Standard Solutions of Cholesterol

Standard solutions of cholesterol are prepared in 0.16 M potassium phosphate buffer, pH 6.50, containing 8% (v/v) of Triton X-100 and 12% (v/v) of ethanol.

Standard Solutions of Triglyceride

Glyceryl trioleate (triolein, Sigma Chemical Co., St. Louis, MO) is used as the standard triglyceride. A 5 mM trioleate dispersion is prepared, with some modification, by the method of Chong-Kit and McLaughlin.[2] Trioleate (900 mg) is placed in a 200-ml conical flask, and 36 ml of Triton X-100 is then added and warmed at 70° for 15 min. After cooling to room temperature, the contents are transferred to a 200-ml volumetric flask and diluted to the mark with 0.15 M NaCl after addition of 100 ml of buffer (0.2 M Tris–HCl, pH 8.0, containing 1.0% in Triton X-100).

Standard Solutions of Phospholipid

Phosphatidylcholine [egg yolk, Type V-E, 100 mg/ml in $CHCl_3$/CH_3OH (9 : 1, v/v); Sigma Chemical Co.] is used as the standard phospholipid. One milliliter of L-α-phosphatidylcholine preparation is placed in a centrifuge tube, and most of the solvent is removed by means of N_2 gas stream at room temperature. Subsequently, after drying under vacuum for at least 30 min, 2.5 ml of 10% Triton X-100 solution is applied to the residues and thoroughly mixed. The mixture is quantitatively transferred to a 50-ml volumetric flask and diluted to the mark with distilled water (2000 mg/liter dispersion).

Ten milliliters of 0.2 M Tris–HCl buffer, pH 8.0, 0.975 ml of 10% Triton X-100 solution, and 2.0 ml of 1.0 M calcium chloride solution are added to 5.0 ml of the 2000 mg/liter phosphatidylcholine solution and diluted to 20.0 ml with distilled water (500 mg/liter dispersion).

Enzyme Immobilization

Alkylamino glass is prepared by treating controlled-pore glass beads (particle size: 40–80, 80–120, and 120–200 mesh, mean pore size: 55.0, 56.9, 72.9, 198.9, and 221.5 nm; Electronucleonics, Analytical Standard AB, Kungsbacka, Sweden) with γ-aminopropyltriethoxysilane (Tokyo

---

[2] R. Chong-Kit and P. McLaughlin, *Clin. Chem.* **20**, 1454 (1974).

Kasei Kogyo Co., Tokyo, Japan) following the procedure developed by Weetall.[3] To 1 g of clean glass beads are added 18 ml of distilled water and 2 ml of γ-aminopropyltriethoxysilane. The pH is adjusted to 3.5 with 6 M HCl. The mixture is then heated for 3 hr on a water bath at 75° and gently stirred intermittently. The glass beads are filtered and washed on a sintered-glass filter with distilled water, then dried in an oven at 115° for at least 4 hr.

The immobilized enzymes used in the experiments described here are prepared by coupling the enzymes to alkylamino glass after activation with a 2.5% glutaraldehyde solution in 0.1 M sodium phosphate buffer adjusted to pH 7.0. The coupling reaction is allowed to proceed in the cold overnight, for the first hour under reduced pressure (see Ref. 3).

Lipase

Lipase (glycerol ester hydrolase, triacylglycerol lipase, EC 3.1.1.3, from *Pseudomonas* sp., twice recrystallized, 2000 units/mg) was obtained from Amano Pharmaceutical Co. (Nagoya, Japan). To 3 g of alkylamino glass, after activation by a 2.5% glutaraldehyde solution, is added 320 mg of lipase dissolved in 15 ml of 0.1 M sodium phosphate buffer, pH 7.0.

Cholesterol Oxidase

Cholesterol oxidase (cholesterol : oxygen oxidoreductase, EC 1.1.3.6, from *Nocardia erythropolis,* 4 units/mg) was obtained from Boehringer (Mannheim, West Germany). To 1 ml of activated sedimented glass is added 0.4 mg of cholesterol oxidase dissolved in 1.5 ml of 0.1 M sodium phosphate buffer, pH 7.0.

Catalase

Catalase (hydrogen-peroxide : hydrogen-peroxide oxidoreductase, EC 1.11.1.6, type C-100, from bovine liver, 30,000–40,000 units/mg) was purchased from Sigma. Activated, sedimented glass beads (0.5 ml) are mixed with 1.0 ml of 0.1 M potassium phosphate buffer, pH 6.5, and 500 μl (23,000 U) of catalase suspension.

Choline Oxidase

Choline oxidase (choline : oxygen 1-oxidoreductase, EC 1.1.3.17, from *Arthrobacter globiformis,* 10.6 units/mg) was obtained from Toyo Jozo Co. To 1.28 g of alkylaminoglass, after activation by a 2.5% glutaraldehyde solution, is added the enzyme mixture: 85.5 mg of choline

---

[3] H. H. Weetall, this series, Vol. 44, p. 139.

oxidase (200 units) and 0.86 mg of catalase (6,440 units) dissolved in 9.0 ml of 0.1 M sodium phosphate buffer, pH 7.0.

Results

Determination of Cholesterol in Standard Solutions and Serum. Calorimetric cholesterol assay based on an enzymatic method involves either the measurement of heat produced by cholesterol oxidation or the detection of decomposition heat from hydrogen peroxide subsequently formed. These reactions are written as follows:

$$\text{Cholesterol} + O_2 \xrightarrow{\text{cholesterol oxidase}} \text{cholest-4-en-3-one} + H_2O_2 \tag{1}$$

$$H_2O_2 \xrightarrow{\text{catalase}} H_2O + \frac{1}{2} O_2 \tag{2}$$

Cholesterol is determined by use of a differential thermistor device, where thermistors are placed at each end of a plastic column with a diameter of 3 mm (i.d.) and a volume of 0.2–0.3 ml of immobilized enzyme. The immobilized cholesterol oxidase is packed in a precolumn (3 × 60 mm) and then used in combination with catalase–thermistor.[4] One-milliliter pulses of the standard solutions are pumped at a flow rate of 1.0 ml/min (0.16 M potassium phosphate buffer, pH 6.50, as carrier solution) through the precolumn containing cholesterol oxidase–glass and then through the enzyme thermistor reactor containing catalase–glass.

A linearity for calibration between temperature response and cholesterol concentration was obtained in the range 0.03–0.15 mM (with a slope of 0.129°/mM). Determination of cholesterol in serum must be performed by extraction of lipids with an organic solvent (hexane). After evaporation to dryness, the above buffer is added to the serum extract and the cholesterol content determined. The extracted serum samples can then be assayed without any nonspecific heat production. The analytical value of this method is therefore limited at this time by involving a discontinuous process.

Determination of Triglycerides in Standard Solutions and Serum. One of the metabolites most successfully studied with the split-flow enzyme thermistor shown in Fig. 1 is triglyceride.[5] The flow system consists of two identical heat exchangers (50 cm of thin-walled stainless steel tube, 0.8 mm i.d.) and two identical columns, one of which contains the immobilized lipase preparation and the other a deactivated preparation (by heating in boiling water for 1 hr). The Tris–HCl buffer (0.1 M, pH 8.0, 0.5% in Triton X-100) is continuously pumped through the enzyme ther-

[4] B. Mattiasson, B. Danielsson, and K. Mosbach, *Anal. Lett.* **9,** 217 (1976).
[5] I. Satoh, B. Danielsson, and K. Mosbach, *Anal. Chim. Acta* **131,** 255 (1981).

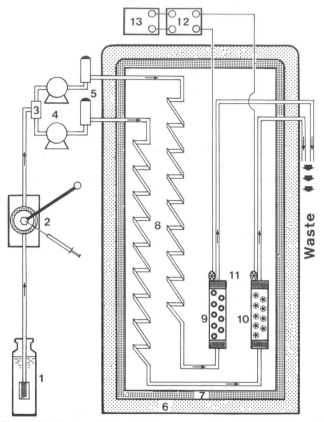

FIG. 1. Schematic diagram of the split-flow enzyme thermistor for triglyceride determination. 1, Buffer reservoir; 2, sample injector; 3, T-type manifold; 4, peristaltic pump; 5, pulse damper; 6, polyurethane foam-insulated casing; 7, aluminum constant-temperature jacket; 8, heat exchanger; 9, immobilized lipase column; 10, reference column; 11, thermistor; 12, Wheatstone bridge equipped with an amplifier; 13, recorder.

mistor by the two peristaltic pumps at a flow rate of 1.5 ml/min in each channel. Standards or serum samples are introduced as short pulses (0.5 ml) via a loop injection valve (Type 50, Teflon Rotary Valve, Rheodyne Inc., Berkeley, CA). The solutions used for calibration are 0.7–5.0 mM glyceryl trioleate, 4.5% in human serum albumin. Three types of controlled-pore glass (CPG) are used for immobilization of the lipase. Table I shows the percentage immobilization and the relative activities of the immobilized enzyme preparations. The amount immobilized is calculated from the absorbance values at 280 nm measured for the enzyme solution before and after coupling. The highest activity is obtained on the largest pore-size glass. The difference in relative activities may be caused by

TABLE I
IMMOBILIZATION OF LIPASE ON VARIOUS TYPES OF
CONTROLLED-PORE GLASS[a]

Type of glass	Mean pore size (nm)	Immobilization yield (%)	Relative activity (%)
CPG-10-700[c]	72.9	98	52
CPG-10-2000[c]	198.9	36	76
CPG-10-2000[d]	221.5	59	100

[a] Modified from Satoh et al., p. 257.[5]
[b] Measured under standard conditions at 30.0° using 1 mM
trioleate as sample, relative to the most active preparation.
[c] Particle size 80–120 mesh.
[d] Particle size 120–200 mesh.

steric hindrance of the large substrate molecules in micellar states. The enzyme immobilized on glass with a pore size of 221.5 nm (particle size 80–120 mesh) was employed in further studies.

The lipoprotein lipase hydrolyzes triglyceride to glycerol and fatty acids as follows:

$$\begin{array}{l} CH_2OOCR \\ | \\ CHOOCR' \\ | \\ CH_2OOCR'' \end{array} + 3H_2O \xrightarrow{\text{lipase}} \begin{array}{l} CH_2OH \\ | \\ CHOH \\ | \\ CH_2OH \end{array} + RCOOH + R'COOH + R''COOH \qquad (3)$$

From measurements in various buffers, it could be deduced that the ester hydrolysis per se produces little heat. In this case a chemical amplification can be achieved by using a buffer with high protonation enthalpy, such as the Tris used here. The time required for a determination is less than 5 min. The calibration graph is linear, with a slope of 0.0014°/mM, up to 5 mM glyceryl trioleate. Standard deviation determined by using 1 mM glyceryl trioleate is found to be ±0.04 mM (RSD 4%) for 10 samples measured within a day. The triglyceride concentration in serum can be directly determined after 2-fold dilution with the previously described Tris buffer up to a concentration of 3 mM. The results agree well with those obtained by the conventional spectrophotometric enzyme method (correlation coefficient of 0.97 for 15 assays).

Determination of Phosphatidylcholine in Standard Solutions and Serum. Phosphatidylcholine is one of the main serum phospholipids. The assay involves the enzymatic hydrolysis of phosphatidylcholine by phospholipase D to liberate choline and phosphatidic acid. The choline liberated is subsequently oxidized, with formation of hydrogen peroxide, by immobilized choline oxidase. The reaction heat evolved by oxidation of choline and decomposition of hydrogen peroxide is monitored by coimmobilized choline oxidase–thermistor and catalase–thermistor.[6] The reaction scheme is as follows:

$$\text{Phosphatidylcholine} + H_2O \xrightarrow{\text{phospholipase D}} \text{phosphatidic acid} + \text{choline} \qquad (4)$$

$$\text{Choline} + 2O_2 + H_2O \xrightarrow{\text{choline oxidase}} \text{betaine} + 2H_2O_2 \qquad (5)$$

$$2H_2O_2 \xrightarrow{\text{catalase}} 2H_2O + O_2 \qquad (6)$$

The heat change in the last two reactions is monitored thermometrically. To 2 ml of the standard solutions is added 0.02 ml of phospholipase D (18 units), and the mixture is allowed to stand for 1 hr at room temperature. Then 0.5-ml pulses of the mixture are pumped at a flow rate of 0.1 ml/min (0.1 M phosphate buffer, pH 8.0, 0.5% in Triton X-100, 50 μM calcium chloride) through the enzyme thermistor reactor containing coimmobilized choline oxidase– and catalase–glass beads (0.6 ml of packed volume). The phosphatidylcholine assay is performed within 5 min. Linearity for calibration is obtained in the range 50–150 mg/liter of phosphatidylcholine (with a slope of 0.02° liters/g). One hundred microliters of fresh human serum is diluted 20-fold with the buffer and then treated as described for the standard solutions. Determinations of phosphatidylcholine in serum by the thermistor method and by a conventional method are in comparatively good agreement (correlation coefficient 0.8 for 60 samples). The choline oxidase–thermistor and catalase–thermistor retained 44% of the initial activity after 1582 assays, including calibrations, during 8 weeks at 30°.

Conclusion

Most serum lipids are solubilized as lipoproteins and therefore have a large aggregate molecular weight. Triglycerides remain mainly as chylomicrons and low density lipoproteins. By the use of rather large pore-size controlled-pore glass, diffusion problems caused by steric hindrance of the large substrate were avoided. Similar approaches using immobilized enzymes coupled with electrochemical devices have been successfully performed.[7,8] In the experiments described the average duration of an analysis was about 5 min, after which the original base-line was obtained. The average precision obtained of ±2% is sufficient for the demands of clinical analysis. Rapid and continuous determination of lipids is possible with the enzyme thermistor. The use of enzyme thermistor devices is

[6] I. Satoh, *Annu. Meet. Agric. Chem. Soc. Jpn.* p. 659 (Abstr. 2S-13) (1985).

[7] I. Satoh, I. Karube, and S. Suzuki, *J. Solid-Phase Biochem.* **2,** 1 (1977).

[8] I. Satoh, I. Karube, S. Suzuki, and K. Aikawa, *Anal. Chim. Acta* **106,** 429 (1979).

simple, reliable, and economical in the biomedical field. Because of these advantages, the enzyme thermistor as a thermal bioanalyzer appears quite promising and very attractive for use in routine clinical analysis. Direct and continuous assays for free cholesterol and esterified cholesterol in serum based on the enzyme thermistor are now under investigation in our laboratory.

[19] Flow Enthalpimetric Determination of Glucose

By Nobutoshi Kiba and Motohisa Furusawa

The development of techniques for enzyme immobilization has made possible the preparation of small enzyme columns with high activity. Efforts have been made to combine the column reactors with various transducers, and many analytical methods based on such combinations have been reported.[1] Immobilized enzyme flow enthalpimetry is based on the combination of a column reactor with a thermistor probe.[2] Most of these methods have the advantage of permitting precise determinations because they are basically end-point or equilibrium assays and, therefore, are less susceptible to errors due to changes in the reaction conditions as well as changes in the concentration of inhibitors and activators. However, immobilized enzyme enthalpimetry based on kinetic assays (batch-type measurements) is less practical because of low analytical sensitivity.[3] On the other hand, end-point assays are more prone to errors arising from nonspecific reactions, even if they are too slow to interfere with a kinetic assay. In order to avoid such errors, high-purity enzymes having high specificity must be used.

A major problem of immobilized enzyme flow enthalpimetry is nonspecific heat due to the interaction between sample and support material. Though a split-flow device developed by Mattiasson et al.[4] has been useful in solving this problem, for common devices such as a single thermistor device and a dual thermistor device (differential type) it is difficult to compensate for interferences in the system by calibration. Thus, the com-

[1] L. B. Wingard, Jr., E. Katchalski-Katzir, and L. Goldstein (eds.), *Appl. Biochem. Bioeng.* **3** (1981).

[2] K. Mosbach and B. Danielsson, *Anal. Chem.* **53**, 83A (1981).

[3] C. L. Cooney, J. C. Weaver, S. R. Tannenbaum, D. V. Faller, A. Shields, and M. Jahnke, *Enzyme Eng.* **2**, 411 (1974).

[4] B. Mattiasson, B. Danielsson, and K. Mosbach, *Anal. Lett.* **9**, 867 (1976).

position of samples, for example, human serum, is quite variable. In order to minimize erroneous heat effects, one should select an optimum support.

The scope of this chapter is limited to the determination of glucose by flow enthalpimetry using immobilized glucose oxidase. Although there have been several reports on the application of flow enthalpimetry to glucose determinations, hexokinase[5] is not suited for a practical method because of its low substrate specificity. Glucose oxidase has high specificity, but a disadvantage is that a linear calibration curve is obtained only up to 0.4 mM, or 0.7 mM if catalase is present, due to the limited solubility of oxygen. Therefore, this method is less useful for practical analysis. In this chapter, we use benzoquinone as the hydrogen acceptor in place of oxygen and determine up to 70 mM glucose.

Experimental Methods

Apparatus. The apparatus includes a pumping system, a thermal detector, an injector, and a strip-chart recorder. The thermal detector is commercially produced by Japan Electron Optics Laboratory (Tokyo, Japan) and has been described in detail elsewhere.[6] The detector consists of two matched, 5-kΩ thermistors (Takara Thermistor, Yokohama, Japan) sealed in thin glass tubes so that the thermistor beads extend into the column. The reference column is packed with 20-mesh solid glass beads and the detection column (glass tube, 3.5 cm length, 1.0 cm o.d., 0.8 cm i.d.) with the immobilized glucose oxidase (1 g). The thermistors form two arms of a Wheatstone bridge responding to changes in temperature due to the heat of reaction, the bridge output being indicated on the chart recorder. A change of the output from the bridge of 1 mV corresponds to a temperature change of 0.1°. The thermal equilibration coil is a Teflon tube 10 m in length with 1.0 mm i.d. The piston pump (Model KHU-W-104) and loop injector (Model KHPUL-130) were purchased from Kyowa Seimitsu Co. (Tokyo, Japan).

Procedure. Glucose oxidase (from *Penicillium amagasakiense,* 110 units/mg; Nagase Biochemicals, Tokyo, Japan) (100 mg) is immobilized on 1 g of the aminoaryl derivative of controlled-pore glass (CPG) beads (Aminoaryl-CPG, 120–200 mesh; Electro-Nucleonics, Fairfield, NJ) diazotized with nitrous acid in 10 ml of 0.1 M potassium phosphate buffer, pH 7.0, at 3–5° for 10 hr. The activity of the enzymes (soluble and immobilized) is determined by measuring H_2O_2 produced by permanganate titra-

[5] L. D. Bowers and P. W. Carr, *Clin. Chem.* **22,** 1427 (1976).
[6] K. Nakagawa, *J. Assoc. Off. Agric. Chem.* **51,** 333 (1968).

tion in a batch method, and the average value for the immobilized enzyme is 5800 units/g CPG.

1,4-Benzoquinone (Tokyo Kasei, Tokyo, Japan) is recrystallized twice from cyclohexane and dried at room temperature in a vacuum for 3 hr. The quinone solutions are prepared in 0.1 M phosphate buffer (KH_2PO_4/Na_2HPO_4, pH 6.0). Black suspended matter develops on standing and is removed by filtration with a membrane filter (pore size 0.45 μm; Toyo, Tokyo, Japan). The stock solution of glucose (1.0 M) is prepared by dissolving β-D-glucose (Tokyo Kasei) in the phosphate buffer, pH 6.0. The solution is allowed to come to anomeric equilibrium (63.5% β-D-glucose) at room temperature. Calibration standards are made by dilution of the stock with the buffered quinone solution.

The buffered quinone solution (45 mM, pH 6.0) is pumped by the double-piston pump at a flow rate of 4.0 ml/min through the thermal equilibration coil into the reference column. After passing the reference thermistor, the flow reaches the loop injector, where 2.0 ml of sample solution is introduced into the flow stream. The sample solution is prepared by dilution of 1.0 ml of sample with 9.0 ml of quinone solution (50 mM) in pH 6.0 buffer. The flow passes through the thermal equilibration coil into the detection column, and then the heated solution passes the detection thermistor and out to waste.

Results

Choice of Support. The magnitude of nonspecific heat effects due to the interaction (adsorption and desorption) between support and sample is dependent on both the sample composition and the kind of support material used. Glass beads (Aminoaryl-CPG, pore size 50 nm, 120–200 mesh; Electro-Nucleonics), polystyrene beads (Amberlite XAD-2; Rohm and Haas, Philadelphia, PA) (powdered in a mortar and sieved to obtain the 120–200 mesh fraction), and polyacrylamide gel (BioGel P-4, 100–200 mesh; Bio-Rad, Richmond, CA) are used as supports and standard serum (Precinorm S; Boehringer Mannheim, Mannheim, FRG) as sample. As shown in Fig. 1, the temperature change due to nonspecific heat with XAD-2 is 4×10^{-3}° (for exothermic adsorption) and 2×10^{-3}° (for endothermic desorption). A similar result is obtained with BioGel. On the other hand, the change for CPG is only 4×10^{-4}°; it is equal to one-fifteenth the value for XAD-2 in magnitude though different in shape. On this point, the CPG is superior to the organic supports. The magnitude of the heat effect for differences in composition of standard sera such as Precinorm S, Precinorm U, and Precilip (Boehringer Mannheim) is in the range of 4×10^{-4} to 8×10^{-4}°. These values correspond to glucose

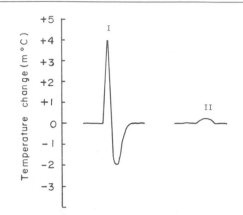

FIG. 1. Response for serum solution using different column packings: I, XAD-2; II, CPG. Serum (Precinorm S) (1.0 ml) is diluted to 10 ml with 50 mM quinone solution (pH 6.0) and filtered through a membrane filter (pore size 0.45 μm). A 2.0-ml portion of the filtrate is injected into the flow line. Flow rate, 4.0 ml/min; column length, 3.5 cm; benzoquinone, 45 mM in buffer (pH 6.0); column temperature, 30.0°.

concentrations in the range of 0.032–0.064 mM by the present method. Ethanol (about 14%) in white wine and dyestuffs in red wine give a relatively large heat effect of about 1×10^{-3}°.

Properties of the Immobilized Glucose Oxidase Column Reactor. The peak height as a function of flow rate and column length was studied over the range 1.0–8.0 ml/min and 1.5–5.5 cm, respectively. The peak height is dependent on both the flow rate and the column length. Figure 2 shows the combined effect of the two factors. When a column of more than 2.5 cm in length is used, the reaction is complete at every flow rate tested, since the peak height increases linearly with an increase in flow rate. The increase in peak height is due to a decrease in heat loss from the column and, though the sample was sufficient, to a slight peak width narrowing. The falloff in response at lower flow rates is due to heat loss and, in addition, to peak width broadening. The phenomenon is characteristic for thermal flow detectors. For a 1.5-cm column, the reaction is incomplete at a flow rate above 2 ml/min because the peak height decreases with an increase in flow rate. At 5.5 cm column length, the peak height, on the whole, is lower than those for shorter columns. This is due to considerable heat loss from the column. Since at 6.0 ml/min the back pressure is relatively high and the flow pulsed, the base-line is unstable.

The influence of the pH of the buffer on peak height and stability of the immobilized enzyme column was studied. Peak height is independent of pH in the range 4.0–7.0. Above pH 8, the peak height decreases gradually

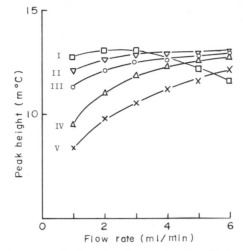

Fig. 2. Effect of flow rate and column length on peak height. Each column is packed with immobilized glucose oxidase. Glucose concentration, 1.0 mM; benzoquinone concentration, 45 mM; column temperature, 30.0°; pH 6.0; column length, cm: I, 1.5; II, 2.5; III, 3.5; IV, 4.5; V, 5.5.

with the number of tests because the stability of CPG is very poor in alkaline solution, and, therefore, the enzyme exfoliates from the surface of CPG beads.[7] This is a serious disadvantage of CPG beads. Furthermore, above pH 8, the quinone solution rapidly turns from yellow to black, resulting in a black precipitate.

Figure 3 shows that the peak height is almost independent of sample volume in the range 1.0–4.0 ml. Even a 4-ml sample does not give saturation (square signal). For smaller samples (<1 ml), precise results can never be obtained, mainly owing to short-term fluctuations in flow rate. A loop injector is preferable to a syringe for reproducible injection of relatively large samples (2.0 ml) into the high flow rate (4.0 ml/min). The larger sample volume gives an increase in peak width and a decrease in sample throughput, but this defect can be removed by application of high flow rate.

A flow rate of 4.0 ml/min, a column length of 3.5 cm, a sample volume of 2.0 ml, and a phosphate buffer of pH 6.0 were chosen as standard conditions. Because of vaporization of quinone, the quinone solution is stable at concentration below 50 mM. Thus, a quinone concentration of 45 mM was chosen. Under these conditions, a linear relationship is obtained

[7] E. P. Plueddemann, "Silane Coupling Agents," p. 227. Plenum, New York, 1982.

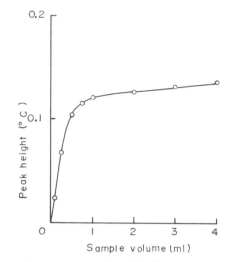

FIG. 3. Effect of sample volume on peak height. Glucose concentration, 10.0 mM; flow rate, 4.0 ml/min; pH 6.0; column length, 3.5 cm; column temperature, 30.0°.

between peak height and D-glucose concentration in the range 0.02–70 mM. The analytical sensitivity is a peak height of 2.5×10^{-4}° in response to 0.02 mM glucose. The rate of analysis is 60 samples per hour. The column lasts for at least 3000 samples.

The quinone concentration affects the peak height. The relative peak height for $1.0 \times 10^{-4}\ M$ glucose is 100, 91.4, 88.3, and 75.5 for a quinone concentration of 0 (that is, free from quinone, only containing oxygen), 9.0×10^{-4}, 9.0×10^{-3}, and $4.5 \times 10^{-2}\ M$, respectively. This is due to an increase in heat capacity of the solution with an increase in quinone concentration.

Applications. The method has been applied to the determination of glucose in serum, wines, soft drinks, beers, and jams.

DETERMINATION OF GLUCOSE IN SERUM. A standard serum (Precinorm S, Boehringer Mannheim) (1.0 ml) is diluted to 10 ml with 50 mM quinone solution in the pH 6.0 buffer and then filtered through a membrane filter (pore size 0.45 μm). A 2.0-ml portion of the filtrate is used as sample solution.

DETERMINATION OF GLUCOSE IN WINES. One milliliter of white wine is boiled for 3 min in a 10-ml evaporation tube in order to evaporate alcohol in the sample, then transferred to a 10-ml standard flask and made up to volume with the 50 mM quinone solution. Red wine (10 ml) is filtered through 5 g of BioGel P-300 (50–100 mesh) to remove colored matter before analysis.

TABLE I
DETERMINATION OF GLUCOSE IN VARIOUS SAMPLES

Sample	D-Glucose (g/100 ml)	CV (%) (n = 7)[a]	Certificate value (g/100 ml)
Serum	0.0995	1.9	0.094–0.111
White wine I	0.231	1.4	0.2
White wine II	0.838	0.51	0.8
Red wine I	0.563	1.7	0.06
Red wine II	0.0844	1.8	0.09
Beer I	0.0260	1.6	0.03
Beer II	0.0361	1.8	0.03
Soft drink I	4.42	0.44	4.4
Soft drink II	4.49	0.98	4.4
Strawberry jam	10.7[b]	0.1	10–11[b]
Apricot jam	9.90[b]	0.1	9–11[b]

[a] CV, Coefficient of variation.
[b] g/100 g.

DETERMINATION OF GLUCOSE IN SOFT DRINKS AND BEER. A 1.0-ml sample is boiled for 1 min in an evaporation tube for degassing, then is transferred to a 10-ml standard flask and made up to volume with 50 mM quinone solution (pH 6.0).

DETERMINATION OF GLUCOSE IN JAM. One gram of sample is dissolved in hot water, followed by dilution to 50 ml with water. Any insoluble material is filtered through a sintered glass disk (pore size 30 μm), and a 1.0-ml portion of the filtrate is diluted to 10 ml with the quinone solution.

The results are in agreement with the certified values (Table I).

Conclusion

The most important thermal condition for the flow enthalpimetric determination is not high precision thermostatting of the detection unit, but the thermal equilibration of the sample solution. In principle, flow enthalpimetry requires a relatively large sample volume, since the sample pulse broadening caused by the thermal equilibration coil influences the signal change with concomitant decrease in the analytical sensitivity. Therefore, the sample volume giving the most sensitive analytical signal is mainly set by the volume of the heat exchanger. An optimum flow rate is determined mainly by a combination of column adiabaticity or column length and the activity of the immobilized enzyme provided a sufficient sample volume is supplied.

A limiting factor of this and other flow enthalpimetry methods is the nonspecific heat effect seen as peak height changes caused by slight differences in sample composition. This is a source of error in the determination. Thus, the reproducibility (coefficient of variation) of this method as shown in Table I is slightly poorer than those of other immobilized enzyme columns with end-point detection. By considering the analytical sensitivity based on the heat of reaction and the magnitude of nonspecific heat, the lower limit of determination is estimated to be 0.1 mM glucose. Therefore, on application of the method to various practical samples satisfactory results are obtained with samples above 1 mM (18 mg/dl) in glucose. The important feature of the method is its practicability and its ease of adaptability for routine analysis.

[20] Use of Hydrogen- and Ammonia-Sensitive Semiconductor Structures in Analytical Biochemistry: Enzyme Transistors

By Fredrik Winquist, Bengt Danielsson, Ingemar Lundström, and Klaus Mosbach

Introduction

In the rapidly growing field of chemical sensors considerable interest has been shown in sensors based on semiconductor technology. Such sensors have important advantages in the biomedical field because of small size, ability to be directly integrated with microelectronics, ruggedness, and the possibility of mass fabrication at low cost. Furthermore, there has been a growing awareness of the potential for combining these sensors with enzymes or other biological systems to provide new analytical tools.

An important group of semiconductor sensors are metal–insulator–semiconductor structures made by the metal oxide semiconductor (MOS) technique, in which the electrical field over the oxide layer controls the capacitance of the structure and the current along the semiconductor surface. Very sensitive detectors for hydrogen have been developed[1] in which the metal gate consists of the catalytically active metal palladium. These Pd–MOS structures are also to some extent sensitive to gases containing hydrogen, e.g., hydrogen sulfide and ammonia. Ammonia sen-

---

[1] I. Lundström, *Sensors Actuators* **1**, 403 (1981).

sitivity can be considerably increased[2,3] if, for example, iridium is deposited as a very thin porous film (nominal thickness 3 nm) over the gate metal. In the ISFET (ion-sensitive field effect transistor), the gate metal is excluded and the insulator is made of an ion-sensitive material (see Caras and Janata, this volume [21]). The Pd–MOS and Ir–MOS sensors differ from the ISFETs since they are used for measurements in the gas phase, while ISFETs are used for measurements in the aqueous phase. In air the lower limit of detection of hydrogen gas is about 0.5 ppmV (volume parts per million) for a Pd–MOS, and the detection limit for ammonia gas with an Ir–MOS is about 1 ppmV.

Background

A hydrogen-sensitive Pd–MOS structure consists of a silicon chip with a silicon dioxide layer (100 nm thick) coated with a thin film (200 nm thick) of a catalytically active metal, palladium. The essential part of the sensor is the palladium gate. Hydrogen molecules in the environment are catalytically dissociated into hydrogen atoms, which are readily adsorbed on the palladium surface. Some of these hydrogen atoms diffuse through the metal film and adsorb on the metal–insulator interface, where they are polarized. The dipole layer of hydrogen atoms thus created causes a voltage drop over the structure, and the voltage drop is related to the hydrogen pressure. The back reaction, when hydrogen is oxidized to water on the palladium surface, is essential to minimize the recovery time of the device. In order to speed up the response and recovery time, the device is normally operated at 100–150°. The elevated temperature also prevents water molecules from sticking to the metal surface. The response time is about 1 min at low concentrations and the sensitivity 1 ppmV or better for hydrogen in air. The chemical reactions on a Pd–MOS structure are illustrated in Fig. 1.

Pd–MOS field effect transistors (FETs) as well as Pd–MOS capacitors have been studied.[1] The hydrogen-induced voltage drop of the metal insulator interface shifts the I_D–V_G characteristics of the transistor and the $C(V)$ curve of the capacitor along the voltage axis (Fig. 2). The catalytic activity of a clean palladium surface in the decomposition reaction of ammonia to nitrogen and hydrogen is low, thus a Pd–MOS device is not very sensitive to ammonia. If the gate oxide in a MOS structure is covered

[2] A. Spetz, F. Winquist, C. Nylander, and I. Lundström. *Proc. Int. Meet. Chem. Sensors,* 479 (1983).

[3] F. Winquist, A. Spetz, M. Armgarth, C. Nylander, and I. Lundström, *Appl. Phys. Lett.* **43,** 839 (1983).

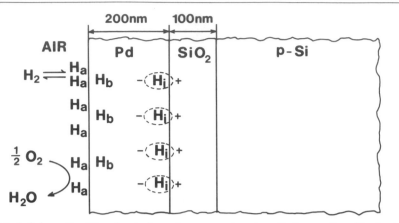

FIG. 1. Schematic of the chemical reactions occurring at a palladium surface. For further details, see text.

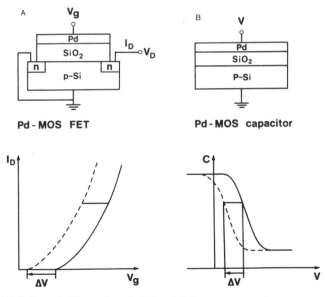

FIG. 2. (A) Schematic illustration of a Pd–MOS field effect transistor and the corresponding I_D–V_G characteristics. The voltage drop, ΔV, due to hydrogen exposure is added to the externally applied voltage. (B) Schematic illustration of a Pd–MOS capacitor. The voltage drop, ΔV, due to ammonia exposure will shift the $C(V)$ characteristics to the left.

by a thin porous layer of iridium, the catalytic activity is increased considerably and thus also the ammonia sensitivity. With an Ir–MOS, ammonia in air can be detected down to 1 ppmV. Furthermore, the hydrogen sensitivity will be decreased, thereby improving the selectivity properties. In contrast with hydrogen-sensitive Pd–MOS structures, the response and recovery times of an Ir–MOS for low gas concentrations (<40 ppmV) are only slightly influenced by the temperature. This means that in practical applications an Ir–MOS can be operated at lower temperatures as well which is advantageous. For the determination of ammonia in aqueous solutions, the structure is normally operated slightly above room temperature (e.g., 35°) in order to avoid condensation of water on the surface of the sensor.

Response Characteristics

The calibration curves of the sensors for steady-state values of the gas concentration are not linear, which is due to the kinetics of adsorption and desorption of gas molecules to the surface of the sensors. Studies of hydrogen gas-sensitive Pd–MOS structures have shown[1] that the relationship between the voltage drop across the sensor, ΔV, and the hydrogen gas partial pressure, P_{H_2}, follows Eq. (1).

$$\Delta V = C_1(P_{H_2})^{0.5} \tag{1}$$

for $P_{H_2} \leq 50$ ppmV, where C_1 is a constant. Corresponding studies of ammonia-sensitive Ir–MOS structures have resulted in a similar equation[2]:

$$\Delta V = C_2(P_{NH_3})^a \tag{2}$$

for $P_{NH_3} \leq 50$ ppmV, where C_2 and a are constants.

The values of the constants C_1, C_2, and a will vary slightly between different sensors, owing to variations in metal film thickness, size of active area, crystallization of the metal film, etc. Each sensor therefore has to be separately standardized. Typical values for the constants are $C_1 = 27$, $C_2 = 24$ (mV/ppmV), and $a = 0.55$.

It has also been shown that if an Ir–MOS is exposed to ammonia at low concentration (≤ 50 ppmV) for a short time, Δt, then the maximum voltage shift during the ammonia pulse is

$$\Delta V = KP_{NH_3}\Delta t \tag{3}$$

where K is a constant.[4] Thus a linear relation between the voltage drop across the component and the ammonia gas concentration is obtained. A

[4] F. Winquist, A. Spetz, I. Lundström, and B. Danielsson, *Anal. Chim. Acta* **164**, 127 (1984).

similar relation is also applicable to hydrogen-sensitive Pd–MOS devices. Normally Δt is below about 60 sec, which is short enough to achieve linearity but long enough to ensure reliable measurements.

Experimental Methods

Instrumentation

Ir–MOS capacitors are made of *p*-type silicon with a 100-nm thermally grown dry oxide layer. Palladium is resistively evaporated through a 1×2 mm^2 T-shaped mask on the oxide to a nominal thickness of 200 nm. Iridium is then evaporated over the upper part of the T-shaped palladium contact to a nominal thickness of 3 nm, and the size of the active area of the sensor is 1.5×2.0 mm^2 (Fig. 3). An aluminum back contact is made prior to the gate metallization. The sensor is mounted on a temperature-controlled sample holder, adjusted to 35°. The shift of the capacitance–voltage characteristics of the capacitor along the voltage axis due to ammonia gas exposure is measured with a constant capacitance controller, connected to an X–t recorder. The temperature-controlled sample holder including the sensor and the constant capacitance controller were built at the laboratory, but can also be obtained from Sensistor AB, Linköping, Sweden. The hydrogen gas-sensitive structures have been of the Pd–MOS FET type, with integrated heater and temperature sensor. The sensor chip (Fig. 4) and the necessary electronic equipment, including a temperature controller, were obtained from Sensistor AB.

Measurements of Gas Dissolved in Aqueous Solution

The Pd–MOS and the Ir–MOS devices are gas sensors and cannot normally be used for direct measurements in aqueous solutions. The separation of gas from the liquid is carried out using a gas-permeable membrane, inserted in a flow–injection system. Owing to the different proper-

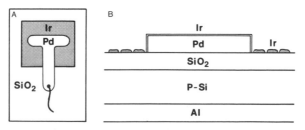

FIG. 3. (A) An Ir–MOS capacitor. The size of the active iridium area is 2.0×1.5 mm^2. (B) Schematic cross section of the capacitor.

Active gate

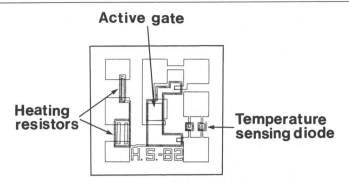

Heating resistors

Temperature sensing diode

H.S.-82

FIG. 4. Pd–MOS field effect transistor with integrated heater and temperature control. The size of the chip is 1.5 × 1.5 mm². The chip is mounted in a TO 18 package.

ties of the two types of sensors, two different flow–injection systems are used. The Ir–MOS capacitor is operated at 35°; the gas-permeable membrane can thus be placed very close to the surface of the sensor. The Pd–MOS FET, however, is normally operated at 150°, and, furthermore, it has bonding wires sticking out from the sensor chip precluding closer attachment of membranes or enzyme layers. In this case, a carrier gas stream leading to the sensor is used.

Experimental Setup for Hydrogen Determinations with Pd–MOS FET. Buffer is continuously pumped through the system at a flow rate of 0.5 ml/min with a multichannel peristaltic pump (Gilson Minipulse, Villiers-le-Bel, France) which is also used to introduce the carrier gas, 2.5 ml/min of air, into the buffer stream. The mixture of buffer and air then enters a cell with a circular (20 mm diameter) gas-permeable membrane made of polytetrafluoroethylene (Fluoropore, Millipore Corp., Bedford, MA, mean pore diameter 0.2 μm). The carrier gas penetrates the gas-permeable membrane and is led to the Pd–MOS FET. A sample injection valve with a 0.5-ml sample loop is inserted in the buffer flow stream. This means that the Pd–MOS FET is exposed to a sample during 1 min, which is within the limit of response linearity. The experimental arrangement is shown in Fig. 5.

Experimental Setup for Ammonia Determinations with Ir–MOS Capacitors. A flow–injection system for ammonia determinations in aqueous solutions is shown in Fig. 6. A peristaltic pump (Microperpex, LKB, Bromma, Sweden) is used to continuously pump buffer at a flow rate of 0.4 ml/min through the system via an injection valve with a 0.2-ml sample loop and a reaction column to a circular (4 mm diameter) Teflon cell, with one wall made of a gas-permeable membrane (SM33, Sartorius Filters, Hayward, CA, mean pore diameter 5.0 μm). The Ir–MOS capacitor is

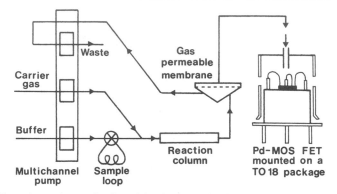

FIG. 5. Experimental setup for flow–injection analysis of hydrogen with a Pd–MOS FET.

placed with the active area of the sensor toward the center of the gas-permeable membrane, with an air gap between the sensor and the membrane of 0.2 mm. Ammonium in the buffer stream will penetrate the gas-permeable membrane in the form of ammonia gas, and be registered by the sensor. After the cell, a 0.5-m long, i.d. 0.2 mm, tubing is attached in order to give a pressure difference over the gas-permeable membrane, thereby increasing the amount of ammonia gas diffusing through the membrane. The sensor is exposed to a 0.2-ml sample for 30 sec.

Hydrogen Gas Determinations

There are many important biochemical routes leading to hydrogen evolution or consumption. Hydrogen gas is produced by many different microorganisms under anaeorobic conditions. Furthermore, considerable

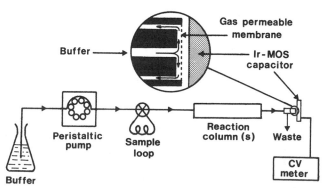

FIG. 6. Experimental setup for flow–injection analysis of ammonia with an Ir–MOS capacitor.

interest has recently been focused on the enzyme hydrogen dehydro-genase (HDH, EC 1.12.1.2) due to, among other things, its capability of reversibly reducing various electron acceptors by molecular hydrogen, including the important coenzymes NAD$^+$ and NADP$^+$, according to Eq. (4). This enzyme can thus be used in assays for these substrates. The

$$\text{NAD(P)}^+ + H_2 \xrightleftharpoons{\text{HDH}} \text{NAD(P)H} + H^+ \tag{4}$$

enzyme used was purified from *Alcaligenes eutrophus* H16, partly follow-ing a procedure described by Schneider and Schlegel.[5]

Purification of HDH. *Alcaligenes eutrophus* H16 (ATCC 17699) is grown autotrophically in a culture medium according to Schlegel *et al.*[6] Cells are cultivated in a 10-liter fermentor at 30° and 80 rpm. The gas mixture contains 80% H$_2$, 10% O$_2$, and 10% CO$_2$. At the end of the growth phase, cells are collected, washed with 50 mM sodium phosphate buffer at pH 7.0 and stored at −30°. One hundred grams of the thawed cell suspension is disrupted by sonication during 15 min, whereafter the prep-aration is centrifuged at 140,000 g for 120 min. The volume of the superna-tant obtained is 75 ml. To this extract, 0.75 ml 2.5% w/v Cetavlon [Ayerst] is added dropwise, and the precipitate is removed by centrifuga-tion at 5000 g for 5 min. The crude enzyme preparation is then fraction-ated with ammonium sulfate by adding solid ammonium sulfate to the solution. Precipitate between 40 and 60% saturation is recovered by cen-trifugation at 5000 g for 5 min, dissolved in 20 ml 20 mM potassium phosphate buffer, pH 7.0, and dialyzed over 12 hr against 2 liters of 20 mM potassium phosphate buffer, pH 7.0.

The enzyme preparation is finally purified by ion-exchange chroma-tography. DEAE-Sepharose, CL-6B (Pharmacia, Uppsala, Sweden) is preequilibrated with 20 mM potassium phosphate, pH 7.0, and packed in a 2.5 × 50 cm column. The enzyme is eluted with a 600-ml linear potas-sium chloride gradient, 0–1 M, in 20 mM potassium phosphate, flow rate 0.4 ml/min. The HDH activity of the fractions is assayed by measuring the initial rate of reduction of NAD$^+$ in 1-cm cuvettes at 340 nm, using 50 mM Tris–HCl buffer, pH 8.0, containing 0.8 mM NAD$^+$ and 0.2 mM NADH, saturated with hydrogen gas. One unit (U) of enzyme activity is defined as the amount of enzyme required to form 1 μmol of NADH per minute at 25°. The pooled, dialyzed fractions from the last purification step contain 27 U/ml or 1 U/mg. Totally 240 U are obtained.

Immobilization of HDH. The enzyme is covalently attached to con-trolled-pore glass (CPG) according to a method described by Weetall and

[5] K. Schneider and H. G. Schlegel, *Biochim. Biophys. Acta* **452,** 66 (1976).
[6] H. G. Schlegel, H. Kaltwasser, and G. Gottschalk, *Arch. Mikrobiol.* **38,** 209 (1961).

Filbert.[7] One milliliter of wet, suspended aminopropyl-CPG, mean pore diameter 200 nm, 80–120 mesh (Corning Glass Works, Corning, NY) is activated by the addition of 10 ml glutardialdehyde (2.5% in 0.1 M sodium phosphate buffer, pH 7.0) and gently shaken for 1 hr. The glass preparation is then carefully washed with twice 10 ml distilled water and twice with the sodium phosphate buffer. To the activated glass beads, 72 U of HDH in 4 ml sodium phosphate buffer is added, and the mixture is gently shaken at 4° for 12 hr. The preparation is then washed twice with 0.5 M sodium chloride, twice with distilled water, and finally twice with sodium phosphate buffer. Out of the 70 U of enzyme applied, an apparent activity of 17 U (22%) is found to be coupled.

The stability of the soluble HDH is poor, as shown by Klibanov and Puglisi.[8] A considerable increase in stability is, however, obtained if the enzyme is immobilized to CPG. A reaction column with immobilized HDH could be used for up to a week with unchanged performance, despite the rather harsh conditions involved in analysis.

Measurements of NAD(P)H and NAD(P)⁺. Using a reaction column containing 1 ml of CPG-immobilized HDH in the flow–injection analysis system shown in Fig. 5, NADH and NAD⁺ can be quantified. For the determination of NADH, the working buffer is 50 mM Tris–HCl at pH 8.0. One-half milliliter samples of NADH at different concentrations are introduced in the system, and the hydrogen gas produced by the enzyme-catalyzed reaction is monitored with a Pd–MOS FET. The calibration curve followed a straight line up to a concentration of 0.5 mM NADH, according to the equation $C_{\text{NADH}} (\mu M) = 0.104 \Delta V$ (mV), where ΔV is the maximum voltage shift obtained from the Pd–MOS FET during the NADH pulse. The lower limit of detection for NADH (defined as $S/N = 3$) is 0.03 mM or 15 nmol. The time for one sample to be analyzed is of the order of 6 min.

This system for NADH determinations can also be used to follow reactions catalyzed by other dehydrogenases. Ethanol is quantified after 10-min incubation in the working buffer containing 0.5 mM NAD⁺ and 0.1 U/ml alcohol dehydrogenase (ADH, EC 1.1.1.1, from yeast, obtained from Sigma Chemical Co., St. Louis, MO) and subsequent determination of the amount of NADH produced. The reaction sequence follows Eq. (5). The principle for the determination of NAD⁺ is based on the reverse

$$\text{C}_2\text{H}_5\text{OH} + \text{NAD}^+ \xrightarrow[\text{CH}_3\text{CHO}]{\text{ADH}} \text{NADH} + \text{H}^+ \xrightarrow{\text{CPG–HDH}} \text{NAD}^+ + \text{H}_2 \qquad (5)$$

[7] H. H. Weetall and A. M. Filbert, this series, Vol. 44, p. 134.
[8] A. M. Klibanov and A. V. Puglisi, *Biotech. Lett.* **2,** 445 (1980).

HDH-catalyzed reaction, where hydrogen gas is consumed. The working buffer is 50 mM Tris–HCl at pH 8.0, containing 0.1 mM NADH. The carrier gas contains 100 ppmV hydrogen gas, and the signal thus generated from the Pd–MOS FET is allowed to stabilize. By introducing NAD$^+$ into the system, a corresponding amount of hydrogen gas will be consumed, and a negative voltage shift, ΔV, will be registered by the Pd–MOS FET. A linear calibration curve with negative slope is obtained for 0.5-ml standard solution samples of NAD$^+$. The plot follows the equation C_{NAD^+} (mM) = $-0.310\Delta V$ (mV) and is linear up to 0.6 mM NAD$^+$. The detection limit for NAD$^+$ is 0.05 mM.

Measurements of Ammonia

Ammonium is a weak acid with a pK_a value of 9.25. Ammonium ions therefore are in equilibrium with ammonia in the pH range 6–12. In the following, ammonia nitrogen concentration will be referred to as the total amount of ammonia nitrogen present, that is, the sum of ammonia gas and ammonium ions.

pH Dependence. In measurements of ammonia nitrogen concentrations in aqueous solutions with the flow–injection system based on the Ir–MOS capacitor, it is the fraction of ammonia gas that is measured. The pH of the solution is therefore an important parameter that must be carefully controlled. Elevated pH will favor ammonia gas formation, and the sensitivity of the system will be correspondingly increased, the maximum sensitivity being obtained for pH values above 11. For the determination of ammonia nitrogen in inorganic samples, such as rainwater and river water, the pH consequently is raised to 12.5. In measurements of biological samples, the pH must not be raised that far, since ammonia gas may be liberated from labile compounds under strong alkaline conditions. For blood samples, the pH must not exceed 8.5.[9]

By combining the flow–injection system with immobilized, ammonia-producing enzymes, many biologically interesting nitrogen-containing compounds can be assayed. Care must be taken, however, when choosing the pH of the working buffer, since the performance of the system is influenced by the pH-dependent parameters of enzyme activity, enzyme stability, and ammonia gas formation.

Ammonia Nitrogen in Aqueous Solutions. The concentration of ammonia nitrogen in aqueous solutions is determined using the flow–injection system shown in Fig. 6, omitting the reaction column. Calibration

---

[9] H. F. Proeless and B. W. Wright, *Clin. Chem.* **19,** 1162 (1973).

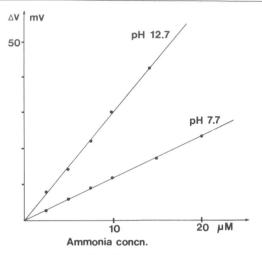

FIG. 7. Calibration graph obtained from 0.2-ml ammonia nitrogen standards at pH 7.7 and 12.7 in the flow–injection system for ammonia nitrogen determinations.

graphs, obtained from 0.2-ml standards at pH 7.7 and 12.7, are linear up to a concentration of 50 μM (Fig. 7). The buffer used in both cases is 50 mM potassium phosphate. The lower limit of detection (defined as $S/N = 3$) is 0.4 μM at pH 7.7 and 0.2 μM at pH 12.7. This flow–injection system has been used for the determination of ammonia nitrogen in various samples, such as rainwater and river water, as well as whole blood, blood serum, and other biological fluids.[4] Samples are diluted prior to analysis with the working buffer to lower the ammonia nitrogen concentration to 5–20 μM. Owing to the high sensitivity of the system, ammonia nitrogen in whole blood can be determined after a 10-fold dilution of a 20-μl sample. Approximately 15 samples per hour can be assayed with a relative standard deviation (RSD) equal to 3.7% ($n = 10$).

Selectivity. The selectivity properties of the system are estimated by the injection of various amines and some other compounds. The response of the Ir–MOS capacitor from 50 μM samples is compared to that obtained from a 50 μM standard solution of ammonium chloride. The measurements are performed using a 50 mM potassium phosphate buffer at pH 7.7, and the results obtained are shown in Table I.

Urea Determinations. Urease (EC 3.5.1.5) catalyzes the decomposition of urea to carbon dioxide and ammonia according to Eq. (6). Immobi-

$$(NH_2)_2CO + H_2O \rightarrow CO_2 + 2NH_3 \tag{6}$$

lized urease in combination with the flow–injection system for ammonia nitrogen determinations will thus provide for an assay method for urea.

TABLE I
SENSOR RESPONSE FOR VARIOUS COMPOUNDS
RELATED TO THE RESPONSE FOR AMMONIA
NITROGEN[a]

Compound	Relative sensitivity (%)
Ammonia nitrogen	100
Methylamine	10
Ethylamine	5
Butylamine	0
Diethylamine	2
Ethanolamine	2
Ethylenediamine	·1

[a] Buffered solution, pH 7.7; ammonia nitrogen
concentration, 50 μM.

The enzyme is immobilized on an epoxy-activated polyacrylic matrix by
the method described by Hannibal-Friedrich et al.[10] Urease from jack
beans (type III, 200 IU, obtained from Sigma), is added to a suspension of
125 mg (dry weight) oxirane acrylic beads (trade name Eupergite C, obtained from Röhm-Pharma GmbH, Darmstadt, FRG) in 4 ml of 50 mM
potassium phosphate buffer at pH 7.0. The mixture is gently shaken at
room temperature for 48 hr and then washed 3 times with 10 ml distilled
water. The suspension is then gently shaken for 12 hr with 4 ml 5 mM
glycine in the potassium phosphate buffer in order to terminate all unreacted epoxy groups in the matrix. Finally, the suspension is washed
twice with 10-ml portions of distilled water, once with 10 ml 0.5 M sodium
chloride, and once with 10 ml of the potassium phosphate buffer. The
volume of the suspended enzyme preparation is 0.31 ml with an apparent
activity of 103 IU. From this preparation, 0.124 ml (40 IU) is packed in a
reaction column (40 mm length, diameter 2 mm) made of Teflon, and the
column is inserted in the flow–injection system for ammonia nitrogen
determinations.

The working buffer is 50 mM Tris–HCl, pH 8.1. The pH optimum for
urease activity is 7.3, but the high enzyme activity of the reaction column
and the low substrate concentration ($\leq$50 μM) will ensure complete substrate conversion. The calibration plot, obtained from 200-μl standard
solutions, is linear up to 40 μM urea and followed the equation C_{urea}
(μM) = 0.47ΔV (mV) (Fig. 8). The lower limit of detection (S/N = 3) is
0.2 μM urea.

The method has been used for the determination of urea in whole

[10] O. Hannibal-Friedrich, M. Chun, and M. Sernetz, *Biotechnol. Bioeng.* **22**, 157 (1980).

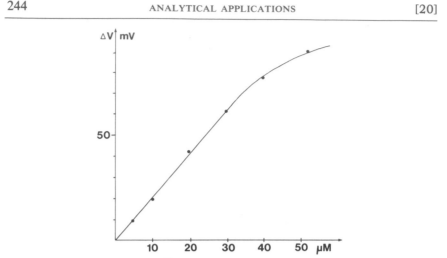

FIG. 8. Calibration graph obtained from 0.2-ml urea standards at pH 8.1. The reaction column, containing 40 U urease, is inserted in the flow–injection system for ammonia nitrogen determinations.

blood and blood serum.[11] The normal concentration range of urea in these specimens is 3.2–6.8 mM; a sample is thus diluted 500-fold before analysis. The precision of the method is 2.3% (RSD, $n = 20$), and 20 samples per hour could be analyzed. Furthermore, the reaction column could be used for over a month with unchanged performance. When not in use, the reaction column was stored at 4° in the working buffer, containing 3 mM sodium azide as a preservative.

L-*Asparagine Determinations.* The method for determination of L-asparagine is based on the enzymatic deamination of L-asparagine by L-asparaginase (EC 3.5.1.1) and subsequent measurement of the ammonia nitrogen formed. In analyses of blood specimens, the assay will be influenced by the normal background of ammonia nitrogen, which is about the same as the L-asparagine level. In these measurements, the ammonia nitrogen is removed by amination of α-ketoglutarate, catalyzed by L-glutamate dehydrogenase (EC 1.4.1.3), prior to the L-asparagine determination. The reaction sequence is shown in Eqs. (7) and (8). L-Asparagin-

$$NH_4^+ + NADH + \alpha\text{-ketoglutarate} \xrightarrow{\text{L-glutamate dehydrogenase}} NAD^+ + \text{L-glutamate} \quad (7)$$

$$\text{L-Asparagine} + H_2O \xrightarrow{\text{L-asparaginase}} \text{L-aspartate}^- + NH_4^+ \quad (8)$$

[11] F. Winquist, A. Spetz, I. Lundström, and B. Danielsson, *Anal. Chim. Acta* **163**, 143 (1984).

ase from *E. coli,* 40 U (Grade VIII, Sigma) and L-glutamate dehydrogenase from bovine liver, 80 U (Type III, Sigma) are immobilized on
oxirane acrylic beads by the same procedure as for urease immobilization,
previously described. The enzymes are applied to each 125 mg (dry
weight) of Eupergite C, and, after the final wash, 0.31 ml of suspended
enzyme preparation of each is obtained. The preparations are stored at 4°
in 50 mM potassium phosphate buffer, pH 7.0, containing 3 mM sodium
azide as a preservative.

The two enzyme preparations are packed in two consecutive Teflon
columns (2 mm i.d., 32 mm length). Each column contains 0.10 ml enzyme-bound beads, carrying an apparent activity of approximately 5 U of
L-asparaginase and 20 U of L-glutamate dehydrogenase, respectively.

The working buffer is 50 mM Tris–HCl, pH 8.2, containing 1 mM
NADH (Grade III, Sigma), 0.5 mM α-ketoglutaric acid, and 3 mM NaN$_3$.
At these substrate concentrations, the L-glutamate dehydrogenase
column is capable of completely removing up to 0.05 mM of ammonia
nitrogen in 0.20-ml samples at a flow rate of the buffer of 0.4 ml/min. This
is sufficient for removing endogeneous ammonia nitrogen in diluted blood
samples.

A calibration graph for L-asparagine is obtained from 0.200-ml standards, linear up to 40 μM L-asparagine and following the equation
$C_{\text{L-asparagine}}$ (μM) = 0.79ΔV (mV). For the determination of L-asparagine in
whole blood, blood serum, and blood plasma, samples are diluted 20-fold
with the working buffer. A 0.200-ml quantity of the dilution is injected
into the flow-through system, and the signal thus generated from the Ir–
MOS capacitor is compared to the calibration graph. Fifteen samples per
hour can be analyzed. After every 20 samples analyzed, the columns and
tubings are washed by passing washing buffer (0.1 M potassium phosphate at pH 7.0, containing 0.8 M sodium chloride) through the system at
a flow rate of 0.4 ml/min over 5 min.

Creatinine Determinations. The method for creatinine is based on the
same principle as the method for the determination of L-asparagine
described above. Creatinine is enzymatically hydrolyzed by creatinine
iminohydrolase (EC 3.5.4.21, creatinine deiminase) and the ammonia
nitrogen thus formed is a measure of the creatinine concentration. The
reaction follows Eq. (9). Creatinine iminohydrolase, 10 U (obtained from

$$\text{Creatinine} + \text{H}_2\text{O} \rightarrow N\text{-methylhydantoin} - \text{NH}_3 \tag{9}$$

Aalto Bio Reagents Ltd., Dublin, Ireland), and L-glutamate dehydrogenase, 80 U (EC 1.4.1.3., from bovine liver, Type III, Sigma), are immobilized on oxirane acrylic beads, following the procedure previously described. The enzyme preparations are packed in two 0.1-ml consecutive

columns. The working buffer is 50 mM Tris–HCl at pH 8.5; the other experimental conditions, including the washing step after every 20 samples, follow the procedure for L-asparagine determination. The calibration graph for 0.200-ml standards follows the equation, $C_{creatinine}$ (μM) = 0.71ΔV (mV), and is linear up to 30 μM creatinine. For the determination of creatinine in whole blood and serum, samples are diluted 25-fold prior to analysis; for the determination of creatinine in urine, specimens are diluted 1000-fold. Fifteen samples per hour can be assayed.

Conclusions

The sensors described in this chapter are hydrogen or ammonia gas-sensitive semiconductor structures. For measurements in aqueous solutions, a continuous flow–injection system utilizing a gas-permeable membrane is used. The phase separation stage is advantageous since the sensor thereby is protected from dirty or viscous solutions, and since the sensor is electrically isolated from the solution there is no need for a reference electrode. Furthermore, the selectivity properties of the flow–injection system for ammonia nitrogen will be improved. The application for the flow–injection systems described can be expanded by introducing reaction columns containing immobilized enzymes. This will provide for a simple and versatile method to follow many hydrogen- or ammonia-producing or -consuming reactions.

Since the sensors are independent of the optical properties of the samples, crude or opaque samples, such as whole blood or fermentation broths, can be directly analyzed. Owing to the high sensitivity of the system, very small sample volumes (1–50 μl) are required.

In contrast with hydrogen gas-sensitive Pd–MOS FETs which are operated at 150°, the ammonia-sensitive Ir–MOS capacitor is operated at 35° for measurements of aqueous solutions. Consequently, the gas-permeable membrane can be placed very close to the surface of the sensor, making the signal response faster. The low temperature of the device also allows enzymes to be placed in close proximity to the structure. In preliminary experiments, an enzyme probe based on the Ir–MOS capacitor for the determination of urea was developed.[11] Urease was thus enclosed between a dialysis membrane and a gas-permeable membrane at the end of a Teflon tubing, with the sensor placed close to the gas-permeable membrane inside the tube.

The impedance of the transistor type of these semiconductor sensors is low, thus the transmission of the signal will be virtually noise free. Furthermore, integrated circuit technology opens up possibilities of designing multisensor chips with several gates of different metals. A corresponding multienzyme probe can be designed, based on an ammonia-sensitive Ir–MOS structure combined with different ammonia-producing enzymes.

The analytical systems, based on hydrogen- or ammonia-sensitive semiconductor structures described in this chapter, offer simple, fast, and sensitive methods for a multitude of biochemically important compounds. Furthermore, this technology involves simple, low-cost instrumentation and can be applied to many fields, such as agriculture, clinical chemistry, environmental control, and fermentation.

[21] Enzymatically Sensitive Field Effect Transistors

By Steve Caras and Jiří Janata

Introduction

Enzymatically coupled field effect transistors (ENFETs) are the only true potentiometric solid-state biosensors made so far.[1-3] In principle they are similar to potentiometric enzyme electrodes and are subject to the same limitations in their lifetime, detection limits, sensitivity, and time response. However, because modern photolithographic techniques can be employed in their fabrication, the parameters which determine their optimum performance can be controlled more closely than is the case with the enzyme electrodes. In this chapter, we will describe the principle of operation of ENFETs, use the glucose-sensitive ENFET as an illustration, and discuss the theoretical limits of these sensors.

A schematic diagram of a dual transistor chip is shown in Fig. 1. The drain current I_D which flows between drain (D) and source (S) is modulated by the potential difference between the p-silicon and the surface of the insulator at the gate 1 and the gate 2, respectively. The surface of the insulator (silicon oxynitride) is pH sensitive. By depositing various ion-selective membranes over this surface, it is possible to make transistors sensitive to different ions.[4] However, to our knowledge an ENFET combining an ion-selective membrane for ions other than H^+ and an enzyme coating layer has not yet been reported; it should not be difficult in principle, however. The theory of operation of ion-sensitive field effect transis-

[1] S. Caras and J. Janata, Anal. Chem. 52, 1935 (1980).
[2] B. Danielsson, I. Lundström, K. Mosbach, and L. Stiblert, Anal. Lett. 12, 1189 (1979).
[3] Y. Hanazato and S. Shiono, Proc. Int. Meet. Chem. Sensors, p. 513 (1983).
[4] J. Janata, "Solid State Chemical Sensors," Chap. 2. Academic Press, New York, 1985.

METHODS IN ENZYMOLOGY, VOL. 137

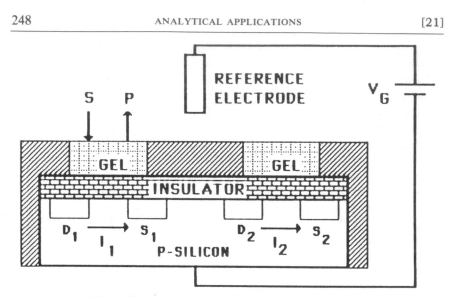

FIG. 1. Schematic diagram of a dual ENFET chip.

tors (ISFETs) has been described elsewhere.[4] For the purpose of this review it suffices to state that there is a functional relationship between the drain current I_D and the activity of the primary ion a_i at the surface of the ion-sensitive gate:

$$I_D = K[(V_G - V_T)V_D + E_{REF} + RT(\ln a_i) - VD^2/2] \qquad (1)$$

for $V_D < V_G - V_T$, where K is a constant which is characteristic of the materials and the geometry of the transistor, V_T is so-called turn-on (or threshold) voltage, and V_D is the voltage between drain D and source C of the transistor. In normal ISFET operation, it is preferable to maintain the drain current constant. In other words, any change in the membrane potential is compensated by the equal and opposite charge of the gate voltage V_G. This is the so-called feedback mode of operation[4] which yields the explicit relationship between the cell voltage and the activity of the measured ion (e.g., the Nernst equation). It is particularly advantageous for so-called equilibrium sensors such as ISFETs.

In contrast, the ENFET is a steady-state sensor. Its response is governed both by the thermodynamic response of the ion-selective membrane to the surface activity of the primary ion and by the diffusion/kinetics properties of the species involved in the enzymatic reaction:

$$S + E \underset{k_{-1}}{\overset{k_1}{\rightleftharpoons}} SE \overset{k_2}{\rightleftharpoons} P + E \qquad (2)$$

The ion I which couples this reaction to the ISFET is related either to the substrate S or, more often, to the product P, for example, PI → P + I. The ionic species which performs this role most frequently is hydronium ion. An example of an ENFET in which hydronium ion and the mobile buffer determine the response of the ISFET and in which the rate constants in Eq. (2) are effectively independent of pH is the penicillin-sensitive transistor described previously.[1] The concentration changes in an arbitrary volume element of the membrane are described by the set of second-order partial differential equations related to the combined transport/kinetic nature of this system. They cannot be solved analytically. Transient changes, steady-state values of concentrations at the transistor surface, and concentration profiles throughout the gel layer surface have been calculated numerically[5] by using the Method of Lines in One Dimension (MOL1D) simulation.

Most enzymatic reactions require a certain pH range for their optimum reaction velocity which is maintained by the buffer. However, this requirement is clearly contradictory to the operation of the pH-sensitive ISFET; if the buffer capacity is too high, the hydronium ions produced by the enzymatic reaction are consumed by the buffer. On the other hand, if the buffer capacity is too low, the velocity of the enzymatic reaction is decreased by self-inhibition, the extent of which depends on the enzyme's pH activity profile. Furthermore, any small changes in the ambient pH are superimposed on the pH change resulting from the enzymatic reaction. In order to avoid this problem, the second transistor gate without the enzyme in the gel is used as a reference. If the drain current from this gate is subtracted from the gate containing the enzyme, the fluctuations of ambient pH, temperature changes, and common noise are eliminated. The response of this sensor then becomes:

$$\Delta I_D = I_{D1} - I_{D2} = K' \ln(a_{H^+})_1 - K' \ln(a_{H^+})_{bulk} \tag{3}$$

A simple circuit used with a differential ENFET probe is shown in Fig. 2.

The partial differential equations describing the case of pH-dependent enzyme kinetics are even more complicated but not different, in principle, from those describing the pH-independent case. They have been again solved numerically under the assumptions (1) there is no unstirred boundary layer at the solution side of the interface; (2) the gel matrix has no buffering capacity, and the total buffer concentration in the gel is constant; (3) the enzyme has no inhibition other than H^+; and (4) the pK_a of the product is lower than the bulk pH and lower than maximum response. The mathematical solution was compared with experimentally obtained

[5] S. Caras, D. Saupe, K. Schmitt, and J. Janata, *Anal. Chem.* **57,** 1917 (1985).

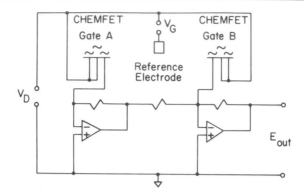

FIG. 2. Simple circuit used with a differential ENFET probe.

results for the response of glucose-sensitive ENFET, which is based on the reaction:

$$\text{Glucose} + 1/2\ O_2 \xrightarrow{\text{glucose oxidase/catalase}} \text{gluconate}^- + H^+ \tag{4}$$

Glucose-Sensitive ENFET

The basic construction of ENFETs sensitive to different substrates is the same, the difference being only in the enzyme immobilized in the gel over the active gate (Fig. 1). As an example, we will use the construction of a glucose-sensitive transistor based on the above reaction [Eq. (5)].

The fabrication of the basic transistor chip is, unfortunately, beyond the capability of many laboratories which may be interested in this research. The fabrication procedure has been described in detail,[6] and the chips are available from the authors' laboratory. Commercial FETs are usually unsuitable for this application because they have not been designed with aqueous application in mind. As the integral part of our devices we have been using a photolithographically patterned encapsulant which defines the depth of the active gate retion. Detailed description of the preparation process is beyond the scope of this chapter, and the reader is referred to the original paper.[7]

The actual enzyme immobilization of the glucose oxidase/catalase system is then performed; the aliphatic amino groups on the enzyme molecule are treated with N-succinimidyl methacrylate (**I**). The derivatized proteins (**II**) are then added (12 mg total protein/ml, containing 50%

[6] R. J. Huber, in "Solid State Chemical Sensors," Chap. 3. Academic Press, New York, 1985.

[7] N. J. Ho, J. Kratochvil, G. F. Blackburn, and J. Janata, *Sensors Actuators* **4**, 413 (1983).

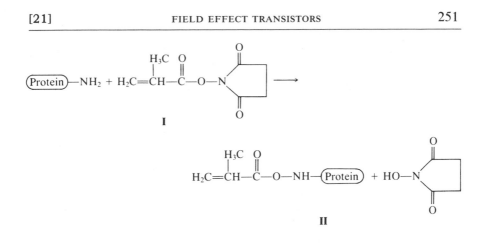

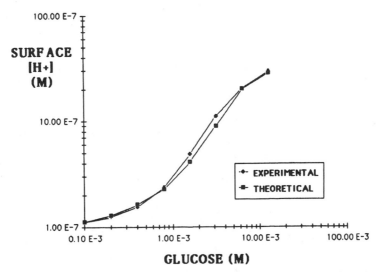

II + acrylamide + bisacrylamide + TEMED + ammonium persulfate $\longrightarrow$

immobilized polyacrylamide gel

glucose oxidase and 50% catalase) to the solution containing 15% of acrylamide and bisacrylamide (in 30:0.8 ratio acrylamide to bisacrylamide),
TEMED (N,N,N',N-tetramethylethylenediamine) and ammonium persulfate. This mixture is applied to the gate of the prepared transistor chip
and allowed to react for 30 min under wet nitrogen. The same procedure is
repeated for the gate 2 except that no enzyme is added. Prior to its use,

FIG. 3. Glucose calibration curve. Conditions: $2 \times 10^{-4} M$ O$_2$, $2 \times 10^{-4} M$ phosphate
buffer, pH 7.

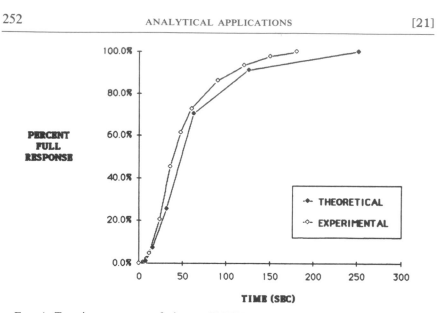

FIG. 4. Transient response of glucose ENFET at a glucose concentration of 1.2×10^{-3} M.

the probe is placed in 0.1 M phosphate buffer (pH 7.2) for approximately 1 hr.

The results of the tests of the glucose ENFET probe are shown in Figs. 3 and 4 together with theoretical curves predicted from the model.

Conclusions

Photolithographic encapsulation allows fabrication of ENFETs with closely defined parameters, particularly the enzyme layer thickness. This enabled us to model both transient and steady-state response of ENFETs with a pH-independent, one-substrate enzymatic system (penicillin)[8] and with the more complicated, two-substrate, pH-dependent kinetics (glucose). The key assumption in our model is that the solution is well stirred so that the concentration profile on the solution side of the membrane/solution interface is flat. Clearly, this condition may not always be satisfied, particularly in biomedical applications of these sensors.

The excellent agreement of the theoretical model with the experiments allows us to extrapolate the theoretical response parameters of these sensors. Although this analysis is done for the ENFET, it is also valid for other potentiometric enzymatic sensors.

[8] S. Caras and J. Janata, *Anal. Chem.* **57,** 1924 (1985).

It is often stated that an enzyme electrode "has or does not have Nernstian response." Since ENFETs and enzyme electrodes are non-equilibrium, rate-dependent sensors, one must be careful in comparing an equilibrium sensor with a rate-dependent (steady-state) sensor. The response of most enzyme electrode systems reported in the literature comes not just from the pH ISFET or electrode but from the diffusion/reaction kinetics within the gel layer as well. Thus, the thermodynamic electrode acts strictly as a chemical transducer, transforming the activity of hydronium ions at the surface to the drain current of the ISFET. The only influence which the electrode has on the diffusion/reaction process is through the boundary condition between the electrode and the gel. Thus, for any rate-based sensor, slopes less than, equal to, or greater than Nernstian are possible. Thus, in the case of pH-based ENFETs, the imposition of the buffer titration curve within the gel layer can cause the maximum slope to differ from 1 pH unit/decade substrate. The slope can depend on the diffusion/reaction kinetics within the gel layer, buffer concentration, and on the gel thickness. For example, Table I shows the effect of buffer and oxygen diffusion coefficients on the maximum slope of the response curve.

The sigmoidal response curve can shift left or right along the substrate concentration axis depending on the buffer diffusion coefficient and buffer concentration. Table II shows the "detection limit," the concentration at which the maximum slope line intersects the glucose concentration axis, for different values of the buffer diffusion coefficient. A range of buffer concentrations will also shift the "detection limit" in the same manner. At very low buffer concentrations, however, the sensor will revert to nearly Nernstian behavior since the buffer capacity is small. At high

TABLE I

MAXIMUM SLOPE FOR VALUES OF THE BUFFER AND O_2
DIFFUSION COEFFICIENT[a]

Diffusion coefficient	Slope (pH units/decade glucose)
$-\log D_{HA}$	
5.3	0.89
5.8	1.2
6.3	0.97
$-\log D_{O_2}$	
5.3	1.2
5.8	1.2
6.3	0.86

[a] Determined by the mathematical model.

TABLE II
DETECTION LIMIT FOR VALUES OF BUFFER
DIFFUSION COEFFICIENT[a]

Diffusion coefficient $-\log D_{HA}$	Detection limit (mM)
5.3	1.0
5.8	0.4
6.3	0.15

[a] Determined by the mathematical model.

buffer concentrations, the buffer capacity is so high there is virtually no response.

Likewise, the maximum response of the glucose sensor depends on the parameters described in the preceding two paragraphs. Table III shows the maximal surface concentration of H$^+$ as a function of buffer and O$_2$ diffusion coefficients. Since the slope and the "detection limit" of these response curves differ, the glucose concentration at which maximal response will occur will differ.

In conclusion let us summarize the salient features of ENFETs. The close control of geometrical parameters allows experimental verification of models of response. The amount of enzyme used per device is minimal (typically 10^{-4} IU) which may be an important consideration for devices utilizing more expensive enzymes. The differential mode of operation allows effective elimination of external interferences. Because of the

TABLE III
MAXIMAL RESPONSE AND GLUCOSE CONCENTRATION AT MAXIMAL
RESPONSE FOR VALUES OF OXYGEN AND BUFFER DIFFUSION
COEFFICIENT[a]

Diffusion coefficient	Glucose concentration at maximal response (mM)	Maximal response (pH$_{surf}$ − pH$_{bulk}$)
$-\log D_{O_2}$		
5.3	70	1.6
5.8	30	1.6
6.3	30	1.7
$-\log D_{HA}$		
5.3	55	1.8
5.8	35	1.6
6.3	7.0	1.1

[a] Determined by the mathematical model.

small area of the enzyme gate, no retaining membrane is needed. This shortens the response time as compared with equivalent enzyme electrodes. Finally, the main goal of solid state chemical sensors: a multichannel device could be realized with ENFETs.

Acknowledgments

Technical assistance of Mrs. D. Petelenz is gratefully acknowledged. This work was supported by a grant from the National Institutes of Health (NIGMS 22952).

[22] Microenzyme Sensors

By Isao Karube and Toyosaka Moriizumi

Introduction

Most organic compounds are determined by spectrophotometry. However, this may require complicated procedures and long reaction times. Electrochemical determination of these compounds has definite advantages as a sample solution does not need to be optically clear. Many biosensors have been developed providing methods for rapid and continuous measurement of various compounds.[1]

In addition, implantable microbiosensors are required in medical fields. Janata et al. have reported an enzyme field effect transistor (FET) which consisted of an immobilized enzyme and a FET[2,3] (see also this volume [21]). Danielsson et al. also described a urea-sensitive device based on a gas-sensing FET (enzyme transistor)[4,5] (see also this volume [20]). We have also reported a preliminary study of a trypsin FET.[6]

In this chapter, two types of microsensors are described. The first one is a urea sensor consisting of immobilized urease and an ion-sensitive field effect transistor (ISFET). The other is a glucose sensor composed of

[1] I. Karube, Biotechnol. Genet. Eng. Rev. 2, 313 (1984).
[2] J. Janata and S. Moss, Biomed. Eng. 11, 241 (1976).
[3] S. Caras and J. Janata, Anal. Chem. 52, 1935 (1980).
[4] B. Danielsson, I. Lundström, M. Mosbach, and L. Stiblert, Anal. Lett. 12, 1189 (1979).
[5] F. Winqvist, B. Danielsson, I. Lundström, and K. Mosbach, Appl. Biochem. Biotechnol. 7, 135 (1982).
[6] Y. Miyahara, S. Shiokawa, T. Moriizumi, H. Matsuoka, I. Karube, and S. Suzuki, Proc. Sensor Symp., 2nd, p. 91 (1982).

immobilized glucose oxidase and an oxygen microsensor fabricated by using an anisotropic silicon etching technique.[7]

Experimental Methods

Materials

Urease (EC 3.5.1.5, from jack beans, 130 units/mg) was purchased from Tokyo Kasei Ind., Japan, and glucose oxidase (EC 1.1.3.4, from *Aspergillus niger,* 191 units/mg) was donated by Amano Pharmaceutical Co., Japan. 1,8-Diamino-4-aminomethyloctane was obtained from Asahi Kasei Ind., Japan. Other reagents were commercially available analytical reagents or laboratory-grade materials.

ISFET

The fabrication procedure of ISFETs is basically the same as that of a metal insulator silicon FET (MISFET).[8] The structure of the ISFET, which was obtained from Kurare Co., Japan, is shown in Fig. 1. The chip was anisotropically etched to be 450 μm wide, 5.5 mm long, and 175 μm thick. The gate insulator of the ISFET consists of a 1000-Å film of thermally grown SiO_2 covered by chemical vapor deposition of a 1000-Å film of Si_3N_4 which is sensitive to H^+ in an electrolyte and protects the device from ion penetration. The channel is 1.2 mm wide and 30 μm long. A typical pH response of the ISFET is linear between pH 3 and 11 with a slope of 56–58 mV/pH unit.

Immobilization of Urease

The Si_3N_4 surface is modified with γ-aminopropyltriethoxysilane (γ-APTES). The ISFET is immersed in a 10% γ-APTES solution (pH 7.0) for 1 hr. Then the enzyme is immobilized on the gate insulator as follows: Triacetylcellulose (250 mg) is dissolved in 10 ml of dichloromethane. Then 100 μl of 50% glutaraldehyde solution and 500 μl of 1,8-diamino-4-aminomethyloctane are added to the reaction mixture. The ISFET chemically modified with γ-APTES is dipped in this mixture for 1 day at room temperature. After reaction it is immersed in 1% glutaraldehyde solution for 1 hr and washed with 10 mM phosphate buffer solution (pH 7.0). Urease is

[7] Y. Miyahara, F. Matsu, T. Moriizumi, H. Matsuoka, I. Karube, and S. Suzuki, *Proc. Int. Meet. Chem. Sensors,* p. 501 (1983).
[8] T. Matsuo and M. Esashi, *Sensors Actuators* **1,** 77 (1981).

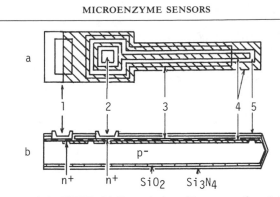

FIG. 1. Structure of the ISFET. (a) Ground plan; (b) cross section. 1, Substrate–source electrode; 2, drain electrode; 3, p-channel stopper; 4, n-diffused region; 5, channel.

covalently immobilized on the surface of the thin organic membrane on the gate insulator. The reaction is performed for 16 hr at 4°. The urease FET is stored in 10 mM phosphate buffer solution (pH 7.0).

Oxygen Microsensor

A miniaturized Clark-type oxygen sensor has been prepared utilizing integrated circuit technology. The structure of the oxygen sensor is shown in Fig. 2. The lower Si wafer is anisotropically etched with 10% KOH solution to make a micropool for the inner electrolyte (1 M KOH solution). The upper Si wafer has a hole (400 × 400 μm) to limit the sensing area. A Teflon membrane was fixed between the upper and lower Si wafers. The fixation of both wafers and the electrical insulation was provided by epoxy resin (EPOTEK H-54, Billerica, MA).

Preparation of Glucose Oxidase Membrane

A triacetylcellulose membrane containing 1,8-diamino-4-aminomethyloctane and glutaraldehyde is prepared as follows: Dichloromethane so-

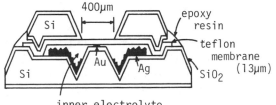

FIG. 2. Structure of the oxygen sensor using two Si wafers (cross section).

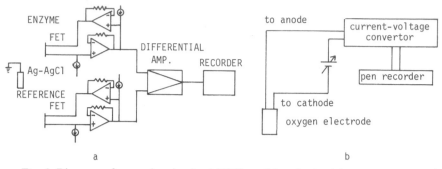

FIG. 3. Diagrams of measuring circuits. (a) Differential mode circuit between enzyme and reference FETs; (b) circuit of measuring the reduction current of the oxygen sensor.

lution containing triacetylcellulose (250 mg), glutaraldehyde (50 μl) and 1,8-diamino-4-aminomethyloctane (500 μl) is cast on a glass plate, and the membrane so prepared is dried for 24 hr at 4°. The membrane is immersed in 1% glutaraldehyde solution for 1 hr, rinsed in distilled water, and immersed in a solution containing glucose oxidase (1 mg/ml in 10 mM phosphate buffer, pH 7.0) for more than 15 hr at 4°.

Procedures

The measurement of urea in solutions is carried out using the differential mode between the enzyme FET and a reference FET. A diagram of the measuring circuit is shown in Fig. 3a. Temperature and ambient pH changes in the solutions are automatically compensated by the differential mode measurement, if the temperature and pH characteristics of both FETs are the same. On the other hand, the current is measued by the circuit shown in Fig. 3b in the case of the oxygen microsensor. All measurements are carried out in a thermostatically controlled bath.

Results and Discussion

Urea Sensor

Urease catalyzes the hydrolysis of urea, and ammonium ions are produced. Therefore, the pH of the solution is changed by the enzymatic reaction, depending on the concentration of urea.

The urea FET, the reference FET, and an Ag–AgCl electrode are placed in 1 ml of 10 mM phosphate buffer (pH 7.0), and then 1 ml of 5 × 10^{-2} g/ml urea solution is added to the phosphate buffer solution. The output voltage rapidly decreases for 2 min and then gradually increases. The initial rapid decrease of the output voltage is due to both the urease reaction and the diffusion of the produced OH^- through the organic membrane on the FET. The following gradual increase is due to the diffusion of the produced OH^-.

The calibration curve of the urease FET in the differential mode is shown in Fig. 4. The voltage is determined at 1 min after injection of a urea solution. Urea in the concentration range from 5×10^{-5} to 1×10^{-2} g/ml could be detected by the sensor.

The lifetime of the urea FET was tested by measuring its response to 1×10^{-3} g/ml urea once a day. The urea FET is stored in 10 mM phosphate buffer (pH 7.0) at 4°. The lifetime of the urea FET is 27 days, with a data fluctuation of ± 3 mV at the output voltage of 20 mV. Therefore, the urease FET can be used for the determination of urea for a long time.

Glucose Sensor

The fundamental characteristics of the oxygen sensor fabricated using the anisotropic etching techniques were examined. The 95% response time was about 12 sec with good reproducibility. The current of the miniaturized oxygen sensor was measured by changing the volume ratio of oxygen and nitrogen gases. The calibration curve shows a good linearity over the range from 0 to 100% oxygen ratios. As glucose oxidase catalyzes oxidation of glucose, oxygen around the enzyme membrane is con-

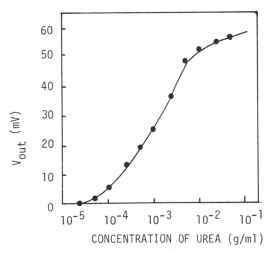

FIG. 4. Calibration curve of the urea sensor. Conditions: 10 mM phosphate buffer (pH 7), 34.9°. The output voltage is determined at 1 min after addition of a sample solution.

sumed depending on the concentration of glucose. Glucose in the range from 0.1 to 2.0 mg/ml could be determined by the glucose microsensor.

In conclusion, the microbiosensors proposed appear to be promising and attractive for the continuous determination of urea and glucose in biological fluids.

Acknowledgment

The authors would like to thank Prof. Shuichi Suzuki, The Saitama Institute of Technology, Mr. Y. Miyahara, Dr. H. Matsuoka for valuable discussions, and Dr. M. Nakamura, Kurare Central Research Laboratory, for supplying ISFETs.

[23] Use of Immobilized Enzyme Column Reactors in Clinical Analysis

By Takashi Murachi and Masayoshi Tabata

Introduction

The principle and usefulness of immobilized enzymes as applied in the form of column reactors to continuous-flow analysis have been described earlier,[1-4] but more recent studies have demonstrated the practicability of the method in routine clinical analyses (see Ref. 5 for review). A column reactor is essentially a minicolumn, usually made of a plastic tubing, which is packed with uniformly sized solid particles. An enzyme or a given number of enzymes are immobilized on the surface of the solid particles which are usually porous glass beads. The enzyme proteins are covalently bound to the alkylamino groups of derivatized glass. An immobilized enzyme reactor thus produced can be inserted into the flow system of an analyzer such as the Technicon type.

In this chapter, the methods for making a minicolumn with enzyme-immobilized glass beads is described, and examples of applications of such column reactors are presented, which include conventional flow analysis as well as flow–injection analysis. These methods are specific for serum samples, but the methods may also be used, with appropriate modi-

[1] O. R. Zaborsky, "Immobilized Enzymes." CRC Press, Cleveland, Ohio, 1973.
[2] L. D. Bowers and P. W. Carr, *Anal. Chem.* **48**, 554A (1976).
[3] K. Mosbach (ed.), this series, Vol. 44.
[4] S. J. Updike and G. P. Hicks, *Nature (London)* **214**, 968 (1967).
[5] T. Murachi, *Enzyme Eng.* **6**, 369 (1982).

fications, for the analysis of urine and other body fluids, as well as to industrial and environmental systems.

Enzyme Immobilization

Principle

Enzyme proteins are coupled to the primary amino groups of commercially available alkylamino glass by two different methods. (1) The enzyme protein and glass beads are allowed to react with a bifunctional reagent, glutaraldehyde, which forms Schiff base linkages both with the primary amino groups (mostly ε-amino groups) of the enzyme protein and with those of the glass beads.[6] (2) The carbohydrate moiety, if present in an enzyme molecule to be immobilized, is first oxidized with periodate to form dialdehyde groups which are then coupled with the primary amino groups of the glass beads through Schiff base formation.[7,8]

Materials

Glucose oxidase (EC 1.1.3.4, type II from *Aspergillus niger,* 15 units/mg) and uricase (EC 1.7.3.3, urate oxidase, grade III from *Candida utilis,* 3.5 units/mg) were obtained from Sigma Chemical Co., St. Louis, MO, and Toyobo Co., Tokyo, Japan, respectively. Glucose oxidase (100 units/mg) and uricase (4 units/mg) from Toyobo were also employed. Alkylamino glass (No. 23908, 20–80 mesh, 550 Å pore size, and No. 23909, 80–120 mesh, 500 Å pore size, both 0.17 mEq functional amino groups/g) was obtained from Pierce Chemical Co., Rockford, IL.

Procedures

Immobilization with Glutaraldehyde. The enzyme is dissolved in 1 ml of 50 mM sodium phosphate buffer, pH 7.0, in a concentration varying from 5 to 20 mg/ml, and the solution is added to 200 mg of glutaraldehyde-treated glass beads. The latter is freshly prepared by incubating the glass beads with 250 mM glutaraldehyde in 50 mM sodium phosphate buffer, pH 7.0, at room temperature for 30 min under reduced pressure to remove air bubbles from the pores, then for 30 min under atmospheric pressure, followed by washing 3 times with distilled water. The enzyme-coupled

---

[6] H. H. Weetall, "Immobilized Enzymes, Antigens, Antibodies and Peptides." Dekker, New York, 1975.

[7] O. R. Zaborsky and J. Ogletree, *Biochem. Biophys. Res. Commun.* **61,** 210 (1974).

[8] P. K. Nakane and A. Kawaoi, *J. Histochem. Cytochem.* **22,** 1084 (1974).

glass beads are thoroughly washed with 50 mM sodium phosphate buffer, pH 7.0, containing 1 M sodium chloride.

Immobilization at the Carbohydrate Moiety. The enzyme (20 mg) is dissolved in 2 ml of 0.3 M sodium bicarbonate, pH 8.1. To the solution is added 0.1 ml of 1% 1-fluoro-2,4-dinitrobenzene in ethanol, and the mixture is gently stirred at room temperature for 60 min after which 1.0 ml of 60 mM sodium periodate is added and mixed for an additional 30 min under protection from light. Then, 0.1 ml of 1.6 M glycerol is added and mixed gently to decompose the excess periodate, and the solution is dialyzed against 10 mM sodium carbonate buffer, pH 9.5, at 4° for 12 hr. To 1.0 ml of the solution of periodate-oxidized enzyme is added 200 mg of alkylamino glass beads, and the coupling reaction is allowed to proceed at pH 9.5 for 120 min at room temperature. The product is thoroughly washed with 50 mM sodium phosphate buffer, pH 7.0, containing 1 M sodium chloride.

Coupling Yield and Assay for Activity. The coupling yield is calculated as percent disappearance of the amount of protein initially added to the reaction mixture for immobilization. The protein concentrations of the enzyme solution before and after the coupling reaction are determined by the method of Lowry *et al.*[9] or by ultraviolet absorption at 280 nm. The enzymatic activity of the immobilized enzyme is determined by a batch method. Thus, a suspension of enzyme-bearing glass beads is rapidly stirred with the substrate, and the supernatant which could be almost instantaneously separated on termination of the reaction is subjected to measurement. For glucose oxidase, the incubation mixture contains 0.1 M sodium acetate buffer, pH 5.6, 0.8 mM 4-aminoantipyrine, 14 mM phenol, 0.2 mM glucose, and horseradish peroxidase (30 μg/ml; EC 1.11.1.7, grade II, 100 units/mg, Boehringer Mannheim–Yamanouchi, Tokyo, Japan), and the mixture is stirred at 25° for 10 min. The increase in absorbance at 500 nm is measured. For uricase, the reaction mixture contains 0.1 mM uric acid and 0.1 M sodium borate buffer, pH 8.6, and the reaction is carried out at 25° for 60 sec. The decrease in absorbance at 293 nm is recorded.

Immobilized Enzyme Column Reactor. The enzyme immobilized to alkylamino glass is packed in a small plastic tubing, 0.5–3.0 mm in inner diameter and 5–40 mm in length. Both ends of the packed column are covered with small pieces of nylon net with lattice of 40 × 40 μm to prevent bed movement that may occur under the pressure pulse from a peristaltic pump. Figure 1 shows two examples of the column reactors.

[9] O. H. Lowry, N. J. Rosebrough, A. L. Farr, and R. J. Randall, *J. Biol. Chem.* **193**, 265 (1951).

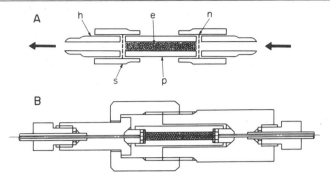

FIG. 1. Schematic representation of immobilized enzyme columns. (A) Homemade column (1.0–2.0 mm i.d. and 5–40 mm length); (B) commercial design (enzyme column size: 3.0 mm i.d. and 20 mm length). e, Immobilized enzyme; h, nipple; n, nylon net; p, plastic tubing; s, silicone rubber tubing. Direction of the flow is from right to left in both A and B.

Conventional Flow Analysis

Apparatus

The standard AutoAnalyzer modules were employed. Sampler II, pump I, 12-inch dialyzer unit with type C membrane, colorimeter with 15-mm tubular flow cell and filter (590 nm), and recorder with chart paper in absorbance units were used.

Procedures [10]

Figure 2 illustrates the setup for the continuous-flow analysis of glucose and uric acid using an immobilized enzyme column reactor integrated into the flow system of an AutoAnalyzer. The sizes of the columns are 1.5 × 20 mm for glucose oxidase and 1.5 × 40 mm for uricase. The reagent baseline on the recorder is adjusted to 0.01 absorbance. Serum samples are directly introduced to the system, while whole blood (for the determination of glucose) is hemolysed with 20-fold dilution beforehand. The analysis is conducted at room temperature.

Results [10]

The analytical data obtained are summarized in Table I. Correlation studies between the present method (Y) and the other method (X) using serum samples from a hospital population gave the results as follows: For glucose in serum, $Y = 0.920X + 3.96$; $r = 0.987$; $n = 100$ (neocuproine

[10] J. Endo, M. Tabata, S. Okada, and T. Murachi, *Clin. Chim. Acta* **95,** 411 (1979).

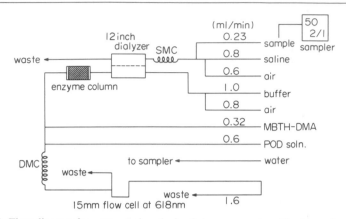

FIG. 2. Flow diagram for automated analysis of glucose or uric acid using an immobilized enzyme column reactor integrated into the AutoAnalyzer I.[10] For the glucose assay, an immobilized glucose oxidase-packed column and 50 mM acetate buffer, pH 5.6, are used; for the uric acid assay, an immobilized uricase-packed column and 50 mM borate buffer, pH 8.6, are used. MBTH, 3-Methyl-2-benzothiazolinone hydrazone; DMA, N,N-dimethylaniline; POD, horseradish peroxidase; SMC and DMC, single and double mixing coils.

method on SMA 12/60[11]). For uric acid in serum, $Y = 0.936X + 0.16$; $r = 0.984$; $n = 68$ (alkaline phosphotungstate method on SMA 12/60[12]). In these equations, r is the correlation coefficient, n is number of assays, and SMA 12/60 denotes a 12-channel Technicon analyzer.

In the present method, uric acid up to 0.20 g/liters did not interfere with glucose assays, and glucose up to 5.0 g/liter also had no interfering effect on uric acid assays. Samples containing bilirubin (0.20 g/liter) and creatinine (0.10 g/liter) did not interfere with either glucose or uric acid assays. However, the presence of 200 mg of ascorbic acid per liter of serum reduced the theoretical glucose values of 0.3 g/liter by 4% and 0.1 g/liter by 21%.

Comments

A minicolumn reactor with an immobilized enzyme can replace conventional chemical reagents or soluble enzyme reagents in an automated flow analysis system. Table II lists successful examples.[10,13-21] The storage stability of the immobilized enzyme column is usually very high: in

[11] D. L. Bittner and M. L. McCleary, *Am. J. Clin. Pathol.* **40,** 423 (1963).
[12] C. P. Paterl, *Clin. Chem.* **14,** 764 (1968).
[13] M. Tabata, C. Fukunaga, M. Ohyabu, and T. Murachi, *J. Appl. Biochem.* **6,** 251 (1984).
[14] T. Murachi, Y. Sakaguchi, M. Tabata, M. Sugahara, and J. Endo, *Biochimie* **62,** 581 (1980).

TABLE I
PREPARATION, PROPERTIES, AND APPLICATION OF IMMOBILIZED GLUCOSE OXIDASE
AND URICASE[a]

Preparation and use	Immobilized glucose oxidase	Immobilized uricase
Yield and activity		
Coupling yield		
%	45–50 (91–95)[b]	70–85
mg enzyme immoblized/g glass beads	13.5–15.0 (18.9–19.0)[b]	10.5–13.0
Specific activity relative to that of soluble enzyme (%)	37 (70)[b]	53
Apparent K_m (M)		
Batch method	5.0×10^{-4}	2.6×10^{-6}
Perfusion method	—	4.8×10^{-6}
Evaluation of analytical data		
Linear range (g/liter)[c]	0–5.0	0–0.15
Coefficient of variation (%)		
Within day	1.5	1.1
	(1.15 g/liter)[d]	(0.0415 g/liter)[d]
Day to day	3.0	2.2
	(1.12 g/liter)[d]	(0.0410 g/liter)[d]
Recovery (%)[e]	93–103	97–108

[a] From Endo et al.[10]

[b] Given in parentheses are values for glucose oxidase immobilized at its carbohydrate moiety.

[c] Substrate concentration in the original serum.

[d] Given in parentheses are mean values assessed by analyzing the same pooled sera in triplicate for 7 days.

[e] Standard solutions of glucose and uric acid to give the respective final concentrations of 1.1–3.4 and 0.07–0.14 g/liter were added to the pooled sera.

most cases the column can be stored in the cold for several months without loss of enzymatic activity. The operational stability is subject to variation depending on the nature of the enzyme immobilized. When used at room temperature, an immobilized glucose oxidase column lost only 20% of the original activity after daily use for a 3-month period with more

[15] M. Tabata, J. Endo, A. Hara, and T. Murachi, *South East Asian Pac. Congr. Clin. Biochem., 1st, 1979,* p. 150 (Abstr.) (1979).

[16] M. Tabata, T. Murachi, and H. Kusakabe, *Congr. Fed. Asian Oceanian Biochem., 3rd,* p. 115 (Abstr.) (1983).

[17] M. Tabata, J. Endo, and T. Murachi, *J. Appl. Biochem.* **3**, 84 (1981).

[18] Y. Sakaguchi, M. Sugahara, J. Endo, and T. Murachi, *J. Appl. Biochem.* **3**, 32 (1981).

[19] M. Tabata, T. Kido, M. Totani, and T. Murachi, *Anal. Biochem.* **134**, 44 (1983).

[20] M. Tabata and T. Murachi, *Biotechnol. Bioeng.* **25**, 3013 (1983).

[21] T. Murachi and M. Tabata, *Biotechnol. Appl. Biochem.* **9**, 303 (1987).

TABLE II
EXAMPLES OF IMMOBILIZED ENZYMES USED FOR CLINICAL ANALYSES

Assay	Enzyme	Source of enzyme	Method of immobilization	Type of enzyme reactor	Reference
Uric acid	Uricase	*Candida utilis*	Glutaraldehyde	Packed bed	10, 13
Glucose	Glucose oxidase	*Aspergillus niger*	Glutaraldehyde	Packed bed	10, 13
	Glucose oxidase	*Aspergillus niger*	Carbohydrate groups	Mixed bed or coimmobilized	14
	Peroxidase	Horseradish	Carbohydrate groups		
Ammonia	Glutamate dehydrogenase	*Proteus* sp.	Glutaraldehyde	Sequential	21
	L-Glutamate oxidase	*Streptomyces* sp.	Glutaraldehyde		
Urea	Urease	Jack bean	Glutaraldehyde	Packed bed or tube	15
	Urease	Jack bean	Glutaraldehyde	Sequential	16
	Glutamate dehydrogenase	*Proteus* sp.	Glutaraldehyde		
	L-Glutamate oxidase	*Streptomyces* sp.	Glutaraldehyde		
Cholesterol (total)	Cholesterol esterase	*Pseudomonas* sp.	Glutaraldehyde	Coimmobilized	17
	Cholesterol oxidase	*Nocardia erythropolis*	Glutaraldehyde		
Glutamate	Glutamate dehydrogenase	Beef liver	Glutaraldehyde	Packed bed	18
Lactate	Lactate dehydrogenase	Rabbit (or hog or beef) heart muscle	Glutaraldehyde	Packed bed	18
Creatinine	Creatinine deiminase	*Corynebacterium lilium*	Glutaraldehyde	Sequential	19
	Glutamate dehydrogenase	*Proteus* sp.	Glutaraldehyde		
Inorganic phosphorus	Pyruvate oxidase	*Pediococcus* sp.	Glutaraldehyde	Packed bed	20

than 2000 assays,[10] whereas an immobilized pyruvate oxidase column became 25% less active within 4 weeks under the similar conditions.[20]

Table II shows that in some cases two enzymes are used in the form of coimmobilized enzymes. This means that the procedures for immobilization were performed with a mixture of two enzymes in one solution. The immobilization of glucose oxidase or uricase with peroxidase and the use of one column having such coimmobilized enzyme glass beads were found to enhance the efficacy of the column reactor compared to the sequential use of the two respective enzyme columns.[14] For total cholesterol assay, the coimmobilization of cholesterol esterase (EC 3.1.1.13) and cholesterol oxidase (EC 1.1.3.6) was essential.[17]

In some other cases, the sequential use of two (or possibly more than two) immobilized enzyme columns is a clue to a problem which otherwise seems to be difficult to solve. For example, when creatinine in serum is to be determined by measuring ammonia produced by the action of creatinine deiminase, the endogenous ammonia in serum has to be removed

$$\text{Creatinine} + H_2O \xrightarrow{\text{creatinine deiminase}} N\text{-methylhydantoin} + NH_3$$

before the sample is introduced into the creatinine deiminase column. The attachment of a precolumn with immobilized glutamate dehydrogenase was found to be ideal for such purposes[19] (Fig. 3).

$$\alpha\text{-Ketoglutarate} + NH_3 + NADPH + H^+ \xrightarrow{\text{glutamate dehydrogenase}}$$
$$\text{glutamate} + NADP^+ + H_2O.$$

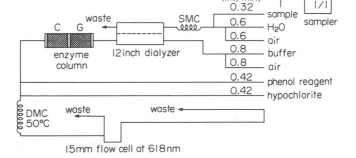

FIG. 3. Flow diagram for the automated analysis of creatinine using an immobilized creatinine deiminase–glutamate dehydrogenase column reactor integrated into an AutoAnalyzer I.[19] Buffer: 10 mM phosphate buffer, pH 7.5, containing 0.4 or 3.2 mM α-ketoglutaric acid and 0.4 or 3.2 mM NADPH. Immobilized enzyme columns: C, creatinine deiminase; G, glutamate dehydrogenase. SMC and DMC, single and double mixing coils.

Flow–Injection Analysis

Principle

Flow–injection analysis (FIA) is a nonsegmented flow system, which may reduce sample volume and shorten assay time.[22–24] An immobilized enzyme column reactor can also be integrated into FIA, and the use of a chemiluminescence detection system in combination has been found to improve greatly the sensitivity of the assay.[13] For example, a FIA system is designed in which hydrogen peroxide is produced according to the following reactions that take place in the respective oxidase columns:

$$\beta\text{-D-Glucose} + O_2 \xrightarrow{\text{glucose oxidase}} \text{D-gluconic acid} + H_2O_2$$

$$\text{Uric acid} + O_2 \xrightarrow{\text{uricase}} \text{allantoin} + CO_2 + H_2O_2$$

The resulting chemiluminescence emitted by the reaction of hydrogen peroxide with a mixture of luminol and potassium ferricyanide is determined by a luminophotometer[13] (Fig. 4).

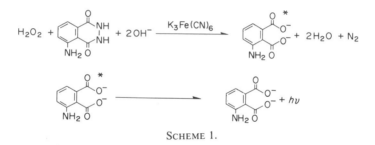

SCHEME 1.

Apparatus

A photomultiplier tube attached to the face of the flow-through cell of 96-μl capacity, 0.7×0.7 mm in dimension and 196 mm in length, is used for chemiluminescence measurements. A sample injector of a rotatory valve form (Fig. 4) or a microsyringe form is used to inject 1 μl of sample. The tubing used is 0.5 mm in inner diameter. A Chromatopack C-R2AX (Shimadzu Corporation, Kyoto, Japan) is used to record the luminescence response and to calculate the area and height of the peak appearing on the chart.

[22] A. Iob and H. A. Mottola, *Anal. Chem.* **52**, 2332 (1980).
[23] J. Ruzika and E. Hansen, *Anal. Chim. Acta* **114**, 19 (1980).
[24] K. K. Stewart, G. R. Beecher, and P. E. Hare, *Anal. Biochem.* **70**, 167 (1976).

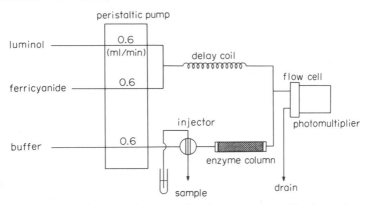

FIG. 4. A flow–injection analysis system for glucose or uric acid using an immobilized enzyme column reactor and chemiluminescence.[13] For glucose assays, immobilized glucose oxidase-packed columns and 10 mM phosphate buffer, pH 7.5, are used; for uric acid assays, immobilized uricase-packed columns and 10 mM phosphate buffer, pH 7.5, and 10 mM borate buffer, pH 8.5, are used.

Reagents

A 7.0 mM luminol stock solution is prepared by dissolving 1.24 g of luminol, 66.01 g of potassium hydroxide, and 61.83 g of boric acid in 1 liter of distilled water, allowing the solution to stand for 3 days to stabilize, and diluting 10-fold with distilled water before use (pH 10.9). A 200 mM potassium ferricyanide stock solution is prepared by dissolving 65.85 g of potassium ferricyanide in 1 liter of distilled water and diluting 10-fold with distilled water before use. The buffers used are 10 mM potassium phosphate, pH 7.5, and 10 mM potassium borate, pH 8.5.

Procedures

An immobilized enzyme column, 1.0 mm in inner diameter and 5 mm (for glucose oxidase) or 20 mm (for uricase) in length, is inserted in the FIA system (Fig. 4). The immobilized glucose oxidase column contains approximately 6.4 mg (wet weight) of glass beads that carries 0.12 mg of the enzyme, and the uricase column approximately 25.6 mg of glass beads with 0.47 mg of the enzyme. The analysis is performed at room temperature and at a speed of 120 samples/hr. The flow rate of luminol, ferricyanide, and the buffer solutions are all 0.6 ml/min. A delay coil of suitable length is introduced to cancel out the luminescence emitted transiently by admixing luminol and ferricyanide solutions only.

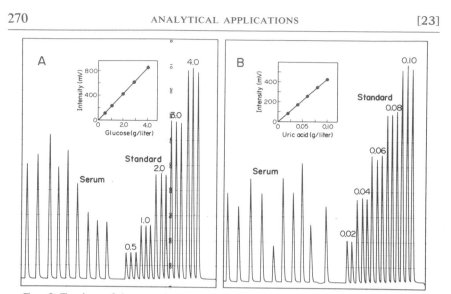

FIG. 5. Tracings of the record and plots for calibration (insets) for (A) glucose and (B) uric acid, in the determination by flow–injection analysis using an immobilized enzyme column reactor.[13] The materials were standard glucose solutions (0.5–4.0 g/liter) and human serum in A, and standard uric acid solutions (0.01–0.10 g/liter) and human serum in B. The rate of injection was 120 samples/hr for the standard glucose or uric acid solutions and 60 samples/hr for human serum.

Results

Figure 5 shows typical chart records of standard and human serum samples obtained for glucose and uric acid.[13] In both cases, linearity of the standard curve was obtained by determining the height as well as the area of the peak appearing on the chart: up to 4.0 g glucose/liter (Fig. 5A) and 0.1 g uric acid/liter (Fig. 5B). With serum samples, however, the peak height data were always 7–8% lower than the data for peak area, probably due to the difference in viscosity between the standard solution and serum samples which may affect the mode of their passage through a compact column.

Using a standard hydrogen peroxide solution, the lower limit of determination was found to be 10 pmol hydrogen peroxide/μl of sample. With pooled human serum samples, the glucose and uric acid assays gave coefficients of 0.9–1.0% for within-day and 1.0–1.5% for day-to-day variations. Ascorbic acid (50 mg/dl) gave only 5–7% negative interference.[13]

Correlation studies between the present method (Y) and the conventional enzymatic and colorimetric methods (X),[25,26] using serum samples

[25] P. Trinder, *Ann. Clin. Biochem.* **6**, 24 (1969).
[26] S. Meites, *Clin. Chem.* **19**, 675 (1973).

from a hospital population, gave the results as follows[13]: $Y = 0.95X - 0.008$; $r = 0.994$; $n = 30$ (for glucose) and $Y = 1.05X - 0.0002$; $r = 0.996$; $n = 25$ (for uric acid).

Comments

An immobilized enzyme column reactor is shown to be applicable not only to a conventional flow analyzer but to an FIA system. The above example combines chemiluminescence detection with the FIA technique and an immobilized enzyme column. One may have a variety of choices for other detection systems as well. With any kind of detection system, one advantage of using FIA is that since the longitudinal dispersion of the sample zone, which is run through the narrow-bore tubing (0.5 mm inner diameter) toward the detector at a high speed, is kept as small as possible, large signals can be obtained with minimum volume of the sample. Another advantage is the rapidness of the performance: only a 10-sec period is required for the system shown in Fig. 4 to complete one assay.[13] Thus, the method can be further applied to other enzymatic analyses that give hydrogen peroxide as a final signal (Table II), and it also should be suitable for a multichannel instrument which has to ensure high sensitivity and compactness.

[24] Use of Immobilized Enzyme Reactors in Flow–Injection Analysis

By MAT H. HO

Introduction

During the past decade, enzymes have become increasingly useful as analytical reagents due to their selectivity and ability to catalyze reactions of substrates at low concentrations. Enzyme-based analyses have been successfully applied to various compounds, and considerable attention has been given to the use of these reagents in flow–injection systems. Since no separation is involved in flow–injection analysis (FIA), the specificity of an assay can be achieved by using an enzyme that is specific for the analyte. Incorporation of the selectivity of an enzyme together with the simplicity, versatility, good precision, high sampling rate, low cost, and automation of FIA provides a useful system for a broad range of

METHODS IN ENZYMOLOGY, VOL. 137

applications, particularly for the analysis of large numbers of samples in clinical laboratories. Unfortunately, the applications of these systems for routine analysis suffer from the cost of enzymes. Purified enzymes suitable for analytical use are expensive, and large quantities of these reagents are consumed in the flow systems, making the cost of the analysis exceedingly high. The limited stability of enzymes in solutions may lead to poor reproducibility. These problems can be overcome by using immobilized enzyme reactors.

This chapter discusses the use of immobilized enzyme reactors in FIA. Several typical applications with different types of reactors, which were developed by other investigators and in our laboratory, will also be presented.

Immobilized Enzyme Reactors in Flow–Injection Analysis

Flow–Injection Analysis (FIA)

FIA is based on the concept of controlled dispersion of a sample zone when injected into a moving, nonsegmented carrier stream.[1] A precisely measured volume of sample is injected into a carrier stream, which may contain necessary reagents such as a cofactor, and then transported through the reactor where immobilized enzymes catalyzed the reaction of substrates to form product(s). The concentration of the analyte can be determined by measuring either the increase in the product concentration or the decrease in the cofactor concentration. Since the reaction is measured before steady-state conditions are established, FIA possesses the potential for high sampling rate, and the effect of interfering species, if any, can be minimized or avoided completely. This technique lends itself to the automation of a wide variety of enzyme-based analytical procedures and has proven to be simple, inexpensive, and highly versatile. Several on-line processes such as dialysis and diffusion can also be easily adapted to FIA, and the system can be interfaced with virtually any detector.

Dispersion is a phenomenon of great importance in FIA. While the sample moves from the injection port to the detector, it disperses into the carrier stream both longitudinally and radially by a combination of laminar flow and molecular diffusion. Dispersion of a sample zone has two negative effects: first, it makes the peak broader and therefore decreases the sensitivity and lowers the possible sampling rate; second, it

---

[1] J. Ruzicka and E. H. Hansen, "Flow Injection Analysis." Wiley, New York, 1981.

increases the dilution of the sample by the carrier, consequently decreasing the concentration to be measured. However, dispersion facilitates mixing of the analyte and reagent necessary for the chemical conversion. This mixing process is always incomplete, but because the dispersion pattern for a given FIA system is perfectly reproducible, FIA yields precise results. The reproducibility is also achieved by controlling sample injection volume and resident time. Dispersion is affected by several factors such as flow rate, reactor dimension and design, manifold dimension, and diffusion coefficient of the analyte. These parameters can be controlled easily to manipulate the dispersion which contributes significantly to the success of FIA. The responses of the detector are recorded in the form of sharp peaks. These peaks reflect both the chemical kinetics of the reaction and the physical dispersion that take place between the injection port and detection point. The peak height or the area under the curve can be used to determine the analyte concentration.

Immobilized Enzyme Reactors

In the design of an immobilized enzyme reactor for FIA applications, two factors that affect the performance of the reactor should be considered: the conversion efficiency of the analyte to product and the dispersion of the sample zone. It is desirable to have a reactor with high conversion efficiency, preferably close to 100%, and low dispersion. High conversion efficiency requires a fast conversion rate and long resident time. However, a long resident time may lead to an increase in dispersion. The conversion rate of a reactor depends on the rate of the enzyme-catalyzed reaction, which depends on the amount and the intrinsic kinetics of the immobilized enzyme, and the rate of mass transfer. In order to have a reactor with high enzyme activity, the total surface area available for enzyme immobilization and the enzyme activity per unit area should be as large as possible. It is therefore necessary to use a highly purified enzyme, a suitable support material, and a suitable immobilization technique. Impurities may compete with active enzyme for binding sites and thus decrease the amount of enzyme per unit surface area. The impurities may also interfere with the assay by decomposing the product, lower the selectivity by forming products which are sensed by the detector, or deactivate the enzyme.

The immobilized enzyme should be stable and concentrated into a small volume. The loading capacity, stability, and K_m value depend on the immobilization method. The glutaraldehyde coupling method has been successfully applied for several enzymes. However, no single coupling procedure is suitable for all enzymes. Since in FIA the resident time is usually not long enough to allow complete conversion and the detector signal seldom reaches the steady-state level, the conversion rate should be as fast as possible in order to obtain high sensitivity. The resident time depends on the desired degree of conversion and dispersion. In practice, these factors should be compromised with each other so that a reactor with adequate conversion efficiency and minimum dispersion can be achieved. The optimization of immobilized enzyme reactors for use in FIA was discussed in detail by Johansson et al.[2]

There are four types of immobilized enzyme reactors that can be used in FIA: open tubular reactors, whisker-walled open tubular reactors, packed-bed reactors, and single bead string reactors.

Open Tubular Reactors. In open tubular reactors, enzymes are immobilized to the inner surface of a nylon, plastic, or glass tubing. The chemical conversion takes place when the analyte and cofactor come into contact with the enzyme on the wall of the tube. Open tubular reactors offer better flow characteristics than packed-bed reactors. Open tubular reactors are usually long in order to provide sufficient enzyme activity and resident time for chemical conversion. The reactors can be coiled to produce secondary flow which will increase radial mass transport and reduce dispersion, thus improving their analytical usefulness. Covalent attachment of the enzyme onto an activated surface of a nylon tube is the most widely used method for immobilization.[3] Although open tubular reactors have been successfully applied in air-segmented continuous-flow systems, they are not well suited for FIA and so far very limited applications have been reported.

Whisker-Walled Open Tubular Reactors. Whisker-walled open tubular columns were developed for chromatography.[4,5] This interesting approach can be applied to immobilized enzyme reactors. The surface area available for enzyme immobilization can be increased significantly by growing whiskers (silicate filaments) on the inner wall of a glass tubing. Gaseous hydrogen fluoride, which can be conveniently generated at high temperature, is used for whisker formation. The silicate surface of the inner wall is then modified with aminoalkylsilyl groups, and the enzyme can be immobilized through glutaraldehyde linkage. Because oxygen may

[2] G. Johansson, L. Ögren, and B. Olsson, *Anal. Chim. Acta* **145,** 71 (1983).
[3] W. E. Hornby and L. Goldstein, this series, Vol. 44, p. 118.
[4] J. D. Schieke, N. R. Comins, and V. Pretorius, *J. Chromatogr.* **112,** 97 (1975).
[5] F. I. Onuska, M. E. Comba, T. Bistricki, and R. J. Wilkinson, *J. Chromatogr.* **142,** 117 (1977).

diffuse in and out of nylon or plastic tubings, a glass reactor is useful when the levels of oxygen consumed or produced are detected with an oxygen electrode. The reactor can also be coiled to improve mass transport and reduce dispersion.

Packed-Bed Reactors. Packed-bed reactors have found applications exclusively in FIA. Enzymes are immobilized onto the activated support, such as controlled-pore glass (CPG), nonporous glass, or Sepharose, and packed into the reactor. CPG is widely used owing to its mechanical strength and large surface area available for enzyme immobilization. Nonporous glass can also be used, although less surface area is available, to avoid peak spreading due to intraparticle diffusion of substrate and product into porous supports. Glass possesses an abundant supply of silanol which can be easily functionalized for covalent coupling. The immobilization procedure involves the preparation of alkylamino-activated glass and the immobilization of enzyme through the glutaraldehyde coupling method. Other materials, such as Sepharose, and other coupling methods can also be used for enzyme immobilization.

If packed-bed and open tubular reactors are compared, it becomes clear that packed-bed reactors are preferred in FIA applications. For two reactors with equal void volumes, packed-bed reactors provide several orders of magnitude more surface area than do open tubular reactors. Packed-bed reactors usually offer better conversion efficiency and lower dispersion. However, with appropriate design, open tubular reactors can be used successfully in FIA for samples which could clog the packed-bed reactors.

Single Bead String Reactors. Recently, single bead string reactors were developed,[6,7] and this interesting concept can be applied to immobilized enzyme–FIA systems. The reactor is constructed by packing a coiled glass or nylon tubing with a single string of glass beads with immobilized enzyme on both the wall of the tube and the beads themselves. This approach offers several attractive characteristics and significantly enhances the performance of the reactor. The dispersion and conversion efficiency in single bead string reactors were found to be much better than those of open tubular reactors with the same dimension, as reported by Mottola for penicillinase.[8] Furthermore, the relatively large support particles used in this type of reactor result in a smaller pressure drop than in packed-bed reactors.

---

[6] J. M. Reijn, W. E. Van der Linden, and H. Poppe, *Anal. Chim. Acta* **123,** 229 (1981).

[7] J. M. Reijn, H. Poppe, and W. E. Van der Linden, *Anal. Chem.* **56,** 943 (1984).

[8] H. A. Mottola, *Anal. Chim. Acta* **145,** 27 (1983).

Applications

Determination of Ethanol with Immobilized Alcohol
  Dehydrogenase Reactor

Alcohol dehydrogenase (EC 1.1.1.1) catalyzes the reaction of ethanol
and nicotinamide adenine dinucleotide (NAD^+) to produce NADH according to Eq. (1). If the concentration of NAD^+ is sufficiently high to

$$CH_3CH_2OH + NAD^+ \underset{}{\overset{\text{alcohol dehydrogenase}}{\rightleftharpoons}} CH_3CHO + NADH + H^+ \qquad (1)$$

keep the reaction pseudo-first order, the concentration of NADH produced is directly proportional to the original concentration of ethanol.
NADH can be detected photometrically, fluorometrically, or amperometrically. In this application, we describe a FIA system for ethanol which
was developed by Kojima et al.[9] This system employs a whisker-walled
open tubular reactor and a photometric detection method. The detailed
procedures for the preparation of whisker-walled enzyme reactor will also
be described.

Reagents. Alcohol dehydrogenase, NAD^+, ethylenediaminetetra-
acetic acid (EDTA), dithiothreitol (DTT), glutaraldehyde (25% aqueous
solution), sodium pyrophosphate, and monobasic sodium phosphate can
be purchased from Sigma (St. Louis, MO). Ethanol standards can be
prepared by diluting the ethanol standards obtained from Sigma with
pyrophosphate buffer (0.1 M, pH 8.5). Phosphate buffers are prepared
from monobasic sodium phosphate and adjusted to the desired pH with
0.2 M sodium hydroxide. Pyrophosphate buffer is prepared from sodium
pyrophosphate and adjusted to the desired pH with 2 M hydrochloric
acid.

Preparation of Reactor. PREPARATION OF WHISKER-WALLED GLASS
COLUMN. The whisker-walled glass column is prepared using the procedure developed by Schieke et al.[4] A coiled Pyrex glass capillary (0.25 mm
i.d., 10 m long) is cleaned thoroughly with an Extran solution (E. Merck,
Darmstadt, FRG), distilled water, and then ethanol. One end of the capillary is connected to a vacuum pump and the other end to a wider bore
glass tubing (5 mm i.d., 1.5 cm long), which is covered by a silicone
rubber septum, using heat-shrinkable Teflon connectors. After the system
is evacuated to 10^{-4} mmHg, the end of capillary attached to the pump is
sealed off by using a microflame. A suitable volume (equivalent to 5% of
the total column volume) of 2-chloro-1,1,2-trifluoroethyl methyl ether
(Columbia Organic Chem., Cassatt, SC) is injected into the capillary

[9] T. Kojima, Y. Hara, and F. Morishita, *Bunseki Kagaku* **32**, E101 (1983).

through the septum, and the end connected to the wider bore tubing is sealed off with a flame. The sealed capillary is then placed in an oven at 400°. At this temperature, the ether vaporized and liberated hydrogen fluoride necessary for whisker formation. In order to obtain an even whisker growth, the fluoroether should be distributed uniformly throughout the capillary and the temperature unchanged during the growth process. After 24 hr, the capillary is removed from the oven, cooled, opened at both ends, and then immediately flushed with hot nitrogen at 200°. A deposit of carbon on the inner surface, which gives the capillary a brownish or blackish appearance, is removed by flushing with oxygen at 450° for 12 hr.

Ammonium hydrogen fluoride can also be used for growing silica whiskers on the inner surface of a glass capillary as described by Onuska et al.[5] A coiled Pyrex glass is sealed at one end with a flame, filled with concentrated hydrochloric acid, sealed at the other end, and then placed in an oven overnight at 80°. The capillary is cooled and opened to remove acid, followed by sequential washes with distilled water, acetone, and diethyl ether. After drying with nitrogen, the capillary is sealed at one end and filled with a 5% (w/v) solution of ammonium hydrogen fluoride (Aldrich Chemical Co., Milwaukee, WI) in methanol. After standing for 1 hr, a uniform nitrogen flow (2 ml/min) is used to remove methanol until a milky film is observed. The other end is then sealed, and the capillary is placed in a furnace at 450° for 3 hr. The capillary is then cooled, opened at both ends (in a hood because of ammonia gas), washed with 20 ml of methanol, and dried with nitrogen.

IMMOBILIZATION OF ENZYME. To prepare the reactor, the whisker-walled capillary is cut to a length of 320 cm, and 50 ml of 2% (v/v) solution of 3-aminopropyltriethoxysilane (Petrach Systems, Bristol, PA) in acetone is circulated through it for 15 min at a flow rate of 8 ml/min. After emptying, the column is allowed to stand at 45° for 48 hr. Alcohol dehydrogenase is immobilized on the aminopropyl-derivatized wall of the glass column by using the glutaraldehyde linkage. A solution of 5% (v/v) glutaraldehyde, prepared by diluting a 25% glutaraldehyde solution with phosphate buffer (0.1 M, pH 7), is circulated through the derivatized glass column at a flow rate of 1 ml/min for 1 hr at room temperature. The tube is then washed 5 times with phosphate buffer (0.1 M, pH 7), 5 times with 0.5 M NaCl, and finally 5 times with phosphate buffer again. A solution of alcohol dehydrogenase (about 10–15 U/ml) in phosphate buffer (0.1 M, pH 8) is circulated through the column for 18 hr at 4°. In order to remove the unbound enzyme, the column is washed extensively with phosphate buffer and then with pyrophoshate buffer (0.1 M, pH 8.5) containing 0.1 M NaCl. When not in use, the reactor is filled with pyrophosphate buffer and stored at 4°. After use, the same washing and storing procedures are repeated.

Flow–Injection System. Figure 1 shows the FIA system for ethanol determination.[9] The system consists of a minipump (Model 45K15GK-A, Kyowa Seimitsu, Japan), a pulse damper, a sample injector (Model VL-611, Japan Spectroscopic, Japan), an immobilized alcohol dehydrogenase reactor, a UV spectrophotometer furnished with a micro flow cell of 1 μl volume and 0.3 mm optical path (Model UVIDEC-3, Japan Spectroscopic), and a recorder. Pyrophosphate buffer (0.1 M, pH 8.5) containing 2 mM NAD$^+$, 1 mM EDTA, and 0.1 mM DTT is used as the carrier solution. The flow rate is maintained at 0.46 ml/min, and 1 μl of sample is injected into the carrier stream each time. Calibration plots are obtained by using the peak heights.

Results. In this application, alcohol dehydrogenase was immobilized on the inner wall of a glass capillary, in which whisker was formed to increase the surface area. Although the calculated and experimental values for the height equivalent of a theoretical plate, H, did not agree well, it was found that the reactor with a larger internal diameter gives a higher H.[9] The H values for the whisker-walled reactor are about 1.2 times higher than those for the non-whisker-walled reactor with the same dimension. However, the peak heights for the former are about 3–5 times higher than that for the latter. The immobilized enzyme reactor gave about 70% of the response in peak heights as compared to those of soluble enzyme using an inactive reactor with same dimension and assay under same conditions (Fig. 2). For the reactor with a 0.25 mm i.d. and a 320 cm length, a residence time of 25 sec was obtained with a flow rate of 0.46 ml/min. The sampling rate of 120 samples/hr can be obtained with a peak resolution of 1.5 as shown in Fig. 2. The K_m value for the immobilized enzyme obtained from the Lineweaver–Burk plots is 15.7 mM as compared to 6.2 mM for soluble enzyme. Ethanol concentrations up to 20 mM can be determined with very good reproducibility. Relative standard deviations of 0.4–3.6% were obtained for different concentrations.

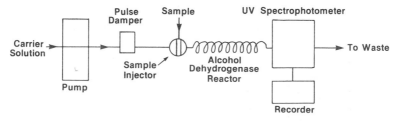

FIG. 1. Flow–injection analysis system for ethanol using the whisker-walled alcohol dehydrogenase reactor. From Kojima *et al.*,[9] with permission of the authors.

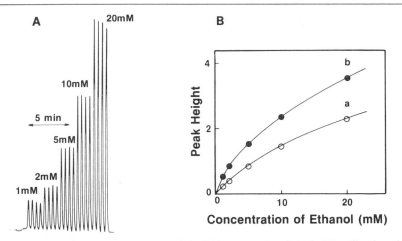

FIG. 2. (A) Typical response curves of the FIA system for ethanol; (B) calibration plots for ethanol using (a) the immobilized enzyme reactor or (b) soluble enzyme with an inactive reactor of the same dimensions as the immobilized enzyme reactor and assay under the same conditions. From Kojima et al.,[9] with permission of the authors.

Determination of Glucose with Immobilized Glucose Oxidase and Peroxidase Reactors

Recently a flow–injection system was developed in our laboratory for the determination of glucose.[10] Two enzymes, glucose oxidase (EC 1.1.3.4) and peroxidase (EC 1.11.1.7), were used to catalyze reactions (2) and (3). The liberated F^-, which can be used for the quantification of

$$\beta\text{-D-Glucose} + O_2 \xrightarrow{\text{glucose oxidase}} \text{gluconic acid} + H_2O_2 \qquad (2)$$

$$H_2O_2 + 4\text{-fluorophenol} \xrightarrow{\text{peroxidase}} F^- + \text{other products} \qquad (3)$$

glucose, is measured by a fluoride ion-selective electrode. In this application, both glucose oxidase and peroxidase are immobilized on CPG to provide packed-bed reactors for glucose analysis.

Reagents. Glucose oxidase (from *Aspergillus niger,* type VII), horseradish peroxidase (type VI), glutaraldehyde (25% aqueous solution), sodium chloride, acetic acid, and sodium phosphate were from Sigma. Sodium fluoride was purchased from Alfa (Danvers, MA), and 4-fluorophenol from Aldrich. A stock glucose solution (10 mg/ml) is prepared from anhydrous β-D-glucose (Sigma) in 1 g/liter benzoic acid (Aldrich) solution and allowed to mutarotate overnight before use. Standard glucose solutions for calibration are prepared by diluting the stock solu-

[10] M. H. Ho and T. G. Wu, manuscript in preparation.

tion with acetate buffer (0.5 M, pH 5). Acetate and phosphate buffers are prepared from glacial acetic acid and monobasic sodium phosphate, respectively, and adjusted to the desired pH with sodium hydroxide.

Preparation of Reactor. Both glucose oxidase and peroxidase are immobilized on aminopropyl-activated controlled-pore glass (AMP-CPG) using the glutaraldehyde coupling method. AMP-CPG (120–200 mesh, 57 nm pore diameter) was obtained from Electronucleonics (Fairfield, NJ). AMP-CPG can also be prepared as follows: One gram of CPG is added to 20 ml of 1% aqueous solution of 3-aminopropyltriethoxysilane, and the pH is adjusted to 3.5 with 6 M HCl; after the reaction is allowed to proceed at 75° for 2 hr, the glass beads are filtered and washed with 20 ml distilled water and then dried in an oven at 115° for 4 hr. AMP-CPG (100 mg) is incubated with 1 ml of 2.5% (v/v) glutaraldehyde in phosphate buffer (0.1 M, pH 7) for 3 hr at room temperature with the first hour under reduced pressure. The activated glass is washed 5 times with distilled water, 5 times with 0.5 M NaCl, and finally 5 times with phosphate buffer. For glucose oxidase immobilization, 300 μl of phosphate buffer (0.1 M, pH 7) containing 134 units of glucose oxidase is added to 30 mg of activated glass. For peroxidase immobilization, the same volume of phosphate buffer containing 800 units of peroxidase is used. The reaction is allowed to proceed at 4° for 18 hr followed by extensive washing with phosphate buffer (0.1 M, pH 7), 0.5 M sodium chloride, and phosphate buffer again. The enzyme-bound glass is packed into two separated reactors, one for glucose oxidase and the other for peroxidase, made of Teflon tubing (2 mm i.d., 25 mm long). These reactors are stored at 4° in pH 7 phosphate buffer when not in use. After use, these reactors should be washed extensively with phosphate buffer before storing.

Flow–Injection System. A schematic representation of the flow–injection system is shown in Fig. 3. Acetate buffer (0.5 M, pH 5) containing 0.5 M sodium chloride, 2.5 mM sodium fluoride, and 15 μM p-fluorophenol is used as a carrier solution and is delivered at a constant flow of 1.7 ml/min using a peristaltic pump (Buchler Instruments, Saddle Brook, NJ). A Rheodyne injection valve (Model 7010, Rheodyne Inc., Cotati, CA) modified with a bypass is used as a sample injector. The fluoride produced by the enzyme reactor, when a glucose sample is injected into the carrier stream, is detected by a combination fluoride ion-selective electrode (Model 96-09, Orion Research Inc., Cambridge, MA) which mounted in a Teflon flow-through cell. The electrode potential is measured with a 901 Orion Ion analyzer and recorded. The sample size is 140 μl. Since fluoride may react with glass, Teflon tubing (0.5 mm i.d.) must be used throughout the system.

Results. The flow–injection system was optimized, and the effects of the sample size, flow rate, and cosubstrate concentrations were investi-

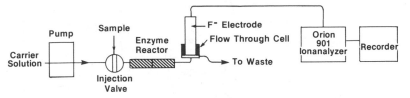

FIG. 3. Flow–injection analysis system for glucose determination using glucose oxidase and peroxidase reactors with potentiometric detection.

gated. At the optimal conditions, linear calibration plots were obtained from 5 to 400 μg/ml of glucose. For 20 replicate analyses of a 300 μg/ml glucose sample, a relative standard deviation of 0.76% was observed. After several months of storage at 4°, the activities of immobilized glucose oxidase and peroxidase were unchanged. These reactors can be used for several weeks without any noticeable reduction in response. For multienzyme systems, such as glucose oxidase and peroxidase in this application, the enzymes can be coimmobilized on the same support or immobilized on separated reactors. The major advantage for using separate reactors is that the immobilized enzymes can be operated and stored under their own optimal conditions. The system is also highly versatile. However, the axial dispersion becomes large if the net length of the reactors is too long. The potentiometric detection method, based on fluoride ion sensing, is highly specific for glucose and free of the interferences usually observed with the amperometric method using a hydrogen peroxide electrode. This system can also be incorporated to several other oxidases to extend the range of applications.

Determination of Ascorbic Acid with Immobilized Ascorbate Oxidase Reactor

In most of the FIA applications using immobilized enzyme reactors, the increase in product concentrations or decrease in some cofactor concentrations is measured by the detector. Another interesting approach has been developed in which the analyte is removed from the flow stream by immobilized enzyme before reaching the detector.[11] This approach was used for the determination of ascorbic acid in brain tissue in the presence of other oxidizable substances such as catecholamines and their metabolites. Ascorbate oxidase (EC 1.10.3.3) can be employed for the determination of ascorbic acid based on Eq. (4). In this application two reactors,

$$\text{L-Ascorbic acid} + 1/2\ O_2 \xrightarrow{\text{ascorbate oxidase}} \text{dehydroascorbic acid} + H_2O \qquad (4)$$

[11] C. W. Bradberry and R. N. Adams, *Anal. Chem.* **55,** 2439 (1983).

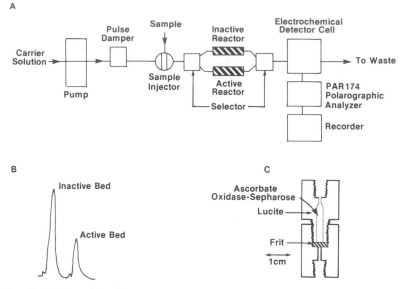

FIG. 4. (A) Flow–injection system for ascorbic acid determination using the ascorbate oxidase reactor; (B) amperometric signals from a mixture of ascorbic acid and other oxidizable substances injected through the inactive and active reactors; (C) detail of reactor design. From Bradberry and Adams,[11] with permission.

one inactive and the other active, were used. If a sample containing ascorbic acid and other electrooxidizable substances is first passed through an inactive reactor, both of them will be detected by an electrochemical detector, resulting in a large peak. If the same sample is now passed over an active reactor, ascorbic acid is selectively removed by immobilized ascorbate oxidase and the detector detects only electrooxidizable substances, resulting in a smaller peak as shown in Fig. 4B. The decrease in the observed signals is proportional to the concentration of ascorbic acid.

Reagents. Ascorbate oxidase was obtained from Boehringer Mannheim (Indianapolis, IN). An ascorbic acid (20 mM) stock solution is prepared by dissolving ascorbic acid (Sigma) in 0.1 M HClO$_4$ and diluting to 1 liter with distilled water. The working standard solutions are prepared by diluting the stock solution with the buffer used as the carrier (0.1 M phosphate buffer, pH 5.6, containing 1 mM EDTA). Glycine, sodium bicarbonate, and sodium phosphate for buffer preparations were obtained from Sigma.

Preparation of Reactor. Ascorbate oxidase is immobilized on the activated Sepharose using the procedure described by Bradberry *et al.*[12]

[12] C. W. Bradberry, R. T. Borchardt, and C. J. Decedue, *FEBS Lett.* **146,** 348 (1982).

CNBr-activated Sepharose 4B can be prepared as described earlier[13] or purchased from Pharmacia Fine Chemicals (Uppsala, Sweden). Three milliliters of CNBr-activated Sepharose gel and 3 ml of sodium bicarbonate buffer (0.1 M, pH 8) containing 510 units of ascorbate oxidase are placed into a small culture tube (1 × 6 cm). The tube is rotated slowly overnight at 4°, and the preparation is rinsed with 200 ml of distilled water on a glass frit over a vacuum. The preparation is returned to the tube, and 2 ml of 1.0 M glycine in sodium bicarbonate buffer (0.1 M, pH 8) is added, followed by rotating for 2 hr at room temperature. The preparation is then washed with 250 ml of distilled water on a glass frit over a vacuum again. The ascorbate oxidase-bound Sepharose is ready for use.

Two matched reactors are prepared from Teflon or Lucite as shown in Fig. 4C.[11] For the active reactor, ascorbate oxidase–Sepharose is loaded into the bed and held by a coarse glass frit located at the bottom of the bed. For the inactivated reactor, ascorbate oxidase–Sepharose is inactivated by boiling for 5 min before being packed into the bed.

Flow–Injection System. A schematic diagram of the flow–injection system is shown in Fig. 4A.[11] This system consists of a pump (Milton Roy minipump, Rainin, Woburn, MA), a damper, a sample injection valve (Rheodyne), an electrochemical detector, and a recorder. The electrochemical flow-through detector, which consists of a glassy carbon working electrode, an Ag/AgCl reference electrode, and a Pt auxiliary electrode, was obtained from Bioanalytical Systems (Lafayette, IN). The glassy carbon electrode is held at +0.8 V versus the Ag/AgCl reference electrode by using a PAR 174 polarographic analyzer (Princeton Applied Research, Princeton, NJ). Other electrochemical detectors such as BAS Model LC 4B (Bioanalytical Systems) can also be used. The current output is recorded on a recorder. Phosphate buffer (0.1 M, pH 5.6) containing 1 mM EDTA is used as carrier.

For calibration, 500 μl of ascorbic acid standards is injected and passed through the inactive reactor, followed by an identical injection through the active reactor. The response peaks of these two injections are recorded, and the difference in peak height (ΔI) is used for construction of a calibration curve. For analyses of ascorbic acid in brain tissues, 500 μl of the pretreated sample is injected into the reactors in the same manner as the standards.

Results. The optimum pH for immobilized ascorbate oxidase was found to be about 5.6,[12] and this pH was used in the FIA. Ascorbic acid up to 400 μM can be effectively removed by the immobilized enzyme reactor. With a pure ascorbic acid standard, a detection limit of 10^{-8} M was observed. The calibration plots for pure standard ascorbic acid and for

[13] P. Cuatrecasas, M. Wilchek, and C. B. Anfinsen, *Biochemistry* **61,** 636 (1968).

ascorbic acid plus dopamine are agreeable within experimental error. Even at a very low concentration of ascorbic acid and an equimolar concentration of dopamine, these two calibration curves never deviate by more than 5%. These results indicate that this system can be used to analyze ascorbic acid in the presence of catecholamines such as dopamine and their metabolites. For replicate analyses of rat brain tissue samples, a relative standard deviation of 1.3% was observed. Since the dispersion in the two beds must be identical in order to have reliable results, the two reactors should be matched.

Determination of NADH with Immobilized Bacterial Bioluminescence Enzymes

Firefly and bacterial bioluminescences have been successfully used to analyze numerous substrates at the picomole level.[14] The sensitivity and specificity of bioluminescence make it attractive for FIA applications. The enzymes NAD(P)H dehydrogenase (FMN) (EC 1.6.8.1) and luciferase (EC 1.14.14.3, alkanal monooxygenase), both isolated from *Beneckea harveyi,* can catalyze reactions (5) and (6). NAD(P)H and FMN

$$NAD(P)H + H^+ + FMN \xrightarrow{\text{NAD(P)H dehydrogenase (FMN)}} NAD(P)^+ + FMNH_2 \qquad (5)$$

$$FMNH_2 + O_2 + decanal \xrightarrow{\hspace{1cm} \text{luciferase} \hspace{1cm}}$$
$$FMN + decanoic\ acid + h\nu \quad (490\ nm) \quad (6)$$

are reduced nicotinamide adenine dinucleotide (phosphate) and flavin mononucleotide, respectively. The intensity of light emitted, which is proportional to the NAD(P)H concentration in the sample, is measured by a photomultiplier tube of a luminometer. By using these two reactions, it is possible to assay NAD(P)H, or any other compounds that can be coupled to the production or consumption of NAD(P)H, with high sensitivity and specificity.

The use of immobilized bacterial bioluminescence enzymes for flow–injection analysis of NADH at picomole levels was developed by Kurkijärvi et al.[15] In this application, NAD(P)H dehydrogenase (FMN) and luciferase are coimmobilized on Sepharose and packed into a reactor. The reactor is placed in front of the photomultiplier tube of a luminometer to provide a bioluminescent detector for NADH.

Reagents. The lyophilized bioluminescence reagent, which contained

[14] M. DeLuca and W. D. McElroy, "Bioluminescence and Chemiluminescence." Academic Press, New York, 1981.
[15] K. Kurkijärvi, R. Raunio, and T. Korpela, *Anal. Biochem.* **125,** 415 (1982).

NAD(P)H dehydrogenase (FMN), luciferase, buffer salts, FMN, and low concentrations of bovine serum albumin (BSA) and DTT, was obtained from LKB-Wallac (Turku, Finland; NADH monitoring reagent 1243 222). NAD(P)H dehydrogenase (FMN) and luciferase can also be purchased from Boehringer Mannheim. Decanal, FMN, NAD(P)H, BSA, DTT, potassium phosphate, and potassium chloride were purchased from Sigma.

Preparation of Reactor. The lyophilized bioluminescence reagent is first dialyzed to remove DTT which has been found to interfere with the immobilization. This reagent is dissolved in 1 ml of distilled water and dialyzed against 2 liters of phosphate buffer (0.1 M, pH 7) containing 20 μM FMN. After dialysis, 0.35 g of lyophilized or 1 ml of swollen CNBr-activated Sepharose is added to the dialysate and shaken gently for 20 hr at 4°. CNBr-activated Sepharose is prepared as described earlier[13] or obtained from Pharmacia. The enzyme-bound Sepharose is then washed with 150 ml of phosphate buffer (0.1 M, pH 7, containing 0.5 mM DTT), 200 ml of another phosphate buffer (0.1 M, pH 7, containing 0.5 mM DTT and 1 M KCl), and finally with 200 ml of the first buffer. The immobilized enzymes are then ready for use. When they are not in use, the immobilized enzymes are suspended in phosphate buffer (0.1 M, pH 7, containing 0.5 mM DTT) and stored at 2° in the dark.

The reactor consists of a glass tube (3 mm i.d., 7 mm o.d., 25 mm long) and two sintered plastic funnels as shown in Fig. 5. Two funnels are tightly fixed to the tube in order to provide a leak-free reactor. The sample holder of the luminometer (LKB model 1250, LKB, Bromma, Sweden) is modified to hold the reactor through its lower funnel and to give a channel for solution outlet. Enzyme bound to Sepharose is packed into the glass tube, and the reactor is placed just in the front of the photomultiplier tube. With this design, the reactor can be removed for storage or refilled with new immobilized enzymes.

Flow–Injection System. Figure 5 shows the diagram of the flow system for NAD(P)H measurement using the immobilized bioluminescence enzymes reactor.[15] The system consists of a pump, a sample injector, an enzyme reactor which is inserted in the measuring chamber of a luminometer, a luminometer, and a recorder. All of these components are similar to those described earlier for other systems, except the reactor and LKB luminometer. Phosphate buffer (0.1 M, pH 7) containing 0.1 M KCl, 10 μM FMN, 0.001% decanal, and 0.5 mM DTT is used as a carrier. This solution is kept in an ice bath and shielded from light. It is stable at least 12 hr at room temperature if kept in the dark. Before the carrier stream enters the pump, a heat-exchanger coil immersed in a water bath at 25° is used to bring the temperature to room temperature. The flow rate is

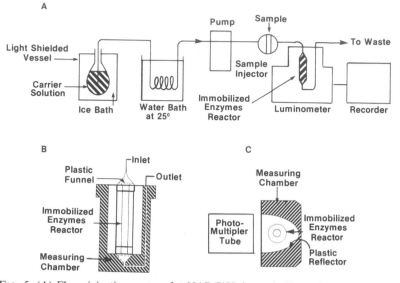

FIG. 5. (A) Flow–injection system for NAD(P)H determination using the immobilized bacterial bioluminescence enzyme reactor; (B) front and (C) top view of the enzyme reactor and bioluminescent detector. From Kurkijärvi et al.,[15] with permission.

maintained at 0.6 ml/min. The precisely measured volume (2–20 μl) of NAD(P)H sample is injected into the carrier stream and transported to the reactor where enzyme-catalyzed reactions occur to produce light. The light emitted is detected by the photomultiplier tube of a luminometer. After use, the reactor is washed with phosphate buffer (0.1 M, pH 7) containing 0.5 mM DTT and 0.1% BSA and stored at 2°. Before use, the reactor is also washed with the same buffer containing 0.1 M KCl instead of BSA and then equilibrated with the carrier solution.

Results. Sepharose was found to be a good support for the immobilization of bacterial luciferase and oxidoreductase.[15–17] By measuring the light emitted, 85% recovery of the activity was found with these two coimmobilized enzymes as compared to soluble enzymes in the same amounts. After 1 month of storage at 2° in phosphate buffer (0.1 M, pH 7) containing 0.5 mM DTT, the activities of these enzymes were unchanged. The reactor can be used for several weeks and up to 400 measurements without any change in sensitivity or accuracy. However, after around 4 days with about 80–100 measurements per day, the peak heights and flow rate began

[16] G. Wienhausen and M. DeLuca, *Anal. Biochem.* **127,** 380 (1982).
[17] L. J. Kricka, G. K. Wienhausen, J. E. Hinkley, and M. DeLuca, *Anal. Biochem.* **129,** 392 (1983).

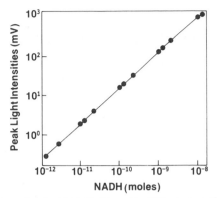

FIG. 6. Calibration plot for NADH. From Kurkijärvi *et al.*,[15] with permission.

to decrease; this decrease may be due to a disruption of the gel matrix. With the same amount of enzymes, the immobilized form allowed us to carry out at least 40 times more analyses during a period of several weeks as compared to soluble enzymes, which are stable for less than a day. Since luciferase may be subjected to product inhibition, the reactor should be extensively washed before storage. The ionic strength of the carrier solution should be high (0.2 M) to prevent the product or other substances from adsorbing on the derivatized Sepharose and therefore increasing the washing time. However, if the ionic strength is greater than 0.5 M, the intensity of emitted light decreases. The system gave good sensitivity over a wide range of linearity as shown in Fig. 6. NADH in the range from 1 pmol to 10 nmol can be determined with a sampling rate of 40 measurements/hr.

This system can be incorporated to several other NADH-dependent enzymes to extend the range of applications. For example, glucose-6-phosphate dehydrogenase or 7β-hydroxysteroid dehydrogenase can be coimmobilized with NAD(P)H dehydrogenase (FMN)/luciferase for the assay of glucose 6-phosphate or primary bile acids, respectively.[17] Recently, an open tubular reactor was developed for air-segmented continuous-flow analysis of NADH.[18] Both NAD(P)H dehydrogenase (FMN) and luciferase were coimmobilized onto the inner wall of a nylon coil, and NADH can be determined in the range 1–2500 pmol. The immobilized enzymes seem to be more stable, and this reactor may also be applied to FIA.

[18] S. Girotti, A. Roda, S. Ghini, B. Grigolo, G. Carrea, and R. Bovara, *Anal. Lett.* **17,** 1 (1984).

[25] Routine Analysis with Immobilized Enzyme Nylon Tube Reactors

By P. V. Sundaram

Immobilized enzymes offer some distinct advantages including cost reduction in the various ingenious applications that they lend themselves to by virtue of a variety of insoluble polymer supports of different chemical structures and physical forms that are available for immobilization. In particular, new analytical devices such as enzyme electrodes, enzyme immunoassay (EIA), and enzyme reactors have become popular for this reason.

In 1970 Sundaram and Hornby[1] showed that enzymes bound to the inside of nylon tubes may be used in flow-through analysis. Then Sundaram *et al.*[2] developed a variety of these "immobilized enzyme nylon tube reactors" (a generic name given by us), for assaying most of the commonly required and clinically relevant analytes such as blood urea,[3] uric acid,[4] glucose,[5] pyruvate and lactate,[6] creatinine and creatine,[7] triglycerides,[8] and cholesterol.[9] Extensive clinical trials including stability tests, cost analysis, and problem solving acceptable in laboratories for routine use were undertaken in the development of these tests.[2]

Principle of Operation

A typical reactor consists of a 1-m-long nylon tube (i.d. 1 mm) wound to form a coil 1 cm in diameter. The enzymes are covalently attached to the inside walls of the tube, and analysis is accomplished by perfusion of the samples separated by air bubbles, i.e., a segmented flow. In routine analysis the reactor is incorporated into the flow system of a Technicon

[1] P. V. Sundaram and W. E. Hornby, *FEBS Lett.* **10,** 325 (1970).

[2] P. V. Sundaram, *Enzyme Microb. Technol.* **4,** 290 (1982).

[3] P. V. Sundaram, M. P. Igloi, R. Wassermann, W. Hinsch, and K.-J. Knoke, *Clin. Chem.* **24,** 234 (1978).

[4] P. V. Sundaram, M. P. Igloi, R. Wassermann, and W. Hinsch, *Clin. Chem.* **24,** 1813 (1978).

[5] P. V. Sundaram, B. Blumberg, and W. Hinsch, *Clin. Chem.* **25,** 1436 (1979).

[6] P. V. Sundaram and W. Hinsch, *Clin. Chem.* **25,** 285 (1979).

[7] P. V. Sundaram and M. P. Igloi, *Clin. Chim. Acta* **94,** 295 (1979).

[8] W. Hinsch, W.-D. Ebersbach, and P. V. Sundaram, *Clin. Chim. Acta* **104,** 95 (1980).

[9] W. Hinsch, A. Antonijevic, and P. V. Sundaram, *J. Clin. Chem. Clin. Biochem.* **19,** 307 (1981).

AutoAnalyzer AA I or AA II. Dialyzed samples flow through the reactor at a predetermined flow rate, and the effluent is analyzed either colorimetrically or, if a cofactor is involved, by absorbance measurements at 340 nm. Reactors are washed, filled with a suitable buffer, and stored at 4° when not in use. Reactor coil length, enzyme concentration in the reactor, specific activity of the immobilized enzyme, flow rate, and in turn the residence time of the substrate in the reactor are factors that influence the turnover of a substrate in addition to pH, temperature, and buffer composition.

Testing and Optimization of Reactor Performance

Individual reactor performance is tested thoroughly before optimization of conditions since very often the kinetic properties such as the pH optima,[4,8,10] apparent K_m, and specific activity of enzymes change on immobilization. Sometimes these changes can be dramatic so that diffusive mass transfer can perturb the system and produce phase changes in the (v) versus (s) progress curve thus leading to a biphasic or triphasic curve.[10] It is critical to ensure that the standard curve for a given method of analysis falls entirely within one of these phases or segments or else routine analysis cannot be automated to give reliably reproducible results.

Routine Analysis

Normally 50–60 assays per hour are carried out in routine automated analysis. Individual flow diagrams for the analysis of the different analytes such as urea, glucose, and cholesterol using reactors containing the respective enzymes are different as shown in the publications.[2-14]

Methods of Immobilization

Nylon tubing (i.d. 1 mm) purchased from Portex Ltd. (Hythe, Kent, UK) is used to make reactors. Unless otherwise specified, reactors are made of 1-m-long tubes cut from the bulk supply, wound around a plastic rod 1 cm in diameter, and fixed in position with adhesive tape. Once the

[10] W. Hinsch and P. V. Sundaram, *Clin. Chim. Acta* **104**, 87 (1980).

[11] P. V. Sundaram, *J. Solid-Phase Biochem.* **3**, 185 (1978).

[12] W. Hinsch, A. Antonijevic, and P. V. Sundaram, *Z. Lebensm.–Unters. Forsch.* **171**, 449 (1980).

[13] W. Hinsch, A. Antonijevic, and P. V. Sundaram, *Clin. Chem.* **26**, 1652 (1980).

[14] W. Hinsch, A. Antonijevic, and P. V. Sundaram, *Fresenius Z. Anal. Chem.* **309**, 25 (1981).

immobilization procedure is complete the coiled tubing may be removed from the plastic rod, which retains its coiled structure. This coiling is essential to ensure turbulence during perfusion of substrates and thus enable a proper mixing of the substrate which thereby facilitates maximum contact of substrate with the immobilized enzyme molecules on the matrix.

Basically two approaches are used in immobilizing enzymes. Either the tube is first partially hydrolyzed before the COOH and NH_2 groups released are further activated to couple the enzyme, or nylon is directly O-alkylated by treatment with dimethyl sulfate (DMS) or triethyloxonium tetrafluoroborate (TTFB). This alkylation produces an imidate derivative of nylon which is very reactive and can be amidinated by reaction with NH_2-bearing compounds. Thus either an enzyme can be directly coupled to the matrix or a spacer may first be coupled followed by an enzyme (Fig. 1).

Enzyme Coupling to Hydrolyzed Nylon

Coiled tubing is filled with 3.6 M HCl, ends sealed after connecting with a piece of soft Tygon tubing, and covered with Parafilm. The tube is incubated in a water bath at 70° for 4–6 min depending on the extent of hydrolysis desired. It is then thoroughly washed starting with warm water and followed by cold water.

Cross-Linking with Glutaraldehyde. Hydrolyzed tube is washed with a $NaHCO_3$ or borate buffer (0.1 M), pH 9.4, and filled or perfused with a freshly prepared 1.25% (v/v) glutaraldehyde solution made up in either of these buffers of choice and allowed to react for 40 min. The tube is then washed with water followed by coupling buffer, usually phosphate buffer (0.1 M), pH 7–8, and filled with the enzyme solution. The average maximum coupling capacity being about 0.1 mg/m tubing by this method, a 2 mg/ml solution of enzyme protein is sufficient provided that the specific activity is around 20 U/mg.[1,15–17]

Cross-Linking with Bisimidates. Hydrolyzed tube is washed with borate buffer (0.1 M), pH 9.4, and then filled with a 4 mg/ml solution of dimethyl adipimidate or suberimidate made fresh. After reaction at room temperature for 2 hr the tube is washed once again with coupling buffer and filled with a 2 mg/ml enzyme solution made in a buffer at pH 6–9. Tris

[15] P. V. Sundaram and D. K. Apps, *Biochem. J.* **161,** 441 (1977).
[16] P. V. Sundaram, *in* "Biomedical Applications of Immobilized Enzymes and Proteins" (T. M. S. Chang, ed.), Vol. 1, p. 317. Plenum, New York, 1977.
[17] P. V. Sundaram, *in* "Enzyme Labelled Immunoassays of Hormones and Drugs" (S. B. Pal, ed.), p. 107. de Gruyter, Berlin, 1978.

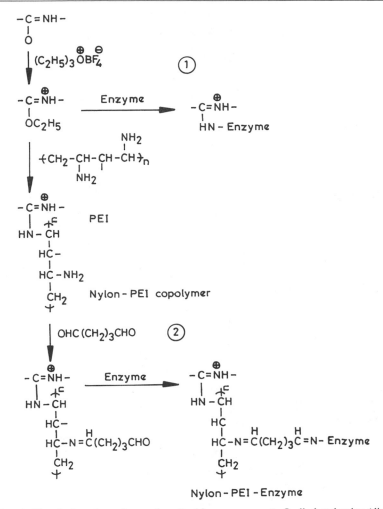

FIG. 1. Chemical rection schemes for attaching an enzyme to O-alkylated nylon (directly) and to nylon–PEI copolymer (indirectly).

buffer is avoided. Above pH 7 coupling is faster but so is the hydrolysis of the imidate. Thus a judicious choice of pH is made. After standing overnight at 4° the tube is washed well with a 0.1 M NaCl solution and then water. The reactor is filled with a suitable buffer and stored.

Coupling to COOH Groups Activated with Soluble Carbodiimides. Hydrolyzed tube is washed well with dry dimethylformamide (DMF) and perfused with 40 mM NHS made in dry DMF for a few minutes after which enough 1-ethyl-3-(3-dimethylaminopropyl)carbodiimide (EDAC) or

1-cyclohexyl-3[2-morpholinoethyl]-carbodiimide metho-p-toulene sulfonate (CMC) weighed out to give a final concentration of 20 mM is added in over a few minutes to the NHS solution ensuring good stirring simultaneously. The mixture is stirred well and perfused through the tube in a closed circuit for 90 min. The tube is then washed quickly with dry DMF and once with coupling buffer before filling with the enzyme solution made up in phosphate buffer, pH 7–8, and left overnight at 4°. The activation is conducted at room temperature. The next day the reactor is washed as usual and stored.

Coupling to Nylon after Alkylation

Nylon yields an imidate derivative on alkylation to which enzymes may be coupled directly or through a spacer such as a diamine, polylysine, or polyethylenimine (PEI). Enzymes are crosslinked to the NH$_2$ groups of the spacer with dialdehydes or bisimidates.

Alkylation with TTFB[18] requires milder conditions than with DMS[19] and is thus preferred to the latter. A 1-m-long coiled tube is filled with a 0.1 M solution of TTFB (supplied by Aldrich Chemical Co., Milwaukee, WI) in dichloromethane and after sealing the ends is allowed to react for 4 min at room temperature. The tube is then emptied into a safety flask by suction and flushed with 50 ml ice-cold methanol followed by ice-cold water. The tube which is now ready for coupling may be filled with an enzyme solution at 2 mg/ml in phosphate buffer (0.1 M), pH 7–9, and allowed to react overnight at 4°, or a spacer molecule is attached. However, kinetics of the coupling process shows that more than 60% couples in 30 min at room temperature.[20]

Spacers are coupled by filling the tube with either 0.1 M hexamethylenediamine or a 4 mg/ml solution of polylysine or a 0.6 M PEI solution (Serva GmbH., FRG, 10,000 MW) made up in NaHCO$_3$ buffer (0.1 M), pH 9.2. After coupling, the tube is washed well with 0.1 M NaCl followed by water and is now ready for coupling enzymes by cross-linking, the same conditions being used as before.[15,20]

Performance Characteristics

The performance characteristics and the statistical parameters of the performance of the various methods evaluated from clinical trials are given in Tables I and II, respectively. Exhaustive information is available

[18] D. L. Morris, J. Campbell, and W. E. Hornby, *Biochem. J.* **147**, 593 (1975).
[19] P. V. Sundaram, *Nucleic Acids Res.* **1**, 1587 (1974).
[20] P. V. Sundaram, *Biochem. J.* **183**, 445 (1979).

TABLE I

Performance Characteristics of Various Immobilized Enzyme Nylon Tube Reactors

Substance analyzed	Enzyme system[a]	Method of analysis	Optimum pH		Average no. of tests	Concentration range	Test of analyzer
			Free enzyme	Immobilized enzyme			
Urea	Urease	Berthelot reaction	7.0	7.0	2000	0–33 mM	AA I
Citrulline	Urease	Ehrlich reagent Different assay	7.0	7.0	—	—	AA I
Uric acid	Uricase	Peroxidase Pap test[b]	7.5–9.5	7.0	4000	20–120 mg/liter	AA II
			—	—	—	—	—
	Uricase/ALDH	NAD[c]	7.0/9.0	8.6	5000	—	AA II
	Uricase/ NADH Peroxidase	NADH	7.0/6.0	7.0	—	—	—
Pyruvate	LDH	NADH	7.0	8.0	4000 (min.)	0–1 mM	AA I/Eppnd 4412
Lactate	LDH/ALT	NAD	7.0/8.0	9.4	2000	0–10 mM	AA I/Appnd 4412
Glucose	Gluc-DH	NAD	7.0–7.6	6.8–7.6	3500	0–4000 mg/liter	AA II
	GOD/ALDH	NAD[c]	5.6/9.0	8.0	—	—	—
	GOD/ NADH Peroxidase	NADH	5.6/6.0	7.0	—	—	—
Glycerol	Gly-DH	NAD	8.5	10.0	3500	0–240 mg/liter	AA II
Triglycerides	Gly-DH	NAD[d]	—	10.0	2000	0–4000 mg/liter	AA II
ATP, PEP, NADH	PK/LDH	NADH	9.0/7.0	8.0	—	—	—
Creatinine	Creatininase/ CK–PK/LDH	NADH	8.0/9.0	8.0 8.5–9.0	—	—	—
Creatine	CK/PK–LDH	NADH	9.0	8.5–9.0	—	—	—
Adenosine, adenine	Adenine deaminase	264 nm	7.0–7.4	7.0	—	—	—
Adenine nucleotides	Alkaline phosphatase/ adenine deaminase	259 nm	10.0/7.0–7.4	8.0	—	—	—
Malathion, Parathion	Parathion hydrolase	400 nm	10.0	8.0–10.5	—	—	—

[a] ALDH, Aldehyde dehydrogenase; ALT, alanine aminotransferase (previously referred to as glutamate pyruvate transaminase, GPT); CK, creatine kinase; Gluc-DH, glucose dehydrogenase; Gly-DH, glycerol dehydrogenase; GOD, glucose oxidase; LDH, lactate dehydrogenase; PK, pyruvate kinase.

[b] PAP Test, peroxidase aminophenazone test.

[c] Catalase and ethanol added in solution.

[d] Lipase and esterase added in solution.

in these tables and only some special features of some of the tests are discussed below.

As already pointed out,[2] reduction in cost of testing is a feature common to all the methods using immobilized enzyme nylon tube reactors. We have employed single enzyme systems for the estimation of urea, uric acid, glucose, and pyruvate using urease,[3] uricase,[4] glucose dehydrogenase (Gluc-DH),[5] and lactate dehydrogenase (LDH)[6] reactors, respectively.

Creatinine and creatine[7] may be estimated using a creatininase–creatine kinase reactor connected to a pyruvate kinase (PK)–lactate dehydrogenase (LDH) reactor. Although all four enzymes may be immobilized in a single reactor, sufficiently high activity of the four enzymes for analysis at the rate of 50–60 per hour was not easily achieved with the enzymes available.

Hitherto triglycerides have been estimated by a chemical method, and now a method using glycerol kinase is commerically available. Our method uses glycerol dehydrogenase which acts on glycerol released from sera after the lipolytic enzymes lipase and esterase have acted on them. The method is very efficient and specific.[8] Immobilization shifts the pH optimum of the enzyme to pH 10.0 and also stabilizes it.[10] This high pH and excess NAD^+ are required for the enzyme to act on glycerol. Approximately 3500 tests are possible with each reactor.

Heterogeneous Multienzyme Systems

A novel approach to analysis is used in a method where yeast aldehyde dehydrogenase (ALDH) is used along with glucose oxidase (GOD), uricase, or cholesterol oxidase and catalase, the last mentioned in solution form, that together make heterogeneous multienzyme systems that are employed in the estimation of glucose,[13] uric acid,[14] or cholesterol.[19] Since these heterogeneous multienzyme systems comprise immobilized enzymes that are separated by an enzyme in solution, transport of substrate and products to and from the polymer surface is involved (Fig. 2), and this could limit the efficient functioning of the systems. However, use of a large excess of catalase and ethanol, which in the presence of H_2O_2 produces CH_3CHO for ALDH to act on, guarantees proper functioning of the systems. In the case of cholesterol, since cholesterol oxidase and cholesterol esterase are unstable on immobilization, the sera are exposed to these enzymes in solution before being passed through an ALDH reactor. In the same way glucose or uric acid may be assayed with an ALDH reactor provided the sample sera are first treated with GOD or

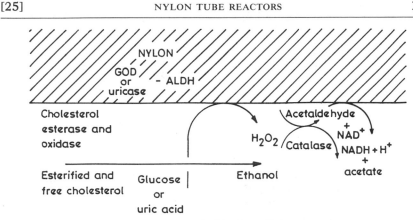

FIG. 2. Reaction schemes of heterogeneous multienzyme systems containing immobilized aldehyde dehydrogenase and other enzymes.

uricase prior to passage through the dialyzer in the AutoAnalyzer. An ALDH reactor is stable for 5000 tests and gives excellent performance (Table II).

Glucose Estimation with the GOD–ALDH Reactor

Having established that the optimum performance of the GOD–ALDH reactor is at pH 7.8,[13] the reactor is assembled as part of the flow circuit of a Technicon AutoAnalyzer II (Fig. 3). Only three different reagent solutions are used in routine analysis of glucose by this system. Buffer A is a potassium phosphate solution (0.1 M), pH 7.8, containing 0.3 g Brij-35 per liter, and Buffer B is the same buffer which in addition contains 1.5 M ethanol, 1 mM EDTA, and 1 mM mercaptoethanol. NAD$^+$ catalase solution is made in phosphate buffer (0.1 M), pH 7.5, containing 5 mM NAD and 1.5×10^6 U of catalase. This method is linear up to 4500 mg/liter glucose.

Uric Acid Estimation with the Uricase–ALDH Reactor

Using similar principles uric acid may be determined with a uricase–ALDH reactor,[14] analysis being carried out in a Tris–HCl buffer, pH 8.6. The flow diagram is slightly modified, and the NAD$^+$/catalase solution contains 10 mM NAD$^+$ and 6×10^6 U of catalase. The method is linear up to 120 mg/liter of uric acid.

TABLE II

STATISTICAL PARAMETERS OF THE PERFORMANCE OF VARIOUS REACTORS

Metabolite	Enzyme reactor	Method compared to	Y	r	CV (%) Within-day	CV (%) Day-to-day
Urea	Urease reactor	DAM (diacetyl monoxime)	0.96X + 1.65	0.998	1.3	3.3
		Urease solution	0.99X + 2.08	0.996		
Uric acid	Uricase reactor	Uricase/Pap solution	0.98X + 0.15	0.978	2.3–8.9[a]	3.1–6.4[b]
		Uricase/MBTH solution	0.87X + 0.69	0.972		
	ALDH reactor + uricase solution	Technicon SMA 12	0.992X − 1.78	0.99	1.5	4.0
	Uricase–ALDH reactor	Uricase/DMA–MBTH	6.95X + 2.17	0.985	1.6–2.9[c]	2.2–3.9[d]
		Technicon SMA12	1.067X + 2.22	0.998		
Glucose	Gluc-DH reactor	HK–G6PDH solution	0.98X + 1.99	0.996	3.2–2.1[e]	3.9–2.8[f]
	GOD–ALDH reactor	HK–G6PDH solution	1.0X − 2.03	0.997	0.6–1.4[g]	2.4–3.3[h]
	GOD solution–ALDH reactor	HK–G6PDH solution	1.02X − 2.45	0.997	1.0–1.1[i]	1.9–3.3[j]
Pyruvate	LDH reactor	LDH solution	1.081X − 2.5	0.997	3.3	5.0
Lactate	LDH–ALT reactor	LDH–ALT solution	0.971X − 0.04	0.993	3.1	4.9
Glycerol	Gly-DH reactor	GK–PK–LDH solution	1.02X + 0.017	0.996	1.7–2.7[k]	3.3–5.5[l]
Triglycerides	Esterase–lipase (solution) Gly-DH reactor	GK–PK–LDH solution	1.02X + 74.2	0.993	3.2–1.6[m]	4.9–3.0[n]
Cholesterol	Cholesterol esterase/cholesterol oxidase solution + ALDH reactor	Technicon SMAC Cholesterol oxidase–PAP	0.99X + 0.05	0.993	1.4–2.6[o]	1.7–3.0[p]

[a] 62–23 mg/liter.
[b] 91–43 mg/liter.
[c] 1.6–2.9 mg/liter.
[d] 2.2–3.9 mg/liter.
[e] 750–2100 mg/liter.
[f] 750–2100 mg/liter.
[g] 1030, 2320, and 840 mg/liter.
[h] 2580–840 mg/liter.
[i] 1030–2320 mg/liter.
[j] 2940–1030 mg/liter.
[k] 0.16–0.7 mM.
[l]
[m] 1040–3400 mg/liter.
[n] 1040–3400 mg/liter.
[o] 1.6–2.4 mM.
[p] 1.7–7.0 mM.

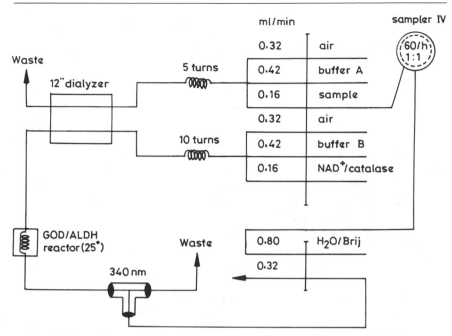

FIG. 3. Flow diagram for the determination of glucose using a GOD–ALDH reactor connected to a Technicon AutoAnalyzer AA II.

Summary

The basic strategy involved in the design, development, and application of immobilized enzyme nylon tube reactors for routine analysis is described in this chapter, touching on some of the attractive features of these methods. Extensive data (Tables I and II) and the references provide details which may be needed based on specific methods.

Acknowledgment

The work described herein was supported by grants from DFVLR and Deutsche Forschungsgemeinschaft. This article was written when the author was a visiting Professor at The Voluntary Health Services Centre, Adyar, Madras, India.

[26] Enzyme Sensors for Fermentation Monitoring: Sample Handling

By Sven-Olof Enfors and Neil Cleland

Introduction

Fermentation control is based on the interaction between the environment and the metabolism of the cell. The methods used to monitor fermentation involve measurement of physical and chemical parameters and application of mathematical models describing the reaction kinetics in relation to the physical and chemical parameters. Development of new sensors and techniques for application of sensors plays a key role in this field. A requirement is that the analysis should be on-line to offer a real-time analysis, thus focusing on sampling systems and *in situ* methods of analysis. The enzymatic analysis technique permits analysis in fermentation broths, like those of sugars and other organic components.

There are two main strategies in this technique: either the sensor is placed outside the reactor and the sample is transported to the sensor ("external enzyme sensor"), or it is placed *in situ* in the reactor. Advantages and disadvantages of either choice will be discussed here.

External Enzyme Sensors

The analyte is transported from the reactor to the sensor by means of a pump in one of three different ways: (I) In a whole-culture flow, (II) in a flow of filtered broth, or (III) in a flow of buffer after diffusion of the analyte from the culture liquid to the buffer. These principles are depicted in Fig. 1. In processes with immobilized cells, where an almost cell-free broth can be withdrawn continuously, method I is preferred. In this case most of the problems of on-line measurement can be avoided because no cell–analyte separation is needed. A system based on this principle has been described by Mandenius *et al.* for sucrose analysis and control in a continuous ethanol process employing an enzyme thermistor.[1]

If the sample flow is not free from cells, the transport tubings are likely to behave as continuous tube reactors where biological reactions during the transport change the sample composition. The kinetics of this reaction can be either of two extremes or of an intermediate type. If the reaction is dominated by growing cells suspended in the sample the expected con-

[1] C. F. Mandenius, B. Danielsson, and B. Mattiasson, *Biotechnol. Lett.* **3**, 629 (1981).

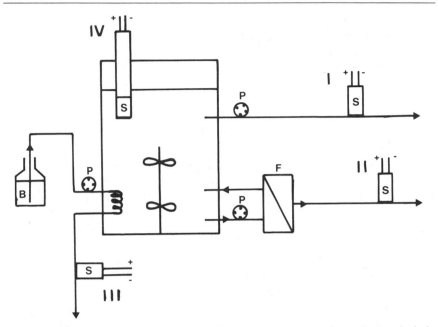

FIG. 1. Alternative methods of sampling to an enzyme sensor for on-line analysis in fermentation broth: (I) whole broth is pumped to the sensor; (II) filtrate from micro- or ultrafiltration is pumped to the sensor; (III) buffer is pumped through a dialysis cell in the fermenter and further to the sensor; (IV) the sensor is introduced into the broth. S, Enzyme sensor; P, pump; D, dialysis cell; F, ultra- or microfilter unit.

centration of an analyte, which is consumed by the cells, can be expressed as

$$S_0 = S + X\tau(\mu/Y + m) \tag{1}$$

where S_0 is the analyte concentration in the reactor, S the observed analyte concentration at the sensor, X the cell concentration, τ the mean residence time of sample in the tubing, μ the specific growth rate of the organisms, Y the yield coefficient (cell/substrate), and m the maintenance coefficient of the cell with respect to the analyte (mainly applicable when the analyte is used as an energy source). A corresponding equation can be deduced for an analyte which is a product of the microbial reaction.

Typical values for $(S_0 - S)/\tau$ for glucose during an *Escherichia coli* fermentation varies from 0.01 g/liter/min in the beginning to several grams per liter per minute at the end of a batch process. This emphasizes the importance of fast sample transport from the reactor to the sensor if this mode of operation is used.

Even if the sample is almost cell free, but nonsterile, cells are likely to grow and attach to the inner surfaces during transport.[2] After some time the microbial film can cause considerable changes as the medium is transported through the tubing. If a plug–flow behavior is assumed the kinetics of the reaction for a consumed analyte (S) is given by

$$\Delta S - K_m \ln(1 - \Delta S/S_0) = V_{max}\tau \qquad (2)$$

where ΔS is the change of analyte concentration during transportation, K_m the Michaelis constant for the reaction, S_0 the analyte concentration in the reactor, V_{max} the maximal reaction rate capacity of the cells during unrestricted conditions, and τ the residence time of sample in the tubing.

K_m is mainly controlled by the diffusion hindrance for the analyte at the film surface. Adhesion of microorganisms to surfaces have been described in several reviews,[2–4] and the metabolic activity of these layers, with thicknesses in the range of tenths of micrometers, can be considerable.[2,5]

Thus, when designing a sample system for reactors with externally located sensors these aspects must be taken into consideration. This technique (I in Fig. 1) is the most straightforward in process monitoring if the reactions described in Eqs. (1) and (2) are negligible. Furthermore, only this technique permits simple calibration by addition of standards.

One way to avoid biological reactions during transport to the sensor is to remove and work with a sterile sample. This can be done either by micro- or ultrafiltration (II in Fig. 1) or by dialysis (III in Fig. 1). If filtration is used precaution must be taken to avoid buildup of a cell layer on the surface of the filter which would work as a biologically active filter during further filtration. The cross-flow filtration technique offers a means to reduce the cell adsorption. However, little information is available on the function of cross-flow filtration in combination with analysis of the permeate. A system of this type—but with a magnetic stirrer instead of cross-flow as a method to reduce the polarization of the filter—has been described.[6] However, no data from process conditions were presented. Unfortunately, ultrafiltration and microfiltration techniques still suffer from considerable concentration polarization of the membrane which causes drift of the flux of components through the membrane.

Dialysis (III in Fig. 1) can also be used to obtain a sterile sample for

[2] B. Atkinson and H. W. Fowler, Adv. Biochem. Eng. 3, 221 (1974).
[3] D. C. Ellwood, J. Melling, and P. Rutter (eds.), "Adhesion of Microorganisms to Surfaces." Academic Press, New York, 1979.
[4] R. C. W. Berkely, J. M. Lynch, J. Melling, P. R. Rutter, and B. Vincent, "Microbial Adhesion to Surfaces." Horwood, Chichester, England, 1980.
[5] I. Nilsson, Ph.D. thesis. University of Lund, Lund, Sweden, 1984.
[6] G. Chotani and A. Constantinides, Biotechnol. Bioeng. 24, 2743 (1982).

transport. In this case the analyte concentration gradient controls the rate of transport of the analyte through the membrane not the pressure gradient:

$$N = [DA(S_0 - S)]/d \qquad (3)$$

where N is the analyte flux through the membrane, d the dialysis membrane thickness, D the diffusion coefficient, A the membrane area, S_0 the concentration of analyte in the reactor, and S the concentration of analyte in the buffer.

The analyte is washed off the dialysis membrane by the buffer and transported to the sensor. The concentration of analyte in the buffer depends also on the buffer flow rate (F):

$$S = N/F \qquad (4)$$

which offers the possibility of extending the linear response range by increasing F. An important advantage of this technique is that the environment at the enzyme sensor can be controlled by selecting a suitable dialysis buffer.

Since there is no pressure gradient which forces materials through the membrane, the tendency to build up a cell layer, as in ultra- or microfiltration, is reduced. However, active attachment and growth of organisms on the membrane can completely change the values of the parameters d and D in Eq. (3), which may require frequent recalibration of the system. This principle was applied for monitoring of glucose during an *E. coli* cultivation.[7]

In Situ Sensors

In situations when sample transport causes problems, the positioning of the sensor in the bioreactor would be an alternative. This has been applied for on-line monitoring of penicillin[8] and glucose.[9] However, some problems arise: (1) The sensor must be sterilizable. (2) The mode of operation does not permit treatment of the sample or conversion of the sample to a suitable environment, with respect to pH, buffer capacity, or oxygen concentration. (3) Recalibration during a process is difficult.

Sterilization. Several methods to sterilize the probe have been suggested: (1) disinfection by chloroform,[8] (2) injection of a suspenion of the enzyme into the autoclavable body of the enzyme electrode after auto-

[7] N. Cleland and S.-O. Enfors, *Anal. Chim. Acta* **163**, 281 (1984).

[8] J. W. Hewetson, T. H. Jong, and P. P. Gray, *Biotechnol. Bioeng. Symp.* **9**, 125 (1979).

[9] N. Cleland and S.-O. Enfors, *Eur. J. Appl. Microbiol. Biotechnol.* **18**, 141 (1983).

claving,[10] and (3) introduction of the enzyme electrode, which is disinfected by, e.g., glutaraldehyde, into an autoclavable housing with an organism-impermeable membrane separating the enzyme from the broth.[9] When using the last principle it is important that the enzyme is also sterile since the sensor signal will start drifting as a result of microbial changes of pH, oxygen tension, etc. when the sensor is subjected to a growth-stimulating sample. This drift is not observed as long as experiments are performed in buffers.[7]

There are several problems caused by the inability to adjust the sample composition when enzyme probes are used *in situ* in the reactor. The two most important will be treated here.

Oxygen Effects. Variations of the dissolved oxygen tension in the sample may infuence the signal from oxygen-dependent sensors. Reduced oxygen tension also limits the linear concentration range.[9] Two principles have been utilized to overcome this limitation. One involves a momentary saturation of the enzyme with oxygen in a matrix of collagen in which oxygen has a high solubility and subsequent analysis according to the dynamic method.[11] However, this principle cannot be used for *in situ* analysis, but it does solve the problem of analysis with oxygen-dependent enzyme electrodes in oxygen-free samples, if the sensor is arranged as an external sensor.

In the oxygen-stabilized enzyme electrode system all oxygen that is consumed by the enzyme reaction is replenished by electrolytic decomposition of water to keep the dissolved oxygen tension in the enzyme probe constant.This is achieved by immobilizing D-glucose oxidase from *Aspergillus niger* (EC 1.1.3.4), catalase of beef liver (EC 1.11.1.6), and BSA on a 8 mm diameter Pt gauze of 24 mesh. The Pt disk is dipped into the polymerizing enzyme solution containing 2.5 mg glucose oxidase (157 units/mg), 2.5 mg catalase (12,925 units/mg), 20 mg BSA, 0.95 ml 25 mM phosphate buffer (pH 7), and 0.1 ml 2.5% glutaraldehyde and then placed in a refrigerator overnight for polymerization. This enzyme disk containing about 0.4 mg glucose oxidase is attached to the surface of an oxygen electrode and covered with a regenerated cellulose dialysis membrane (24 Å, Union Carbide). The reduction of the oxygen electrode signal caused by glucose is used to control electrolytic decomposition of water at the Pt gauze surface at a rate which maintains the dissolved oxygen tension at the enzyme surface constant. This principle is shown in Fig. 2.[9,12]

A glucose electrode of this type has been used to control glucose in

[10] S.-O. Enfors and H. Nilsson, *Enzyme Microb. Technol.* **1**, 260 (1979).
[11] J. L. Romette, B. Froment, and D. Thomas, *Clin. Chim. Acta* **95**, 249 (1979).
[12] S.-O. Enfors, *Enzyme Microb. Technol.* **3**, 29 (1981).

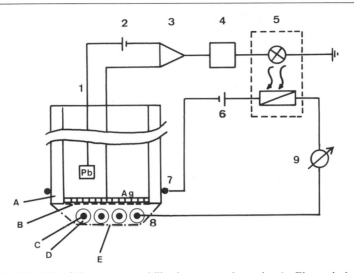

FIG. 2. Principle of the oxygen-stabilized enzyme electrode. A, Electrode house; B, membrane of oxygen electrode; C, Pt wire gauze; D, immobilized enzyme; E, electrode membrane (a dialysis membrane); 1, oxygen electrode anode/cathode; 2, reference voltage; 3, amplifier; 4, proportional, integrating, and derivating (PID) controller; 5, current controller; 6, electrolysis voltage source; 7, electrolysis cathode; 8, electrolysis anode (= C); 9, microampere meter (= enzyme electrode signal).

glucose-fed batch cultures.[9] By introducing the oxygen stabilization principle the linear response range can be increased considerably, and the effect of variations in oxygen concentration of the sample is reduced. A special feature of the oxygen-stabilized electrode is that the oxygen concentration of the enzyme can be set at any value (up to saturation), and can be utilized to further extend the lower response range.[9]

When this sensor operates in an oxygen concentration that is higher than the ambient oxygen concentration, some oxygen is lost by diffusion from the enzyme to the sample. A correlation equation, which can be applied on-line, has been developed[13]:

$$S = (I - k_D\Delta DOT)k_G^{-1} \qquad (5)$$

where S is the analyte concentration (g/liter), I the electrode signal (μA), k_D the diffusion constant (μA/DOT unit), ΔDOT the dissolved oxygen tension difference between enzyme and sample (arbitrary units), and k_G the electrode sensitivity to glucose at $\Delta DOT = 0$ (μA liters/g). k_G is

[13] S.-O. Enfors and N. Cleland, in "Chemical Sensors" (T. Seiyama, K. Fueki, J. Shiokawa, and S. Suzuki, eds.), p. 672. Elsevier, Amsterdam, 1983.

determined from the response to a known addition of glucose to the reactor, and k_D is determined from the response to a stepwise change in the dissolved oxygen tension which can be applied at any time during the process.

pH Effects. Problems related to variations in pH and buffer capacity of the broth during *in situ* monitoring have been studied mainly during penicillin analysis.[8,10,14] The main interactions can be summarized as follows: The pH-based enzyme probe responds to a change of pH which is caused by the enzyme reaction. The extent of pH change depends not only on the rate of liberation (or uptake) of protons in the enzyme reaction but also on the buffer capacity which itself is a function of pH.

The possibility of deriving a correlation function for on-line analysis in a penicillin process has been discussed.[13] According to the literature data, the sensitivity of a penicillin electrode depends on the buffer capacity described by a hyperbolic function [Eq. (6)][8,15]:

$$(k_{pc})^{-1} = a\beta - b \tag{6}$$

where k_{pc} is the electrode sensitivity (pH units/mM), β is the buffer capacity of the broth (mM/pH unit), and a and b are constants. a has been determined to 2080, and b was 2.3 in corn steep liquor-based penicillin production medium.[15]

Since β can be evaluated from pH control data during the process, a correlation for changes in buffer capacity can be obtained:

$$S = \Delta pH_e/[aB/(\Delta pH - b)] \tag{7}$$

where ΔpH_e is the differential pH between enzyme and broth (i.e., the electrode response), ΔpH the pH response of the broth after an addition of B mM H^+ (or OH^-) during a pH control operation.

Externally Buffered in Situ Sensors. Table I shows some characteristics of externally located and *in situ* sensors. The main problems with the *in situ* sensor are calibration and the necessity to operate at ambient conditions. The latter problem can be overcome by using external sensors since the sample can be changed by the choice of buffer (III, Fig. 1) or by utilizing the flow–injection analysis[16] principle for analysis according to method I or II.

The "externally buffered enzyme electrode"[17] was developed to combine the advantages of *in situ* operation with selection of buffer type and

[14] H. Nilsson, K. Mosbach, S.-O. Enfors, and N. Molin, *Biotechnol. Bioeng.* **20,** 527 (1978).
[15] S.-O. Enfors and N. Molin, *Proc. Eur. Congr. Biotechnol., 1st,* p. 35 (1978).
[16] J. Růžiča and E. H. Hansen, *Anal. Chim. Acta* **114,** 19 (1980).
[17] N. Cleland and S.-O. Enfors, *Anal. Chem.* **56,** 1880 (1984).

TABLE I
CHARACTERISTICS OF THE SAMPLING METHODS DESCRIBED (FIG. 1)

	Sampling principle			
Characteristic	Direct flow (I)	Filtration (II)	Dialysis (III)	*In situ* (IV)
Calibration	Easy	Only the enzyme, not the membrane		Difficult
Optimization of enzyme environment	By adding an additional flow	By selecting buffer		External buffering
Transport reactions	Can be estimated according to Eqs. (1) and (2)	By growth on membranes; must be checked by calibration		
Sample consumption	Considerable	Negligible		

sample dilution factor obtained with external sensors. The design of this electrode is shown in Fig. 3. It is a refinement of the oxygen-stabilized glucose electrode described above. The enzymes are immobilized as mentioned previously and the electronic circuits shown in Fig. 2 are utilized for the externally buffered electrode. To facilitate mixing in the diffusion chamber of the electrode, two nylon net spacers (15 mesh, Monyl HD,

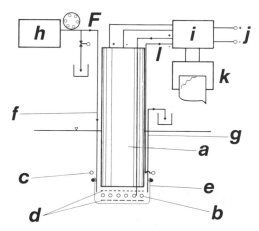

FIG. 3. Main parts of the externally buffered enzyme electrode: a, oxygen electrode; b, Pt gauze with immobilized enzymes; c, Pt coil (cathode); d, nylon nets; e, dialysis membrane; f, incoming buffer stream; g, buffer effluent; h, buffer reservoir; i, PID controller; j, reference potential; k, recorder; I, electrolysis current; F, buffer flow. From Cleland and Enfors,[17] by permission of the American Chemical Society.

ZBF, Zürich, Switzerland) are placed on each side of the enzyme net as shown in Fig. 3. The buffer is fed to and from the enzyme chamber through 1 mm (i.d.) stainless steel syringes connected to Tygon tubing. Low-pulsing pumps (e.g., Multiperpex, LKB, Bromma, Sweden) are used to feed the buffer at rates varying from 0.004 to 1.15 ml/min. Formation of air bubbles in the buffer flow system can create problems in long-term experiments. In that case the buffer should be air-saturated in its reservoir at 60° which reduces the oxygen content to about half that at 25°.

With this electrode the linear response range for glucose analysis can be extended from about 0–10 g/liter at low flow rates to about 0–150 g/liter at high flow rates as demonstrated in Fig. 4. This is of course an effect of the dilution of the intraelectrode concentration of glucose.

Another advantage of the externally buffered electrode is its reduced sensitivity to detrimental composition of sample. This has been most clearly demonstrated by continuous operation at pH 2.0, an environment that irreversibly inactivates an *in situ* glucose electrode of conventional design.[17] The intraelectrode pH is the parameter that determines the stability in such an environment, and therefore the performance of the electrode is strongly dependent on the capacity and flow rate of the buffer.

The stability of the externally buffered enzyme electrode is strongly dependent on the stability of the buffer flow rate. Drifting at a rate of 5%

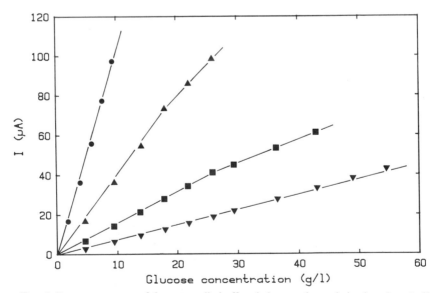

FIG. 4. Response curves of the externally buffered glucose electrode in phosphate buffer at different flow rates: (●) 0.042 ml/min, (▲) 0.095 ml/min, (■) 0.18 ml/min, and (▼) 0.28 ml/min. From Cleland and Enfors,[17] by permission of the American Chemical Society.

per 24-hr period is common and mainly caused by corresponding drift of the flow.

Calibration

Recalibration of sensors to correct for changes in the permeability of the analyte through the membrane between enzyme and sample (*in situ* sensors, method IV) or between sample and buffer flow (methods II and III) remains a major obstacle for on-line analysis. The problem is also common for other electrochemical sensors such as oxygen and pH electrodes used in fermentation processes. It is a well-known fact that the problems can be reduced by proper choice of electrode materials and position of the electrode in the reactor. A general solution to the calibration problem of *in situ* sensors would be to utilize the "retractable probe," a principle utilized by Ingold AG (Zürich, Switzerland) for their pH electrodes.[18] This technique was developed to enable exchange of electrodes during an operation, but it could be utilized for recalibration of *in situ* probes.

[18] H. Bühler and W. Ingold, *Process Biochem.* **11**, 19 (1976).

[27] Enzyme Thermistors for Process Monitoring and Control

By Carl Fredrik Mandenius and Bengt Danielsson

The usefulness of the enzyme thermistor as an on-line sensor for monitoring of bioreactors has been demonstrated in several publications.[1–5] In these studies monitoring was often combined with control of the substrate and/or the product concentrations in the reactors used. Continuous moni-

[1] B. Danielsson, E. Rieke, B. Mattiasson, F. Winquist, and K. Mosbach, *Appl. Biochem. Biotechnol.* **6**, 207 (1981).

[2] C. F. Mandenius, L. Bülow, B. Danielsson, and K. Mosbach, *Appl. Microbiol. Biotechnol.* **21**, 135 (1985).

[3] B. Danielsson, B. Mattiasson, R. Karlsson, and F. Winquist, *Biotechnol. Bioeng.* **21**, 1749 (1979).

[4] C. F. Mandenius, B. Danielsson, and B. Mattiasson, *Acta Chem. Scand.* **34B**, 463 (1980).

[5] C. F. Mandenius, B. Danielsson, and B. Mattiasson, *Biotechnol. Lett.* **3**, 629 (1981).

toring of glucose and cellobiose in a dialyzate from a cellulosic hydrolysis was performed with an on-line enzyme thermistor arrangement,[1] sucrose[2] and lactose[3] conversion in plug–flow enzyme reactors were monitored and controlled by continuous measurement of glucose, and ethanol fermentation with immobilized yeast was controlled by monitoring the sucrose substrate.[4,5] Examples will be given here of sucrose/glucose monitoring of sucrose hydrolysis and cellobiose monitoring of a cellulose hydrolyzate.

Monitoring and Control of Sucrose Hydrolysis

Using on-line sampling methods presented in this volume, biosensor monitoring systems can be designed for glucose and sucrose analysis with enzyme thermistors. Figure 1 shows an enzyme reactor with invertase immobilized to Sepharose. The reactor effluent is monitored continuously with a glucose and a sucrose enzyme thermistor. The thermistor signals are used to control the influent sucrose concentration by setting the mixing ratio of two peristaltic pumps, one for concentrated sucrose and the other for buffer solution.

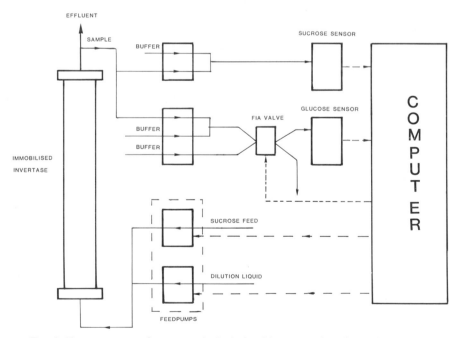

FIG. 1. Enzyme reactor for sucrose hydrolysis with enzyme thermistors for glucose and sucrose on-line monitoring and control. (From Ref. 2.)

Sucrose Monitor

The effluent sample is withdrawn from the bioreactor in a thin silicone tubing (0.5 mm i.d.) and at a high flow rate (10 ml/min) to a degasser cup for removal of gas and to speed up the sampling. A flow of 0.1 ml/min is collected from the degasser for sucrose determination. This stream is diluted with 0.1 M citrate buffer (pH 4.6), using a flow rate of 0.9 ml/min delivering a stream with flow rate of 1.0 ml/min. The stream is continuously introduced into an enzyme thermistor. The enzyme thermistor contains a 0.9-ml column filled with invertase immobilized to aminoalkylated controlled-pore glass (CPG) by glutardialdehyde coupling according to Weetall.[6] Thirty milligrams dialyzed invertase (β-fructofuranosidase, EC 3.2.1.26, grade VI, from bakers' yeast, 200 U/mg, Sigma Chemical Co., St. Louis, MO) is coupled to 1 ml aminoalkylated CPG (40–80 mesh, 1350 Å pore size, Corning Co., Corning, NY) activated by 2.5% glutardialdehyde. The coupling is performed in 0.1 M phosphate buffer (pH 7.0) at room temperature for 4 hr. With a 10-fold dilution of the original sample, it is possible to analyze sucrose up to 0.3 M and resulting in a temperature response of 0.080°.

Glucose Monitor

Glucose is determined by correspondingly collecting, a sample stream from another cannula inserted in a degasser, also using a flow rate of 0.1 ml/min, and followed by a 10-fold dilution of 0.1 M sodium phosphate (pH 7.0). Subsequently the resulting stream fills a 50-μl loop of a 6-port sampling valve (type 50 Teflon rotary valve, Rheodyne, Cotati, CA). Buffer solution flows through the other channel of the valve and is continuously conducted to a second enzyme thermistor. This enzyme thermistor contains a 0.9-ml column with 25 mg glucose oxidase (EC 1.1.3.4, from *Aspergillus niger,* 200 U/mg, Sigma) and 0.10 ml catalase (EC 1.11.1.6, from beef liver, 800,000 U/ml, Sigma) coimmobilized to CPG as described above. Samples from the 50-μl loop are injected every 5 min and give a 0.050° response at 0.3 M glucose.

Enzyme Reactor

The invertase used in the enzyme reactor is coupled according to the tresyl chloride method.[7] Thirty milliliters of Sepharose beads (CL-6B, Pharmacia, Uppsala, Sweden) is activated with 1 ml of 2,2,2-trifluoro-

---

[6] H. H. Weetall, this series, Vol. 44, p. 134.
[7] K. Nilsson and K. Mosbach, *Biochem. Biophys. Res. Commun.* **102,** 449 (1981).

ethanesulfonyl chloride (tresyl chloride, Fluka AG, Buchs, Switzerland) under conditions described. After thorough washing, 0.5 g of invertase (from bakers' yeast, 300 U/mg, Boehringer Mannheim, FRG; dialyzed against 0.1 M phosphate buffer, pH 7.0) in phosphate buffer (0.1 M, pH 7.5) is added and coupled for 5 hr at room temperature. After washing with phosphate buffer, the beads are packed in a 50-ml column.

Computer

A personal computer (ABC 80, 16K ROM, Luxor AB, Motala, Sweden) was used for operating the injection valve and pumps and performing the control algorithm. The programs used were written in real-time BASIC.

The reproducibility of the monitors was checked by carrying out predetermined concentration changes directly introduced into the monitors or after passage of the enzyme reactor. Figure 2 shows an example from such an experiment where it is obvious that the monitors are able to follow the changes of that particular periodicity. Some inadequacy is

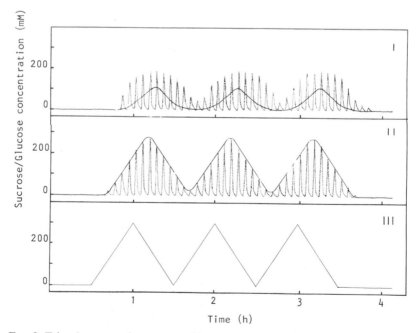

FIG. 2. Triangle-wave pulses generated by computer-controlled flow rate changes in the two pumps shown in Fig. 1. (I) Enzyme thermistor signals for glucose (·····) and sucrose (——) after passage of the enzyme reactor. (II) Signal before entry of the reactor. (III) Generated concentration profile. (From Ref. 2.)

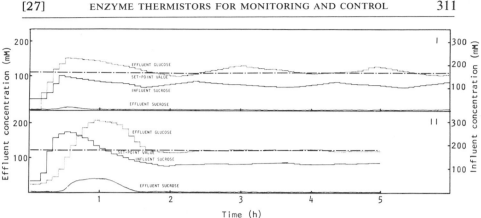

FIG. 3. Control of sucrose conversion using the setup in Fig. 1. The effluent is controlled from the glucose monitoring at the outlet of the reactor using a PI (proportional and integral) controller algorithm to adjust the feeding concentration of sucrose. (From Ref. 2.)

shown in following the concentration change in the inflection region due to the extended fall time of the sensor. Rapid changes in concentrations can therefore be difficult to resolve.

Control experiments can be performed in various ways. Figure 3 shows an example where the effluent glucose concentration is maintained at a preset value by adjusting the influent sucrose concentration from the glucose signal. Other examples of control are given in Ref. 2.

Monitoring of Cellobiose of a Cellulose Hydrolyzate[1]

Figure 4 shows an extension of the glucose enzyme thermistor as described above for cellobiose monitoring which uses a β-glucosidase precolumn. The setup is useful for following degradation of cellulose where estimation of both glucose and cellobiose is desirable. By converting the cellobiose of the sample to glucose in this precolumn prior to the enzyme thermistor, the total amount of glucose residues in the sample is determined. A subsequent determination before passing the β-glucosidase column gives the original amount of glucose in the sample.

Immobilization

β-Glucosidase is immobilized as described above. Twenty milligrams enzyme (EC 3.2.1.21, from sweet almonds, 40 U/mg, Boehringer Mannheim, FRG) is dialyzed against 0.2 M potassium phosphate (pH 7.4) and added to 1.5 ml activated glass beads. The slurry is agitated for 48 hr at 4°. A glass tubing, 5 mm in diameter, is filled with the slurry. Glucose oxidase and catalase are coimmobilized as described above or in Ref. 1.

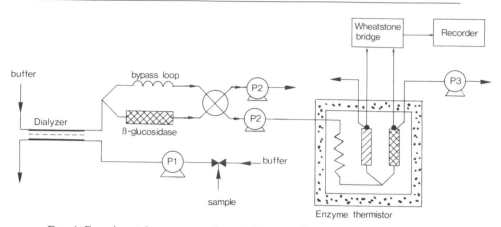

FIG. 4. Experimental arrangement for cellobiose monitoring with an enzyme thermistor with glucose oxidase/catalase and a precolumn with β-glucosidase. The sample is continuously or discontinuously fed to the dialyzer unit and transferred to the on-line buffer stream for the precolumn and the enzyme thermistor. (From Ref. 1.)

Setup

With the setup shown in Fig. 4 the sample stream, obtained from the dialyzer unit, is bypassed in a separate flow line to the enzyme thermistor, thereby generating a continuous temperature signal for its glucose content (see Fig. 4). By switching a 6-port valve (Rheodyne) the sample stream is directed through the β-glucosidase column and passed to the enzyme thermistor. The continuous temperature signal then increases to a higher temperature level if cellobiose is present in the sample. Note that two glucose residues are formed per cellobiose, thus doubling the temperature signal.

The crude hydrolyzate is dialyzed continuously in a Technicon 24 in. dialyzer equipped with a premount dialysis membrane, type C. Samples are dialyzed as discrete pulses or continuously against 50 mM sodium citrate and 0.1 M sodium phosphate (pH 5.5) with thymol which is also passed through the enzyme thermistor. A split-flow enzyme thermistor (see Danielsson, this volume [16]) with a dual column arrangement is advantageous here since variations in the crude sample composition may occur and give rise to nonspecific heat effects of the glass support.

Figure 5 shows the thermal response for cellobiose concentration in the 10–40 mM range and the nonlinearity of the curve due to the transport of the dialysis. Standard deviation of the crude samples was ±5. Additions of known glucose and cellobiose concentrations to fermentation samples were accounted for by 98–102%.

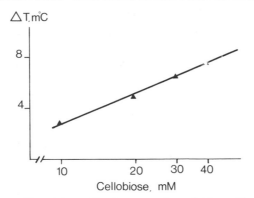

FIG. 5. Enzyme thermistor calibration curve for cellobiose. (From Ref. 1.)

Thermistor Signals for Control Purposes

For a computer it is irrelevant whether the signals are obtained from an enzyme thermistor device or any other sensor device; its demands are the same. The signal must have a reasonably short response time and should have a moderate level of noise. However, depending on the dynamics of the controlled process, relatively long response times can sometimes be accepted and noisy signals can be analogously and/or digitally filtered.

The fastest possible concentration change that can be expected in the process to be controlled must be estimated. If the sensor response time is not short enough to cover these changes, a control algorithm capable of compensating for long delays in the sensor signal must be found. Such algorithms are extensively described in control theory. Examples of algorithms used with enzyme thermistors can be found elsewhere.[8,9]

Computers are also useful for other purposes in continuous monitoring, e.g., evaluation of peak heights of discretely injected samples. This type of sample introduction is simpler to evaluate since it provides baseline checking at each injection. Furthermore, a computer for process control usually samples the signal at regular intervals which should be synchronized with the injection interval time. On the other hand, if disturbances or noise levels are so high that several values are necessary for mean value calculation, then a continuous signal is preferable.

Sometimes there is a pronounced need for rapid response biosensors,

[8] J. P. Axelsson, P. Hagander, C. F. Mandenius, and B. Mattiasson, in "Modeling and Control of Biotechnical Processes" (A. Halme, ed.). Pergamon, New York, 1982.

[9] C. F. Mandenius, B. Mattiasson, J. P. Axelsson, and P. Hagander, *Biotechnol. Bioeng.* **29,** 941 (1985).

e.g., for the detection of the formation of toxic by-products in a bio-process where the toxic compound can rapidly affect the bioactivity. Flow stream analyzers can seldom respond within the necessary time for this kind of rapid change. In such cases *in situ* biosensors are preferred, although the same accuracy cannot always be obtained.

The long-term stability of enzyme thermistors may be affected during extended control periods. This can be checked relatively easily by re-peated daily calibrations. Again, a system with an injection valve is pre-ferred where standard calibration solutions are a part of the monitoring setup, and where the computer itself corrects the last-made calibration run.

Sampling Procedures

In process control or any other form of control application, it is usu-ally necessary to use automatic sampling procedures. With crude samples these procedures must also comprise automatic treatment of the sample. This is a serious problem (see Enfors and Cleland, this volume [26]) which remains largely unsolved, especially where sterile operation is required. Undoubtedly, it is hampering any wider use of biosensors in fermentation process monitoring. For example, automatic centrifugation and sample filtration are not always easy to accomplish in small systems on an analyt-ical scale. However, modern membrane technology such as ultrafiltration and tangential flow filtration can be automatically and continuously ac-complished.

The limited operational range of enzymes and other active biomolecu-lar preparations as well as the sensitivity of the transducer device used often make it necessary to either enrich or dilute the sample continuously before introducing it to the biosensor. It may also be necessary to adjust pH, to transfer the sample to a buffer medium containing cofactors for the enzyme in use, or to add bacterostatic agents.

Thus, there are at least five operational steps involved in continuous monitoring before a sample can be analyzed in a biosensor: (1) collection of the sample, (2) purification of the sample, (3) enrichment or dilution of the sample, (4) adjustments of the sample medium, and (5) introduction of the sample. Some simple ideas and solutions for performing these steps are given.

Pretreatment of the Sample

Sampling from a bioreactor is easily performed as long as the risk of or sensitivity to contamination is not severe. Plastic tubing or stainless steel tubes equipped with coarse filters in connection with peristaltic pumps are straightforward setups. It is important to insert the sampling probe in positions representative of the entire bioreactor. In large reactors it is sometimes necessary to install several sampling probes and to avoid badly agitated positions.

When sterility demands make it necessary to use a closed bioreactor system, special equipment is needed for automatic on-line sampling. Briggs and Gardiner[10] have designed a dialysis probe for on-line sampling where a buffer stream passes a spiral groove enclosed by a dialysis membrane (available from Leeds & Northrup, North Wales, PA). This device is immersed in the medium to be analyzed, whereby the analyte diffuses through the membrane to the buffer stream and is simultaneously diluted to a degree determined by the flow rate of the buffer. The analyte-carrying stream can subsequently be conducted to any flow stream analyzer. To diminish the risk of contamination, Valentini and Razzano (1982) have suggested letting the buffer stream pass a UV radiator before entering the bioreactor.[11]

Dialysis Probe

Another device makes use of a strong tangential flow, generated by an impeller very close to the membrane, in order to prolong the operational time of the dialysis membrane.[12] Figure 6 shows this dialysis probe in cross section. The steel shaft fixes the probe to the fermenter top. The probe is made of stainless steel, Teflon, and polycarbonate materials and is thus autoclavable *in situ* (construction details can be obtained from Department of Biotechnology, University of Lund, Lund, Sweden).

Using an impeller speed of 600 rpm, fouling of the membrane surface was almost completely prevented in a cellulose slurry of 15% (w/v) during 5-day periods. The choice of membrane is a crucial point in order to obtain a suitable transfer through the membrane. Ultrafiltration membranes (cutoff 30,000, polysulfone, Millipore) give a 2% transfer of the bulk concentration that surrounded the probe with a buffer flow rate of 1 ml/min. Using the same membrane type but with a cutoff of 1,000,000, more than 55% is transferred to the buffer stream.

This sampling probe can be connected to any flow stream analyzer, such as enzyme electrode, enzyme thermistor, or flow–injection (FIA) systems, with enzyme reactors. For example, the probe can be connected

K. Briggs and W. Gardiner, U.S. Patent 3,830,106 (1974).
[11] L. Valentini and G. Razzano, *in* "Modelling and Control of Biotechnical Processes" (A. Halme, ed.), pp. 253–258. Pergamon, New York, 1982.
[12] C. F. Mandenius, B. Danielsson, and B. Mattiasson, *Anal. Chim. Acta* **163,** 463 (1984).

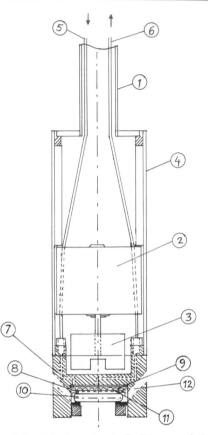

FIG. 6. Cross section of the dialysis probe. 1, Stainless steel shaft; 2, autoclavable dc motor; 3, permanent magnet; 4, stainless steel housing; 5 and 6, incoming and outgoing dialysis solution; 7, polycarbonate membrane holder; 8, membrane; 9, membrane support with spiral groove; 10, magnetic stirrer bar; 11, thin Teflon washer; 12, channels for air escape and improved circulation. (From Ref. 12.)

to a FIA system for glucose analysis where glucose oxidase and peroxidase microcolumns are used for formation of a colored complex. Using a sample injection volume of 40 μl and 30,000 cutoff membrane in the probe, the concentration range of glucose analysis is extended up to 150 g/ liter. Without this dilution system glucose oxidase is limited, for measuring glucose, to concentrations less than 150 mg/liter.

Figure 7 shows the influence of a particulate suspension of cellulose, in a wide concentration range. A particulate suspension apparently does not affect the transfer through the membrane.

Thus, it is advisable to estimate the degree of change in the medium

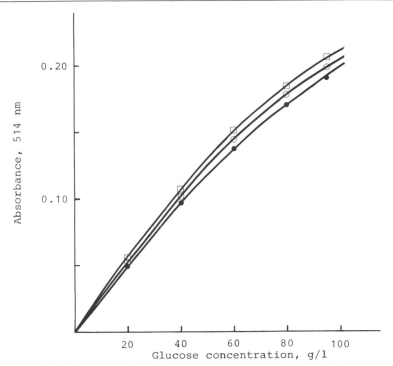

FIG. 7. Calibration curve from glucose measurements made in cellulose suspensions using the dialysis probe. (From Ref. 12.)

during the measurement as well as calibrating the probe in a medium which is representative of the measured media. Injection of standard solutions into the buffer stream after the passage of the probe is recommendable for checking the response of the sensor. Calibration of *in situ* probes is another problem and can be avoided by using a membrane (coarse enough) permitting constant and complete transfer (100%) of low molecular weight components.

Examples of Devices for Sample Treatment

Fermentation samples can often contain CO_2 or air which has to be removed so as not to influence analysis. Especially when using peristaltic pumps for transporting the sample, pressure drops at certain points along the flow line easily result in gas formation from saturated solution with CO_2/air.

Example 1. Relatively large gas volumes can be removed by using a degasser device of the kind shown in Fig. 8a, where gas bubbles escape

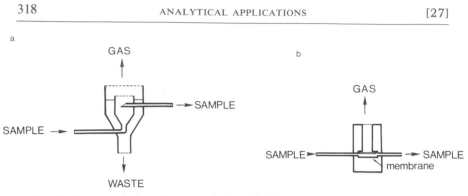

FIG. 8. Degasser devices for removal of gas bubbles in flow streams (a) for large gas volumes (b) for small gas volumes.

from a funnel fed from the bottom with the sample stream.[13] The sample is further transported by withdrawing it from the funnel with a cannula inserted through the wall of the funnel a small distance apart from the bottom of the inlet, in order to prevent the bubbles from passing near the inlet of the cannula. It is usually necessary to use a fast flow through the funnel. A flow of 10 ml/min gives a holdup of about 10 seconds, whereby fast transients are obtained when the sample concentration changes. This will of course also reduce the sampling time. The device can be constructed in plastic or stainless steel.

Example 2. When gas formation is moderate, a gas-permeable membrane tubing or a device as shown in Fig. 8b can suffice to remove unwanted gas bubbles. Using a PTFE membrane (Fluoropore filter, pore size 1 μm, Millipore) inserted in a membrane holder open to air with the membrane in an upright position, a short channel of 5 mm is enough to remove bubbles at a flow rate of 1 ml/min (Fig. 8b). The device is constructed of plastic materials.

Example 3. An example of enrichment of the samples is given in Ref. 14 for use with cyanide samples. Cyanide is mixed with sulfuric acid to lower the pH to 1 and introduced into a channel with a porous membrane (Mitex, 5 μm, Millipore). Due to HCN gas formation at low pH, the samples cross the membrane in gas phase to a NaOH stream (pH 10) on the other side. By using a lower flow rate of the alkali stream, an enrichment is obtained which is related to the relative flow rate difference between the two streams. An enrichment of 5–8 times was measured with a flow rate ratio of 1 : 40. A similar procedure can be used with ammonia and other volatile, alkaline components (see Winquist *et al.*, this volume [20]).

[13] C. F. Mandenius, B. Danielsson, and B. Mattiasson, *Biotechnol. Lett.* **3,** 629 (1981).
[14] C. F. Mandenius, L. Bülow, and B. Danielsson, *Acta Chem. Scand.* **37B,** 739 (1983).

[28] Portable Continuous Blood Glucose Analyzer

By HAKAN HAKANSON

Introduction

In recent years blood glucose measurements have become increasingly important, especially in the treatment of diabetic patients, where pump infusion[1] or pancreas transplantations demand careful supervision. Even in the treatment of type II diabetes, great improvements could be made with continuous measurement of blood glucose.[2]

In other clinical applications as well, the continuous measurement of glucose has become more important. In open-heart surgery large doses of insulin are administered to improve the condition of the heart.[3-5] Continuous measurement of the glucose level is necessary in order to prevent the surgical patient from lapsing into hypoglycemia.

To facilitate the taking of these measurements, we have developed an automatic analyzing system. This system is a revised version of the somewhat larger system used previously.[6]

General Principles of the Analyzer System

Figure 1 shows the basic principle of our analyzing system. Heparin is fed through a pump to the outer channel of a double lumen catheter. The same pump then draws blood and heparin into the inner lumen of the catheter. The blood and heparin are forced through a dialyzer to a waste bag. On the other side of the dialyzer membrane saline solution is being drawn. Glucose from the blood passes through the membrane to the dialysis solution and then to the enzyme reactor, where the glucose is consumed by immobilized glucose oxidase. This causes a change in the oxygen level of the solution, which is measured by an oxygen-sensitive

[1] I. Lager, P. Loennroth, H. von Schenck, and U. Smith, *Br. Med. J.* **287,** 1661 (1983).
[2] B. Hagander, B. Scherstén, N.-G. Asp, G. Sartor, C.-D. Agardh, J. Schrezenmeir, H. Kasper, B. Ahren, and I. Lundqvist, *Acta Med. Scand.* **215,** 205 (1984).
[3] S. Svensson, E. Berglin, W.-O. R. Ekroth, I. Milocco, F. Nilsson, and G. William-Olsson, *Cardiovasc. Res.* **18,** 697 (1984).
[4] F. Nilsson, B. Bake, E. Berglin, W.-O. R. Ekroth, J. Holm, I. Milocco, S. Svensson, J. Waldenstroem, and G. William-Olsson, *J. Parenter. Enteral. Nutr.* **2,** 159 (1984).
[5] O. Reikeraas, P. Gunnes, D. Soerlie, R. Ekroth, R. Jordes, and O. D. Mjoes, *Eur. Heart J.* **6,** 451 (1985).
[6] U. Nylén, T. Lindholm, H. Thysell, D. Heinegaard, A. Qvarnstroem, H. Hakanson, L. Ohlsson, and C. Gullberg, *Artif. Organs* (1979).

METHODS IN ENZYMOLOGY, VOL. 137

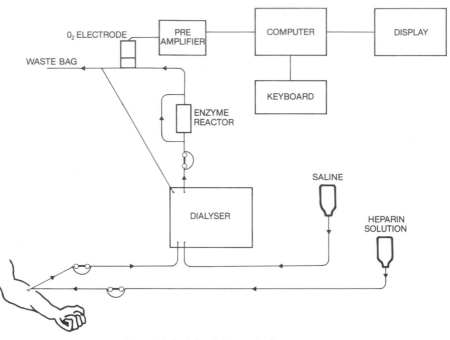

FIG. 1. Principle of the analyzing system.

electrode. The reference level of oxygen in the solution is measured by bypassing the enzyme reactor. After passing the enzyme reactor the fluid is discarded into the waste bag.

The electrical signal from the oxygen electrode is fed through a preamplifier to a computer. The computer converts the electrical signal to a glucose concentration, expressed in millimoles per liter, or milligram per deciliter, which is presented on the display. Simultaneously the system can either plot data or a diagram of the measurement. Up to 20 hr of data can be stored for printout when required.

Description

Fluid System

A pump tubing (Elkay, USA) with 0.25 mm inner diameter is used for the heparin flow. This gives a flow of about 3–4 ml of heparin solution per hour. The heparin (Loevens 5000 IU/ml, Sweden) concentration is 20,000 IU in 100 ml saline solution. The double-lumen catheter is Venflon, (Viggo, Sweden) (Fig. 2), and blood and heparin are drawn back with a

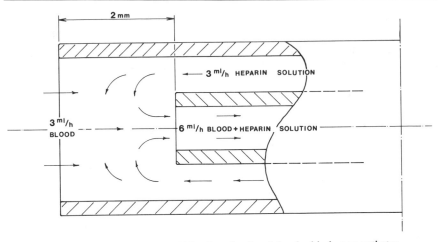

FIG. 2. Flow of heparin and blood at the tip of the double-lumen catheter.

tubing of approximately 0.38 mm inner diameter, resulting in a blood consumption of about 3–4 ml per hour.

The dialyzer has a cuprophane membrane (thickness 13.5 μm, Enka, FRG). The exposed area of the dialyzer is a 350-mm-long and 1.5-mm-wide channel giving an effective area of about 5 cm^2 (Fig. 3). Saline solution is drawn on the other side of the membrane. A tubing of 1.14 mm inner diameter is used, enabling a flow of around 60 ml/hr. Thus continuous measurement is possible with a 500-ml bag for more than 8 hr.

Before reaching the enzyme reactor the fluid passes a degassing filter (Gortex L10 415, USA). The reactor is 30.5 mm long and has an inner diameter of 4 mm. Pore glass from ElectroNucleonics Inc. (Fairfield, NJ)

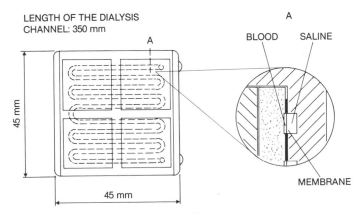

FIG. 3. The dialyzer.

with a mean pore diameter of 55 nm is used as carrier material. The glass is washed, silanized, and activated with glutaraldehyde solution. Thereafter 1100 units/ml of glucose oxidase (EC 1.1.3.4, type V, from *Aspergillus niger,* Sigma Chemical Co. St. Louis, MO) is coupled. The reactor is stored in a solution of sodium azide. The oxygen electrode (Beta Research Lab., Sweden) is a silver–platinum electrode with a Teflon membrane.

Measurements take place every sixtieth second. The measurement is followed by automatic check of the oxygen reference level which lasts for 30 sec.

The blood flow can be switched off using a mechanical valve. A calibration solution (5.5 m*M* Farmacevtiska Centralen, Soedersjukhuset, Sweden) is then mixed with the heparin. The complete system can be calibrated in this way. To facilitate handling all disposable parts are placed together in a molded polycarbonate casette (Fig. 4).

Electrical System

Figure 5 shows the basic block diagram of the electronic system. An INTEL 8031 microprocessor with 8-kByte PROM and 2-kByte RAM is used. The processor monitors the temperature sensors (Craftemp, Sweden), the valves, and the pumps. Furthermore, the processor controls alarm conditions for low battery voltage, glucose measurements, temperature, and pump speed regulation out of range. The program automatically starts up the measurement procedure and informs the 16-character

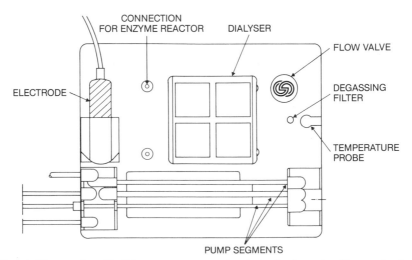

FIG. 4. The casette with dialyzer pump segments, electrode, degassing filter, and valve.

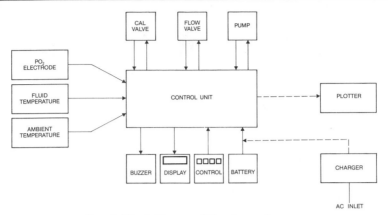

Fig. 5. Block diagram of the electronic system.

liquid crystal display (Hitachi, Japan) of the next step. The machine also performs automatic calibration.

A dc motor (Philips Model 4322010, Holland) is used as the main pump motor together with a gear box (Ritomex Model 349047 (10 : 1), Sweden). The valve motors (Minimotors Model 1516E0045, Switzerland) are also dc motors. The power is supplied by seven nickel/cadmium batteries each with a capacity of 1.8 A hr (Sanyo, Japan). A conventional charger reloads the battery pack to full capacity within 14 hr.

For printing data or plotting curves an Itoh CX 4800 plotter (Japan) is used. Figure 6 shows the complete apparatus.

In Vitro Characteristics of the Analyzing System

Laboratory tests have been performed to verify the data on the system. The results are tabulated below.

Parameter	Value
Measuring range	0–20 mM (0–3.6 g/liter), ±20%
Accuracy	±7% or ±0.2 mM, whichever has the greatest value
Response time	~4 min
Linearity	±5%
Blood consumption	3.5 ml/hr ±20%
Time of operation (off-line)	~10 hr with fully charged batteries
Dimensions	310 × 220 × 130 mm
Weight	3.5 kg fluids excluded, 4.2 kg fluids included for 8 hr of measurements

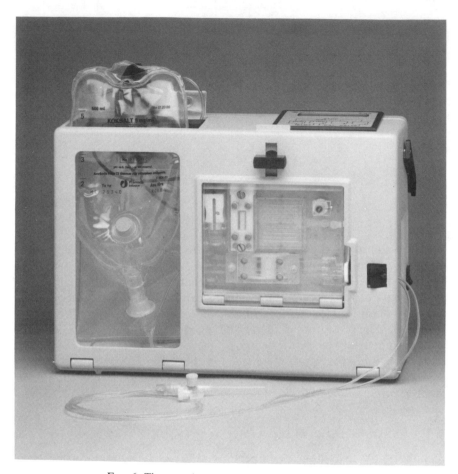

FIG. 6. The complete continuous analyzing system.

Clinical Results Obtained with the Analyzer

Comparative measurements with other methods have been carried out. An *in vitro* test on diabetic patients with over 200 measuring points between 4.8 and 20 mM gave a correlation factor of 0.97 between our method and the hexokinase method (manual method: Kit No. 124330, Boehringer Mannheim, FRG; automated method: Abbot Dichromatic Analyzer 100 with Kit. No. 6082-02). This is also verified by later investigations.[7] Figure 7 shows *in vivo* measurements from a diabetic patient who

---

[7] B. Hagander, Thesis, VI. Lund, Sweden, 1987.

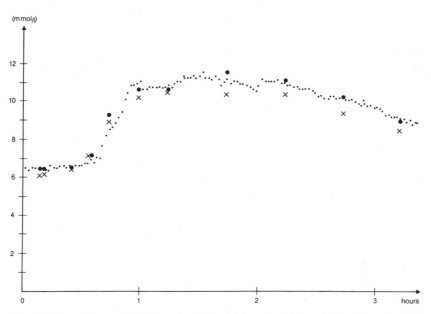

FIG. 7. Record from a diabetic patient after intake of a light meal. Key: ($\cdots\cdots$), AMG; (X), hexokinase method; (●), automated method.

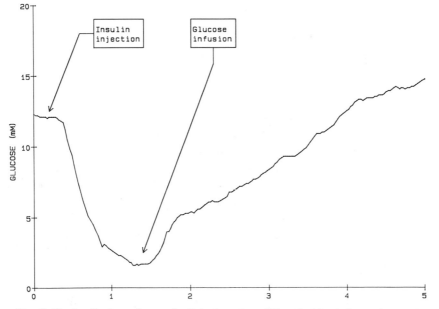

FIG. 8. The insulin dependence of a diabetic patient. When the blood glucose has sunk to the hypoglycemic level, glucose infusion is started.

had just eaten a light meal. Other methods of determining the glucose level are also shown. Observe the rapid increase of glucose.

Discussion

As the glucose oxidase method is very selective in comparison with the hexokinase method, which also measures lactate and fructose, care must be taken when comparing these methods. Furthermore, the response time of our system is only about 4 min, which is significant if the glucose level is fluctuating. The response time can be decreased by shortening the tubing to the patient.

The advantages of a system like ours are that it is easy to handle, it has a short response time, and it can be checked or recalibrated during the measurements. These features allow the physician to get much more information from the investigation and also to make it safer (see Fig. 8). However, it is important that the tubings connected to the patient are not obstructed as this will cause deviations from expected values.

[29] Needle-Type Glucose Sensor

By Motoaki Shichiri, Ryuzo Kawamori, and Yoshimitsu Yamasaki

Principle of Measurement

Using glucose oxidase, glucose concentration can be measured either by determining the production of hydrogen peroxide or by determining the consumption of oxygen. Glucose sensors which measure both determinants are available for laboratory use.[1,2] Techniques for enzyme immobilization have made it possible not only to stabilize the activity of the sensor but also to extend its lifetime. So far glucose sensors can detect glucose concentrations only in the low range, and their outputs are affected by changes in oxygen tension. Thus, they need a supplementary flow circuit which dilutes samples with buffer solution and stabilizes oxygen tension.

---

[1] S. J. Updike, M. C. Shults, and M. Busby, *J. Lab. Clin. Med.* **93,** 519 (1979).
[2] A. H. Clemens, P. H. Chang, and R. W. Myers, *Horm. Metab. Res., Suppl. Ser.* **7,** 22 (1977).

In attempts to measure glucose concentrations in blood or tissue fluids with a glucose sensor by inserting it directly into veins or subcutaneous tissue, where glucose concentrations are high and the oxygen tension is rather low, application of a biomembrane with selective permeabilities for glucose and oxygen seems to be crucial. The application of a hydrophobic membrane which is more freely permeable to oxygen than glucose might solve the oxygen limitation problems and extend the linearity of glucose upward. The authors have prepared a needle-type glucose sensor measuring hydrogen peroxide, since this sensor does not need a reference oxygen sensor and the sensor output is not affected by fluctuations in oxygen tension.[3]

Experimental Methods

Preparation of a Needle-Type Hydrogen Peroxide Electrode

A hydrogen peroxide electrode is prepared according to the method described by Hagihara et al.[4] modified as follows: the tip of a platinum wire (diameter 0.2 mm, length 4 cm) is melted in an oxygen–natural gas flame to form a small bulb (diameter 0.3–0.7 mm) then sealed into a soft glass capillary by melting in the flame. The tip of the electrode is polished with fine sand paper (#2000) until the platinum surface (anode) is uncovered.

Using a stainless steel tube (i.d. 0.4 mm, length 2 cm) as a cathode, silver plating is carried out in a solution containing 0.5% (w/v) silver cyanide, 6% (w/v) potassium cyanide, and 0.15% (w/v) potassium carbonate at a current density of 1.6 A/dm² for 30 sec and then in a solution containing 3.6% silver cyanide, 10% potassium cyanide, and 4.5% potassium carbonate at a current density of 0.1 A/dm² for 3 hr. Then the tube is kept in 3% ammonium solution and washed with distilled water. This silver-plated tube serves as the cathode of the electrode. Finally the platinum–glass anode is inserted and fixed tightly by heating with flame.

Coating with Immobilized Glucose Oxidase Membrane

The electrode tip is dipped into 1% (w/v) cellulose diacetate (Eastman Kodak Co., Rochester, NY) dissolved in 50% acetone–50% ethanol solu-

[3] Y. Yamasaki, Med. J. Osaka Univ. 35, 25 (1984).
[4] B. Hagihara, F. Ishibashi, N. Sato, T. Minami, Y. Okada, and T. Sugimoto, J. Biomed. Eng. 3, 9 (1981).

tion for 5 sec and then exposed to acetone vapor for 5 min. These procedures are repeated twice. The tip is then dipped into 2.5% (w/v) cellulose diacetate dissolved in 50% acetone–50% ethanol solution for 30 sec. Then 0.2 μl of glucose oxidase solution, in which 50 mg of glucose oxidase (EC 1.1.3.4; from *Aspergillus niger,* type II, 17300 U/g, Sigma Chemical Co., St. Louis, MO) is dissolved in 1 ml of distilled water, is dropped onto the electrode tip, the dipped end being kept upward. For the immobilization of glucose oxidase, 0.1 μl of 2% glutaraldehyde solution (Wako Pure Chemical Industries Ltd., Japan) dissolved in distilled water is dropped onto the electrode tip. The electrode is kept in air for 2 hr at 25° and then is exposed to acetone vapor for 5 min at 25°. The tip is dipped into 2% polyurethane (Japan Erastran Co., Japan) dissolved in tetrahydrofuran (Wako Pure Chemical Industries) for 2 sec followed by drying in air. Then the tip is dipped into 15% (w/v) polyurethane in 50% tetrahydrofuran–50% dimethylformamide (Wako Pure Chemical Industries) for another 10 sec. The needle-type glucose sensor (Fig. 1) thus prepared is stored in the refrigerator (4°) until it is used.

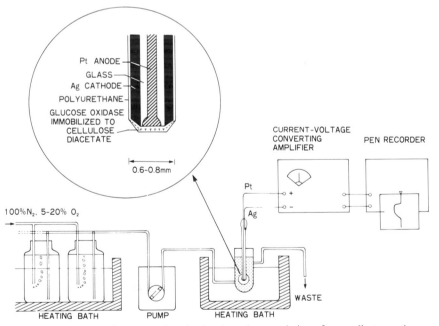

FIG. 1. Apparatus for measuring the *in vitro* characteristics of a needle-type glucose sensor. A schematic structure of the glucose sensor is shown in the enlargement.

In Vitro Characteristics of the Glucose Sensor

Apparatus for Measurement and Methods

A needle-type glucose sensor loaded with a polarizing voltage of +0.6 V supplied by a lithium battery (9 V, Matsushita Electric Industries Ltd., Japan) is connected to the current–voltage converting amplifier (POG-200A, Unique Medical Co., Ltd., Japan), which amplifies a current of 1 nA to a voltage of 100 mV. A pen recorder (VP6621A, Matsushita Communication Industrial Co., Ltd., Japan) is connected to the amplifier in order to record sensor outputs. These circuits are shown in Fig. 1.

The characteristics of the sensor are tested in 0.9% NaCl solution containing 7% (w/v) bovine albumin (Fraction V, Miles, Elkhart, IN) with varying glucose concentrations in a temperature-, flow rate-, and oxygen tension-controllable chamber as shown in Fig. 1. The output current of the sensor is calibrated initially after a stabilization period of at least 10 min.

Drift and Noise Range of Measurement

The drift of the base-line and noise range of the sensor are expressed as a percentage change of the sensor output in response to 100 mg/100 ml glucose solution. The base-line drift was $0.8 \pm 1.3\%$ per 24 hr, and the noise range was $0.3 \pm 0.4\%$. The residual current against glucose-free saline solution was $1.3 \pm 0.6\%$ (Table I).[5,6]

Dose Response against Glucose

The dose–response pattern and rapidity of the sensor output in response to the alteration in glucose concentration are measured by infusing solutions with 0–500 mg/100 ml glucose at 37°. As shown in Fig. 2, the sensor output responds well to changes in glucose concentrations, and the rapidity in response shown by $t_{90\%}$ was 16.2 ± 6.2 sec. Linear response is obtained in the range 0–500 mg/100 ml (Fig. 2).

Effect of Temperature and Oxygen Tension

The temperature coefficient measured by changing the temperature of the solution from 33 to 42° is $2.3 \pm 1.0\%/1°$. The current dependency of oxygen tension is checked by admitting a varying oxygen/nitrogen gas

[5] M. Shichiri, R. Kawamori, Y. Yamasaki, N. Hakui, and H. Abe, *Lancet* 2, 1129 (1982).
[6] M. Shichiri, R. Kawamori, Y. Goriya, Y. Yamasaki, M. Nomura, N. Hakui, and H. Abe, *Diabetologia* 24, 179 (1983).

TABLE I
CHARACTERISTICS OF THE NEEDLE-TYPE GLUCOSE SENSOR
in Vitro[a]

Test	Performance
Residual current (%)	1.3 ± 0.6
Base-line drift (%/24 hr)	0.8 ± 1.3
Noise range (%)	0.3 ± 0.4
Output generated to 100 mg/100 ml glucose (nA)	1.2 ± 0.4
Range of glucose concentrations producing a linear dose–response pattern (mg/100 ml)	0–500
$t_{90\%}$ response time (sec)	16.2 ± 6.2
Temperature coefficient (%/1°)	2.3 ± 1.0

[a] Results are shown as mean ± SD for 15 sensors. The drift of the base-line, noise range, and temperature coefficient are expressed as a percentage change of the sensor output in response to 100 mg/100 ml glucose solution.

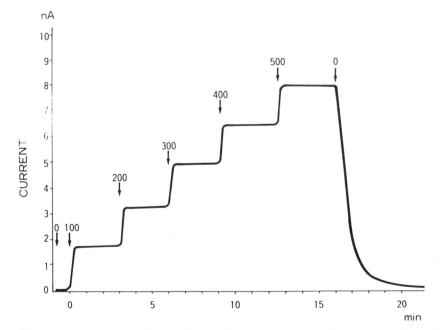

FIG. 2. Output currents of a needle-type glucose sensor to various concentrations of glucose.

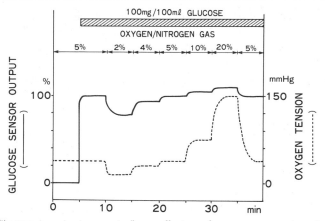

FIG. 3. Changes in output current of a needle-type glucose sensor against the change in oxygen tension of the infusate. Oxygen tension is monitored by an oxygen electrode.

mixture to the solution and by monitoring the oxygen tension (15–150 mmHg) with an oxygen sensor. The output current in response to 100 mg/ 100 ml glucose concentration increases by only 0.1% per 1 mmHg (Fig. 3).

Life Expectancy

The life expectancy of the glucose sensor is examined in the chamber by continuous recirculation of a solution containing glucose (100 mg/100 ml) at 37°. Each sensor is equilibrated in this solution for 2 hr, and output currents are continuously recorded for 7 days without calibration. During continuous monitoring *in vitro,* the output current gradually decreases to 76.2 ± 6.9% of the initial value at 7 days after the initiation of the monitoring.

In Vivo Characteristics of the Glucose Sensor

Apparatus for in Vivo Measurement and Methods

For *in vivo* monitoring, a needle-type glucose sensor is connected to a current–voltage converting amplifier device (4 × 5 × 2 cm in size) which has been constructed by using a CMOS operation amplifier (ICU 7613, Intersil Inc., Mansfield, OH). The polarizing voltage for the glucose sensor is supplied by the lithium battery built in the device. A pen recorder is connected to the amplifier to monitor the sensor output.

Each sensor's output is calibrated with a standard glucose solution in which 200 mg of glucose is dissolved in 100 ml of sterilized 0.9% NaCl solution maintained at 37°. Then a glucose sensor is inserted by means of an indwelling needle (gauge #18) into the jugular vein or subcutaneous tissue of healthy and diabetic dogs, or into subcutaneous tissue of the forearm in healthy and diabetic human volunteers. The sensor output is compared with blood glucose concentrations simultaneously measured by a bedside-type artificial endocrine pancreas system.[7,8] Then oral glucose loads are administered.

Noise Range of in Vivo Measurements

The noise range at *in vivo* monitoring by inserting the sensor into subcutaneous tissue in generally anesthetized and unanesthetized normal dogs is $1.3 \pm 0.5\%$ ($n = 5$) and $3.1 \pm 0.8\%$ ($n = 5$), respectively. Strenuous muscular exercise in dogs produces noise in the range up to 13.4% of outputs.

Response of the Glucose Sensor for Blood Glucose

The output of the sensor when kept in the jugular vein of dogs (Y) relates to the results of intravenous glucose monitoring by a bedside-type artificial endocrine pancreas system (X) according to ($Y = 0.98X + 2$; $r = 0.998$; $n = 92$). A significant relationship also exists between the glucose concentrations obtained by the needle-type glucose sensors in subcutaneous tissue (Y) and the blood glucose concentrations (X) determined by the bedside-type monitoring system in dogs ($Y = 0.85X + 3$; $r = 0.956$; $n = 144$). In human volunteers, a high correlation ($Y = 0.79X + 17$; $r = 0.96$; $n = 115$) is also observed.

In Vivo Sensor Output Dependency on Oxygen Tension

In dogs, a reference oxygen electrode is also inserted into the subcutaneous tissue 2–3 cm from the glucose sensor to monitor background oxygen tension. After monitoring the base-line for more than 30 min, the dogs inhale 100% nitrogen gas or 95% oxygen plus 5% carbon dioxide gas. The reference oxygen electrode shows fluctuations in subcutaneous tissue oxygen tension in the range of 27–72 mmHg. However, the glucose sen-

---

[7] M. Shichiri, R. Kawamori, and H. Abe, *Diabetes* **28,** 272 (1979).
[8] R. Kawamori, M. Shichiri, M. Kikuchi, Y. Yamasaki, and H. Abe, *Diabetes* **29,** 762 (1980).

sor shows stable output regardless of oxygen tension changes, and the output is consistent with monitored blood glucose concentrations.

Life Expectancy on an in Vivo Basis

In order to examine changes in the characteristics of sensors inserted into subcutaneous tissue during continuous monitoring, both "relative" output current and "relative" response time of the sensors are determined as follows: The "relative" output of the sensor kept in subcutaneous tissue for 3 days is calculated by comparing the sensor output with simultaneously measured blood glucose levels. The "relative" response time of the sensor is calculated from the time lag between the rise in blood glucose and the rise in sensor output after meal intake. After three days, the "relative" sensor output decreases to 73.5% of the initial level, and the "relative" response time increases up to 13.5 min. *In vitro* characteristics of the sensor determined after removal show a 23% reduction in output current and a 14-sec delay in response (Table II).

TABLE II

CHANGES IN *in Vitro* AND *in Vivo* "RELATIVE" CHARACTERISTICS OF THE NEEDLE-TYPE GLUCOSE SENSOR AFTER 1–3 DAYS' CONTINUOUS TISSUE GLUCOSE MONITORING

Characteristic	Time after application			
	Just after	1 day	2 days	3 days
In Vivo				
"Relative" output current generated to 100 mg/100 ml glucose (%)	100	93.4 ± 4.4	82.0 ± 4.4	73.5 ± 3.4
"Relative" response time to blood glucose (min)	5.1 ± 2.2	7.3 ± 11.9	9.0 ± 0.7	13.5 ± 1.5
	Before application			3 days
In vitro				
Residual current[a] (nA)	1.0 ± 0.4			1.4 ± 1.2
Output current generated to 100 mg/100 ml glucose (nA)	2.2 ± 0.5			1.7 ± 0.1
$t_{90\%}$ (sec)	29.0 ± 6.0			43.0 ± 6.0
Linearity in dose response to glucose concentration (mg/100 ml)	0–500			0–500
Base-line drift (%/24 hr)	1.1 ± 1.0			1.3 ± 1.4

[a] Residual current denotes the output current of the electrode in the absence of glucose.

Clinical Applications of the Needle-Type Glucose Sensor

The application of the needle-type glucose sensor has taken two paths. The first has been the application in a blood glucose monitoring system. The sensor can measure glucose concentration in discretely obtained whole blood, and it can monitor glucose concentrations on a nonvascular basis by measuring subcutaneous tissue glucose concentrations as mentioned beforehand.

The second has been the application to a closed-loop glycemic control system. The wearable artificial endocrine pancreas system consists of a needle-type glucose sensor, a microcomputer system, a 2-roller pump-driving system for insulin and glucagon infusions, and lithium batteries, packed into a small unit (12 × 15 × 6 cm in size, 400 g in weight). Glycemic control in five hospitalized insulin-dependent diabetic patients with the wearable artificial endocrine pancreas has been attempted. In these trials, a sensor is inserted into the subcutaneous tissue of the forearm, and insulin and glucagon are infused intravenously. Perfect glycemic control was established during continuous blood glucose regulation up to 3 days with a single sensor and up to 7 days by replacing the sensor with a new one. Comparison of the glycemic regulation by a wearable closed-loop control system in these five patients with that obtained by multiple subcutaneous insulin injections or open-loop continuous subcutaneous insulin infusion regimes shows a superiority of the closed-loop regulation in glycemic control.[9]

[9] M. Shichiri, R. Kawamori, N. Hakui, Y. Yamasaki, and H. Abe, *Diabetes* **33**, 1200 (1984).

[30] Automated TELISA Procedure for Process Monitoring

By Staffan Birnbaum, Leif Bülow, Bengt Danielsson, and Klaus Mosbach

The synthesis of valuable proteins by recombinant prokaryotic or eukaryotic cells in the laboratory as well as in industrial-scale biotechnological processes requires rapid, sensitive, specific, and automated means of determination for the optimization and control of the production system. This can be achieved by combining the enzyme-linked immunosorbent assay (ELISA) technique, described earlier in this series, with the enzyme thermistor as detector for the enzyme activity to form a thermo-

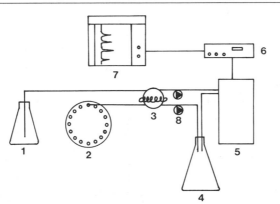

FIG. 1. Schematic diagram of the automated TELISA system consisting of a PBS reservoir (1), an autosampler (2), an HPLC valve (3), a waste reservoir (4), an enzyme thermistor (5), a Wheatstone bridge (6), a recorder (7), and a dual channel pump (8).

metric enzyme-linked immunosorbent assay (TELISA).[1-3] Below, we illustrate an automated TELISA system for the analysis of insulin which we have employed to monitor the production of human proinsulin in fermentation broth by *Escherichia coli*.[4] For other examples of the TELISA technique see also the chapter by Mattiasson [60] on affinity immobilization in this volume.

Apparatus

The experimental design of the analytical apparatus is shown in Fig. 1. The autosampler contains three test tubes for each fermentation sample analyzed. The first test tube contains the fermentation supernatant or the standard sample mixed with an equal volume of the insulin–peroxidase conjugate (typically a 1:20 dilution of the stock solution). The second tube contains the substrate for the peroxidase reaction, 2 mM H_2O_2, 14 mM phenol, and 0.8 mM 4-aminoantipyrine in phosphate-buffered saline (PBS) (50 mM sodium phosphate, 0.15 M NaCl, pH 7.4). The third tube contains 0.2 M glycine–HCl, pH 2.2. An HPLC valve with a 0.5-ml loop is employed to regulate sample injection into the thermistor. The HPLC valve is rotated by an electric motor. The autosampler and the HPLC valve are controlled by a preset timer. The enzyme thermistor contains a

[1] E. Engvall, this series, Vol. 70, p. 419.
[2] B. Danielsson, B. Mattiasson, and K. Mosbach, *Appl. Biochem. Bioeng.* **3**, 97 (1981).
[3] B. Mattiasson, C. Borrebaeck, B. Sanfridson, and K. Mosbach, *Biochim. Biophys. Acta* **483**, 221 (1977).
[4] S. Birnbaum, L. Bülow, B. Danielsson, and K. Mosbach, *Anal. Biochem.* **158**, 12 (1986).

1-ml column housing the immobilized antibodies. A highly sensitive thermistor is situated at the outlet of the column. The column and thermistor are thermostatted at 25° within an insulated aluminum block. The heat registered by the thermistor is converted by a Wheatstone bridge and appears as peaks on the recorder.

Alternatively, a programmable controller can be employed. We have used a Hizac Model D-28 Hitachi programmable controller in combination with a motor-driven sampling valve, solenoid valves for selection of substrate and washing agents, and a modified sample changer from a Technicon AutoAnalyzer in which the timer control was replaced by a relay operated from the programmable controller to actuate the sample changing mechanism. Various parameters, such as washing time and substrate volume, can then be easily changed. With the sample changer omitted, this system can be used for automatic sampling from a process stream as well. In that case it is recommended to include a suitable filter in the sample stream to remove any cells present.

Preparation of Immobilized Antibodies

Guinea pig antibodies to porcine insulin are initially affinity purified against beef insulin. Beef insulin (11.6 mg) is covalently coupled to 2.5 ml Sepharose 4B activated with 15 μl tresyl chloride as described in this series.[5] Guinea pig anti-porcine insulin serum is applied to a column containing the insulin-bound Sepharose 4B. The column is washed with PBS until the A_{280} reading is 0.0. Subsequently, the antibodies are eluted with 0.2 M glycine–HCl, pH 2.2, with reverse flow, mixed immediately with an equal volume of 0.5 M NaHCO$_3$, and dialyzed overnight at 4° against 0.2 M NaHCO$_3$. The purified antibodies (0.2 mg/ml) are then coupled to Sepharose 4B (1 ml/g wet gel) activated with tresyl chloride (6 μl/g wet gel).[5]

Preparation of Insulin–Peroxidase Conjugate

Beef insulin is conjugated to horseradish peroxidase by a modified version of the method of Wilson and Nakane.[6] Horseradish peroxidase is dissolved in 4 ml H$_2$O at a concentration of 4 mg/ml. Add 0.8 ml of freshly prepared sodium metaperiodate (21.4 mg/ml) to the horseradish peroxidase solution and stir for 30 min at room temperature. Dialyze the perox-

[5] K. Nilsson and K. Mosbach, this series, Vol. 135 [3].
[6] M. B. Wilson and P. K. Nakane, in "Immunofluorescence and Related Staining Techniques" (W. Knapp, K. Holubar, and G. Wicks, eds.), p. 215. Elsevier/North-Holland, Amsterdam, 1978.

idase–periodate solution against 1 mM sodium acetate buffer, pH 4.4, at 4° overnight. Beef insulin (8 mg) is dissolved in 2 ml 50 mM HCl, 2 ml 50 mM NaOH is added, and the solution is dialyzed against 10 mM sodium bicarbonate buffer, pH 9.5, overnight at 4°. After dialysis, the pH of the peroxidase–aldehyde solution is adjusted to 9–9.5 with addition of 50–200 μl 0.2 M sodium bicarbonate buffer (pH 9.5). The dialyzed insulin solution is then immediately added, and the mixture is allowed to stir at room temperature for 2 hr. The insulin–peroxidase solution is subsequently dialyzed against 10 mM sodium bicarbonate buffer containing 0.5 mg/ml NaBH$_4$ for 2 hr at 4°. The conjugate is then dialyzed overnight at 4° against PBS. Finally, the preparation is chromatographed on a Sephacryl S-200 (35 × 2.5 cm) with PBS, and 3-ml fractions are collected (0.25 ml/min); the absorbances at 403 and 280 nm are measured, and the fractions containing the desired conjugate peak are pooled. The conjugate is stored at 4° until used.

Assay Procedure

The contents of the first tube on the autosampler is pumped (2.4 ml/min) into the injection loop. After 20 sec the valve is rotated, and the contents of the loop are pumped (0.6 ml/min) into the enzyme thermistor unit where the conjugate and the insulin or proinsulin molecules compete for the antibody sites. After 2 min the valve returns to the fill position. An

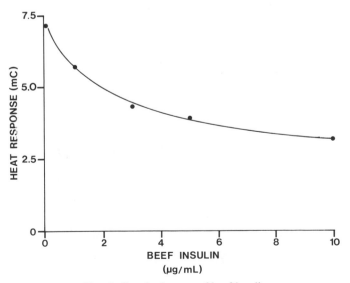

FIG. 2. Standard curve of beef insulin.

additional 2 min elapse before the autosampler advances to the next tube, containing the substrate, which is injected in a similar manner. As the substrate passes the immobilized antibody-bound conjugate, catalysis occurs, and the heat generated is registered by the thermistor and appears on the recorder. Thus, less heat is generated as more free insulin or proinsulin is present in the standard or sample (see Fig. 2). Finally, the autosampler advances to the third tube, containing the low pH glycine buffer, which removes the bound (pro)insulin and conjugate from the immunosorbent and thereby regenerates the immobilized antibody column for subsequent assays. PBS is used as the running buffer throughout the system.

Comments

Sample determination takes roughly 7 min, and the entire assay cycle takes 13 min. TELISA methods for the analysis of albumin, gentamicin, and monoclonal antibodies have also been described.[3,7,8]

[7] B. Mattiasson, K. Svensson, C. Borrebaeck, S. Jonsson, and G. Kronvall, *Clin. Chem.* **24,** 1770 (1978).
[8] O. W. Merten, manuscript in preparation.

[31] Solid-Phase Optoelectronic Biosensors

By CHRISTOPHER R. LOWE and MICHAEL J. GOLDFINCH

A biosensor comprises two principal components: a biological system which recognizes the analyte and responds to its concentration with an overall physicochemical change, and the transducer element which converts this change into an amplifiable and processible electrical signal.[1–3] The biologically responsive component normally comprises an enzyme, lectin, antibody, organelle, whole cell, or tissue slice immobilized in close proximity to a suitable transducer. Ideally, the biorecognition system should be highly specific for the analyte of interest and thereby discriminate between the target analyte and other materials with related structures present in the sample. The biological system should also respond to the analyte over the concentration range found in the potential samples

[1] C. R. Lowe, *Trends Biotechnol.* **2,** 59 (1984).
[2] C. R. Lowe, *Biosensors* **1,** 3 (1985).
[3] N. C. Foulds and C. R. Lowe, *Bioessays* **3,** 129 (1985).

and generate a physicochemical change which can be detected with a suitable transducer. The electrochemical transducer may take one of a number of forms, including a metal or semiconductor electrode,[4] an ion-selective or gas-sensing electrode,[5] a thermistor,[6,7] a piezoelectric crystal,[8] a transistor,[9] a conductance device,[1] or an optical or optoelectronic device.[10,11] The transducer should ideally possess the following attributes: be responsive to the physicochemical change occasioned by the biorecognition system with a moderately fast response time (0–60 sec), be capable of miniaturization, and be sturdy, reliable, and calibratable.

A simple and relatively inexpensive device which meets most of these requirements is the solid-phase optoelectronic biosensor.[12,13] Triphenylmethane dyes such as bromocresol green and bromothymol blue are covalently attached to a transparent cellophane membrane and sandwiched between a light source, a red light-emitting diode, and a detector system, a silicon photodiode with integral amplifier. Adsorption of human serum albumin to the dyed membrane buffered to pH 3.8 causes a yellow to blue–green color change in the membrane which is monitored by a fall in the output voltage of the photodiode detector system. In addition, the optoelectronic biosensor may be exploited to determine enzyme substrates by coimmobilizing a pH-sensitive dye and appropriate enzymes which generate or consume protons. This chapter describes the synthesis of the dye analogs and membranes as well as construction and operation of the solid-phase optoelectronic sensor.

Synthesis of Dye Analogs

Reaction of the triphenylmethane dyes, bromocresol green and bromothymol blue (Fig. 1), with reduced glutathione at pH 6.5–7.0 generates the corresponding glutathionyl conjugates in yields of approximately 10–15%.[12–14] Typically, bromocresol green (BDH Chemicals, Poole, UK; 840 mg, 1.4 mmol) and reduced glutathione (1840 mg, 6.0 mmol) are dissolved

[4] C. R. Lowe, *FEBS Lett.* **106**, 405 (1979).

[5] G. H. Fricke, *Anal. Chem.* **52**, 295R (1980).

[6] K. Mosbach, B. Danielsson, A. Borgerud, and M. Scott, *Biochim. Biophys. Acta* **403**, 256 (1975).

[7] B. Danielsson, L. Buelow, C. R. Lowe, I. Satoh, and K. Mosbach, *Anal. Biochem.* **117**, 84 (1981).

[8] J. Hlavay and G. G. Guilbault, *Anal. Chem.* **50**, 965 (1978).

[9] B. Danielsson, I. Lundström, K. Mosbach, and L. Stiblert, *Anal. Lett.* **12**, 1189 (1979).

[10] M. T. Flanagan, *Electron. Lett.* **20**, 968 (1984).

[11] J. R. North, *Trends Biotechnol.* **3**, 180 (1985).

[12] M. J. Goldfinch and C. R. Lowe, *Anal. Biochem.* **109**, 216 (1980).

[13] M. J. Goldfinch and C. R. Lowe, *Anal. Biochem.* **138**, 430 (1984).

[14] A. G. Clark and S. T. Wong, *Anal. Biochem.* **92**, 290 (1979).

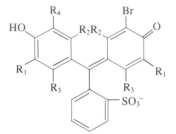

FIG. 1. Structures of triphenylmethane dyes and their glutathione conjugates. Bromocresol green: R^1, Br; R^2 H; R^3, CH_3; R^4, Br. Bromothymol blue: R^1, $CH(CH_3)_2$; R^2, CH_3; R^3, H; R^4, Br. Bromocresol green–glutathione conjugate: R^1, Br; R^2, H; R^3, CH_3; R^4, —$SCH_2CHCONHCH_2COO^-$. Bromothymol blue–glutathione conjugate: R^1,

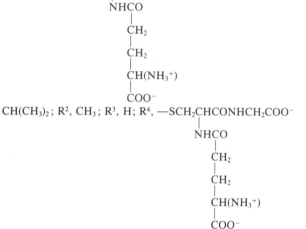

in distilled water (90 ml) and the pH adjusted to 6.5–6.6 with 0.880 specificity gravity NH_4OH. The solution is incubated at 40° for 48 hr, degassed under reduced pressure for 1 hr and then rotary evaporated to dryness at ≤60°. The residue is dissolved in distilled water (30 ml), acetic acid is added to give a pH of 3.0, and ethyl acetate (~70 ml) added to give a final ethyl acetate : water ratio of 70 : 30 (v/v). Unreacted dye is extracted with ethyl acetate, and the procedure is repeated 5 times prior to taking to dryness by rotary evaporation. The residue is dissolved in n-butanol/ acetic acid/water (4 : 1 : 5, by volume; upper phase, 20 ml) and applied to a column (60 × 240 mm) of silica gel (Kieselgel 60 PF$_{254}$; E. Merck, Darmstadt, FRG) equilibrated with the same solvent system. Fractions (10 ml) are collected at a flow rate of 30 ml/hr and sustained with a hand bellows.

The column separates the mixture into three principal components:

unreacted bromocresol green (R_f 0.71),[15] an unknown mauve product (R_f 0.58), and finally the yellow–green pH-sensitive bromocresol green–glutathione conjugate (R_f 0.34). Unreacted glutathione remains tightly adsorbed to the surface of the silica column. The product (R_f 0.34) is recovered in approximately 10% overall yield and is repeatedly rotary evaporated at $\leq 60°$ and redissolved in water until all traces of acetic acid are removed. The yellow product becomes dark blue in water of pH ∼5 and is finally lyophilized.

The procedure for the synthesis of the bromothymol blue–glutathione conjugate is similar to that described above except that the ethyl acetate extraction stage is omitted. The yellow adduct is recovered in approximately 15% overall yield.

Dye–Conjugate Characterization

The dye–glutathione conjugates are formed by the nucleophilic displacement of the aryl halide by the thiol function of reduced glutathione. The glutathione conjugate of bromocresol green has a slightly reduced molar absorption coefficient (E_m 42,550 liters mol^{-1} cm^{-1}) and a slightly higher pK_a (5.1) than that of the parent dye (E_m 45,700 liters mol^{-1} cm^{-1}, pK_a 4.7). In contrast, the glutathionyl conjugate of bromothymol blue displays a markedly lowered molar absorption coefficient (E_m 6,615 liters mol^{-1} cm^{-1}) and higher pK_a (8.3) than the unmodified dye (E_m 42,460 liters mol^{-1} cm^{-1}; pK_a 7.1). The R_f values on silica gel TLC, the positive reaction to ninhydrin, the spectral properties, the solubilities, and the capacity to couple to activated polysaccharides all confirm the structures assigned to the conjugates and shown in Fig. 1.

Preparation of Bromocresol Green Membranes[12]

Visking tubing (Size 5, 24/32 in., Medicell International Ltd., London) (3 g dry weight/10 ml water) is exhaustively washed with distilled water and activated with CNBr (140–200 mg/g dry weight) for 30 min at 12–15° and pH 10.8 ± 0.2. The activated membrane is thoroughly washed with distilled water and 0.1 M ice-cold NaHCO$_3$–Na$_2$CO$_3$ buffer, pH 9.3, prior to the addition of the bromocresol green–glutathione conjugate (114 μmol/3 g activated membrane). The coupling mixture is tumbled slowly for 18 hr at 0–4°, and free dye conjugate is removed by extensive washing with distilled water, 0.1 M HCl, 8 M urea, 1% Triton X-100, and finally water. The concentration of the membrane-bound dye is determined by

---

[15] Determined by TLC on DC Fertigfolien F1500 LS 254 Kieselgel plates in n-butanol/acetic acid/water (4 : 1 : 5 by volume, upper phase).

hydrolysis of a known weight of membrane in 1 M NaOH for 1 hr at 35°
and measurement of the absorbance at λ_{max}.

Preparation of Enzyme–Dye Membranes

The dyed enzyme membranes are prepared by cutting cellophane
Visking tubing (Size 5, 24/32 in.) (1 g) into single strips, boiling in distilled
water (100 ml) containing EDTA (1 mmol), washing, and activating with
1-chloro-2,3-epoxypropane (0.13 mol) in 1.0 M NaOH (130 ml) containing
sodium borohydride (100 mg, 2.64 mmol) for 24 hr at 22°. The activated
membrane is washed thoroughly with water and 0.1 M NaHCO$_3$–Na$_2$CO$_3$
buffer, pH 11.5, and incubated with 1,3-diamino-2-hydroxypropane (3
mmol) in the same buffer for 24 hr at 22°. The aminohydroxypropylated
membrane contains approximately 720 μmol amino groups per gram wet
weight.

Penicillinase–bromocresol green membranes are prepared by incubating a small square (1 cm^2) of the aminohydroxypropylated membrane in a
miniature petri dish with penicillinase (penicillin amido-β-lactam hydrolase, β-lactamase, EC 3.5.2.6, from *Bacillus cereus*; 2 mg, 71 units), 1-
ethyl-3-(3-dimethylaminopropyl)carbodiimide hydrochloride (50 mg,
0.26 mmol), water (100 μl), and bromocresol green–glutathione conjugate
(70 μl, 0.3 μmol) for 2 hr at 0°. The membrane is finally washed with
distilled water.

A similar procedure is used for the preparation of glucose oxidase–
bromocresol green membranes.[13] The urease–bromothymol blue membrane is prepared by a two-stage procedure. Aminohydroxylpropylated
membrane is incubated at 2° for 2 hr with sodium phosphate buffer, pH 6.3
(100 mM, 100 μl), bromothymol blue–glutathione conjugate (6 μmol), and
glutaraldehyde (5 μl, 0.6%, v/v, final concentration). The membrane is
washed with distilled water and the phosphate buffer prior to incubating
with urease (urea amidohydrolase, EC 3.5.1.5, from *Canavalia ensifor-
mis*; 10 mg, 641 units) for 1 hr at 2° in sodium phosphate buffer, pH 6.3
(100 μl, 100 mM), containing glutaraldehyde (5 μl, 0.6%, v/v, final concentration). The membrane is exhaustively washed to remove unbound
dye and enzyme.

Construction of the Optoelectronic Sensor

The optoelectronic sensor comprises a small sample cell, 8 mm diameter and 4 mm deep, cut into a Lucite sheet (50 × 50 mm, 5 mm thick) with
2 mm diameter feed and out-flow holes and closed on the open side with a
Lucite sheet 2 mm thick. The transparent dyed membrane (10 × 10 mm) is
squeezed between the two blocks. The light source, a red light-emitting
diode (LED) (Siemens; high brightness, 12° half-viewing angle, Type LD

S2-C), and the detector system, a silicon photodiode with integral amplifier (RS Components Type 308-067), are located on either side of the flow cell, held in position by further Lucite sheets (5 mm thick) and accurately aligned with metal studs and bolts. The whole assembly is mounted vertically in an aluminum light-proof box with entry and exit points for the sample. The LED is powered with a dual-output stabilized power supply unit (Farnell Electronic Components Ltd., Leeds, UK; Type LT 30/1) through an external series current-limiting resistor (470 Ω, 11 W). The voltage across the LED is typically 1.885 V. The photodiode amplifier circuitry is also powered by the stabilized power supply unit, and the output voltage (V_o) is monitored on a Pederson Model 310-106 chart recorder and on a Sinclair Model DM450 4.5 digit multimeter across a 10 kΩ–1% load resistor. The schematic layout of the components of the optoelectronic biosensor is shown in Fig. 2.

Operation of the Optoelectronic Biosensor

For the measurement of human serum albumin, the output voltage (V_o) of the detector is initially set at 7.000 V by adjusting the stabilized power supply when 20 mM citrate buffer pH 3.8 is pumped through the flow cell at 4 ml/min. The albumin sample in the same buffer is pumped through, the new V_o recorded, 8 M urea pumped through, and the membrane reequilibrated with the citrate buffer, pH 3.8, ready for the next sample.

Measurement of metabolite levels is effected by pumping the sample, usually made up to 3.0 ml with 1 mM sodium phosphate buffer, pH 7.0, through the flow cell at 10 ml/min. The flow is stopped and the change in output voltage (ΔV_o) per unit time recorded. The system is regenerated by pumping through fresh buffer solution until V_o returns to its initial value.

Solid-Phase Albumin Sensor

The triphenylmethane dye, bromocresol green, and its glutathione conjugate bind to human serum albumin at pH 3.8 with a substantial increase in absorbance at 620 nm and hence a visible color change from yellow through green to blue.[12] This color change encountered when albumin binds to immobilized bromocresol green forms the basis for the optoelectronic biosensor by sandwiching the dyed membrane between a red LED (λ_{max} 630–633 nm) and a silicon photodiode/amplifier. Figure 3 illustrates the standard curve for human serum albumin over the concentration range 0–45 mg/ml. The change in output voltage is approximately linear within the range 5–35 mg/ml and corresponds to 13 mV per mg/ml increment in the albumin concentration, although at concentrations above 35 mg/ml a progressive saturation is observed. After regeneration of the membrane with 8 M urea and returning the membrane to the starting

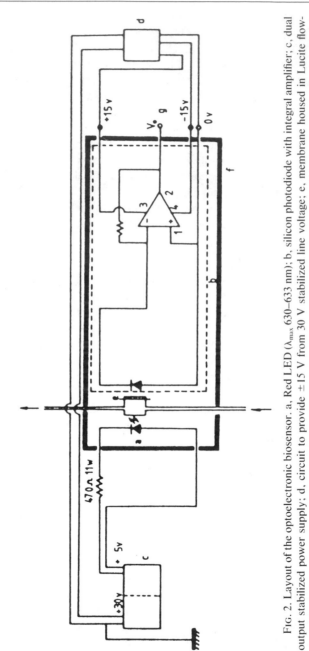

FIG. 2. Layout of the optoelectronic biosensor. a, Red LED (λ_{max} 630–633 nm); b, silicon photodiode with integral amplifier; c, dual output stabilized power supply; d, circuit to provide ±15 V from 30 V stabilized line voltage; e, membrane housed in Lucite flow-through cell; f, aluminum light-proof box; g, output voltage (V_o) monitored on a Sinclair DM450 4.5 digit multimeter. Reproduced with permission from Ref. 12.

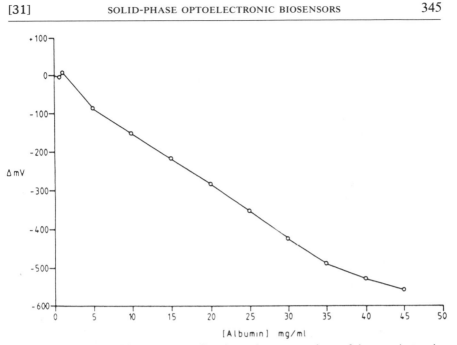

FIG. 3. The effect of human serum albumin on the output voltage of the optoelectronic albumin sensor. The ordinate is expressed as a change (in millivolts) of the output voltage (V_o) from its initial value of 7.000 V in the absence of albumin. Reproduced with permission from Ref. 12.

buffer, the output voltage reverts to its initial value of 7.000 V. Each measurement/regeneration cycle can be completed in less than 8 min. In 10 separate cycles with a standardized albumin concentration of 10.8 mg/ ml, the observed output voltage was 164.20 ± 2.35 mV (SD).

pH Response of the Optoelectronic Sensor

The pH responses of the membrane-immobilized triphenylmethane dyes are measured by monitoring the output voltage (V_o) of the photodiode/amplifier as a function of pH. The initial V_o is set at 7.000 V when 100 mM sodium acetate buffer, pH 3.3, is pumped through the flow cell. Figure 4 shows the membrane response when buffers of pH 3.3–6.8 are pumped through the flow cell. The bromocresol green membrane, containing approximately 2 μmol dye/g dry weight, changes color from yellow to blue–green and causes a 3–4 V fall in output voltage from the detector, with an apparent pK_a value of 5.1. Similarly, the bromothymol blue membrane, containing approximately 10 μmol dye/g dry weight, changes from yellow to blue over the pH range 6–10 with an apparent pK_a

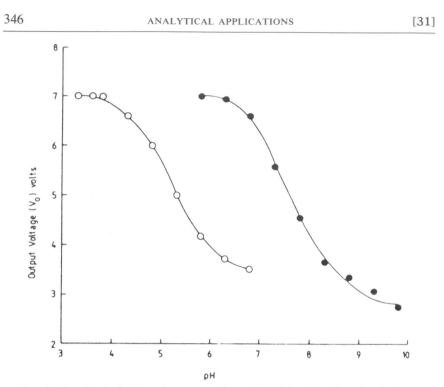

FIG. 4. The effect of pH on the output voltage (V_o) of the photodiode/amplifier of the optoelectronic sensor with a bromocresol green (○) and bromothymol blue (●) membrane. Reproduced with permission from Ref. 13.

of ~7.5 and with a 3–4 V change in V_o. Both dyed membranes generate an output voltage change of 1.6–2.0 V per pH unit around their respective apparent pK_a values and thus permit pH values in these regions to be determined to at least three decimal places. This sensitivity to pH forms the basis for the enzyme-loaded optoelectronic biosensors.

Optoelectronic Enzyme Biosensor

An optoelectronic sensor containing a membrane comprising coimmobilized bromocresol green and penicillinase responds to the substrate, penicillin G, in the concentration range 0–20 mM. The output voltage change is linear within the range 0.5–5 mM but appeared to increasingly saturate at concentrations above 10 mM (Fig. 5). The apparent Michaelis constant (K_m) deduced from a double-reciprocal plot of (ΔmV min^{-1})$^{-1}$ versus [penicillin G]$^{-1}$ is 12.9 mM. At a concentration of 5 mM penicillin G, the sensor produces a response equivalent to approximately 30 mV

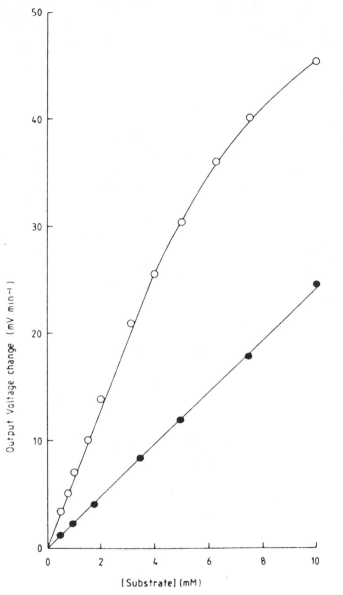

FIG. 5. The response of the penicillinase optoelectronic sensor to penicillin G (○) and ampicillin (●). Bromocresol green–penicillin membrane; V_{LED} 1.8848 V; V_o = 2.500 V at time zero; 24 ± 2°. Reproduced with permission from Ref. 13.

min^{-1}, with V_o set to 2.500 V and the LED voltage (V_{LED}) to 1.8848 V.[16] In eight separate measurement/regeneration cycles of the sensor, at a standardized penicillin G concentration of 10 mM, the ΔV_o min^{-1} was 45.50 $\pm$ 0.73 (SD), a variation of less than $\pm 2\%$.

The penicillinase/bromocresol green sensor also responds to penicillin analogs that are catalytically hydrolyzed by *B. cereus* β-lactamase. For example, the sensor responds linearly to both ampicillin and cephaloridine in the concentration range 0–10 mM, but with ΔV_o min^{-1} values reduced to 53 and 29%, respectively of that with penicillin G. In contrast, penicillinase-resistant antibiotics such as methicillin, vancomycin, and cloxacillin produce no detectable response at 10 mM.

Enzyme optoelectronic biosensors also respond to reactions, such as that catalyzed by urease, which consume protons. The urea-responsive biosensor is constructed with a coimmobilized urease/bromothymol blue (pK_a ~7.5) membrane which responds over the pH range 6.5–9.5 with a 3 V change in V_o. The urease biosensor responds to urea concentrations in the range 0–40 mM, linearly up to 10 mM, and typically with a ΔV_o min^{-1} of approximately 125 mV min^{-1} at 10 mM urea and a variation of $\pm 4\%$ or less.

Conclusions

The binding of bromocresol green to albumin is widely used for the estimation of this analyte in clinical laboratories since it is relative insensitive to competition with bilirubin, fatty acids, and drugs.[14] The optoelectronic technique of monitoring the dye–protein interaction is reagentless, simple, inexpensive, and extremely reproducible. Similarly, coimmobilization of pH-sensitive dyes with proton-consuming or -producing enzymes to a thin transparent membrane permits the technique to be extended to the measurement of enzyme substrates. The operational responses of the enzyme optoelectronic biosensor compare very favorably with conventional enzyme electrodes,[17] enzyme-linked field effect transistors,[18] and enzyme thermistors.[6,7] However, the optoelectronic device has advantages in that it can be readily constructed in fiber optic format[19] and thus have a high degree of mechanical flexibility combined with small size, low cost, and disposable construction.

---

[16] The rate of change of output voltage (V_o min^{-1}) could be increased linearly with LED voltage. For example, doubling the V_{LED} from 2.000 V to 4.000 V almost exactly doubled V_o min^{-1}.

[17] L. D. Bowers, *Trends Anal. Chem.* **1,** 191 (1982).

[18] S. Caras and J. Janata, *Anal. Chem.* **52,** 1935 (1980).

[19] M. J. Goldfinch and C. R. Lowe, unpublished observations.

[32] Optical Fiber Affinity Sensors

By JEROME S. SCHULTZ and SOHRAB MANSOURI

Introduction

Recent developments in the technologies related to optical fibers, membranes, and molecular biology have made possible a new generation of miniature biosensors which are sensitive, selective, and stable. As a result of developments in the communications industry, single optical fibers are available with excellent transmission characteristics and a whole array of couplers, light sources, filters, and light detectors.

The recent proliferation of commercially available, hollow-fiber membranes is of particular interest to biosensor development. Because of their small dimensions (of the order of tenths of a millimeter) and selective permeability properties, hollow fibers provide a "protective" environment which can respond rapidly to external chemical changes.

We have utilized these technologies for the development of a new, generic type of biosensor that can be adapted to a wide range of assays.[1] Initial evaluation of this "affinity" optode sensor concept was made by constructing a glucose sensor based on the use of concanavalin A as a receptor site.[2-4]

The principle of detection is based on the reversible competitive binding of glucose and a high-molecular-weight fluorescein-labeled (FITC) dextran for sugar-binding sites on concanavalin A (Con A), which is immobilized on the interior surface of a hollow dialysis fiber. The configuration is shown in Fig. 1. The amount of free (unbound) FITC-dextran is related to the concentration of glucose by the following reversible reactions:

$$\text{Glucose + Con A–membrane} \rightleftharpoons \text{glucose–Con A–membrane} \qquad (1)$$
$$\text{(free)} \qquad\qquad\qquad \text{(bound)}$$

$$\text{FITC-dextran + Con A–membrane} \rightleftharpoons \text{FITC-dextran–Con A–membrane} \qquad (2)$$
$$\text{(free)} \qquad\qquad\qquad\qquad \text{(bound)}$$

The interior glucose concentration is the same as in the external medium because it freely permeates through the hollow fiber. The dextran molecular weight is chosen so that it is retained within the transducer

[1] J. S. Schultz, U.S. Patent 4,344,438 (1982).
[2] J. S. Schultz and G. Sims, *Biotechnol. Bioeng. Symp.* **9**, 65 (1979).
[3] J. S. Schultz, S. Mansouri, and I. J. Goldstein, *Diabetes Care* **5**, 245 (1982).
[4] S. Mansouri and J. S. Schultz, *Bio/Technology* **2**, 885 (1984).

METHODS IN ENZYMOLOGY, VOL. 137

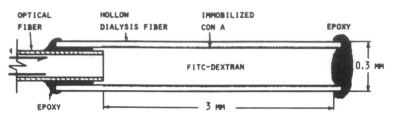

FIG. 1. Schematic diagram of the affinity sensor transducer element.

chamber, and the amount of free dextran is measured by the single optical fiber fluorometer which is inserted into the lumen of the hollow fiber. The field of view of the optical fiber is limited to the central region of the hollow dialysis fiber, and, thus, bound FITC-dextran on the membrane Con A sites is not detected.

Component Characteristics

Optical Fibers

The feasibility of optical fibers as "wires for light" is based on the fact that optical energy will be efficiently transported through a transparent material if the size of the waveguide transverse to the direction of light propagation is of the order of the dimension of the wavelength of light or more. Only light rays traveling at a small angle to the axis remain confined to the waveguide, and a transverse gradient in refractive index at the walls serves to reflect the light and trap the energy. This principle has been applied to the development of optical fibers as waveguides by creating a symmetrical annular structure where the index of refraction of the inner core is greater than that of the outer coating. Figure 2 illustrates three forms of optical fibers that behave as waveguides. Generally, the cost of the fibers increases as one proceeds from top to bottom.

Optical fibers have some properties that are very interesting for sensor development. They provide a capability both to bring light energy to remote locations and also to monitor optical signals from remote locations. The inherent minute size of optical fibers almost automatically leads to miniaturization of sensors based on optical effects.

Some of the characteristics of optical fibers that are particularly relevant for sensor applications will be reviewed here. For a more extensive analysis of optical fibers the reader should refer to introductory texts.[5]

[5] J. C. Palais, "Fiber Optic Communications." Prentice-Hall, Englewood Cliffs, New Jersey, 1984.

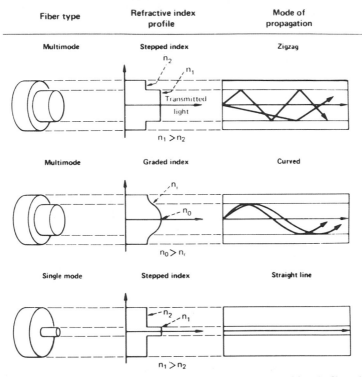

FIG. 2. Types of optical fibers commercially available. Top: Multimode fibers have the advantage that they are inexpensive, large numerical apertures are available, and they are easy to connect. A disadvantage is that, in a pulsing mode, there is large dispersion for different wavelengths of light. Bottom: Monomode fibers have the advantage that there is little dispersion with wavelength, and small numerical apertures are available. Disadvantages are that a laser light source is required and there are no good connectors for these fibers. Middle: The graded index fiber has properties inbetween the other two.

Since one of the most important applications of optical fibers is to act as a conduit of light, the attenuation of light in an optical fiber as a function of wavelength gives an indication of the useful range of the optical spectrum that is available. A typical absorption spectrum for a glass fiber is shown in Fig. 3. Transmission of light is highest at the longer wavelengths, i.e., in the infrared, and least in the ultraviolet. However, it should be noted that the transmission figures are given in decibels per kilometer since most optical fibers are manufactured for the communications industry, where distances of the order of kilometers are the norm. For analytical purposes, typical transmission distances are of the order of meters, and in these circumstances the absorption effect may be ignored over most of the spectrum. Referring to Fig. 3, for 1 m of fiber the light loss is about

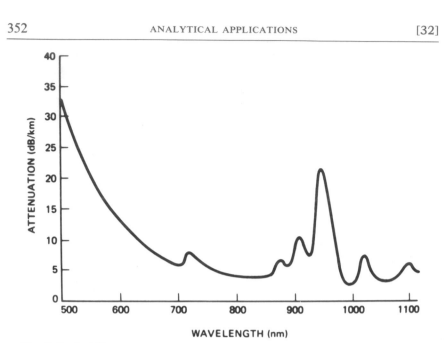

WAVELENGTH (nm)

FIG. 3. Typical fiber attenuation curve as a function of wavelength of light. QSF-A grade material, *NA*, 0.18 made by quartz and silica.[6]

0.1% at 800 nm and 1% at 500 nm. Plastic optical fibers adsorb light to a much larger extent (at 800 nm there is a 10% loss for a meter length) so that they are not as effective for communication purposes, but they still are candidate materials for sensors.

If a pulse of light, covering a range of wavelengths, is passed through an optical fiber, there is a slight spatial separation of the different wavelength bands due to the variation of index of refraction of the glass with wavelength. Thus, for long fibers (of the order of kilometers), one obtains a temporal spectrophotometer effect since the longer wavelengths will reach the end of the fiber earlier. Also, the pulse width will be broadened on the order of 20 nsec.

Some operational considerations for the use of optical fibers in sensors are the need for light sources, connectors, and detectors. Fortunately, the expanding applications of optical fibers for the communications industry has promoted a number of commercially available, light-emitting diodes (LEDs), laser-injection diodes, photodiodes, and couplers to meet these assembly needs.[6]

Modes of Operation. There are a number of ways of utilizing optical fibers for sensor applications. In general, for use to measure chemicals,

[6] E. D. Lacy, "Fiber Optics." Prentice-Hall, Englewood Cliffs, New Jersey, 1982.

light has to be coupled from the optical fiber to an external medium containing the chemical species that are to be measured. Light from a fiber can be coupled to external media either from the end of the fiber or from the lateral surface of the fiber.

Optical fibers have an effective acceptance angle and emission angle for light which is determined by the refractive indices of the core material and external fluid. This critical angle (α) is related to the numerical aperture (NA) of the fiber by Eq. (3):

$$NA = N_0 \sin \alpha \tag{3}$$

where N_0 is the refractive index of the fluid outside the fiber. Figure 4a shows the field of view of optical fibers for typical numerical apertures of commercially available fibers.

This view factor characteristic of optical fibers has important implications for sensor design. If a single fiber design is used, only the space within the cone volume is optically sampled. Thus, one can achieve a spatial separation of detection if some of the chemical species are preferentially distributed to either the central volume or the other space. This property of spatial view separation is one critical element of the affinity biosensor to be described in detail below.

The light energy density within the central cone is highest at the fiber interface and falls off rapidly with distance from the end of the fiber. Figure 5 shows the relative fluorescence detected as a function of distance from the end of the optical fiber when the excitation light beam comes from the same fiber; it is clear that most of the response is obtained within 2 mm of the end of the optical fiber. Thus, the approximate dimensions of the optically sampled volume of the device depicted in Fig. 1 are 0.3 mm diameter by 2 mm in length.

If colorimetric light absorption measurements are desired, light can be redirected back into a fiber by the use of fine particle scatters within the sample volume. When two optical fibers are employed for absorption measurements, one for the light source and the other to measure transmitted light, the geometry of the sensor and sample volume will be dominated by numerical aperture considerations. If the fibers are aligned side by side, only a small portion of the fields of view of the two fibers will overlap (see Fig. 4b).

Finally, if a portion of the cladding is removed from the optical fiber, then a portion of the light energy enters the fluid sample through the circumferential surface of the core fiber (Fig. 4c). This evanescent field falls off exponentially from the surface of the fiber, and, for practical purposes, the volume sampled is within 1000 Å of the surface. Thus, surface-adsorbed films in the surrounding fluid will dominate the response

a

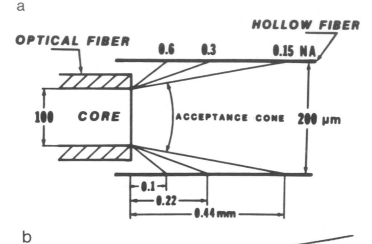

OPTICAL FIBER

HOLLOW FIBER

0.6   0.3   0.15 NA

100   CORE   ACCEPTANCE CONE   200 µm

0.1
0.22
0.44 mm

b

INPUT CORE

DETECTION FIELD

OUTPUT CORE

BLIND VOLUMES

c

EVANESCENT RADIATION

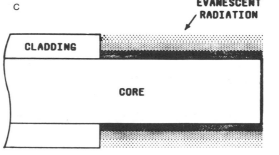

CLADDING

CORE

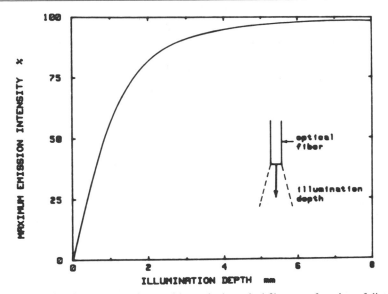

FIG. 5. Relative fluorescence detected by a single optical fiber as a function of distance from the end of the optical fiber.

if the adsorbed species have optical properties, e.g., fluorescence or adsorption.

Components for Optical Sensors. A wide variety of optical fibers are commercially available in both production and experimental quantities. Some of the properties and sources of optical fibers are given in Table I.

In addition to these basic optical fiber elements, many companies manufacture a variety of optical cables containing from 1 to 18 or more optical filaments. A key requirement for the fabrication of optical fiber devices is the availability of connectors or couplers. Ideally, when light intensities are being measured, the loss in a connector should be negligible, or at the very least a constant amount. Unfortunately, at present, this component of the optical train has the least level of reproducibility. The technical problem is that individual fibers must have "mirror smooth" ends and must be aligned with 2° and displaced less than 5% of the core diameter. Because of the small diameter of single mode fibers (5 μm), the last requirement has defeated attempts to produce inexpensive couplers for these fibers.

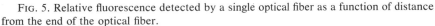

FIG. 4. Light interaction patterns between optical fibers and the surrounding medium. (a) Field of view as a function of numerical aperture. (b) Overlapping field of view of two adjacent optical fibers. (c) Region of light penetration from the lateral surface of an optical fiber without cladding.

TABLE I

TYPICAL OPTICAL FIBERS MANUFACTURED FOR COMMUNICATION SYSTEMS[a]

Manufacturer	Product	Core/clad material	Core diameter (μm)	Fiber diameter (μm)	Numerical aperture
ITT	GS-02-1	Glass	50	125	0.25
Times Fiber Communications, Inc.	G5054	Glass	50	125	0.16
Corning	1051	Glass	63	125	0.20
Corning	SDF	Glass	100	140	0.30
Quartz Products Corp.	QSF-A-200	Glass/plastic	200	400	0.22
DuPont	PFX-S120R	Glass/plastic	200	600	0.4

[a] From Schultz *et al.*[3]

Ordinary lenses can be used to focus light into an optical fiber, but special cylindrical lenses have particularly favorable characteristics for use with optical fibers. The basic problem is that most light sources (e.g., arcs, filaments) have dimensions larger than the diameter of the core, and thus slight movements of the light source can result in drastic variations in the amount of light entering the optical fiber. Figure 6 shows the optical setup that we found useful for this type of work.

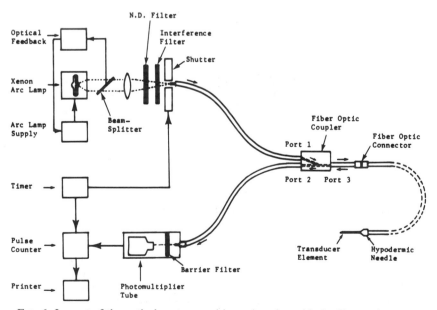

FIG. 6. Layout of the optical system used in conjunction with the fiber optic sensor.

Miniature light sources, such as LEDs and solid-state lasers, will eventually allow miniaturization of the entire sensor device. However, at the moment, the optimum wavelength ranges of the devices available are in the infrared, which do not match the optimum wavelengths for typical fluorescent dyes such as fluorescein or rhodamine.

Membrane Characteristics

Membranes have an important role in the fabrication of a biosensor. As mentioned earlier, membranes serve to isolate the chemical elements of the transducer from the external environment. The particular characteristics of membranes for this application are differential permeability, suitability as a matrix for immobilization of various chemicals, and availability in the form of hollow fibers.

Since the biosensor we have developed requires the free passage of the analyte through the membrane with complete retention of the competing analyte, membranes with sharp size cutoff characteristics are best. Since most membranes are not homoporous, there is no one characteristic molecular weight cutoff. Figure 7 shows the permeability properties of a typical commercial membrane[7] with a cutoff molecular weight of about 2000. For example, if the analyte was of molecular weight 2000, for sensor use the competing analyte should have a molecular weight at least 30 times greater, i.e., 60,000. Table II lists some commercially available hollow fibers, materials, and typical permeabilities.

In our biosensor the receptor site is immobilized in the hollow fiber membrane to spatially separate the bound competing analyte from the field of view of the optical fiber. Thus, membranes must either be of materials that can be activated for chemical immobilization of the receptor species or be of asymmetric structure with the rejecting surface on the outer surface and an open spongelike structure on the luminal side. In the latter situation the receptor species could be ultrafiltered from the lumen into the internal spongy region and then immobilized in place by various cross-linking agents.

Receptors

The primary property of interest in biosensors is the selectivity of the device, and this in turn is a property of the specificity of the receptor moiety chosen for the sensor. There is a large variety of biomacromole-

[7] Enka AG, "Cuprophan Hollow Fibers." Enka AG, Wuppertal, Federal Republic of Germany.

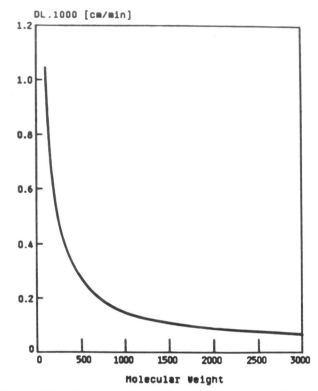

FIG. 7. Permeability of Cuprophan, a dialysis membrane, as a function of permeant molecular weight.[7]

cules that have receptor activity: the most obvious are enzymes, antibodies, membrane-binding proteins, and nucleic acids.

The key requirements for this application are (1) totally reversible binding, (2) rapid equilibration (of the order of minutes or less), (3) no destruction of the analyte or competing analyte, (4) high differential selectivity (binding constants) for analytes of similar structure, and (5) long-term stability.

Due to the enormous amount of work that has been done in developing antibodies to a large variety of haptens, this class of biomolecules is an obvious first choice for finding receptor candidates for drugs, metabolites, and hormones. However, most of the antibodies selected for immunoassay purposes have very high binding constants for their analytes, and this may be the result of very slow dissociation rates. Thus, antibodies to be used in biosensors should be screened for high dissociation rates. Monoclonal antibodies appear to be more suitable from this point of view.

TABLE II
HOLLOW FIBERS USED FOR HEMODIALYSIS MEMBRANE[a]

Membrane	Dimensions of dry fiber (μm)		Polymer and description	Manufacturer	Glucose permeability (10^4 cm/sec)
	Internal diameter	Wall thickness			
Enka B2-AH* (Cuprophan)	230	20	Cellulose produced via cuprammonium intermediate	Enka AG, FRG.	2.9
Cordis Dow	200	30	Cellulose regenerated from acetate	Cordis Dow	2.3
Cordis Dow 3500	210	40	Cellulose acetate	Cordis Dow	
Asahi PAN-15	200	50	Anisotropic polyacrylonitrile	Asahi Medical, Japan	
Amicon PMD	215	60	Anisotropic polysulfone	Amicon Corp.	4.0

[a] From Schultz et al.[3]

Construction of a Prototype Glucose Sensor

Receptor (Con A) Immobilization. Hollow Cuprophan dialysis fibers obtained from Enka are prepared for immobilization by first removing the isopropyl myristate and other preservatives by washing in dry methanol. Afterward, the fibers are dried and stored in N_2 prior to use. In order to control the exposure and concentration of the various solutions used in the immobilization procedure, the two ends of a single fiber, about 15 cm long, are sealed with epoxy cement into the tips of two 23-gauge needles. The various solutions are perfused through the lumen of the fiber with an infusion pump. Throughout all operations the fiber is maintained immersed in appropriate buffer solutions with gentle stirring.

In general the procedure is to partially oxidize the cellulose fibers with periodate to form dialdehyde groups. Then a 1,6-hexanediamine spacer arm is attached, which also serves to provide free amino groups. Subsequently, the receptor protein (Con A) is attached by a glutaraldehyde coupling sequence. Finally, the preparation is stabilized by reduction of all free aldehyde groups with sodium borohydride. The conditions for each step are summarized in Table III. In all steps the external bathing solution consists of the same buffer solution as is used for the infused reagents except for the Con A immobilization step. Here the external medium is a 0.01% glutaraldehyde solution which permeated through the membrane into the fiber and promoted cross-linking of free Con A molecules to the immobilized Con A, resulting in a thin gel layer of Con A.

Fabrication of the Sensing Element. A piece of Con A-bound Cuprophan dialysis fiber about 2 cm long is placed in a petri dish filled with 15 μg/ml FITC-dextran, 70,000 MW (Sigma, St. Louis, MO), in phos-

TABLE III

REACTION STEPS FOR IMMOBILIZING CON A IN A SINGLE HOLLOW FIBER

Reaction step	Reactant	Concentration	Medium	pH	Reaction time (hr)
Oxidation	Sodium periodate	50 mM	Water	Neutral	5
Spacer addition	1,6-Hexanediamine	1%	Phosphate buffer	8	2
Spacer coupling	Glutaraldehyde	2%	Phosphate buffer + NaCl	7.2	0.75
Immobilization	Con A	1.0 mg/ml	Phosphate buffer + NaCl	7.2	6
Reduction	Sodium borohydride	1%	0.5 M Phosphate	8	1

phate-buffered saline (PBS), pH 7.4. A syringe with a 20-gauge needle is employed to push the dextran solution through the fiber to ensure complete equilibration between FITC-dextran bound to Con A inside the fiber and the FITC-dextran in solution.

The fabrication of the sensing element is performed under a low power dissecting microscope. The fiber optic end is brought through a hypodermic needle (prepared as shown in Fig. 8) into the FITC-dextran solution, and the hollow fiber is slid about 5 mm over the optical fiber. The unit is lifted out of the solution, and a small amount of cyanoacrylate adhesive (Eastman 910, Eastman Kodak, Rochester, NY) is quickly applied to the entire circumference of the optical fiber/hollow fiber interface. The amount of adhesive is minimized by applying it with a thin needle (0.2 mm diameter). The hollow fiber is then resubmerged. The adhesive in contact with water polymerizes and forms an impermeable seal. The optical fiber with attached hollow fiber is then withdrawn into the needle and aligned as in Fig. 8. The free end of the hollow fiber is cut near the tip of the needle by dissecting scissors. The hollow fiber, once again, is lifted out of the solution, the adhesive is quickly applied to the free end of the hollow fiber to provide a seal, and the completed transducer is resubmerged in the solution.

There are two primary factors in determining the concentration of FITC-dextran in the transducer. First, the concentration of FITC-dextran should be substantially higher than the background fluorescence in the medium (in plasma, the background fluorescence is equivalent 0.03 μg/ml FITC-dextran). Second, the total amount of dextran should be less than the amount of immobilized Con A in the transducer in order to maximize the ratio of bound to free dextran in the absence of glucose. The solution used to infuse the hollow fiber had a concentration of 15 μg/ml FITC-dextran, which resulted in a total concentration (free plus bound FITC-dextran) of approximately 100 μg/ml or 1.5×10^{-6} M. This total concentration of FITC-dextran satisfies both requirements: it is at least 30 times

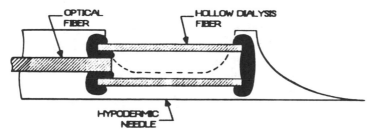

FIG. 8. Assembly details of a sensor transducer element inside a hypodermic needle.

greater than the detection limit of the fluorescence and about 7 times less than the estimated Con A concentration of 10^{-5} M in the dialysis fiber.[8]

Description of the Optical System.[4] Figure 5 schematically shows the optical system developed for the sensor. The excitation light is transmitted into the transducer element by means of the optical coupler and the optical fiber. Free FITC-dextran in the illumination field of the optical fiber becomes excited and emits fluorescent light. A portion of the fluorescence enters the optical fiber and is directed by the coupler into the light detector. A barrier filter mounted in front of the detector blocks the back-scattered source radiation, allowing the detection of the fluorescence only.

The optical system consists of a number of components: a light source, suitable optical filters, a light detector, signal processing electronics, and a readout. All of these components are contained in a single support module. The transducer element is extended back approximately 1 m and is connected to the support module through the optical fiber. Specific design details concerning these components are presented below.

LIGHT SOURCE. The necessary requirements for the light source are its illumination stability, ability to focus the light down to a small spot size, and adequate intensity at 490 nm wavelength (peak absorption of FITC). These requirements are met with a low power xenon arc lamp (75 W, Osram) with a small luminous area (0.25 × 0.5 mm). The lamphouse (Stabilarc 250, Ealing, Newport Beach, CA) is equipped with an optical feedback controller to minimize "arc wander" effects.

LIGHT DETECTOR. A detection capability of extremely small fluorescence intensities with a reasonable signal-to-noise ratio is the main requirement of the light detector. This is met with a 16-stage photomultiplier tube (FW130, ITT, Fort Wayne, IN) having a 0.36-cm-diameter sensitive area on its cathode. The cathode sensitivity is maximum at 420 nm and dropped to 75% of its maximum at the 520 nm peak of the fluorescein emission spectrum. The housing for the photomultiplier tube (PMT) has a built-in photon-counting amplifier–discriminator module (ADH-1400-120, Products for Research, Danvers, MA). The PMT is operated at 1800 V, using a 0–3000 V power supply (6516 A, Hewlett-Packard, Corvallis, OR). A 5 V regulated power supply (LZD 21, Lambda, Melville, NY) drives the amplifier–discriminator module. The output from the PMT housing is fed to a pulse counter (Model 5845, Data Precision, Danvers, MA). Counts registered by the counter are recorded on a thermal printer (NP-7A, Gulton, East Greenwich, RI).

[8] S. Mansouri, Ph.D. thesis. University of Michigan, Ann Arbor, Michigan, 1983.

SHUTTER AND OPTICAL FILTERS. The output from the light source focused onto the tip of the optical fiber which is connected to port 1 of the coupler. The optical fiber is held by a 5-axis fiber optic positioner (FP-2, Newport Research, Fountain Valley, CA). A linear variable neutral density fiber (density ranged from 0.04 to 3.0, Ealing), an interference filter (peak wavelength 488 nm, Edmund Scientific, Barrington, NJ), and an electronic shutter (Model 844, Newport Research) are positioned between the lens and the optical fiber. The interference filter is taped over the 6-mm-diameter aperture of the shutter. All of the optical components, including the light source, are mounted on a 36-inch-long optical rail (URL-36, Newport Research). A barrier filter (OG 515, Americal Optical, Buffalo, NY) is positioned in the PMT housing between the photocathode and the optical fiber connected to port 2 of the coupler.

TIMER. The timer controls the shutter and the counter. In every measurement, the timer signals the shutter to open and, at the same time, resets and starts the counter. When the counting period ($\sim$2 sec) is completed, the timer closes the shutter and stops the counter, and the count is printed. This process is automatically repeated every 2 min.

OPTICAL FIBER AND COUPLER. In this work no attempt was made to find an optimal optical fiber. The SDF Corning optical fiber (100 μm core and 140 μm cladding, Corning Glass Works, Corning, NY) was selected, based on its large core/cladded diameter ratio, small fiber diameter (small enough to be easily inserted into a dialysis hollow fiber), and relatively low numerical aperture (*NA* 0.3).

A three-port fiber optic directional coupler (TC3-B, Canstar, Scarborough, ON) provides a separation between the excitation and the emission lights. This is an improvement over the initial design in which a dichroic beam splitter was used instead of the coupler.[3] The incorporation of the optical coupler not only simplifies the optical system, but also increases the detection efficiency of the fluorescence. Approximately 20% of the excitation light entering the optical fiber at port 1 is directed into the transducer element (excluding connection losses), and of the fluorescent light reaching the coupler 75% is directed to the light detector. Ports 2 and 3 of the coupler are terminated using AMP optical connectors (Connector No. 227285-4, AMP, Harrisburg, PA). Port 2 is connected to the PMT housing. At port 1, about 5 cm of the fiber jacket is stripped, and the fiber is cut to produce a clean, flat end. The optical fiber is then installed onto the fiber optic positioner. The attachment of the optical fiber to the transducer element is through 1 m of the SDF fiber which is enclosed in a black rubber tube (1 mm o.d.) for most of its length. An AMP optical connector is assembled at one end, the fiber at the other end is cut to produce a clean flat end, and subsequently the sensing element is fabricated at this end as mentioned before.

Performance Tests of the Sensor[4]

The characteristics of the sensor were tested in 50 mM phosphate buffer containing 150 mM NaCl with varying glucose concentrations. The assay solutions in which the sensing element is immersed are covered with a cardboard box to eliminate room light effects.

In phosphate buffer the signal corresponding the fluorescence of free dextran in the sensing element increases with glucose concentration as shown in Fig. 9. The response is linear from 0.5 to 4 mg/ml glucose concentration. Above 4 mg/ml there is a change in slope, and the response levels off due to saturation of Con A sites with glucose.

Figure 9 also indicates variation in the performance of the sensor with time. During a 15-day period in which many other experiments were also performed with this sensor, including about 10 calibrations with other sugars, exposure to plasma, and mechanical stresses of stirring, the maximum fluorescence intensity was reduced approximately 22%. Most of the reduction, however, occurred during the initial 5 days (20%), and there was little change afterward. In the range of interest for blood glucose

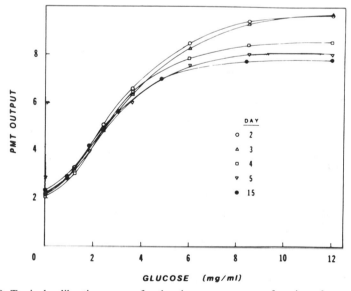

FIG. 9. Typical calibration curves for the glucose sensor as a function of sensor age.

levels (0.5–4 mg/ml), the drift in the calibration curve was less pronounced and did not exceed about 15% in 15 days.

Plasma samples from patients receiving glucose tolerance tests were obtained from the Diabetes Center at the University of Michigan, Ann Arbor. The pH of the samples was adjusted to 7.4 by solid KH_2PO_4, and the glucose level in some of the samples was artificially increased by the addition of glucose powder. Glucose concentrations were determined (Beckman Astra-8 Glucose Analyzer) and subsequently measured by the sensor which had been previously calibrated with standard glucose solutions. The results of the sensor were compared to those of the reference method.

The relation between plasma glucose concentrations determined by the sensor and by a Beckman glucose analyzer is shown in Fig. 10. The slope of 1.066 indicated a small proportional error, the S_y (standard error of estimate in the y direction) implied a random error of 0.13 mg/ml, and the y intercept suggested a negligible constant error of −0.08 mg/ml. The

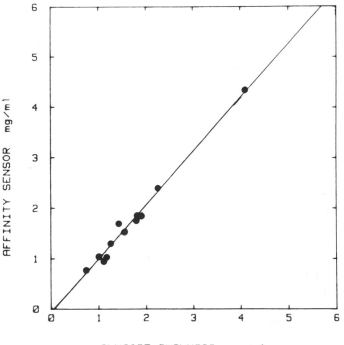

FIG. 10. Comparison of sensor readings and chemical assays for glucose content of blood plasma.

correlation coefficient was 0.991 and, therefore, did not suggest any significant difference between the two sets of values. It appears that the sensor measures plasma glucose accurately in the range of 0.5–4 mg/ml, and the other plasma components do not interfere with the measurement.

[33] Surface-Oriented Optical Methods for Biomedical Analysis

By HANS ARWIN and INGEMAR LUNDSTRÖM

Introduction

Two important features of optical methods are their nondestructive character and the fact that high surface sensitivity can be achieved without the necessity of a vacuum. These properties facilitate the application of optical methods in biomedical analysis, because most molecules and reactions of interest in this area are confined to the liquid phase. Thus, optical techniques offer possibilities to study biochemical reactions either after completion, *ex situ*, in air (end-point measurements) or kinetically, *in situ*.

We will limit our discussion to optical methods in the visible and near-ultraviolet photon energy range, that is, wavelengths between 200 and 800 nm. The major drawback is the poor identification capability, which is basically due to the absence of sharp and well-separated optical structures in the optical response of the organic material under study. Therefore, in this context it is necessary to use light probes mainly for detection and only rarely for identification. The implication is that we must know what we are measuring.

Our objective is the characterization and determination of the amount of adsorbed organic material on a solid substrate surface. Quantification is the most important task from a clinical point of view. Our goal can be, for example, to measure the amount of antibodies adsorbed on an antigen-covered surface or the rate of enzymatic degradation of a thin preadsorbed organic layer.

There are a number of currently used or suggested surface-oriented optical methods suitable for biomedical analysis.[1] Some are indirect, like staining with Coomassie brilliant blue[2] or detecting changes in wettability

[1] B. Ivarsson and I. Lundström, *CRC Crit. Rev. Biocompat.* **2**, 1 (1986).
[2] A. L. Adams, M. Klings, G. C. Fisher, and L. Vroman, *J. Immunol. Methods* **3**, 227 (1973).

by water condensation.[3,4] Direct optical methods often involve specially prepared solid surfaces. Giever[5] made use of surfaces coated with thin indium films for which changes in light scattering occur due to antigen–antibody complex formation. Adams et al.[2] and later Sandström et al.[6] have shown that changes in interference colors of thin layers can be used to qualitatively monitor antigen–antibody reactions occurring on the surface of the layers (see also the subsequent contributions by Jönsson et al. [34] and Mandenius et al. [35]). Direct optical properties are generally more attractive because they are easier to implement in routine work and often comprise fewer steps in laboratory procedure. This facilitates understanding and thereby increases the confidence in the methods.

This work is, however, not intended to be a review, and we will restrict our description to important and promising direct optical methods based on reflection of polarized light. After a short introduction on reflection of light, we will describe principles, instrumentation, and applications of ellipsometry, reflectometry, and surface plasmon resonance techniques. It should be noted that we have selected illustrative examples mainly from research in our laboratory, which of course is only a partial representation of the activities in the field.

Reflection Techniques

In this section, we give some background to aid in understanding the different reflection techniques used to determine optical properties of materials. In general, the optical properties of a sample can be determined if the properties of both incident and reflected waves are known. Reflectometry and ellipsometry are two techniques for obtaining this information.

In reflectometry one measures the change in intensity due to reflection. The general formula relating the intensity of the incoming (I_i) and reflected (I_r) beams is

$$R = I_r/I_i \qquad (1)$$

where the quantity R is called reflectance. Two important parameters in R are the angle of incidence and the state of polarization of the incoming light.[7]

In ellipsometry, the change in state of polarization is the important

[3] H. Elwing, Ph.D. thesis. Gothenburg Univ., Sweden, 1980.
[4] B. M. Wikström, H. Elwing, and Å. J. R. Möller, Enzyme Microb. Technol. 4, 265 (1982).
[5] I. Giever, J. Immunol. 110, 1424 (1973).
[6] T. Sandström, M. Stenberg, and H. Nygren, Appl. Opt. 24, 472 (1985).
[7] J. D. Jackson, "Classical Electrodynamics." Wiley, New York, 1975.

quantity, and an ellipsometric measurement provides us with the ratio

$$\rho = \tan \psi \, e^{i\Delta} = r_p/r_s \tag{2}$$

where $\tan \psi$ and Δ are the amplitude and phase, respectively, of the complex reflectance ratio, ρ. r_p and r_s are the complex reflectances for light polarized parallel (p) and perpendicular (s) to the plane of incidence, respectively.

We can now make some general remarks about the relative advantages and disadvantages of ellipsometry and reflectometry. Ellipsometry is strictly a nonnormal incidence technique, whereas reflectometry can be performed at either normal or nonnormal incidence. Reflectometry deals with intensities, whereas ellipsometry deals with intensity-independent complex quantities. Because more complicated quantities are involved, an ellipsometer is necessarily more complicated than a reflectometer. The most serious drawback of ellipsometers is the general requirement of transmission through optical elements (e.g., polarizers). Therefore, such measurements are limited to those wavelength ranges where good-quality transmitting elements are available.

Ellipsometry is unquestionably the more powerful for a number of reasons. First, two parameters instead of one are independently determined in any single measurement operation. Second, ellipsometric measurements are relatively insensitive to intensity fluctuations of the source, temperature drifts of electronic components, and macroscopic roughness. Third, accurate reflectance measurements are difficult, in general requiring double-beam methods. In contrast, ellipsometry is intrinsically a double-beam method where one polarization component serves as amplitude and phase reference to the other. Finally, ρ explicitly contains phase information, which makes ellipsometry generally more sensitive to surface conditions.

Ellipsometry

Principles

In ellipsometry, the complex reflectance ratio [Eq. (2)] is measured. In our case, we are interested in determining properties of thin surface films, and other system parameters are assumed to be known. In an ideal three-phase model (ambient–thin film–substrate) the complex reflectances in Eq. (2) are functions of all optical parameters.[8] We are here mainly concerned

[8] R. M. A. Azzam and N. M. Bashara, "Ellipsometry and Polarized Light." North-Holland, Amsterdam, 1977.

with determination of the thickness and refractive index of a thin surface film. Other parameters are assumed to be known.

For very thin films adsorbed on conventional substrates, it is usually found that only one of the two ellipsometer parameters Δ and ψ changes due to formation of a thin surface film or that the changes in Δ and ψ are correlated so that only one effective datum is available. In general, the film refractive index is then assumed to be known, and the film thickness remains as the only unknown and is calculated from Eq. (2). The sensitivity is so high that changes in film thicknesses of fractions of an angstrom can be resolved.

In biomedical analysis, because we are dealing with macromolecules of nanometer dimensions, it is advantageous to convert the more or less nonphysical "average thickness" to surface coverage, θ, or adsorbed amount, Γ, in terms of mass per unit area ($\mu g/cm^2$). A simple empirical conversion formula has been deduced by de Feijter et al.[9]:

$$\Gamma = \frac{0.1d(N_1 - N_0)}{dn/dc} \quad \text{mg/m}^2 \tag{3}$$

where dn/dc is the refractive index increment which is of the order 0.18–0.19 ml/g, d is the thickness (nm), and N_1 and N_0 are the refractive indices of the material in the film and of the ambient, respectively.

Other more elaborate theories have also been developed.[10] The validity of ellipsometer data has also been verified by radioactive labeling techniques.[11] The observed linear relationship between the ellipsometer parameter Δ and film thickness or, alternatively, the surface concentration, Γ, has been confirmed.

Instrumentation

An ellipsometer consists of a light source, a polarizing system, a test surface, an analyzer, and a detector (Fig. 1). The light source provides a collimated beam of monochromatic light and is usually a laser or a mercury or xenon lamp with a filter. The polarizing system and the analyzer are high quality optical devices like polarizing prisms and retarders.

A variety of different configurations of ellipsometer systems has been suggested,[12] but the majority of published results comes from classical null ellipsometers or from photometric ellipsometers of rotating ana-

[9] J. A. de Feijter, J. Benjamins, and F. A. Veer, *Biopolymers* **17**, 1759 (1978).
[10] P. A. Cuypers, J. W. Corsel, M. P. Janssen, J. M. M. Kup, W. T. Hermens, and H. C. Hemker, *J. Biol. Chem.* **258**, 2426 (1983).
[11] U. Jönsson, M. Malmqvist, and I. Rönnberg, *J. Colloid Interface Sci.* **103**, 360 (1985).
[12] P. S. Hauge, *Surf. Sci.* **96**, 108 (1980).

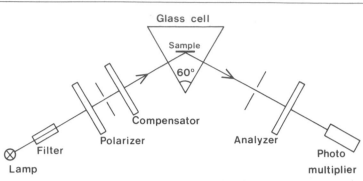

FIG. 1. Optical components in a null ellipsometer. The polarizer, analyzer, and compensator can be rotated. The compensator is normally set at $C = -45°$, and the polarizer and analyzer are rotated until the light at the photodetector is extinguished. The angular positions of the polarizer and the analyzer are the ellipsometer outputs.

lyzer[13] type. Photometric ellipsometer systems are not yet commercially available. They are technically rather complicated and find their main use in more fundamental research projects. Null ellipsometers (Fig. 1) of different types are supplied by Gaertner Scientific Co. (Chicago, Illinois) and Rudolph Research (Fairfield, NJ). Modern computerized instruments are straightforward to use. Necessary software for data interpretation are supplied with the instrument. Some instruments are of the push-button type which means that the measurement procedure is reduced to the sequence: insert the sample, push a button, wait 30 sec, and read the result. The result is often presented in film thickness (nm) or in adsorbed amount per unit area ($\mu g/cm^2$).

A comparison ellipsometer, the Isoscope (Sagax Instrument, Sundbyberg, Sweden) is now available.[14] This instrument is based on ellipsometric comparison between a reference surface and the sample surface. If these two surfaces are optically equivalent, the light through the instrument is totally extinguished. By using a suitable reference surface, e.g., with a gradient in film thickness, one then obtains extinction of light only at spots where the sample surface is equal to the reference surface. This technique is very attractive because thin layers, such as surface adsorption profiles, can be visualized. The operator can actually look in an eyepiece and see the thickness variations in films of nanometer dimensions.

[13] D. E. Aspnes and A. A. Studna, *Appl. Opt.* **14,** 220 (1975).
[14] M. Stenberg, T. Sandström, and L. Stiblert, *Mater. Sci. Eng.* **42,** 65 (1980).

Applications

Since the original works by Rothen,[15] Vroman and Lukosevicius,[16] and Trurnit[17] ellipsometry has been applied to a substantial number of investigations in biochemical and biomedical analysis. A recent review on fundamental protein adsorption is given by Ivarsson and Lundström.[1] Most measurements have been done on immunological[18,19] and enzymatic[16] systems, but as far as we are aware, ellipsometry has not been applied routinely in laboratory assays in the area of biomedical analysis.

Here we demonstrate the application of ellipsometry using two examples which are closely related to questions in clinical research. In Fig. 2, the interaction of serum with an antibody-covered silicon surface is studied by combining ellipsometry and the diffusion in gel (DIG) technique.[20] The antigen–antibody system is human serum albumin (HSA) (AB Kabi, Stockholm, Sweden) and rabbit anti-HSA (Behringwerke AG, Marburg-Lahn, FRG).

The test surfaces are prepared by immersing silicon wafers, 3.5 cm in diameter and 0.2 mm thick (Wacker-Chemie, Munich, FRG), for 15 sec in 3% solution of dichlorodimethylsilane, $Si(CH_3)_2Cl_2$, in trichloroethylene and then rinsed in trichloroethylene. This treatment gives a hydrophobic surface which spontaneously adsorbs HSA when immersed in HSA 100 mg/liter in 0.15 M NaCl for 30 min. The HSA-coated silicon wafers are placed in the bottom of 4.5-cm Petri dishes. Then 1% agarose (Pharmacia Fine Chemicals, Uppsala, Sweden) in 0.15 M NaCl at 56° is poured on the plates to form an agar layer 2 mm thick. A 30 mm long and 2 mm wide trough is cut in the agar layer on each plate (Fig. 2). The trough is filled with anti-HSA diluted 1/4 in 0.15 M NaCl. After 24 hr diffusion in a humid atmosphere, the agar layer is rinsed off and a new gel layer is cast on the silicon plates. Troughs 6 mm wide and 25 mm long are cut in the agar perpendicular to the initial trough. These troughs are filled with complement-rich serum or other reagents and incubated 1.5 hr at 37°. The gel is then rinsed off the plates with distilled water and the surfaces dried with nitrogen.

Ellipsometric measurements are performed spotwise at different dis-

[15] A. Rothen, *Rev. Sci. Instrum.* **16**, 26 (1945).
[16] L. Vroman and A. Lukosevicius, *Nature (London)* **204**, 701 (1964).
[17] H. J. Trurnit, *Arch. Biochem.* **51**, 176 (1954).
[18] C. Mathot, A. Rothen, and J. Casals, *Nature (London)* **202**, 1181 (1964).
[19] G. Poste and C. Moss, *Prog. Surf. Sci.* **2**, 139 (1972).
[20] H. Elwing, C. Dahlgren, R. Harrison, and I. Lundström, *J. Immunol. Methods* **71**, 185 (1984).

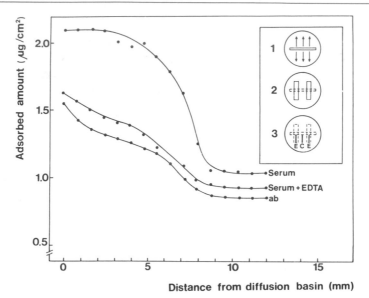

FIG. 2. Interaction of human serum with an anti-HSA-covered surface. The inset demonstrates the experimental technique. Antibodies are allowed to diffuse from a trough in an agar gel on an antigen-coated silicon wafer (1). After removal of the gel, a new gel is formed on the surface, and rectangular troughs in the gel are filled with serum containing complement (2). After removal of the gel and drying of the wafer, the amount of organic material is ellipsometrically determined along the solid lines indicated in (3). The curve marked ab is the control (or the antibody profile on top of the antigen layer), measured along C in (3). The curve marked serum is obtained (along E) when fresh human serum is incubated in the experimental trough. An addition to the amount of adsorbed organic material is detected. The middle curve is obtained when fresh human serum plus 10 mM EDTA is used in another experimental trough (but on the same wafer). Only a slight increase of adsorbed amount of organic material is detected. The antigen layer on the methylated silicon wafer corresponds to about 0.8 μg/cm^2. After Elwing et al.[20]

tances from the edge of the antibody basin either within the reagent basins (E = experimental) or outside (C = control) (Fig. 2). The implications of the results in Fig. 2 are discussed in Ref. 20.

Figure 3 demonstrates a similar experiment performed with an Isoscope.[21] Silicon wafers are made hydrophobic by incubation in dichlorodimethylsilane as described above. Bovine serum albumin is adsorbed to the surface (100 μg/ml in saline for 2 hr), and an agarose gel (1%) is poured over the antigen-coated surface in a 2 mm thick layer. A longitudinal well (3 × 20 mm) is punched out with a razor blade and filled with anti-BSA antiserum containing 2 mg/ml of specific antibodies. The diffusion is

[21] M. Stenberg and H. Nygren, J. de Phys. C10 (Suppl. 12), C10-83 (1983).

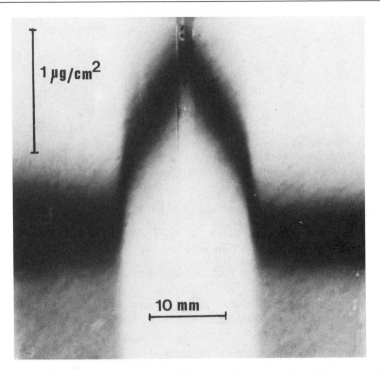

Fig. 3. Isoscope analysis of the lateral diffusion profile obtained by diffusion of antibodies over an antigen-coated surface. The dark areas represent the surface concentration profile. The region in the middle is the 3-mm-wide well from which lateral diffusion takes place. The vertical bar shows the scale of the surface concentration.

terminated by removal of the serum, and the gel and the plate are rinsed in water and dried. The rapid thickness increase at the advancing front indicates a diffusion rate-limited surface reaction, followed by a concentration-dependent reaction. Further details are found in Ref. 21. Note that the Isoscope result is a photograph of what the operator can see in the eyepiece.

Reflectometry

Principles

In reflectometry, the intensity change due to reflection of light [Eq. (1)] is measured. However, reflectometry is in general rather insensitive to surface conditions, except in certain cases. One such case is reflection of light polarized parallel to the plane of incidence at an angle of incidence

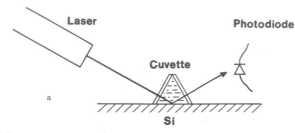

FIG. 4. (a) Instrumental setup for a simple reflectometer. See text for details. (b) Flow-through cell for reflectometry mounted on a HeNe laser. The silicon substrate is pressed against the cell and constitutes one of the walls in the sample compartment.

equal to the so-called Brewster angle of a dielectric substrate. The Brewster angle, ϕ_B, is defined by

$$\tan \phi_B = N_s/N_a \tag{4}$$

where N_s and N_a are the refractive indices of the substrate and ambient, respectively.

In the more general case of a transparent ambient and an absorbing (nondielectric) substrate, the substrate refractive index is a complex number, $N_s = n_s + ik_s$. In this case, the reflectance has a minimum at the so-called pseudo-Brewster angle. For highly polarizable materials, with low absorption (n_s high and k_s low), this minimum is very pronounced.[8] Furthermore the reflectance is very sensitive to the presence of overlayers.[22] The absolute magnitude of the reflectances is low, but the relative change due to an overlayer is large, especially if a substrate with high refractive index, like silicon, is used.

Instrumentation

Figure 4a shows a simple experimental setup.[23] This low-cost instrument consists of a HeNe laser, a glass cell for studies in liquids, and a photodiode as detector. The laser can be replaced with a lamp, a simple filter, and a Polaroid filter. The polarized laser (or the Polaroid filter) is rotated to give *p*-polarized light, and the components are mounted at an angle of incidence equal to the pseudo-Brewster angle. For a water–silicon interface this angle is 72.2° at a wavelength of 632.8 nm. For normalization, a second photodiode (not shown in Fig. 4) measures the light intensity of the source. The instrument can be made very compact by

[22] H. Arwin and I. Lundström, *Anal. Biochem.* **145,** 106 (1985).
[23] S. Welin, H. Elwing, H. Arwin, I. Lundström, and M. Wikström, *Anal. Chim. Acta* **163,** 263 (1984).

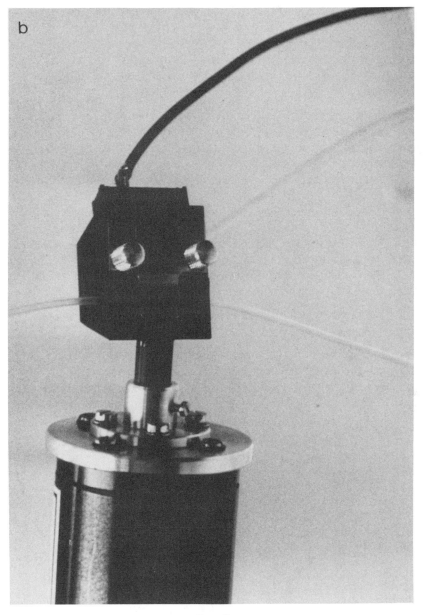

Fig. 4b.

mounting the cell directly on the laser and the detector directly on the cell. The cell can be of the flow-through type (Fig. 4b), whereby the liquid sample under test is pumped into the cell and comes in contact with the test surface which is pressed over an opening in the cell.

Applications

The Brewster-angle reflectometry technique has the same general area of application as ellipsometry. We here demonstrate its use in studies of tryptic digestion of human immunoglobulin (IgG). Silicon surfaces are made hydrophobic by silanization as discussed above. A test surface is mounted in the cell which is filled with 0.5 g/liter of human immunoglobulin, IgG (Kabi Vitrum, Stockholm, Sweden), dissolved in 0.15 M sodium chloride (37°). A rapid uptake of protein occurs as can be seen in Fig. 5. Then a rinse (R) in saline is done, and, finally, 1 g/liter trypsin (from bovine pancreas, Sigma Chemical Company, St. Louis, MO) is added to the cell. The initially adsorbed monolayer of IgG is clearly digested by the trypsin. The rate of initial decrease in signal (layer thickness) can be taken as a measure of the enzymatic activity.

Surface Plasmon Resonance

Surface plasmons are collective oscillations of the conduction electrons in the surface of a metal. A surface plasmon can be demonstrated by light in the surface of a thin metal film on a glass prism at a certain angle of

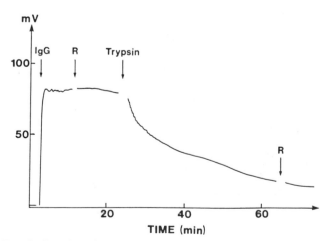

FIG. 5. Tryptic digestion of IgG adsorbed on a silicon surface. Arrows indicate addition of reagents as explained in the text.

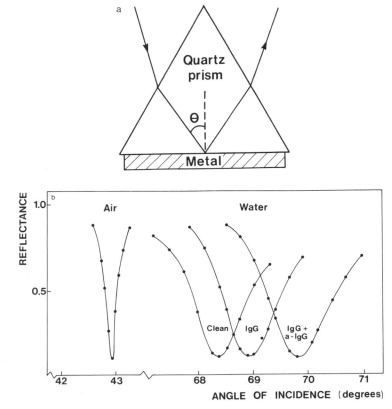

FIG. 6. (a) Principle of surface plasmon generation. The metal film can be evaporated on a glass slide which is pressed against a prism. (b) Surface plasmon resonance curves for a 600 Å silver film in air and in aqueous solutions. On adsorption of an IgG layer, the curve shifts to a greater angle of incidence. Additional adsorption of anti-IgG gives rise to a further shift.

incidence of the light.[24] This is a resonance phenomenon, and the angle depends very strongly on the medium outside the metal film and on adsorbates at the metal surface. The principle of surface plasmon generation and some experimental results[25] are shown in Fig. 6.

Figure 7 demonstrates the use of surface plasmon resonance for the kinetic study of an antigen–antibody reaction on a surface. In this case, the change in reflected light intensity close to the reflectance minimum for the antigen-covered surface is followed and the change later translated into a shift of the resonance angle.

[24] H. Raether, *Phys. Thin Films* **9**, 145 (1977).
[25] B. Liedberg, C. Nylander, and I. Lundström, *Sensors Actuators* **4**, 299 (1983).

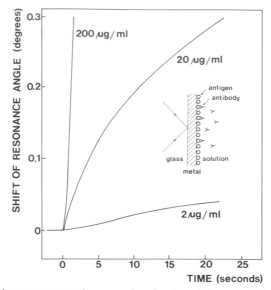

FIG. 7. Shift in resonance angle versus time for three different anti-IgG concentrations. The shift is measured as an increase of the reflected light intensity. The inset illustrates the antibody adsorption.

The surface plasmon technique has not yet been applied extensively for biosensing purposes. It appears, however, to be very easy to implement. The sensitivity is equal to or better than that for ellipsometry. It is probably one of the optical techniques we will see further developed in the near future.

Discussion

The direct optical methods discussed here offer several advantages for both research and clinical applications in biomedical analysis. One of the most important features is that they can be applied nondestructively *in situ* for kinetic studies of biochemical reactions. Ellipsometry already is established as an accurate and versatile tool in basic research, while reflectometry and surface plasmons are still under development. Because ellipsometry is a more general technique, it will probably be more widely used. In clinical applications, reflectometry and surface plasmons will probably play a more important role in the future. Ellipsometry is more complicated, and most instruments are expensive and contain moving parts, which are subject to wear.

It has been stated that "at its present state of development, the ellipso-metric technique is semiquantitative."[26] This statement is based on the observed semilogarithmic relation between number of adsorbed mole-cules and their concentration in bulk. The accuracy in determination of bulk concentration of, e.g., antibodies will therefore be reduced consider-ably even though the amount of adsorbed antibodies (film thickness) can be determined very precisely and reproducibly. However, this is not a principle limitation and can be circumvented by using other assay proce-dures, such as end-point titration, where instead the detection limit of the method determines the ultimate accuracy.

The drawback discussed above applies only to the classical type of measurement where the bulk concentration of a substance is of primary interest. However, in many problems in modern biomedical analysis, other aspects of macromolecules are more important, and parameters like conformation, activity, binding capacity, specificity, and general function are of more relevance. New analytic methods must therefore be devel-oped. Surface-oriented optical methods will play an important role. The diffusion profiles in Figs. 3 and 4 are examples of such "alternative" applications.

So far we have only discussed principles and instrumentation of opti-cal methods based on reflection of light at a solid/liquid or solid/air inter-face. However, a successful application of these methods is based on the ability to prepare reproducible surfaces. No matter how accurate the instrument is, the overall accuracy of the assay relies on the fact that different sample test surfaces have identical properties, which, e.g., can be antibody affinity or specificity depending on application. In most cases, surface specificity is required, and surface preparation therefore has at least two critical steps. The first step is to produce a reproducible substrate surface. Silicon and various metals like gold and chromium have been used extensively. Silicon has several advantages: (1) Its surface chemistry is well known because of the extremely well-developed pro-cesses in the semiconductor industry. (2) The silicon surface can routinely be polished close to atomic flatness, which is very favorable when adsorp-tion of macromolecules is to be studied. (3) Silicon also has a very large polarizability, which implies high sensitivity because the coupling factor between film thickness and instrument output is large.

There are two different ways of preparing reproducible surfaces: ei-

[26] M. A. Genshaw, in "Clinical Immunochemistry: Principles of Methods and Applications" (R. C. Boguslaski, E. T. Maggio, and R. M. Nakamura, eds.), p. 221. Little, Brown, Boston, 1984.

ther one cleans an "as grown" surface to obtain an intrinsically clean substrate, or one modifies the surface in a controlled way. The latter is probably the best approach from the standpoint of stability. A very commonly used method is silanization, which is discussed in detail in this volume.[27] The second step in sample surface preparation is the coupling or adsorption of specific macromolecules such as antigens, receptors, or enzyme substrates to the solid substrate. Special precautions have to be taken in order to retain the biological function of the macromolecules.

Finally we want to comment on a few points which appear to form a communication gap between biochemists, physicians, and physicists. The first is the suspicion that optical methods have to rely on intensity interference and thus can only resolve dimensions of the order of the wavelength of the light. Two of the methods discussed here are based on changes in the phase of the field components in thin films. For example, in the ellipsometric principle, phase changes due to film formation right at the reflection spot are converted to a geometric angle in terms of the (angular) direction of polarization measured, e.g., with a polarizer. High resolution can now be understood if we study the case with light incident at the Brewster angle of a film. Then the p-wave component does not "see" the film (total refraction), while, for simplicity, we suppose that the s-wave component is completely reflected at the ambient/film interface. Thus the p-wave component travels approximately $2d$ farther, resulting in a relative s–p phase shift, $\Delta\theta = 4\pi nd/\lambda$. This phase is converted to a geometric angle, which we can measure with an accuracy of 0.001°. We then find that thickness variations of $d = \lambda\Delta\theta/4\pi n \approx 0.02$ nm can be resolved ($\lambda = 500$ nm, $n = 1.5$). Similar arguments apply to reflectometry.

The second problem concerns the way in which data are presented. Usually ellipsometric data are interpreted in optical models in which a thickness is defined. The result is then presented in this form which can be very misleading because we are dealing with discrete macromolecules adsorbed in the form of a discontinuous "film" on a surface. The way out of this dilemma is to use conversion formulas like Eq. (3) to calculate surface concentration in units of $\mu g/cm^2$ instead. This procedure does not introduce any approximations because, as long as the size of the macromolecules is much smaller than the wavelength of the light, the surface "film" constitutes an effective medium, which in first approximation can be described by its thickness and density. With this limit ($d \ll \lambda$), the light probe cannot resolve the detailed microstructure.

---

[27] U. Jönsson, M. Malmqvist, G. Olofosson, and I. Rönnberg, this volume [34].

Acknowledgments

We would like to thank H. Elwing, M. Stenberg, and S. Welin for providing us with source material, and all our collaborators and research students in the biomedical analysis field for their work. Our research on the development of methods in biomedical analysis is supported by grants from the Swedish Board for Technical Development.

[34] Surface Immobilization Techniques in Combination with Ellipsometry

By Ulf Jönsson, Magnus Malmqvist, Göran Olofsson, and Inger Rönnberg

The rapid development of bioanalytical methods and biosensors involving immobilized enzymes or antibodies focuses interest on the interaction of biomolecules at inorganic surfaces. It is also of importance to use efficient, gentle, and durable immobilization techniques and that the techniques are suited for large-scale production of sensor devices. The function of several surface-sensitive sensors based on electrochemical,[1,2] optical,[3,4] and gravimetric[5] methods requires a proper immobilization of biomolecules to inorganic surfaces. Most such surfaces do not carry suitable chemically reactive groups. Such groups may, however, be introduced by chemical surface modification such as silanization. This process is commonly performed by deposition of the silane from a liquid phase.[6] Several difficulties, such as nonuniformity of the deposited silane film thickness, codeposition of polymeric silane particles, and hydrolytic removal of the deposited film, occur using this method[7-11] and are not

[1] N. Yamamoto, Y. Nagasawa, S. Shuto, H. Tsubomura, M. Sawai, and H. Okumura, *Clin. Chem.* **26,** 1569 (1980).

[2] J. Janata and R. J. Huber, in "Ion-Selective Electrodes in Analytical Chemistry" (H. Freiser, ed.), Vol. 2, p. 107. Plenum, New York, 1980.

[3] K. I. Lundström, H. R. Arwin, E. Rieke, G. Sielaff, and N. Hennrich, European Patent Application EP-A-073980 (1980).

[4] C. Nylander, B. Liedberg, and T. Lind, *Sensors Actuators* **3,** 79 (1982/1983).

[5] J. E. Roederer and G. J. Bastiaans, *Anal. Chem.* **55,** 2333 (1983).

[6] H. W. Weetall, this series, Vol. 44, p. 136.

[7] W. D. Bascom, *Macromolecules* **5,** 792 (1972).

[8] D. F. Untereker, J. C. Lennox, L. M. Wier, P. R. Moses, and R. W. Murray, *J. Electroanal. Chem.* **81,** 309 (1977).

[9] I. Haller, *J. Am. Chem. Soc.* **100,** 8050 (1978).

[10] E. P. Plueddemann, "Silane Coupling Agents." Plenum, New York, 1982.

[11] C. H. Lochmüller, A. S. Colburn, M. L. Hunnicutt, and J. M. Harris, *Anal. Chem.* **55,** 1344 (1983).

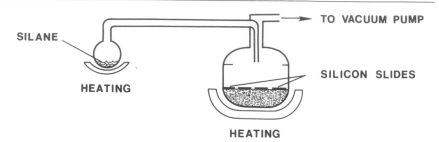

FIG. 1. The silanization apparatus used for chemical vapor deposition of silane.

acceptable for optimum performance of the surface-sensitive measuring devices.

Vapor phase silanization can overcome some of these difficulties. To obtain aqueous-stable silane films the silanization should be performed at a temperature suitable for reaction between the silane and the surface. We report here on a chemical vapor deposition technique for alkoxysilanes. The technique is a further development of that described by Mittal and O'Kane.[12] The main deviation from that method is that the slides to be silanized together with the silanization apparatus are kept at a temperature suitable for the reaction between the alkoxysilane and the surface.

The technique is exemplified by the covalent binding of protein A to silanized silica surfaces and the interaction of the immobilized protein A with immunoglobulins. The result is evaluated by *in situ* ellipsometry, an optical technique based on reflection of polarized light from a surface.

Chemical Vapor Deposition

Single crystalline silicon (Wacker Chemitronic, FRG) covered by spontaneously formed silica is cut into 3×3 mm slides and first washed in $NH_4OH-H_2O_2-H_2O$ ($1:1:5$) at $80°$ for 5 min, rinsed in water, then washed in $HCl-H_2O_2-H_2O$ ($1:1:5$) at $80°$ for 5 min and again thoroughly rinsed in water. The water used is double distilled. This treatment renders hydrophilic slides, which are a necessary requirement for a successful chemical vapor deposition.

The slides are blown dry in nitrogen gas and immediately transferred to the apparatus shown in Fig. 1 which, except for the silane-containing vessel, is kept at $90 \pm 1°$ by means of a surrounding air-circulating oven. The silane is kept around $0°$ by means of an ice/water slurry-filled Dewar

[12] K. L. Mittal and D. F. O'Kane, *J. Adhes.* **8**, 93 (1976).

vessel. The modification begins with evacuating the apparatus to 0.5–1 N/m^2. The temperature of the silane-containing vessel is slowly increased by allowing the Dewar vessel temperature equilibrate with the oven. The silane, (3-[(2-aminoethyl)amino]propyl)trimethoxysilane (Union Carbide, USA), is now slowly vacuum distilled for 24 hr.

After vacuum distillation of the silane the temperature is increased to 160° for a further 12 hr. The system is then allowed to cool to room temperature and slowly filled with dust-free air. The silanized slides are kept in ethanol until use.

Further details concerning the chemical vapor deposition technique may be found in Ref. 13. The silanized surfaces are characterized by ellipsometry, contact-angle measurements, and scanning electron microscopy, revealing smooth, aqueous-stable, and reproducible films of monolayer character.[14]

Immobilization of Biomolecules to Silica Modified by Chemical Vapor Deposition of Silanes

The surface amino groups introduced by silanization may be used for covalent binding of biomolecules either directly utilizing, e.g., carbodiimide as a coupling reagent or indirectly as an anchor for further surface syntheses. Such syntheses may be used to introduce protected thiol groups on the surface. After reduction these groups may be used for covalent binding of proteins via thiol–disulfide exchange reactions utilizing the bifunctional reagent N-succinimidyl-3-(2-pyridyldithio)propionate (SPDP) (Pharmacia, Uppsala, Sweden). This immobilization technique eliminates the risk of cross-linking and homopolymerization, thus giving monolayer coverage in close contact with the surface. Furthermore, the degree of SPDP substitution on the proteins can be controlled and therefore also the maximum number of covalent bonds to the solid surface.

The proposed chemical reaction steps are shown in Fig. 2. Pyridyl disulfide groups are introduced to aminosilanized silica surfaces by reaction with 5–10 mM SPDP in 0.1 M sodium phosphate, 0.1 M NaCl, and 1 mM EDTA, pH 7.5, for 30 min. After rinsing the slides are reduced with 100 mM dithiothreitol (Fluka, Buchs, Switzerland) for 30 min. The slides are rinsed again and incubated for 8–16 hr with 100 μg/ml of protein A

[13] M. Malmqvist, G. Olofsson, and U. Jönsson, Swedish Patent 8,200,442-5 (1982).
[14] U. Jönsson, G. Olofsson, M. Malmqvist, and I. Rönnberg, *Thin Solid Films* **124**, 117 (1985).

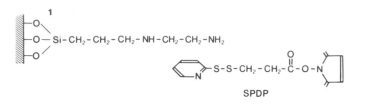

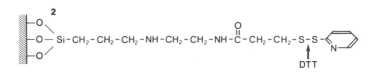

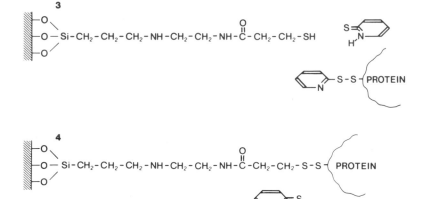

FIG. 2. Proposed chemical reactions on the aminosilanized silica surface. In (1) this surface is reacted with the SPDP reagent. The SPDP-reacted surface is then reduced with dithiothreitol (DTT) (2). In (3) the reduced surface is reacted with SPDP-modified protein thus covalently attaching the protein to the surface (4).

(Pharmacia) modified with SPDP according to Carlsson *et al.*[15] The degree of SPDP modification is on average 2–3 SPDP/protein A molecule. A higher degree of modification was found to decrease the interaction of immobilized protein A with immunoglobulin G.

    To avoid transfer of impurities from the air–liquid interface and also to ascertain a proper rinsing after the various incubation steps, all handling of the slides takes place with a continuous back-flow of buffer through a glass filter funnel thereby causing an overflow of buffer from the funnel. All reactions are carried out at room temperature.

[15] J. Carlsson, H. Drevin, and R. Axén, *Biochem. J.* **173,** 723 (1978).

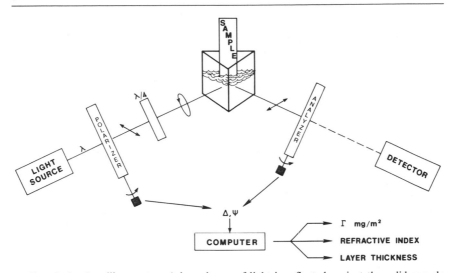

FIG. 3. *In situ* ellipsometry. A laser beam of light is reflected against the solid sample surface at an angle of incidence of 60°. Before reaching the surface the light passes through a polarizer and a quarter-of-wavelength plate. After reflection the light passes through a second polarizer and is detected by a photodiode. The polarizers are automatically rotated in a way that the resulting light intensity reaching the detector is kept at a minimum. The position of the polarizers gives the angles Δ and ψ, which may be translated into amount of adsorbed or interacting proteins on the surface.

Interaction of Immobilized Protein A with Immunoglobulin G

The immobilized protein A is expected to bind immunoglobulin G (IgG) from several species in its Fc part.[16] To elucidate the possibilities of utilizing a properly immobilized molecule in bioanalytical sensor applications, slides with immobilized protein A are incubated in 50 μg/ml human serum albumin (Sigma, St. Louis, MO, rinsed on protein A–Sepharose to remove trace impurities of human IgG) together with 1–30,000 ng/ml of rabbit IgG (Dako, Denmark) for 8 hr. These slides are compared with unmodified protein A adsorbed onto aminosilanized slides.

All results are evaluated by *in situ* ellipsometry (Fig. 3). For further information concerning the instrumental setup and theory, see Jönsson *et al.*[17] and Azzam and Bashara,[18] respectively. The amount of interacting

[16] G. Gyka, V. Ghetie, and J. Sjöqvist, *J. Immunol. Methods* **57,** 227 (1983).
[17] U. Jönsson, B. Ivarsson, I. Lundström, and L. Berghem, *J. Colloid Interface Sci.* **90,** 148 (1982).
[18] R. M. A. Azzam and N. M. Bashara, "Ellipsometry and Polarized Light." North-Holland, Amsterdam, 1977.

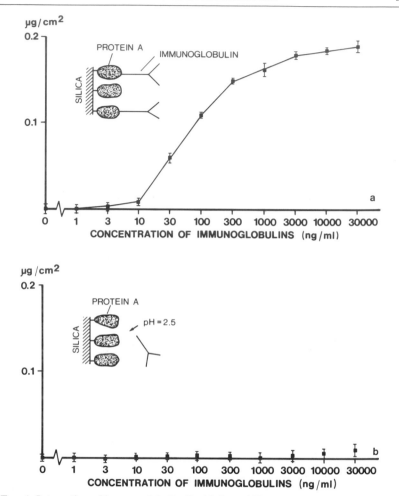

FIG. 4. Interaction of immunoglobulin G with immobilized SPDP-modified protein A on silica surfaces. The results are evaluated by *in situ* ellipsometry. In (b) the same slides as in (a) are transferred to a solution of pH 2.5 and again measured in the ellipsometer.

IgG is determined by simultaneous ellipsometric and radiometric measurements.[19]

Figure 4 (a) and (b) shows the interaction of IgG (at the indicated concentrations) with immobilized protein A on silica surfaces. The bars indicate mean ± maximum deviations from three different slides at each concentration. In Fig. 4 (b) the same slides as in (a) are transferred to 0.1

[19] U. Jönsson, M. Malmqvist, and I. Rönnberg, *J. Colloid Interface Sci.* **103**, 360 (1985).

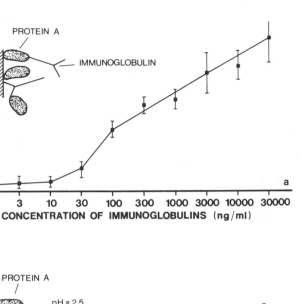

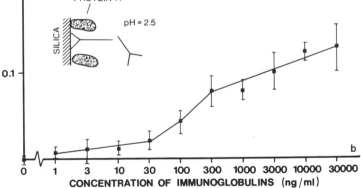

FIG. 5. Interaction of immunoglobulin G with adsorbed unmodified protein A on silica surfaces. In (b) the same slides as in (a) are transferred to a solution pH 2.5 and again measured in the ellipsometer.

M glycine buffer, pH 2.5, for 5 min and again measured in the ellipsometer. The IgG is eluted, and the protein A surfaces can be used in another experimental run. The amount of irreversibly bound IgG, however, gradually increases, and the number of experimental runs with the same type of results as given in Fig. 4 is limited to about three. No nonspecific interaction with human serum albumin could be detected although the concentrations of this protein are far above the concentration of IgG.

In Fig. 5 (a) and (b) the same experiments as in Fig. 4 are repeated.

The main difference, however, is that unmodified protein A is adsorbed onto the silica surfaces. These experiments indicate that adsorbed protein A is exchanged by IgG from solution and/or the IgG is adsorbed into "holes" in the protein A film due to molecular packing of the protein A molecules on the surface.

The experiments described here stress the importance of attaining stable and reproducible biomolecular layers for use in bioanalytical applications. In the protein system tested, protein A–IgG interactions, it was found that covalently attached protein A molecules show excellent reproducibility, stability, and biospecific interaction as compared with adsorbed protein A. By lowering the pH, the surfaces could be regenerated, thus making possible the development of nondisposable sensors. Finally we would like to emphasize that *in situ* ellipsometry, combined with well-defined surfaces, is a fast and sensitive method for the study of biomolecular interactions.

[35] Coupling of Biomolecules to Silicon Surfaces for Use in Ellipsometry and Other Related Techniques

By Carl Fredrik Mandenius, Stefan Welin, Ingemar Lundström, and Klaus Mosbach

The formation of a biomolecular layer on a silicon surface can easily be followed by ellipsometry. In this chapter, examples of the use of ellipsometry for the study of biomolecular interactions between receptor–cell and enzyme–coenzyme are given. By coupling one of the interacting components to the silicon surface the binding of the other can be monitored specifically and sometimes also quantitatively.

The immobilization techniques employed all include formation of a hydrocarbon layer on top of the silicon oxide present on all silicon surfaces. By binding with established immobilization methods to this hydrocarbon layer, biomolecules such as coenzymes, antibodies, or other active proteins can be attached to the silicon surface.

It is important that the properties of the Si surface should not affect the biomolecules, leading, for instance, to denaturation, and that nonspecific adsorption which might impede any biospecific binding does not occur. The introduction of the hydrocarbon layer significantly limits nonspecific adsorption and thus protects the coupled biomolecules.

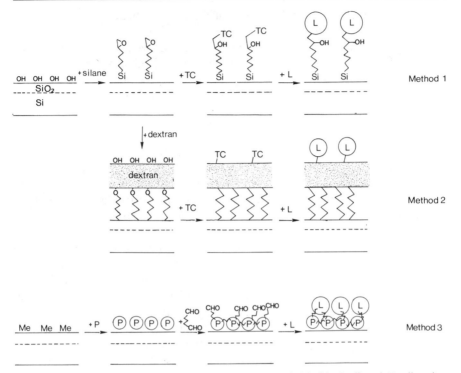

FIG. 1. Coupling methods to silicon wafers. TC, Tresyl chloride; L, ligand; P, albumin; Me, methyl.

The three methods described here (Fig. 1) may be used as alternatives to the methods described by Jönsson *et al.* in this volume.[1]

Methods for Immobilization of Silicon Wafers

Materials

The following materials are used: 2-glycidoxypropyltrimethoxysilane (epoxysilane, Z6040, Dow Chemical Co.), 2,2,2-trifluoroethanesulfonyl chloride (tresyl chloride, Fluka AG, Buchs, Switzerland), human serum albumin (Kabi Diagnostics, Stockholm, Sweden), dichlorodimethylsilane, triethylamine, hydrogen fluoride, glycine, ethanolamine, pyridine, glutardialdehyde, Tris (Merck, Darmstadt, FRG), and dextran (MW 70,000 and 500,000 g/mol, Pharmacia Fine Chemicals, Uppsala, Sweden). Labo-

---

[1] U. Jönsson, U. Malmqvist, G. Olofsson, and I. Rönnberg, this volume [34].

ratory quality acetone and toluene are dried by means of a molecular sieve (4 Å) before use.

Silicon wafers can be purchased from Wacker Chemitronics (Burghausen, FRG) (diameter 50 mm, thickness 0.35 mm, crystal orientation 1–1–1). The NAD analog described below can be obtained from Sigma Chemical Co. (St. Louis, MO).

Method 1. Coupling to Silane-Coated Si Wafers[2]

Alkylation of Silicon Wafers. Silicon wafers are treated with 3 ml ammonia (25%) and 3 ml hydrogen peroxide (30%) in 15 ml distilled water at 80° for 5 min each to cleanse the surface. This is followed by treatment in 3 ml concentrated hydrogen chloride and 3 ml peroxide (30%) in 18 ml distilled water at 80° for 5 min in order to render the surfaces hydrophilic. The wafers are then rinsed with distilled water, acetone, and dried toluene and are hung in clips in a round-bottomed flask containing 200 ml carefully dried toluene. The toluene is gently stirred, and 8 ml 100% 2-glycidoxy-propyltrimethoxysilane is slowly added with a pipet. The flask is then sealed, equipped with a reflux cooler, and adapted to a dry nitrogen stream to exclude all traces of moist air. Triethylamine (160 μl) is finally added to initiate the alkylation reaction, and the mixture is boiled for 3 hr. The treated wafers are consecutively transferred to hot toluene, acetone, diethyl ether, and water and are stored at room temperature until required.

Activation with Tresyl Chloride and Coupling of Biomolecules. The epoxy groups on the alkylated silicon wafers are hydrolyzed with HCl/water, pH 1.5, by heating at 80° for 30 min. The hydroxyl groups thus formed are reactive and can be activated with tresyl chloride (for details of the tresyl method see Ref. 3, Vols. 135 and 136, and elsewhere in this volume) and 0.2 ml pyridine in 20 ml dry acetone gently stirred for 10–20 min on an ice-bath. The wafers are then rinsed with solutions of acetone: a 1 : 1 mixture of acetone and 1 mM HCl; aqueous 1 mM HCl; and finally doubly distilled water. The tresyl-activated wafers are then transferred to the immobilization buffer. The tresylate groups are displaced by biomolecules containing amino or sulfhydryl groups, thus forming covalent linkages between the biomolecules and the wafers. Proteins are immobilized at 10° for at least 12 hr by gentle stirring at appropriate pH and concentrations (e.g., 1 mg/ml).

[2] C. F. Mandenius, S. Welin, B. Danielsson, I. Lundstrom, and K. Mosbach, *Anal. Biochem.* **137,** 106 (1984).
[3] K. Nilsson and K. Mosbach, *Biochem. Biophys. Res. Commun.* **102,** 449 (1981).

Method 2. Coupling to Dextran-Coated Silicon Wafers[4]

Silanization. Silicon chips are treated as described above or in chromosulfuric acid for 15 min to obtain a hydrophilic surface. The chips are placed in a well-dried vacuum chamber, together with pure epoxysilane in a separate cup, and evacuated. The chamber is heated to 160° for 1 hr whereby the silane is deposited on the chips. These are subsequently washed with distilled water and dried by blowing with nitrogen gas.

Dextran Coating. Dextran polymers are dissolved in doubly distilled water at a concentration of 20 wt%. Epoxidized chips are covered with the solution for 20 hr in a sealed moist chamber at room temperature and subsequently heated to 130° for 1 hr. It is important to avoid drying of the dextran solution during the heating procedure. Extensive washing with distilled water for 15 min and nitrogen blowing are used to achieve a clean dextran-modified surface.

Immobilization of Biomolecules. The dextran-coated chips are activated with tresyl chloride dissolved in pyridine (carefully dried by means of molecular sieve). One-tenth milliliter tresyl chloride is added in 20 ml pyridine for each treatment taking place in a beaker, stirred and cooled to 4° in an ice bath. Thoroughly dried chips are immersed in the solution for 15 min without stirring. Immediate washing in water and blowing with nitrogen is repeated until a clean surface is obtained.

The solution containing dissolved biomolecules is allowed to form a droplet of 10 μl on a glass slide, and the tresyl chloride-activated chip is placed upside down over the drop for coupling. After 20 hr in a moist atmosphere at room temperature, excess liquid is rinsed off and the chips blown in nitrogen and stored at 4°.

Method 3. Coupling to Albumin-Coated Silicon Wafers[2]

Silicon chips are cleansed and made hydrophilic as described above and then treated with 10% dichlorodimethylsilane in trichloroethylene in order to obtain a strong hydrophobic surface. The chips are then incubated with an albumin solution (1 mg/ml in 20 mM sodium phosphate, pH 7.5) for 2 hr. The albumin layer thus formed is used as a matrix for subsequent covalent coupling of biomolecules using glutardialdehyde. This is accomplished by first treating the chip with 2.5% glutardialdehyde solution (in 20 mM sodium phosphate, pH 6.8) for 25 min at 5°, stirring gently. This is followed by inactivation of unreacted glutardialdehyde groups with gly-

---

[4] C. F. Mandenius, K. Mosbach, S. Welin, and I. Lundström, *Anal. Biochem.* **157,** 157 (1986).

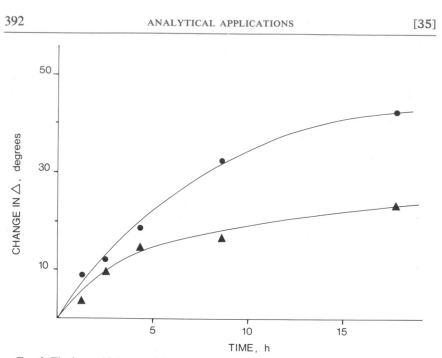

FIG. 2. The layer thickness of *S. aureus* cells (expressed as change in Δ) affinity bound to an IgG wafer, as a function of incubation time. Two cell concentrations applied are shown, 3 μg/ml (●) and 5 μg/ml (▲). (From Ref. 2.)

cine for 80 min (0.1 M, pH 7.5) and subsequent washing with urea (4 M) to remove nonspecifically bound biomolecules.

Applications

Using an automatic ellipsometer (Auto El-II, Rudolph Research, Fairfield, NJ), silicon chips treated as described have been monitored in various affinity binding studies. The basis of the ellipsometry technique is given elsewhere in this volume.[5] For example, immunoglobulin G (IgG) was coupled according to Method 1. Such IgG chips were incubated in suspensions of *Staphyloccccus aureus* cells. *Staphylococcus aureus* contains protein A in the cell membrane and can thus bind to the coupled IgG. The chips were measured intermittently during a 30-hr period at two levels of cell concentrations. Figure 2 shows the time course of the experiment. The curves show a pronounced dependence on cell concentration. Using a intermediate incubation time of 4.5 hr, chips were incu-

[5] H. Arwin and I. Lundström, this volume [33].

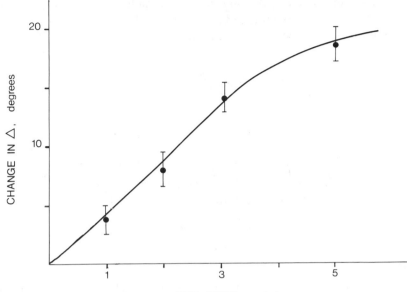

FIG. 3. The average thickness of the layers of *S. aureus* cells affinity bound to an IgG wafer, as a function of cell concentration. The IgG solution was incubated for 4.5 hr. (From Ref. 2.)

bated in cell concentrations of 1, 2, 3, and 5 μg/ml cells. A linear response was thereby observed when measured with the ellipsometer (Fig. 3). For the significance of ellipsometry parameters, see the chapter by Arwin and Lundström.[5]

Control experiments with *Staphylococcus* cells lacking protein A (*S. albus*) gave only minor binding to the surface (less than one-twentieth that of *S. aureus*).

An NAD analog was coupled to wafers coated with dextran (Method 2). The affinity binding of alcohol (ADH) and lactate dehydrogenase (LDH) were studied in a reflectometer, a simplified version of an ellipsometer.[5] ADH bound rapidly to the NAD chips and could subsequently be biospecifically eluted from the chips by addition of a solution of 2 mM NAD and 2 mM pyrazole allowing formation of the competing ternary complex.

In conclusion, these three methods are all applicable to ellipsometry studies using silicon wafers. Methods 1 and 3 should be used when immobilizing proteins. Method 2 is useful for immobilizing small biomolecules, such as coenzymes, but is also an alternative for protein immobilization.

Additional applications of the methods could include other analytical devices with biomolecules attached to silicon or glass supports. For example, similar methods are used in an integrated optical biosensor presented
recently.[6]

[6] M. Seifert, K. Teifenthaler, K. Heuberger, W. Lukosz, and K. Mosbach, *Anal. Lett.* **19,**
205 (1986).

[36] Construction of Dry Reagent Chemistries: Use of Reagent Immobilization and Compartmentalization Techniques

By BERT WALTER

Introduction

As in many technical fields, techniques in clinical analysis have undergone many advances in the last few decades. The basic needs in clinical
analyses are unambiguous analyte-specific assays that provide both identification of sample components and their concentration levels. The importance of this is self-evident since most substances analyzed are part of
a multicomponent biological fluid (e.g., urine, serum). Advances in enzyme and immunochemical assay techniques provided ideal systems for
component analysis in biological fluids, serum and, to a lesser extent,
urine being the most frequently analyzed fluids. In adapting a procedure
for clinical analysis, a premium is placed on its accuracy and convenience
to the user. With this in mind, great efforts have been devoted to developing miniature integrated analytical elements for the quantitative analysis
of serum analytes. These devices integrate several conventional analytical steps into a single element.[1] This includes separation steps as well as
several self-contained chemical reactions. The development of integrated
dry reagent chemistries can be likened to the evolution of miniature electronic devices where many components used in the past are currently
integrated into microscopic circuits.

Dry reagent chemistries open a new era in user convenience. The user
needs only apply the sample to the element to initiate an analysis; the
analysis is usually rapid, taking only a few minutes. No reagent preparations or analyte separations from sample components are required. By

[1] B. Walter, *Anal. Chem.* **55,** 498A (1983).

virtue of their stability and discrete formats, these elements allow both low and high volume testing to be cost effective. These devices are easy to store, requiring less space than reconstituted conventional wet chemistry reagents, and they are easily disposable after use.

One of the earliest dry reagent elements described made use of glucose oxidase–peroxidase chemistry to analyze for glucose in urine.[2] The results were semiquantitative in nature since the color generated by the element was compared with color blocks to estimate glucose concentrations. With the introduction of instrumentation, quantitation became feasible.[3] Owing to the physical configuration of these elements, reflectance spectroscopy and, where appropriate, front-face fluorescence are used to monitor chemical and physical events. Although reflectance spectroscopy has been around for a long time, its popularity as a tool for routine analysis has increased with the advent of dry reagent chemistries. This chapter will attempt to describe the basic features of dry reagent chemistry elements, the multiplicity of scientific disciplines involved in element construction, the use of compartmentalization to segregate incompatible components involved in cascading reactions (enzyme, immunochemical, and organic), and the use of reflectance photometry in monitoring assays.

Basic Features of Dry Reagent Chemistries

Most dry reagent chemistries are designed to be self-contained analytical devices. Each element may have several functional zones that are introduced as single layers or combined into one layer during construction. Regardless of the number of layers present, all dry reagent elements have a support function, a reflectance function, and an analytical function as illustrated in Fig. 1. Some elements may also contain a sample-spreading function.

The support function usually is a layer of material that serves as the foundation for the dry reagent element. It consists of a transparent thin rigid plastic or plasticlike material. The support layer may also be opaque and contain the reflective function.

By design, as well as convenience, dry reagent chemistries are monitored by either reflectance photometry or front-face fluorescence. This necessitates the presence of a reflective surface. This surface reflects to a detector light emitted or light not absorbed by the chemistry of the element. A reflective function is usually constructed by introducing reflective (or scattering) centers into a dry chemistry element layer. Commonly

---

[2] E. C. Adams, Jr., L. E. Burkhardt, and A. H. Free, *Science* **125**, 1082 (1957).
[3] E. L. Mazzaferri, R. R. Lanese, T. G. Skillmann, and M. P. Keller, *Lancet* **1**, 331 (1960).

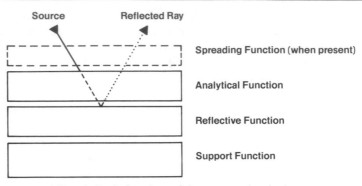

FIG. 1. Basic functions of dry reagent chemistries.

used materials include pigments such as TiO_2, $BaSO_4$, MgO, and ZnO, reflective materials such as metals and foils, or fibrous materials such as paper and fabrics. The primary requirement of a reflective material is that its absorbance of electromagnetic radiation in the area of interest be negligible.

The analytical function of a dry reagent chemistry is often the most complex. It usually consists of multiple chemical reactions as well as specific physical functions. This may include separation zones, masking zones to hide specific components generated during an analysis, trapping zones to immobilize specific chemical species, and reactive zones where specific chemical reactions take place. These zones may be found as distinct layers in an element or integrated in various combinations into a single layer. Hence a layer may contain several functions. Layers constituting the analytical function are commonly constructed with film-forming polymers, fibrous materials, or nonfibrous porous materials.

Where present, the sole purpose of a sample spreading function is to rapidly spread a sample laterally after application.[4] This feature mediates a uniform sample volume per unit volume of the element as illustrated in Fig. 2. Substances commonly used in construction of spreading functions include fabrics, membranes, and paper.

Monitoring of Dry Reagent Chemistries

Chemical reactions in dry reagent elements are usually monitored by diffuse reflectance photometry[5] and, where appropriate, by front-face

[4] E. P. Przybylowicz and A. G. Millikan, U.S. Patent 3,992,158 (1976).
[5] G. Kortüm, "Reflectance Spectroscopy: Principles, Methods, Applications." Springer-Verlag, New York, 1969.

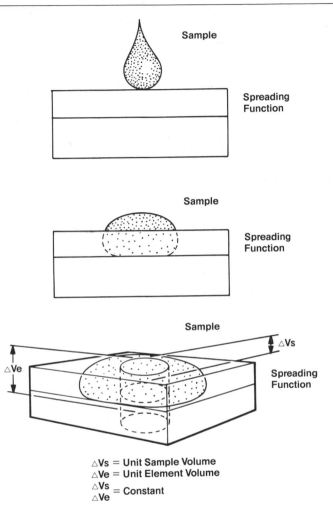

$\triangle Vs$ = Unit Sample Volume
$\triangle Ve$ = Unit Element Volume
$\dfrac{\triangle Vs}{\triangle Ve}$ = Constant

FIG. 2. Spreading function.

fluorescence.[6] Since light scattering is common during photometric measurements of dry reagent chemistries, reflectance photometry offers a distinct advantage over transmittance. For reflectance measurements, an analytical element can be illuminated by direct lighting as illustrated in Fig. 3A or by diffuse lighting. In both direct and diffuse lighting, two kinds of reflection must be considered. One is specular reflection, which is the

[6] A. J. Pesce, C. G. Rosen, and T. L. Pasby, "Fluorescence Spectroscopy." Dekker, New York, 1971.

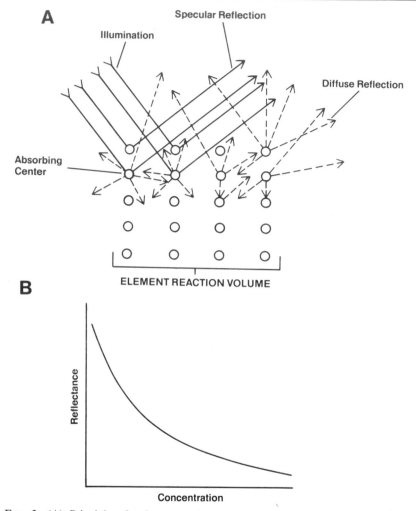

A

Specular Reflection

Illumination

Diffuse Reflection

Absorbing Center

ELEMENT REACTION VOLUME

B

Reflectance

Concentration

FIG. 3. (A) Principle of reflectance photometry. (B) Relation of concentration to reflectance measurements [see Eq. (1)]. Reprinted with permission from V. Marks and K. G. M. M. Alberti (eds.), "Clinical Biochemistry Nearer the Patient." Churchill-Livingstone, Edinburgh, Scotland, 1984.

mirrorlike reflectance from the surface where the angle of incidence is equal to the angle of reflectance. This reflection is of limited value in monitoring the chemistries of dry reagent elements. The second is diffuse reflection, which is a reflection from within the reagent layers of the analytical element. This is the predominant reflection mode of interest. Diffuse reflection is not a surface phenomenon but the result of light interacting with various chemical and physical factors in the layers of the

TABLE I

ALGORITHMS TO LINEARIZE REFLECTANCE MEASUREMENTS
WITH CONCENTRATION[a]

Kubelka–Munk equation	Williams–Clapper equation
$C\alpha K/S = \dfrac{(1 - R)^2}{2R}$	$C = \beta(D_T - D_\beta)$
C = concentration K = absorption coefficient S = scattering coefficient R = % reflectance/100	$D_T = -0.194 + 0.469D_R + \dfrac{0.422}{1 + 1.179e^{3.379D_R}}$ $D_R = \log R$ C = concentration β = reciprocal absorptivity D_T = transmittance density D_β = blank density D_R = reflectance density R = reflectance

[a] Reprinted with permission from V. Marks and K. G. M. M. Alberti (eds.), "Clinical Biochemistry Nearer the Patient." Churchill-Livingstone, Edinburgh, Scotland, 1984.

analytical and reflective functions of the element. These factors include the absorption, transmission, and scattering properties of the illuminated material.

When the reaction compartment of an element, where a chromophore is either generated or degraded, is illuminated at a suitable wavelength, the amount of diffuse light recovered with the aid of the reflective function is a measure of the progress of the reaction. The commonly used expression for reflectance [Eq. (1)] describes the amount of diffuse light reflected

$$\%R = (I_s/I_r)R_r \tag{1}$$

from the analytical element relative to a known standard, where I_s, I_r, and R_r represent, respectively, the reflected light from the sample, the reflected light from the standard, and the percent reflectivity of the standard. Percent reflectance measurements are comparable to transmittance measurements where the relation to concentration is not linear, as illustrated in Fig. 3B. To make the reflectance measurements useful and convenient, several algorithms are available to linearize the relation of $\%R$ versus analyte concentration. The two most notable are the Kubelka–Munk[7] and the Williams–Clapper equations[8] (Table I). The specific algorithm used in an analysis depends on the nature of illumination, the reflectance characteristics of the dry reagent element, and the geometry of the instrument.

[7] P. Kubelka and F. Munk, Z. Tech. Phys. 12, 593 (1931).

[8] F. C. Williams and F. R. Clapper, J. Opt. Soc. Am. 43, 595 (1953).

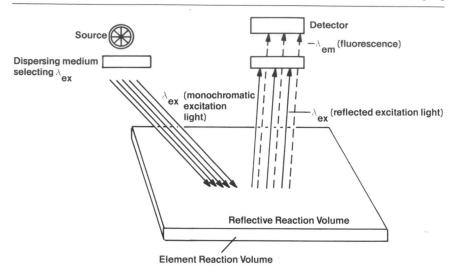

FIG. 4. Principle of front-face fluorescence measurements. Reprinted with permission from B. Walter, *Anal. Chem.* **55,** 498A (1983). Copyright © 1983, American Chemistry Society.

Front-face fluorescence can be used to monitor fluorescence, either generated or destroyed, within an analytical element.[9,10] The general principle of front-face fluorescence is illustrated in Fig. 4. Irradiating light is passed through a filter, or a light-dispersing device such as a monochromator, to select the appropriate wavelength of excitation light. With the aid of the reflective function, a portion of the fluorescence and irradiating light is reflected toward the detection system. A filter or monochromator in the detection apparatus segregates the fluorescence from the irradiating light, thus permitting the transmission of fluorescence light only. Unlike reflectance, measured fluorescence is linear with fluorophore concentrations in the absence of self-quenching.

Construction of Dry Reagent Chemistry Elements

Layer Construction

For most purposes, layers of dry reagent elements are constructed by trapping materials in the layer matrix.[11] Materials range from small inor-

[9] J. C. W. Kuan, H. K. Y. Lau, and G. G. Guilbault, *Clin. Chem.* **21,** 67 (1975).
[10] W. E. Howard III, A. Greenquist, B. Walter, and F. Wogoman, *Anal. Chem.* **55,** 878 (1983).
[11] J. P. Comer, *Anal. Chem.* **28,** 1748 (1956).

ganic molecules to large molecular weight enzymes. Although covalent linking of components to the matrix may be necessary in certain cases, the technique is rarely used. Two approaches are most commonly used in constructing element layers. These involve the simultaneous trapping of components and matrix formation and the trapping of components in preformed matrices.

Simultaneous Component Entrapment and Matrix Formation. Film casting techniques are commonly used to trap system components and simultaneously form a layer matrix. The technology of thin-film casting has been developed and advanced by industries involved in plastics packaging, electronics, and photographic materials.[12-14] The photographic industry has developed techniques for casting thin film layers that contain specific functional chemistries.[15,16] A familiar example is that of color film, which may contain as few as 3 chemistry layers or as many as 12 in the case of instant color films.[17] Each layer serves a specific physical or chemical function in image development and stabilization. In like fashion, layers can be constructed in building dry reagent elements. The main differences are that these layers entrap specific chemistries and functions unique and necessary for a specific analysis.

A variety of natural and synthetic polymers are used for constructing element layers. The primary requirement is that the polymer form a porous film or be adaptable to membrane formation.[18] Gelatin is a natural polymer commonly used in layer construction because of its emulsifying action as well as other chemical and physical attributes.[19-21] Enzymes and other reagents can be dissolved or suspended in gelatin emulsions and easily coated onto a support material. Other polymers found useful in layer construction include polyacrylamide,[22] poly(vinyl alcohol) hydroxypropylmethylcellulose, methylcellulose,[23] α-methylglutamate N-carbo-

[12] B. M. Deryagin and S. M. Levi, "Film Coating Theory." Focal Press, London, 1964.

[13] H. F. Payne, "Organic Coating Technology," Vol. 2. Wiley, New York, 1961.

[14] S. C. Zink, "Coating Processes—Encyclopedia of Chemical Technology" (D. F. Othmer, ed.), Vol. 6, p. 386. Wiley, New York, 1979.

[15] T. H. James, "The Theory of the Photographic Process." Macmillan, New York, 1977.

[16] A. Rott and E. Weyde, "Photographic Silver Halide Diffusion Processes." Focal Press, London, 1972.

[17] C. C. Van de Sande, *Angew. Chem., Int. Ed. Engl.* **22,** 191 (1983).

[18] R. R. Myers and J. S. Long, "Film Forming Compositions," Vols. 1, 2, and 3. Dekker, New York, 1967, 1968, and 1969, respectively.

[19] I. Tomka, J. Bohonek, A. Spühler, and M. Ribeand, *J. Photogr. Sci.* **23,** 97 (1975).

[20] A. Veis, "Macromolecular Chemistry of Gelation," p. 396. Academic Press, New York, 1964.

[21] B. E. Tabor, "Photographic Gelatin," p. 83. Academic Press, London, 1972.

[22] G. D. Jones, U.S. Patent 2,504,074 (1950).

[23] R. L. Davidson and M. Sitting, "Water Soluble Resins." Reinhold, New York, 1968.

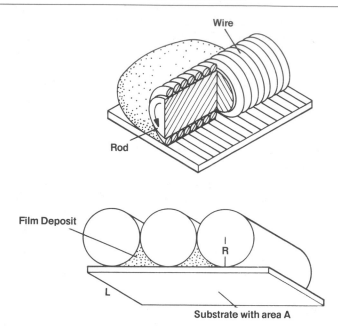

Unit Volume Deposited = dV = (1 − π/4) 2R²dL

Film Thickness Deposited = dV/dA

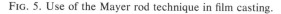

FIG. 5. Use of the Mayer rod technique in film casting.

benzyloxylysine copolymer,[24] poly(vinyl acetate), agarose, alginate, and carrageenan.[25] The choice of polymer for layer construction is a function of porosity needs, compatibility with required chemistries, and any additional analytical functions required in the layer.

One of the crucial parameters which must be controlled during film casting is film thickness. Commercially, this parameter is controlled by coating equipment specifically designed to cast a uniform layer thickness.[14] Hence, a variety of coating head configurations are available for precision casting of specific materials at desired thicknesses; however, for rapid, convenient, and low cost coating at the laboratory bench, the Mayer rod technique is most practical.[26] This device consists of a cylinder with a tightly wrapped wire in helical configuration as shown in Fig. 5. The rod is used to pull the solution across the surface of a receiving

[24] H. W. Wood, *Nature (London)* **2,** 106 (1958).
[25] S. C. Sharma, *Food Technol.* **Jan.** (1981).
[26] H. L. Weiss, "Coating and Laminating Machines." Converting Technology Co., Milwaukee, Wisconsin, 1977.

TABLE II
RELATION BETWEEN COATING THICKNESS AND
WIRE DIAMETER OF MAYER ROD

Wire diameter (μm)	Wet coating thickness (μm)
50	5.37
100	10.73
150	16.08
200	21.45
250	26.82
300	32.18
350	37.54
400	42.91
450	48.28
500	53.65

material. The thickness of the resulting coat is controlled by the diameter of the wrapped wire. The conversion of Mayer rod wire diameter to expected coating thickness is summarized in Table II for several rods.

The use of film casting techniques can be illustrated by constructing a film layer that traps a nitrite-sensitive chemistry.[27] The coating solution has the following composition:

N-(1-Naphthyl)ethylene diamine dihydrochloride	5.6 mmol
Sulfanilamide	1.5 mmol
Trichloroacetic acid	1.8 mmol
Poly(vinyl acetate)	20.0 g
Ethyl acetate	100 ml

All components are dispersed, and the mixture is applied to a polyester sheet to a thickness of 500 μm. The resulting composition is dried at 40°. Application of a nitrite solution to the dry layer results in the formation of a red color. The color intensity is proportional to the nitrite concentration. The quantities of reagents available during analysis are controlled by their concentration in the casting medium, the thickness of the film wetted by the sample, and the solubility in the sample.

Component Entrapment in Preformed Matrices. The most widely used alternative approach in layer construction involves trapping components in a preformed matrix by saturation techniques. In this approach, the matrix of choice is saturated with a solution of desired components or reagents. Commonly used preformed matrices include paper, woven fabrics, and a variety of porous membranes. The use of this technique can be

[27] H. G. Rew, P. Rickman, H. Wielinger, and W. Rittersdorf, U.S. Patent 3,630,957 (1971).

illustrated by constructing a layer that entraps a glucose-sensitive chemistry.[28] The saturation solution has the following composition:

Glucose oxidase (200 U/mg)	40 mg
Horseradish peroxidase (100 U/mg)	5 mg
o-Tolidine	100 mg
Ethanol (50% aqueous)	10 ml

Whatman filter paper is saturated with the solution and allowed to dry at 40°. The resulting layer is affixed onto a support with double-adhesive tape. Application of glucose solutions result in the formation of a blue–green color. The rate of color production is proportional to the glucose concentration. The availability of reagents during analysis is controlled by the concentration of reagents in the saturation solution, the thickness of the matrix, the porosity of the matrix, the absorptivity of the solution by the matrix fibers, and the solubility of the reagents in the applied sample.

Integration of Element Functions

A dry reagent chemistry is designed to simplify a multi-step analytical procedure for the user. The sole function of the element is to convert an analyte in a sample to a specific quantifiable material. To accomplish this, all the physical and chemical functions needed for an analysis must be integrated into one element. Thus, the primary need in element construction is to integrate all the required functions into one small device.

Figure 6 diagrams the two basic approaches employed to integrate functions in element construction. With coating techniques, desired functions as well as incompatible chemistry compounds can be introduced as separate layers. Hence, the element is constructed by coating a series of layers on top of each other. The alternative approach is to introduce incompatible reagents and functions by a multistep saturation process. By carefully choosing the solvents, interactions between components can be prevented. The solvents of successive saturations are chosen to prevent dissolution of reagents deposited by previous saturations. This approach allows the construction of single-layer elements that otherwise would require multilayer coating. In some cases, a combination of the two approaches are employed. The final configuration of a dry reagent chemistry element is dependent on the mode of construction and the specific analytical requirements. The examples that follow illustrate the basics of element construction.

Element Construction by Multisaturation. Single-layer elements prepared by multiple saturations have been described for the analysis of

[28] K. H. Kallies, German Patent 1,240,884 (1971).

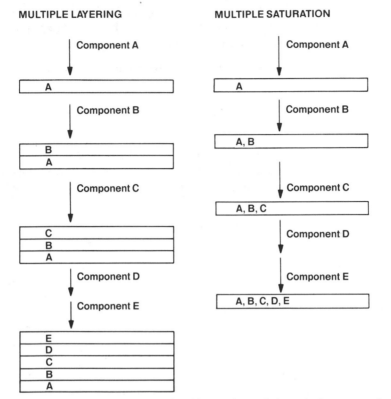

FIG. 6. Comparison of multilayer and multisaturation techniques in dry reagent element construction.

several therapeutic drugs. The approach can be illustrated by the preparation of an element for the quantitation of gentamicin.[29] The chemistry, a competitive-binding assay based on a substrate-labeled fluorescent immunoassay,[30] specific for gentamicin can be summarized in Eqs. (2) and (3)

$$Ab + G + \beta\text{-Gus} \rightarrow Ab:G + Ab:\beta\text{-Gus} \qquad (2)$$
$$\beta\text{-Gus} + enzyme \rightarrow \beta\text{-Gus (f)} \qquad (3)$$

where G and β-Gus, representing the drug (gentamicin) and its respective fluorescence-quenched conjugate (β-galactosylumbelliferone–sisomicin), compete for a limited quantity of antibody (Ab) binding sites. The unbound β-Gus is treated with an enzyme (β-galactosidase) to remove the quenching group, and the fluorescent conjugate β-Gus (f) can be moni-

[29] B. Walter, A. C. Greenquist, and W. E. Howard III, *Anal. Chem.* **55,** 873 (1983).
[30] J. Burd, R. Carrico, M. Fetter, R. Buckler, R. Johnson, R. Boguslaski, and J. Christner, *Anal. Biochem.* **77,** 56 (1977).

tored. The resulting fluorescence is proportional to the gentamicin concentration. The analytical element is prepared by saturating Whatman 31ET filter paper with the following aqueous solution:

Antisera to gentamicin	250	ml/liter (binding capacity 2 μmol)
β-Galactosidase	104	U/liter
Bicine buffer	0.36 M	
MgCl$_2$	0.05 M	
pH	8.3	

After drying, the conjugate is introduced into the same layer via a second saturation using an organic solvent to prevent its interaction with the antibody and enzyme. The composition of the solution is summarized below:

β-Galactosylumbelliferone–sisomicin	$2 \times 10^{-6}\ M$
Solvent	Toluene

After drying, the reagent layer is fixed onto a support material and cut into a desirable dimension. The various functions can be identified in the final element as shown in Fig. 7A. In the dry state, the element contains in one compartment a conjugate in the presence of antibody and a developing enzyme where interactions have been prevented. On sample application, reagents are reconstituted and a competitive binding assay initiated. The paper fibers act as a spreading function, a reflective function, as well as provide the analytical function. The fluorescence generated on sample application is proportional to the gentamicin concentration.

Element Construction by Multilayers. The alternative to multiple saturations is to construct a device by multilayers. All element functions are introduced in successive layers where incompatibilities are avoided. This can be exemplified by the construction of a multilayer element for cholesterol analysis.[31,32] A cross section of such an element is shown in Fig. 7B. The detection chemistry is summarized by Eqs. (4)–(6). Element con-

$$\text{Cholesterol esters} \xrightarrow[\text{esterase}]{\text{cholesterol}} \text{cholesterol + fatty acids} \qquad (4)$$

$$\text{Cholesterol} + O_2 \xrightarrow[\text{oxidase}]{\text{cholesterol}} \text{4-cholestene-3-one} + H_2O_2 \qquad (5)$$

$$H_2O_2 + \text{indicator (reduced)} \xrightarrow{\text{peroxidase}} H_2O + \text{indicator (oxidized)} \qquad (6)$$
$$\text{(colorless)} \qquad\qquad\qquad\qquad\qquad \text{(color)}$$

[31] C. T. Goodhue, H. A. Risley, R. E. Snoke, and G. M. Underwood, U.S. Patent 3,983,005 (1976).
[32] G. M. Dappen, P. E. Cumbo, C. T. Goodhue, S. Y. Lynn, C. C. Morganson, B. F. Nellis, D. M. Sablanskas, J. R. Schaeffer, R. M. Schubert, R. E. Snoke, G. M. Underwood, C. O. Warburton, and T.-W. Wu, *Clin. Chem.* **28,** 1159 (1982).

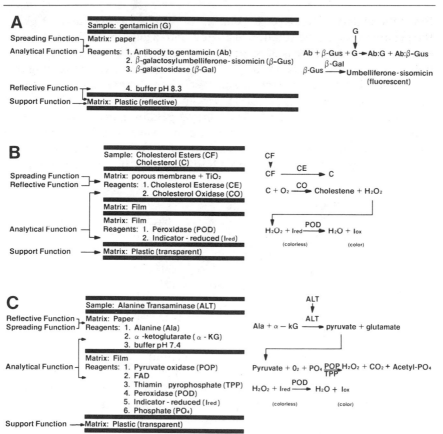

FIG. 7. (A) Cross section and chemistry of Ames (Miles Laboratories, Inc.) dry reagent strip for serum gentamicin analysis constructed by multiple saturations. (B) Cross section and chemistry of an element for cholesterol analysis constructed by multilayer techniques. (C) Cross section and chemistry of an element for serum ALT analysis constructed by a combination of film casting and saturation techniques.

struction is initiated by first coating the peroxidase chemistry onto a support. The composition of the solution is summarized below:

Gelatin	14% (w/v)
Peroxidase	4.6×10^5 U/liter
Triton X-100	1.8% (w/v)
4-Methoxy-1-naphthol	0.5% (w/v)
5,5-Dimethyl-1,3-cyclohexanedione	0.15% (w/v)
Phosphate buffer	0.1 M
pH	6.4

Since the matrix of the layer is gelatin, the solution is kept at 45° to keep the gelatin molten. The solution is coated onto a gelatin-subbed polyethylene support to a thickness of 50 μm. The coating is allowed to cool and gel before drying at low temperatures (<30°). A spacer layer consisting of a 1% (w/v) n-isopropylacrylamide and 1% Triton X-100 is coated on top of the first layer to a thickness of 15 μm and allowed to dry. Since many cholesterol esters will not penetrate into a gelatin matrix, the remaining chemistry is applied in a layer with greater porosity. This is accomplished by incorporating the chemistry into the element spreading layer. The emulsion composition for the spreading layer has the following composition:

Cellulose acetate	9.7	g
TiO$_2$	64.5	g
Lipase M (15 U/mg)	1.1	g
α-Chymotrypsin	21.2	g
Triton X-405	30.0	g
Cholesterol oxidase	405	Units
Chloroform	30	ml
Acetone	25	ml
Xylene	15	ml

The emulsion is prepared by dissolving cellulose acetate in chloroform before the other two solvents are added. The remaining ingredients are added, and the mixture is emulsified. This emulsion is coated on top of the spacer layer to a thickness of 100 μm. After drying, the material is cut into a desirable dimension. As with the previous examples, all the basic functions can be identified. Upon application, the sample is spread by the spreading layer. The freely diffusing cholesterol esters are hydrolyzed, and the resulting cholesterol is oxidized to generate H$_2$O$_2$. Owing to its low molecular weight, H$_2$O$_2$ readily diffuses into the gelatin layer to initiate the peroxidase reaction.

Element Construction by Hybrid Methods. Integrated elements can also be constructed by the combined use of coating and saturation techniques. This can be exemplified by the construction of element for alanine transaminase (ALT, alanine aminotransferase) analysis.[33] The cross section of such an element is shown in Fig. 7C. The chemistry is summarized by Eqs. (7)–(9). The pyruvate oxidase and peroxidase reactions are intro-

$$\text{Alanine} + \alpha\text{-ketoglutarate} \xrightarrow{\quad\text{ALT}\quad} \text{pyruvate} + \text{glutamate} \tag{7}$$

[33] H. Katsuyama, Y. Amano, A. Konlo, and S. Asuka, German Patent DE 32,067,231 (1982).

$$\text{Pyruvate} + \text{phosphate} + O_2 \xrightarrow[\text{thiamin pyrophosphate}]{\text{pyruvate oxidase}} H_2O_2 + CO_2 + \text{acetyl phosphate} \quad (8)$$

$$H_2O_2 + \underset{\text{(colorless)}}{\text{indicator (reduced)}} \xrightarrow{\text{peroxidase}} H_2O + \underset{\text{(color)}}{\text{indicator (oxidized)}} \quad (9)$$

duced by a coating procedure. The composition of the solution is summarized below:

N,N-Bis(β-hydroxyethyl)-m-toluidine	0.2% (w/w)
4-Aminoantipyrene	0.3% (w/w)
Gelatin	16.7% (w/w)
Pyruvate oxidase	2.7×10^4 U/liter
FAD	$2.5 \times 10^{-4}\%$ (w/w)
Peroxidase	1×10^5 U/liter
Thiamin pyrophosphate	0.1% (w/w)
$MnCl_2$	0.1% (w/w)
$Na_2HPO_4 \cdot 12H_2O$	3.6% (w/w)
$NaH_2PO_4 \cdot 2H_2O$	1.0% (w/w)
p-Isononylphenoxypolyglycido ether (Surfactant 10G)	0.13% (w/w)
pH	7.4

The molten gelatin solution (45°) is coated onto Gel Bond at 100 μm thickness and dried at room temperature. Since ALT does not penetrate the gelatin matrix, the components for the transamination reaction are introduced in a layer that facilitates free diffusion of the enzyme. This is accomplished by saturating filter paper (200 μm thick) with the following solution:

α-Ketoglutarate	0.15% (w/w)
L-Alanine	3% (w/w)
$Na_2HPO_4 \cdot 12H_2O$	2.2% (w/w)
$NaH_2PO_4 \cdot 2H_2O$	0.6% (w/w)
Polyacrylamide	0.3% (w/w)
Surfactant 10G	0.1% (w/w)
pH	7.4

After drying, the paper layer is laminated onto the film layer by moistening the film surface and placing the paper on top. The system is allowed to dry before it is cut into a desired configuration. On sample application the paper matrix facilitates sample spreading and free diffusion of ALT. The resulting pyruvate readily diffuses into the gelatin layer to initiate the pyruvate oxidase and peroxidase reactions. On sample application a linear relation is established between rate of change in reflectance density and ALT levels.

Survey of Dry Reagent Chemistries in Clinical Analysis

Dry reagent chemistries have been described for the analysis of a variety of blood constituents. These include metabolites, enzymes, electrolytes, hormones, and therapeutic drugs. A partial list is summarized in Table III. With the exception of electrolytes, nearly all the analyses depend on enzyme chemistry including immunochemical assays. A brief survey of element structures will serve to illustrate how physical functions and chemical reactions used in conventional multistep procedures have been integrated in the construction of dry reagent devices. These examples will illustrate how reactions in dry reagent elements can be compartmentalized and how end products are shunted to other compartments for further reaction. In its final form, each element provides a complete analytical procedure.

Metabolites

The oldest example of an integrated dry reagent chemistry for the quantitative analysis of a metabolite is the Dextrostix reagent strip (Ames Division, Miles Laboratories, Inc.) for whole blood glucose analysis. The cross-section of the element is illustrated in Fig. 8A. The detection chemistry is the well-known glucose oxidase–peroxidase procedure. Approximately 50 μl of whole blood is applied to the surface of the element (0.5 $\times$ 1 cm) where plasma glucose is separated from red blood cells by the element matrix. After a 1-min reaction time, the red blood cells are removed by washing, and the developed color is monitored from above with a Glucometer reflectance photometer and translated to glucose concentrations. The basic functions of the element are easily identifiable. The element foundation is the support function constructed of a transparent plasticlike material. The analytical function consists of a film membrane cast on top of a paper matrix that contains the detection chemistry. The membrane excludes red blood cells from the detection chemistry and quantitatively meters a plasma volume. The paper matrix also provides the reflective function. The Dextrostix strip has reduced blood glucose analysis from a multistep laboratory process to a simple procedure that allows diabetics to routinely monitor their own blood glucose at home.

Elements with similar functions but with different configurations are also available. For illustrative purposes, the cross sections of two elements, developed by Boehringer Mannheim Corporation (BMC)[34] and

---

[34] Boehringer Mannheim GmbH, British Patent 1,464,359 (1974).

TABLE III
ANALYSES FOR WHICH DRY REAGENT CHEMISTRIES
HAVE BEEN REPORTED[a]

Category	Analyte
Metabolites	Glucose
	Blood urea (BUN)
	Creatinine
	Bilirubin
	Cholesterol
	Triglycerides
	Uric acid
Enzymes and proteins	Total protein
	Albumin
	Hemoglobin
	Lactate dehydrogenase
	Aspartate aminotransferase
	Creatinine kinase
	Alanine aminotransferase
	Amylase
	Alkaline phosphatase
	γ-Glutamyltransferase
Electrolytes	Sodium
	Potassium
	Chloride
	Ammonium
	Carbon dioxide
	Calcium
Hormones and	Theophylline
therapeutic drugs	Gentamicin
	Amikacin
	Tobramycin
	Carbamazepine
	Phenytoin
	Primidone
	Tryroxine
	Insulin
	Phenobarbital
	Quinidine

[a] Reprinted with permission from V. Marks and
K. G. M. M. Alberti (eds.), "Clinical Biochemis-
try Nearer the Patient." Churchill-Livingstone,
Edinburgh, Scotland, 1984.

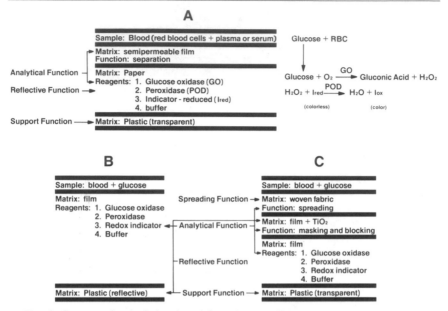

FIG. 8. Cross section and chemistry of (A) Ames (Miles Laboratories, Inc.) Dextrostix reagent strip for blood glucose analysis. Cross sections of (B) BMC and (C) Fuji elements for blood glucose analysis.

Fuji Photo Film Co., Ltd.,[35,36] are shown in Fig. 8B and 8C, respectively. In the BMC element, the porous film matrix which excludes the red blood cells also contains the detection chemistry. After a specified reaction period, the cells are removed by wiping with an absorbing material (e.g., cotton ball), and the color produced during the reaction is analyzed. As with Dextrostix strips, the reflective zone is part of the support layer, and the element is monitored from above.

Instead of removing red blood cells, the Fuji element makes use of a masking layer. This layer excludes the cells, serum, and platelets from the reaction layer. The masking layer prevents the optical interference of red blood cells with instrumental measurements. After the sample is spread by the spreading layer, which consists of a woven thin fibrous web with an active surface, the glucose diffuses through to the reaction layer which contains the detection chemistry. The chemical reaction is monitored from below through the transparent support layer.

[35] N. Kobyashi, M. Tuhata, T. Akui, and K. Okuda, *Clin. Rep.* **16**, 484 (1982).
[36] A. Ohkubo, S. Kamei, M. Yamanaka, F. Arai, M. Kitajima, and A. Kondo, *Clin. Chem.* **27**, 1287 (1981).

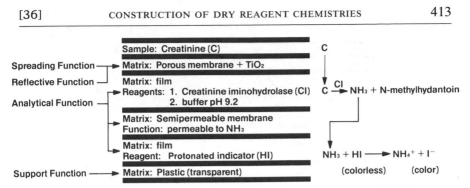

Fig. 9. Cross section and chemistry of Eastman Kodak's Ektachem slide for serum creatinine analysis.

An element making use of a semipermeable membrane to separate a gaseous product is exemplified by the Ektachem (Eastman Kodak) slide for creatinine analysis.[37,38] A cross section of the element is illustrated in Fig. 9. The element consists of a transparent support material, a layer containing the spreading and reflective functions, and an analytical function having two reagent layers separated by a semipermeable membrane. A 10-μl sample of undiluted serum is applied to the reflective-spreading layer to meter a uniform reaction volume. As the sample enters the first reagent layer, the creatinine is converted to ammonia and N-methylhydantoin in a reaction catalyzed by the enzyme creatinine deiminase (CI). The semipermeable membrane acts as a barrier to hydroxyl ions, and allows the diffusion of gaseous ammonia into the second reagent layer where the ammonia deprotonates a pH indicator. The color developed in this process is monitored by reflectance from below the carrier. Since the generated dye cannot diffuse across the membrane, the second reagent layer behaves as a trap for immobilizing the end product.

Enzymes

Construction of dry reagent chemistries for blood enzyme analysis presents new levels of complexity since enzymes are too large to readily diffuse through most conventional matrices. In addition, many enzyme analyses require coupling of multistep reactions which are frequently catalyzed by other enzymes. Some dry reagent element matrices have a large

[37] M. W. Sunberg, R. W. Becker, T. W. Esders, J. Figueras, and C. T. Goodhue, *Clin. Chem.* **29,** 645 (1983).

[38] R. W. Spayd, B. Bruschi, B. A. Burdick, G. M. Dappen, J. N. Eikenberry, T. W. Esders, J. Figueras, C. T. Goodhue, D. D. LaRassu, R. W. Nelson, R. N. Rand, and T.-W. Wu, *Clin. Chem.* **24,** 1343 (1978).

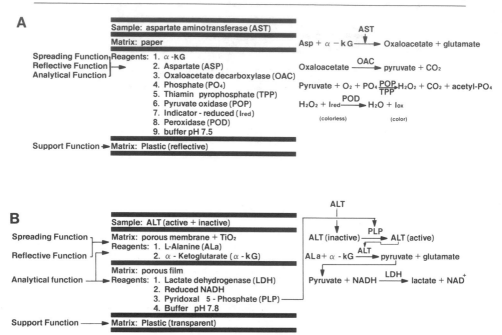

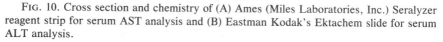

FIG. 10. Cross section and chemistry of (A) Ames (Miles Laboratories, Inc.) Seralyzer reagent strip for serum AST analysis and (B) Eastman Kodak's Ektachem slide for serum ALT analysis.

open lattice which allows free diffusion of macromolecules, whereas others have a lattice that is impermeable to macromolecules. Both configurations can be used in constructing elements for enzyme analysis. An example of a dry reagent element with an open lattice matrix is represented by the Ames Seralyzer reagent strip (Miles Laboratories, Inc.) for aspartate aminotransferase (AST).[39] The cross section of the element is shown in Fig. 10A. The element consists of a paper matrix fixed onto a reflective support. On application, the sample spreads into the reagent zone by capillary action. The chemistry couples an oxaloacetate decarboxylase (OAC) reaction, a pyruvate oxidase (POP) reaction, and a peroxidase reaction (POD) to the oxaloacetate produced during the transamination. The dry reagent strip requires 30 μl of solution derived from the 3-fold dilution of a serum specimen (10 μl of undiluted serum is required per assay).

A dry reagent chemistry with a matrix lattice that is impermeable to macromolecules is exemplified by the Ektachem slide (Eastman Kodak)

[39] B. Walter, L. Berreth, R. Co, and M. Wilcox, *Clin. Chem.* **29,** 1267 (1983).

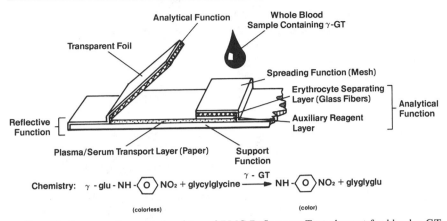

FIG. 11. Cross section and chemistry of BMC Refloquant Test element for blood γ-GT analysis.

for the analysis of alanine aminotransferase (ALT).[40] A cross section of the element is illustrated in Fig. 10B. The format consists of a transparent support layer coated with a reagent layer and a reflective spreading layer. The chemistry is partitioned between these layers. On application of a 10 μl volume, the sample rapidly diffuses into the porous spreading layer. As the sample penetrates the gelatin layer, pyridoxal 5-phosphate (PLP) diffuses into the spreading layer to activate ALT. ALT then catalyzes the transamination reaction between L-alanine and α-ketoglutarate. The resulting pyruvate then diffuses into the gelatin reagent layer where NADH is oxidized to NAD^+. Since the large molecular weight of ALT prevents its diffusion into the gelatin layer, the total chemistry must be mediated by diffusion of low molecular weight substances, such as PLP and pyruvate.

A recent report by BMC describes a dry reagent chemistry for the analysis of an enzyme in whole blood. Figure 11 illustrates the cross section of the Refloquant Test element for γ-glutamyltransferase (γ-GT) analysis in whole blood.[41,42] The blood sample is applied to the protective mesh which acts as a spreading layer. The sample then enters a matrix containing glass fibers where the red blood cells are separated from the serum or plasma by chromatographic principles. The process takes 15–30 sec. The resulting sample diffuses through the bottom paper matrix and

[40] J. N. Eikenberry, W. F. Erickson, K. J. Sandord, J. W. Sutherland, R. S. Vickers, and M. S. Weaver, *Annu. Meet. Am. Assoc. Clin. Chem., Anaheim, California, August 8–13, 1982.*

[41] E. W. Bush, *Kongr. Lab. Med., Vienna, April 19–23, 1983.*

[42] F. Stahler, *Clin. Biochem.: Princ. Pract., Plenary Lect. Symp. South East Asian Pac. Congr. Clin. Biochem.,* 2nd, 1981, p. 297 (1982).

comes in contact with the reagent layer to start the chemical analysis where the reaction is monitored through the transparent foil. Again, all the basic features of a dry reagent chemistry elements are present. This device demonstrates the ability of dry reagent devices to separate macromolecules from cells.

Electrolytes

Dry reagent chemistries are also available for serum electrolytes. Although elements using ion-selective electrochemistries have been described,[43,44] devices based on colorimetric responses are also known.[45] Recently, a dry reagent chemistry for serum potassium analysis based on a colorimetric assay has been described (Ames Division, Miles Laboratories, Inc.)[46] A cross section of the element is illustrated in Fig. 12A. The element consists of an organic medium containing a potassium-specific ionophore and a dye molecule. On applying a potassium sample, the ionophore mediates a K^+-H^+ exchange between the aqueous and organic phase. On deprotonation, the absorption band of the dye shifts from 460 nm to 640 nm. The accuracy of the element can be seen in its correlation with gravimetric methods, as illustrated in Fig. 12B. Potassium analyses in serum samples also correlate well with flame photometry (F) (element $= 0.988F + 0.092$ mM K^+) and ion-selective electrodes (ISE) (element $= 0.970ISE + 0.112$ mM K^+). This device is a good example of how organic/aqueous partition chemistry may be employed for electrolyte analysis.

Immunochemical Analysis

Dry reagent chemistries have also been described for the immunochemical analyses of therapeutic drugs and hormones. In all cases, the detection chemistry is based on an enzyme assay. Instrumentation and elements for fluorescence immunoassay of therapeutic drugs have been described previously.[10,29] More recently, dry reagent chemistries have been developed (Ames Division, Miles Laboratories, Inc.) for therapeutic drug analysis that are based on an apoenzyme reactivation immunoassay (ARIS).[47] The cross section of the element is shown in Fig. 13A. ARIS is a

[43] H. G. Curme, K. Babaoglu, B. E. Babb, C. J. Battaglia, D. J. Beavers, M. J. Bogdanowicz, J. C. Chang, D. S. Daniel, S. H. Kim, T. R. Kissel, J. R. Sandifer, P. N. Schnipelsky, R. Searle, D. S. Secord, and R. W. Spayd, Clin. Chem. 25, 1115 (1979).

[44] P. Costello, N. P. Kubasik, B. B. Brody, H. E. Sine, J. A. Bertsch, and J. P. O'Souza, Clin. Chem. 29, 129 (1983).

[45] S. C. Charlton, A. Zipp, and R. L. Fleming, Clin. Chem. 28, 1857 (1982).

[46] D. Wong, G. Frank, S. Charlton, P. Hemmes, A. Lau, R. Fleming, J. Atkinson, and E. Makowski, Clin. Chem. 30, 962 (1984).

[47] P. A. Rupchock, R. G. Sommer, and A. C. Greenquist, Clin. Chem. 30, 1014 (1984).

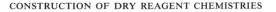

A

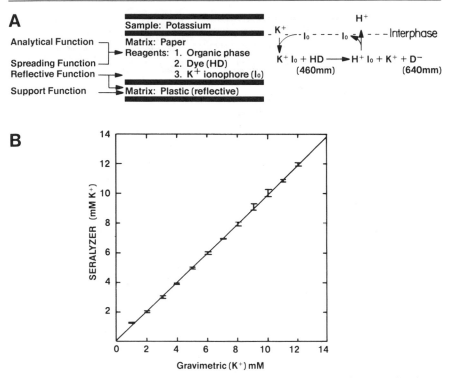

FIG. 12. (A) Cross section and chemistry of Ames (Miles Laboratories, Inc.) Seralyzer reagent/strip for serum potassium analysis. (B) Relation of gravimetric analysis for K^+ and Ames dry reagent strip analysis.

colorimetric immunoassay in which a drug and the drug–FAD conjugate compete for limited number of antibody (Ab) binding sites. Unbound conjugate reactivates apoglucose oxidase. The reconstituted enzyme activity is then proportional to drug concentration in the sample. Figure 13B illustrates kinetic profiles of strip color generation as a function of theophylline concentration, and Fig. 13C illustrates the proportionality between the rate of color generation and drug concentration.

The cross section of an element described by Fuji Photo Film Co., Ltd.[48] for the detection of the hormone thyroxine is illustrated in Fig. 14. This is one of the few examples where an assay component is covalently bound to a layer matrix. On applying the sample, the thyroxine (T_4) is mixed with the conjugate (thyroxine–peroxidase, POD-T_4). As the solution diffuses through the immobilized antithyroxine antibody (Ab) zone,

---

[48] S. Nagatoma, Y. Yasuda, N. Masuda, H. Makiachi, and M. Okazaki, European Patent Application 81,108,365.8 (1981).

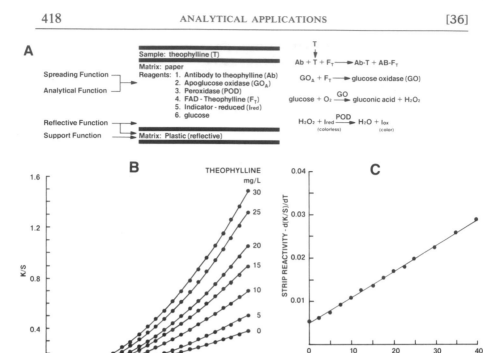

FIG. 13. (A) Cross section and chemistry of Ames (Miles Laboratories, Inc.) Seralyzer reagent strip for serum theophylline analysis. (B) Kinetic profile of reacting strips where the change in the reflectance function K/S is monitored as a function of time for several levels of theophylline. (C) Relation of theophylline concentration to the rate of change in the reflectance function $[d(K/S)/dt]$.

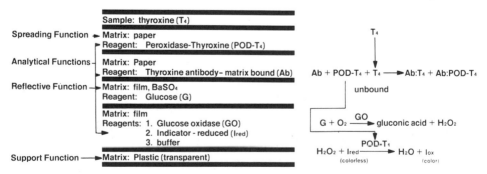

FIG. 14. Cross section and chemistry of a Fuji (Fuji Photo Co., Ltd.) element for thyroxine analysis.

TABLE IV

COMPARISON OF DRY REAGENT CHEMISTRIES WITH REFERENCE METHODS[a]

System (source)	Chemistry	Comparison system	N	Regression equation	Correlation coefficient	$S_{y,x}$	Reference[b]
Seralyzer (Ames)	Glucose	Hexokinase	191	$y = 0.99x + 2$	0.990	11 mg/dl	(1)
	BUN	Berthelot	194	$y = 0.98x + 0.9$	0.950	3.4 mg/dl	(1)
	Bilirubin	Jendrassik-Groff	278	$y = 1.04x + 0.1$	0.980	0.33 mg/dl	(1)
	Uric acid	Uricase	223	$y = 0.92x + 0.57$	0.980	0.42 mg/dl	(1)
	Creatinine	Jaffe	162	$y = 1.00x + .11$	0.990	0.4 mg/dl	(2)
	Creatinine kinase	Rosalki	87	$y = 0.98x + 5.5$	0.998	16 IU/liter	(2)
	LDH	LDH UV	154	$y = 1.01x - 4.3$	0.970	28 IU/liter	(2)
	Cholesterol	Cholesterol oxidase	114	$y = 1.01x + 3$	0.970	22 mg/dl	(2)
	AST	Modified Karmen UV method	95	$y = 1.00x + 0.3$	0.991	9.3 IU/liter	(3)
	ALT	Modified Wroblewski UV method	74	$y = 0.98x + 1.7$	0.994	5.1 IU/liter	(4)
	Potassium	Flame photometer	80	$y = 0.99x + 0.09$	0.993	0.15 mM	(5)
Ektachem (Eastman Kodak)	Glucose	SMA II	162	$y = 0.97x - 36.7$	1.00	36.4 mg/liter	(6)
	Uric acid	SMA II	184	$y = 0.98x - 2.4$	0.98	2.4 mg/liter	(6)
	Triglyceride	CentrifiChem	149	$y = 1.11x + 61.8$	0.98	124.0 mg/liter	(6)
	Albumin	SMA II	215	$y = 1.11x - 6.4$	0.81	3.2 g/liter	(6)
	Amylase	ACA	62	$y = 1.2x + 15.4$	0.98	16.7 U/liter	(6)
	Calcium	SMA II	203	$y = 0.92x + 8.4$	0.90	2.6 mg/liter	(6)
	Sodium	C800	101	$y = 1.07x - 9.3$	0.91	2.7 mM	(6)
	Potassium	C800	177	$y = 1.03x - 0.1$	0.98	0.1 mM	(6)
	Chloride	C800	180	$y = 1.03x - 4.1$	0.96	1.8 mM	(6)
	CO_2	C800	191	$y = 0.92x + 0.1$	0.76	2.2 mM	(6)
Reflotron (Boehringer Mannheim)	Hemoglobin	Cyanmethemoglobin	200	$y = 1.12x - 0.61$	0.984	—	(7)
	Cholesterol	Cholesterol oxidase	214	$y = 0.95x + 4.5$	0.977	—	(7)
	γ-GT	Eppendorf	90	$y = 0.96x + 9.8$	0.994	—	(7)
Dry Chem (Fuji)	Glucose	Glucoroder-S	124	$y = 0.99x + 3.2$	0.993	8.8 mg/dl	(8)

[a] Reprinted with permission from V. Marks and K. G. M. M. Alberti (eds.), "Clinical Biochemistry Nearer the Patient." Churchill-Livingstone, Edinburgh, Scotland, 1984.

[b] References: (1) L. Thomas, W. Plischke, and G. Storz, Ann. Clin. Biochem. **19**, 214 (1982); (2) Seralyzer Products Package Inserts, Ames Division, Miles Laboratories; (3) B. Walter, L. Berreth, R. Co, and M. Wilcox, Clin. Chem. **29**, 1267 (1983); (4) B. Walter, R. Co, and E. Makowski, Clin. Chem. **29**, 1168 (1983); (5) D. Wong, G. Frank, S. Charlton, P. Hemmes, A. Lau, R. Fleming, J. Atkinson, and E. Makowski, Clin. Chem. **29**, 498 (1983); (6) C. L. Kadinyer and R. T. O'Kell, Clin. Chem. **29**, 498 (1983); (7) E. W. Bush, Kong. Lab. Med., Vienna, April 19–23, 1983; (8) N. Kubyashi, M. Tuhata, T. Akui, and K. Okuda, Clin. Rep. **16**, 484 (1982).

the free thyroxine and conjugate are partitioned between the matrix-bound antibody and free solution. The free conjugate continues to diffuse through the reflective layer into the detection layer where a glucose oxidase-based peroxidase assay is initiated. The rate of color development is proportional to the free conjugate concentration which, in turn, is proportional to the thyroxine concentration in the sample. Elements for immunochemical detection of other analytes have been described by both Eastman Kodak Co.[49] and Fuji Photo Film Company Ltd.[50]

Performance

The success of any new technology depends on its correlation to established methodologies. Table IV summarizes comparisons between dry reagent chemistries and conventional solution chemistries made by several manufacturers.[39,41,51,52] In most cases the correlation is better than 0.95, and the slopes of the correlations are close to unity.

Summary

The development of dry reagent systems has provided convenience to the user as well as devices that are more versatile and suitable for a variety of analyses. Most dry reagent chemistries are usually less than 7 cm^2 by 300 μm thick packaged as discrete test devices. This reduces spoilage of unused reagents. Sample volumes needed for analysis are usually in the range of 3–30 μl, with 10 μl most commonly used. The use of such small volumes makes these devices suitable for neonatal and geriatric patients where large sample volumes are not often available. Hence, 150 μl of serum (approximately 300 μl blood) is sufficient for at least 15 different analyses on a sample. Dry reagent chemistries are easy to store, readily available for use, and disposable. Only application of a sample is needed to start an analysis.

[49] R. S. Zon and S. F. David, European Patent Application 79,303,004.8 (1979).
[50] N. Hiratsuka, Y. Mihara, N. Masuda, and T. Miyazako, U.S. Patent 4,337,065 (1982).
[51] L. Thomas, W. Plischke, and G. Storz, *Ann. Clin. Biochem.* **19,** 214 (1982).
[52] C. L. Kadinyer and R. T. O'Kell, *Clin. Chem.* **29,** 498 (1983).

[37] Mass Spectrometry Combined with Immobilized Cells and Enzymes

By JAMES C. WEAVER

Introduction

A mass spectrometer (MS), which can measure small quantities of volatile substances, can be combined with immobilized enzymes or cells in order to measure nonvolatile substances. A MS can be thought of as a programmable extension of a membrane-covered oxygen electrode (Clark electrode). Just as suitable immobilized enzymes (oxidases) can be combined with the Clark electrode in order to create an enzyme electrode, any enzyme which catalyzes a reaction involving volatile reactants can be considered for use with a MS. Immobilized cells can also be combined with an MS. In this way, measurement based on the consumption of volatile substrates and release of volatile products can be exploited.

Immobilized Enzymes and a Mass Spectrometer

The basic arrangement of the immobilized enzyme region within the MS interface is shown in Fig. 1.[1,2] It is important to distinguish two cases: (1) The enzyme (or cells) catalyzes a reaction wherein a volatile substrate (e.g., O_2) is consumed, and it is the volatile reactant which is measured by the MS. (2) The enzyme (or cells) catalyzes a reaction wherein a volatile product (e.g., CO_2) is liberated, and it is the volatile reactant which is measured by the MS. In general, it appears that there are more enzyme-catalyzed reactions with suitable volatile products than there are with suitable volatile substrates.[3]

Volatile Substrate Consumption

Substrate measurement requires only that the volatile substrate be intercepted by the layer of immobilized enzyme, so that the flux of volatile substrate reaching the MS decreases as the enzyme-catalyzed rate

[1] J. C. Weaver, M. K. Mason, J. A. Jarrell, and J. W. Peterson, *Biochim. Biophys. Acta* **438,** 296 (1976).
[2] J. C. Weaver, *in* "Biomedical Applications of Immobilized Enzymes and Proteins" (T. M. S. Chang, ed.), Vol. 2, p. 207. Plenum, New York, 1977.
[3] T. E. Barman, "Enzyme Handbook," Vols. 1 and 2 (1969) and Supplement I (1974). Springer-Verlag, Berlin.

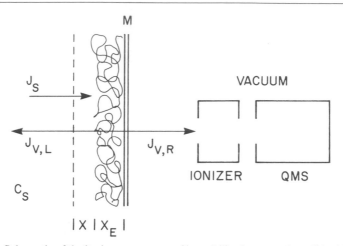

FIG. 1. Schematic of the basic arrangement of immobilized enzyme (or cells) within a MS interface. M is a gas-permeable membrane, vacuum is the very low pressure gas region within a MS. The MS (in this example a QMS) consists of an ionizer and a mass filter. The enzyme (or cells) are immobilized within a layer of thickness X_E, while X is an unstirred layer. The case of volatile product liberation is shown. J_S is the substrate flux to the enzyme from the sample, where substrate is present at concentration C_S. The volatile product flux has two components: (1) a "signal flux" to the right, $J_{V,R}$, which enters the MS, and (2) a "lost flux" to the left, $J_{V,L}$, which enters the sample.[1,2]

increases. This is analogous to the use of oxidases with the Clark electrode, wherein the diffusion-limited flux of oxygen to the electrode decreases because of immobilized oxidase when nonvolatile substrates are present, and oxygen is supplied at known levels.

Volatile Product Liberation

Product measurement requires that more consideration be given to the location of the immobilized enzyme with respect to the gas-permeable membrane. Volatile product originates within the immobilized enzyme layer, but is removed by diffusion in two directions: (1) toward the gas-permeable membrane of the MS interface, and (2) away from the membrane (toward the sample).[1] In short, the product flux is divided between the MS and the sample. Because the flux into the MS alone contributes to the signal, a significant fraction of the product flux should enter the MS. Thus, attention must be paid to the relative permeabilities of the immobilized enzyme region, any unstirred layers and the membrane interface of the MS.

Immobilization of Enzyme or Cell within the MS Interface

A typical MS interface membrane configuration is shown in Fig. 2. Either enzyme or cells can be immobilized near the membrane. Because the gas-permeable membrane of the MS interface is usually silicone based, an entrapment immobilization method is convenient. Gas-permeable membranes with a fibrous backing onto which enzyme can be immobilized are not as widely available as those without.[1] Any of the entrapment methods used with enzyme electrodes or cell electrodes can be

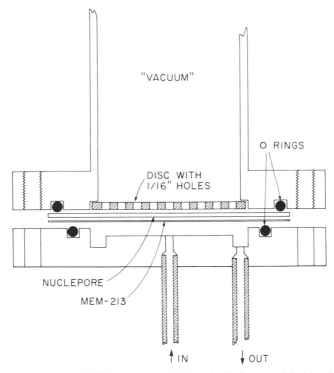

FIG. 2. Membrane-based MS interface which is suitable for use with either immobilized enzymes or cells, with enzyme or cells immobilized within or upstream from the interface.[8,9] The interface is constructed from stainless steel, with solder or epoxy used to attach 0.159-cm outer diameter stainless steel tubes. The Viton O rings provide both liquid and vacuum seals. The membrane is usually silicone based (see text). Mechanical support for the membrane is provided by a 1-μm Nuclepore filter and a stainless steel disk with 0.159-cm holes. The vertical separation between the membrane and the underlying stainless steel is 0.01 cm, except at the outer collection annulus which has a 0.159 × 0.159 cm cross section. The membrane area exposed to the flowing aqueous solution is 1.3 cm^2.

considered.[4-7] The thickness of the immobilized layer usually dominates the measurement response time.

Immobilization of Enzymes or Cells Upstream from the MS Interface

Immobilization is most readily accomplished if the enzyme or cells are located upstream from the MS interface in a flow-through configuration. For cells, a reversible immobilization procedure is very simple: Cells are injected into a thin chamber which is separated from a flowing stream by a porous membrane.[8] The membrane retains the cells, but permits diffusive transport of substrates to and products away from the entrapped cells. Removal of the reversibly immobilized cells is readily accomplished by flusing the chamber with water. For example, two stainless steel slabs with flat surfaces are held together, clamping a porous membrane between the slabs.

A machine shop with a milling machine is employed to mill flat channels into two flat pieces of 3/8-inch-thick stainless steel plate. A 1/16-inch end mill is used to cut a 5-inch-long shallow (4 mil $\approx$ 100 μm deep) channel with flat sides and semicircular ends. Care is paid that the two channels exactly overlap when the two pieces of stainless steel are held together. Then 0.1065-inch-diameter holes (drill size 36) are drilled 1 inch apart around the perimeter of both pieces while they are clamped together. The orientation of the two pieces is marked with a pair of scratches, and a 1/8-inch pry hole is ground on the side of piece one so that insertion and twisting by a screwdriver can be used to forcefully separate the two pieces after use. The holes in one piece are enlarged to 0.1495 inch diameter (6-32 bolt clearance holes; free fit drill size 25), while the holes in the mating piece are taped for 6-32 stainless steel bolts. At this point the reversible immobilization chamber can be repeatedly assembled and disassembled.

Four stainless steel tubes 1/16 inch outer diameter are used to provide inlets and outlets to the two shallow chambers. Four communication holes of diameter 0.055 inch are drilled at the ends of the channels. These holes are enlarged to 0.065 inch from the outside of the pieces, such that

[4] K. Mosbach (ed.), this series, Vol. 44.
[5] G. G. Guilbault, "Handbook of Enzymatic Methods of Analysis." Dekker, New York, 1976.
[6] P. W. Carr and L. D. Bowers, "Immobilized Enzymes in Analytical and Clinical Chemistry: Fundamentals and Applications." Wiley, New York, 1980.
[7] T. Seiyama, K. Fueki, J. Shiokawa, and S. Suzuki (eds.), "Chemical Sensors." Elsevier, Amsterdam, 1983.
[8] J. C. Weaver, C. R. Perley, F. M. Reames, and C. L. Cooney, Biotechnol. Lett. 2, 133 (1980).

the 0.065 inch diameter extends approximately half way through the 3/8-inch-thick pieces. Four 1/16-inch stainless steel tubes with Swagelock connectors attached to one end are inserted into the holes and bonded by epoxy, soft solder, or brazing. The bonding material has only a small area exposed to aqueous solutions. Epoxy is least strong mechanically, but should be used whenever there is concern that no heavy metals be exposed to the aqueous solution. The four tubes emerge perpendicular to the stainless steel pieces. In order to provide strain relief, and thereby protect the bonding of the tubes, each tube is bent back and secured by a small cable clamp fastened by one of the 6-32 bolts.

A dialysis membrane with molecular weight cutoff of about 50,000 is clamped between the two pieces, thereby forming two chambers which are separated by approximately 2 cm^2 of membrane. The clamped membrane also serves as a gasket at the perimeter; occasionally a light application of silicone grease is used if leaks at the perimeter occur. In order to place a dialysis membrane between the chambers, the membrane is immersed in water and stretched over a wooden frame (such as is used for embroidery). The two stainless pieces are brought together with the taut membrane held between, and the stainless screws inserted (a needle is used to pierce the taut membrane at each screw location). The screws are tightened partially in turn, working around the perimeter. After several passes, the membrane is taut and tightly clamped between the two stainless pieces, ready for use.

Tube connection is accomplished easily by the Swagelock connectors. One chamber has the sample stream passed through it and on to the MS interface. The other is connected to a syringe outfitted with a Swagelock connector, so that suspensions of either cells or enzymes can be immobilized by entrappement within the 100-μm-deep chamber following sample injection using the syringe. Removal of enzymes or cells is accomplished by repeated flushing in combination with brief exposures to high concentrations of HCl to kill cells or inactivate susceptible enzymes.

Because the immobilized cell chamber and MS interface can be spatially separated, it is possible to exploit physiochemical changes (e.g., temperature increase, pH decrease for volatile acids) which increase the volatility of dissolved volatile compounds. The immobilized cell chamber is maintained at biological conditions, while the MS interface is heated (e.g., 50°) or maintained at an extreme pH (e.g., $1 \leq pH \leq 12$).[9] For pH alteration it appears desirable to use a volatile acid or base, e.g., formic acid if acetic, acrylic, or propionic acid are measured (Weaver and Naber, unpublished data).

[9] J. C. Weaver and J. H. Abrams, *Rev. Sci. Instrum.* **50**, 478 (1979).

Device Materials

Wherever possible stainless steel should be used. Compared to glass, plastics, epoxies, etc., stainless steel neither takes up nor releases significant quantities of volatile species. This is well recognized in vacuum technology, wherein the release of very small amounts of volatile species is devastating. The practices of vacuum technology are extremely useful in selection of materials and the methods of fabrication of all of the apparatus contacting aqueous solutions.[10–12] A particularly convenient approach to plumbing is based on the use of stainless steel tubing and Swagelock connectors. Malleable stainless tubing with an outer diameter of 1/16 inch (0.159 cm) is available with inner diameters from 0.0178 to 0.130 cm from general commercial suppliers such as Tubesales (Chicago, IL) and chromatography equipment suppliers such as Alltech Associates Inc. (Deerfield, IL).

MS Interface

The combined properties of the MS interface and the dissolved volatile species determine whether a particular dissolved volatile compound can be measured at reasonably low concentrations. An interface for use with flowing aqueous solutions is shown in Fig. 2.[9] The mechanically supporting disk and Nuclepore filter (Nuclepore Corp., Pleasanton, CA) do not contain critical design features. However, the gas-permeable membrane is critical. Water permeation into the MS is tolerable, but not desired. Because of the high concentration of water in aqueous solutions, even a low water permeability results in a large water flux into the MS. This causes water to be the dominant species in the background gas of the MS, and governs the operating pressure of the MS. Background gas causes scattering within the MS, such that there is a low level interference of water with the measurement of any other volatile compounds. Any variations in the water flux couple directly into the signals associated with volatile compounds.

The requirement that the membrane be highly permeable to a volatile compound implies that a dissolved volatile compound must partition significantly into the membrane, and that the membrane must both be thin and allow a high rate of diffusion. Silicone-based membranes are good

---

[10] S. Dushman and J. M. Lafferty (eds.), "Scientific Foundations of Vacuum Technique," 2nd Ed. Wiley, New York, 1962.

[11] G. L. Weissler and R. W. Carlson (eds.), "Vacuum Physics and Technology." Academic Press, New York, 1979.

[12] J. F. O'Hanlon, "A User's Guide to Vacuum Technology." Wiley, New York, 1980.

choices. They have a relatively high permeability to many volatile compounds in comparison to their water permeability. However, silicone membranes can vary widely in mechanical strength. For example, a proprietary membrane, MEM-213 (General Electric, Schenectady, NY), is preferred over dimethylsilicone because its mechanical strength is a factor of 300 greater than dimethylsilicone whereas its permeability is smaller only by a factor of 2 to 3.

Measurement of dissolved volatile compounds is affected by the flow rate through the MS interface.[9] Flow variations alter the mass transport of volatile molecules into the MS and directly alter the signal. Although purposeful modulation of the flow with a well-controlled pattern could be an advantage, allowing lock-in techniques to be employed, typical peristaltic pumps do not appear to possess a sufficiently well-controlled pumping pattern. For this reason quiet, steady pumps such as gravity feed or gas overpressure are generally used. Syringe pumps are not good choices, because their short-term motion is irregular.

Heating the interface (e.g., to 50°) increases the vapor pressure of dissolved volatile analytes and of water, usually with an improved signal-to-noise ratio, which is beneficial.[9,13] Use of several MS interfaces with one MS is also feasible. By placing a vacuum quality valve between each interface and the main chamber of the vacuum system of the MS, multiplexing can be used. Computer control of the valves allows association of signals with individual MS interfaces.

Mass Spectrometers

General properties of the various types of mass spectrometers have been described by others.[14–16] The quadrupole mass spectrometer (QMS) appears best suited for use with immobilized enzymes and cells. Key features of the QMS are as follows:

1. The ability to computer control the measured ion (m/z, mass to charge ratio) with a linear control voltage. The integrated peak area, corrected for background, is analogous to the output current

[13] J. P. Mieure, G. W. Mappes, E. S. Tucker, and M. W. Dietrich, in "Identification and Analysis of Organic Pollutants in Water" (L. H. Keith, ed.), p. 113. Ann Arbor Science, Ann Arbor, Michigan, 1976.

[14] G. R. Waller (ed.), "Biochemical Applications of Mass Spectrometry." Wiley (Interscience), New York, 1972.

[15] J. R. Chapman, "Computers in Mass Spectrometry." Academic Press, London, 1978.

[16] G. R. Waller and O. C. Dermer (eds.), "Biochemical Applications of Mass Spectrometry," Suppl. 1. Wiley, New York, 1980.

of an oxygen electrode, and comprises the signal by which a volatile compound is measured.[17,18]

2. The ability to use "selected ion monitoring," wherein only selected peaks (m/z) in the mass spectrum are measured, and to switch rapidly (microseconds) between selected peaks. The speed of the QMS allows several volatile compounds to be measured at each MS interface within a few seconds.

3. The ability to easily alter which peaks are measured using software. This allows different volatile compounds to be measured on different days, if the type of enzyme or cell is changed.

Overall, the QMS can be regarded as a flexible, rapidly responding sensor whose specificity can be altered by computer programming. Because of the relatively high cost and complexity of any mass spectrometer, the most attractive applications will involve multiplexing, such as multiple monitoring of fermenters.

Vacuum Systems

A vacuum system which contains a MS used for measurements with enzymes or cells has special requirements. These include the following:

1. The vacuum system must be capable of tolerating a high water input for long periods (hours to weeks).

2. The pumps of the vacuum system must pump water well for long periods. This can be accomplished by using cryopumps (which pump water extremely well), provided that they are used in alternating pairs. Turbomolecular pumps are a reasonable alternative, while diffusion pumps are less desirable.

3. The entire vacuum system, including pressure gauges and vacuum pumps, must be sufficiently sturdy that occasional rupture of the gas-permeable membrane is well tolerated. An all stainless steel vacuum system is generally required. Both cryopumps and turbomolecular pumps recover well from the large loading of water, salts, and other materials which enter the MS vacuum system on membrane rupture. Diffusion pumps are generally less forgiving, requiring a long bake-out of the diffusion pump oil.

4. The entire vacuum system and MS should be heated, preferably to 50–100°, to minimize adsorption of volatile compounds to the sur-

[17] E. Pungor, Jr., C. R. Perley, C. L. Cooney, and J. C. Weaver, *Biotechnol. Lett.* **2,** 409 (1980).
[18] E. Pungor, Jr., E. J. Schaefer, C. L. Cooney, and J. C. Weaver, *Eur. J. Appl. Microbiol. Biotechnol.* **18,** 135 (1983).

faces within the vacuum system and MS. Significant adsorption leads to hysteresis, wherein large magnitude signals decay slowly.

Criteria for Volatile Compounds

Although each candidate volatile compound must be individually tested for measurement while dissolved in aqueous solution, some guides are available. Compounds generally regarded as gases often have published solubilities, either directly as solubilities or as a Henry's Law constant.[19,20] Less volatile species such as alcohols, ketones, and low molecular weight acids usually have published boiling points, with lower boiling points often corresponding to higher volatility in the dissolved aqueous phase. As a guide, liquid compounds with boiling points of less than 150° are worth pursuing as candidates. In the case of simultaneous multiple measurements, considerable thought should be given to the compounds' fragmentation patterns in the mass spectrometer.

Stable isotopes such as ^{13}C or ^{2}H can also be distinguished by a MS, such that dissolved volatile species containing different isotopes can also be measured.[21] In cases where appropriately stable isotope-labeled compounds are available, the measurement of a volatile compound containing an isotope of low natural abundance can greatly reduce background effects. For example, use of an immobilized decarboxylase with a substrate labeled with ^{13}C at the appropriate position results in the liberation of $^{13}CO_2$. The MS can readily measure $^{13}CO_2$ at m/z 45 in the presence of larger amounts of $^{12}CO_2$ which has its primary mass peak at m/z 44. The natural abundance of the selected isotope must also be considered, because it affects the degree of background reduction. For example, ^{13}C is present as about 1% of total carbon, so that the $^{13}CO_2$ background is about two orders of magnitude lower than the $^{12}CO_2$ background. There also exist stable isotopes of hydrogen, nitrogen, oxygen, and sulfur which are two to four orders of magnitude lower than the more abundant isotopes.

Computer Considerations

Real time computation is used to both control a QMS and to analyze the resulting data.[15,17,18,21] Computer-generated control voltages govern the use of selected ion monitoring, such that a user can enter which ion

[19] W. Braker and A. L. Mossman, "Matheson Gas Data Book." Matheson Gas Products, East Rutherford, New Jersey, 1971.
[20] C. D. Hodgman, R. C. Weast, and S. M. Selby (eds.), "Handbook of Chemistry and Physics," 38th Ed. Chemical Rubber Co., Cleveland, Ohio, 1956.
[21] E. Pungor, Jr., A. M. Klibanov, C. L. Cooney, and J. C. Weaver, Biomed. Mass. Spectrom. 9, 181 (1982).

peak (m/z ratio) is to be measured, and the computer will generate the appropriate control signals based on previous calibrations. Real time analysis of the data is realistic because the mass peak area S_i (the ith signal) is related to the jth dissolved volatile compound concentration C_j by a linear equation [Eq. (1)]. Here the $A_{i,xj}$ are coefficients determined in a calibrat-

$$S_i = A_{i,1}C_1 + A_{i,2}C_2 + \cdots + A_{i,j}C_j + \cdots + A_i(0) \qquad (1)$$

ing run, and $A_i(0)$ is the background contribution to the ith mass peak. Because the relationship is linear, a real time computer-generated solution for the corresponding concentrations is possible. However, there is some evidence that the mass transport of some chemical species can interact.[22,23] In such cases, the coefficients $A_{i,xj}$ can no longer be treated as constant over a wide range of concentrations. Even in this case, however, useful MS measurements of dissolved products of cells were measured.

[22] P. Doerner, J. Lehman, H. Piehl, and R. Megnet, *Biotechnol. Lett.* **4**, 557 (1982).
[23] J. W. Lorenz, B.S. thesis. Massachusetts Institute of Technology, Cambridge, Massachusetts, 1983.

[38] Bioelectrocatalysis: Conductive and Semiconductive Matrices for Immobilized Enzymes

By S. D. Varfolomeev

Introduction

The progress made in the field of enzymology and immobilized enzymes has led to the idea that highly efficient and selective enzyme catalysts can be employed to promote electrochemical processes. As shown by investigations in recent years, immobilized enzymes can catalyze electrode processes, including the stages of molecular ionization and electron transport to an electron-conducting matrix. The term bioelectrocatalysis has been suggested to define the phenomenon of enzyme-catalyzed electrochemical reactions.[1-6]

[1] I. V. Berezin, S. D. Varfolomeev, A. I. Jaropolov, V. A. Bogdanovskaya, and M. R. Tarasevich, *Dokl. Akad. Nauk SSSR* **225**, 105 (1975).
[2] S. D. Varfolomeev, A. I. Jaropolov, I. V. Berezin, M. R. Tarasevich, and V. A. Bogdanovskaya, *Bioelectrochem. Bioeng.* **4**, 314 (1977).
[3] S. D. Varfolomeev and I. V. Berezin, *J. Mol. Catal.* **4**, 387 (1978).
[4] I. V. Berezin and S. D. Varfolomeev, *Appl. Biochem. Bioeng.* **2**, 259 (1979).

The fact that enzymes can operate as good catalysts in electrode processes provides a considerable incentive for bioelectrocatalytic studies.[6-10] By way of example, the reader is referred to Ref. 5. Ordinarily, immobilization of enzymes on carriers produces samples containing 10–100 mg of the protein per gram of the carrier. The "median" molecular mass of the protein, 3×10^4 g/mol corresponds to an enzyme active site concentration of 3×10^{-7}–3×10^{-6} mol/g. If this kind of sample is used for an electrode operating under kinetic conditions, the current densities should be 3–30 A/g. These are substantially superior values to those obtained in "classical" nonenzymatic electrochemical reactions. Coupled with the highest specificity and selectivity, as well as with the capacity for an infinitely wide range of organic reactions, the electrode-immobilized enzymes set the stage for a new generation of electrochemical processes.

Electron Transfer between the Enzyme Active Site and the Electrode

Electron transfer between the enzyme active site and the electrode constitutes the most complex problem of bioelectrocatalysis. Several effective ways are known to "populate" the enzyme active sites with electrons (or electron vacancies)[1-6]:

1. Electron transfer is effected with the aid of a diffusionally mobile low-molecular-weight electron carrier, the mediator. The sequence of the process is illustrated in Eq. (1), where E and E^0 are the oxidized and

$$S + E \rightarrow P + E^0, \qquad E^0 + M \rightarrow E + M^0, \qquad M^0 \xrightarrow{\text{electrode}} M \pm e^- \qquad (1)$$

reduced forms of the enzyme active site, M and M^0 the oxidized and reduced forms of the mediator, S the substrate, and P the product of the enzymatic reaction. The mediator mechanism of electron transport is fairly widely used in electrochemical enzymatic reactions (for review, see Ref. 7).

2. A direct electrocatalytic transfer of electrons is possible between the electrode and the enzyme active site. This is the most attractive

[5] I. V. Berezin, S. D. Varfolomeev, and M. V. Lomonosov, *Enzyme Eng.* **5,** 95 (1980).

[6] H. A. O. Hill and I. J. Higgins, *Philos. Trans. R. Soc. London, A* **302,** 267 (1981).

[7] S. D. Varfolomeev and I. V. Berezin, *in* "Advances in Physical Chemistry: Current Developments in Electrochemistry and Corrosion" (J. M. Kolotyrkin, ed.), p. 60. Mir, Moscow, 1982.

[8] E. V. Plotkin, I. J. Higgins, and H. A. O. Hill, *Biotechnol. Lett.* **3,** 187 (1981).

[9] L. S. Suzuki, I. Karube, T. Matsunaga, S. Kuriyana, N. Suzuki, T. Schirogami, and Y. Takamura, *Biochimie* **62,** 353 (1980).

[10] L. B. Wingard, C. H. Schaw, and J. F. Castner, *Enzyme Microbiol. Technol.* **4,** 137 (1982).

proposition for bioelectrocatalytic processes. This electron transport mechanism has been determined for a limited number of electrocatalytic systems: for laccase, responsible for four-electron reduction of oxygen,[11] for peroxidase in hydrogen peroxide reduction,[12] and for hydrogenase operating in the mode of electrochemical ionization or hydrogen formation.[13,14] Electron transfer between the electrode and the acceptor (donor) group of the active site is known as electron tunnelling.[7,15]

3. An electrocatalytic coupling of enzymatic and electrode processes can be implemented by enzyme incorporation in organic semiconductors.[5,16–18] In this case, electron transfer between the enzyme active center and the semiconductor domains takes place with subsequent electron transport to the electrode.

Below we shall consider some of the bioelectrocatalysis techniques.

Oxygen and Hydrogen Enzymatic Electrodes, Carbon Matrix for Enzyme Immobilization, Direct Transport of Electrons between the Enzyme Active Site and the Electrode

Electroreduction of oxygen is one of the most formidable problems in "classical" electrochemistry. It is common knowledge that the equilibrium redox potential of the pair O_2/H_2O, equal to 1.23 V,[19] is established on specially pretreated platinum and in extra pure solutions. Oxygen exchange currents on the platinum are insignificant and equal 10^{-11} A/cm^2.

By using laccase, immobilized on different electron-conductive matri-

[11] I. V. Berezin, V. A. Bogdanovskaya, S. D. Varfolomeev, M. R. Tarasevich, and A. I. Jaropolov, *Dokl. Akad. Nauk SSSR* **240,** 615 (1978).

[12] A. I. Jaropolov, V. Malovik, S. D. Varfolomeev, and I. V. Berezin, *Dokl. Akad. Nauk SSSR* **240,** 1399 (1979).

[13] A. I. Jaropolov, A. A. Karyakin, I. N. Gogotov, N. A. Zorin, S. D. Varfolomeev, and I. V. Berezin, *Dokl. Akad. Nauk SSSR* **274,** 1434 (1984).

[14] A. I. Jaropolov, A. A. Karyakin, and S. D. Varfolomeev, *Vestn. Mosk. Univ., Khim.* **24,** 523 (1983).

[15] A. I. Jaropolov, T. K. Suchomlin, A. A. Karyakin, S. D. Varfolomeev, and I. V. Berezin, *Dokl. Akad. Nauk SSSR* **260,** 1192 (1981).

[16] S. D. Varfolomeev, S. O. Bachurin, I. V. Osipov, K. V. Aliev, I. V. Berezin, and V. A. Kabanov, *Dokl. Akad. Nauk SSSR* **240,** 615 (1978).

[17] I. V. Berezin, S. D. Varfolomeev, B. E. Davidov, G. V. Mavrenkova, and I. V. Tisachnaya, U.S.S.R. Patent 707,244 (1978).

[18] K. V. Aliev, I. V. Berezin, S. O. Bachurin, S. D. Varfolomeev, and V. A. Kabanov, U.S.S.R. Patent 707,245 (1978).

[19] All potential values are given with respect to the hydrogen electrode in the same solution.

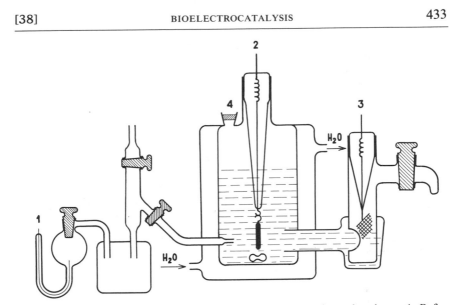

FIG. 1. The thermostatted electrochemical cell for a three-electrode scheme. 1, Reference electrode; 2, enzyme electrode; 3, auxiliary electrode (platinum net); 4, membrane for taking gas samples.

ces, especially carbonic carriers, one can obtain electrodes which actively reduce oxygen. In a similar fashion, electrodes with immobilized hydrogenase have been obtained, capable of ionizing a hydrogen molecule at an equilibrium potential.[20]

Enzymes are immobilized on carbon carriers by using the adsorption method. An electrode formed from carbon black with a specific surface of 50 m²/g is immersed in an enzyme solution with a concentration of 10^{-6}– 10^{-5} M and kept at room temperature for 24 hr. The enzyme adsorption process is completed within this period of time. Laccase and hydrogenase adsorption on carbon black is practically irreversible.

It is most convenient to test the electrocatalytic properties of immobilized enzymes by using a thermostatted electrochemical cell equipped with a special gas blow tap and a membrane for taking gas samples (Fig. 1). All the electrochemical measurements are taken according to a three-electrode scheme with the use of a potentiostat. A steady potential of the working electrode is maintained with respect to the subsidiary one, and the current registered. The dependence of the current on the enzymatic electrode potential is a criterion of the efficiency of the electrode process.

[20] V. A. Bogdanovskaya, S. D. Varfolomeev, and M. R. Tarasevich, U.S.S.R. Patent 877,980 (1979).

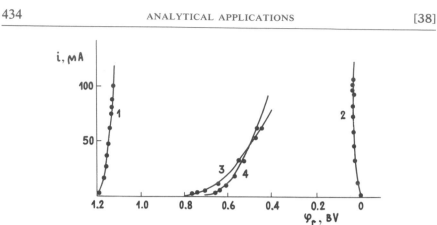

Fig. 2. Current versus potential (polarized curves) for electrodes responsible for electro-reduction of oxygen with the aid of immobilized laccase (curve 1) and hydrogen electrooxi-dation with the aid of immobilized hydrogenase (curve 2). Curve 3 describes the oxygen reduction process in the absence of immobilized laccase, and curve 4, that in the presence of immobilized laccase, inhibited by 1 mM NaN$_3$. Oxygen electroreduction conditions: pH 3.1, 50 mM sodium phosphate, 50 mM sodium acetate, 0.1 M potassium nitrate, 25°. Hydrogen electrooxidation conditions: pH 7.1, 50 mM sodium phosphate, 0.1 M sodium acetate, 0.1 M potassium chloride, 25°. The electrodes are made of carbon black, and the immobilization is performed by irreversible absorption of the enzymes from the solution directly to the elec-trode.

The methods for electrochemical kinetics investigation are given else-where.[21]

In neutral or weakly acidic solutions the electroreduction of oxygen on carbon electrodes proceeds extremely slowly. This is expressed in con-siderable overvoltage values (Fig. 2). The use of an electrode with ad-sorbed laccase even from very dilute solutions (10^{-9} M) involves a signifi-cant acceleration in oxygen electroreduction. This is revealed as high rates of the electrochemical reaction (high values of the specific current) at the potentials, when no process, in fact, takes place in the absence of an enzyme.

These effects do not depend on what kind of electrode is employed. Electrochemical measurements were taken on electrodes made of carbon black, pyrographite, vitreous carbon, or gold. A maximum potential value, +1.207 V, close to the equilibrium potential of an oxygen elec-trode, was attained on carbon black electrodes, stored in a laccase solu-tion (10^{-5} M) for 24 hr.

[21] J. Kuta and E. Yeager, in "Techniques of Electrochemistry" (E. Yeager and A. J. Salkin, eds.), Vol. 1, p. 151 (Russ. ed.) or p. 137 (Engl. ed.). Wiley (Interscience), New York, 1972.

The enzymatic nature of electrocatalysis was proved by specific inhibition of the process by fluoride and azide ions (Fig. 2), by heat inactivation of the enzymes, and by comparison of the pH dependence of the electrocatalytic effects and the catalytic activity in the oxidation of ferricyanide by oxygen.

Experimentally the stationary potential of the electrode depends on the partial pressure of oxygen and pH of the solution. To elucidate the nature of the stationary potential of an electrode with immobilized laccase, dependence of φ_{st} on the partial pressure of oxygen and on pH was investigated. It was found that φ_{st} (pH) constitutes 10–12 mV, and φ_{st} (pO_2) is 60 mV. These magnitudes are close to the coefficients in the Nernst equation for the system O_2/H_2O. Consequently, the electrochemical process on an electrode with immobilized laccase is determined by a reaction of four-electron reduction of oxygen to water:

$$O_2 + 4\,e^- + 4\,H^+ \rightarrow 2\,H_2O \tag{2}$$

An enzyme exerts significant catalytic effects on oxygen reduction. Acceleration estimates (comparison of catalytic currents in the presence or absence of an immobilized enzyme) produce values of 10^6–10^9.

If hydrogenase, an enzyme activating molecular hydrogen, is immobilized on carbon or metal materials, a hydrogen enzymatic electrode can be obtained, close in its characteristics to a classical highly dispersed platinum electrode.[13,14] In a hydrogen atmosphere the electrode (vitreous carbon, pyrographite, or carbon black) produces the same hydrogen oxidoreduction potential as the best platinum electrodes. Fig. 2 shows the current–potential dependence for such model electrodes with immobilized hydrogenase which promotes the equilibrium:

$$H_2 \rightleftharpoons 2\,H^+ + 2\,e^- \tag{3}$$

The use of theoretical and experimental methods of electrochemistry as applied to immobilized enzymes operating in the bioelectrocatalytic mode furnishes an insight into enzyme-induced catalysis. Under experimental conditions, the use of electrochemical kinetics methods involves an investigation of the dependence of current on potential for a given electrode with an immobilized enzyme at different concentrations of substrate, enzyme, and hydrogen ion. This investigation was carried out on an oxygen electrode with immobilized laccase and on a hydrogen electrode with immobilized hydrogenase.[13]

In the case of an oxygen electrode, the dependence of current on potential was studied under potentiostatic conditions for different concentrations of laccase immobilized on the carbon; current densities versus oxygen and hydrogen ion concentrations were likewise studied. Figure 3

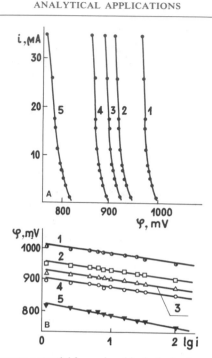

FIG. 3. (A) Current versus potential for carbon black electrodes with immobilized laccase at different pH values: 1, 3.08; 2, 4.0; 3, 4.5; 4, 5.0; 5, 6.05. Conditions: 0.1 M potassium nitrate, 50 mM sodium phosphate, 50 mM sodium acetate, 25°. (B) Linear dependence of the data in (A) in semilog coordinates.

plots current density versus potential in semilog coordinates at different pH values. Empirically, the following dependence was established:

$$i \approx [E][O_2][H^+]^2 \exp[\varphi/(26 \pm 3)] \qquad (4)$$

where [E] is the enzyme surface concentration, [O_2] the oxygen concentration, φ the potential value, and the 26 ± 3 values are given in millivolts.

Theoretical analysis of the reaction mechanism involves a search of the kinetic scheme in conformity with Eq. (4). Details of the theoretical interpretation of the experimental data on electrochemical kinetics of immobilized enzymes are explained in Refs. 6 and 20. The results of the experimental and theoretical analyses are summed up in the scheme:

$$E + O_2 \overset{\text{fast}}{\rightleftharpoons} EO_2, \qquad EO_2 + e^- \rightleftharpoons EO_2^-$$

$$EO_2^- + H^+ \overset{\text{fast}}{\rightleftharpoons} EO_2^-H^+, \qquad EO_2^-H^+ + H^+ \rightleftharpoons EO_2^-H_2^{2+}$$

$$EO_2^-H_2^{2+} + 2\,e^- \xrightarrow{\text{limiting step}} EO^- + H_2O$$

According to the data obtained, the mechanism of immobilized-laccase electrocatalysis includes the rapid step of oxygen addition, the rapid electrochemical step of transfer of an electron from the electrode to the active site acceptor group, two rapid equilibrium steps of proton acceptance, and the slow, limiting step of the synchronous transfer of two electrons to the active site. These reactions are followed by rapid steps of oxygen reduction at the active site with formation of a second molecule of water.

Enzymes in Organic Semiconductors

Organic polymer semiconductors constitute a large class of potential carriers for bioelectrocatalysts. The electric conductivity of semiconductor polymers can change over a wide range (10^{-15}–10^4 ohm^{-1} cm^{-1}) and approach the electroconductivity of metals. At least two classes of organic semiconductors are of interest in terms of enzyme immobilization:

1. Polymers with a system of conjugated bonds having a long conjugation chain. Noted for comparatively high conductivity, these polymers are electronically heterogeneous systems where polyconjugated zones with "metallic" conductivity are separated by dielectric sections. Electron transfer across the dielectric sections determines the overall barrier for electron transport.

2. Polymers with a charge transfer complex (CTC). Ion–radical pairs are the limiting case of strong CTCs. The conduction mechanism is determined by ion–radical disproportionation in "packs" consisting of electron donors and acceptors.

Tetracyanoquinodimethane (TCNQ) salts are a classical example of conductors with the ion–radical mechanism of conduction. Conduction in these systems is usually attributed to CTC formation between TCNQ molecules.

Thermally Modified Poly(acrylonitriles)

To obtain a polymeric semiconductor, the poly(acrylonitrile) samples are treated at 300 or 850° under anaerobic conditions for a week. The electroconductivity of the samples depends on the treatment temperature. Increased temperature at the anaerobic treatment of the material results in increased electric conductivity. The carriers are chemically modified by oxidation with strong nitric acid (introduction of nitro and oxo groups) and by boiling for 48 hr, and then washed clean with water. Hydrogenase was isolated from the bacterium *Thiocapsa roseopersicina* by ion-ex-change chromatography on DEAE-cellulose.

Hydrogenase is immobilized according to the following procedure: a 50 ml solution of the enzyme (12 mg protein/ml) in a phosphate buffer (pH 7.8, 2.0 M phosphate) is supplemented by 3 g of the thermally treated poly(acrylonitrile) and stirred for 1 hr. To reduce Schiff base formation in the process, 100 mg of $NaBH_4$ is added, and the suspension is stirred for another hour. The preparation thus obtained is separated by centrifuga-tion and washed 3 times with buffer solution to remove excessive immobi-lized enzyme. The carrier capacity is up to 150 mg per gram of the ther-mally treated poly(acrylonitrile).[16] Such a heterogeneous catalyst can be used as a component of composite electrode materials.

Semiconductive Polymer Carriers Based on Complex Salts of Tetracyanoquinodimethane

Immobilized enzymes are prepared by coprecipitating polycations with enzymes in the presence of a semireduced TCNQ salt. An enzyme is immobilized in high-conductivity (up to 10^{-2} ohm^{-1} cm^{-1}), water-perme-able polymers.

The following polymers (**I–III**) were used as polycations:

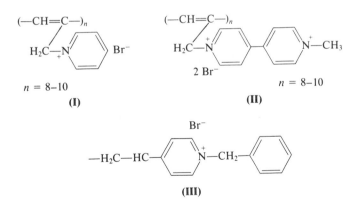

Polycation synthesis methods are given in Ref. 22. Methyl dipyridyl, a specific hydrogenase substrate, operates as the cation in the case of poly-mer **II**.

The polymer and LiTCNQ are taken in proportions so as to obtain a

[22] V. A. Bogdanovskaya, S. D. Varfolomeev, M. R. Tarasevich, and A. I. Jaropolov, *Elec-troanal. Chem.* **104,** 393 (1979).

1:2 molar ratio of z-LiTCNQ to polymer link (z is the number of charges on a polymer link). In an experiment involving polymer **II**, a solution of 90 mg polypropargylpyridinium bromide in 4 ml water is supplemented with 6 ml of hydrogenase solution (10 mg protein/ml) and stirred for 15 min; then, while continuing the stirring, 160 mg LiTCNQ in 20 ml of distilled water is added. As a result of this procedure, protein is coprecipitated in a complex with a polycation and LiTCNQ. The mixture is stirred for another 15 min, and thereafter the precipitate with the immobilized enzyme is separated by centrifugation and washed with buffer solution.

The capacity of the carriers obtained is up to 500 mg of the protein per gram of carrier. The properties of the enzymes (hydrogenase, glucose oxidase, and peroxidase) immobilized in polycation–TCNQ complex salts are described in Refs. 23–25.

"Electron Sponge" Effect in Semiconductive Matrix–Enzyme Systems

The electron-conducting semiconductive matrix operates as an electron donor (acceptor) for the active site of an enzyme. In the course of electrochemical reactions with the matrix (e.g., the polymer reduction) one can observe the "electron sponge" effect—a reversible oxidation or reduction of the matrix—and processes of enzymatic catalytic transformation of electricity carriers.[23,24]

The suspension of the semiconductor with the enzyme immobilized on it is polarized in a test cell with the aid of an extra electrode. On cutting off the polarizing potential, the formation of molecular hydrogen caused by the active site of hydrogenase is observed. The process can be controlled (i.e., hydrogen evolution ceased) with the aid of an enzyme inhibitor.

The kinetics of hydrogen evolution on initial electroreduction of the matrix with the immobilized enzyme was studied in the semiconductor polymer matrix-hydrogenase system. Just as in the case of enzyme immobilization on thermally treated poly(acrylonitrile), a time-dependent process of catalytic evolution of hydrogen to the gaseous phase is observed when the enzyme is incorporated in the polycation–TCNQ complex. Here an intermediary, electroreduced polymer substrate acts as electron donor. The rate and the amount of hydrogen formed depend on the quantity of enzyme in the polymer.[26]

[23] V. A. Kabanov, K. V. Aliev, and J. Richmond, *J. Appl. Polym. Sci.* **19,** 1275 (1975).

[24] S. O. Bachurin, S. D. Varfolomeev, I. V. Tisyachnaya, B. E. Davidov, G. B. Marzenkova, and I. V. Berezin, *Dokl. Akad. Nauk SSSR* **253,** 370 (1980).

[25] S. D. Varfolomeev, S. O. Buchurin, and A. Nagui, *J. Mol. Catal.* **9,** 223 (1980).

[26] S. D. Varfolomeev and S. O. Bachurin, *J. Mol. Catal.* **27,** 305 (1984).

The limiting amount of hydrogen evolved per active site of the enzyme indicates about 40 domains per active site of the enzyme (assuming that each domain can accept one extra electron).[27]

The semiconductor–enzyme systems involve several consecutive processes, namely (1) interdomain electron exchange, (2) transfer of electrons to the active site (reduction of the active site), and (3) formation of a product of reduction by the enzyme active site. The investigations into the kinetics of hydrogen liberation after a preliminary polarization of the matrix showed that the enzymatic step is the rate-determining one. The processes of interdomain exchange and electron transfer to the active sites of the enzyme are considerably more rapid.

[27] S. D. Varfolomeev and S. O. Bachurin, *J. Mol. Catal.* **27,** 315 (1984).

Section II

Medical Applications

Editor

Klaus Mosbach

[39] Overview

By KLAUS MOSBACH

The section on medical applications of immobilized enzymes and cells has been structured as follows: The contribution by Chang [40] is meant to serve as a general introduction. Aspects of extracorporeal shunt systems are covered in contributions [41]–[45]; more specifically, adsorption of antibodies is covered in paper [41], immunoadsorption of plasma lipoproteins in paper [42], immobilized L-glutaminase and asparaginase in papers [43] and [44], and immobilized protein A in paper [45].

Contributions [46]–[48] deal with the use of immobilized enzymes in medical instrumentation. The use of immobilized heparin lyase for blood deheparinization [46], clinical application of urokinase [47], and immobilized enzymes used for biomaterials in surgery [48] are some of the applications presented. Two contributions deal with the use of enzymes in medicine for thrombolytic therapy [49] and enzyme replacement therapy [50]. Finally, paper [51] focuses on the use of immobilized cells, i.e., microencapsulated pancreatic islets as an artificial pancreas. See also Vol. 135, Chap. [33], mentioning similar studies [e.g., K. Nilsson and K. Mosbach, *FEBS Lett.* **118,** 145 (1980)].

The purpose of this section is not only to provide selected methodological techniques, but also to act as a stimulus for further research. In addition, the reader is referred to Vol. 44 in this series and to Vols. 112 and 149 (K. J. Widder and R. Green, eds.) which covers enzyme targeting and related topics. One aspect of increasing interest involves the use of polyethylene glycol-modified enzymes. In the area of targeting, a paper on plasmin immobilized to magnetic beads should be mentioned.[1] An additional aspect worth mentioning involves recent attempts to prepare an artificial liver support.[2]

Medical applications of immobilized enzymes and cells have great potential, but from a clinical point of view they are still at an early stage of development. Because the contributors to this section come from various disciplines and because of the variety of topics presented, this section may appear heterogeneous in character, but I am sure all contributions will provide useful information.

---

[1] K. Mosbach and U. Schröder, *FEBS Lett.* **102,** 112 (1979).
[2] T. Akimoto *et al., Artif. Organ* **14,** 249 (1985).

[40] Medical Applications of Immobilized Proteins, Enzymes, and Cells

By T. M. S. CHANG

Introduction

Discussion of the medical applications of immobilized proteins, enzymes, and cells[1-9] usually emphasizes methods of enzyme immobilization and areas of possible application. Since extensive *in vivo* evaluations are being carried out in this area of research, methodologies for *in vivo* applications will also be covered in this chapter.

About 20 years ago immobilized enzymes in the form of artificial cells containing catalase, asparaginase, or urease were shown to be effective for replacing hereditary enzyme deficiency in acatalasemic mice[10]; for suppressing the growth of lymphosarcoma in mice[11]; or for decreasing system urea levels in animals.[12,13] Since that time, despite extensive efforts by many researchers using all available immobilization technologies, progress toward the actual large-scale clinical application of immobilized enzymes and cells has been very slow.[1-9] This does not negate the earlier promise of the potential medical applications of immobilized enzymes and cells but shows the complexities and problems associated with this type of approach. Although proteins, enzymes, and cells are potentially much more specific, powerful, and useful, there are also major problems related to immunogenicity, toxicity, and needs for targeting. Furthermore, there

[1] T. M. S. Chang, "Artificial Cells." Thomas, Springfield, Illinois, 1972.
[2] T. M. S. Chang (ed.), "Biomedical Applications of Immobilized Enzymes and Proteins," Vols. 1 and 2. Plenum, New York, 1977.
[3] T. M. S. Chang (ed.), "Microencapsulation Including Artificial Cells." Humana, New York, 1984.
[4] I. Chibata, "Immobilized Enzymes." Wiley (Interscience), New York, 1978.
[5] J. S. Holcenberg and J. Roberts (eds.), "Enzymes as Drugs." Wiley (Interscience), New York, 1981.
[6] J. S. Holcenberg, *Annu. Rev. Biochem.* **51**, 795 (1982).
[7] K. J. Widder and R. Green, eds., this series, Vol. 112, p. 1.
[8] T. M. S. Chang, *Int. J. Biomater., Artif. Cells Artif. Organs* **15**, 1 (1987).
[9] T. M. S. Chang, *Appl. Biochem. Biotechnol.* **10**, 5 (1984).
[10] T. M. S. Chang and M. J. Poznansky, *Nature (London)* **218**, 242 (1968).
[11] T. M. S. Chang, *Nature (London)* **229**, 117 (1971).
[12] T. M. S. Chang, *Science* **146**, 524 (1964).
[13] T. M. S. Chang, *Trans. Am. Soc. Artif. Intern. Organs* **12**, 13 (1966).

METHODS IN ENZYMOLOGY, VOL. 137

are also problems related to the availability of suitable enzyme systems, the need for multienzyme system with cofactor requirements, and the viability of cell cultures. Recent progress in biotechnology is beginning to solve some of these problems. As a result an increasing number of researchers are seriously investigating the medical applications of immobilized proteins, enzymes, and cells. Since many of the applications are described elsewhere in this volume, we will discuss other examples not presented herein.

Inborn Errors of Metabolism

A number of years ago we reported the successful use of microencapsulated catalase for enzyme replacement in acatalasemic mice with congenital deficiency of the enzyme catalase.[10] Whereas repeated injections of heterogenous catalase resulted in anaphylactic shocks, microencapsulated catalase did not result in immunological reactions and continued to carry out its functions.[1,10,14] However, this demonstration 15 years ago did not result in routine clinical application in this and other types of inborn errors of metabolism. There are many problems to be solved before immobilized enzymes and cells can be used routinely. Three of the most important problems are (1) the availability of enzymes, (2) the requirement for multienzyme systems in many conditions, and (3) the need for targeting to specific organs or intracellular locations in many types of congenital enzyme defects. Extensive effort has been carried out to solve these problems.

One of the most intensively studied areas involves the use of liposomes for the microencapsulation of enzymes.[15] Thus it has been reported that liposomes containing sulfatide, phosphatidylcholine, and cholesterol may be incorporated into the central nervous system.[16] Surface incorporation of antibodies onto liposomes has also been used in targeting.[15] Other modifications include surface attachment of lectins, glycoproteins, ligands, and others. Liposomes have the advantage of being biodegradable; however, they appear to enhance immune response to the entrapped protein. This area has been reviewed in detail elsewhere.[17]

Red blood cells have also been used to microencapsulate enzymes by

[14] M. J. Poznansky and T. M. S. Chang, *Biochim. Biophys. Acta* **334,** 103 (1974).
[15] G. Gregoriadis, *in* "Enzyme Replacement Therapy of Lysosomal Storage Diseases" (J. M. Tager, J. M. Hooghwinkel, and W. T. Daoms, eds.), p. 131. North-Holland, Amsterdam, 1974.
[16] M. Naoi and K. Yagi, *Biochem. Int.* **1,** 591 (1980).
[17] G. Gregoriadis, "Drug Carriers in Biology and Medicine." Academic Press, New York, 1979.

hemolysis and resealing.[18] Rh antibody-coated human red blood cells containing β-glucosidase have been used for targeting in Gaucher's disease.[17] Other immobilization techniques have also been extensively investigated,[8] including the cross-linkage of enzymes with proteins[19,20] or polymers.[21,22] Other approaches involve the modification of the enzyme by selective removal of carbohydrates, coupling of the recognition marker, and selection of a specific isoenzyme.[23,24]

Multienzyme systems enclosed within artificial cells have been used to convert ammonia or urea into amino acids.[25–29] In our recent studies, artificial cells can convert urea or ammonia into essential amino acids, such as leucine, isoleucine, and valine.[30] Enzymes in the urea cycle have also been immobilized to carry out reactions in the urea cycle.[31]

By using artificial cells containing bacterial phenylalanine ammonia-lyase, we have solved the problem of the availability of the human enzyme system and the requirement for cofactor recycling.[32] By administering these systems orally we have also solved the problem of *in vivo* accumulation, resulting in ease of administration.[33,34] We have carried out randomized control studies in rats with phenylketonuria[33,34] and found that oral administration in the phenylketonuria rat model for 7 days resulted in the lowering of system phenylalanine levels from the control group of 331.4 ± 26.4 mg/dl to the treated group of 82.7 ± 7.0 mg/dl ($p < 0.001$).[33,34] The level in the treated group is not significantly different from that of normal rats (33.6 ± 29.3 mg/dl).

[18] G. Ihler and R. Glew, *in* "Biomedical Applications of Immobilized Enzymes and Proteins" (T. M. S. Chang, ed.), p. 219. Plenum, New York, 1977.

[19] T. M. S. Chang, *Biochem. Biophys. Res. Commun.* **44,** 1531 (1971).

[20] M. J. Poznansky, *J. Appl. Biochem. Biotechnol.* **2,** 41 (1984).

[21] A. Abuchowski and F. F. Davis, *in* "Enzymes as Drugs" (J. S. Holcenberg and J. Roberts, eds.), p. 367. Wiley (Interscience), New York, 1981.

[22] R. L. Foster and T. Wileman, *J. Pharm. Pharmacol.* **31** (Suppl.), 37P (1979).

[23] G. A. Grabowski and R. J. Desnick, *in* "Enzymes as Drugs" (J. S. Holcenberg and J. Roberts, eds.), p. 167. Wiley (Interscience), New York, 1981.

[24] G. Gregoriadis and M. F. Dean, *Nature (London)* **278,** 603 (1981).

[25] T. M. S. Chang, *Enzyme Eng.* **5,** 225 (1980).

[26] Y. T. Yu and T. M. S. Chang, *Enzyme Microb. Technol.* **4,** 327 (1982).

[27] T. M. S. Chang, this series, Vol. 136, p. 67.

[28] E. Ilan and T. M. S. Chang, Appl. Biochem. Biotechnol. **13,** 221 (1986).

[29] H. P. Wahl and T. M. S. Chang, *J. Mol. Catalysis* **39,** 147 (1986).

[30] K. F. Gu and T. M. S. Chang, *Int. J. Biomater., Artif. Cells Artif. Organs* **15,** 297 (1987).

[31] N. Siegbahn and K. Mosbach, *FEBS Lett.* **137,** 6 (1982).

[32] L. Bourget and T. M. S. Chang, *Appl. Biochem. Biotechnol.* **10,** 57 (1984).

[33] L. Bourget and T. M. S. Chang, *FEBS Lett.,* in press (1985).

[34] L. Bourget and T. M. S. Chang, *Biochim. Biophys. Acta* **883,** 432 (1986).

Asparaginase and Other Enzymes in Chemotherapy

Since the first publication by Broome,[35] extensive research has been carried out using asparaginase and other enzymes which can degrade essential amino acids required by tumor cells.[5,6] The use of these enzymes is associated with problems related to toxicity, immunogenicity, and duration of action. Initial attempts at immobilization involved the successful use of microencapsulated asparaginase to suppress the growth of lymphosarcoma in mice.[11] Because of the importance of cancer treatment, extensive study has been carried out since then by many groups using all available immobilization approaches.[1–8,21,22,36–41] Details of the different immobilization approaches have been discussed under the section "Inborn Errors of Metabolism" in this chapter. Asparaginase itself can also be modified.[6] For example, it can survive longer in circulation by deamination, acylation, and carbodiimide reactions with free amino groups. Certain poly(amino acids) have also been incorporated into the enzyme to prolong half-life and decrease immunogenicity.[6] Poly(N-vinylpyrrolidine) conjugated to β-D-N-acetylhexosaminidase A increased survival and decreased immunogenicity.[6]

Detoxification

Many of the detoxifying functions in the body are carried out by enzymatic reactions; as a result it is not too surprising that research has been carried out to investigate the use of immobilized enzymes in detoxification. It is now clearly established that the earlier high expectations of immobilized enzymes in detoxifying applications have been overshadowed by the much more rapid advances in the use of artificial cell-immobilized adsorbents for detoxification.[1,8,42–48] Microencapsulated activated

[35] J. D. Broome, *J. Exp. Med.* **118**, 99 (1963).

[36] T. M. S. Chang, *Enzyme* **14**, 95 (1973).

[37] E. D. SiuChong and T. M. S. Chang, *Enzyme* **18**, 218 (1974).

[38] T. Mori, T. Tosa, and I. Chibata, *Biochim. Biophys. Acta* **321**, 653 (1973).

[39] L. D. S. Hudson, M. B. Fiddler, and R. J. Desnick, *J. Pharmacol. Exp. Ther.* **208**, 507 (1979).

[40] G. Schmer and J. S. Holcenberg, *in* "Enzymes as Drugs" (J. S. Holcenberg and J. Roberts, eds.), p. 385. Wiley (Interscience), New York, 1981.

[41] D. A. Cooney, H. H. Weetall, and E. Long, *Biochem. Pharmacol.* **24**, 503 (1975).

[42] T. M. S. Chang, J. F. Coffey, P. Barre, A. Gonda, J. H. Dirks, M. Levy, and C. Lister, *Can. Med. Assoc. J.* **108**, 429 (1973).

[43] T. M. S. Chang, J. F. Coffey, C. Lister, E. Taroy, and A. Stark, *Trans. Am. Soc. Artif. Intern. Organs* **19**, 87 (1973).

[44] T. M. S. Chang, *Clin. Toxicol.* **17**, 529 (1980).

[45] M. C. Gelfand, J. F. Winchester, J. H. Knepshield, K. M. Hansen, S. L. Cohan, B. S. Stranch, K. L. Geoly, A. C. Kennedy, and G. E. Schreiner, *Trans. Am. Soc. Artif. Intern. Organs* **23**, 599 (1977).

charcoal when used in extracorporeal circulation in hemoperfusion is much more effective than standard hemodialysis or any presently available immobilized enzyme system for the removal of many drugs. This is particularly so in the case of poisoning. Many of the commonly used medications in suicidal overdose can be very effectively removed by hemoperfusion at a level close to the blood flow rate.[8,42-48] As a result, hemoperfusion using artificial cells containing activated charcoal has now replaced standard hemodialysis as the treatment of choice for many types of poisoning. However, this approach is not as specific as enzyme systems. Where it is important to have specificity there is still an important place for immobilized enzyme systems.

Chronic Renal Failure

Renal failure results in the inability of the body to excrete waste metabolites, electrolytes, and water. The standard treatment is hemodialysis. Studies on patients demonstrated that 100 g of artificial cells containing activated charcoal can remove uremic metabolites and toxins much more efficiently than the hemodialysis machine.[1,8,46-51] Hemoperfusion in series with dialysis[49-55] has been successfully used to cut down the time required for hemodialysis and also for the treatment of uremic complications.[46-55] A recent crossover control clinical trial shows that 8 hours/week hemoperfusion–hemodialysis is as effective as 12 hours/week hemodialysis alone.[55] A second-generation artificial kidney has been developed consisting of 100 g of artificial cells containing activated char-

[46] V. Bonomini and T. M. S. Chang (eds.), "Hemoperfusion" (Contributions to Nephrology Series). Karger, Basel, Switzerland, 1982.

[47] S. Sideman and T. M. S. Chang (eds.), "Hemoperfusion: I. Artificial Kidney and Liver Support and Detoxification." Hemisphere, Washington, 1980.

[48] T. M. S. Chang and N. Nicolaev, eds., Int. J. Biomater., Artif. Cells Artif. Ogans 15, 1 (1987).

[49] T. M. S. Chang, E. Chirito, P. Barre, C. Cole, and M. Hewish, Trans. Am. Soc. Artif. Intern. Organs 21, 502 (1975).

[50] T. M. S. Chang, Kidney Int. 10, S305 (1976).

[51] T. M. S. Chang, Clin. Nephrol. 11, 111 (1979).

[52] J. F. Winchester, M. T. Apiliga, J. M. MacKay, and A. C. Kennedy, Kidney Int. 10, 315 (1976).

[53] S. Stefoni, L. Coli, G. Feliciangeli, L. Beldrati, and V. Bonomini, Int. J. Artif. Organs 3, 348 (1980).

[54] A. M. Martin, T. K. Gibbins, T. Kimmit, and F. Rennie, Dial. Transplant. 8, 135 (1979).

[55] T. M. S. Chang, P. Barre, and S. Kuruvilla, Trans. Am. Soc. Artif. Intern. Organs 31, 572 (1985).

coal in series with a small ultrafiltrator.[49,56] This way uremic metabolites, water, and sodium chloride can all be controlled. Oral adsorbents can be used to remove potassium and phosphate. The only additional step required before this second-generation artificial kidney is completed is the need to remove urea. Unfortunately, adsorbents available at present do not have sufficient capacity for urea.

Artificial cells containing urease can be used in extracorporeal blood circulation to rapidly convert urea into ammonia.[13] However the amount of ammonium adsorbent required is too large for extracorporeal blood recirculation in patients. This principle has been successfully adopted for use in dialyzate regeneration where a large amount of adsorbent can be used.[57] The use of microencapsulated urease together with ammonium adsorbent for oral administration[1,58,59] resulted in a significant lowering of systemic urea levels in the rat. This has been developed further[60] and tested clinically in patients.[61] Initial clinical trials in patients demonstrated that this approach is effective in lowering the urea level in chronic renal failure patients; however the volume required for ingestion needs to be decreased.[61] We have recently carried out further study and analysis and demonstrated that with proper design of the immobilized system it should be possible to decrease the volume required.[62] This promising approach using immobilized urease may become the one additional step required to complete the second-generation artificial kidney based on hemoperfusion and ultrafiltration. Instead of using ammonia adsorbent, we have prepared artificial cells containing a multienzyme system of urease, glutamate dehydrogenase, glucose, dehydrogenase, and a transaminase.[25-29] This way each artificial cell can convert urea to ammonia which is then converted to glutamic acid and other amino acids. The cofactor required is recycled by glucose dehydrogenase. More recently, by using artificial cells with another multienzyme system and cofactor recycling, we can convert urea into essential amino acids (leucine, isoleucine, and valine.)[30]

[56] T. M. S. Chang, E. Chirito, P. Barre, C. Cole, C. Lister, and E. Resurreccion, *Artif. Organs* **3**, 127 (1979).

[57] A. Gordon, A. J. Lewin, M. H. Maxwell, and R. Martin, in "Artificial Kidney, Artificial Liver and Artificial Cells" (T. M. S. Chang, ed.), p. 23. Plenum, New York, 1978.

[58] T. M. S. Chang and S. K. Loa, *Physiologist* **13**, 70 (1970).

[59] T. M. S. Chang, *Kidney Int.* **10**, S218 (1976).

[60] D. L. Gardner, R. D. Falb, B. C. Kim, and D. C. Emmerling, *Trans. Am. Soc. Artif. Intern. Organs* **17**, 239 (1971).

[61] C. Kjellstrand, H. Borges, C. Pru, D. Gardner, and D. Fink, *Trans. Am. Soc. Artif. Intern. Organs* **27**, 24 (1981).

[62] E. A. Wolfe and T. M. S. Chang, *Int. J. Artif. Organs* **10**, 43 (1987).

Liver Failure

Hemoperfusion using artificial cells containing activated charcoal can result in the temporary recovery of consciousness of grade IV coma patients.[63-65] When treated in the earlier grades of acute fulminant hepatic failure there is a 70% long-term recovery rate as compared to 30% in control groups.[66-68] However this is only useful in acute liver failure. In order to complete the artificial liver support system for use in chronic hepatic failure like cirrhosis, many other metabolic functions of the liver in addition to detoxification have to be supplemented.

Metabolic disturbances in amino acids is very marked in liver failure, with elevations of aromatic acids like tyrosine and phenylalanine. We have microencapsulated tyrosinase in artificial cells. These cells are then used in an extracorporeal system for hemoperfusion in galactosamine-induced fulminant hepatic failure rats.[69,70] Hemoperfusion results in significant lowering of the tyrosine level. We have also prepared artificial cells containing phenylalanine ammonia-lyase to effectively remove phenylalanine *in vitro* and also *in vivo*.[32-34] Another waste metabolite to be removed is ammonia. We have carried out studies where artificial cells containing multienzyme systems of glutamate dehydrogenase, glucose dehydrogenase, and a transaminase have been used to convert ammonia to amino acid with the required cofactor recycled by glucose dehydrogenase.[25-29] Further research has led to the use of artificial cells to convert ammonia into essential amino acids of leucine, isoleucine, and valine.[30] Other groups have used other methods of immobilizing multienzyme systems. An exciting approach is the immobilization of the enzymes required for the urea cycle.[31] Microencapsulation within artificial cells with hepatic organelles to carry out the required enzymatic functions has also been studied.[71] Attempts to supplement the detoxifying function have also been

[63] T. M. S. Chang, *Lancet* **2,** 1371 (1972).
[64] R. Williams and I. M. Murray-Lyon, "Artificial Liver Support." Pitman, London, 1975.
[65] M. C. Gelfand, J. F. Winchester, J. H. Knepshield, S. L. Cohan, and G. E. Schreiner, *Trans. Am. Soc. Artif. Intern. Organs* **24,** 239 (1978).
[66] T. M. S. Chang, C. Lister, E. Chirito, P. O'Keefe, and E. Resurreccion, *Trans. Am. Soc. Artif. Intern. Organs* **24,** 243 (1978).
[67] T. M. S. Chang, *in* "Artificial Liver Support" (G. Brunner and F. W. Schmidt, eds.), p. 126. Springer-Verlag, Berlin, 1981.
[68] A. E. S. Gimson, S. Brande, P. J. Mellon, J. Canalese, and R. Williams, *Lancet* **2,** 681 (1982).
[69] C. D. Shu and T. M. S. Chang, *Int. J. Artif. Organs* **4,** 82 (1981).
[70] Z. Q. Shi and T. M. S. Chang, *Trans. Am. Soc. Artif. Intern. Organs* **28,** 205 (1982).
[71] Z. Y. Yuan and T. M. S. Chang, *Int. J. Artif. Organs* **9**(1), 63 (1986).

analyzed using enzymes extracted from liver cells and immobilized in Sepharose.[72] A recent review is available.[73]

Microencapsulated Cell Cultures in Diabetes Mellitus and Other Applications

Artificial cells can be used to microencapsulate biological cells, and we have proposed the use of this approach for cells like islet cells, hepatocytes, and other cells for *in vivo* replacement to avoid immunological rejection.[1,74,75] This has now been supported by Sun *et al.* who have microencapsulated islet cells.[76,77] They have shown that intraperitoneal injection can successfully maintain diabetic rats for over 12 months. The microencapsulated islet cells respond to glucose and secrete the required amount of insulin to maintain a suitable glucose level in diabetic rats. Microencapsulation of hybridoma cell cultures have also been used for the large-scale production of monoclonal antibodies and interferon.[78] Our recent studies show that microencapsulated hepatocyte culture can increase the survival of acute liver failure rats.[79]

Immunoadsorbents

We have complexed albumin to collodion-coated activated charcoal to improve blood compatibility and to allow albumin to facilitate the transport of loosely bound protein molecules.[1,59] Terman found that this albumin–collodion–charcoal system could also be used to remove albumin antibodies in a perfusion system.[80] He later replaced albumin with other types of antigens or antibodies for immunosorbents in perfusion.[80] He found in preliminary studies that plasma perfusion with protein A–collodion-activated charcoal decreased the size of breast carcinoma in pa-

[72] G. Brunner and F. W. Schmidt (eds.), "Artificial Liver Support." Springer-Verlag, Berlin, 1981.

[73] T. M. S. Chang, *Sem. Liver Dis. Ser.* **6,** 148 (1986).

[74] T. M. S. Chang, Ph.D. thesis. McGill University, Montreal, Quebec, 1965.

[75] T. M. S. Chang, F. C. MacIntosh, and S. G. Mason, *Can. J. Physiol. Pharmacol.* **44,** 115 (1966).

[76] F. Lim and A. M. Sun, *Science* **210,** 908 (1980).

[77] A. M. Sun, G. M. O'Shea, and M. F. A. Goosen, *J. Appl. Biochem. Biotechnol.* **10,** 87 (1984).

[78] Damon Corp., "Bulletin on Tissue Microencapsulation." Damon Corp., Needham Heights, Massachusetts, 1981.

[79] H. Wong and T. M. S. Chang, *Int. J. Artif. Organs* **9,** 335 (1986).

[80] D. S. Terman, *in* "Sorbents and Their Clinical Applications" (C. Giordano, ed.), p. 470. Academic Press, New York, 1980.

tients.[81] However, we have pointed out that this way the immobilized protein A may be released into the circulation acting as a slow release system.[82] Protein A has also been immobilized by covalent linkage to Sepharose for hemoperfusion in patients by groups.[83]

A synthetic immunoadsorbent has been prepared for the removal of blood group antibodies anti-A and anti-B. However, the problems of blood compatibility and release of particulates has limited its application. We found that by using an ultrathin coating of collodion and albumin, a synthetic immunoadsorbent could be made that was blood compatible and would not release particles so that it could be used for hemoperfusion.[84] Clinical trials have been conducted on patients to remove anti-A and anti-B blood group antibodies before bone marrow transplantation.[85]

Blood Substitutes

Immobilization technology has also been used in the search for a red blood cell substitute. Outside the red blood cell, hemoglobin is converted in the circulation to a dimer and removed rapidly. We have earlier demonstrated that hemoglobin could be cross-linked by bifunctional agents to polyhemoglobin.[1,12] This resulted, however, in an increase in oxygen affinity so that oxygen is not readily released for use when required. Benesch *et al.*[86] demonstrated that pyridoxalation of hemoglobin decreases oxygen affinity. There is a renewed interest at present in the use of polyhemoglobin.[87] Work being carried out[87–91] has shown that peridoxylated hemoglobin can be cross-linked to soluble polyhemoglobin which can reversibly carry oxygen. In this form the polyhemoglobin can survive in the circulation so that 3 hours after intravenous injection 77% still remains in the circulation as compared to 25% for free hemoglobin.[88–90] It has also been shown that this type of polyhemoglobin is effective in resuscitating rats with hemorrhagic shock.[92] Unlike fluorocar-

[81] D. S. Terman, J. B. Young, W. T. Shearer, and Y. Daskal, *N. Engl. J. Med.* **305,** 1195 (1981).
[82] T. M. S. Chang, *N. Engl. J. Med.* **306,** 936 (1982).
[83] I. M. Nilsson, S. Jonsson, S. B. Sundquist, A. Ahlberg, and S. E. Bergentz, *Immunology* **14,** 38 (1981).
[84] T. M. S. Chang, *Trans. Am. Soc. Artif. Intern. Organs* **26,** 546 (1980).
[85] W. I. Bensinger, D. A. Baker, C. D. Buckner, R. A. Clift, and E. D. Thomas, *N. Engl. J. Med.* **304,** 160 (1981).
[86] R. E. Benesch, R. Benesch, R. D. Renthal, and N. Maeda, *Biochemistry* **11,** 3576 (1972).
[87] T. M. S. Chang, *Trans. Am. Soc. Artif. Intern. Organs* **26,** 354 (1980).
[88] P. Keipert, J. Minkowitz, and T. M. S. Chang, *Int. J. Artif. Organs* **5,** 383 (1982).
[89] P. Keipert and T. M. S. Chang, *Trans. Am. Soc. Artif. Intern. Organs* **29,** 329 (1983).
[90] P. Keipert and T. M. S. Chang, *Appl. Biochem. Biotechnol.* **10,** 133 (1984).
[91] R. B. Bolin, R. P. Geyer, and G. J. Nemo (eds.), *Adv. Blood Substitute Res.* (1983).

bons,[91] cross-linked hemoglobin is biodegradable in the body after use. Repeated injection of homologous polyhemoglobin is not immunogenic.[93,94] Extensive research is being carried out using pyridoxalated polyhemoglobin or microencapsulated hemoglobin as blood substitute.[93-96]

In Vivo Evaluations

As shown in the above discussions, extensive studies have already been carried out on the use of different types of immobilization techniques. Perhaps one of the next major steps should be to study the different possible routes of *in vivo* actions of the immobilized systems. For eventual clinical application, the routes used in experimental animal studies may have to be greatly modified. The following factors are extremely important in considering any system for actual clinical application.

Mode of Action

If the mode of action of the immobilized enzymes and cells requires rapid interactions with body fluids, then one of the best routes is to use extracorporeal blood circulation. Intraperitoneal administration is another alternative although this is usually used in experimental investigations in animals rather than for actual clinical applications. If the metabolites can diffuse freely across the intestinal tract, another possible approach is to use oral ingestion of immobilized enzymes and cells. In controlled release, intramuscular or subcutaneous injection could be carried out. The greatest problem is encountered when the immobilized enzyme or cells require localization in specific organs, cells, or intracellular organelles.

Biocompatibility

Depending on the mode of action and therefore the route of administration, the requirements for biocompatibility are different. In extracorporeal blood circulation the immobilized system has to be blood compatible, nontoxic, and should not release any harmful material into the circulation. In systems involving intraperitoneal injection, intramuscular injection, or subcutaneous injection, the material also has to be biocompatible, sterile,

[92] P. Keipert and T. M. S. Chang, *J. Biomater., Artif. Organs Med. Devices* **13**, 1 (1985).
[93] C. M. Hertzman, P. E. Keipert, T. M. S. Chang, *Int. J. Artif. Organs* **9**, 179 (1986).
[94] T. M. S. Chang and R. Varma, *Int. J. Biomater., Artif. Cells Artif. Organs* **15**, 443 (1987).
[95] T. M. S. Chang, *Int. J. Biomater., Artif. Cells Artif. Organs* **15**, 323 (1987).
[96] T. M. S. Chang and R. Geyer, eds., "Blood Substitutes." Dekker, New York, 1988.

and nontoxic, in addition to being noncarcinogenic and nonimmunogenic. Oral administration involves the least requirements for biocompatibility.

Accumulation

Immobilized enzymes and cells implanted by intraperitoneal injection, intramuscular injection, subcutaneous injection, or intravenous injection are retained for some time in the body. Since the duration of action of immobilized enzymes and cells is not indefinite, it is necessary to repeat the injections at various intervals. This may not be a problem in feasibility studies in animals. However, accumulation of immobilized material in the body could be a major problem in actual clinical applications if the treatments have to be repeated over a long period of time. Biodegradable polymers like poly(lactic acid), poly(glycolic acid), or cross-linked protein could solve the problem. However, proteins, enzymes, and cells which have been immobilized would be exposed and left behind when the biodegradable materials are removed. It is important to assure that this does not give rise to allergic and immunological reactions. To solve the problem of introduction of material into the body one could use the extracorporeal route, for instance, extracorporeal blood circulation, in those cases where this is applicable. The immobilized materials, having carried out their functions extracorporeally, can be disconnected and removed with no problems of accumulation. Another approach to solve the problem of *in vivo* accumulation is the use of oral administration in those few cases where this route is applicable.

On the basis of the above factors, one of the following approaches could be selected.

Methodology

Intraperitoneal Administration

The intraperitoneal route of administration is a convenient way for investigating the *in vivo* action of immobilized enzymes and cells on substances present in body fluids. While this approach is very convenient and useful for initial feasibility studies in animals, the peritoneal cavity is an extremely sensitive area. Intraperitoneal introduction of foreign materials in humans may result in peritonitis due to reaction to foreign materials or infections. Furthermore, chronic fibrotic reaction to the long-term intraperitoneal introduction of foreign material could result in intestinal obstruction and other major problems. Thus in most cases, the peritoneal route should be used only as a convenient initial feasibility study in animals, to be followed by other approaches for actual clinical applications.

Although large animals such as dogs and primates can be used, this approach is most convenient in introducing large amounts of immobilized proteins, enzymes, and cells into smaller animals like mice and rats. The procedure in small animals is very simple and can be summarized as follows. When the particles are about 50 μm in diameter, they can be injected as a suspension through an 18- or 20-gauge needle. Care should be taken to make sure that the injection is not made into the subcutaneous tissue or into the lumen of the gastrointestinal tract. Two millimeters of suspension can be introduced into each rat without causing any problems. With particles much larger than 50 μm diameter, one has to use larger needles. In these cases, it may be necessary to suture the hole made by the larger bore needle in smaller animals like rats and mice. Small flexible particles smaller than 8 μm are removed by the lymphatic system from the peritoneal cavity. Larger particles are retained in the peritoneal cavity to act on metabolites equilibrating across the peritoneal membrane. However, unless the material is very biocompatible, it will be encased by a fibrous capsule because of foreign body reactions.

This is a convenient approach which has been used in the initial investigation of many of the earlier studies employing immobilized enzymes and cells. These involved the use of microencapsulated urease as a model system,[1,12,74,75] microencapsulated catalase for replacement of enzyme deficiency or acatalsemia,[1,10,14] microencapsulated asparaginase for tumor suppression separation,[11,36–38] microencapsulated islet cells for diabetes mellitus,[76,77] and microencapsulated hepatocytes for acute liver failure.[79]

Extracorporeal Peritoneal Recirculation

Extracorporeal peritoneal recirculation attempts to make use of the intraperitoneal approach without introducing any foreign materials into the peritoneal cavity. Immobilized enzymes or cells are retained in a small column; 2 ml sterilized peritoneal dialysis fluid is injected into the peritoneal cavity of each anesthetized mouse. The peritoneal fluid is continuously recirculated through the extracorporeal column and returned to the peritoneal cavity. At the completion of the recirculation the extracorporeal immobilized enzymes and cells are discarded or stored for later use. This approach has been used successfully in acatalasemic mice for the enzymatic removal of peroxides by artificial cells containing catalase.[1,10]

Extracorporeal Blood Recirculation

Immobilized proteins, enzymes, and cells are retained in an extracorporeal column.[13] Blood from the animal is recirculated through the column by means of a pump. This technique can be used conveniently in animals as small as rats and in acute studies can be carried out in anesthetized animals with femoral artery and vein cannulation. In cases where repeated extracorporeal circulation is required, one can use chronic cannulation so that repeated extracorporeal circulation can be carried out in fully conscious animals. A number of techniques can be used depending on the size of the animal. When using rats two approaches are possible. Using the jugular vein and carotid artery to form an arteriovenous shunt, the cannula is exteriorized near the top of the head, and the tubings are attached permanently to the top of the cage. This results in restrictions in the mobility of the animal. Another approach is to use the femoral artery and vein and exteriorize the cannulae through the tail veins.[97] This is much more convenient since the animal, when not being treated, can be freely mobile.

In the case of extracorporeal blood circulation one does not introduce foreign material into the body. However the immobilized enzymes and cells have to be blood compatible, sterile, and not release any harmful material. Artificial cell-immobilized adsorbents have already become a routine procedure in clinical practice in patients for the treatment of poisoning,[42–48] chronic renal failure,[46–56] and removal of aluminum.[98] With the large amount of clinical experience already available for immobilized adsorbents, clinical applications of immobilized enzymes and cells could be easily developed. Experimental investigation of immobilized enzymes and cells includes the use of immobilized urease,[13] asparaginase,[40,41] tyrosinase,[8,69,70] heparinase,[99] and other enzymes.[1–8]

Oral Administration

In some cases, the metabolites to be acted on can equilibrate readily from the bloodstream into the lumen of the gastrointestinal tract. In these cases oral administration of immobilized enzymes is possible. This is one of the most convenient approaches. A gastric tubing can be easily introduced into the stomach of an unanesthetized rat and materials inserted by means of a syringe. This approach has been used for the removal of urea using artificial cell-immobilized urease and ammonia adsorbent in rats[1,58–62] and the recent use of artificial cell-immobilized phenylalanine ammonia-lyase in phenylketouria rats.[32–34] In humans, oral route of administration is the most convenient and least problematic, since it will be equivalent to drinking a suspension or taking a few pills.

[97] Y. Tabata and T. M. S. Chang, *J. Artif. Organs* **6,** 213 (1982).
[98] T. M. S. Chang and P. Barre, *Lancet* **2,** 1051 (1983).
[99] R. Langer, P. J. Blackshear, T. M. S. Chang, M. D. Klein, and J. S. Schultz, *Trans. Am. Soc. Artif. Intern. Organs* **32,** 639 (1986).

Intravascular Injection

Immobilized enzymes and cells can be injected intravenously into the systemic circulation, e.g., through foreleg veins in dog, ear veins in rabbits, and tail veins in rats and mice. Chronic cannulations could also be used for long-term intermittent administration. Most intravenously injected particulate matters larger than 2 μm in diameter are filtered out in the pulmonary circulation. Most of the smaller particles which pass through the pulmonary circulation are removed by the reticuloendothelial system particularly the liver and spleen. Immobilized enzymes which have been injected intravenously localize in the reticuloendothial system.[1,2,5-8,15-18] Immobilized enzymes or proteins in the soluble form such as cross-linked hemoglobin,[87-96] enzymes cross-linked to albumin,[20] or enzymes cross-linked to soluble polymer[21,22] can survive much longer in the circulation. The use of magnetic microcapsules[13] allows the material to be localized by externally applied magnetic forces. Intraarterial injections have been used to localize material in specific organs.

Intramuscular or Subcutaneous Injection

Intramuscular or subcutaneous injection routes are more suitable in cases where the immobilized system is used as a depot for slow release of immobilized materials like hormones and drugs.

Discussion

The first *in vivo* studies carried out about 20 years ago demonstrated the possible medical application of immobilized enzymes and cells[1,10-13]; however, many different approaches are now available.[1-8] Further progress toward actual routine clinical applications will depend on careful design of *in vivo* experiments which can be adapted for use in clinical applications. The clinical uses of artificial cells containing adsorbents have developed rapidly so that, although this technique started much later than immobilized enzymes, it is already in routine clinical use.[41,42] Experience gained can be applied to future large-scale clinical applications of immobilized proteins, enzymes, and cells. With the recent rapid progress in biotechnology, the required proteins, enzymes, and cells will be more readily available. Many factors will contribute to the future possibilities of clinical application of immobilized proteins, enzymes, and cells. Artificial cells or microencapsulated enzyme in the intestine can act effectively on plasma substrates which can enter the intestine.[100] This may lead to an easy route for routine applications of immobilized enzymes.

[100] L. Bourget and T. M. S. Chang, *J. Biomat. Artif. Cells Artif. Organs* (in press).

[41] Extracorporeal Systems for Adsorption of Antibodies in Hemophilia A and B

By Christian Freiburghaus, Sten Ohlson, and Inga Marie Nilsson

Introduction

Congenital hemophilia is a disorder inherited as a sex-linked recessive trait which is expressed only in males, but is transmitted by female carriers, who are not bleeders themselves. Statistically there are about 70 hemophiliacs in a population of one million. The disease begins in early childhood with an abnormal tendency to bleeding. The two types, hemophilia A and hemophilia B, depend on whether the defect is caused by a deficiency of factor VIII or IX. The severity of the disease is determined by the ability of the patient to synthesize the inadequate coagulation factor. Nowadays there are commercial preparations of enriched and purified human coagulation factors, derived from donor plasma, available for the treatment of hemophiliacs. During treatment with factor concentrates a small number of the severe hemophiliacs develop antibodies, so-called inhibitors, directed against the deficient coagulation factor. If the antibodies occur in high titer, administration of factor concentrates for prophylaxis is impossible, and the treatment of this group becomes very complicated. This group also includes a small number of individuals who have developed an antibody against factor VIII without being hemophiliacs. The origin of this antibody against factor VIII is unknown, but the bleeding manifestations are similar to those seen in severe hemophilia.

Different models for treatment of the acute situations involving inhibited patients have been presented. There are alternatives like activated prothrombin complex concentrates[1,2] and porcine factor VIII,[3] but the mechanisms and clinical effects of these treatments will not be discussed here. This chapter describes the *in vivo* elimination of inhibitors by two different extracorporeal affinity chromatography systems. The first is a group-specific system in which immunoglobulins are adsorbed to matrix-bound protein A in a continuous two-column plasma system. In the second system antibodies to factor IX are specifically adsorbed to purified factor IX covalently linked to macrobeads of agarose in an extracorporeal whole-blood system.

[1] J. M. Lusher, *Prog. Clin. Biol. Res.* **150,** 227 (1984).
[2] J. A. Penner, *Prog. Clin. Biol. Res.* **150,** 291 (1984).
[3] P. B. A. Kernoff, *Prog. Clin. Biol. Res.* **150,** 207 (1984).

Group-Specific System

Protein A is a bacterial cell wall protein found in some strains of *Staphylococcus aureus*. One of the most widely studied strains is Cowan type I which has been estimated to carry protein A equal to 80,000 binding sites of IgG per organism.[4] Some methicillin-resistant strains of *S. aureus* produce extracellular protein A. One of these strains, A676, is often used in the production of protein A. Protein A has a molecular weight of 42,000 and an isoelectric point of 5.1. Although the protein reacts with immunoglobulins from many mammalian species, it reacts chiefly with the Fc portion of human IgG_1, IgG_2, and IgG_4. IgA, IgE, and IgM also react with protein A to a limited extent.

Cultivation of S. aureus A676[5]

Strain A676 of *S. aureus* can be cultivated in different nutrient media, but a high yield of bacteria does not always correspond with a large production of protein A.

Three hundred grams of trypticase soy broth (170 g trypticase peptone, 30 g phytone peptone, 50 g NaCl, 25 g K_2HPO_4, 25 g dextrose) is added to a 10-liter fermentation vessel. The pH is adjusted to 7.3, and 400 g glucose is added as an energy source. Vitamins (20 mg thiamin, 40 mg nicotinic acid) are added as well as trace elements (200 mg $MgSO_4$, 100 mg $MnSO_4$, 64 mg $FeSO_2$, 65 mg citric acid). Polypropylene glycol may be used as an antifoaming agent. The glucose, vitamin, and trace element solutions are sterilized separately and then added to the autoclaved medium before inoculation of *S. aureus* A676.

Cultivation is carried out in a stirred aerated fermenter to an early stationary phase (8–10 hr). The medium is then heat inactivated at 90° for 10 min and centrifuged at 2000 *g* for 30 min to remove the cells.

Isolation of Protein A

Various methods for purifying protein A include ion-exchange and gel chromatography.[6] However, with the introduction of affinity techniques on IgG gel significant advances have been made.

Affinity Chromatography on IgG Gel.[7] The supernatant from *S. aureus* is applied to IgG–Sepharose 4B (200 ml, 20 cm²) at a rate of 600 ml/hr. After washing with 50 m*M* phosphate buffer, 0.5 *M* NaCl, and 0.02%

[4] G. Kronvall, P. G. Quie, and R. C. Williams, Jr., *J. Immunol.* **104**, 273 (1970).

[5] P. Landvall, *J. Appl. Bacteriol.* **44**, 151 (1978)

[6] J. Sjöqvist, B. Meloun, and H. Hjelm, *Eur. J. Biochem.* **29**, 572 (1972).

[7] R. Lindmark, J. Movitz, and J. Sjöqvist, *Eur. J. Biochem.* **74**, 623 (1977).

NaN$_3$, pH 7.4, until A_{280} reaches 0.05, protein A is eluted with 0.1 M glycine–HCl, pH 3.0. Fractions containing protein A are pooled, neutralized, and desalted in water. The protein A solution is freeze-dried or stored at $-20°$. The recovery of protein A is 33 mg/liter of supernatant.

Analysis of Protein A

To detect protein A during cultivation, purification, and immobilization, a number of analytical procedures based on the physiochemical and immunological characteristics of protein A has been developed. These are only briefly discussed here, and the interested reader is referred to the original literature references. Protein A can be quantified according to its ability to interact with immunoglobulins using several different immunological procedures such as Mancini immunodiffusion with dog serum,[8] rocket electrophoresis with IgG for various species,[9] and chromatography with immobilized IgG. In the 250–280 nm range, the UV spectrum of protein A shows a characteristic appearance which is partly due to its lack of tryptophan. Only the phenylalanyl and tyrosyl residues contribute to the UV response here. Standard polyacrylamide gel electrophoresis procedures as described by Weber and Osborn[10] with or without sodium dodecyl sulfate are of value when studying the presence of contaminants in protein A preparations.

Selection of Matrix

Protein A can be linked to various matrices using different techniques. For clinical applications, the choice of matrix must be guided by the surface area available and criteria such as stability, biocompatibility, bead size, and reproducibility. Different types of gels and resins satisfy these criteria, e.g., derivatized silica and bead agarose, two matrices which are used for commercial applications of immobilized protein A. In our work we have selected bead agarose (bead size 50–120 μm) to which protein A was covalently coupled using the CNBr method.

Cyanogen Bromide Activation. Because of the toxic nature of cyanogen bromide, necessary safety precautions must be taken including satisfactory ventilation in a hood. Two hundred milliliters of sedimented Sepharose gel (4B or 6MB) is suspended in water (100 ml). Cyanogen bromide (18 g) is dissolved in 400 ml of water and subsequently added to the Sepharose solution. Acetonitrile may be used to increase the solubil-

[8] G. Mancini, O. Carbonara, and J. F. Heremans, *Immunochemistry* **2**, 235 (1965).
[9] C.-B. Laurell, *Scand. J. Clin. Lab. Invest.* **29** (Suppl. 124), 21 (1972).
[10] K. Weber and M. Osborne, *J. Biol. Chem.* **244**, 4406 (1969).

ity of cyanogen bromide. The gel suspension is stirred at room temperature, and the pH is maintained at 11 for 10 min. The activated gel is then filtered off and washed with 5 liters of cold water.

Coupling of Protein A to CNBr-Activated Agarose. Protein A (600 mg) is dissolved in 0.1 M NaHCO$_3$, 0.5 M NaCl, pH 8.3. The activated gel (~200 ml) is mixed with protein A solution and stirred "end-over-end" for 2 hr at 21°. The gel is washed with coupling buffer, and the remaining active CNBr groups are then blocked with 1 M ethanolamine, pH 8.0, for 2 hr at 21°. Finally the adsorbent is washed alternatively in coupling buffer and acetate buffer (0.1 M, pH 4) containing NaCl (0.5 M).

Columns

A normal preparation of protein A–Sepharose gives a binding capacity of 20 mg IgG/ml gel. In the extracorporeal removal of IgG 50–60 g must often be eliminated from the plasma. It is both impractical and very expensive to use a single column for the total removal, so two smaller columns are run alternatively with intermittent regeneration.

A multiple adsorption system, the so-called plasma regeneration technique, has been developed by Gambro AB and used in the treatment of hemophiliacs who have developed inhibitors.[11] In this system two 100-ml columns are used. They are packed under aseptic conditions and thereafter tested for sterility and pyrogenicity. The columns are stored at 4° with the addition of 0.1% Merthiolate (thimerosal).

Clinical Setup

The clinical setup (Fig. 1) consists of a plasma separator and a computerized adsorption system. Plasma may be obtained with either a plasma filter or a cell centrifuge. In the treatment of hemophiliacs the cell centrifuge is used.

Venous blood is drawn with a flow of up to 50 ml/min. The blood is mixed with sodium citrate (0.13 M, ACD) at a ratio of about 1:10. The citrate is added as an anticoagulant and to prevent the activation of complement in the centrifuge and on the gel. After separation in the cell centrifuge, the plasma is pumped over to the adsorption system at a rate of 10–25 ml/min. The plasma is passed over the first column until the column is saturated with IgG. The plasma is then switched over to the second column while the first column is rinsed and the IgG eluted with a pH gradient (pH 7.2–2.4). The column is neutralized before the new

[11] I. M. Nilsson, S. Jonsson, S.-B. Sundqvist, Å. Ahlberg, and S.-E. Bergentz, *Blood* **58,** 38 (1981).

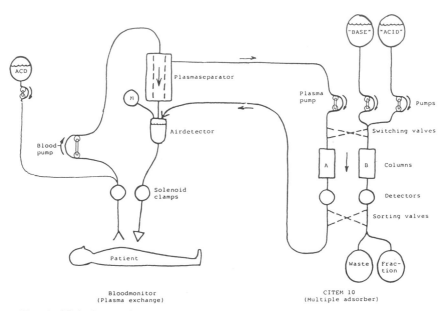

FIG. 1. Clinical setup for a group-specific system using the Gambro multiple adsorber (CITEM 10). (Designed by L.-Å. Larsson.)

plasma is applied. The eluting fluids from the columns are controlled by continuous pH and UV detectors. The data from these detectors aid the computer to identify the fluids as plasma, waste, or fraction. A series of sorting valves guides the fluids in the right direction. The treated plasma is then mixed with the blood cells and returned to the patient. The time required for a treatment depends on the inhibitor titer and the IgG level; 6–12 hr is the average treatment time. At high titer, a 2-day session including a night's rest for equilibrium of intra- and extravascular IgG is also an alternative.

Case Report[12]

One case has been reported of a 16-year-old boy with severe hemophilia A and a high titer of antibodies against factor VIII in need of treatment for severe joint bleeding. The patient had an initial inhibitor titer of 129 Bethesda inhibitor units (BIU) per ml (this means that 1 ml of plasma inactivates about 40 times more factor VIII than is present in 1 ml of normal plasma). In patients with more than 10 BIU/ml, it is not possi-

[12] I. M. Nilsson, S.-B. Sundqvist, and C. Freiburghaus, *Prog. Clin. Biol. Res.* **150,** 225 (1984).

ble to neutralize the antibodies by giving only factor VIII concentrates. A total of 6.4 liters of his plasma was treated, and his inhibitor decreased to 5.7 BIU/ml while his total IgG declined from 20 to 7 g/liter. Factor VIII concentrates were then given to neutralize the remaining antibodies and to increase factor VIII to a hemostatic level, e.g., 50% of normal. Cyclophosphamide and intravenous γ-globulin are often given after the treatment as an attempt to prevent the resynthesis of the inhibitor. This has been successful in 5 of 8 cases.

Specific Adsorption

Coagulation factor IX is a glucoprotein with a molecular weight of 55,000. It is a vitamin K-dependent protein with a concentration of less than 10 μg/ml in normal plasma. There are many commercial factor IX concentrates available for the treatment of hemophilia B. These concentrates include the other vitamin K-dependent coagulation factors II, VII, and X as well, and the factor IX content is often only 1–2% of the total protein. A more purified factor IX concentrate is required for the specific adsorbent system.

Purification of Factor IX[13]

Frozen plasma is thawed at 0°. Any precipitate is filtered off. Ethanol is added to the plasma fraction (8% final concentration of ethanol), and the precipitate is removed by filtration. Two hundred milliliters DEAE-Sephadex A-50 (wet weight) is added to the ethanol–plasma solution (6 liters), and the mixture is stirred for 30 min at 0°. After filtration and washing of the gel with 0.3 M NH$_4$HCO$_3$, pH 8.0, the factor IX fraction is eluted with 0.75 M NH$_4$HCO$_3$, pH 8.0. This fraction is freeze-dried and redissolved in 50 mM Tris, 20 mM citrate, 0.1 M NaCl, pH 7.4. The material is then applied to the heparin–Sepharose 4B column (5 × 27 cm) in the above buffer, and elution is achieved by performing a linear gradient of NaCl (0.15–2 M) in 50 mM acetate, 20 mM citrate, pH 5.0. The factor IX-containing fractions are pooled (250 ml) and rechromatographed on heparin–Sepharose 4B as described above. The factor IX material is concentrated with (NH$_4$)$_2$SO$_4$ (55%, pH 5.0) and subsequently applied to a Sephadex G-200 column in 50 mM Tris, 10 mM citrate, 0.1 M NaCl, pH 7.4. As the final step, the factor IX peak is once again applied to a DEAE-Sephadex A-50 column and eluted with a NaCl gradient (0.05–0.55 M) in 50 mM Tris, 20 mM citrate, pH 7.9. The purification is estimated to be 12,200-fold, with a recovery of 22%.

[13] L.-O. Andersson, H. Borg, and M. Miller-Andersson, *Thromb. Res.* **7**, 451 (1975).

Choice of Matrix

When dealing with specific adsorption, only a small portion of the patient's IgG is removed, and there is no need for regeneration of the columns during the treatment. Macrobeads of agarose are well known for their ability to allow cells to pass between the beads. Sepharose 6MB is the matrix chosen for the whole-blood adsorption system. The factor IX is coupled to Sepharose using the CNBr method described above. The gel has a binding capacity of 2100 BIU/ml gel and is packed in two 62.5-ml columns under aseptic conditions, tested for sterility and pyrogenicity, and preserved with 0.1% Merthiolate.

Clinical Setup

The whole-blood system is controlled by an immunotherapy monitor, ITM 10 (Gambro AB, Lund, Sweden). The apparatus (Fig. 2) consists of two pumps, one for blood and one for the addition of sodium citrate. The monitor is also equipped with pressure gauges and air detectors for the protection of the patient as well as of the columns.

Venous blood is drawn at a rate of approximately 20 ml/min and citrated directly. While the blood passes over the column, it is depleted of its antibodies against factor IX. After adsorption the blood is retransfused to the patient. The extracorporeal volume of the system is only 200 ml.

Case Report[14]

The first patient treated was a 10-year-old boy with severe hemophilia B and a high titer inhibitor. The patient was severely handicapped, bound to a wheelchair, and incapable of a normal posture or walking. Orthopedic correction was required because of a contracture in his left knee joint. The patient had an initial inhibitor titer of 120 BIU/ml, his body weight was 31 kg, and his blood volume was estimated to be approximately 2400 ml.

On 2 consecutive days, 4 hours per day, approximately double the patient's blood volume was passed over the columns. The inhibitor titer declined from 120 to 21 BIU/ml the first day. After equilibrium of the intra- and extravascular fluids his inhibitor was found to be 27 BIU/ml the following morning. After another 4 hr of treatment his inhibitor was down to 7.2 BIU/ml. The antibodies were then neutralized, and substitution therapy was given to achieve normal hemostasis. After successful correction, the boy maintained normal hemostasis for another 7 days before his

---

[14] I. M. Nilsson, C. Freiburghaus, S.-B. Sundqvist, and H. Sandberg, *Plasma Ther. Transfus. Technol.* **5**, 127 (1984).

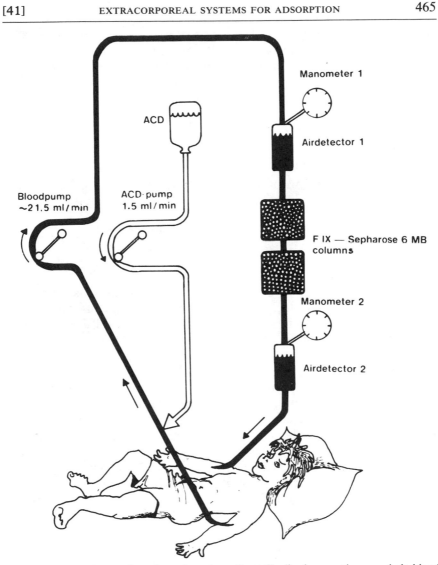

FIG. 2. Clinical setup for adsorption of specific antibodies in a continuous whole-blood system.

inhibitor returned. His walking improved tremendously after 2 weeks of intensive physiotherapy.

Conclusion

These adsorption techniques are both safe and rapid. With citrate added to the blood there are no signs of activation of the coagulation, fibrinolytic or complement system, nor is there any sign of hemolysis. The field of extracorporeal adsorption will expand with the development and purification of new adsorbents, like factor VIII, and the use of monoclonal antibodies for removal of harmful substances as well as for harvesting desirable components from the blood.

[42] Low Density Lipoprotein-Pheresis: Selective Immunoadsorption of Plasma Lipoproteins from Patients with Premature Atherosclerosis

By T. S. PARKER and J. F. STUDEBAKER

Introduction

Kilogram quantities of immunoadsorbents with defined specificity can be produced with the techniques of modern immunology and protein chemistry. The power and versatility of these immunoadsorbents can be brought to bear on human disease through a new technology called extracorporeal immunoadsorption. In its simplest form blood or plasma is pumped from a vein or artery through an extracorporeal loop that includes an immunoadsorption device and back into a second blood vessel until most or all of the targeted substance has been removed. The amount of pathogen or toxin removed during each immunoadsorption procedure is determined by the volume of plasma treated. This in turn is decided by adjusting the rate of plasma flow through the immunoadsorption columns and time of treatment. The long-term, time-averaged lowering depends on the rate of rebound and the interval between treatments.

To the basic scientist, extracorporeal immunoadsorption provides an experimental system that lies between perfused organ models and the intact animal. It can be used to hold the concentration of a plasma protein at some fixed low level as a "metabolic clamp" to study regulation. Or it can be used as a "metabolic sink" to completely deplete a given protein from the plasma to characterize nascent secretory proteins and the cell biology of protein secretion or uptake and catabolism. To the physician, the specificity of immunoadsorption offers a means of selectively lowering the plasma concentration of pathogenic proteins without the need for the plasma replacement solutions that are required in plasma exchange.

Here we discuss LDL-pheresis, the selective immunoadsorption of atherogenic plasma lipoproteins from patients, as an example of the use of immunoadsorbents in medicine. Atherosclerosis accounts for over 50% of all deaths (stroke and heart disease combined) in the United States and other developed countries. Atherosclerosis risk increases with the amount of plasma cholesterol carried on low density lipoprotein (LDL). The atherogenic LDL contain apoprotein B and transport cholesterol from the liver to peripheral cells. Antiatherogenic high density lipoprotein (HDL) contain apoproteins A-I and A-II and accept excess cholesterol from peripheral cells for reverse transport to the liver. LDL-pheresis is used to treat patients with severe hypercholesterolemia in the hope that selective lowering of LDL cholesterol by immunoadsorption of apoprotein B-containing lipoproteins will halt the progression of atherosclerosis and that the antiatherogenic HDL that remain in the circulation can promote regression of preexisting disease.

Stoffel and co-workers in Cologne, Germany introduced continuous LDL-pheresis first in the pig[1] and later in man.[2] Since that time, over 20 patients have undergone a total of more than 500 LDL-pheresis treatments either in Cologne or at our facility in New York.[3,4] This makes LDL-pheresis the most commonly used form of extracorporeal immunoadsorption. The equipment and procedures that have evolved out of this experience are described below.

Description of the System

The components of our LDL-pheresis system are shown in Fig. 1. They accomplish four separate tasks: plasma separation, immunoadsorption, regeneration of the immunoadsorbent, and flow control for blood, plasma, and regeneration buffers. Plasma separators are necessary because most solid supports cannot be safely used with whole blood. Contact with foreign surfaces can activate any of several reactive systems in blood: i.e., the cellular immune system (marcophages and leukocytes), complement, or the coagulation system (platelets and clotting factors). Bioactive factors that are generated in these reactions could harm patients if carried along in the returning plasma stream.

Two immunoadsorption columns, each capable of binding 10–12 g of LDL, are committed to the exclusive use of each patient. Under normal operating conditions, 2 to 3 "columns" are required to remove the 15–30 g of LDL that is present in the bloodstream of a typical patient with homozygous familial hypercholesterolemia (HmFH). Virtually unlimited

[1] W. Stoffel and V. Demant, *Proc. Natl. Acad. Sci. U.S.A.* **78,** 89 (1981).
[2] W. Stoffel, H. Borberg, and V. Greve, *Lancet* **2,** 1005 (1981).
[3] H. Borberg, W. Stoffel, and K. Oette, *Plasma Ther. Transfus. Techol.* **4,** 459 (1983).
[4] S. Saal, T. Parker, B. Gordon, J. Studebaker, L. Hudgins, E. H. Ahrens, and A. Rubin, *Am. J. Med.* **80,** 583 (1986).

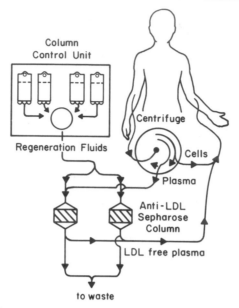

FIG. 1. Physical layout of the LDL-pheresis system. The principal components shown, listed in the sequence of blood flow are as follows: cell-separating centrifuge, immunoadsorption columns, tubing manifold and column control unit (regeneration buffers from left to right: saline, acid glycine, phosphate-buffered saline, and saline), lines returning LDL-depleted plasma and blood cells to the patient, and lines directing desorbed LDL and washes to waste.

immunoadsorption capacity is achieved by means of a "column control unit" that isolates each column when its binding capacity is saturated and regenerates the immunoadsorbent by pumping series of buffers [saline, acid glycine, phosphate-buffered saline (PBS), and saline] through the column at high speed. During the 1-year lifetime that we have arbitrarily set for our immunoadsorption columns, they undergo 100–150 adsorption-desorption cycles in the course of 50–60 LDL-pheresis procedures without significant loss of binding capacity. Not shown in Fig. 1, but discussed below, are the bactericidal buffers and aseptic procedures that have been developed to maintain the immunoadsorption columns in a sterile and pyrogen-free state.

Materials

Continuous Cell Separators

Plasmapheresis is currently performed with a variety of cell-separating devices including continuous cell centrifuges (Fenwal, IBM Biomedical, now COBE Laboratories), hollow fiber filters, and flat membrane cartridges produced by several manufacturers (COBE, Fresenius, Asahi, Kanegafuchi, and others). The membranes pass fewer platelets into the plasma stream but give lower plasma flow at equivalent whole blood flow rates than the centrifugal separators. On the other hand, the smaller dead-volume of the membrane and hollow fiber devices is an advantage for work with small laboratory animals.

We use an IBM 2997 cell-separating centrifuge. We have attempted to make this description of our LDL-pheresis system apply as generally to other separators as possible. The adjustments needed to accommodate the system to other separators are small. Because considerable experience is required to safely carry out plasmapheresis, we recommend that each center use the equipment which it knows best. We further recommend that immunoadsorption centers be organized as collaborative efforts between a basic science laboratory and a clinical apheresis unit.

Suppliers. Single-stage, 2 channel/seal assembly (972679-176); tubing kit (972679-104); pump tube loop (975578-309); and waste bag, 2 liters (975578-312) are available from COBE Laboratories, 1201 Oak St., Lakewood, CO 80215.

Immunoadsorption Columns

Immunoadsorption columns consist of a glass shell, a flow spreader assembly, Luer-lock fittings, and a tubing assembly that connects the inlet and outlet in a closed loop for storage. Designs for each of these parts are illustrated in Fig. 2. The barrel shape of the shell assures an even distribution of flow through the bed at the high flow rates (~250 ml/min). High flow rates are achieved by keeping the gel bed short (~6 cm) and the diameter wide (9.5 cm i.d.). The medium/course porosity of the glass frit permits high flow rates at low pressure with good retention of Sepharose gel and gel fragments.

Newly made columns are cleaned with chromic acid, then washed with water and dried at 120°. Care is needed to avoid heating these columns too rapidly so that they do not crack at the joint between the frit and glass shell. The frit and inside surfaces of all glass (but not plastic) parts are then treated with a 10% solution of Surfasil (Pierce Chemical Co., Box 117, Rockford, IL) in hexane.

The flow spreader is assembled as shown in Fig. 2. Holes are cut for the ports at the bottom of the spreader with a punch made from 18-gauge needle. The fit of the spreader adapters should be tested. Teflon sleeves (T 29/42, ACE Scientific, East Brunswick, NJ) can be used to ensure a good seal. The assembled column, spreader and tubing loops can be wrapped, autoclaved separately and stored. We recommend that several spares be kept on hand as replacements.

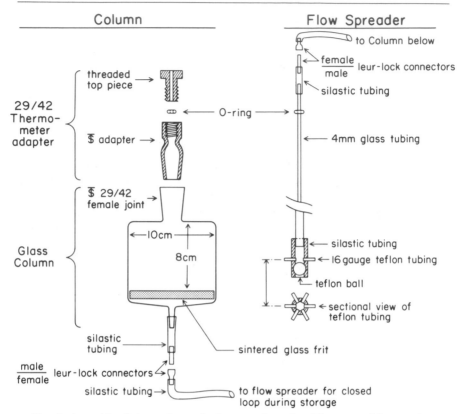

FIG. 2. Assembly of glass columns for immunoadsorption. Three assemblies are shown: glass shell, flow spreader, and storage loop. In addition to the specifications shown, we recommend that the sintered frit be ground to fit the column tube (~95 mm) and that the gap between the frit and the base be limited to <4 mm. The inlet and outlet fittings must be selected to match the endpieces of the particular plasma-separating device used.

Parts and Suppliers. COLUMN. Suppliers are as follows: Pyrex glass tubing and joints, Corning Glass Works (Corning, NY); sintered glass frits (10 cm, RH-1000S, medium coarse), R and H Filter Company Inc. (Box 99, Rt. 1, Rehobeth Beach, DE); ℥ 29/42 thermometer adapter, SGA Scientific Inc. (Bloomfield, NJ); Dow Corning medical-grade, silastic tubing (outlets: 1/8 in. i.d., 1/4 in. o.d.), Surgical Specialties Instrument Co. (2124 Baldwin Ave., Crofton, MD); Luer-lock connectors (nylon or Teflon), Hamilton Co. (P.O. Box 10030, Reno, NV).

FLOW SPREADER. Pyrex tubing (4 × 150 mm), silastic tubing at bottom (3/16 in. i.d., 3/8 in. o.d. × 4 cm), Teflon tubing (16-gauge, 1 cm lengths), and teflon balls (1/4 in.) are obtained from Small Parts Inc. (6910 NE Third Ave., Miami, FL).

STORAGE LOOP. Silastic tubing (0.078 in. i.d., 0.125 in. o.d. × 50 cm) and female Luer-lock connectors are obtained from the suppliers listed above.

Immunoadsorption Gel

Special care must be taken to avoid chemical and microbiological contamination in preparing antibody and immunoadsorbents for use *in vivo*. A "preservative solution," containing sodium azide and chloroform, is used to control microbial growth, and aseptic technique is practiced from the collection of immune serum onward. Methods for the isolation and immobilization of monospecific anti-LDL from sheep immune serum have been described by Stoffel *et al.*[1,2] or are adequately discussed elsewhere in this volume. Briefly, a narrow density cut (1.030–1.050 g/ml) of human LDL, isolated by sequential flotation in the ultracentrifuge,[5] is used to immunize sheep and to prepare LDL–Sepharose columns. Monospecific, sheep, anti-human LDL antibody is purified by adsorption/desorption on these columns and dialyzed against 10 mM NaCO$_3$ (pH 7.5). Considerable amounts of IgM precipitate from some preparations during the dialysis. The antibody solution is centrifuged (or membrane sterilized) to remove particulate matter and brought to a protein concentration of approximately 10 mg/ml by concentration over a PM10 pressure dialysis membrane (Amicon Corp., Bedford, MA). Sepharose CL-4B is washed thoroughly, fined, and activated with cyanogen bromide (50 g/500 ml of Sepharose).

Regeneration Buffers and Preservative Solution

Because most laboratory distilled water contains pyrogens, intravenous (iv)-grade saline is used to prepare all solutions. The glass bottles and their rubber-stoppered tops serve as autoclavable containers. Unless specified otherwise, chemicals are reagent grade.

Acid Glycine. Stock glycine buffer is prepared as a 10 times concentrated solution and diluted as needed by injecting 110 ml into each 1-liter plastic bag of iv-grade saline. Thus the 2000 mM stock solution becomes a ~200 mM glycine (pH 3) working solution. For 10 liters of concentrated stock, transfer the contents of nine 1-liter bottles of saline to a large beaker and add 1501.25 g of glycine (ultrapure grade; BioRad, 2200 Wright Ave., Richmond, CA 94804). Add concentrated reagent grade HCl to bring the pH to 2.9. Allow the solution to cool, then adjust the pH again and bring the volume to 10.0 liters. Return the solution to the original 1-liter bottles. Replace the rubber top and insert a 23-gauge needle in each

---

[5] R. Havel, H. Eder, and J. Bragdon, *J. Clin. Invest.* **34**, 1345 (1955).

to act as a pressure vent during sterilization. Cover with aluminum foil and autoclave around 40 min on slow exchange. Remove the needles and cover with an alcohol pad and foil. Store at room temperature.

Phosphate-Buffered Saline. The stock PBS buffer is prepared as a 2.5 times concentrated solution meant to be added to 1-liter plastic bags of saline in the ratio of 55 ml to 1 liter. In this way the 473 mM stock becomes ~25 mM potassium phosphate working solution. To prepare 10 liters transfer the saline from nine bottles to a large beaker. Add 20 g of KCl, 100 g of anhydrous KH_2PO_4, and 1080 g of $Na_2HPO_4 \cdot 7H_2O$. Bring the pH to 7.4 with 10 N KOH. Adjust to a final volume of 10.0 liters with saline. Autoclave and store as described above.

10% Sodium Azide. Add 15 g of reagent grade NaN_3 (Eastman Kodak) to a 150-ml bottle of saline (wear plastic gloves and work in the fume hood). Shake carefully to dissolve and store in the refrigerator at 4°.

Preservative Solution. Add 5 ml of 10% NaN_3 and 4 ml of distilled $CHCl_3$ to a 1-liter bottle of pyrogen-free saline. Shake the bottle until all of the $CHCl_3$ dissolves and use immediately.

Column Control Unit and Tubing Set

A manually operated column control unit can be assembled from standard medical-grade three-way stopcocks and iv line-extension sets as shown in Fig. 3. The essential elements are (1) connectors that match the tubing set of the cell-separating device; (2) a safety bypass that, when activated, takes the immunoadsorption columns out of the extracorporeal loop by shunting plasma directly into the return line. This is shown in Fig. 3 as the connection between the inlet and outlet and clamps at points marked by **I**; (3) connections to the column for switching between plasma and regeneration lines; (4) outlet connections from the columns to either the plasma return line or to waste; and (5) a manifold for switching through the sequence of regeneration buffers.

We use the plasma line from the IBM 2997 as an inlet and the plasma pump tube as the outlet. Thus we operate our immunoadsorption columns under negative pressure. The warning signs of underpressure, either collapse of the plastic tubing or the appearance of bubbles in the gel bed, are easily noted and less troublesome than the popping of tubing joints and squirting of plasma that can occur accidently in positive pressure systems. The flow into junction points, marked by circles in Fig. 3, may be controlled manually by means of clamps and stopcocks. It is also possible to automate the procedure to some extent by mounting the tubing and columns on a frame and controlling the flow with solenoid-actuated pinch valves. The operator still selects the proper state for the columns (process, switch, or regeneration), but the combination of valves appropriate

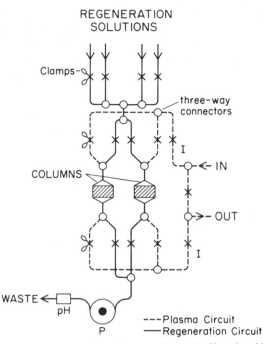

FIG. 3. Column control unit tubing diagram. Symbols: X, solenoid pinch valves or clamps; O, three-way stopcocks or connectors; I, location of the clamps used to isolate the system; P, high-speed, variable-control pump; pH, flow-through pH monitor.

for that state is determined by the instructions carried in a read-only memory within the controller. Experimental versions of such devices, with their associated tubing sets, have been built independently in Cologne and New York and used successfully.[3,4]

Parts and Suppliers. If an integrated tubing set is assembled for operation with solenoid-actuated pinch valves, it should be made of tubing which passes the Class VI requirements for medical plastics of the U.S. Pharmacopoeia.[6] We have found PVC tubing with a durometer of 78 ± 5 occludes well when a valve is closed and reopens cleanly when released. Y Connectors in the sets were obtained from Lifemed (Compton, CA).

High-Speed Pumps

Two variable speed peristaltic pumps, each capable of delivering fluid at rates up to 250 ml/min, are needed. One pump is used in the column control unit to match the flow rate in the fresh immunoadsorption column

[6] U.S. Pharmacopoeial Convention, "The Pharmacopoeia of the United States," 20th revision, pp. 725–734. U.S. Pharmacopoeia Convention, Rockville, Maryland, 1979.

to that of the plasma pump during the switchover (see "Switch from Column 1 to 2" in the procedures section below). After the switch over this pump is connected to the saturated column and operated at high speed for the regeneration sequence. A second high speed pump is kept in a separate room where it is used to handle the preservative solution.

pH Indicator Cartridge

The pH can be monitored with a flow-through indicator cartridge that contains pH-sensitive indicator beads. The beads are made by a suspension polymerization procedure that was originally developed for phenol red by Peterson et al.[7] and modified by us[8] to incorporate bromphenol red which has a better transition pK. The polyacrylamide beads are held within a nylon mesh bag (80 mesh), which is in turn contained within a piece of transparent PVC tubing of i.d. 5/8 in., o.d. 13/16 in. and length approximately 2-3/4 in. Transparent ABS (acrylonitrile–butadiene–styrene) end caps of the appropriate size are then solvent welded to the two ends of the tubing.

LDL-Pheresis Procedure

Preparation of New Immunoadsorption Columns

Charging and Storage of Immunoadsorption Columns. Charging and all subsequent operations involving open columns are carried out using careful aseptic technique in a laminar flow hood. Each column is charged with 400 ml of immunoadsorption gel. Newly packed columns are stored in preservative solution for a minimum of 48 hr to assure sterility. The columns are flooded with preservative solution by venting air through the thermometer adapter. This adapter is used only as a vent. When air must be allowed to enter the column to displace the preservative solution, it is introduced through the filters of the iv sets that connect the saline bottles to the tubing manifold.

Scrubbing and Coating. Each new column is run through 10 regeneration cycles (1 liter each of the buffers: saline, acid glycine, PBS, and saline) to desorb most of the uncoupled or poorly coupled protein that remains associated with the Sepharose after the cyanogen bromide procedure. Next a mixture of 250 ml of 6% human albumin (Immuno AG, Vienna, Austria) and 250 ml of saline is pumped into the column and left

---

[7] J. Peterson, S. Goldstein, and R. Fitzgerald, *Anal. Chem.* **52,** 864 (1980).
[8] J. Studebaker and J. DeBroy, *J. Clin. Apheresis* **2,** 235 (1985).

there for 2 hr to allow the albumin to bind and inactivate any reactive sites within the immunoadsorption gel. After washing out the albumin (saline, 2 liters), the column is flooded with preservative solution and stored at 4° until needed.

Bioconditioning. The immunoadsorption columns are exposed to plasma to "condition" them for *in vivo* use. A standard plasma exchange is carried out, replacing a total of 2 liters of plasma with 2 liters of a mixture of albumin and saline (1 : 1) per exchange. Each column is perfused with 1 liter of the harvested plasma. After a regeneration cycle to desorb any bound LDL, the columns are flooded with preservative solution and stored. Once this step is completed the column is committed to the exclusive use of the plasma donor.

Bioconditioning is repeated.

Allergy Screen. The immunoadsorption columns are connected to the column control and cell separator units as in an immunoadsorption procedure. Plasma (1 liter) is processed through each column and collected, but none is returned to the patient. Instead the processed plasma is replaced with human albumin as in Bioconditioning above. When the procedure is completed, the cell-separating device is disconnected, and a 25-ml aliquot of processed plasma from each column is administered iv to the patient at half-hour intervals. The patient is monitored closely for changes in vital signs or allergic reactions.

On-Line Immunoadsorption

Preservative Washout. On the day of each LDL-pheresis procedure the preservative solution is washed out of the immunoadsorption columns by passing 6 liters of saline through each column. The last part of this wash is collected and sent to the laboratory to be tested for pyrogens and microbes. It is good practice to carry out all procedures involving the preservative solution in a separate "preparation" room.

Prime. The cell separator is tubed and primed as for a routine plasmapheresis according to the manufacturer's instructions. The plasma outflow line is then detached from its pump and connected to the inflow of the column unit. Flow from the columns is directed back to the patient via the plasma pump-tube of the cell separator. The four 1-liter bags of regeneration solution (in order: saline, acid glycine, PBS, and saline) are hung in position and connected to the appropriate tubing. The tubing lines of the regeneration system and the cell separator are primed with saline.

Vascular Access and Anticoagulation. Patient access is obtained via appropriate needles (intake, 17-gauge Terumo; return, 19-gauge butterfly) placed one in each antecubital area. Anticoagulation is achieved by administration of heparin as a bolus of 40 units/kg followed with an infusion of ACD-A/heparin solution (5000 units/500 ml ACD-A) at rates of 2.8, 2.6, 2.4, and 2.4 ml/min for columns 1, 2, 3, and 4, respectively.

For patients who are unable to tolerate the above regimen, ACD-A alone may be used. At the start of the procedure the plasma and anticoagulant pumps are adjusted to provide a ratio of 16 ml of blood/ml of anticoagulant (note that the absolute citrate flow should not exceed 3 ml/min). During the LDL-pheresis procedure, the ratio should be gradually decreased to 20:1 depending on citrate (paresthesia) tolerance.

Process Column 1. Whole blood is pulled from the patient to the cell separator, and plasma is drawn from the separator into the columns. Gradually the saline prime is displaced from the system, and plasma begins to return to the patient. A plasma flow rate of ~25 ml/min delivers enough LDL to saturate the capacity of a column in 30–50 min. Saturated columns develop an orange color from the carotenes carried within the hydrophobic core of the bound LDL.

Switch from Column 1 to 2. Plasma is directed to column 2, and the speed of the waste pump is adjusted to match that of the plasma pump which continues to pull LDL-free plasma from column 1. For the next 10–15 minutes a transition takes place. The plasma in column 1 is displaced from it by saline entering at the top. This plasma is returned to the patient. At the same time, the saline prime in column 2 is displaced by incoming plasma and directed to waste. When this transition has been completed, the outflow from column 1 is switched to waste and that from column 2 is switched to the plasma return line.

Regenerate Column 1 and Process Column 2. Now the first column is regenerated at high speed while plasma is processed in column 2.

Switch from Column 2 to Column 1. When the second column is saturated with LDL (and the first column has been regenerated) the switching procedure described above ("Switch from Column 1 to 2") is repeated this time changing the plasma flow to column 1 and regenerating column 2.

Return Plasma and Stop. The steps from "Process Column 1" to "Switch from Column 2 to Column 1" are repeated until the desired cholesterol lowering is obtained or a predetermined volume of plasma is processed. The LDL-pheresis procedure is terminated by switching the blood intake line to saline. Blood and plasma are purged from the system and returned to the patient. A typical LDL-pheresis procedure lasts 3 hr with small variations due to differences in the plasma flow rate.

Column Storage. After each LDL-pheresis the immunoadsorption columns are moved to the "prep room," flooded with 1 liter of preservation solution and stored at 4°.

Results and Discussion

LDL-pheresis has proven to be a safe and remarkably effective procedure for selectively lowering plasma LDL cholesterol levels. There have been no serious complications in over 250 procedures carried out in six patients over the 2-year period from 1982 to 1984. Complications, such as citrate toxicity or minor bleeding at the venipuncture sites, occur no more frequently with immunoadsorption than they do with plasmapheresis. Potential complications unique to immunoadsorption, such as pyrogen reactions or sepsis, have not occurred.

Sensitive ELISA immunoassays for sheep IgG have detected the shedding of small quantities of immobilized protein. Routine allergy testing by the "skin prick" method has shown that one of our six patients is sensitized to the immunoadsorption columns. On his forty-seventh treatment, this patient began to experience "shaking chills" and withdrew from our clinical study of LDL-pheresis.

Retrospective immunoassay of frozen plasma samples for human antibody against sheep IgG demonstrated that this patient (and no other) began LDL-pheresis with a preexisting sensitivity to sheep IgG. Moreover, the titer of human anti-sheep antibody increased with exposure to the anti-LDL immunoadsorption columns. No other patient has developed "shaking chills" or had a positive skin test. Nevertheless, we are investigating the mechanism of shedding and subsequent sensitization in the hope that these can be understood and prevented in the future.

An attractive feature of LDL-pheresis is that the degree of LDL-lowering, achieved by immunoadsorption, is limited only by the amount removed at each procedure and the frequency of treatment. Both of these variables are under the control of the investigator. In the example shown in Fig. 4, immunoadsorption of 2.5 plasma volumes removed ~74% of the plasma cholesterol, primarily as very low density lipoprotein (VLDL) and LDL, with only a small and transient lowering of HDL. In longer treatments, we have removed up to 93% of the LDL to achieve posttreatment levels as low as 12 mg/dl.

Analysis of the material desorbed from a washed immunoadsorption column shows that 93% of the bound cholesterol is carried on apoprotein B-containing lipoproteins (VLDL and LDL) and only 7% is found in the HDL fraction. Apoprotein B accounted for 50% of the bound protein; albumin and fibrin made up most of the remainder. HDL apoproteins accounted for less than 4% of the total. By these criteria the immunoadsorption columns are performing with the degree of specificity expected of them.

Some nonspecific loss of plasma proteins does occur. Most of this

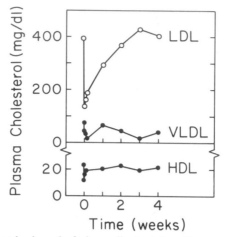

Fig. 4. Lowering and rebound of plasma lipoproteins. Changes in plasma lipoprotein cholesterol achieved by LDL-pheresis of a 9-year-old, female patient with HmFH. Apoprotein B-containing lipoproteins (VLDL and LDL) are selectively removed.

protein is lost from the void spaces of the system during the various washes that take place in the regeneration sequence. These losses are usually less than 10%. Weekly LDL-pheresis does not deplete plasma of proteins other than apoprotein B. In fact just the opposite has taken place in cases of HmFH. The abnormally low HDL levels commonly seen in these patients seem to be corrected by bringing the plasma LDL concentration into the normal range. Time-averaged plasma HDL levels increased from ~20 to ~45 mg/dl as the plasma LDL concentration was lowered 70% by weekly LDL-pheresis.[9]

Although these are the very changes in plasma lipoproteins that would be expected to arrest the progression of atherosclerosis or promote regression of existing atherosclerotic lesions, the ability of LDL-pheresis to achieve these goals remains to be proven. Clinical trials to test this question are planned or underway in Cologne and New York. Nevertheless, we believe that the experience gained so far will prove useful to any who might wish to apply immunoadsorption to other medical problems or to take advantage of it to study regulation of the synthesis, secretion, and clearance of plasma proteins.

[9] T. Parker, B. Gordon, S. Saal, L. Hudgins, E. H. Ahrens, Jr., and A. Rubin, *Proc. Natl. Acad. Sci. U.S.A.* **83**, 777 (1986).

[43] Enzyme Reactors for the Removal of Amino Acids from Plasma

By Gottfried Schmer and Wayne L. Chandler

Introduction

This chapter deals with the preparation of enzyme reactors for the removal of amino acids from plasma as a potential form of tumor treatment. Certain amino acids such as L-asparagine, L-glutamine, and L-tryptophan have been recognized as important cancer nutrients, the removal of which can lead to remissions in tumors.[1–3] Since the so-called antitumor enzymes are derived from bacterial or fungal sources, immunological responses are observed after parenteral administration.[4] In an attempt to prevent immunological sensitization, extracorporeal enzyme reactors were developed containing antitumor enzymes covalently bound to a water-insoluble, polymeric surface. However, with the successful preparation of enzyme reactors, new problems arose, notably a decrease in the stability of the immobilized enzyme, an appreciable change in the substrate affinity, and difficulty in sterilization of the devices.

Problems Associated with Enzyme Reactors

Change in Substrate Affinity

Enzymes bound to water-insoluble matrices often show a decrease in substrate affinity reflected by an increase of the K_m apparent.[5–7] L-Asparaginase covalently bound to a Dacron vascular prothesis and exposed to a back-and-forth flow of 100 strokes/min, showed an increase in K_m from 1.2×10^{-5} M for free asparaginase to 2.0×10^{-3} M for bound enzyme.[6] L-

[1] J. M. Hill, J. Roberts, and E. Loeb, *JAMA, J. Am. Med. Assoc.* **202,** 882 (1967).

[2] A. S. D. Spiers and H. E. Wade, *Br. Med. J.* **1,** 1317 (1976).

[3] J. Roberts, F. A. Schmid, and H. J. Rosenfeld, *Cancer Treat. Rep.* **63,** 1045 (1979).

[4] R. G. Peterson, R. E. Handschumacher, and M. S. Mitchell, *J. Clin. Invest.* **50,** 1080 (1971).

[5] G. J. H. Melrose, *Rev. Pure Appl. Chem.* **21,** 83 (1971).

[6] D. A. Cooney, H. H. Weetall, and E. Long, *Biochem. Pharmacol.* **24,** 503 (1975).

[7] J. P. Allison, L. Davidson, A. Gutierrez-Hartman, and G. B. Kitto, *Biochem. Biophys. Res. Commun.* **47,** 66 (1972).

Asparaginase bound to nylon tubing showed an increase of the K_m apparent to 3.0×10^{-3} M with a substrate flow rate of 1.8 ml/min, but the K_m decreased to 6.7×10^{-4} M when the substrate flow was increased to 4.3 ml/min. This improvement of apparent substrate affinity by higher flow rates indicates a diffusion-controlled enzyme–substrate interaction at the interface. Enzyme reactors should therefore be subjected to the maximum blood or plasma flow possible to improve their performance. L-Asparaginase coupled to Dacron fibers by γ-aminopropyltriethoxysilane and glutaraldehyde showed a strong dependence of the K_m apparent on flow rate. High flow rates through this system greatly improved the apparent substrate affinity.[8] L-Asparaginase bound to nylon tubing exhibiting a polyionic surface, showed a K_m apparent of 2.1×10^{-5} M under comparable flow conditions to those described above.[9] This points to another important issue in the design of enzyme reactors, namely, the composition of the matrix as a more or less favorable microenvironment for the insolubilized enzymes.

Stability of the Immobilized Enzyme

The cyanogen bromide activation method of binding, which has been used widely for the insolubilization of enzymes, has decreased in importance in the synthesis of enzyme reactors since the isourea bonds formed between the enzyme and the matrix are hydrolyzable.[10] This makes enzyme leakage an appreciable problem, especially in the presence of strong nucleophiles.[11,12] To alleviate this problem cyanuric chloride activation has been introduced, which results in a stable, nonhydrolyzable bond.[13,14]

Matrix-bound enzymes exposed to blood or plasma *in vivo* exhibit similar problems of stability. L-Asparaginase bound to Dacron and implanted into dogs for 10 days showed a loss in enzyme activity.[6] L-Asparaginase coupled to Dacron wool and enclosed in a cartridge leaked 7×10^{-4} IU/ml of blood.[8] It is assumed that the main problem of enzyme reactor stability *in vivo* is caused by the digestion of the insolubilized enzymes by plasma enzymes, notably, plasmin.[15] Succinylation of bound L-asparaginase blocks the amino groups essential for tryptic cleavage and increases the *in vivo* resistance of the enzyme reactor when used on sheep.[15]

[8] R. Y. C. Ko and L. S. Hersh, *J. Biomed. Mater. Res.* **10**, 249 (1976).
[9] C. Horvath, A. Sardi, and J. S. Woods, *J. Appl. Physiol.* **34**, 181 (1973).
[10] J. Porath and R. Axen, this series, Vol. 44, p. 19.
[11] J. Lasch and R. Koelsh, *Eur. J. Biochem.* **82**, 181 (1978).
[12] J. Schnapp and Y. Shalitin, *Biochem. Biophys. Res. Commun.* **70**, 8 (1976).
[13] F. Dumler, P. R. Singh, C. E. Jackson, *et al.*, *ASAIO J.* **4**, 70 (1981).
[14] T. H. Finlay, V. Troll, M. Levy, A. J. Johnson, and L. T. Hodgins, *Anal. Biochem.* **87**, 77 (1978).

Sterilization

Recent progress has been made in sterilizing enzyme reactors and preserving at least an appreciable part of the enzyme activity. Enzyme reactors containing a matrix derived from biological materials, such as red cell ghosts or fibrin plates, are conveniently sterilized by 0.02% sodium azide in 0.15 M NaCl, 40 mM Tris–HCl, pH 7.4. A disadvantage of this method is the requirement of an extensive wash prior to *in vivo* use in experimental animals because of the high toxicity of azide. Good results have been achieved in the sterilization of enzyme reactors based on the hollow fiber artificial kidney (HFAK) principle. Sterilization of an L-asparaginase reactor was carried out by exposing the reactor to γ radiation for 72 hr at room temperature (total dose of 2.5×10^3 rads). Only 25% of the original enzymatic activity was lost during sterilization.[16] The same enzyme reactor system was sterilized using an ethylene oxide/Freon 12 mixture (12:88 v/v) with 50% humidity for 4 hr at 45° (1.0 kg/cm² gas pressure). With this method the enzyme reactor maintained 37% of its original biological activity after sterilization.[16] An arginase enzyme reactor system was sterilized with an ethylene oxide/freon mixture (12:88) and 50% humidity for 6 hr at 35° with a gas pressure of 1.0 kg/cm². The sterilized material was then stored for 1 day at room temperature and subsequently aerated for 10 days at room temperature. These conditions were sufficient to sterilize 1.0×10^4 *Bacillus subtilis* var. *niger*. Sterilized in this way, reactor dialyzers maintained 70% of their original enzyme activity.[17]

Methodology Involved in the Preparation of Enzyme Reactors[17a]

Because of the importance of the matrix for the activity of the insolubilized enzymes as described above, this chapter will discuss three preparations of enzyme reactors: (1) preparation of enzyme reactors derived from biological materials such as fibrin plates; (2) preparation of enzyme reactors from commercially available devices such as artificial kidneys (dialyzers); and (3) preparation of enzyme reactors as part of a plasmapheresis system (i.e., "on-line" plasmapheresis–enzyme reactor system).

[15] G. Schmer, L. N. Rastelli, M. B. Dennis, J. C. Detter, G. von Sengbusch, and H. D. Lehmann, *Trans. Am. Soc. Artif. Intern. Organs* **28,** 374 (1982).

[16] L. Callegaro and E. Denti, *Intern. J. Artif. Organs* **6/S-1,** 107 (1983).

[17] V. Rossi, A. Malinverni, and L. Callegaro, *Intern. J. Artif. Organs* **4,** 102 (1981).

[17a] The general remarks regarding the Synthesis of Enzyme reactors as well as specific comments of their *in vivo* and *in vitro* efficiency were taken from the chapter Enzyme Reactors: Achievements, Problems, Future Perspectives. G. Schmer, W. L. Chandler (International Symposium 1987 Therapeutic Plasma Exchange and Selective Plasma Separation, Schattauer Verlag, Stuttgart 1—West Germany) with the kind permission of the publisher.

Enzyme Reactors Derived from Biological Materials

L-Asparaginase–Fibrin Reactor. This procedure is a modification of the method of Schmer et al.[18,19]

Reagents

Solution A: 0.15 M NaCl, 20 mM imidazole, pH 7.35
Solution B: 0.15 M NaCl, 100 U/ml bovine thrombin, 20 mM imidazole, pH 7.35
Solution C: citrated bovine or human plasma, heat defibrinated at 57° for 3 min, 0.1 M CaCl$_2$, 20 U/ml bovine thrombin

Procedure. Lyophilized 75% clottable bovine fibrinogen (Pentex) is dissolved in solution A to a concentration of 30 mg/ml. The fibrinogen solution is activated by adding 0.05 ml of solution B to 10 ml of the fibrinogen solution and immediately spread on a 10 × 10 cm roughed glass plates (Elphor). The fibrin-coated plate is incubated for 15 min at 25° allowing the fibrin to firmly polymerize. The plate is then turned over, supported at the edges, and the opposite side grafted in the same manner. The plates are then placed in solution C (which contains activated factor XIII) at 25° for 4–6 hr to cross-link the fibrin. After stabilization the plates are washed with solution A until the absorbance at 280 nm is less than 0.04. The remaining fibrin amino groups are activated by incubating the plates in solution A containing 10% (v/v) glutaraldehyde for 2 hr. This step also makes the fibrin resistant to fibrinolysis. The plates are again washed extensively in solution A and grafted with Escherichia coli asparaginase by incubation in solution A containing 1500 IU of L-asparaginase per plate (200 cm^2). Unbound asparaginase is washed off the plates with solution A until no trace of the enzyme can be detected. The completed asparaginase–fibrin reactor consists of 20 parallel fibrin plates enclosed in a polyacrylic cubicle and stored in solution A with 0.02% sodium azide added.

Efficiency. L-Asparagine (2800 μmol/min/m^2 of surface) was degraded in vitro using a 1.0 mM L-asparaginase solution in 0.025 M sodium phosphate, pH 7.0, with a flow rate of 100 ml/min. In healthy sheep a drop in plasma L-asparaginase from 53 nmol/ml plasma to 15 nmol/ml plasma could be achieved after extracorporeal blood circulation through the unit for 6 hr. After 50 hr of in vivo exposure no decrease in enzyme activity was noted. The enzyme reactor is stable for 2 years, if stored in solution A

[18] G. Schmer, J. Bisbroek, M. B. Dennis, and J. C. Detter, Trans. Am. Soc. Artif. Intern. Organs 26, 129 (1980).
[19] G. Schmer, L. N. Rastelli, M. L. Newman, M. B. Dennis, and J. S. Holcenberg, Intern. J. Artif. Organs 4, 96 (1981).

containing 0.02% sodium azide, without losing any enzyme activity. No
L-asparaginase leakage was detected using an ultrasensitive radioassay
for L-asparaginase that could detect enzyme concentrations as low as
1.0×10^{-5} IU/ml.[20]

Enzyme Reactors Derived from Hollow Fiber Artificial Kidneys (HFAK)

Enzymes can be covalently bound to solid polysaccharide supports by
activating the surface of the support. Covalent binding of the enzyme
significantly reduces nonspecific enzyme absorption. This section
presents several different methods of surface activation and binding.

Immobilization of L-Asparaginase by Cyanuric Chloride Activation[13,14,21]

Device

C-DAK Artificial Kidney 2.5D (Cordis-Dow Corp., Miami, Florida)

Reagents

Cyanuric chloride (99%, Aldrich)
L-Asparaginase (*E. coli,* Merck, Sharp and Dohme; "Elspar").
Solution A: 0.1 *M* phosphate buffer, pH 7.4.

Procedure. The dialyzer is washed with deionized water and treated
with 2.5 liters of 0.15 *M* NaOH overnight at 4° by continuous circulation.
After washing with deionized water, the dialyzate compartment is rinsed
sequentially with 1 liter of 10%, 25%, and 50% dioxane followed by a 400
ml wash with 0.2 *M* triethylamine in 50% dioxane. A freshly prepared
solution of 5 g of cyanuric chloride in 125 ml of dioxane is then made up to
200 ml with deionized water, and 7.3 ml of triethylamine (0.25 *M*) is added
slowly as the solution is circulated through the dialyzate compartment.
The first 30 ml of effluent is discarded, and the remaining solution recirculated for 2 hr. The dialyzer is then washed with 1 liter each of 50%, 25%,
and 10% dioxane, and 1 liter of solution A. L-Asparaginase (3000 IU) in
120 ml of solution A is recirculated through the dialyzate compartment at
room temperature for 24 hr. The dialyzer is then washed with 2 liters of
solution A, and excess reactive groups are blocked with 500 ml of 1 *M*
ethanolamine in solution A. The dialyzer is again washed with 2 *M*

---

[20] D. A. Cooney and H. A. Milman, *Biochem. J.* **129,** 953 (1972).
[21] G. Mazzola and G. Vecchio, *Intern. J. Artif. Organs* **3,** 120 (1980).

guanidine-HCl and solution A. Enzyme reactor dialyzers are stored filled with solution A containing 0.2% sodium azide.

Immobilization of L-Asparaginase by Glutaraldehyde Cross-Linkage[21,22]

Device

50 Cuprophan hollow fibers (Cl 1M, Enka Glanzstoff), 8–25 cm long with an internal to external compartment volume ratio of 1:3–4

Reagents

0.25% glutaraldehyde, phosphate buffer, pH 8.0, bovine serum albumin

L-Asparaginase (Sigma, Grade V)

Solution A: 0.1 M phosphate buffer, pH 8.0

Procedure. A solution of 5 mg of L-asparaginase in 3 ml of phosphate buffer, pH 8.0, is ultrafiltered from the jacket of the device into the lumen of the fiber. When all the solution is ultrafiltered, the glutaraldehyde solution is pumped into the lumen for 3 min and then washed free with phosphate buffer. This procedure does not change the transport characteristics of membrane.

Efficiency. Since the L-asparaginase has been bound to the outer surface of the hollow fiber device, a relatively high K_m apparent of 1.0×10^{-3} M was observed by recycling L-asparagine through the internal compartment of the device (blood compartment side). The advantage of binding the enzyme to the outer surface of the fibers is the certain prevention of an immune response toward the enzyme. This sort of device is severely limited in that it can only hydrolyze freely dialyzable, non-protein-bound amino acids. The K_m apparent for L-asparagine in this system is 200 times higher than that of the free enzyme, which reflects the dependence of the system on the route of solute transfer through the membrane.

The enzyme system described above ("Immobilization of L-Asparaginase by Cyanuric Chloride Activation") is the first one to be tested in a clinical setting.[13] Three patients were treated using this device: one patient with acute lymphoblastic leukemia, one patient with poorly differentiated lymphocytic lymphoma, and one with acute rejection episodes after transplantation of a cadaver kidney. In the first two cases, L-asparagine in the blood decreased to undetectable levels after 3 hr of extracorporeal therapy. One patient, who had a history of allergic reactions following

[22] C. Horvath and B. A. Solomon, *Biotechnol. Bioeng.* **14,** 885 (1972).

parenteral administration of L-asparaginase, showed no immunological reaction during the extracorporeal enzyme therapy. Both tumor patients showed only a transient remission. The patient with the lymphoblastic leukemia did not seem to respond to the therapy after 1 week of treatment (i.e., no reduction in blood L-asparagine levels). A similar observation has been noted in sheep.[23] L-asparagine levels in the blood increased to 60–120 nmol/ml followed by a increase in blast counts. After an initial remission, the patient with lymphoma developed increasing blast counts in the peripheral blood with a subsequent blast crisis, in spite of low serum L-asparagine levels. The third patient showed a decrease in L-asparagine levels, but renal allograft function did not improve.

Preparation of an Arginase–Hollow Fiber Artificial Kidney[17]

Reagents

Solution A: 0.1 M NaHCO$_3$, pH 8.3

Procedure. The inner surface of a 1.3 m^2 Cuprophan hollow fiber hemodialyzer (Spiraflow SD-1) is oxidized using a solution of 21 mg/ml sodium metaperiodate for 60 min at 25°. This is followed by 250 ml of 1.0 M 3,3′-diaminodipropylamine, pH 5.0, which is circulated through the inner chamber overnight at 25°. The pH of the circulating amine solution is then adjusted to pH 9.0 with solid sodium carbonate, and the hemodialyzer is cooled to 4°. Solid sodium borohydride is added in two aliquots of 1.25 g to the solution (final concentration 10 mg borohydride/ml) and allowed to circulate for 4 hr. The dialyzer is then washed with 10 liters of 1.0 M NaCl and stored overnight in the same solution at 4°. After washing with 10 liters of water, 250 ml of 2.5% glutaraldehyde in water is circulated through the fibers for 60 min at 25°. The inner chamber is washed with 10 liters of water at 60° and equilibrated with solution A. The dialyzer is then filled with 25 ml of solution A containing 1 mg/ml arginase (1250 IU activity) and incubated overnight at 25°. Finally, the dialyzer is washed with 1.0 M glycine, pH 7.4, and stored, filled with 10 mM Tris–HCl, pH 7.4, containing 5 mM MnCl$_2$ until used.

The hemodialyzer containing the immobilized arginase is sterilized prior to use with an ethylene oxide/freon 12 mixture 12 : 88 for 6 hr at 35° in 50% humidity (gas contact pressure 1.0 kg/cm^2). The sterilized material is stored for 1 day at 25° and subsequently aerated for 10 days at 25°.

Immobilized enzyme activity can be estimated using a circulating solution of 0.1 M arginine in 10 mM Tris–HCl, pH 7.4, at a flow rate of 170 ml/

[23] P. Edman, V. Nylen, and I. Sjoholm, *J. Pharmacol. Exp. Ther.* **225**, 164 (1983).

min. The enzymatic activity is determined by following the amount of urea formed as a function of circulation time.

Efficiency. Beef liver arginase bound to the inner surface (blood surface) of the dialyzer was able to metabolize 250 μmol of L-arginine/min at 37°. The authors noted that the extracorporeal use of this reactor was well tolerated in sheep and that the device exhibited good stability. Precise data were not given. Although several tumor lines which are sensitive to arginine deprivation have been described, it appears that this type of arginase is of minor importance as an antitumor enzyme because of its high K_m.

Preparation of Enzyme Reactors Containing L-Phenylalanine Ammonia-Lyase[24,25]: Method a—Hollow Fiber Reactor

Principle. The enzyme L-phenylalanine ammonia-lyase is absorbed but not covalently bound to the walls of asymmetric hollow fibers incorporated into a multitubular enzyme reactor cartridge.

Device. Asymmetric hollow fibers (PM10, Amicon) with a molecular weight cutoff of 10,000 and inner and outer diameters of 0.5 and 0.8 mm, respectively, are surrounded by a Plexiglass sheath (Rohm and Haas) and sealed with epoxy resin (T640, Amicon).[24]

Reagents

L-Phenylalanine ammonia-lyase (PAL, P & L Biochemicals), purified according to Fritz *et al.*[26] Concentrated PAL solution = 20 U/ml solution A.
Solution A: 0.1 *M* phosphate buffer, pH 7.2

Procedure. Fill the outer compartment of the dialyzer with a concentrated PAL solution. Ultrafilter the solution through the hollow fibers by suction of the fiber input. The enzyme will be deposited in the porous shell near the fiber membrane. Repeat this procedure a total of 5 times. This results in approximately 100 U of PAL being absorbed in the hollow fibers. The outer compartment ports are sealed, and the lumen of the fibers is filled with solution A containing 0.1% sodium azide. Cap the ends of the reactor and store at 5°.

Method b—Nylon Tube Reactor

Principle. PAL is covalently bound to nylon tubing which is then incorporated into a multitubular enzyme reactor.

[24] H. Peterson, C. Horvath, and C. M. Ambrus, *Res. Commun. Chem. Pathol. Pharmacol.* **20,** 559 (1978).
[25] C. L. M. Ambrus, J. L. Ambrus, C. S. Horvath, H. Peterson, S. Sharma, C. Kant, E. Mirand, R. Guthrie, and T. H. Paul, *Science* **201,** 837 (1978).
[26] R. R. Fritz, D. S. Hodgins, and C. W. Abell, *J. Biol. Chem.* **251,** 4646 (1976).

Device. A 5-m-long nylon tube (Polymer Corp.), having inner and outer diameters of 0.8 and 1.2 mm, respectively, after binding of enzymes, are cut into 27-cm lengths and mounted 15 at a time into cartridges as described by Peterson *et al.*[24]

Reagents

PAL as listed above
Glutaraldehyde (Aldrich)
Solution A: 0.1 *M* phosphate buffer, pH 7.2 (as above)
Solution B: 0.2 *M* phosphate buffer, pH 7.8

Procedure. A layer of polyethyleneamine is covalently bound to the inner wall of the nylon tube using the method of Horvath.[22] About 20 ml of solution B containing 25% glutaraldehyde is circulated through the tube using a peristaltic pump at a flow rate of 1 ml/min for 60 min. The tube is then washed with 500 ml of distilled water at a flow rate of 3 ml/min and subsequently purged with dry nitrogen. The tube is placed in a water bath at 4°, and 5 ml of concentrated PAL is circulated through the tube at a flow rate of 1 ml/min for 6 hr. The tube is then washed with 500 ml of distilled water and stored as described above in solution A containing sodium azide.

To test the activity of the reactor, circulate 100 ml of L-phenylalanine (0.1–10 m*M*) in 50 m*M* phosphate buffer, pH 7.2, through the enzyme reactor and the flow cell of a spectrophotometer at a flow rate of 40 ml/min, 25°. The progress of phenylalanine breakdown is monitored by observing the increase in absorbance at 290 nm.

Efficiency. The L-phenylalanine ammonia-lyase enzyme reactors exhibited a K_m apparent of 5.0×10^{-4} *M*, which is identical to the K_m of the free enzyme. This indicates that no diffusional effects were involved. Typical reaction rates for the disappearance of phenylalanine at flow rates ranging from 30 to 80 ml/min and substrate concentrations from 0.1 to 10 m*M* phenylalanine, were 500–800 μmol/min for the multitubular hollow fiber method and 10 to 150 μmol/min for the nylon tube reactor. Dogs with artificially high phenylalanine levels (raised by diet plus administration of *p*-chlorophenylalanine, a phenylalanine inhibitor), showed a drop in the levels of this amino acid to 18% of pretherapy levels. No enzyme leakage was reported.[25]

L-Asparaginase in Acrylic Microspheres in a Hollow Fiber Dialyzer[23]

Reagents

Solution A: 5 m*M* phosphate buffer, pH 7.4, 6% (w/v) acrylamide, 2% (w/v) *N,N'*-methylenebisacrylamide

Solution B: chloroform–toluene (1 : 4 v/v), 2.5 mg/ml Pluronics F-68 detergent (Wyandotte Chemicals)

Solution C: 500 mg/ml ammonium peroxydisulfate.

Procedure. L-Asparaginase, with an activity of 40,000 IU, is dissolved in 20 ml of solution A which is then flushed with nitrogen gas to remove oxygen. Similarly, 800 ml of solution B is freed of oxygen by nitrogen flushing. The L-asparaginase solution is combined with 0.4 ml of solution C and immediately homogenized with 800 ml of deoxygenated solution B using an Ultra Turrax TP 18-10 homogenizer. Polymerization is started by adding 4.0 ml of N,N,N',N'-tetramethylethylenediamine to the organic phase. The microparticles are isolated by centrifugation, washed several times with 5 mM phosphate buffer, pH 7.4, and suspended in 150 mM NaCl.

Microparticles containing L-asparaginase are installed on the outer surface of a Gambro FH 202 hemodialyzer. A solution of sterile physiologic saline containing the microparticles is circulated in a closed circuit under a slight pressure through the outer chamber of the dialyzer. To install the microparticles, the outlet to the hollow fibers is opened allowing the saline and microparticles to flow into asymmetric pores of the fibers from the side with the wider pore opening. The microparticles, which cannot pass through the smaller opening of the pores in the hollow fibers, are partially trapped within the fiber membrane. After preparation of the hemofiltration units and before use, the dialyzers are checked for contamination and particle leakage by pumping the eluate from the inside of the hollow fibers through a 0.22-μm membrane filter. The filter is then analyzed for L-asparaginase activity and bacterial contamination.

Reactor activity can be estimated by circulating 50 μM L-asparagine in 20 mM phosphate buffer, pH 7.4, containing 150 mM NaCl through the unit. Samples are withdrawn for analysis of L-asparagine content before circulation and at approximately 15 min intervals thereafter.

Efficiency. This system is one of the most potent devices to convert L-asparagine to L-aspartic acid. More than 95% of the aspartic acid in a 5-liter solution of 50 μM L-asparagine in buffer was metabolized in 1 hr using an enzyme reactor containing 2000 IU of L-asparaginase. Equally impressive are the *in vivo* data in healthy sheep showing total L-asparagine depletion after 1–2 hr of extracorporeal treatment with a blood flow of 150–200 ml/min through the reactor. The authors make the important statement that repeated extracorporeal treatments in sheep decreased the efficiency of the system in obtaining low asparagine levels probably due to an increase in resynthesis of L-asparagine by the action of L-asparagine synthase. This fact points to the limiting factor for the use of extracorporeal tumor therapy, where a prolonged decrease of the respective amino acids is necessary.

Plasmapheresis–Enzyme Reactor System

The Enzyme Reactor Unit[15,27]

Principle. Asparaginase or glutaminase is covalently bound to glutaraldehyde-activated L-lysylcellulose and then succinylated to create a nonhydrolyzable enzyme reactor.

Reagents

> Asparaginase (*E. coli,* Merck; "Elspar") or glutaminase (*Acinetobacter*)[28]
> Isocyanate cellulose (granular, Gambro)
> Glutaraldehyde (Aldrich)
> Bovine albumin (RIA grade, Sigma)
> Solution A: 0.2 M sodium acetate, pH 5.5
> Solution B: 20 mM sodium phosphate, pH 7.35, 150 mM NaCl

Procedure. One hundred grams of dry isocyanate cellulose is suspended in 500 ml of solution A containing 10% L-lysine. The suspension is stirred gently at 4° for 24 hr and then placed in a column and washed extensively with 1 M NaCl in solution A until the ninhydrin reaction becomes negative. After a final wash with solution A, wet L-lysylcellulose is mixed 1 : 1 with 25% glutaraldehyde, the pH is adjusted to 5.5, and the solution is gently stirred for 24 hr at 4°. The above wash cycle is repeated to remove unreacted glutaraldehyde. The activated gel is then incubated in solution A containing 2 mg of albumin with 100 IU of asparaginase or 1 mg of glutaminase per milliliter of wet gel. The reaction is allowed to proceed for 24 hr at 4°. The resulting enzyme–albumin gel is then extensively washed with solution A, 2.0 M NaCl in solution A and 50 mM sodium borate, pH 8.5, in that order, and finally equilibrated in solution B. The gel can be stored in this form at 4° with 0.02% sodium azide added.

To further stabilize the gel, it can be succinylated. One hundred grams of wet albumin–enzyme gel is suspended in 100 ml of solution B and vigorously stirred. Ten grams of finely ground succinic anhydride and 20 gm of anhydrous Na_2CO_3 are simultaneously and rapidly added to this suspension. After 5 min the succinylated enzyme cellulose is washed over a funnel with solution B and stored in solution B containing 0.02% sodium azide at 4°.

The coarse enzyme granules are packed into a cartridge consisting of a Travenol HFAK device, the fibers of which have been removed. The bottom of the cartridge is closed by a polystyrene filter (Glasrock Products Inc., Fairburn, GA) with an average pore size of 25 μm.

[27] M. B. Dennis, G. Schmer, N. L. Rastelli, and J. C. Detter, *Am. Soc. Artif. Intern. Organs* **29**, 744 (1983).
[28] J. R. Roberts, J. S. Holcenberg, and W. C. Dolowy, *J. Biol. Chem.* **247**, 84 (1972).

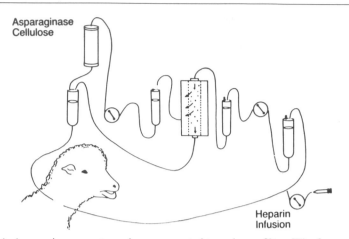

FIG. 1. Asparaginase reactor column connected to a plasma filter. [The figure was taken from *Trans. Am. Soc. Artif. Intern. Organs* **28,** 377 (1982), with the kind permission of the publisher (J. B. Lippincott, Hagerstown, MD 21740).]

Plasmaphersis is carried out by using a polypropylene plasma filter (Plasmaflux P-2, Fresenius, Homburg/Saar, FRG). The principle of the plasmaphersis enzyme reactor system is shown in Fig. 1. Blood derived from a jugular–carotid shunt (sheep or dog) is pumped through the plasma filter. The plasma filtrate is passed through the cartridge containing the L-asparaginase or L-glutaminase and then returned to the animal through the venous blood line.

Efficiency. The enzyme reactor system containing only L-asparaginase decreased the L-asparagine levels in the plasma filtrate of sheep from an average of 63 to 5 nmol/ml with a concomitant drop in L-glutamine from 290 to 158 nmol/ml due to the 2–5% glutaminase activity present in *E. coli* asparaginase. Systemic L-asparagine levels dropped to less than 10% of pretreatment levels within 6 hr of extracorporeal treatment.

The same enzyme reactor system containing L-asparaginase and L-glutaminase was tested in dogs with lymphoma.[27] L-Asparagine dropped to undetectable levels in one dog, remaining low throughout the 3 weeks of treatment. L-Glutamine never dropped below 50% of its pretreatment values. This dog exhibited a good remission; however, it relapsed after 3 months and was then resistant to further treatment. No enzyme leakage was detected (lower limit of detection was 1.0×10^{-5} IU/ml).

Summary and Outlook

The development of an enzyme reactor system for cancer treatment is still in its infancy, although considerable progress has been made toward stable, nonhydrolyzable, and highly active systems. The main problem in the future to be dealt with is the rapid rebound phenomenon of L-asparagine and L-glutamine which greatly reduces the value of extracorporeal tumor treatment as a sole method.[27] It will be necessary to combine extracorporeal enzyme treatment with the administration of amino acid analogs, which either decrease the utilization of their respective amino acid or decrease its synthesis rate. The combination of a tryptophan-degrading enzyme administered parenterally along with a tryptophan analog has shown great promise in cancer therapy.[29]

[29] J. Roberts, H. Rosenfeld, and B. Dunn, *Proc. Am. Assoc. Cancer Res. Am. Soc. Clin. Oncol.* **22,** 218 (Abstr.) (1981).

[44] Dialysis Membranes Containing Asparaginase Entrapped in Microparticles

By Peter Edman, Ulf Nylén, and Ingvar Sjöholm

During the last decade, interest in immobilized therapeutic enzymes has increased considerably. L-Asparaginase which is used in the treatment of acute lymphoblastic leukemia (ALL), has been used as model enzyme in many cases.[1,2] In several systems, the immobilized form has been superior to soluble enzyme especially when considering the duration of the effect *in vivo.*[3]

However, the immunological reactions which are inevitably associated with the use of foreign proteins/enzymes *in vivo* are strengthened in many immobilized systems, e.g., in microspheres. Therefore, attempts have been made to overcome the immunological reactions and at the same time maintain the advantages of an immobilized system. Immobilizing enzymes on carriers used in an extracorporeal system is one way to avoid the above mentioned problems.[4,5] An extracorporeal system in which the enzyme is immobilized in polyacrylamide microparticles and installed in the outer part of a hemofilter membrane having asymmetric pores, is an

[1] T. M. S. Chang, this series, Vol. 44, p. 201.
[2] E. D. Neerunjun and G. Gregoriadis, *Biochem. Soc. Trans.* **4,** 133 (1976).
[3] P. Edman and I. Sjöholm, *J. Pharmacol. Exp. Ther.* **211,** 663 (1979).
[4] D. Sampson, L. S. Hersh, D. Cooney, and G. P. Murphy, *Trans. Am. Soc. Artif. Intern. Organs* **18,** 54 (1972).
[5] A. Jackson, H. R. Halvorson, J. W. Furlong, K. O. Lucast, and J. D. Shore, *J. Pharmacol. Exp. Ther.* **209,** 271 (1979).

example of a device which prevents direct contact between blood and enzyme.[6]

Preparation of Microparticles Containing Asparaginase

L-Asparaginase (370 mg, corresponding to 40,000 IU) is dissolved together with 1.2 g acrylamide and 0.4 g N,N'-methylenebisacrylamide in 20 ml 5 mM phosphate buffer, pH 7.4. Oxygen is removed from the solution prior to, during, and after homogenization by bubbling nitrogen through the solution. After addition of ammonium peroxydisulfate (100 μl of a solution of 0.5 g/ml) the water phase is poured into a mixture consisting of 800 ml toluene–chloroform (4:1, v/v) and 0.5 g of a detergent, Pluronics F-68. The mixture is then homogenized to produce a water-in-oil emulsion with a mechanical homogenizer, e.g., an Ultra Turrax TP 18–10. After formation of a stable emulsion, N,N,N',N'-tetramethylethylenediamine (TEMED, 2 ml) is added to induce polymerization of the droplets to small microparticles. The speed of the homogenizer is reduced when TEMED is added. One minute after the addition of the accelerator, the homogenizer is removed. The suspension is then stirred magnetically under nitrogen for 10–20 min to complete the polymerization. The suspension is centrifuged, and the organic phase is discarded. The microparticles in the water phase are washed several times with buffer or physiological saline. Generally, the mean diameter of the polyacrylamide microparticles is less than 0.5 μm as determined by scanning electron microscopy.

When the particles are to be used for *in vivo* studies or are to be stored for long times, aseptic techniques have to be used. All solutions, which can not be heat sterilized, are then filtered through a bacteria-retentive filter (pore diameter 0.22 μm) and the equipment wiped with alcohol. Sterile physiological saline supplemented with penicillin (1 mg/ml) and gentamicin (2.5 μg/ml) can be used during the washings. All the manipulations are carried out in a clean room equipped and built for aseptic work and when possible in a laminal air flow (LAF) bench.

Analytical Methods

Assay of L-asparaginase activity is performed according to the method of Yellin and Wriston.[7] The sample is mixed with 0.1 M sodium tetraborate, pH 8.0, to obtain a volume of 1.0 ml. After addition of L-asparagine (0.5 ml of a 40 mM solution in borate buffer), the mixture is incubated at 37° for different time periods. The reaction is stopped with 0.5 ml of

[6] P. Edman, U. Nylén, and I. Sjöholm, *J. Pharmacol. Exp. Ther.* **225,** 164 (1983).
[7] T. O. Yellin and J. C. Wriston, Jr., *Biochemistry* **5,** 1605 (1966).

30% (w/w) trichloroacetic acid (TCA). After centrifugation at 4° and 1000 g, the supernatant (1.0 ml) is withdrawn and diluted with distilled water (8.0 ml). Nessler's reagent (1.0 ml) is then added, and after incubation at 22° for 10 min the resulting color is measured photometrically at 500 nm.

L-Asparagine and L-aspartic acid assays are performed according to the method of Cooney et al.[8] In brief, fresh serum (1.0 ml) is carefully mixed with 5% TCA (1.0 ml), and after centrifugation at 4° the supernatant is extracted with ether to remove TCA. NADH (7.1 mg), α-ketoglutaric acid (7.3 mg), and aspartate aminotransferase (58 IU) are dissolved in 5 ml of 0.5 M Tris buffer, pH 8.0. The sample (0.5 ml) diluted in 0.5 ml Tris buffer, is added to the NADH solution. The adsorbance is measured at 340 nm. When a stabel level is established, 50 μl of a malate dehydrogenase solution (20 IU/ml) is added, and the decrease of the adsorbance is followed for up to 30 min or to the time when a new stable value is obtained. The difference between the starting and end values is used to calculate the concentration of L-aspartic acid. L-Asparaginase (0.1 IU) is then added, and the adsorbance measured each 5 min until a stable value is reached again. From the changed absorbance, the concentration of L-asparagine in the sample can be determined with relevant standards.

Preparation of the Extracorporeal Dialysis Unit

Microparticles containing L-asparaginase are installed in the outer surface of the hollow fibers in a commercial hemofilter (Gambro hemofilter, FH 77). The hollow fiber membranes have asymmetric pores and are semipermeable with a cutoff at about 30,000 daltons. The diameter of the pores is about 1 μm on the outside and less than 0.01 μm on the inner surface. Small L-asparaginase microparticles (<1 μm in diameter) are suspended in buffer and placed in the outer chamber of the hemofilter. The medium is forced cautiously through the membrane with a slight overpressure. The particles will then be immobilized within the pores of the membrane. The outlet of the hemofilter is open during the installation, and the performance of the reactor is controlled by filtering the eluate through a 0.22-μm membrane filter. The filter is tested for the absence of enzyme activity and/or bacterial contamination. Before the enzyme unit is used, 5 liters of sterile physiological saline is pumped through the unit to remove any antibiotics.

In Vitro and in Vivo Use of the Extracorporeal Unit

The clearance of asparagine in the reactor is self-evidently dependent on the amount of enzyme installed in the hollow fibers and on the flow

[8] D. A. Cooney, R. L. Capizzi, and R. E. Handschumacher, Cancer Res. **30**, 929 (1970).

through the unit. As can be seen in Table I, the clearance value increases with the amount of L-asparaginase in the reactor, when tested *in vitro* by recirculation of 5 liters of 50 μM L-asparagine in phosphate buffer, pH 7.4. The concentration of asparagine corresponds to that normal in sheep. With a flow rate of 200 ml/min maximal clearance is obtained with 2000 IU of the enzyme in the reactor. This means that the substrate is totally cleared during one passage through the hemofilter, i.e., converted to L-aspartic acid. The clearance is improved further when the flow is increased to 300 ml/min as seen in the table. With the reactor containing 2000 IU of L-asparaginase and the flow rate 200 ml/min, the buffer solution (5 liters) is completely cleared of L-asparagine in about 1 hr.

The reactor has been used in sheep and in a clinical trial with a patient. The *in vivo-* experiments in sheep showed that the L-asparagine concentration in the blood was reduced to 0 after 2 hr when a reactor containing 2000 IU of enzyme and a flow rate of 150–200 ml/min was used. However, the initial concentration in the blood was partly restored after 1 day (see Fig. 1). This rebound phenomenon was enhanced when the treatment was repeated. Almost no effect on the plasma L-asparagine level was seen after 7 daily perfusions, probably due to increased resynthesis of L-asparagine in the liver by induction of L-asparagine synthase.

TABLE I

Conversion of L-Asparagine to
L-Aspartic Acid[a]

Amount of enzyme (IU)	Flow (m/min)	Clearance (m/min)	K_{el}[b]
500	200	90	0.018
1000	200	115	0.023
1500	200	161	0.032
2000	200	200	0.040
500	300	134	0.027
2000	300	268	0.054

[a] L-Asparaginase in microparticles is installed in hemofiltration units consisting of 7000 asymmetric hollow fibers with a total membrane area of 1.2 m². Five liters of 50 μM L-asparagine in sodium borate buffer, pH 7.4, is circulated through the reactor at 22°. The clearance and the rate constant for the conversion of L-asparagine to L-aspartic acid are determined at different flows.

[b] K_{el}, rate constant.

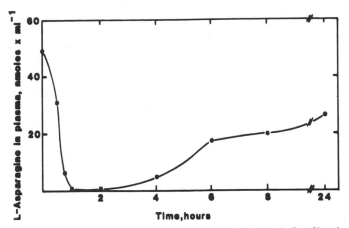

FIG. 1. Plasma level of L-asparagine in a sheep, F10B, during and after filtration for 2 hr through an enzyme reactor containing 2000 IU of L-asparaginase. The blood flow rate through the unit was 150–200 ml/min.

Similar effects were seen in a patient with lymphoblastic leukemia. During the first treatments, the L-asparagine concentration was reduced to almost zero but was restored to normal values the following day. This rebound phenomenon was remarkably strong and later treatments in a 10-day program had no effect on the plasma concentration of L-asparagine.

Comments

The system described is an efficient enzyme reactor primarily due to the combination of an immobilized particle-bound enzyme and a hemo-filtration unit consisting of hollow fibers with asymmetric membranes. The immobilization of the enzyme in microparticles of polyacrylamide without covalent coupling means that the stability of the enzyme is improved with retained biological activity.[9,10] The loss of enzymatic activity from the particles is small, and the tight inner membrane of the hollow fibers prevents the microparticles from escaping into the circulating medium. The particles are at the same time small enough (0.2–0.5 μm in diameter) to penetrate deep into the membrane and stick to the walls of the pores, since the pores are asymmetric with outer openings around 1 μm in diameter. The enzyme will consequently be concentrated in the fiber membranes close to the circulating medium, resulting in a higher reactor efficiency owing to the short distance between substrate and en-

[9] B. Ekman, C. Lofter, and I. Sjöholm, *Biochemistry* **15,** 5115 (1976).
[10] B. Ekman and I. Sjöholm, *J. Pharm. Sci.* **67,** 693 (1978).

zyme. The enzyme will still not be in contact with the medium, which is an advantage in many applications.

The enzyme reactor has been tested with L-asparaginase, which has been used in the clinic for many years for the treatment of acute lymphoblastic leukemia. A clinical summary[11] has shown its usefulness, particularly in combination with other antineoplastic drugs, to induce prolonged remission duration in asparagine-dependent T-cell-related leukemias. Our own limited experience from one patient indicates, however, that asparagine synthase may be markedly induced in long-term treatment programs as a result of hemofiltration through the L-asparaginase reactor, thus neutralizing the asparagine degradation at the end of the treatment period. It seems necessary, therefore, to combine the use of the reactor with other therapy aiming at decreasing or preventing the effects of the enzyme induction or to limit its use to patients, in which the induction is less pronounced. The technique has recently also been suggested to be used to remove bilirubin from blood with bilirubin oxidase.[12]

[11] J. F. Holland and T. Ohnuma, *Cancer Treat. Rep.* **65,** 123 (1981).
[12] A. Lavin, C. Sung, A. M. Klibanov, and R. Langer, *Science* **230,** 543 (1985).

[45] Preparation of Protein A Immobilized on Collodion-Coated Charcoal and Plasma Perfusion System for Treatment of Cancer

By DAVID S. TERMAN

Introduction

Tumoricidial responses in dogs with spontaneous anaplastic mammary carcinoma were observed after treatment with plasma that was perfused over protein A-bearing *Staphylococcus aureus* which were immobilized in a microporous membrane filtration system.[1] Using an identical perfusion system and canine tumor model Holohan and colleagues at the National Cancer Institute confirmed these findings,[2] and subsequently several others observed similar effects in various other canine and rodent cancer model systems.[3-7] Because protein A appeared to be one of the

[1] D. S. Terman, T. Yamamoto, M. Mattioli, G. Cook, R. Tillquist, J. Henry, R. Poser, and Y. Daskal, *J. Immunol.* **124,** 795 (1980).
[2] T. V. Holohan, T. M. Phillips, C. Bowles, and A. Diesseroth, *Cancer Res.* **42,** 3663 (1982).
[3] S. Sukumar, B. Zbar, N. Terata, and J. J. Langone, *J. Biol. Resp. Modif.* **3,** 303 (1984).
[4] P. D. Cooper and G. R. Masinello, *Int. J. Cancer* **32,** 737 (1983).

tumoridicial materials, we subsequently immobilized commercial protein A in a collodion–charcoal matrix and used this system to treat dogs and humans with mammary cancer.[8-10] Tumoricidial effects and tumor regressions were observed with this system.[8-10]

The method for preparation of protein A immobilized in collodion–charcoal (PAAC) used for preclinical and clinical studies is given herein.[11] Initially, the system involved on-line perfusion of host plasma over PACC using a plasma–cell separator. However, it became apparent during human studies that similar tumoricidial effects could be produced by direct infusion of host plasma which was perfused over PACC.[9] Objective regressions of human breast adenocarcinoma were noted after treatments which were associated with toxicity of fever, chills, nausea, and hypotension.[9-12] Using a modified protein A–collodion–charcoal (PACC) system, tumor regressions were also noted by Bertram *et al.* in human breast and brain cancer which were associated with similar toxicity.[13-15] Recent analysis of clinical data suggests that maximum antitumor effects were invariably associated with constitutional symptoms of fever, chills, and rigors.[9,10,12-15]

As clinical and basic studies have progressed, several important biomolecules have been identified which are either contaminating commercial protein A preparations or present in plasma which has been perfused over PACC.[16-18] Recent studies have shown that commercial protein A

[5] B. R. Gordon, R. E. Matus, S. D. Saal, G. E. MacEwen, A. I. Hurvitz, K. H. Stenzel, and H. L. Rubin, *JNCI, J. Natl. Cancer Inst.* **70**, 1127 (1983).

[6] F. R. Jones, L. H. Yoshida, W. C. Ladiges, and M. A. Kenny, *Cancer* **46**, 625 (1980).

[7] W. T. Liu, R. W. Engleman, L. Q. Trang, K. Hau, R. A. Good, and N. K. Day, *Proc. Natl. Acad. Sci. U.S.A.* **81**, 3516 (1984).

[8] D. S. Terman, T. Yamamoto, R. Tillquist, J. F. Henry, G. L. Cook, and W. T. Shearer, *Science* **209**, 1257 (1980).

[9] D. S. Terman, J. B. Young, W. T. Shearer, J. C. Ayus, D. Lehane, C. Mattioli, R. Espada, J. F. Howell, T. Yamamoto, H. Zaleski, L. Miller, P. Frommer, L. Feldman, J. Henry, R. Tillquist, G. Cook, and Y. Daskal, *N. Engl. J. Med.* **303**, 1195 (1981).

[10] I. Daskal, C. Mattioli, J. Kao-Jen, and D. S. Terman, *Cancer Res.* **44**, 2225 (1984).

[11] D. S. Terman, Canadian Patent 415,005 (1985).

[12] J. B. Young, J. C. Ayus, L. K. Miller, R. R. Miller, and D. S. Terman, *Am. J. Med.* **75**, 278 (1983).

[13] J. H. Bertram, J. C. D. Hengst, and M. S. Mitchell, *J. Biol. Resp. Modif.* **13**, 235 (1984).

[14] J. H. Bertram, S. M. Grunberg, I. Shulman, M. Apuzzo, D. Boquiren, and M. Mitchell, *Cancer Res.* **45**, 4486 (1985).

[15] D. S. Terman and J. Bertram, *Eur. J. Clin. Oncol.* **21**, 1115 (1985).

[16] J. Balint, Jr., Y. Ikeda, J. J. Langone, W. T. Shearer, Y. Daskal, K. Meek, G. Cook, J. Henry, and D. S. Terman, *Cancer Res.* **44**, 734 (1984).

[17] D. S. Terman, *CRC Crit. Rev. Oncol. Hematol.* **4**, 103 (1985).

[18] J. J. Langone, C. D. Bennet, and D. Terman, *J. Immunol.* **133**, 1057 (1984).

preparations are contaminated with trace, but biologically significant, amounts of staphylococcal enterotoxins.[17] In addition, it has now been shown that plasma perfused over PACC contains anaphylatoxins[18] and conjugates of protein and IgG with molecular weights ranging from 600,000 to 2,000,000.[16] Our present thought is that these products contribute significantly to the antitumor effects and the toxicity of the procedure.

Additional studies have demonstrated mitogenic activity in plasma perfused over PACC in patients responding to perfusion therapy, whereas those patients whose perfused plasma was not mitogenic failed to show objective responses or constitutional reactions.[13,14] PACC-perfused plasma has also been shown to stimulate human granulocyte chemiluminescence and ^{45}Ca release.[17]

Our present hypothesis is that the above constellation of bioreactive products released or formed in plasma after contact with PACC results in antitumor effects via activation of multiple immune and inflammatory systems in vivo.[17] Staphylococcal enterotoxins, in very small quantities, are powerful interleukin[19,20] as well as interferon inducers[21] and T cell mitogens.[22,23] They produce hemodynamic changes in monkeys identical to those which we observed with the PACC perfusion system.[9,12,24] Protein A is a well-known immunocyte mitogen[25,26] and is capable of augmenting natural killer cell and antibody-dependent cellular cytotoxicity (ADCC) activity.[27] Anaphylatoxins are known to have immunopotentiating activity.[28] They may stimulate leukotriene release[29] and augment interleukin I production,[30] accounting, in part, for the bronchopulmonary toxicity observed during perfusions.[9,12] Synchronous activation of these systems by the above-mentioned powerful bioreactants may result in maximal antitumor effects whereas partial or weak activation may yield minimal clinical effects.

[19] T. Ikejima, C. A. Dinarello, D. M. Gill, and S. M. Wolff, J. Clin. Invest. **73**, 1312 (1984).
[20] J. Parsonnet, R. K. Hickman, D. D. Eardley, and G. B. Pier, J. Infect. Dis. **151**, 514 (1985).
[21] E. M. Smith, A. M. Johnson, and J. E. Blaylock, J. Immunol. **130**, 773 (1983).
[22] M. P. Langford, G. J. Stanton, and H. M. Johnson, Infect Immun. **22**, 62 (1978).
[23] N. J. Poindexter and P. M. Schlievert, J. Infect. Dis. **151**, 65 (1985).
[24] W. R. Beisel, Toxicon **10**, 433 (1972).
[25] T. Sakane and I. Green, J. Immunol. **120**, 302 (1978).
[26] S. Romagnani, R. Baigiotti, G. Guidizi, F. Almerigogna, A. Alessi, and M. Ricci, J. Immunol. **132**, 566 (1984).
[27] W. J. Catalona, T. L. Ratliff, and R. E. McCool, Nature (London) **291**, 77 (1981).
[28] M. G. Goodman, D. E. Chenoweth, and W. O. Weigle, J. Exp. Med. **156**, 912 (1982).
[29] N. P. Stimler, M. K. Bach, C. M. Bloor, and T. E. Hugli, J. Immunol. **128**, 2247 (1982).
[30] M. Rola-Pleszczynski and I. Lemaire, J. Immunol. **131**, 801 (1983).

Numerous *in vitro* studies have demonstrated that *Staphylococcus aureus* Cowan 1 (SAC) or staphylococcal enterotoxins may potentiate the immunocyte-activating and differentiative properties of various endogenous lymphokines. For example, SAC-induced B-cell differentiation and proliferation are augmented in the presence of small quantities of interleukin 2.[31,32] Moreover, SAC will synergize with soluble B-cell growth factors or mitogen-activated T cells to induce B-cell differentiation to immunoglobulin-secreting activity.[33,34] Furthermore, staphylococcal enterotoxins will potentiate T-cell proliferation in the presence of small amounts of interleukin 1.[19,20,35] While natural killer (NK) cell function is significantly augmented by interferon, levels of cytotoxicity are even greater following pretreatment with enterotoxin A.[36]

Staphylococcus protein A and enterotoxins are capable of inducing lymphocyte mitogenesis and production of several lymphokines from immunocytes and macrophages. T cells will differentiate to a cytotoxic state in the presence of soluble staphylococcal protein A and enterotoxins.[37–39] Moreover, mononuclear cells are activated to produce interleukin 1 and interferon by staphylococcal protein A and enterotoxins.[19,20,22,27,40,41] Indeed, synergy between staphylococcal enterotoxins and additional bacterial products for interferon production by mononuclear cells has been demonstrated.[40] Finally, staphylococcal enterotoxins, SAC or soluble protein A, and additional staphylococcal products are known to be powerful immunocyte mitogens.[22,42–47] Thus, it is clear that various staphylococcal products working alone or together with endogenous lymphokines

[31] T. Teranishi, T. Hirano, B. Lin, and K. Onoue, *J. Immunol.* **33**, 3062 (1984).

[32] R. J. M. Falkoff, L. P. Zhu, and A. S. Fauci, *J. Immunol.* **129**, 97 (1982).

[33] O. Saiko and P. Ralph, *J. Immunol.* **127**, 1044 (1981).

[34] T. Hirano, T. Teranishi, B. Lin, and K. Onoue, *J. Immunol.* **133**, 798 (1984).

[35] J. Parsonnet, Z. A. Gillis, and G. B. Pier, *J. Infect. Dis.* **154**, 55 (1986).

[36] I. Kimber, J. Bakacs, and M. Moore, *Clin. Exp. Immunol.* **54**, 39 (1983).

[36a] Drying may be a very crucial step. In a modification of the above procedure, Bertram dries the charcoal for up to 7 days in a desiccator attached to low-pressure suction (J. Bertram, personal communication; Ref. 14).

[37] T. Kashahara, H. Harad, K. Shiorri-Nakaho, M. Imain, and T. Sano, *Immunology* **42**, 175 (1981).

[38] T. Zehavir-Willner and G. Berke, *J. Immunol.* **137**, 2682 (1986).

[39] M. Sumiya, S. Kano, and F. Takaku, *Int. Arch. Allergy Appl. Immunol.* **67**, 66 (1982).

[40] M. Tsujimoto and N. Higashi, this series, Vol. 119, p. 93.

[41] M. DeLey, J. VanDamme, and A. Billiau, this series, Vol. 119, p. 88.

[42] D. L. Peavy, W. H. Adler, and R. T. Smith, *J. Immunol.* **105**, 1453 (1970).

[43] A. S. Kreger, G. Cuppari, and A. Taranta, *Infect. Immun.* **5**, 723 (1972).

[44] N. R. Ling, K. Spicer, and N. Williamson, *Br. J. Hematol.* **11**, 421 (1969).

[45] A. A. Forsgren, A. Svedjelund, and H. Wigzell, *Eur. J. Immunol.* **6**, 207 (1981).

may augment significantly the activity of various immune and inflammatory systems. In purified form, these substances would seem to have great potential for tumor immunotherapy, particularly if utilized in *ex vivo* tissue-culture systems. In this fashion, various biological activities may be monitored and quantitated prior to *in vivo* administration to tumor-bearing hosts.

It should be noted that the protein A produced by most manufacturers is presently made for *in vitro* use and is not intended to serve as a pharmaceutical. Most preparations emerge with few or no quality controls, no determination of contaminating staphylococcal components, and little or no information regarding biological function. Recently, manufacturers have indicated that since 1980 there have been multiple changes in the fermentation step in production of protein A in an effort to increase the yield.[48] Such changes were likely to influence the quality of the protein A as well as the contaminating elements in these preparations.

The relative role of protein A in the antitumor effects and toxicity is unknown at this point. Indeed, it is now likely that additional staphylococcal products leaching from PACC after plasma perfusion are of equal if not greater importance in the antitumor effects. Therefore, it is recommended that investigators proposing to use these reagents for clinical or experimental studies bear in mind the heterogeneous composition of commercial protein A and that they institute their own quality controls on these preparations before embarking on clinical trials. In addition to standard microbiological and pyrogen testing of the PACC system at various phases in the production process, we recommend further testing of protein A for enterotoxin quantity. The plasma which has passed over PACC should also be tested for the presence of protein A–IgG conjugates, anaphylatoxins, and mitogenic activity.

In its present state of development, the procedure described in this chapter is still highly experimental. It should be used for clinical trials with great caution and only after appropriate preclinical animal studies and approval by human review boards. It is essential that proper commercial protein A is chosen since batches may vary with respect to the quantity of additional staphylococcal products and toxins.[43,46,49,50] Crude preparations of protein A, prepared by the lysostaphlin digestion method or by isolation of secreted products from a mutant staphylococcal strain, have

[46] B. Petrini and R. Mollby, *Infect. Immun.* **31,** 952 (1981).
[47] A. Romagnani, A. Amadori, M. G. Giudizi, R. Biagiotti, F. Maggi, and M. Ricci, *Immunology* **35,** 471 (1978).
[48] M. Inganass (Pharmacia, Piscataway, New Jersey), personal communication (1984).
[49] A. W. Bernheimer and L. I. Schwartz, *J. Gen. Microbiol.* **30,** 455 (1963).
[50] M. Rogolsky, *Microbiol. Rev.* **43,** 320 (1979).

been used for clinical studies and are likely to contain important additional staphylococcal components which are as yet unclassified but which will need to be identified and assayed. Assessment of enterotoxins in commercial protein A is a critical quality control. A simple radioimmunoassay method which we have employed for detection of enterotoxins in protein A preparations is given in the Appendix.

Critical determinants of clinical efficacy with this system appear to be the quantity and quality of products released from PACC or formed in plasma after contact with PACC.[17] The quantity of protein A leaching from PACC appears to be influenced by the amount immobilized, the extent of the washing procedure, and the volume and dilution of plasma perfused over it.[17] In short, it appears that larger quantities of immobilized protein A and more undiluted plasma perfused over PACC result in larger amounts of leached protein A. Details of preparation and use of the PACC perfusion system employed in two successful trials are given herein and in previous publications.[9,12,14] However, until standardized preparations of commercial protein A are developed, it will be important for each investigator to establish conditions of safety and efficacy for each batch utilized in preclinical and clinical studies. A recommended battery of tests that might be employed for each batch of commercial protein A would be as follows: (1) polyacrylamide gel electrophoresis; (2) assessment of Fc binding activity of protein A; (3) quantitative assessment of contaminating staphylococcal enterotoxins in protein A (see Appendix for method); (4) quantitative assessment of contaminating staphylococcal α-toxins in protein A[32]; (5) assessment of capacity of plasma perfused over PACC to generate C3a anaphylatoxins [using commercially available kits (Upjohn Co.)]; and (6) assessment of mitogenic activity of effluent plasma perfused over PACC column.[14]

Based on presently available information, protein A preparations with a maximum of 0.05% contamination with enterotoxin B would be acceptable for clinical trials. For clinical studies, we would recommend beginning with 0.5–5 mg of immobilized protein A perfused with 100 ml of autologous plasma and administered to the host at flow rates of 5–20 ml/min. Dose and flow rate escalation might proceed at 5 mg and 5 ml/min increments, respectively, depending on the presence of tumor necrosis and limiting host toxicity. Detailed description of host toxicity is given in a previous study.[12] The absence of host constitutional systems associated with treatments would suggest that the system is inactive and should be reevaluated with particular reference to the quantities of enterotoxins in the protein A preparations.

[51] P. J. McConahey and F. J. Dixon, *Int. Arch. Allergy Appl. Immunol.* **29**, 185 (1966).

Procedure 1: Charcoal Washing

Materials

Activated charcoal, Fisher activated coconut charcoal, 6–14 mesh,
  #5-685-B, Fisher Scientific (1 4-lb can)
Tyler sieves, Tyler U.S. Standard Sieve Series,
  Sieve #6, #8323-R10 (1),
  Sieve #16, #8323-R28 (1),
    A. H. Thomas Co.
Sterile water for irrigation, 1-liter bottles, #2F7114 Travenol
Tripod stand, 11 cm i.d. #15-300B, Fisher Scientific (1)
Tri-pour beakers, 1 liter #B2722-1 L, American Scientific Products
  (3)
Vacuum desiccator, #08-631A, Fisher Scientific (1)
Desiccator cabinet, #08-645-6, Fisher Scientific (1)
Indicating Drierite absorbent, #07-578-3A, Fisher Scientific (1)
Percolator funnel, #13-391A, Fisher Scientific (1)
Tygon tubing, #14-169-9H, Fisher Scientific (1)
Erlenmeyer flask, 1 liter, 10-047F, Fisher Scientific (1)
Autoclave tape
Cloth drapes for sterile wrapping

Method

1. Set up washing apparatus (see Fig. 1).
2. Weigh out approximately 400 g of charcoal.
3. Place #6 sieve on top of the #16 sieve.
4. Over a waste receptacle, pour the charcoal into the #6 sieve.
5. Shake the sieve. Charcoal particles larger than #6 mesh will remain in the upper sieve. Particles smaller than #16 mesh will pass through the bottom sieve.
6. Place the charcoal in the 1-liter Tri-pour of the washing apparatus. Place the #16 sieve on top of the Tri-pour.
7. Begin washing the charcoal by pouring sterile water in the percolator funnel.
8. Shake the Tri-pour occasionally to release trapped air.
9. Wash the charcoal with 25 liters of sterile water.
10. Transfer the charcoal as a slurry to another Tri-pour. Have the water level at least 1 inch above the charcoal.
11. Place the Tri-pour into the vacuum desiccator. Apply a vaccum of 600 mmHg for 30 min. This draws trapped air and charcoal fines from the interstices of the charcoal particles.

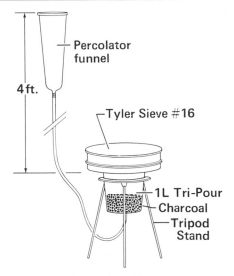

FIG. 1. Charcoal washing apparatus.

12. Return the charcoal to the washing apparatus and wash the charcoal with another 5 liters of sterile water.
13. As a slurry, pour the charcoal into the Erlenmeyer flask, covering the charcoal with at least 1 inch of water.
14. Autoclave at standard temperature and pressure (STP) for 30 min.
15. Return the charcoal to the washing apparatus and wash with another 25 liters of sterile water.
16. Place the charcoal into the #16 sieve, allow to drain, and cover with a cloth drape. Place the charcoal in a drying oven at 66° and allow it to dry for 48 hr with open exhaust.[36a]
17. After the drying period, allow the charcoal to cool and weigh out 30-gm aliquots.
18. Wrap the aliquots in cloth drapes and seal with autoclave tape. Label the tape with the date washed, charcoal weight, and the initials of the washing technician.
19. Autoclave the charcoal aliquots at STP for 45 min.
20. Place the aliquots in a drying oven at 66°. Allow to dry for 48 hr.
21. After the drying period, place the charcoal aliquots in the desiccator cabinet for storage.
22. The final aliquot from step 20, which is less than 30 g, is to be used for pyrogen testing.

Procedure 2: Coating Charcoal with Purified Protein A

Materials

Prepackaged Sterile Items
  30 g washed charcoal (see Procedure 1)
  5-ml volumetric pipets (2)
  5-ml syringe (1)
  1-ml syringe (1)
  20-gauge needles
  0.22-μm Millex filters, #SLG02505, Millipore (2)
  Sterile water, 1 liter, #2F7114 Travenol (1)
  Sterile gloves (1 pair)
Items Sterilized by Investigator
  Forceps (1)
  Stir bar (1)
  Applicator sticks, #01-340, Fisher Scientific (4)
  400-ml Tri-pour beaker, #B2722-400, Scientific American Products
    (1)
  8 × 10 inch Pyrex baking dish (1)
  100-ml graduated cylinder (1)
  Cloth drapes (2)
Nonsterile Items
  Protein A, 5 mg, Pharmacia (1 vial)
  Absolute pure ethyl alcohol, 100%
  Ether (ethyl) anhydrous, MAL 0848-1, Scientific American Products
    (1 lb)
  Collodion, U.S.P., MAL 4560, Scientific American Products (1 pint)
  Tris(hydroxymethyl)aminomethane, #T-395, Fisher Scientific
  NaCl, #S-671, Fisher Scientific
  HCl, 34% #A-144, Fisher Scientific
  Mechanical stirrer

Method. The charcoal coating procedure should be carried out at least
1 day prior to patient perfusion. This procedure is to be carried out in a
class 100 room with an externally exhausted fume hood with a capability
of 804 cubic feet per minute (CFM) and the sash open 1 foot. Aseptic
technique is observed throughout the procedure.

  1. The fume hood should be cleaned before the coating procedure is
     carried out. All surfaces are first washed with bleach. The sash is
     then lowered to the 1 foot position, and the hood is purged for 5
     min with the exhaust fan turned on. All surfaces are then washed
     with 7% alcohol and purged in the same manner for 5 min.

2. Prepare 0.15 M Tris, 0.19 M NaCl solution, pH 7.5. The Tris–NaCl solution is to be used within 12 hr of preparation.

3. Using the 5-ml syringe, inject 5 ml of the Tris–NaCl solution into the protein A vial, filtering the solution through a 0.22-μm Millex filter. Gently shake the vial until the contents are dissolved. This yields a 1 mg/ml concentration.

4. Withdraw 1 ml of the Tris–protein A solution from the vial using the 1-ml syringe.

5. Attach a 0.22-μm Millex filter to the syringe.

The following steps of the procedure are carried out in the fume hood with the exhaust fan turned on.

6. Place the 400-ml Tri-pour beaker with stir bar on the mechanical stirrer. Turn on the mechanical stirrer to a slow rate of speed.

7. Dispense into the beaker 4.1 ml of absolute alcohol using a 5-ml volumetric pipet. Next dispense 4.1 ml of collodion using another 5-ml volumetric pipet. (Note: The alcohol is a solvent for the collodion and must be dispensed into the beaker first.)

8. Measure 40 ml of ether in the graduated cylinder and add this to the alcohol and collodion.

9. Dispense the desired quantity of 0.22-μm filtered Tris–protein A solution into the alcohol–collodion–ether solution.

10. Allow the solution to stir for 2 min, turn off the mechanical stirrer, remove the stir bar, and add 30 g of washed charcoal to the solution.

11. Immediately begin gently mixing the charcoal using applicator sticks to coat uniformly. The sash of the fume hood should be lowered as much as possible to increase the air flow to aid in the evaporation of the liquid phase.

12. Continue gentle mixing, bringing the charcoal on the bottom to the surface. Continuous mixing is necessary to prevent charcoal particle aggregation.

13. As the alcohol and ether evaporate, the collodion forms a thin membrane on the charcoal surface which entraps the protein A.

14. The charcoal should be dry in approximately 15 min. Once dry the charcoal particles will no longer stick to the beaker walls. The charcoal particles will be light grey in color as opposed to the dark black of untreated charcoal.

15. Place the dry charcoal on the Pyrex baking dish, distributing it evenly across the bottom. Check for aggregated particles, removing any present.

16. Cover the Pyrex dish with a sterile cloth drape and permit any further ether fumes to evaporate overnight (10–16 hr).

17. The coated charcoal may be wrapped in sterile drapes and stored in a desiccator cabinet until future use.

Procedure 3: Protein A–Collodion–Charcoal Cartridge Assembly, Loading, and Washing

Materials. All items are wrapped and sterile.
PACC cartridge components (see Fig. 2)
Cartridge top (1)
PACC cartridge gasket (1)
Screens (2)
Spacer ring (1)
Cartridge bottom (1)
Reservior (1)
Reservior release tool (2)
O ring gasket (1)
Wing nuts (4)
Protein A-coated charcoal, 30 g
Accessory Perfusion Pack #491067 (Texas Medical Products, Inc., 10940 S. Wilcrest, Houston, TX 77099), contents:
  Tube item #1, female connector
  Tube item #2, male connector
  Female–female connector
Plasma transfer sets, 24 inches, Fenwal #4C2240 (3)
Transfer pack, 2 liters, Fenwal #4R2041 (5)
Solution administration set, Travenol #2C0017s (1)
400-ml disposable beaker, Tri-pour beaker, #B27722-400 American Scientific Products (1)
Applicator sticks Fisher Scientific Co. #01-340 (5)
Normal saline, 1 liter (9)
Sterile gloves (1 pair)
Tubing clamp (1)

Method. This procedure is to be performed in a class 100 room. Aseptic technique is observed throughout the procedure.

The rehydrated charcoal may then be loaded into a variety of containers. In our preclinical studies, we employed a cylindrical column 10 cm in height, 3 cm in diameter. The rehydrated charcoal should be loaded as a slurry into a column which contains a *fluid* bed. As the charcoal is gently loaded into the column, the particles will gravitate through the fluid and

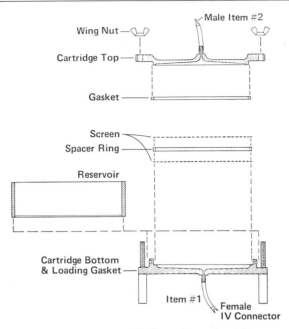

Fig. 2. Protein A–collodion–charcoal cartridge.

settle into position. The fluid level of the column may be adjusted as loading ensues by dispelling small amounts through an opening in the bottom of the column controlling flow with a screw clamp.

While a chromatography column may be adequate, we employed a "flat bed" column for human studies. Details of loading of this column are given below (see also Fig. 3):

1. Place the coated charcoal into the 400-ml Tri-pour beaker.
2. Attach the administration set to the 1-liter normal saline bag.
3. Add 300 ml normal saline to the beaker. Allow the charcoal to hydrate for 30 min. During this period, occasionally mix the charcoal gently with applicator sticks to allow air bubbles on the charcoal to escape.
4. See Figs. 2 and 3. Connect tube item #1 to the cartridge bottom.
5. Place the bottom screen into the cartridge bottom followed by the spacer ring. Attach the reservior to the cartridge bottom. Be sure to fully engage the O ring gasket.
6. Connect the administration set to tubing item #1 and fill the cartridge with saline until the reservior is approximately one-third full.

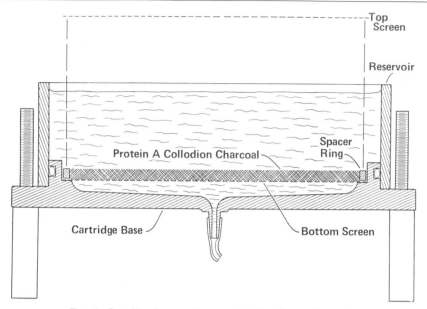

Fig. 3. Cartridge base and reservoir in loading configuration.

7. Pour the hydrated charcoal as a slurry into the cartridge bottom screen. Gently distribute the charcoal evenly across the bottom screen using an applicator stick. The charcoal should be inside the spacer ring.
8. Clamp tubing item #1. Disconnect the administration set and connect a plasma transfer set. Remove a transfer pack from its container and place a knot in the line to close it. Insert the coupler of the transfer set into the transfer pack.
9. Open the clamp and allow the saline to drain just to the level of the charcoal, clamping the line to stop saline from being drained completely. Gently redistribute the charcoal if necessary.
10. Remove the reservior. Place the upper screen over the charcoal resting on the spacer ring.
11. Place the PACC cartridge gasket around the outer edge of the cartridge top. Connect item #2 of the Accessory Perfusion Pack to the cartridge top.
12. Place the cartridge top onto the cartridge bottom. (Check the gasket circumstances for uniformity. A perfect seal is imperative.)
13. Put the wing nuts onto the bolts and hand tighten snugly. Tighten opposite wing nuts in sequence to ensure uniform sealing.
14. Remove the transfer set and transfer pack from tube item #1, and

connect to tube item #2 using the female–female connector provided in the Accessory Perfusion Pack.

15. Connect the administration set to tube item #1.
16. Open all clamps and allow saline to flow into the cartridge assembly to purge air.
17. The cartridge must be fully purged of air before continuing.
18. Proceed with washing of the PACC cartridge using 8 liters of normal saline. Saline may be run in by gravity at approximately 80 ml/min.
19. Observe the PACC cartridge for leaks. If any are present, tighten the wing nuts further. If leaks continue, abort the procedure and begin again using new materials and equipment.
20. The washed transfer packs are saved for microbiological testing.
21. Upon completion of washing, clamp tube item #1 and disconnect the administration set.
22. Connect tube item #1 and #2 together.
23. The PACC cartridge is ready to be transferred to the patient perfusion area.

A recent modification of the above procedure was described by Bertram *et al.* and used to treat 15 patients, with good clinical results.[13,14] Bertram *et al.* used a 50-ml syringe as a column to house the PACC and emptied the column of saline before plasma perfusion.[14] Prior to plasma perfusion, Bertram *et al.* washed the protein A–charcoal with 500 ml of sterile saline compared to 8 liters by our method.[14] Bertram pooled all perfused plasma and then administered it to patients. An additional technical point in Bertram's protein A–charcoal preparation was a considerably longer drying time (1 week of desiccation) of the washed and autoclaved charcoal before being used for protein A immobilization.

Procedure 4: Assembly of Pump and Filter

Procedures 4 and 5 are optional. Use of the filter system described below may not be necessary. It may be possible to perfuse plasma directly through the cylinder or flat-bed column without first perfusing through the filter. Recent studies suggest that the choice of systems will depend on the potency of the perfused plasma ascertained by *in vitro* biological testing.

Materials. All items are wrapped and sterile.
   Filter housing: 5 parts, requires cleaning and sterilization after each use; contents:
   Knurled lock rings (2)
   Housing cylinder (1)

Top end cap (1)
Bottom end cap (1)
Filter housing gaskets (2 large, 1 small)
Nonsterile Items
  Precision peristaltic pump (1)
  Power outlet board (1)
  Cart (1)
  Ring stand (1)
  Filter housing clamp (1)
  Drip chamber clamp (1)
Items Provided by Investigator
  Patient Perfusion Pack (Texas Medical Products), #411009, contents:
    Y-type blood solution recipient set
    Pump chamber tubing
    Two-way stopcock, Milex filter
    Extension tubing
    Drip chamber line with medication site
    Female–female connector
  Normal saline, 8 liters
  Plasma transfer sets, 24 inches, Fenwal #4R2041 (4)
  Sartorius filter (Sartorius Filters, Inc., 26575 Corporate Avenue, Hayward, CA 94545) (1)

Method. Aseptic technique is observed throughout the procedure.

1. See Fig. 4. Place one large filter housing gasket onto the bottom end cap and place the housing cylinder onto this, locking with ring.
2. Insert filter into housing, sealing completely.
3. Place small filter gasket onto central filler of top end cap. Place second large housing gasket in place.
4. Insert top end cap into housing cylinder and lock with ring.
5. Place filter housing into clamp on ring stand.
6. Remove Y-type set from patient perfusion package and close all clamps. Connect set to 1-liter normal saline bag.
7. Thread pump chamber through perfusion clamp. Connect downstream female connector to filter inlet. Connect upstream connector to Y-type set.
8. Place stopcock into filter vent and insert Millex filter into stopcock.
9. Insert extension tubing into filter outlet. Connect extension set to a transfer set. Remove a transfer pack from its container and place

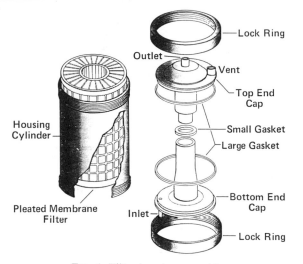

FIG. 4. Filter housing assembly.

a knot in the line to close it. Insert the coupler of the transfer set into the transfer pack.

10. Open the Y-type set clamps to allow saline to flow, keep the second Y-clamp closed.
11. Make sure the stopcock on the air vent is open.
12. Turn on the pump and adjust to 2 ml/min. Saline will begin to fill the filter housing.
13. Close the stopcock when saline appears.
14. Saline will then exit the filter to the transfer pack.
15. Wash the filter with 7 liters normal saline. The pump flow rate can be increased to 100 ml/min.
16. Observe the filter housing and connections for leaks.
17. Replace transfer packs as they fill, saving the final pack for microbiological testing.

Procedure 5: Final Stage Assembly and Priming of Filter and Cartridge

Materials

Drip chamber line and female–female connector from Patient Perfusion Pack (see Fig. 5).
2-liter transfer pack (1)
Normal saline, 1 liter

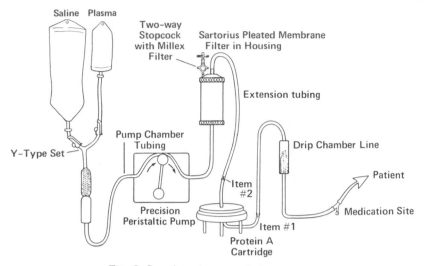

FIG. 5. Complete plasma perfusion system.

Sterile gloves (1 pair)
Small carpenter's bubble level (1)
Precision peristaltic pump (American Instrument Co.)

Method

1. Close the extension set from the filter outlet.
2. Connect this to tube item #2 of the PACC cartridge assembly.
3. Connect the drip chamber line to tube item #1.
4. Attach an empty transfer pack to the drip chamber line using the female–female connector.
5. Open all clamps on all lines and turn on pump at 20 ml/min.
6. Momentarily invert the drip chamber until the saline level reaches the mesh filter, then reinvert. This fills the drip chamber to the correct level. Place the drip chamber into the drip chamber clamp.
7. Wash the entire system with 1 liter normal saline. On completion turn the pump control fully counterclockwise to OFF position.
8. Place the bubble level on top of the PACC cartridge and check to see if it is level, adjusting the PACC cartridge if necessary. The PACC cartridge must be level if the plasma is to make uniform contact with the charcoal.
9. Disconnect the transfer pack and make the connection to the patient intravenously (iv).
10. The transfer pack is saved for microbiological testing.

Procedure 6: Patient Treatment

Method

1. Thaw patient plasma in a 37° water bath.
2. Connect the plasma bag to the second coupler of the Y-type set.
3. Close the clamp to the saline bag and open the clamp to the plasma bag.
4. Make sure all lines are open and check the iv connection to the patient.
5. Begin the perfusion by slowly, increasing to a final flow rate of 3–20 ml/min.
6. When the plasma bag empties, close the clamp to the plasma bag and open the clamp to the saline bag.
7. Flush 700 ml of normal saline through the system at the same flow rate as plasma was perfused.
8. Throughout the perfusion observe the system for leaks.
9. On completion of the normal saline flush, turn off the pump and close the drip chamber line.
10. Disconnect the drip chamber line from the patient and connect a slow normal saline drip to the patient iv to maintain patency.

It should be noted that tumoricidal responses in humans were also noted with an on-line plasma perfusion system.[9] The above-described off-line system may be directly connected to the plasma line of a conventional plasma–cell separator. In this mode, the host is initially connected to a plasma–cell separator device, and whole blood is partitioned into formed elements and plasma. After 15 min of such separation, to include formation of a large buffy coat layer, the plasma line may be opened to perfusion of the PACC column at flow rates of 10–20 ml/min. Approximately 100–200 ml of plasma may be perfused over the column by this method for each treatment.

Procedure 7: Cleaning and Sterilization of Equipment

Materials

Ultrasonic cleaner
Sterile distilled water, 10 liters
Alconox solution, 2%, in sterile distilled water

Method

1. Discard all tubing and the filter.
2. Disassemble filter housing and PACC cartridge assembly to their

component parts. Discard charcoal unless it is to be saved for testing.

3. Rinse all items in sterile distilled water.
4. Wash parts in an ultrasonic cleaner in a distilled water–Alconox solution for 10 min.
5. Thoroughly rinse all parts in sterile distilled water.
6. Dry all parts; a drying oven may be used, 66°, 3–4 hr.
7. Sterile wrap all items. Label as to contents and date.
8. Ethylene oxide sterilization is preferred but all items except the Sartorius filter may be steam autoclaved (121°, 15 psi, 15 min, fast exhaust). All items must be dry before storage.

Appendix: Radioimmunoassay for Detection of Enterotoxins in Commercial Protein A

The dilutions of rabbit antisera specific for enterotoxins A, B, C, E, and F required to precipitate 50% of ^{125}I-labeled enterotoxins ranged from 1 : 100 to 1 : 72,900. Dilutions of secondary antibody (IgG fraction of sheep anti-rabbit IgG Fc fragment) required to precipitate 50% of the rabbit IgG were determined and ranged from 1 : 2 to 1 : 16. For performance of the assay, enterotoxins are then iodinated by the chloramine-T technique.[34] Labeled enterotoxins (1.0 ml, 20,000–30,000 cpm), normal rabbit serum (0.1 ml, 1 : 100 dilution) or primary antibody dilution (0.1 ml), and test sample (0.1 ml) or enterotoxin standard (0.1 ml) are preincubated for 30 min at 37° in a volume of 0.9 ml with buffer (150 mM sodium chloride, 5 mM veronal, 10 mM EDTA, and 0.1% gelatin, pH 7.2). Test samples and standard enterotoxin solutions are diluted 1 : 3 to a final dilution of 1 : 27. After preincubation, secondary antibody is added, the mixture incubated for 30 min at 37° and an additional 12 hr at 4°. Precipitates are collected by centrifugation at 25,000 rpm for 30 min at 4°. Supernatants are decanted and tubes dried. ^{125}I replicates are counted in a Beckmann 5500 γ counter. A standard curve for each enterotoxin is prepared over a range of 0.133 ng to 100 ng/0.1 ml. Control samples containing normal rabbit serum instead of rabbit antienterotoxin sera (primary antibody) are used to determine nonspecific precipitation.

To determine whether the SpA present in the preparations or human albumin could alter the test results, HPLC-purified commercial protein A (5 μg/50 ml) or human albumin is added to ^{125}I-labeled enterotoxin A (SEA) and SEA-specific antisera in the preincubation phase. Control tubes receive ^{125}I-labeled enterotoxins and SEA-specific antisera without commercial protein A. Secondary antibody is added, the tubes are centrifuged, and precipitates are collected as described.

Our results have shown that the addition of protein A albumin does not alter the quantitation of SEA since binding values are comparable in control samples regardless of whether protein A or albumin was added. Percent inhibition is calculated from the formula:

$$\%\text{Inhibition} = 100\% - \frac{\text{precipitated cpm}}{\text{maximum precipitated cpm}} \times 100$$

The quantity of detectable enterotoxin in the unknown sample is derived from a standard inhibition curve in which known quantities of enterotoxins are added.

Acknowledgments

Supported by a grant (9R01 CA 28441) and a Research Career Development Award (5K04 AI 00302) from the National Institutes of Health.

[46] Immobilized Heparin Lyase System for Blood Deheparinization

By HOWARD BERNSTEIN, VICTOR C. YANG, CHARLES L. COONEY, and ROBERT LANGER

Extracorporeal devices perfused with blood (e.g., artificial kidney, pump oxygenator) have been an important part of clinical medicine for many years. These devices all rely on the patient being heparinized to prevent clotting in the device. However, heparinization of the patient often leads to hemorrhagic complications. With the prospect of longer perfusion times with machines such as the membrane oxygenator, the drawbacks of systemic heparinization are increased. Although a number of approaches have been tested to solve this problem, control of heparin levels in the blood remains a serious threat to human safety.

We have suggested a novel approach that would permit full heparinization of blood entering any extracorporeal device but which would enable enzymatic elimination of the heparin before the blood is returned to the patient. This approach consists of placing a blood filter containing immobilized heparinase (heparin lyase)—which degrades heparin into small saccharides—at the effluent of the extracorporeal device. We report here on the methods of heparinase production, purification, immobilization, and reactor design. A section on the assay procedure is provided initially, and we conclude with a brief summary of some of the results obtained.

Assays

Protein is measured by the method of Bradford[1] or Lowry et al.[2] Amino acid concentrations are measured by the method of Hill et al.[3] consisting of reverse-phase separation and fluorescence detection of the amino acids as their o-phthaldialdehyde/ethanethiol derivatives. A high-performance liquid chromatograph (HPLC) (Waters Associates) consisting of two Model 600 A pumps, a Model 660 Solvent Programmer, and a U6K injector was used. The Model FS-970 Liquid Chromatographic Fluorometer was from Schoeffel Instrumental Corp. The reversed-phase column was a μBondapak C_{18} (Waters Associates).

Heparinase activity is followed in vitro in two ways: (1) the disappearance of heparin, and (2) the appearance of heparin degradation products. The former is measured by the metachromatic shift of Azure A from blue to red in the presence of heparin, whereas the latter is measured by the increase in ultraviolet absorption at 232 nm due to the formation of α,β-unsaturated uronides. Because contaminating enzymes such as glycuronidase can act on degradation products, resulting in a loss of chromophores measured, the UV 232 nm assay is utilized only on relatively pure preparations of heparinase.

Reagents

0.25 M sodium acetate, 2.5 mM calcium acetate, adjusted to pH 7 with acetic acid (assay mixture).

Substrates. Heparin (from porcine intestinal mucosa, 157 USP units/mg, Hepar International, Columbus, Ohio), heparan monosulfate (kindly supplied by Dr. Cifonelli of Wyler Children's Hospital, Chicago), Chondroitin 4- and 6-sulfates, dermatin sulfate, and hyaluronic acid (Sigma) are prepared at 25 mg/ml in assay mixture, with the exception of hyaluronic acid, which is prepared at 7 mg/ml owing to the high viscosity of this compound.

Azure A Assay. Fifty microliters of enzyme solution, 50 μl of assay mixture, and 50 μl of substrate solution are combined and incubated at 37°. At various times, 10 μl of sample is added to 10 ml of Azure A (Fisher Scientific, prepared at 0.02 mg/ml in distilled water and adjusted to pH 7).

---

[1] M. M. Bradford, Anal. Biochem. 72, 248 (1976).
[2] O. H. Lowry, N. J. Rosebrough, A. L. Farr, and R. J. Randall, J. Biol. Chem. 193, 265 (1951).
[3] D. W. Hill, F. H. Waler, T. D. Wilson, and J. D. Stuart, Anal. Chem. 51, 1338 (1979).

The change in absorbance is measured at 620 nm and compared with a standard curve of 0–8.3 mg of heparin/ml of assay broth.[4]

UV 232 nm Assay. One hundred microliters of enzyme solution, 500 μl of assay mixture, and 300 μl of substrate solution are combined and incubated at 37°. At various times, aliquots of 50–70 μl are withdrawn in duplicate and quenched in 1.5 ml 0.03 N HCl. The increase in absorbance at 232 nm is measured.[5] A molar extinction coefficient of 5,100 cm^{-1} M^{-1}[6] is used to determine the amounts of α,β-unsaturated uronides formed. One unit of enzyme activity is defined as the amount of enzyme which degrades 1 mg of heparin per hour (international units are not used due to difficulty in relating metachromatic shift of the dye to moles of bonds broken).

For the immobilized enzyme, the support material is assayed for heparinase activity using a modified 232 nm assay. To 1.5 ml of a 25 mg/ml heparin solution (0.25 M sodium acetate, pH 7.4), 300 μl of gel suspension (1 part bead : 2 parts buffer) is added. The mixture is well agitated in a water bath and 100-μl aliquots in duplicate are withdrawn at time intervals of 0, 10, and 20 min and quenched in 1.5 ml of 0.03 N HCl. The increase in absorbance at 232 nm is determined.

For studies of heparinase activity in blood or *in vivo*, normal plasma is prepared from citrated whole blood samples (9 : 1 v/v whole blood to 3.8% by weight trisodium citrate) by centrifugation at 3000 g for 20 min at 4°. After centrifugation, the top 1.5 ml of plasma is pipetted off, stored in polypropylene vials on ice, and assayed within 2 hr. The loss of heparin's biological activity through the action of heparinase is tested by a number of available assays including aPTT,[5] which involves the measurement of heparin in plasma by measuring the clotting time, and the chromogenic Factor Xa assay, which determines the inhibition of factor Xa cleavage of an artificial substrate by heparin.[5] By using these two assays which measure different facets of heparin's biological activity, the possibility of artifacts is reduced.

Activated Partial Thromboplastin Time (aPTT). A test tube containing 100 μl of Activated Cephaloplastin Reagent (Dade Diagnostics Inc., Aguada, Puerto Rico) prewarmed to 37° is incubated with 100 μl of the plasma sample for 2 min. One hundred microliters of 20 mM CaCl$_2$ is then added and after 25 sec, a platinum innoculating loop is drawn through the mixture until a clot is formed, and the time is recorded. The aPPT assay is linear in the range of heparin concentrations of 0–0.6 units/ml.

[4] A. C. Grant, R. J. Linhardt, G. Fitzgerald, J. J. Park, and R. Langer, *Anal. Biochem.* **137**, 25 (1984).

[5] R. Linhardt, A. Grant, C. L. Cooney, and R. Langer, *J. Biol. Chem.* **257**, 7310 (1982).

[6] A. Linker and P. Hovingh, *Biochemistry* **11**, 563 (1972).

Factor Xa Inhibition. To a test tube, 800 μl of 0.2 *M* Tris buffer, pH 7.4, 100 μl of antithrombin (1.0 IU/ml, Kabi Diagnostica, Stockholm, Sweden), and 100 μl of the plasma sample are added. To three test tubes (two test and one blank tube) are transferred 200-μl aliquots of the test sample. These tubes are incubated for 2 min, and to the test tube is added 100 μl of Factor Xa (7 nkat/ml, Kabi Diagnostica); after 30 sec, 200 μl of artificial substrate S-2222 (0.75 mg/ml in distilled water, Kabi Diagnostica) is added. After an additional 3 min, 300 μl of 50% acetic acid is added to the test and blank tubes. To the blank tube, 300 μl of distilled water is added. The absorbance of the test samples are read against the blank at 405 nm. The assay is linear up to 0.8 units/ml.

Production of Heparinase

Analytical Determinations. Samples of *Flavobacterium heparinum* to be assayed for heparinase are centrifuged at 4° for 10 min at 12,800 g. Cell pellets are immediately sonicated at a cell protein concentration of 8 mg/ml as previously described[7] and frozen for later assays. Cell pellets are resuspended in 5 ml potassium phosphate buffer (0.15 *M*, pH 7.0). Heparinase activity was measured using an Azure A dye assay.[7]

Microorganisms. *Flavobacterium heparinum* is a small soil bacterium, gram-negative, rod shaped (1.0 × 0.3 μm), nonmotile, and non-spore forming. We use a strain obtained from Dr. Alfred Linker (Veterans Administration Hospital, Salt Lake City, UT). The microorganism is grown at 23° or at room temperature. The culture is stored on agar slants of defined medium as previously described.[7,8] Media for culture storage are the same as for shake flask cultivation (described below) except that the glucose concentration is 4.0 g/l.

Shake Flask Cultivation. Experiments are conducted in 0.5-liter sidearm baffled shake flasks with 50 ml of medium. The "low-sulfate" medium consists of 10^{-4} *M* trace salts ($CaCl_2 \cdot 2H_2O$, $FeCl_2 \cdot 2H_2O$, $CuCl_2 \cdot 2H_2O$, $Na_2MoO_4 \cdot 2H_2O$, $CoCl_2 \cdot 6H_2O$, $MnCl_2 \cdot 4H_2O$) and the following compounds (g/liter): glucose (sterilized separately), 8.0; NH_4Cl, 2.0; K_2HPO_4, 2.5; Na_2HPO_4, 2.5; $MgCl_4 \cdot 6H_2O$ (sterilized separately), 0.5; L-histidine (Sigma), 0.2; L-methionine (Sigma), 0.2; P-2000 antifoam, 4 drops/liter. The low-sulfate medium does not contain any heparin. Glucose and magnesium salts are autoclaved separately and added to the

[7] P. M. Galliher, C. L. Cooney, R. Langer, and R. J. Linhardt, *Appl. Environ. Microbiol.* **41**, 360 (1981).
[8] P. M. Galliher, R. J. Linhardt, L. J. Conway, R. Langer, and C. L. Cooney, *Eur. J. Appl. Microbiol. Biotechnol.* **15**, 252 (1982).

sterilized medium; L-histidine and L-methionine are added after sterilization by filtration through 0.22-μm Millipore syringe filters. Trace salts are dissolved and stored at 25 times concentrated at pH 2.0, with 0.5 ml/liter of concentrated H_2SO_4 or HCl. After sterilization, the pH of both media are 7.4 ± 0.2.

A 4–6% inoculum in exponential growth is used. Cultures are incubated at room temperature on a 2.5 cm stroke shaker at 200 rpm. Growth is measured by turbidity in a Klett-Summerson colorimeter with a red filter (#66); the conversion factor is 250 Klett units per g/liter dry cell weight. Growth is allowed to continue until 400–500 Klett units is achieved during which time the pH falls to 5.0 ± 0.2. Flasks are then harvested for enzyme assays, or used to innoculate other flasks or a fermentor. In all low-sulfate medium experiments, deionized distilled water is used, and innocula are transferred several times in low-sulfate medium to avoid sulfate contamination from the agar. A trace analysis of the low-sulfate medium done by Galbratih Laboratories Inc. (Knoxville, TN) indicated that sulfur contamination of the medium was $4.5 \times 10^{-5} M$ sulfur in the form of sulfate and sulfite.

Fermentor Cultivation. A 2-liter fermentor (with a 1.5-liter working volume) is equipped with controlled agitation, aeration, pH, temperature, and dissolved oxygen. The fermentor media are the same as above, except that KH_2PO_4 (1.0 g/liter) and NaH_2PO_4 (1.0 g/liter) are substituted for K_2HPO_4 and Na_2HPO_4. The fully equipped fermentor is autoclaved with the basal salt medium at pH 5 for 30 min at 121°. Amino acids are sterilized by filtration as described above. Aeration and agitation are controlled to maintain a dissolved oxygen level of about 50% of air saturation. Temperature is controlled at 23 ± 1° and pH at 7.0 ± 0.2 with NH_4OH (12%) addition.

Production of Heparinase in Low-Sulfate Medium. Growth of *F. heparinum* and heparinase production in the low-sulfate medium is carried out in a 1.5-liter (working volume) fermentor (Fig. 1). Initial nutrient concentrations are 10 g/liter for glucose, 0.2 g/liter each for L-histidine and L-methionine, and 0.5 g/liter for NH_4Cl. At 33 hr, more glucose (5 g/liter) is added. L-Histidine and L-methionine (0.1 g/liter each) are added at 35 hr. The innoculum consists of a culture grown in the low-sulfate medium.

There is an initial lag phase, during which cells synthesize heparinase. Just before exponential growth, at a specific growth rate of 0.22 hr^{-1}, heparinase activity falls; during the rapid exponential growth phase, both heparinase specific and volumetric activities decrease. At 0.75 g DCW/liter (dry cell weight/liter), growth slows to a specific growth rate of 0.10 hr^{-1} and later, at 2 g DCW/liter, to 0.04 hr^{-1}. As the growth rate de-

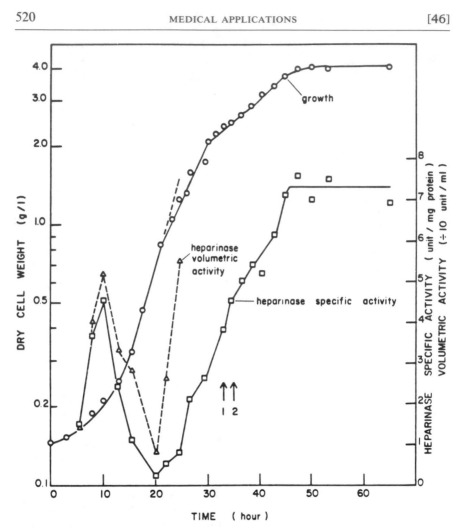

FIG. 1. Heparinase production by growth of *F. heparinum* in low-sulfate medium in a 1.5-liter fermentation. Arrows 1 and 2 show the point of addition of glucose and amino acids, respectively.

creases, glucose and amino acids are added. At the onset of slow growth, production of heparinase starts and continues until the stationary phase. Heparinase specific activity reaches a maximum value of 7.5 units/mg protein and remains stable during the stationary phase. Higher specific activity of heparinase can be achieved by growth in the same medium with or without additional sulfate if heparin (1 g/liter) is added as an inducer.[7] However, the use of heparin is very expensive and is undesirable in a large-scale production situation.

Purification of Heparinase

In addition to heparinase, *Flavobacterium heparinum* contains a variety of other catalytic enzymes such as chondroitinases, hyaluronidase, and heparitinase which act on heparin-like glycosaminoglycans, as well as sulfatases and glycuronidases which act on the heparin degradation products. Although heparinase can be purified to homogeneity by successive column chromatographic procedures with an increase of the specific activity of 300-fold,[9] both the overall mass yield (0.003%) and the total activity recovery (0.8%) are too low to satisfy the high demand for blood deheparinization studies. Moreover, these procedures are time consuming and are not well suited for the preparation of large quantities of the enzyme. The major objective in the enzyme purification has therefore been focused on developing a method conducive to rapidly producing large quantities (grams) of catalytically pure enzyme (i.e., free of the other catalytic enzyme contaminants described above). This has been accomplished by converting two successive chromatographic procedures, hydroxylapatite and QAE–Sephadex, from a column technique to a batch process.

Preparation of the Crude Cell Extract. One hundred grams (dry weight) of flavobacterial cells is suspended in 2 liters of cold 10 mM sodium phosphate buffer (pH 6.8) and then homogenized at 5,000 psi for 8–10 min (about 3–4 passages) through a Gaulin Model 15 M homogenizer thermostattically controlled at 4° by circulating ice water. To the homogenate, 7.5 g of protamine sulfate is added. The cell debris together with the precipitated nucleic acids are removed by centrifugation at 4°, 9,000 g for 15 min. The supernatant thus obtained is defined as the crude cell extract.

Hydroxylapatite Purification. Two liters of the crude cell extract (containing a protein concentration of about 30 mg/ml) is batch loaded onto 240 g of hydroxylapatite (BioGel HTP, Bio-Rad) preequilibrated with 10 mM sodium phosphate buffer (pH 6.8). The hydroxylapatite-bound protein is then washed stepwise with 2 liters of buffers of increasing ionic strength. The solutions used for the sequential, stepwise elutions are prepared by combining 10 mM sodium phosphate buffer (pH 6.8) with 0.25 M sodium phosphate, 0.5 M NaCl buffer (pH 6.8) according to a ratio (v/v) of 8/0, 7/1, 6.5/1.5, 6/2, 5.5/2.5, 5/3, 4/4, 3/5, 2/6, 1/7, and 0/8, respectively. The hydroxylapatite particles are then removed by centrifugation at 3,000 g for 3 min, and the supernatant is assayed for heparinase activity and protein concentration.

[9] V. C. Yang, R. Linhardt, H. Bernstein, C. L. Cooney, and R. Langer, *J. Biol. Chem.* **260,** 1849 (1985).

Heparinase activity elutes over a broad range of ionic strength ($I = 0.22-0.42$), with an activity maximum occurring at $I = 0.29$ (the 6/2 wash). Enzymatic assays indicated that this preparation is not catalytically pure and contains low amounts of activities attributed to the other catalytic enzyme contaminants. The preparation is stable when stored either lyophilized or frozen at $-20°$, and it maintains more than 90% of activity after storage in the freezer for 6 months.

QAE–Sephadex Purification. Two liters of the enzyme preparation (containing a protein concentration of 0.2–0.4 mg/ml), obtained from the hydroxylapatite chromatography in the washing steps ranging from 6.5/1.5 to 5.5/2.5, is dialyzed for 4 hr against 100 volumes of 10 mM phosphate buffer (pH 6.8) and then added to 10 g of QAE–Sephadex A-50 (Pharmacia). The mixture is gently shaken at $4°$ for 10–15 min, and the gel particles are removed by centrifugation at 3,000 g for 3 min. The supernatant is assayed for heparinase activity and protein concentration.

In spite of the fact that heparinase is a basic protein with an isoelectric pH of 8.5,[9] anion-exchange chromatography based on a negative adsorption on QAE–Sephadex is used after the hydroxylapatite purification for several reasons. First, the predominant fraction (>70%) of the protein in the hydroxylapatite preparation is acidic (as judged by isoelectric focusing), and can be removed in a single operation. Second, heparinase binds to the cation-exchanger strongly, possibly leading to difficulties in subsequently eluting it. Third, the use of negative adsorption avoids tedious washing steps required by conventional ion-exchange chromatography. After QAE–Sephadex purification, the specific activity is enriched 3- to 4-fold over that obtained from the hydroxylapatite purification, and a total activity recovery as high as 90% is found.

Table I summarizes the results of the large-scale heparinase purification. With the adaption of batch procedures, gram quantities of heparinase can be purified in 4–5 hr. The enzyme is purified 25-fold over the cell homogenate. The specific activity of the QAE-purified enzyme is

TABLE I

PURIFICATION OF HEPARINASE

Purification step	Total protein (g)	Specific activity (units/mg)	Total activity (units/10^3)	Recovery of activity (%)
Cell homogenate	60	10	600	
Crude extract	40	13	520	86.7
Hydroxylapatite	4.8	75	360	60
QAE–Sephadex	1.2	240	288	48

about 250 units per mg of protein. The recovery of heparinase activity is about 50%. In addition, the resulting enzyme preparation is nearly free of the catalytic enzyme contaminants. The relative activities of these contaminating enzymes, as compared to the heparinase activity (100%) on the basis of absorbance change at 232 nm per hour, are as follows: chondroitinase B 0%, chondroitinase AC <4%, hyaluronidase <4.2%, heparitinase <4.5%. Over a period of 1 hr, less than 2.5% of the sulfate groups is desulfated by sulfatases, and only 1–2.8% of the unsaturated uronide chromophores is degraded by glycuronidase.

Immobilization of Heparinase

In order for heparinase to work effectively in removing heparin and not to produce any immunological problems, we felt that it should be covalently immobilized to a support material. This facilitates enzyme recovery and increases stabilization. Our studies involved (1) attempting to understand what coupling techniques, support materials, and reaction conditions are most effective in immobilizing heparinase, (2) using a model system to explore the feasibility of immobilized heparinase with respect to stability and activity recovery. Many immobilizations have been tried. The supports giving the highest level of immobilized activity include the following: CNBr-activated Sepharose 4B (50–91% activity immobilized), tresyl chloride-activated Sepharose (25% activity immobilized), Sephadex (5–56%), and cellulose from hollow fiber hemodialyzers (4%); carbodiimide- or active ester-activated CM–Sephadex (4%), CM–cellulose (1%), and polyacrylamide (PAN) (36%); and epoxy-activated (oxirone) acrylic beads (1%).

It appears that macroporous supports, possibly because of their large surface area, increase the levels of total activity that can be recovered on immobilization. On the other hand, highly negatively charged supports result in very low levels of immobilized activities even when a large amount of protein is immobilized. The reduced activity observed may be due to the electrostatic repulsion between the support and the highly negatively charged substrate, heparin.

The support material that yielded the highest activity retention is Sepharose 4B. The gel is made up of spherical particles with diameters in the swollen state of 60–140 μm. However, because of its low mechanical strength, some of the limitations of the 4% agarose gel are (1) the nonautoclavibility of the gel posing sterilization problems and (2) the highly deformable nature of the gel resulting in the inability to withstand high operating pressures and in gel deterioration under high shear conditions. These problems became apparent after extensive testing of the beads in a sheep animal model at high blood flow rates (250 cm^3/min.). Beads were found downstream of the reactor in the filter trap before the animal which is normally used to prevent blood clots from entering the animal. When the beads were examined using high power light microscopy, numerous fractures were observed. The surfaces of the beads were no longer smooth, and the particle-diameter size distribution had shifted to a lower mean value. The recovery of beads was always less than 20% of that initially primed to the reactor.

A search for an agarose-based support that is mechanically stronger yet at the same time permits high retention of heparinase activity was initiated. The strength of the beads was increased in two independent ways: first, by increasing the percentage of agarose in the gel and, second, by cross-linking the beads with 2,3-dibromopropanol. Studies were conducted with Sepharose CL-4B (Pharmacia), Sepharose CL-6B (Pharmacia), BioGel 8% agarose (Bio-Rad), and BioGel 8% agarose cross-linked with 2,3-dibromopropanol to determine whether heparinase could be immobilized with a high degree of activity retention and if the beads could mechanically withstand the *in vivo* operating conditions. For Sepharose CL-4B, Sepharose CL-6B, and non-cross-linked BioGel 8%, bead fracturing with resultant leakage from the reactor was comparable to that observed with Sepharose 4B. After perfusion, the bead recovery rates approached 50–60% and the beads had a lower mean particle diameter. However, with the BioGel 8% cross-linked agarose (150–300 μm diameter), the recovery was 100% with no significant change in particle size.

Preparation of Cross-Linked 8% Agarose. BioGel 1.5 (500 ml, 8% agarose, Bio-Rad) is washed with distilled water to remove preservatives and then resuspended in 500 ml of 1 *M* sodium hydroxide. The mixture is heated to 50°, and 10 ml of 2,3-dibromopropanol (Aldrich) is added. The reaction is continued at 50° for 1 hr, and after that an additional 10 ml of the cross-linking agent is added. The reaction is allowed to continue for an additional hour, and the gel is removed by filtration. The gel is resuspended in 500 ml of 1 *M* sodium hydroxide, and the above procedure is repeated for a second time. The agarose is then filtered and washed with 15–20 liters of distilled water. A sample of known volume of the gel is withdrawn, autoclaved for 30 min, and the volume of gel after the autoclaving procedure is determined. If less than 100% of the gel has survived the autoclaving procedure, the cross-linking step is repeated. The beads are stored at 4° in physiological saline containing 0.02% sodium azide.

Gel Activation with Cyanogen Bromide (CNBr). The 8% cross-linked agarose (100 ml) is washed with distilled water to remove the azide preservative. The beads are resuspended in 100 ml of chilled distilled water and 200 ml of chilled 2 *M* sodium carbonate. To the agarose suspension, 40 ml of CNBr in acetonitrile is added (1 g CNBr/ml of acetonitrile), and the mixture is vigorously stirred in a fume hood for 5 min. The reaction mixture is quickly filtered and washed with 1000 ml of chilled distilled water and then 500 ml of chilled 1 mM HCl. A 150-mg sample of suction-dried beads is assayed for the degree of activation according to the method of Kohn and Wilchek.[10] If the gel contains between 10–20 μmol of cyanate esters/g of resin, it is suitable for immobilizing heparinase.

Coupling Procedure. One hundred milliliters of the gel is washed with 500 ml of 0.1 M NaHCO$_3$, 0.5 M NaCl, pH 8.3, and suctioned-dried. The beads are mixed immediately with 400 ml of enzyme which contains 40 mg heparin/mg protein. The mixture is placed on an orbital shaker, and coupling proceeds for 20–24 hr at 4°. The suspension is filtered, resuspended in 300 ml of NaHCO$_3$ buffer (0.1 M, pH 8.3) containing 0.5 M NaCl and 0.2 M lysine, and incubated at 4° for 12 hr to block remaining cyanate esters. The beads are washed to removed noncovalently attached protein with 750 ml of chilled buffer containing 0.25 M sodium phosphate, 0.5 M NaCl, pH 7.0, and then with 750 ml of chilled buffer containing 72.5 mM sodium phosphate, 125 mM NaCl, pH 7.0, and are resuspended in this buffer. The washes are collected to assay for both protein concentration and enzymatic activity. About 85–95% of protein is immobilized with a retention of 50–90% of the heparinase activity.

Reactor Fabrication and Priming

Two 3/4 inch holes are drilled on either side of the outlet port of a Bentley AF-1025 arterial blood filter. The ends of a 20 inch piece of silastic tubing (Cole Palmer #6418) are inserted into the two holes and sealed under vacuum with silicone RTV gel. The tubing serves as a source of agitation to maintain the gel in a fluidized state. The inlet of the filter is connected to a peristaltic pump via silastic tubing, and the system is vented as sterile physiological saline (Abbott Labs.) is pumped through the system. After all air has been purged, the immobilized heparinase bead slurry is slowly pumped into the reactor. The reactor is ready to be hooked up to the animal.

In Vivo Testing Procedure

The reactors are tested in a manner as close as possible to the clinical situation of kidney dialysis using sheep as an animal model. Blood access is provided by arterial–venous shunts installed in the carotid artery and the jugular vein. A 100-liter Travenol tank dialysis unit with a negative

[10] J. Kohn and M. Wilchek, *Enzyme Microb. Technol.* **4,** 161 (1982).

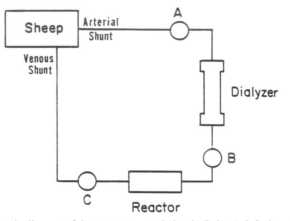

FIG. 2. Schematic diagram of the extracorporeal circuit. Point A, infusion site of heparin. Point B, sampling port before reactor. Point C, sampling port after reactor.

pressure converter is used. From the animal the blood passes through the arterial tubing, the artificial kidney, a venous tubing line with thrombus trap, the reactor, and back to the animal through a small venous tubing equipped with a thrombus trap. Heparin is continuously infused into the bubble trap located before the artificial kidney using a syringe infusion pump. The animal is secured to the sideboard of a wooden platform via two belt straps and is conscious throughout the whole procedure.

An intravenous injection of heparin (100 units/kg) is administered through the venous shunt to systemically heparinize the animal. The arterial tubing is hooked to the animal's arterial shunt, and the blood pump is started at a flow rate of 250 ml/min. The saline used to prime the system is discarded except for the last 50 ml. Blood samples are withdrawn through two sampling ports, one before and one after the reactor. A schematic of the system is shown in Fig. 2. At selected time intervals, two 3-ml whole blood samples are drawn in EDTA for complete blood cell counts and in sodium citrate for platelet counts and for the heparin assays. The samples for heparin determination are kept on ice. Cell counting is done on a Coulter counter while platelet counts are performed manually using a light microscope.

Results and Discussion

Properties of Immobilized Heparinase. The activity retained based on activity bound is between 50–90%. The beads are stored in sterile phosphate-buffered saline (Abbott Labs.) at 4° until they are used. Heparinase immobilized to agarose via cyanogen bromide has an enhanced thermal stability. The half-life of deactivation at 4, 25, and 37° is increased to 3600, 125, and 15 hr, respectively, compared to free enzyme values of 125, 30, and 1 hr. These stability results indicate that the immobilized enzyme should be adequately stable under conditions of storage and operation. The immobilized enzyme exhibits an activity profile broadened over a larger temperature range than the free enzyme and a shift in activity maximum from 30 to 37°.

In Vivo Results. Preliminary studies have shown that a conventional blood filter containing immobilized heparinase can remove over 90% of heparin's anticoagulant activity in dogs within minutes.[11] However blood flow rates were at most 50 ml/min, considerably less than the 250–300 ml/min expected in the clinical situations of chronic kidney dialysis or pediatric extracorporeal membrane oxygenator. The objectives of the more recent work have been to design and test a reactor which is capable of degrading clinically used amounts of heparin at flow rates of up to 250 ml/min. For scale-up purposes sheep were selected since comparable blood volumes and flow rates to those in humans are attainable.

The reactor is being tested for efficiency, hemocompatibility, and safety. Studies have been conducted with AF-1025 blood filters (American Bentley Laboratories) containing 50–75 ml of 8% cross-linked agarose beads loaded with 0.4–0.6 mg/ml of protein. Single pass conversion of heparin has varied between 15 and 45% as determined by both the aPTT clotting assay and the Factor Xa chromogenic assay. No hemolysis is observed over the 1-hr runs that are currently being done. The white blood cells drop to 40% of their initial value within the first 20 min of the treatment and return to 90–100% by the end of the experiment, similiar to what is observed under normal kidney dialysis. The platelet values drop to 40–50% of their initial values for very active reactors while only to 60–80% for less active reactors. The pressure drop across the reactor starts off at 10–20 mmHg and rises to 40–80 mmHg by the end of the experiment.

A number of toxicology studies have been conducted. In the studies with dogs in which the blood was exposed to the immobilized heparinase filter for 90 min, blood was drawn from the dogs 1, 2, and 5 months after the experiments, and no antibodies to heparinase were detected. In addition, products of the enzymatically degraded heparin were tested for cytotoxic immunogenic effects on *Salmonella typhimurium*. No toxicity or mutagenicity was observed even with concentrations 1000 times in excess of those anticipated clinically.[12]

[11] R. Langer, R. Linhardt, S. Hoffberg, A. K. Larsen, C. L. Cooney, D. Tapper, and M. Klein, *Science* **217,** 719 (1982).
[12] H. Bernstein, V. C. Yang, D. Lund, M. Randhawa, W. Harmon, and R. Langer, *Kidney Int.* **32,** 452 (1987).

Studies have also been done to examine the rate at which the heparin fragments are eliminated from the body. If the heparin fragments produced by the heparinase filter during extracorporeal treatment remain in the body for prolonged time period, the potential value of the blood filter would be limited. However, studies using both anticoagulant tests and radiolabeled heparin and heparin fragments showed that heparin fragments are excreted at a much more rapid rate than parent heparin in normal rats and in nephrectomized rats (nephrectomized rats are used because kidney dialysis is one of the targets for the filter).[13] In addition, histopathological studies have been conducted to examine the fate of a variety of different organs in rats when given heparin and heparin products. Although the addition of heparin causes bleeding in the lungs of those rats given heparin, no toxic effects were observed as judged by histopathology to the lung or any other organs dosed with identical amounts of heparin degradation products.[14]

Future studies will be directed toward developing purer heparinase using efficient procedures amenable for scale-up. This will be approached both by fermentation using genetic engineering to develop improved bacterial strains and by adapting the purification approaches that are currently being used to make completely pure heparinase.[9] We will also be examining other linkages to combine heparinase to support materials to obtain the most stable bond possible. In addition, other support materials may also be examined. A major focus will be on developing the most optimal reactor for the flow rates that we are currently considering (250 ml/min). We may also attempt to develop systems that can operate at even higher flow rates such as those used in open heart surgery. Finally, extensive animal testing, examining in detail such issues as reactor efficiency, complement activation, and antibody formation, will be conducted.

An immobilized heparinase filter could be used in a variety of clinical situations. For example, it might be used at the end of a clinical procedure to eliminate heparin without the toxic effects of heparin-neutralizing substances such as protamine. Alternatively it might be used continuously in such situations such as kidney dialysis, continuous arteriovenous hemofiltration, or extracorporeal membrane oxygenators, to prevent high levels of heparin from ever entering the patient. Blood filters as large as 2 liters have been used at the effluent of extracorporeal devices to remove microemboli. Heparinase might be bonded to the biomaterials of these filters. Unlike many proposed applications of immobilized enzymes, heparinase could be used in procedures in which blood must enter the extra-

[13] A. K. Larsen, S. Hetelkidis, and R. Langer, *J. Pharmacol. Exp. Ther.* **231,** 373 (1984).
[14] A. K. Larsen, P. M. Newberne, and R. Langer, *Fund. Appl. Toxicol.* **7,** 86 (1986).

corporeal circulation anyway and where existing biomaterials already interface with blood at the desired locations (the end of the extracorporeal circuit). Thus the heparinase reactor may require no additional apparatus or invasive procedure.

Finally, the heparinase filter may also be a prototype for other enzymatically catalyzed removal systems. The advantage of using enzymes are that they are inexpensive and the products are continuously removed, therefore causing the catalyst to be regenerated (this would not be the case if an antibody or ion-exchange resin were used). Thus, the opportunity to have small efficient reactors using enzymes is possible. The heparin removal system may be a useful prototype for designing reactors to remove other undesirable substances. Examples of these include bilirubin, amphetamines, barbiturates, digitalis, and low density lipoprotein. It is hoped that the methods developed here, while useful for the heparinase filter in their own right, may also serve as useful guidelines for the kinds of studies that could be done for other immobilized enzymes systems to be used in blood purification.

Acknowledgments

This work was supported by National Institutes of Health Grant GM 25810. We thank Dr. Robert Linhardt, Dr. Michael Klein, Dr. David Tapper, Parrish Galliher, Edith Cerbelaud, Margaret Flanagan, Annette Larsen, Cynthia Zannetos, Arthur Grant, Gerald Fitzgerald, and Steven Hoffberg.

[47] Clinical Applications of Urokinase-Treated Material

By TAKESHI OHSHIRO, MOU CHUNG LIU, JUNICHI KAMBAYASHI, and TAKESADA MORI

The need for antithrombogenic material has recently increased in the medical field, and various materials have been developed to meet this requirement. However, they are not always satisfactory when used for long periods of time *in vivo*, because of poor antithrombogenic properties. In fact, Formaneck et al.[1] reported that thrombus formation in catheters in the Seldinger method reached 50% or more after only a few minutes, and we[2] found that the rate of thrombus formation in nylon filament retained in inferior caval vein of rats for 5 min was 91.7%. Thrombus

---

[1] G. Formaneck, R. S. Frech, and K. Amplatz, *Circulation* **41,** 833 (1970).

[2] T. Ohshiro and G. Kosaki, *Artif. Organs* **4,** 58 (1980).

formation is caused not only by the material used but also by the blood circulated, because blood has a high coagulability in contact with foreign material. Factor XII (contact factor) is readily activated by a negative charge, the coagulation system comes into action, and fibrin is formed. Taking this into account, new materials are being manufactured by selecting a polymer which is physicochemically stable.

How does the circulating blood maintain fluidity in a blood vessel? It is the reason why the coagulation system and the fibrinolysis system are balanced under the dynamic equilibrium. The epithelial layer of the vessel becomes a trigger for accelerating the coagulation system and a trigger for accelerating the fibrinolysis system, so that the fibrin monomer, precipitated on the epithelial layer, is simultaneously lysed. In this phenomenon, urokinase plays an important role. This is why we have utilized urokinase-treated material in our study of antithrombogenic medical material. The initial experiment on immobilized urokinase was reported by Kusserow and Larrow[3] in 1972. We[4] have also done basic experiments since 1975 and we[5] began clinical applications in 1978. The basic experiments are summarized as follows: (1) Immobilized urokinase possesses fibrinolytic activity. (2) Immobilized urokinase does not lose the fibrinolytic activity even after storage. (3) Immobilized urokinase neither liberates nor splits easily from a carrier. (4) Effective enzyme activity is approximately 30%, as compared to that of soluble urokinase. (5) Increase of binding sites results in increase in immobilization amount of urokinase. (6) The binding form on the surface of the carrier comprises 20–30% covalent binding and 70–80% stable ionic binding. (7) Urokinase-treated material exhibits antithrombogenic ability *in vivo* and could be an excellent antithrombogenic medical material. This chapter includes clinical applications of urokinase-treated material.

Materials

One material is an ethylene–vinyl acetate copolymer (Evatate), which is used for intravenous catheters, 1.5 mm in diameter. Another material is polyvinyl chloride (PVC), which is used for drainage tubes, 9 mm in diameter.

Immobilization Procedure of Urokinase

Urokinase is immobilized as shown in Fig. 1. The surface of Evatate is subjected to saponification and aminoacetalization followed by peptide

[3] B. K. Kusserow and R. Larrow, *Circulation, Suppl.* **2**, 54 (1972).
[4] T. Ohshiro, K. Mukai, and S. Koh, *J. Med. Enzymol.* **1**, 72 (1975).
[5] T. Ohshiro, A. Takahashi, and K. Mukai, *Artif. Organs (Jpn.)* **7**, 210 (1978).

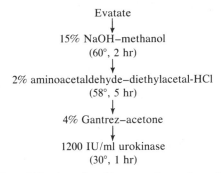

FIG. 1. Immobilization of urokinase on the surface of Evatate.

linking of urokinase via maleic anhydride–methyl vinyl ether copolymer (Gantrez) which is also linked.

The immobilization procedure is as follows: Evatate is (1) etched by 15% NaOH–methanol solution at 60° for 2 hr, (2) dipped in 2% aminoacetaldehyde–diethylacetal-HCl solution at 58° for 5 hr, (3) dipped in 4% Gantrez–acetone solution, (4) dipped in 1200 IU/ml of urokinase solution at 30° for 1 hr, and (5) dried after washing with 1 M NaCl. Polyvinyl chloride is linked with urokinase via Gantrez which is membranously covered. Urokinase immobilized on the surface of materials increases in quantity in accordance with the concentration of urokinase solution. Figure 2 shows this relationship in a polyvinyl chloride tube. Immobilization reaches a plateau (22.8 ± 0.9 IU/cm^2 of tube) in 5000 IU/ml of urokinase.

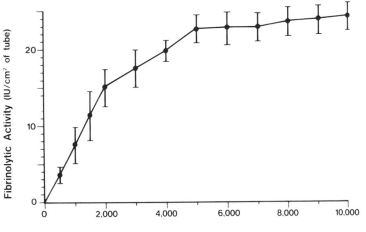

FIG. 2. Relationship between fibrinolytic activity of tube and concentration of urokinase.

Assay of Urokinase Activity

Urokinase activity is measured by the MCA method,[6] shown in Fig. 3. Peptide–MCA (5.3 mg, 10.6 μmole) is dissolved in 1 ml of dimethyl sulfoxide, and the solution is diluted to give a final concentration of 0.1 mM, using 50 mM Tris–HCl buffer, pH 8.0, containing 100 mM NaCl and 10 mM CaCl$_2$. Soluble or immobilized urokinase is added to 1 ml of the peptide–MCA solution and incubated at 37° for 10 min. Urokinase hydrolyzes a peptide–MCA substrate, releasing 7-amino-4-methylcoumarin (AMC) which is highly fluorescent. The enzyme reaction is stopped by adding 1.5 ml of 17% acetic acid. The rate of AMC is measured fluorometrically with excitation at 380 nm and emission at 460 nm, and the amount is estimated photometrically at 370 nm. The fibrinolytic activity of urokinase-treated material was 31.6 ± 4.2 IU/10 cm of tube in the catheter and 9.3 ± 1.1 IU/cm^2 of tube in the drain. The residual fibrinolytic activity after use is expressed by percentage of initial fibrinolytic activity.

Stability of Immobilized Urokinase

It has been reported that the immobilized enzyme is stable. Here, we investigated the stability of immobilized urokinase.

Stability in Storage

Urokinase-treated Evatate was stored in NaCl solution at 37 and 4°. The fibrinolytic activity was measured by MCA method after storage. The result is shown in Fig. 4. Urokinase-treated Evatate maintained high fibrinolytic activity after 6 months regardless of the temperature. Immobilized urokinase was nearly stable during storage. On the other hand, soluble urokinase completely lost fibrinolytic activity after 20 days at 37°. When immobilized urokinase was stored at low temperature, low oxygen, and low moisture, the stability markedly increased.

Stability against Urokinase Inhibitor, Antiurokinase Antibody, and Cephalothin Sodium

It has been reported that the fibrinolytic activity of soluble urokinase is inactivated under the dose–response of urokinase inhibitor, antiurokinase antibody, and cephalothin sodium (CET, cephem antibiotics). We investigated the influence of these agents on the fibrinolytic activity of soluble and immobilized urokinase. Figure 5A is the result on urokinase inhibitor

---

[6] M. Zimmerman, E. C. Yurewicz, and G. Patel, *Anal. Biochem.* **70,** 258 (1976).

Peptide–MCA,
dimethyl sulfoxide,
Tris–HCl buffer, pH 8.0
+
Soluble urokinase
or
immobilized urokinase

| 37°
| 10 min

17% acetic acid

| Spectrofluorometer
| excitation   380 nm
| emission   460 nm

7-Amino-4-methylcoumarin

FIG. 3. MCA method to measure urokinase activity. Peptide–MCA: Glutaryl-glycyl-arginyl-methylcoumarin 7-amide.

(Merck) which was prepared from human placenta. Urokinase inhibitor (100 IU) was mixed to 50 IU of soluble and immobilized urokinase. This experimental sample was incubated at 37° for a definite time (10, 30 min; 1, 2, 3, 6 hr) and measured by the MCA method. The fibrinolytic activity of soluble urokinase was no longer exhibited after 1 hr, while that of immobilized urokinase was maintained at approximately 70%.

Figure 5B is the result on antiurokinase antibody (Green Cross Corporation) which was prepared from rabbit serum, immunized with purified human urinary urokinase. Antibody titer was not less than 2× by the

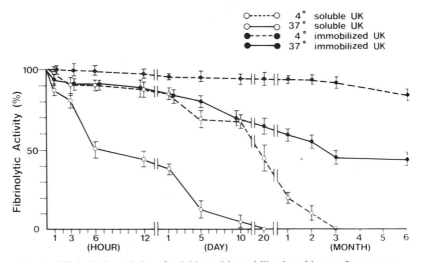

FIG. 4. Fibrinolytic activity of soluble and immobilized urokinase after storage.

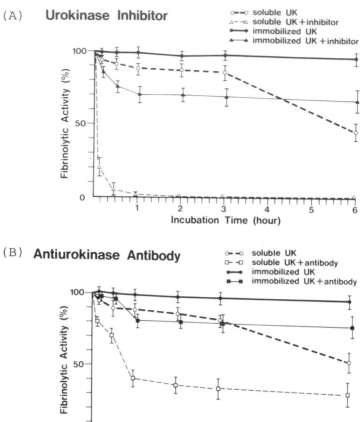

(A) Urokinase Inhibitor

o――o soluble UK
△--△ soluble UK + inhibitor
●――● immobilized UK
▲――▲ immobilized UK + inhibitor

Fibrinolytic Activity (%)

Incubation Time (hour)

(B) Antiurokinase Antibody

o――o soluble UK
□--□ soluble UK + antibody
●――● immobilized UK
■――■ immobilized UK + antibody

Fibrinolytic Activity (%)

Incubation Time (hour)

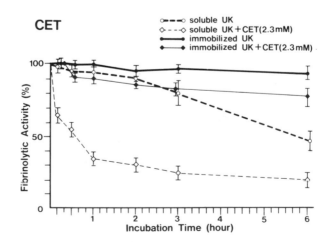

(C) CET

o----o soluble UK
◇----◇ soluble UK + CET(2.3mM)
●――● immobilized UK
◆――◆ immobilized UK + CET(2.3mM)

Fibrinolytic Activity (%)

Incubation Time (hour)

micro-Ouchterlony technique when dissolved in 1 ml water. Twofold anti-urokinase antibody was mixed with 50 IU of soluble and immobilized urokinase. This sample was incubated and measured as above. The fibrinolytic activity of soluble urokinase decreased to 40–30% after 1–6 hr, while that of immobilized urokinase remained at 80–75%.

Figure 5C is the result on cephalothin sodium (Shionogi Corporation) which is one of the cephalosporins. Cephalothin sodium (2.4×10^{-3} mmol) was mixed with 50 IU of soluble and immobilized urokinase. This sample was incubated and measured as above. The fibrinolytic activity of soluble urokinase decreased to 35–20% after 1–6 hr, while that of immobilized urokinase stayed at 90–80%.

Figure 6 shows the influence of a combination of these agents on the fibrinolytic activity of soluble and immobilized urokinase. The experiment was performed at 37° after 1 hr incubation. The residual fibrinolytic activity of soluble urokinase which was incompletely depressed by antiurokinase antibody or cephalothin sodium decreased to the lowest level using this combination of urokinase inhibitor, while that of immobilized urokinase was reduced to approximately 75–70%, which corresponds to a further decrease of 10–15%, due to the influence of the urokinase inhibitor. The combination of urokinase antibody and cephalothin sodium had no influence on the fibrinolytic activity of soluble and immobilized urokinase. The influence of the three agents combined is almost similar to the influence of urokinase inhibitor alone.

Stability against Gastric Juice, Pancreatic Juice, and Bile

Figures 7A, 7B, and 7C are the results of gastric juice, pancreatic juice, and bile on soluble and immobilized urokinase. Immobilized urokinase was relatively stable against these digestive enzymes, compared to soluble urokinase.

Hypothesis Regarding the Stability of Immobilized Urokinase

The reason why immobilized urokinase is stable under rigorous conditions is unknown, but it can be suggested by the following hypotheses.

Hypothesis on Storage. The stability in storage of immobilized enzymes has already been evaluated in the study of immobilized trypsin by A. Bar-Eli and E. Katchalski.[7] If this assumption could be applied to

[7] A. Bar-Eli and E. Katchalski, *Nature (London)* **188,** 856 (1960).

FIG. 5. Influence of (A) urokinase inhibitor, (B) antiurokinase antibody, and (C) CET on fibrinolytic activity of soluble and immobilized urokinase.

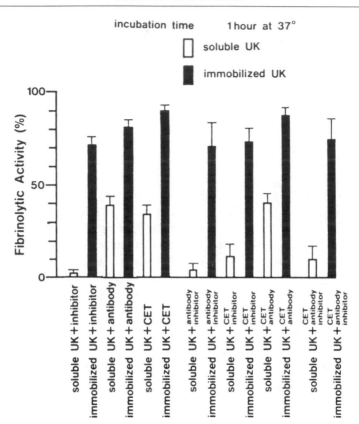

FIG. 6. Influence of various agents on fibrinolytic activity of soluble and immobilized urokinase.

immobilized urokinase, we may say that the amino group in the lysine residue of a urokinase molecule is protected by immobilization, resulting in the prevention against proteolysis by lysine-specific protease and against autolysis by urokinase itself.

Hypothesis on Agents. The following three dogma can be proposed on this hypothesis. (1) The carrier physically blocks the binding reaction of agents. (2) Ion charge changes to the direction of disadvantage in binding of agents and their receptors. (3) Chemical structure changes to the direction of protection in binding of agents and their receptors. Of them, we accept the third hypothesis, as shown schematically in Fig. 8. The hypothesis concerning chemical structure explains the stability of immobilized urokinase against various agents such as urokinase inhibitor, anti-urokinase antibody, and CET. From our experiment, it is clear that half of

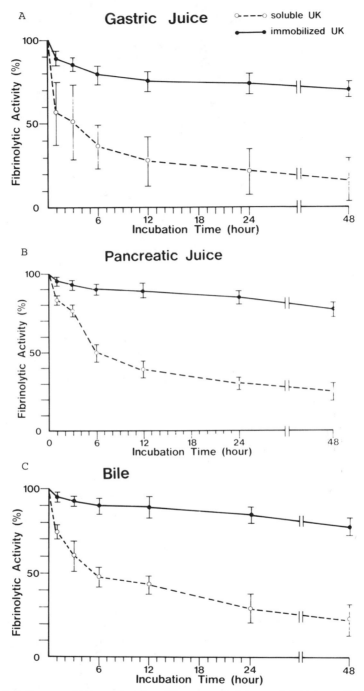

FIG. 7. Influence of (A) gastric juice, (B) pancreatic juice, and (C) bile on fibrinolytic activity of soluble and immobilized urokinase.

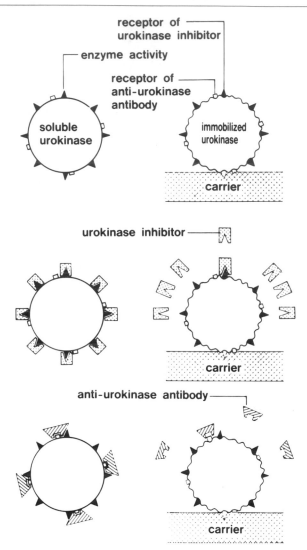

FIG. 8. Hypothesis regarding the stability of immobilized urokinase.

the receptors of urokinase inhibitor interfere with the binding reaction of antiurokinase antibody and its receptors. Antibody binding sites are not identical with or are far away from enzyme activity sites. Inhibitor binding sites are considered identical to or closely adjacent to enzyme activity sites. One of the advantages in immobilized urokinase is its insusceptibility to the influence of urokinase inhibitor.

Antithrombogenicity of Urokinase-Treated Tubes

Chandler Loop Method

The Chandler loop method tests the antithrombogenicity of a uro-kinase-treated tube. The tube is filled with blood and rotated until a thrombus is formed on the inside of the tube. Two types of tubes are used and one of the following is added: physiological saline, $4\times$ antiurokinase antibody, 200 IU urokinase inhibitor, or 4.8×10^{-3} mM CET. As a control 100 IU of soluble urokinase is added to an untreated tube; this corresponds to the fibrinolytic activity of experimental group. After these pre-treatments, each tube is filled with 0.2 ml of 3.8% citric acid, 1.8 ml fresh whole blood, and 0.2 ml 0.25 M CaCl$_2$.

Figure 9 shows the result of the Chandler loop method. In the control group using untreated tubes, the thrombus formation time was about 20 min in the saline series and about 25–45 min in the antibody, inhibitor, or CET series. In the experimental group using urokinase-treated tubes, the thrombus formation time exceeded 45 min in tubes of the saline, antibody, inhibitor, or CET series, and the same result was obtained when only soluble urokinase was added in untreated tube. Fibrin degradation prod-

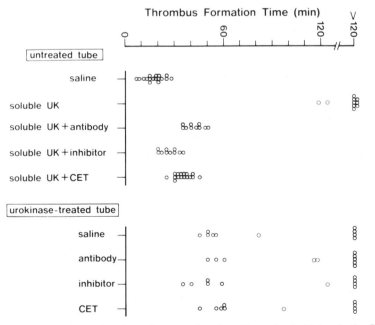

FIG. 9. Thrombus formation time of untreated and urokinase-treated tubes in the Chandler loop method.

uct (FDP) remained within normal levels in series with the shortened thrombus formation time, whereas it increased more than 50 $\mu g/ml$ in series with delayed thrombus formation time. It is supposed that the increase of FDP resulted from the fibrinolytic activity of immobilized or soluble urokinase. Soluble urokinase is partially antagonized by urokinase inhibitor, antiurokinase antibody, or CET, whereas immobilized urokinase is hardly affected.

Mechanism of Antithrombogenicity

We[8] interpret the antithrombogenicity mechanism for urokinase-treated tubes as follows. When blood is in contact with the surface of tube, factor XII is activated and fibrin manomer is formed on the surface. Plasminogen which has an affinity for fibrin gathers and adheres densely around fibrin filament and is converted to plasmin by the immobilized urokinase. Plasmin which has fibrinolytic ability can immediately degrade the contiguous fibrin filament. As a result of this process, we guess that the urokinase-treated tube exhibits the antithrombogenic property.

The mechanism of this antithrombogenicity is characterized as follows: (1) Antithrombogenic ability of urokinase-treated tubes is obtained by the localized fibrinolytic phenomenon which occurs at the contact point of tube with fibrin. Accordingly, the antithrombogenic ability is not affected by antiplasmin which is abundantly present in blood. (2) This fibrinolytic phenomenon is extremely similar to the property of the endothelia of blood vessels. (3) This fibrinolytic phenomenon is of localized fibrinolysis, not of generalized fibrinolysis.

Clinical Application

As the possibility[9] that urokinase-treated material had effective antithrombogenic ability was established, we utilized urokinase-treated Evatate tube for intravenous catheters and urokinase-treated polyvinyl chloride tube for drainage tubes.

Use of Intravenous Catheters

The catheter, 1.5 mm in diameter and 40 cm in length, had a fibrinolytic activity of 31.6 ± 4.2 IU/10 cm of tube (38.8–26.4 IU/10 cm of tube) and was inserted in the superior caval vein through the subclavian vein. We used 70 catheters, which were utilized for total parenteral nutrition in

---

[8] T. Ohshiro, in "Biocompatible Polymers, Metals, and Composites" (M Szycher, ed.), p. 275. Therm. Elect. Co., New York, 1983.
[9] V. L. Gott, J. D. Whiffen, and G. C. Dutton, *Science* **142**, 1297 (1963).

50 patients after surgery and in 20 terminally ill patients. Fever due to catheterization was noted only in 5 patients, but there was no correlation between thrombus formation and pyrexia.

The fibrinolytic activity of urokinase-treated Evatate tubes after catheterization is shown in Fig. 10. The residual fibrinolytic activity gradually decreased with a prolonged catheterization period, but it still showed the activity of approximately 4.1 IU/10 cm of tube after 1 week and 0.4 IU/10 cm of tube after 3 weeks. The six catheters, which were covered with adhered thrombus, are marked with an asterisk in Fig. 10. Out of them, 5 catheters gave a value of less than 1 IU/10 cm of tube, and accounted for 19.2% of 26 catheters whose fibrinolytic activity was less than 1 IU/10 cm of tube. On the other hand, 44 catheters gave a value above 1 IU/10 cm of tube and only one (2.3%) was associated with thrombus formation. From these results, the thrombus formation seems to correlate not with the catheterization period, but with the fibrinolytic activity.

Use of Drainage Tubes

The drain, 28–20 French (i.e., 1 French = 0.33 mm) in size, had a fibrinolytic activity of 9.3 ± 1.1 IU/cm² (10.9–7.8 IU/cm²) and was placed

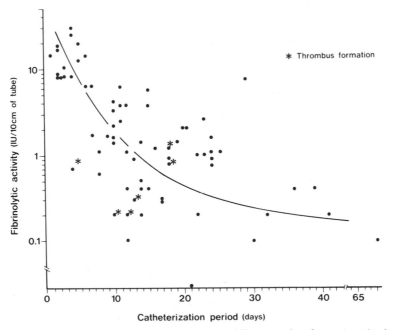

FIG. 10. Fibrinolytic activity of urokinase-treated Evatate tube after catheterization.

in the thoracic or abdominal cavity. Of 100 drains used, the results indicated that there was no thrombus formation in 84 drains and that in the other 16 drains exhibiting thrombus formation occlusions were noted in 4 drains (25.0%). On drainage, thrombus formation was noted in 8 of 43 (18.6%) in thoracic drainage, 6 of 48 (12.5%) in abdominal drainage, and 2 of 9 (22.2%) in other drainage. Based on drain size, thrombus formation was noted in 6 of 59 (10.2%) in 28 French, 4 of 14 (28.6%) in 24 French and 6 of 27 (22.2%) in 20 French. The average drainage period for 16 drains was 7.5 ± 5.6 days, and there was no significant difference, compared to 6.9 ± 4.3 days in total.

The fibrinolytic activity of a urokinase-treated polyvinyl chloride tube after drainage is shown in Fig. 11. The residual fibrinolytic activity decreased with the prolonged drainage period: approximately 5 IU/cm^2 of tube after 3 days, 1.5 IU/cm^2 of tube after 7 days, and 0.6 IU/cm^2 of tube after 2 weeks. The 16 drains in which thrombi were formed are marked with an asterisk in Fig. 11. The residual fibrinolytic activity of them ranged from the maximum 5.4 IU/cm^2 of tube to the minimum 0.2 IU/cm^2

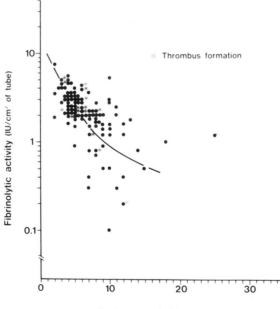

FIG. 11. Fibrinolytic activity of urokinase-treated polyvinyl chloride tube after drainage.

of tube. Thrombus formation correlates with neither the drainage period nor the residual fibrinolytic activity.

Fibrinolytic Activity after Application

The fibrinolytic ability of urokinase-treated tubes abruptly decreased after clinical application contrary to the stability *in vitro*. This phenomenon may be explained by the binding form in the immobilization. Urokinase is supposed to bind to the surface of carrier via physical adsorption, ionic binding, and peptide linking. The many factors in the blood may partially account for the segregation or destruction of urokinase, even those immobilized in the forms of stable ionic binding and tight peptide linking. This phenomenon seems to be related to the following factors: (1) physiological constant temperature (37°); (2) the presence of urokinase inhibitor in serum (α_2-plasmin inhibitor, α_2-macroglobulin, etc.); (3) leakage of protease from cell (peptidase, amidase, etc.); and (4) other biological factors. We are making an effort at present to establish the strongest binding form.

Comparative Study

This study on the investigation of intravenous catheter for total parenteral nutrition was performed with the cooperation of our multiinstitutions

TABLE I
BACKGROUND IN COMPARATIVE STUDY

	Total parenteral nutrition	
	UK-treated EVA group	Untreated PVC group
Case	71	80
Age (year)	57.7 ± 12.8	58.8 ± 12.9
Sex (male/female)	42/29	47/32
Catheter size (16 gauge)	63/71 (88.7%)	59/80 (73.8%)
Catheter insertion site		
Subclavian vein	64/71 (90.1%)	71/80 (88.8%)
Catheterization period (day)	21.3 ± 15.6	19.1 ± 15.0
Infusion volume (ml/day)	2170 ± 410	2090 ± 390
Infusion method		
Filter	43/71 (60.6%)	40/80 (50.0%)
Air needle	59/71 (83.1%)	60/80 (75.0%)
Three sides cock	68/71 (95.8%)	78/80 (97.5%)

TABLE II
RESULT OF COMPARATIVE STUDY

	UK-treated EVA group	Untreated PVC group	Difference
Total days of catheterization (day)	1511	1526	
Completion after planned infusion	55/71 (77.5%)	53/80 (66.3%)	
Thrombus on catheter	11/71 (15.5%)	23/80 (28.8%)	$p < 0.1$
Occlusion in catheter	5/71 (7.0%)	9/80 (11.3%)	
Period to occlusion (day)	27.4 ± 15.4	8.8 ± 4.6	$p < 0.01$
Catheter sepsis	1/71 (1.4%)	3/80 (3.9%)	
Sepsis ratio per 1000 days	0.7	2.0	
Positive bacterial culture	0/64 (0%)	4/80 (6.3%)	$p < 0.05$

to compare 71 catheters of urokinase-treated Evatate catheter group (UK-treated EVA group) with 80 catheters of untreated polyvinyl chloride catheter group (untreated PVC group).

Background. Table I shows the background between two groups. There was no major difference in patient conditions, catheter conditions, catheter insertion site, and catheterization periods, which was 21.3 ± 15.6 days in the UK-treated EVA group and 19.1 ± 15.0 days in the untreated PVC group.

Result. Table II shows the results of comparative study.

The total days of catheterization was 1511 and 1526 days, respectively. The catheter, which was removed after a planned infusion schedule, was 55/71 (77.5%) versus 53/80 (66.3%), adhered thrombus observed; on the surface was 11/71 (15.5%) versus 23/80 (28.8%), $p < 0.1$, complete occlusion recognized; in the hole was 5/71 (7.0%) versus 9/80 (11.3%). The period to occlusion was 27.4 ± 15.4 days in UK-treated EVA group and 8.8 ± 4.6 days in untreated PVC group, presenting a significant intergroup difference ($p < 0.01$). Thus, the former is evaluated to be superior to the latter in antithrombogenic ability. The catheter sepsis was diagnosed in 1/71 (1.4%) versus 3/80 (3.9%). One patient in the former was a 48-year-old male with esophageal varicose vein rupture, who was fevered on the ninth day and normalized after the removal of the catheter. Three patients in the latter group were a 74-year-old male with gastric cancer (sixty-eighth day), a 34-year-old female with meningitis (sixth day), and a 71-year-old male with rectal cancer (seventeenth day). The positive bacterial culture of the catheter tip was 0/64 (0%) versus 4/63 (6.3%), $p < 0.05$.

Superiority of Urokinase-Treated Material

In the materials, immobilized antithrombogenic substances, the heparin-treated tubes reported by Gott *et al.*[9] and Idezuki *et al.*,[10] have already been used in medical clinics. In heparin-treated tube, anticoagulant is tentatively immobilized on tube, gradually released from tube, and finally lost, while in urokinase-treated tube, fibrinolytic enzyme is semipermanently immobilized, fairly retained on the tube, and effectively worked. These differences result in their merits and demerits. Judging from the viewpoint of antithrombogenic material, immobilized urokinase is preferred to immobilized heparin, because the former has a similar function to catalyst. The clinical applications of antithrombogenic medical apparatuses will probably be further developed in future and be put to practical use.

[10] Y. Idezuki, H. Watanabe, and M. Hagiwara, *Trans. Am. Soc. Artif. Intern. Organs* **21,** 282 (1975).

[48] Application of Immobilized Enzymes for Biomaterials Used in Surgery

By Satoshi Watanabe, Yasuhiko Shimizu, Takashi Teramatsu, Takashi Murachi, and Tsunetoshi Hino

With a view to producing biomedical materials for permanent substitution of organic defects in reconstructive surgery, we have developed a composite material made from collagen and a synthetic polymer, which possesses high tissue compatibility suitable for clinical use.[1,2] We are now investigating the replacement of various organs with this new material experimentally and clinically, including reconstruction of the trachea, chest wall, and diaphragm in the field of thoracic surgery.[3] With the development of techniques for binding enzymes to insoluble supports, immobilized enzymes have recently been utilized in various types of med-

[1] Y. Shimizu, R. Abe, T. Teramatsu, S. Okamura, and T. Hino, *Biomater., Med. Devices, Artif. Organs* **5,** 49 (1977).
[2] Y. Shimizu, Y. Miyamoto, T. Teramatsu, S. Okamura, and T. Hino, *Biomater., Med. Devices, Artif. Organs* **6,** 375 (1978).
[3] Y. Shimizu, Y. Miyamoto, T. Teramatsu, T. Hino, U. Shibata, and S. Okamura, *in* "Biomaterials 1980" (G. D. Winter, D. F. Gibbons, and H. Plenk, Jr., eds.), p. 745. Wiley, New York, 1982.

ical applications, especially in clinical analysis and also in therapeutic medicine.

This collagen–synthetic polymer composite material was applied as a support for the immobilization of enzymes in order to establish their biological functions on the surface of the material, and enzymes were successfully bound to the collagen membrane layer by activation of carboxyl groups.[4–6] Trypsin and urokinase were chosen with the intention of adding proteolytic (antiinflammatory) and fibrinolytic (thromboresistant) activities, respectively, to the surface properties of the composite material. Lysozyme and a peptide antibiotic, polymyxin B,[7] were bound onto the material for the purpose of producing bacteriolytic and antibacterial biomaterials to prevent serious problems caused by bacterial infection from occurring when artificial organs and biomaterials are implanted into the body. Here we describe the novel method of producing this enzyme-bound collagen–synthetic polymer composite material, and we discuss the enzymatic characterization and some *in vivo* experiments made with the material carrying immobilized enzymes.

Materials and Methods

Polyethylene (PE) and polypropylene (PP) were employed as synthetic polymer supports. PE film, PE monofilament, and PP mesh (Marlex Mesh, for medical use) were obtained from commercial sources. Bovine skin collagen, which was treated with protease in order to remove telopeptides and reduce its antigenicity to a minimum, was purchased from Nippi Co. (Tokyo). The other chemical reagents were obtained from the usual commercial sources.

Coating of the synthetic polymer with collagen was performed according to the method devised by Okamura and Hino.[8] Active radicals are first generated on a surface of a synthetic polymer by glow discharge (50 watt, 0.5 torr in air for 30 min) or spark discharge (3-cm-long spark for 15 min), and the polymer surface is coated with 1% collagen solution (pH 3). This collagen–synthetic polymer composite material is neutralized in ammonia gas and dried. The quantity of collagen coating is 1–5 mg/cm² surface area of the synthetic polymer. The collagen membrane layer is then

[4] S. Watanabe, H. Kato, Y. Shimizu, T. Teramatsu, J. Endo, T. Murachi, and T. Hino, *Artif. Organs* **3** (Suppl.), 200 (1979).

[5] S. Watanabe, Y. Shimizu, T. Teramatsu, T. Murachi, and T. Hino, *J. Biomed. Mater. Res.* **15**, 553 (1981).

[6] S. Watanabe and T. Teramatsu, *Enzyme Eng.* **6**, 459 (1982).

[7] D. R. Storm, K. S. Rosenthal, and P. E. Swanson, *Annu. Rev. Biochem.* **46**, 723 (1977).

[8] S. Okamura and T. Hino, U.S. Patent 3,808,113 (1974) and U.S. Patent 3,955,012 (1976).

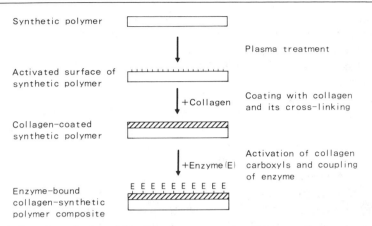

FIG. 1. Procedures for immobilization of enzymes onto collagen–synthetic polymer composite material support.

cross-linked with ethanol containing 2.5% glutaraldehyde for 10 min to increase its acid stability, then washed with ethanol several times and dried (Fig. 1).

Immobilization of enzymes onto the collagen layer was performed according to two methods (Fig. 1).

Coupling of Enzymes after Activation of the Collagen Membrane Layer by the Azid Method.[9,10] The first step in this method is acidic methylation of the collagen layer in methanol containing 0.2 N HCl for 1 week, followed by reaction with 1% hydrazine hydrate for 12–15 hr and formation of acyl azid in a mixture of 0.5 M NaNO$_2$ and 0.3 N HCl for 5 min at 0°. After removal of all excess reagents, composite material is dipped into an enzyme solution in borate or Tris buffer, pH 8.8, which ensures spontaneous coupling of the enzyme at 0° for 2 hr. Finally the enzyme-bearing composite material is washed several times and stored in 50 mM Tris buffer, pH 8, at 4°.

Coupling of Enzymes after Activation of the Collagen Membrane Layer by Preparing the Active Ester Intermediate with a Water-Soluble Derivative of Carbodiimide.[11] The active ester intermediate of the collagen membrane layer is prepared by reacting it with 0.1 M N-hydroxysuccinimide and 0.1 M 1-ethyl-3-(3-dimethylaminopropyl)carbodiimide hy-

[9] P. R. Coulet, J. H. Julliard, and D. C. Gautheron, *Biotechnol. Bioeng.* **16,** 1055 (1974).
[10] P. R. Coulet and D. C. Gautheron, *in* "Analysis and Control of Immobilized Enzyme Systems" (D. Thomas and J. P. Kernevez, eds.), p. 165. North-Holland, Amsterdam, 1976.
[11] P. Cuatrecasas and I. Parikh, *Biochemistry* **11,** 2291 (1972).

drochloride in distilled water stirred gently for 2 hr. After removal of excess reactants by rinsing, enzyme is coupled by immersing the composite material in an enzyme solution of 60 mM phosphate saline buffer, pH 7.4, at 4° for 4 hr with mild agitation. This is followed by extensive washing in 60 mM phosphate saline buffer, pH 7.4, and 1 M KCl, to remove loosely bound enzyme. Finally, the enzyme-coupled composite material is stored in 60 mM phosphate saline buffer pH 7.4, at 4°.

Trypsin (from bovine pancreas) was purchased from Boehringer Manheim, and urokinase (from human urine, 35,700 IU/mg protein) was kindly provided by Kanebo Pharmaceuticals, Ltd. (Tokyo). The esterase activity of trypsin is determined by measuring the initial rate of hydrolysis of a synthetic substrate, α-N-benzoyl-L-arginine ethyl ester (BAEE), using a pH stat assembly (Radiometer, Copenhagen, Denmark).[12] Urokinase activity for hydrolysis of a synthetic substrate, α-N-acetylglycyl-L-lysine methyl ester (AGLME, Protein Foundation, Osaka, Japan) is also determined by the pH-stat method.[13]

The esterolytic reaction of the enzymatic composite material is initiated by dipping it into the reaction mixture and is stopped simply by removing it. The specific activities of the enzymatically active composite material are expressed as nanomoles substrate transformed per minute per square centimeter surface area of the material. The yields of immobilized trypsin and urokinase are calculated from these specific activity values.

The caseinolytic activity of trypsin is measured by the casein digestion method described by Laskowski.[14] The fibrinolytic activity of urokinase is determined by a modified fibrin–agarose plate method.[15] The inhibitory effect of blood plasma on the fibrinolytic activity of immobilized urokinase is examined on a fibrin–agarose plate by adding human, canine, and rabbit plasma directly to the urokinase-bound PE monofilament composite material, which has been cut off and placed on the plate. The *in vivo* stability of the fibrinolytic activity of immobilized urokinase is investigated by inserting urokinase-bearing PE monofilament composite material into a rabbit blood vessel (from the jugular vein to the superior vena cava).

Lysozyme (hen egg white) was purchased from Sigma Chemical Co. (St. Louis, MO), and polymyxin B sulfate (activity level of 10,000 unit/mg) was obtained from Pfizer Taito Corp. (Tokyo). The activity of lyso-

[12] K. A. Walsh and P. E. Wilcox, this series, Vol. 19, p. 31.
[13] F. C. Capet-Antonini and J. Tamenasse, *Can. J. Biochem.* **53,** 890 (1975).
[14] M. Laskowski, this series, Vol. 2, p. 26.
[15] P. L. Walton, *Clin. Chim. Acta* **13,** 680 (1966).

zyme for hydrolysis of the cell walls of *Micrococcus lysodeikticus* ATCC 4698 is measured by an agarose plate method which we have devised. Agarose plates are made from equal volumes of 0.8% agarose and 0.09% dried cell suspension of *M. lysodeikticus* dissolved in 60 m*M* phosphate buffer (pH 6.2), which are mixed at 45° and cooled. After incubating the plates with test enzyme at 26° for 40 hr, the diameter of bacteriolytic zones is measured. Pieces (5 × 5 mm) of lysozyme-bearing PP mesh composite material are placed directly on the agarose plate. Ten-microliter standard dilutions of lysozyme (0.1–10 μg/ml of 60 m*M* phosphate buffer, pH 6.2) and 5 × 5 mm pieces of the composite material without bound lysozyme are also applied. Antibacterial activity of polymyxin B against *Bordetella bronchiseptica* ATCC 4617 is measured by the disk-agar diffusion method with a modified Mueller–Hinton medium.[16] After 40-hr incubation at 30°, the diameter of the growth inhibition zones is measured. We implant 1 × 1 cm pieces of this lysozyme- or polymyxin B-bearing PP mesh composite material into the dorsal subcutaneous tissue of rabbits (Japan Whites), and investigate histological reactions of the specimens which are extirpated at various intervals. Hematological examinations are also performed in the course of implantation.

Results and Discussion

Trypsin-Bound Composite Material.[4,5] When the trypsin concentration in the coupling solution was 1.3 mg/ml, the yield of immobilized trypsin, calculated from the specific activity values, was 2.4 μg protein/cm² for the PE film composite material and 4.9 μg protein/cm² for the PP mesh composite material. The fact that the yield for the PP mesh composite material was about twice that for the PE film composite material must be attributed to the difference in the nature of their surfaces, as the surface area of the PP mesh composite material is larger than that of the PE film composite material. These composite materials were also found to possess sufficient hydrolytic activities with respect to a high molecular weight substrate casein. The trypsin activity retained on the PE film composite material increased in proportion to the trypsin concentration in the coupling solution until maximum activity was reached at about 10 mg trypsin/ml.

Urokinase-Bound Composite Material.[4,5] When the urokinase concentration in the coupling solution was 1.3 mg/ml, the yield of immobilized urokinase, calculated from the specific activity values, was 60 IU/cm² for

---

[16] S. M. Finegold and W. J. Martin, "Diagnostic Microbiology," 6th Ed. Mosby, St. Louis, Missouri, 1982.

the PE film composite material and 120 IU/cm² for the PP mesh composite material. PE film and PP mesh composite materials carrying immobilized urokinase produced fibrinolytic zones in proportion to the size of the composite materials. The yield of the bound enzyme was considered to be approximately 10% of the total enzyme protein in the coupling solution when estimated from the change of the enzyme concentration in the routine course of immobilization. The catalytic activity of the immobilized enzyme per unit surface area was sufficiently high to demonstrate that the enzyme had been bound onto the surface of the collagen membrane layer. The Michaelis constants (K_m values) obtained for AGLME were 1.0×10^{-3} M for soluble urokinase and 2.0×10^{-3} M for immobilized urokinase, showing that the apparent affinity for the substrate was decreased by immobilization but the intrinsic kinetic properties of the immobilized urokinase were not significantly affected. Immobilization resulted in decreased pH sensitivity over the range 7.0–9.0 with no shift in pH optimum. The relative activity of immobilized urokinase increased with increasing temperature in the same manner as soluble urokinase.

The esterase activity of immobilized urokinase, when stored at 4° in the buffer solution, was found not to change over a period of 10 months during which the same enzymatic preparations were used repeatedly. Both the esterolytic and fibrinolytic activities of the urokinase-bearing composite material were stable on disinfection treatment with antibiotic saline solution (10 mg/ml of ampicillin or cephaloridine). Human, rabbit, and canine blood plasma showed certain different inhibitory effects on the fibrinolytic activity of the immobilized urokinase. When the urokinase-bearing composite material was applied to rabbit blood vessel, its *in vivo* fibrinolytic activity gradually decreased, but about 80% of its initial activity was maintained after 3 hr.[4,5]

Lysozyme-Bound Composite Material.[6] The bacteriolytic activity of lysozyme-bearing PP mesh composite material against *M. lysodeikticus* was equivalent to that of about 10 μg/ml of soluble lysozyme when the lysozyme concentration in the coupling solution was 10 mg/ml. The activity decreased to about 50% of its initial level during 6 months of storage, but some activity still remained for about 19 months after preparation. The lysozyme-bound composite materials were adapted to the subcutaneous tissue of a different species, rabbit, after prolonged implantation, although various histological reactions were seen around them in the early stage of implantation. There were no remarkable differences between the composite materials with immobilized lysozyme and without immobilized lysozyme in terms of the inflammatory reaction and in the thickness of the surrounding connective tissue layer. Hematological examination showed no abnormal findings.

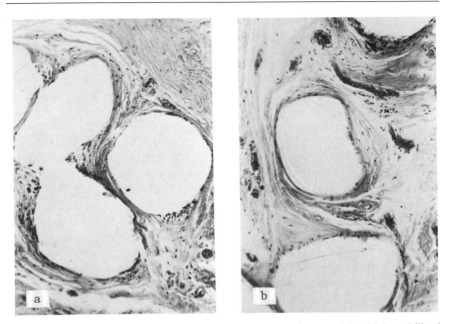

FIG. 2. Histological findings of collagen–PP mesh composite materials with immobilized polymyxin B (a) and without immobilized polymyxin B (b), implanted in rabbit subcutaneous tissue for 5 months (hematoxylin–eosin stain, ×100).

Polymyxin B-Bound Composite Material.[6] The yield of immobilized polymyxin B, calculated from the antibacterial activity value, was 97 units/cm^2 for the PP mesh composite material when the polymyxin B concentration in the coupling solution was 10 mg/ml. When the polymyxin B-bound composite material was dipped into human serum for 30 min, then washed with 60 mM phosphate buffer, pH 7.4, sufficient antibacterial activity remained. The antibacterial activity of immobilized polymyxin B was also maintained during 6 months of storage. Although some inflammatory reactions were found soon after implantation, the polymyxin B-bearing composite material was later adapted to the subcutaneous tissue, as shown in Fig. 2. Side effects, such as renal and neural toxicity, associated with soluble polymyxin B have never been seen in the course of implantation.

Enzyme–collagen–synthetic polymer composite materials may have wide application as biofunctional materials for biomaterials and artificial organs. For example, we have made a urokinase-bound poly(vinyl chloride) tube composite material for intrathoracic drainage, a lysozyme-bound PP mesh composite patch for organic defects, and a polymyxin B-bound silicone tube composite material for urethral catheters.

[49] Immobilized Enzymes for Thrombolytic Therapy

By V. P. Torchilin, A. V. Maksimenko, and A. V. Mazaev

At present, thrombolytic therapy is based on the knowledge of the molecular mechanisms of fibrinolysis regulation *in vivo* (see, for example, Wiman and Collen[1]). Thrombolytic enzymes are becoming more important and more widely used in practical applications of therapy. Thrombolytic enzyme therapy is based on two principal approaches: (1) the use of an enzyme (or its zymogen) which is responsible for fibrin degradation in physiological conditions when an enzyme (or zymogen) deficiency is present in the blood, and (2) the use of proteolytic enzymes capable of specific transformation of the appropriate zymogen into the fibrin-degrading enzyme when normal quantities of the zymogen are present in the blood but transformation into the active enzyme proceeds too slowly. A combination of both approaches is also promising. Thrombolytic enzyme therapy can be useful for the treatment of deep vein thrombosis, acute pulmonary embolism, arterial occlusion, and myocardial infarction.[2]

Thrombolytic Enzymes[3]

Plasminogen and Plasmin (EC 3.4.21.7).[4] Human plasminogen is a single chain glycoprotein (carbohydrate content ~2%) with a molecular weight of about 90,000. In the native form plasminogen has an N-terminal glutamic acid (so-called Glu-plasminogen) and can be easily converted to Lys-plasminogen, containing N-terminal lysine, valine or methionine. Both zymogens are proteolytically converted to the active form, plasmin, which is a trypsinlike enzyme, consisting of two chains bound together by two S–S bridges and containing His-602, Asp-645, and Ser-740 at the active site. Plasmin from human blood has limited use in clinics, and the therapeutic enzyme is usually obtained by trypsin activation of purified porcine plasminogen.

[1] B. Wiman and D. Collen, *Nature (London)* **272,** 549 (1978).
[2] F. Duckert, *Semin. Thromb. Hemostasis* **10,** 87 (1984).
[3] P. J. Gaffney and S. M. Balkuv-Ulutin (eds.), "Fibrinolysis: Current Fundamental and Clinical Concepts." Academic Press, London, 1978.
[4] K. C. Robbins and L. Summaria, this series, Vol. 45, p. 257.

Urokinase (EC 3.4.21.31, Plasminogen Activator).[5,6] The serine protease urokinase can be isolated from human urine or kidney tissue culture and exists in two molecular forms (molecular weights around 33,000 and 53,000). Urokinase is very selective in its action on proteins, the conversion of plasminogen to plasmin being the only reaction catalyzed by the enzyme, but it is similar to trypsin and plasmin in reactions with low molecular weight synthetic substrates (the most widely used substrate of urokinase is *N*-α-acetyl-L-glycyl-L-lysine methyl ester, AGLME).

Tissue Plasminogen Activator (TPA).[7,8] TPA is also a serine protease with molecular weight around 68,000, which is synthesized in endothelial cells and present in most or all human tissues. TPA has some homology with urokinase and plasminogen and demonstrates high affinity for fibrin. The physiological role of TPA is the activation of plasminogen to plasmin in the region of thrombus formation, where the activity of TPA is greatly enhanced by the presence of fibrin.

Streptokinase (EC 3.4.24.4).[2,9] Streptokinase is the neutral proteinase from *Staphylococcus aureus* which catalyzes the conversion of plasminogen (the only substrate for streptokinase) to plasmin. It is now known that streptokinase itself does not catalyze proteolytic cleavage of plasminogen, but it forms complexes with circulating plasmin or plasminogen, which in turn act as catalysts of plasmin formation. The molecular weight of streptokinase is 45,000–47,000. The preformed native or acylated complex of streptokinase with plasminogen can also be used successfully as a thrombolytic agent.[10]

Some other enzymes, e.g., brinase from *Aspergillus oryzae,* can be used for thrombolytic therapy, but their limited availability does not permit us to draw more definite conclusions.

Limitations in the Use of Thrombolytic Enzymes

It should be emphasized, however, that the everyday clinical use of thrombolytic enzymes is still limited by a number of factors. Urokinase, plasmin, and TPA are very expensive and are produced only on a small

---

[5] G. H. Barlow, this series, Vol. 45, p. 239.

[6] R. Paoletti and S. Sherry (eds.), "Thrombosis and Urokinase." Academic Press, London, 1977.

[7] D. C. Rijken and D. Collen, *J. Biol. Chem.* **256,** 7035 (1981).

[8] F. Bachmann and I. E. K. O. Kruithof, *Semin. Thromb. Hemostasis* **10,** 6 (1984).

[9] F. J. Castellino, J. M. Sodetz, W. J. Brockway, and G. E. Siefring, Jr., this series, Vol. 45B, p. 244.

[10] K. C. Robbins, L. Summaria, R. C. Wohl, and W. R. Bell, *Thromb. Haemostasis* **50,** 787 (1983).

scale. Streptokinase, which is the most widely used activator of fibrinolysis, is isolated from the broth of β-hemolytic streptococci and demonstrates high nonspecific toxicity and antigenicity. The general drawbacks to using these thrombolytic enzymes are their quick inactivation under physiological conditions and their fast clearance from the body, which considerably increases the consumption of the enzyme in the course of treatment; the high sensitivity of these enzymes to the action of various endogenous proteases and natural inhibitors; the impossibility of obtaining high local drug concentration without increasing its total concentration (which is often undesirable or even impossible); and the pronounced depressive action of thrombolytic enzymes on the coagulative component of hemostasis. The drawbacks listed often produce the following clinical manifestations usually accompanying thrombolytic enzyme therapy: the necessity of prolonged and continuous intravenous administration, which is possible only in stationary conditions; frequent allergic reactions; the achievement of thrombolysis effect on the background of pronounced hemostasis destabilization, which often leads to hemorrhagic and re-thrombotic complications (in 80% of cases hemorrhagic complications are registered; in 30% rethrombosis; 30% of cases are accompanied with allergic reactions[11,12]); and difficulties in controlling and in carrying out combined therapy with heparin.

Role of Immobilization

It becomes evident that the problems listed above can be resolved to a great extent by the use of modified, stabilized, or immobilized enzymes.[13,14] Present techniques yield enzyme preparations possessing increased stability in the circulation toward thermoinactivation, pH-inactivation, and the action of endogenous inhibitors and proteases. Modified (immobilized) enzymes also demonstrate decreased toxicity and antigenicity. At the same time enzyme catalytic properties and therapeutic activity remain unchanged. Even more, the use of proper modifying agents (carriers) or immobilization conditions yields preparations possessing more than one useful therapeutic activity and capable of specific recogni-

[11] A. Sahara and J. E. Datten, *J. Cardiovasc. Med.* **5**, 793 (1980).
[12] E. I. Chazov and A. P. Golikov, *Kardiologiya* **21**, 10 (1980).
[13] V. P. Torchilin, A. V. Mazaev, E. V. Il'ina, V. S. Goldmacher, V. N. Smirnov, and E. I. Chazov, *in* "Future Directions for Enzyme Engineering" (L. B. Wingard, Ir, I V Berezin, and A. A. Klyosov, eds.), p. 219. Plenum, New York, 1980.
[14] T. M. S. Chang (ed.), "Biomedical Applications of Immobilized Enzymes and Proteins," Plenum, New York, 1977.

tion of an affected zone (for a review see, for example, Larionova and Torchilin[15] and Torchilin[16]).

Carriers Which Can Be Used for Thrombolytic Enzyme Immobilization

Medical requirements considerably limit the number of synthetic or natural carriers or modifiers potentially fit for the stabilization of therapeutic enzymes. The carriers used should be highly pure and homogenic, cause no irritating, toxic, or carcinogenic effects on the organism, undergo biodegradation quickly enough, give no toxic products on biodegradation, and should be completely removed or metabolized after having fulfilled their therapeutic function. At present three types of polymeric carriers can be conditionally selected for therapeutic enzyme immobilization: (1) natural inert carriers (mainly polysaccharides, such as soluble or insoluble reactive derivatives of dextran, cellulose, or starch; but also proteins, like albumin, and other compounds); (2) synthetic, inert, biocompatible, soluble or insoluble polymers or copolymers (e.g., polymers containing biodegradable bonds in the main chain or in side chains); and (3) natural or synthetic physiologically active carriers, which can facilitate or improve the action of the immobilized enzyme (heparin, specific antibodies, synthetic sulfo-containing polymers).

Two possible ways of creating therapeutic (thrombolytic) immobilized enzymes can be outlined in a very general form.[13,16] If an enzyme is meant for prolonged circulation in blood or if it must of necessity be present in different organs and tissues of an organism, it appears reasonable to create water-soluble stabilized enzymes capable of prolonged circulation in the organism. It is also desirable that the immunogenicity of such preparations be low. When it is preferable for the enzyme to be localized in the organism during the treatment of local lesions (e.g., thrombosis, atherosclerotic deposits, and tumors) and its presence in other organs and tissues is unnecessary or even undesirable, one should synthesize biocompatible and biodegradable enzyme derivatives, that is, microparticles, which can be localized by means of standard methods and can be degraded there over a period of time, while continuously releasing the active agent into the local surroundings. As a result, an opportunity arises to create high local concentrations of an enzyme while utilizing small doses. This increases the therapeutic effect and reduces unfavorable side effects on the organism.

[15] N. I. Larionova and V. P. Torchilin, *Khim–Farm. Zh.* **4,** 21 (1980).
[16] V. P. Torchilin, *in* "Targeted Drugs" (E. Goldberg, ed.), p. 127. Wiley, New York, 1983.

Biodegradable (Slow Soluble) Preparations of Thrombolytic Enzymes

Biodegradable preparations developed by us[17,18] differ from typical slow-release systems, because in our case an enzyme, which is covalently bound to the solubilized fragment of the carrier, is released into the surroundings, and the rate of the carrier biosolubilization determines the time of local drug depot existence. To obtain such preparations, cross-linked insoluble polysaccharides can be used (Sephadex, Macrodex, etc.) which on partial oxidation form appropriate aldehyde derivatives capable of slow dissolution in water at physiological pH values, the rate of solubilization depending on the degree of modification. At the same time protein covalent immobilization can proceed via aldehyde groups.

Immobilization of Plasmin on Aldehyde–Sephadex (AS).[18,19] Twenty grams of Sephadex G-25 (Pharmacia) is suspended in 160 ml of 0.1 M acetate buffer, pH 4, and supplemented with 17.3 g of HIO_4. The reaction is performed in the dark at room temperature for different time intervals. The reaction is stopped by washing with 5% hyposulfite (dithionite) or with 0.1 M KCl in 0.1 M HCl. Then the product is repeatedly washed with distilled water and dried with acetone. The degree of oxidation (expressed as the quantity of aldehyde groups per 100 dextran units) is determined based on the iodine value of the preparation.[17] The granulometric composition of the carrier is controlled microscopically. The AS fraction with a granular size up to 30 μm (60–80 μm in the swollen state) is isolated by sieving (Fritsch Analysette, FRG). The rate of AS solubilization in 0.15 M phosphate buffer, pH 7.4, is measured spectrophotometrically by following the increase in absorbance at 220 nm due to carbonyl groups. As a result AS samples can be obtained with an aldehyde content up to 60 carbonyl groups per 100 dextran units (48 hr of oxidation) and with a solubilization time of about 3 hr (30 mg of AS in 100 ml of 0.15 M phosphate buffer, pH 7.4, 37°, gentle shaking). Under analogous conditions Sephadex with a smaller content of cross-linkages gives products with a similar degree of oxidation, but the time of solubilization is decreased.

For immobilization 30 mg of plasmin (fibrinolysin, Koch-Light, England), with a 60% concentration of active enzyme as determined by titration with *p*-nitrophenyl-4-guanidinobenzoate,[20] is added at 4° to a

[17] V. P. Torchilin, A. S. Bobkova, V. N. Smirnov, and E. I. Chazov, *Bioorg. Khim.* **2,** 116 (1976).

[18] V. P. Torchilin, E. G. Tischenko, V. N. Smirnov, and E. I. Chazov, *J. Biomed. Mater. Res.* **11,** 223 (1977).

[19] E. I. Chazov, A. V. Mazaev, V. P. Torchilin, B. S. Lebedev, E. V. Il'ina, and V. N. Smirnov, *Thromb. Res.* **12,** 809 (1978).

[20] T. Chase, Jr., and E. Shaw, *Biochemistry* **8,** 2212 (1969).

freshly prepared suspension of 100 mg of AS (solubilization time ~3 hrs at 37°) in 5 ml of 50 mM phosphate buffer, pH 8.2, and the reaction proceeds overnight (carrier solubilization at 4° is negligible). Nonbound enzyme is washed off by successive washing with ice-cold phosphate buffer. The reaction proceeds with Shiff base formation between AS aldehyde groups and ε-amino groups of protein lysine residues. To make protein–carrier bonds nonhydrolyzable Shiff bases (and simultaneously nonreacted carbonyls) are reduced by the addition of excess NaBH$_4$ at 4°. After 1 hr incubation NaBH$_4$ is destroyed with HCl at pH 4, and the preparation is washed with phosphate buffer, water, and a water–acetone mixture. Finally the preparation is dried with acetone and sterilized with UV light. The preparation can be stored for years. According to the data from spectrophotometric analysis (at 280 nm on complete solubilization of the preparation) and the protein determination by the Lowry method, approximately 120 mg of the enzyme is bound per 1 gram of carrier. The activity of the bound enzyme is around 70% in the insoluble state and about 80% in solution after complete solubilization as determined by following the initial rate of hydrolysis of a specific plasmin substrate, N-benzoyl-L-arginine ethyl ester, in the cell of a TTT-1c pH-stat (Radiometer, Copenhagen, Denmark) at 25°, pH 7.5 in 0.1 M KCl. Initial substrate concentration is 5×10^{-3} M.

An alternative procedure of protein immobilization in biodegradable microspheres of polyacryldextran is given by Edman et al.[21]

Water-Soluble Immobilized Thrombolytic Enzymes

Plasmin Immobilization on Soluble Dextran. To avoid the use of a toxic compound (like CNBr) for dextran activation, reactive groups are introduced in dextran molecules by partial oxidation.[22,23] For this purpose 10 g of soluble dextran (molecular weight 35,000–50,000, NBC, USA) and 8.5 g of HIO$_4$ are dissolved in 50 ml of 0.1 M phosphate buffer, pH 4, and reacted together for 16 hr at 25°. The reaction is stopped by the addition of 5% hyposulfite solution. Aldehyde–dextran (AD) obtained (22 aldehyde groups per 100 dextran units according to the iodine value) is precipitated with methanol, dried with acetone, and stored in a refrigerator in the dark.

For immobilization, 100 mg of plasmin (Koch-Light) is added to a solution of 0.5 g of AD in 20 ml of 50 mM phosphate buffer, pH 8.2, and

[21] P. Edman, B. Ekman, and I. Sjöholm, *J. Pharm. Sci.* **69,** 838 (1980).

[22] V. P. Torchilin, I. L. Reizer, V. N. Smirnov, and E. I. Chazov, *Bioorg. Khim.* **2,** 1252 (1976).

[23] E. I. Chazov, V. N. Smirnov, V. P. Torchilin, I. M. Tereshin, B. V. Moskvichev, G. M. Grinberg, A. Z. Skuya, and G. I. Kleiner, U.S. Patent 4,446,316 (1984).

the reaction is allowed to proceed overnight at 4°. The Shiff bases formed are reduced with NaBH₄, any excess of which is destroyed by adding diluted hydrochloric acid to pH 4. Immobilized plasmin is separated from low molecular weight impurities and from nonreacted dextran and protein by gel chromatography on Sepharose 4B (K-16 column, Pharmacia, Uppsala, Sweden). The molecular weight of the conjugate obtained is around 130,000 as determined by gel chromatography. The degree of binding is about 95%, which corresponds to binding of 190 mg of protein per gram of AD. Plasmin activity (with N-benzoyl-L-arginine ethyl ester as substrate, see above) is preserved at 85%. The product is dialyzed against 1000 volumes of distilled water overnight at 4° and then freeze-dried and stored in a refrigerator. The specific fibrinolytic activity is approximately 22 units/mg of the preparation. The product is a yellowish powder with no taste or odor and is readily soluble in water and 0.9% isotonic solution.

Urokinase Immobilization on Acrylamide–Acrylic Acid Copolymer.[24] The synthesis, structure, and properties of biocompatible and bio- (enzymatically) degradable soluble and insoluble synthetic polymers are reviewed by Ringsdorf,[25] Kopeček and Rejmanova,[26] Hoffman,[27] and Drobnik and Rypaček.[28] Water-soluble copolymers of acrylamide and acrylic acid are obtained by radical copolymerization of monomers in the presence of 0.01% (wt) ammonium persulfate in aqueous solution.[29] To obtain copolymers with various concentrations of acrylic acid units the composition of a monomeric mixture of acrylamide and acrylic acid is calculated using the constants of copolymerization $r_1 = 1.38$ and $r_2 = 0.36$, respectively.[30] Copolymerization is carried out at 60° during 4 hr; total monomer concentration in the reaction mixture is 5% (wt/wt). The copolymers are precipitated by methanol and dried with acetone. The composition of copolymers is determined by infrared spectroscopy based on the adsorbance of the carboxylic groups (1726 cm⁻¹). The measurements are made on the UR-20 spectrophotometer (GDR).

Immobilization of urokinase (Abbot, USA, low molecular weight enzyme) on the copolymer is performed using as a coupling reagent 1-ethyl-

[24] A. V. Maksimenko, V. P. Torchilin, V. N. Smirnov, and E. I. Chazov, U.S. Patent 4,349,630 (1982).

[25] H. Ringsdorf, *J. Polym. Sci.* **51,** 135 (1975).

[26] J. Kopeček and P. Rejmanova, *in* "Controlled Drug Delivery" (S. D. Bruck, ed.), Vol. 1, p. 81. CRC Press, Boca Raton, Florida, 1983.

[27] A. S. Hoffman, *in* "Macromolecules" (H. Benoit and P. Rempp, eds.), p. 321. Pergamon, Oxford, 1982.

[28] J. Drobnik and F. Rypaček, *Adv. Polym. Sci.* **57,** 1 (1984).

[29] V. P. Torchilin, E. G. Tischenko, and V. N. Smirnov, *J. Solid-Phase Biochem.* **2,** 19 (1977).

[30] D. Ham (ed.), "Copolymerization." Wiley Interscience, New York, 1966.

3-(3-dimethylaminopropyl)carbodiimide (EDC). For this purpose 40 mg of a copolymer with a molecular weight of about 80000 and a 3% content of acrylic acid units are dissolved in 10 ml of 0.1 M phosphate buffer, pH 8.3. Then in the cold (4°), 5 mg of EDC is introduced in this solution, and after 10 min of incubation 5 mg of urokinase is added as a solution in 1 ml distilled water. The reaction proceeds overnight at 4°, and nonbound enzyme is separated by gel filtration on a Sephadex G-150 column (60 × 2.5 cm). Elution is performed with 0.1 M phosphate buffer, pH 8.0 at elution rate 0.7 ml/min. Gel filtration is controlled spectrophotometrically by measuring the eluent absorbance at 280 nm (Turner UV spectrophotometer, USA) and by measuring catalytic activity of eluent aliquots with a pH-stat TTT-1c (Radiometer), using the specific urokinase substrate AGLME, (Sigma, St. Louis, MO). Catalytic activity is determined by potentiometric titration of AGLME enzymatic hydrolysis reaction at 25°, pH 7.5, 0.1 M KCl, 10 mM substrate. The degree of binding is 85%; 81% of the catalytic activity is preserved. The product (immobilized urokinase) is dialyzed against 1000 volumes of distilled water overnight at 4°, lyophilized, and stored in a refrigerator.

New Generation of Immobilized Thrombolytic Enzymes

To increase the therapeutic efficacy of thrombolytic enzymes, they can be modified with different substances which themselves possess useful physiological activity or increase the therapeutic potential of the modified enzyme. On the other hand, compounds capable of specific recognition of affected zones (thrombus) can be used as carriers for enzyme (thrombolytic enzyme) modification.

Urokinase Modification with Heparin.[31] Taking into consideration the fact that thrombolytic therapy is usually performed along with heparinization, one can use heparin to modify thrombolytic enzymes.[32–34] Other advantages of heparin are its low antigenicity, high chain rigidity, and high content of reactive groups.

Modification of urokinase (Abbot, low molecular weight enzyme, specific activity 3125 IU/mg) with heparin (Koch-Light) is performed with the previous activation of heparin carboxyl groups with EDC. For this purpose 5 mg of EDC is added to 4 ml of 50 mM phosphate buffer, pH 8.3, containing 8 mg of heparin at 4°, and the mixture is incubated with stirring

[31] A. V. Maksimenko and V. P. Torchilin, *Thromb. Res.* **38,** 277 (1985).
[32] V. P. Torchilin, E. V. Il'ina, Z. A. Streltsova, V. N. Smirnov, and E. I. Chazov, *J. Biomed. Mater. Res.* **12,** 585 (1978).
[33] V. P. Torchilin, *Bioorg. Khim.* **4,** 566 (1978).
[34] E. V. Il'ina, V. P. Torchilin, V. N. Smirnov, and E. I. Chazov, U.S.S.R. Patent 671,285 (1979).

for 30 min. Then 2 ml of urokinase solution (250,000 IU) in the same buffer is added to this solution, and incubation continues for 5 hr at 4°. Then the preparation of modified urokinase is isolated on a Sepharose 4B column (K-16, Pharmacia) equilibrated with solution of 50 mM phosphate buffer, pH 8.3, and lyophilized. The degree of binding (determined as in the case of urokinase immobilization on soluble copolymer) is 94%; urokinase activity preservation is 86%.

Urokinase Modification with Fibrinogen.[35] To increase the affinity of thrombolytic enzymes toward thrombus material they can be immobilized on fibrinogen. Attachment of urokinase (Green Cross, Japan, high molecular weight enzyme, specific activity 480 IU/mg) to fibrinogen (Calbiochem, San Diego, CA) is performed via the "spacer," 1,2-dodecamethylenediamine. Preliminarily fibrinogen carboxyl groups are activated with EDC. A fibrinogen solution (1.3 mg/ml) in distilled water is supplemented with 100-fold molar excess of EDC at pH 4.5 (HCl), and the mixture is incubated for 10 min at 20°; then a 100-fold molar excess of diamine in 0.1 M phosphate buffer, pH 8.3, is added to the mixture at 4°, and incubation continues for 5 hr. The product is dialyzed for 6 hr at 4° against distilled water, and then against 0.1 M phosphate buffer, pH 8.3 (product I). To activate urokinase carboxyl groups, the enzyme (24,000 IU) is dissolved in 2 ml of distilled water and dialyzed at 4° for 2 hr against distilled water, then for 2 hr against 50 mM phosphate buffer, pH 8.3. A 100-fold molar excess of carbodiimide is added, and the solution is incubated at 4° for 20 min (product II). The binding reaction is performed by combining product I and product II and subsequent incubation for 18 hr at 4°. The preparations of urokinase, covalently bound to fibrinogen, are isolated by gel chromatography on Sepharose 4B or ultrafiltration (Amicon XM-100 filter). The degree of binding and activity preservation determinations are performed as described above.

Attachment of urokinase to fibrinogen via a spacer binds 90% of the enzyme to protein matrix. After lyophilic drying the derivative appears as a white floccular powder readily soluble in water. The molecular mass of the derivative is 360,000–440,000 Da according to gel chromatography data on Sepharose CL-6B with protein markers. The enzyme content in the preparation is 10–20%; the activity preserved is around 50%.

Some General Remarks

One should realize that all the methods described are just typical examples, and each of them can be applied with minor variations to other

---

[35] A. V. Maksimenko and V. P. Torchilin, *Thromb. Res.* **38,** 289 (1985).

representative thrombolytic enzyme groups. Thus, urokinase and streptokinase can be immobilized on oxidized Sephadex; soluble activated dextran or heparin can be used as carriers for plasmin and streptokinase immobilization, etc.

The proper choice of immobilization conditions in all the cases described leads to obtaining immobilized thrombolytic enzymes possessing, in comparison with native ones, the following properties: (1) increased stability in physiological conditions; (2) decreased clearance rate from the circulation; and (3) decreased antigenicity. At the same time the catalytic properties of enzymes in the immobilized state practically do not change.

It is also necessary to mention that some problems arise in the determination of fibrinolytic (thrombolytic) activity of immobilized thrombolytic enzymes. If the activity of native and immobilized enzymes toward low molecular weight substrates can be easily compared in homogenous medium, in the case of immobilized enzymes the use of traditional methods of fibrinolytic activity determination (see, for example, Refs. 3, 4, 5, and 9) is impossible. An apparent decrease in immobilized enzyme fibrinolytic activity exists because under experimental conditions the increase in the molecular weight of the enzyme–carrier conjugate leads to an increase in the diffusional limitation of the penetration of immobilized enzymes into a fibrin clot.

In order to "equalize" experimental conditions for native and immobilized enzymes we developed a determination of the rate of fibrin clot lysis.[24,36] For this purpose 0.2 ml thrombin solution (4 mg/ml) is added to 0.5 ml fibrinogen solution (10 mg/ml) in 0.1 M phosphate buffer, pH 7.4, and the mixture is kept for 1 hr at room temperature. During this time a fibrin clot is formed. It is placed into a plastic tube 1.0–1.5 cm in diameter with the bottom made of a permeable microwell net. The bottom of the tube is washed with 15 ml of 0.1 M phosphate buffer solution, pH 7.4, and plasminogen is dissolved (0.33 mg/ml) (control) in this solution, along with preparations of native and modified thrombolytic enzymes taken in a ratio which provides equal catalytic activity toward low molecular weight specific substrates. Dissolution of fibrin clot is measured spectrophotometrically following the increase in optical density of the washing solution at 280 nm with time. The ratio of the solution optical density growth ΔA_{280} (as a result of fibrinolysis of the clot and transition of its solubilized fragments through the net into the solution) to the period of time t during which it occurred characterizes the rate of the fibrinolysis process in this system. This parameter, graphically determined as the tangent of a straight-line inclined angle in coordinates ΔA_{280} versus t, together with a

[36] V. P. Torchilin, E. V. Il'ina, and E. G. Tischenko, U.S.S.R. Patent 824,053 (1981).

time value for complete clot dissolution, is the criterion for the efficiency of the fibrinolysis process.

Experimental Study of Biodegradable Preparation of Immobilized Plasmin

For animal experiments 10–15 kg dogs are used. Left and right femoral arteries are isolated in morphine–barbamyl anesthetized animals. Isolation is carried out under regional anesthetization with 0.5% novocaine solution, and during the experiment arteries are punctured every 20 min with novocaine. Both arteries are shunted with a system of ball-flowmeter to measure the blood flow. The blood flow is registered on Mingograph-34. Artery branches are ligated, and the measurements of the arterial blood flow are made. Animals are heparinized (1000 units of heparin per kilogram weight). The amount of thrombotic mass which is administered to each artery depends on the size of the latter and varies from 1 to 3 g. The red thrombus is prepared from the venous blood of the same animal which had been taken 1 hr before the experiment. In order to prepare the red thrombus (agglutinated erythrocytes, coagulated fibrinogen, and precipitated proteins) a blood sample which had been taken with a dry glass syringe is stored for 1 hr at 20°. Embolization is performed by means of a polyethylene catheter with inner diameter of 4 mm. The inner diameter of embolized arteries is 3–4 mm. Thirty minutes after the thrombus administration into right femoral artery, 10 mg of preparation (prepared as described above; time of complete solubilization about 3 hr in *in vitro* conditions, 0.15 M phosphate buffer, pH 7.4, 37°) containing around 1000 IU of plasmin immobilized on oxidized Sephadex is administered by means of a polyethylene catheter which under X-ray control is introduced directly to the surface of thrombus.

In preliminary experiments it was shown that intraarterial administration of carrier granules in the same doses but without enzyme has no effect on blood flow and perfusion pressure. In all 6 experiments performed we observed the complete lysis of the thrombus with a recovery of blood flow (see Table I). Table I shows that the blood flow was restored only in the artery where immobilized plasmin was administered. No restoration of the blood flow was noticed in nontreated artery.

If plasmin has been used normally, i.e., after intravenous administration, the dose required for the solubilization or lysis of a similar thrombus would be equal to 100,000 units. It is evident that the effective amount of the immobilized enzyme is lower by 2 orders. Even under conditions of regional infusion of plasmin the dose of continuously administered native enzyme required for the lysis of the thrombus was equal to 15,000–20,000 units.

TABLE I
BLOOD FLOW IN DOG FEMORAL ARTERIES

Artery	Animal no.	After thrombosis	1 hr after immobilized plasmin administration
		% of the initial flow	
Treated	1	10	110
	2	28	130
	3	15	88
	4	20	113
	5	18	114
	6	25	140
Untreated	1	17	12
	2	20	30
	3	7	21
	4	8	0
	5	17	22
	6	17	26

Some Results of Clinical Studies of Streptokinase Immobilized on Oxidized Dextran ("Streptodekaza")[37]

Beginning in 1980 immobilized streptokinase was produced in the USSR on an industrial scale. It is used for the treatment of acute myocardial infarction, acute pulmonary artery thromboembolism, periferal arterial and deep vein thrombosis, and hemophthalmia.[38–42] In comparison with the native enzyme, Streptodekaza has a prolonged lifetime in the circulation (up to 80 hr) and causes practically no complications (see Table II). Instead of continuous intravenous drop-by-drop infusion of the drug, a single injection of the entire therapeutic dose can be given in an ambulance or even to a patient who is not hospitalized.

[37] E. I. Chazov, V. N. Smirnov, V. P. Torchilin, I. M. Tereshin, and B. V. Moskvichev, F.R.G. Patent DE 3,032,606 C2 (1984).
[38] V. P. Torchilin, Y. I. Voronkov, and A. V. Mazaev, *Ter. Arkh.* **54,** 21 (1982).
[39] E. I. Chazov, V. N. Smirnov, L. A. Suvorova, A. V. Suvorov, A. V. Mazaev, Y. I. Voronkov, and V. P. Torchilin, *Kardiologiya* **21,** 18 (1981).
[40] E. I. Chazov, A. V. Mazaev, V. P. Torchilin, and V. N. Smirnov, *Klin. Med. (Moscow)* **58,** 51 (1980).
[41] V. N. Smirnov, V. P. Torchilin, A. V. Mazaev, L. A. Suvorova, and Y. I. Voronkov, *Ukr. Biokhim. Zh.* **55,** 311 (1983).
[42] E. I. Chazov, R. A. Gundorova, A. D. Romaschenko, V. P. Makarova, A. V. Mazaev, Y. I. Voronkov, and V. P. Torchilin, *Vestn. Oftal'mol.* **4,** 61 (1982).

TABLE II

Comparison of the Clinical Efficacy of Streptokinase/Heparin and Streptodekaza/Heparin Administration in the Treatment of Patients with Acute Myocardial Infarction

	% of cases	
Clinical criterion	Streptokinase/ heparin (n = 37)	Streptodekaza/ heparin (n = 30)
Relief of pain syndrome	63.9	83.8
Fast ECG dynamics (first 24 hr)	45.7	67.6
Prolonged course and secondary infarction	13.3	5.4
Cardiac tamponade	13.3	2.7
Circulatory insufficiency (%)		
On admission	42.8	59.4
Disappearance on 14th day	20.0	51.3
Specific complications		
Hemmorrhagia	71.4	—
Thromboembolism	14.3	5.4
Allergic reactions	26.6	—

Streptodekaza is available in two commercial forms: vials containing 1,500,000 IU of streptokinase activity (for the treatment of cardiovascular diseases) and vials containing 100,000 IU (for the treatment of ophthalmological diseases). In case of cardiovascular thromboses the therapeutic dose is 3,000,000 IU (in some cases 6,000,000 IU). The whole dose is administered intravenously as a single infusion into 20–40 ml of physiological solution (the total time of the infusion is ~5 min) in 1 hr after the injection of a test dose (3,000,000 IU) in 2 ml of physiological solution. Because of the complete absence of hemorrhagic complications, heparin therapy can be started 6–12 hr after Streptodekaza administration. In the case of hemophthalmia total therapeutic doses (30,000–50,000 IU) in 0.2–0.3 ml of physiological solution can be administered locally, subconjunctivally, or intravitreously.

The clinical efficacy of Streptodekaza in the treatment of myocardial infarction can be seen from Table II. Analogous data are obtained in the treatment of other pathologies mentioned above. On treatment with Streptodekaza the increase in fibrinolytic and activator activity is pronounced and prolonged (3–10 days), the drug influence on the coagulative chain of the hemostasis is minimal, i.e., the increase in blood fibrinolytic activity is not accompanied by "streptokinase-like" depression of blood coagulative properties. In contrast to the treatment with streptokinase, there is no sharp increase in blood inhibitory activity; acid-labile inhibitors either disappear totally or are changed into activators. Streptodekaza therapy does not cause an increase in the number of platelets and does not influence their aggregation. The direct effect of the preparation on fibrinogen is not observed, and sharp variations which are characteristic of the therapy with the native enzyme are absent. The degradation products of the fibrinogen–fibrin complex are present in a lesser amount and disappear much faster as compared to treatment with the native enzyme.

Appendix

Some new data on the synthesis of immobilized thrombolytic enzymes have been published.

Polyethylene glycols of various molecular weights have become rather popular polymeric carriers. Urokinase has been immobilized on polyethylene glycols with molecular weights of 200–20,000[43] and streptokinase on polymers with molecular weights of 2000, 4000, and 5000.[44] Rajagopalan et al.[44] have carried out a detailed study of streptokinase–polyethylene glycol conjugate kinetic behavior in the reaction of plasminogen activation. It was shown that the use of carbonylimidazole-activated polymers gave products exhibiting practically unchanged activity. However, it is interesting to note that the molecular weight of polyethylene glycol has a noticeable impact on conjugate properties: streptokinase derivatives have somewhat different catalytic properties, stability, ability to interact with plasmin, etc. This result points to the necessity of optimizing properties of immobilized enzymes, not only according to the type of a carrier, but also according to its molecular weight. Modified enzymes, irrespective of the carrier molecular weight, demonstrate sharply decreased ability for the interaction with antibodies against native streptokinase. In addition, immobilized enzyme remains in the blood longer than does native enzyme.

Great attention is now being given to obtaining targeted enzymatic drugs. Drug targeting permits one to decrease the therapeutic dose of an enzyme and, thus, the side reactions accompanying its administration. In addition to urokinase binding to fibrinogen (which has already been mentioned), serious attempts are being made to bind thrombolytic enzymes with antibodies against thrombi components. The most natural target seems to be fibrin. Despite the structural similarity between fibrin and fibrinogen, monoclonal antibodies against fibrin have been obtained and

---

[43] K. Shimizu, T. Nakahara, and T. Kinoshita, U.S. Patent 4,495,285, Jan. 22, 1985.
[44] S. Rajagopalan, S. Gonias, and S. V. Pizzo, *J. Clin. Invest.* **75,** 413 (1985).

then used for the synthesis of urokinase–antifibrin antibody conjugate.[45,46] The binding was performed via SPDP. In *in vitro* experiments, the authors have shown that the fibrinolytic activity of antibody-immobilized urokinase is 100-fold higher than in the case of the native enzyme.

Low-molecular-weight urokinase has also been immobilized on the plasmin heavy chain via disulfide bridges.[47,48] The preparation obtained has been shown to be 10-fold more active than the native enzyme in the treatment of experimental thromboembolism in rabbits.

The immobilization of thrombolytic enzymes on modified thrombin (on modification, thrombin maintains its ability to recognize and to bind platelets, but loses completely its clotting ability) also gives preparations increased affinity toward thrombi.[49]

An interesting approach also seems to be the immobilization of thrombolytic enzymes on ferromagnetic carriers, which can be concentrated in the target zone by external magnetic fields.[50,51] In these experiments, different enzymes (plasmin, streptokinase, urokinase) and different magnetic carriers (magnetic Sephadex, polysaccharide-coated fine ferromagnetic particles, etc.) have been used. The immobilization usually proceeds via the activation of carriers with CNBr or periodate. Animal experiments (the lysis of the experimental thrombus in carotid arteries of dogs) have shown very high thrombolytic activity in the preparations described.

[45] K. Y. Hui, E. Haber, and G. Matsueda, *Science* **222,** 1129 (1983).
[46] K. Y. Hui, E. Haber, and G. Matsueda, *Thromb. Hemost.* **54,** 524 (1985).
[47] Y. Nakayama, W. Miyazaki, and M. Shinahara, U. S. Patent 4 545 988, Oct. 8, 1985.
[48] Y. Nakayama, M. Shinahara, T. Tani, T. Kawaguchi, T. Furuta, T. Izawa, H. Kaise, W. Miyazaki, and Y. Nakano, *Thromb. Hemost.* **56,** 364 (1986).
[49] A. V. Maksimenko, A. N. Rusetsky, and V. P. Torchilin, *Bull. Exp. Biol. Med.* (Russian) **103,** 35 (1987).
[50] V. P. Torchilin, M. I. Papisov, and V. N. Smirnov, *J. Biomed. Mater. Res.* **19,** 461 (1985).
[51] V. P. Torchilin, M. I. Papisov, N. M. Orekhova, A. A. Belyaev, A. D. Petrov, and S. E. Raginov, *Hemostasis,* in press.

[50] Soluble Enzyme–Albumin Conjugates: New Possibilities for Enzyme Replacement Therapy

By Mark J. Poznansky

Enzymes, by virtue of their exquisite specificities, low general toxicity, and ability to function under physiological conditions, might be expected to behave as ideal drugs. The potential use of enzymes as a means of therapy in a range of genetic and metabolic diseases, many of them manifested as specific enzyme deficiencies or defects, has long been anticipated but never properly realized. Except in the treatment of a number of digestive disorders, enzymes have not been widely used as therapeutic agents. Several serious drawbacks have limited the common use of enzymes in medicine: (1) Availability of the enzyme in a sufficiently pure and nontoxic form. (2) Biodegradation of administered enzyme due to proteolysis and heat inactivation. (3) Immunological reactivity of the administered enzyme, usually a foreign protein, often bacterial. (4) Delivery of enzyme to appropriate and specific sites of action—tissues, cells, and often intracellular organelles.[1]

Our laboratory is concerned with attempts to make enzymes more amenable to use in medicine in order to circumvent some of the stated limitations. The concept of packaging enzymes in order to protect and/or deliver the enzyme has been proposed by a number of investigators using semipermeable microcapsules or artificial cells, lipid vesicles or liposomes, intact cells, and synthetic or biopolymers (see Ref. 1 for a recent review). All of these solutions both solve and create problems especially as they relate to the carriers' interactions with the reticuloendothelial system (RES), the body's primary defense mechanism against particulate foreign materials.

Our approach has been to use albumin, a "natural" plasma protein, to immobilize enzymes in a soluble conjugated form in an effort to produce a stable nonimmunogenic and potentially targetable enzyme product. Thomas and colleagues[2] first described the preparation of cross-linked preparations of albumin and either uricase or L-asparaginase using the bifunctional reagent, glutaraldehyde, as the cross-linking agent. We have extended these initial physicochemical studies to a number of different cross-linking protocols (see below) and another eight different enzymes. We have also examined the immunological properties of a number of different conjugates using both heterologous and homologous albumins.[3] The possibility of conjugating specific "targeting agents" (poly- or monoclonal antibodies against cell surface-specific antigens or cell surface-specific ligands including a number of different polypeptide hormones) to the enzyme–albumin conjugate has been established with excellent results in tissue culture experiments and promising *in vivo* data.[4] The problems associated with multiple barriers and intravenous administration of such enzyme–carrier systems is discussed.

[1] M. J. Poznansky and R. L. Juliano, *Pharmacol. Rev.* **36,** 277 (1984).
[2] B. Paillot, M. H. Remy, D. Thomas, and G. Broun, *Pathol. Biol.* **22,** 491 (1974).
[3] M. J. Poznansky, *Pharmacol. Ther.* **21,** 53 (1983).
[4] M. J. Poznansky, R. Singh, B. Singh, and G. Fantus, *Science* **223,** 1304 (1984).

Materials and Methods

All chemicals and reagents were purchased from Sigma Chemicals (St. Louis, MO). Enzymes were purchased from Sigma [uricase (urate oxidase), α-glucosidase, L-asparaginase, catalase] and Boehringer Mannheim (α-glucosidase, cholesterol esterase, superoxide dismutase, glucose-6-phosphatase) and used without further purification except dialysis to standardize salt conditions during cross-linking and to wash the enzyme preparations of low molecular weight contaminants (6000 and less). α-1,4-Glucosidase from human placenta was prepared by ammonium sulfate precipitation and affinity chromatography using Sephadex G-100 (the sugar residues on the gel apparently function as a natural affinity site for the enzyme) according to established procedures.[5] The resultant product (as a result of maltose elution of the column) shows two peaks by SDS–PAGE eluting with apparent molecular weights of 58,000 and 66,000, values which suggest two subunits since the enzyme elutes with an approximate MW of 120,000 by molecular sieve chromatography. Enzymes and enzyme–albumin conjugates are assayed by standard procedures except for experiments examining enzyme activity in subcellular fractions where ultrasensitive micromethods have to be adopted where possible.

Chemical Cross-Linking

There are no specific *a priori* rules set to the cross-linking procedures, and a certain degree of trial and error is required in order to maximize (1) cross-linking, (2) enzyme recovery, and (3) enzyme stabilization. We have utilized three different cross-linking agents (see Ref. 1 for a recent review): glutaraldehyde to cross-link protein molecules by primary amino groups; 1-ethyl-3-(3-dimethylaminopropyl)carbodiimide (EDCI), a water-soluble carbodiimide, to cross-link a free carboxy group on one protein with a free amino group on the other; and periodate oxidation using sodium periodate to effect a cross-link between a sugar residue on one protein (a glycoprotein) with a primary amino group on another.[1] Using these procedures we have no assurance that intermolecular cross-linking does not occur as well, and this may be the source of loss in enzyme activity. In all cases we attempt to protect the active site of the enzyme during the cross-linking step by including an excess of substrate in the medium and by enhancing the possibility that the site is occupied (so that it cannot be cross-linked) by decreasing the turnover rate of the enzyme. This may be accomplished by carrying out the reaction at 4° and by leaving out essential cofactors during the cross-linking step. Following cross-linking the final conjugate may be further purified by either one or a

[5] T. deBarsy, P. Jacqaemin, P. Devos, and H.-G. Hers, *Eur. J. Biochem.* **31**, 156 (1972).

combination of dialysis, molecular sieve chromatography, HPLC, or affinity chromatography. The following represent examples of the three different methods of cross-linking we have used to produce enzyme–albumin and enzyme–albumin–targeting agent complexes.

Method 1. Here we describe the straightforward cross-linking of α-1,4-glucosidase (from *Aspergillus*) with a 20-fold molar excess of human serum albumin using glutaraldehyde to form a Schiff base cross-link between primary amino groups. One milligram of enzyme is mixed with 10 mg of albumin in 3 ml phosphate-buffered saline (PBS) (100 mM NaCl, 5 mM Na$_2$HPO$_4$, pH 6.8) and 10 mg maltose at 4°; 50 μl of glutaraldehyde (grade I 25%, v/v) is added with stirring, and the reaction is allowed to proceed for 3 hr at 4°. Fifty milligrams of glycine is added to stop the reaction, and the resultant product is dialyzed against PBS containing 1% (w/v) glycine using dialysis tubing with an exclusion limit of 14,000–15,000 MW for 24 hr with several changes of dialysis fluid. The product may be further purified by any number of chromatographic procedures or by ultrafiltration using Amicon XM300 to filter away small molecular weight conjugates (<300,000) or unreacted enzyme and albumin. We do not regularly reduce the resultant Schiff base body with sodium borohydride or sodium cyanoborohydride since this does not alter the yield or stability of the conjugate.

Method 2. The object of this procedure is to cross-link insulin to α-glucosidase–albumin conjugates in order to direct the enzyme to cells rich in insulin receptors. The enzyme–albumin conjugate is prepared according to Method 1 (above), but other methods of conjugation may also be used. Ten milligrams of insulin in 2 ml PBS is reacted with 100 μl of 25% (v/v) glutaraldehyde for 1 hr at room temperature. This represents a 156-fold molar excess of glutaraldehyde and minimizes the incidence of insulin–insulin or intramolecular cross-linking. The reaction mixture is then dialyzed for 24–48 hr against PBS with at least 4 washes using a dialysis tubing with an exclusion pore size 1,000–2,000 Da. This assures that no free glutaraldehyde is present. The 10 mg of insulin with one glutaraldehyde molecule ostensibly linked to at least one of insulin's three available free amino residues is then added to 20 mg of the enzyme–albumin conjugate in a total of 5 ml PBS (glycine-free) containing 25 mg of maltose. The reaction is allowed to proceed for 2 hr at 4°, after which the mixture is dialyzed against PBS for 24–48 hr with numerous washings using a dialysis tubing with a molecular weight cutoff of 12,000–14,000. This would be expected to get rid of most unreacted insulin.

The resultant enzyme–albumin–insulin may be further purified by gel chromatography. Using an insulin affinity column (antiinsulin antibodies bound to protein A–Sepharose) it is possible to separate enzyme–albumin conjugates with and without attached insulin. Radioimmunoassays for insulin performed on the resultant conjugates indicated a minimum of 30 insulin molecules per enzyme–albumin conjugate, considering an average MW of 8×10^5 for the latter. We have not validated the insulin radioimmunoassay quantitatively for insulin in the conjugated form so this value must be considered a lower limit. With that we calculate an average molar ratio of 1 : 10 : 30 for enzyme–albumin–insulin.

Method 3. The object in this procedure is to attach an intact immunoglobulin (IgG) molecule to the enzyme–albumin polymer. L-Asparaginase–albumin is prepared exactly as described for α-glucosidase in Method 1, using glutaraldehyde as the cross-linking agent. The prepared conjugate (20 mg) is then reacted in 3 ml of PBS with 5 mg of IgG in the presence of 35 mg of sodium periodate and allowed to react for 3 hr at 4°. The Schiff bases formed between the opened sugar residue on the IgG molecular and primary amino residues on the enzyme–albumin conjugate are stabilized by the addition of 30 mg of sodium cyanoborohydride. The conjugate is then dialyzed overnight against PBS. The enzyme–albumin–antibody complex can then be separated by gel chromatography on Sepharose 4B from free IgG or from enzyme–albumin. Using L-asparaginase and producing enzyme–albumin conjugates according to Method 1 (with ultrafiltration with an Amicon XM300 membrane), the average molecular weight was 8.2×10^5 with 90% of the protein eluting with molecular weights between 6×10^5 and 1.1×10^6. Following cross-linking of the IgG molecule to the conjugate the approximate molecular weight increased to 10.2×10^6 suggesting that on average 1.5 antibody molecules are conjugated to each enzyme–albumin complex giving a final calculated molar ratio of 1 : 10 : 1.5, enzyme–albumin–antibody.

We chose to use periodate oxidation for cross-linking the antibody to the conjugate since the sugar residues are apparently located on the Fc portion of the IgG molecule and we are more concerned with maintaining the integrity of the Fab portions which realize the antibody's specificity. Attempts to utilize carbodiimides as cross-linking agents were generally not as productive with much greater loss of enzyme activity possibly due to intermolecular cross bridging between adjacent free carboxyl and amino groups. Alterations in cross-linking conditions (time, temperature, protein, and/or cross-linking reagent concentrations) yielded almost linear changes in the conjugate size.

Results

The data represented in the tables attempt to show how enzymes can be modified to function as more effective therapeutic agents. Table I demonstrates the increased resistance of the enzyme conjugate to pro-

TABLE I

STABILITY OF ENZYME–ALBUMIN CONJUGATES

Enzyme preparation	Conditions	$T_{1/2}$
L-Asparaginase	37° + 5 U trypsin	30 min
L-Asparaginase–albumin	37° + 5 U trypsin	6 hr
α-1,4-Glucosidase (yeast)	37° + 5 U trypsin	10 min
α-1,4-Glucosidase–albumin	37° + 5 U trypsin	180 min
α-1,4-Glucosidase (human placenta)[a]	37° + 5 U trypsin	≫10 hr
α-1,4-Glucosidase–albumin	37° + 5 U trypsin	≫10 hr
α-1,4-Glucosidase	45°	60 min
α-1,4-Glucosidase–albumin	45°	240 min
α-1,4-Glucosidase plus albumin[b]	45°	95 min

[a] Placental α-1,4-glucosidase in both free and albumin-conjugated forms show no alteration in enzyme activity over a 6-hr period at 37° in the presence of 20% fetal calf serum.

[b] Enzyme and albumin coincubated at 45° without benefit of cross-linking.

teolytic degradation or heat denaturation when compared to an equivalent amount of free enzyme with or without free (unconjugated) albumin. A protein environment in and of itself without benefit of chemical cross-linking may also serve to stabilize enzyme activity. The extent of recovery of enzyme activity following the cross-linking step is largely a function of the procedures used and the extent to which attempts are made to maximize recovery of enzyme activity. In our hands using glutaraldehyde (Method 1) as the cross-linking agent we have achieved recoveries ranging from 48% for cholesterol esterase–albumin conjugates to 100% for superoxide dismutase–albumin conjugates. An initial concern was that the cross-linking procedure would limit access of certain substrates to the active site of the enzyme due to steric hindrance. This was shown to be a minor factor in comparing the ability of α-glucosidase and α-glucosidase–albumin conjugates to reduce artificial (and small) substrates, p-nitrophenylglucoside or maltose or natural (and large) substrate, glycogen.

Table II demonstrates the altered plasma clearance of enzyme introduced in free and albumin–conjugate forms.[6] Superoxide dismutase represents the most dramatic change allowing for a significant presence of the enzyme–albumin conjugate in the plasma as a function of time compared to the half-life of the free enzyme which can be measured in seconds.[7] Similar increases have been possible with all enzymes attempted thus far.

[6] M. J. Poznansky and D. Bhardwaj, *Can. J. Physiol. Pharmacol.* **58**, 322 (1980).

[7] K. Wong, L. G. Cleland, and M. J. Poznansky, *Agents Actions* **10**, 231 (1980).

TABLE II
PLASMA CLEARANCE OF ENZYME–ALBUMIN
CONJUGATES

Enzyme preparation	Circulation half-life
α-1,4-Glucosidase (yeast)	2.5 hr
α-1,4-Glucosidase–albumin	12 hr
α-1,4-Glucosidase–albumin–insulin[a]	5 hr
Superoxide dismutase (hog liver)	1–2 min
Superoxide dismutase–albumin (1 : 10)	16 hr
Superoxide dismutase–albumin (1 : 20)	4 hr
Uricase (hog liver)	3 hr
Uricase–albumin	12 hr

[a] Mole ratio, enzyme : albumin : insulin = 1 : 10 : 60.

Attaching a targeting agent, such as a monoclonal antibody, to the conjugate can further increase the rate of clearance of the conjugate from the circulation.

Table III represents a summary of work demonstrating the potential of these procedures to decrease the immunogenicity of normally highly immunogenic foreign protein molecules. The initial work from our laboratory[8,9] has now been verified by others[10] and suggests that the use of a molar excess of homologous albumin conjugates to the foreign enzyme masks the antigenic determinants on the enzyme providing the organism a recognition of only a self protein. We know less about the intracellular processing of the enzyme–albumin conjugate and the possible exposure of antigenic sites following proteolysis within the cell, but we have no evidence that antibody formation occurs even with repeated injections in mice over a period of months.

One of the primary limitations to enzyme replacement therapy remains the question of how to target the enzyme or enzyme–carrier complex to appropriate sites of substrate accumulation. In Pompe's disease or Type II glycogenosis, glycogen accumulates as a result of α-1,4-glucosidase deficiency in lysosomes of most cells. The disease is fatal due to continued lysosomal storage and cellular dysfunction in liver, spleen, and muscle, death usually occurring by the age of 2 or 3 due to cardiac and

[8] M. H. Remy and M. J. Poznansky, Lancet 2, 68 (1978).
[9] M. J. Poznansky, M. Shandling, M. A. Salkie, J. Elliott, and E. Lau, Cancer Res. 42, 1020 (1982).
[10] T. Yagura, Y. Kamisaki, H. Wada, and Y. Yamamura, Int. Arch. Allergy Appl. Immunol. 64, 11 (1981).

TABLE III

IMMUNOGENICITY[a] OF ENZYME–ALBUMIN CONJUGATES

Enzyme preparation	Immune response
Uricase (hog liver)	+++
Uricase–albumin (1 : 10)	−
α-1,4-Glucosidase (human placenta)	+++
α-1,4-Glucosidase–albumin (1 : 10)	−
Superoxide dismutase (bovine)	+++
Superoxide dismutase–albumin (1 : 5)	+
Superoxide dismutase–albumin (1 : 10)	−
L-Asparaginase (E. coli)	+++
L-Asparaginase–albumin (1 : 5)	+
L-Asparaginase–albumin (1 : 10)	−

[a] Immunogenicity was determined as described in Refs. 8 and 9. Both immunodiffusion and radioimmunoassays were used to detect the presence or absence of antibodies.

respiratory muscle dysfunction. There is some indication that hepatic tissue glycogen might be reduced by simple infusion of the enzyme, the liver and particularly cells of the RES being the natural site of clearance of foreign protein.[11–13] No evidence for reduction of glycogen in the more serious site, muscle tissue, has been presented. We have demonstrated that insulin[4] covalently conjugated to either enzyme or enzyme–albumin conjugates is capable of binding to and being internalized by the insulin receptor by a process which resembles receptor-mediated endocytosis. Table IV demonstrates that the process results in the degradation of cholesterol ester stored within lysosomes of fibroblasts from a patient with cholesterol ester storage disease.

This is an important demonstration of the possibility of introducing a deficient enzyme into a cell and into a particular intracellular storage vesicle resulting in the degradation of an accumulating substrate. Not only can we realize an effective enzyme replacement in cells with a particular enzyme deficiency, but we make use of the cells own protein transport system (the insulin receptor) to direct the administered enzyme to the appropriate site. By piggybacking the enzyme–albumin conjugate into the cell by the process of receptor-mediated endocytosis we are using the

[11] M. J. Poznansky and D. Bhardwaj, Biochem. J. 196, 89 (1981).
[12] A. Abuchowski and F. F. Davis, in "Enzymes as Drugs" (J. S. Holcenburg and J. Roberts, eds.), p. 367. Wiley Interscience, New York, 1981.
[13] F. S. Furbish, C. J. Steer, N. L. Krett, and J. A. Barranger, Biochim. Biophys. Acta 673, 425 (1981).

TABLE IV
ENZYME REPLACEMENT[a] IN CHOLESTEROL ESTERASE-DEFICIENT FIBROBLASTS

Enzyme preparation	Cholesterol ester (μg/mg cell protein)	Cholesterol ester/ cholesterol
1. Cholesterol + esterase	36.1	0.690
2. Cholesterol + esterase–albumin	36.3	0.681
3. Cholesterol esterase–albumin–insulin	12.2	0.198
4. Cholesterol esterase–insulin	12.1	0.204
5. Insulin	36.9	0.705
6. Cholesterol esterase + free insulin	36.0	0.652
7. Cholesterol esterase–insulin + 500× excess of free insulin	35.8	0.621
Control cholesterol esterase-deficient fibroblasts	37.4	0.710
Control normal fibroblasts	4.8	0.104

[a] Confluent fibroblasts (normal) or cholesterol esterase-deficient fibroblasts were incubated for 4 hr at 37° in the presence of various enzyme preparations. Two micrograms of enzyme was added per 2×10^6 cells. Preparations 1–7 use cholesterol esterase-deficient fibroblasts; 3–6 contain 4 μg of insulin either free or conjugated to enzyme; 7 contains 2 mg of unconjugated insulin.

cell's own mechanism for directing exogenous proteins into the cell for processing, often via the lysosome.

While Table IV demonstrates effective enzyme replacement in enzyme-deficient fibroblasts established in tissue culture, one important barrier must still be dealt with, and that is the endothelial barrier. Can the targetable enzyme–carrier complex be designed so as to use a physiological pathway to leave the circulation and traverse the specific endothelial barrier to reach underlying target tissue? Unfortunately little is yet known of the chemical specificity of the barrier or of the mechanism which might allow large proteins to traverse.

Acknowledgment

We are grateful to Damyanti Bhardwaj for her ongoing contributions, to Jennifer Halford and Deanna Moores for their technical support, and to the Medical Research Council of Canada.

[51] Microencapsulation of Pancreatic Islet Cells: A Bioartificial Endocrine Pancreas

By Anthony M. Sun

Since the pioneering work of Chang and others,[1-4] numerous methods have been developed for immobilizing biologically active materials such as enzymes. The next obstacle was to develop a method for the microencapsulation of living cells within a semipermeable membrane which would permit the passage of nutrients and oxygen, but not of cells or high molecular weight substances. However, attempts to synthesize semipermeable microcapsules which are biocompatible with human and animal tissues have met with very little success until recently. The *in vivo* survival times of transplanted microencapsulated tissue or cells have typically been less than 2–3 weeks. This limitation severely restricts the applicability of the microencapsulation procedure to the treatment of diseases such as diabetes[5] which could benefit from cell transplantation.

It has been established that syngeneic islet transplantation can prevent or reverse the retinal and renal complications of diabetics. However, the immunogenicity of islet cells remains a major obstacle to the use of this therapeutic approach. A solution to this problem may be provided by the introduction of a physical, semipermeable barrier between the transplanted islets and the immune system of the host.

Studies in our laboratory have demonstrated that capsules made of an alginate–poly(L-lysine) (PLL)–alginate membrane survive *in situ* for nearly 1 year after transplantation,[6,7] indicating that this membrane is biocompatible. When isolated rat pancreatic islets were microencapsulated in this membrane and implanted into diabetic rats[7] and mice,[8] the diabetic state was reversed; in some cases, for the life span of the recipient.

With our encapsulation method, greater than 90% of the islet cells retain their viability and functional properties. However, current studies

---

[1] T. M. S. Chang, *Science* **146,** 524 (1964).

[2] K. Mosbach and R. Mosbach, *Acta Chem. Scand.* **20,** 2807 (1966).

[3] T. M. S. Chang, F. C. MacIntosh, and S. G. Mason, *Can. J. Physiol. Pharmacol.* **44,** 115 (1966).

[4] V. Hackel, J. Klein, R. Megret, and F. Wagner, *Eur. J. Appl. Microbiol.* **1,** 291 (1975).

[5] F. Lim and A. M. Sun, *Science* **210,** 908 (1980).

[6] Y. F. Leung, G. M. O'Shea, M. F. A. Goosen, and A. M. Sun, *Artif. Organs* **7,** 208 (1983).

[7] G. M. O'Shea, M. F. A. Goosen, and A. M. Sun, *Biochim. Biophys. Acta* **804,** 133 (1984).

[8] G. M. O'Shea and A. M. Sun, *Diabetes* **35,** 943 (1986).

to optimize some encapsulation parameters such as the purity and viscosity of the sodium alginate, the viscosity average molecular weight ($\bar{M}_v$) of the PLL, the alginate–PLL reaction time, and PLL concentration are critical in the preparation of improved microcapsules.

Materials

Pancreatic Islets. Rat or dog islets were isolated by the collagenase-digestion technique and either hand-picked or purified through discontinuous Ficoll gradients as described previously.[9,10] Briefly, pancreatic tissue is dressed and perfused with Hank's balanced salt solution. The minced tissue is then digested for 12–15 min at 37° with collagenase (12–15 mg/4 ml of tissue). Islets can be handpicked from the digest with the aid of a dissecting microscope, or the tissue is mixed with 25% Ficoll in a centrifuge tube. Three Ficoll concentrations (23, 20, and 11%) are then layered above this suspension. Centrifugation is carried out for 10 min at 800 *g*. The islets are harvested from the interface of the 23 and 20% Ficoll layers. The cells can either be microencapsulated immediately or cultured for 1–3 days prior to encapsulation.[11]

Sodium Alginate Solutions. To prepare sodium alginate solutions, 3.0 g of sodium alginate (Kelco Gel L.V.) is sprinkled or sifted slowly into 100 ml of distilled water with stirring. Then 100 ml of 1.8% NaCl solution is added and well mixed. The mixture is then centrifuged at 30,000 *g* for 1 hr at 4°. The supernatant is sterilized by filtering through a Nalgene filtration unit (0.2 μm) and stored at 4°. The 0.15% sodium alginate solution is prepared by diluting the 1.5% solution 10-fold with physiological saline.

Calcium Chloride Solutions. To prepare 1.1% calcium chloride solution, 11.0 g of $CaCl_2$ (anhydrous) is dissolved in 100 ml of distilled water. The 0.55% and 0.28% $CaCl_2$ solutions are made by progressive 1 : 1 dilutions of the stock solution with physiological saline.

2-(N-Cyclohexylamino)ethanesulfonic Acid (CHES). To prepare the stock solution, 2.0 g of CHES is dissolved in 100 ml of 0.6% NaCl, and the pH is adjusted to 8.2 with 1 *N* NaOH. Five milliliters of the stock solution is added to 95 ml of the 1.1% $CaCl_2$ solution to make the 0.1% (w/v) CHES solution.

Poly(L-lysine) Solution (PLL). Fifteen milligrams of PLL (Sigma) is dissolved in 30 ml of saline.

[9] P. E. Lacy and M. Kostianovsky, *Diabetes* **16,** 35 (1967).
[10] A. M. Sun, B. J. Lin, and R. E. Haist, *Can. J. Physiol. Pharmacol.* **51,** 175 (1973).
[11] A. M. Sun, G. M. Healy, I. Vacek, and H. G. MacMorine, *Biochem. Biophys. Res. Commun.* **79,** 185 (1977).

Sodium Citrate Solutions. Trisodium citric acid (3.23 g) is dissolved in 100 ml of distilled water to which 100 ml of saline is added and mixed (55 m*M* sodium citrate solution).

Culture Medium. Medium CMRL-1969 is supplemented with fetal calf serum (7.5%) and gentamicin (20 mg/mL) for tissue culture.

Procedure for Microencapsulation

A syringe pump with a 10-ml syringe connected to a special air jet is used for making sodium alginate islet droplets.

For each preparation, 2000–3000 islets in 0.2 ml of saline is mixed gently with 2 ml of 1.5% sodium alginate, transferred to the 10-ml syringe, and connected to the air jet. The distance from the tip of the air jet needle to the surface of the collecting fluid is set at precisely 4 cm. The syringe pump and air flow are then turned on to extrude sodium alginate droplets containing islets into 50 ml of the 1.1% $CaCl_2$ solution in a beaker. During extrusion, the islets are kept in suspension by gently rotating a small magnet inside the 10-ml syringe. After the extrusion process is completed, the spherical calcium alginate gel droplets are transferred to a 50-ml polystyrene test tube with a conical bottom, and allowed to settle before withdrawing the supernatant down to 5 ml using a vacuum aspirator. The gel droplets are washed once with 30 ml of 0.55% $CaCl_2$, once with 0.28% $CaCl_2$, and then suspended in 25 ml of 0.1% CHES solution for 3 min. After aspirating the CHES solution, the capsules are washed with 1.1% $CaCl_2$ and suspended in 25 ml of 0.05% (W/V) poly(L-lysine) for 6 min. After further washing with CHES, $CaCl_2$, and saline, the microcapsules are incubated in 0.15% sodium alginate for 4 min and washed with saline. The capsules are then suspended in 10 ml of 55 m*M* sodium citrate solution for 5 min. The final product is washed twice with saline and once with medium CMRL-1969, and then transferred to Falcon flasks for incubation at 37° until required for *in vitro* and *in vivo* studies.

In Vitro and *in Vivo* Assessment of Capsules

For gross assessment, the finished microcapsules can be examined under an inverted microscope (see Fig. 1). A perfect spherical shape is important to the *in vivo* survival time of the capsules. Evaluation of the surface finish, wall thickness, and membrane uniformity can be done by scanning electron microscopy.[12] The results should show smooth interior

---

[12] M. F. A. Goosen, G. M. O'Shea, H. M. Gharapetian, S. Chou, and A. M. Sun, *Biotechnol. Bioeng.* **27**, 146 (1985).

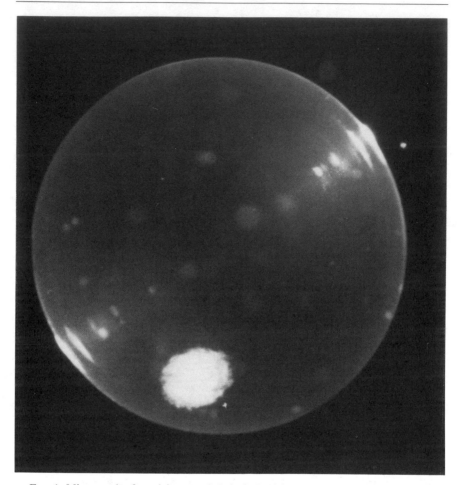

FIG. 1. Micrograph of an alginate–poly(L-lysine)–alginate capsule containing a rat pancreatic islet of Langerhans. The diameter of the capsule is 700 μm and the wall thickness 4 × 5 μm.

and exterior capsular surfaces, with a membrane thickness of 4–5 μm. Histochemical staining of paraffin sections with aldehyde thionine is used to evaluate the viability and integrity of the microencapsulated islets. A normal degree of glucagon and insulin granulation should be observed in the encapsulated islet cells.

Insulin secretion from microencapsulated islets in response to glucose challenge can be measured by (1) long-term culture and (2) perfusion experiments using dual chambers, as described by Lacy *et al.*[13] In general, the insulin secretion from the microencapsulated islets into the culture media or the perifusate is comparable in quantity with that from control, unencapsulated islets. In the perifusion experiments, a biphasic response of insulin release from both groups of islets is observed when the glucose concentration is raised from 50 to 300 mg/dl.

Detailed descriptions of *in vivo* tests of the efficacy and biocompatibility of microencapsulated islets have been published elsewhere.[6,7,8,14] Briefly, diabetes is induced in animals with streptozotocin. Microencapsulated islets suspended in saline are implanted into the peritoneal cavity of diabetic animals, using a cannula attached to a 10-ml syringe. When 4.5×10^3 encapsulated islets are implanted in diabetic rats normoglycemia is restored in recipient animals within 2 days, and maintained for more than 3 months. In our laboratory, one animal remained normoglycemic until sacrificed, 780 days posttransplantation. Xenografts of encapsulated rat islets reversed diabetes in mice for a mean of 80 days.

Free intact capsules were recovered from transplant recipient animals and were shown by histochemical studies to contain viable islet cells. Furthermore, when placed into culture, the retrieved islets secreted insulin in response to a glucose challenge.

Variations

Variations in membrane properties among the different preparations are difficult to eliminate. However, the variations can be minimized by optimizing several important microencapsulation parameters. The strength and the permeability of the capsule membrane depend on the viscosity average molecular weight $\bar{M}_v$ of the PLL used in the encapsulation procedures; the lower the $\bar{M}_v$ of PLL, the less permeable the capsules. The alginate–PLL contact time is also a factor which affects the strength of the capsule membrane. The second coating of sodium alginate on alginate–PLL capsules plays a key role in the long-term durability of the capsules during *in vivo* studies, although the *in vitro* stability is not affected. In previous reports microcapsules prepared with alginate–PLL–polyethyleneimine all failed to survive for a long period of time in animal studies.[15]

[13] P. E. Lacy, E. H. Finke, S. Conant, and S. Naber, *Diabetes* **25,** 484 (1976).
[14] A. M. Sun, G. M. O'Shea, and M. F. A. Goosen, *in* "Biocompatible Polymers, Metals and Composites" (M. Szycher, ed.), p. 929. Technomic, Lancaster, Pennsylvania, 1983.
[15] A. M. Sun, G. M. O'Shea, and M. F. A. Goosen, *Appl. Biochem. Biotechnol.* **10,** 87 (1984).

Summary

It was about two decades ago that Chang proposed the use of microencapsulated islets as artificial beta cells. By using alginate–poly(L-lysine)–alginate membranes, biocompatible, durable capsules containing viable islet cells can be produced which are impermeable to cells and effector molecules of the immune system, thus providing a total protection to transplanted islets against rejection. The capsule wall contains 93% (w/w) water and can be classified as a hydrogel. Many hydrogels have gained general acceptance as being biocompatible materials. Microencapsulation of pancreatic islets for use as an artificial endocrine pancreas would not only obviate the need for immunosuppressive therapy but also has the potential to prevent the long-term complications of diabetes. Furthermore, the microencapsulation technique can be applied to other types of cells to produce antibodies or enzymes, and to treat a whole range of diseases requiring endocrine replacement therapy.

Section III

Novel Techniques for and Aspects of Immobilized Enzymes and Cells

Editor

Klaus Mosbach

[52] Overview

By K. Mosbach

This section includes a number of contributions on topics not presented in the preceding sections. The paper on enzyme stabilization by Monsan and Combes [53] serves as review and is followed by chapters by Shaked and Wolfe [54] and Martinek and Torchilin [55]. Since this review was written much additional emphasis has been given to the aspect of enzyme stabilization in organic solvents at normal and elevated temperatures and its application. The reader will also find a number of related papers in Vol. 136, Section II on Immobilized Enzymes/Cells in Organic Synthesis. The article by Mozhaev, Berezin, and Martinek (Vol. 135 [53]) also touches on the aspect of enzyme stabilization.

The aspect of cell permeabilization with its potential usefulness in facilitating in- and out-diffusion of chemicals to be biologically transformed is discussed in the chapter by Felix [58].

Two contributions deal with long-term stability of immobilized cell systems (Häggström and Förberg [56] and Bülow *et al.* [57]). In the latter chapter more specifically the use of immobilized genetically engineered cells is discussed. The interested reader should also read an article dealing with this important aspect of long-term stability by Klein and Wagner [*Enzyme Eng.* **8,** *Ann. N.Y. Acad. Sci.* **501,** 306 (1987)].

The preparation of cells coentrapped with enzymes is described in the chapter by Hahn-Hägerdal [59]. Apart from this important aspect, mention should be made to other applications in which coimmobilization and more specifically coentrapment can be utilized beneficially. The reader is referred to some relevant publications on this topic. For *in situ* generation of oxygen, algae have been coentrapped with the microorganism of interest [e.g., P. Wikström, E. Szwajcer, P. Brodelius, K. Nilsson, and K. Mosbach, *Biotechnol. Lett.* **4,** 153 (1982); P. Adlercreutz, O. Holst, and B. Mattiasson, *Enzyme Microb. Technol.* **4,** 395 (1982)]. Alternatively MnO_2 or carbon has been coentrapped [P. Brodelius, K. Nilsson, and K. Mosbach, *Appl. Biochem. Biotechnol.* **6,** 293 (1981)]. Further, media containing O_2- and/or CO_2-enriching solvents have been used [P. Adlercreutz and B. Mattiasson, *Eur. J. Appl. Microbiol. Biotechnol.* **16,** 165 (1982); A. Leonhardt, E. Szwajcer, and K. Mosbach, *Appl. Microbiol. Biotechnol.* **21,** 162 (1985)]. Although slightly peripheral to the subject I cannot help referring to the interesting work on immobilized systems, more specifically on entrapped hemoglobin with potential as an artificial lung investigated by Bonaventura. [*Proc. Int. Conf. Enzyme Eng. Helsingör, Denmark, 8th,* p. 45 (Abstr.) (1985)]. Alternative "immobilization techniques" are given by Mattiasson [60, 61] presenting both affinity immobilization and bioconversions in two-phase systems.

Some more unusual but interesting applications of immobilized enzyme/cell systems are then given dealing with the aspect of biochemical energy conversion by immobilized photosynthetic bacteria (Karube and Suzuki [62]) and the use of immobilized cytochrome *P*-450 enzyme (King *et al.* [63]). Linko and Linko [64] give a useful example of the production of enzymes by immobilized cells, an area with potential for not only enzyme but also hormone production. In addition, the scope and further potential of the application of immobilized cells in water purification is discussed by Bryers and Hamer [65]. The preparation and retention of a sterile immobilized system is not discussed in this volume. However, within Vols. 135–137 the reader will find information in a number of topics, e.g., in Section III, Vol. 136, on the industrial use of enzymes/cells. Admittedly the collection of titles is rather heterogeneous. So is the structure of the various contributions, but hopefully all this will, in addition to providing methodological know-how, act as a stimulus for further research.

[53] Enzyme Stabilization by Immobilization

By Pierre Monsan and Didier Combes

Introduction

Developments in biological catalyst applications (isolated enzymes, organelles, whole cells) require continuous reactors to be used in many cases. Such operations call for biocatalyst immobilization by inclusion, adsorption, cross-linking, or covalent grafting, techniques which have been described in the present work. Moreover, one of the key parameters with regard to the feasibility of a biotechnological process is the combination of efficiency and stability of the biocatalyst used.[1–7] Immobilization

[1] O. R. Zaborsky, "Immobilized Enzymes." CRC Press, Cleveland, Ohio, 1973.
[2] I. Chibata, "Immobilized Enzymes: Research and Development." Halsted, New York, 1978.
[3] R. D. Schmid, *Adv. Biochem. Eng.* **12**, 41 (1979).
[4] V. P. Torchilin and K. Martinek, *Enzyme Microb. Technol.* **1**, 74 (1979).
[5] A. M. Klibanov, *Adv. Appl. Microbiol.* **29**, 1 (1983).
[6] V. V. Mozhaev and K. Martinek, *Enzyme Microb. Technol.* **6**, 50 (1984).

very often results in a greatly increased resistance to various denaturation factors: extreme pH and temperature values, high ionic strength, denaturating reagents, proteases, etc. Another approach to stabilization is to modify the biocatalyst microenvironment by the use of additives such as salts, polyols, sugars, and various polymers.[3–7] It has thus been possible to obtain enzyme preparations with a highly increased resistance to denaturation factors, e.g., certain protease preparations used in the manufacture of detergents. We shall now take a closer look at the possibilities of biocatalyst stabilization, on the one hand by immobilization, and on the other hand by immobilization plus additives, with the help of a few significant examples.

Stabilization by Immobilization

Enzyme Inclusion

Direct comparison of the operational stability, i.e., in the presence of a continuous supply of substrate, of an immobilized enzyme with that of a free enzyme has been possible in the ultrafiltration reactor. For example, we have taken invertase (β-D-fructofuranosidase, EC 3.2.1.26), which hydrolyses sucrose into glucose and fructose, using a clay, bentonite, as immobilization support.

Preparation of Immobilized Invertase.[8] One hundred grams of bentonite is suspended in 200 ml of dioxane containing 5 g cyanuric chloride (Fluka). The suspension is stirred for 90 min at 20°. The bentonite–cyanuric chloride (BCC) complex is separated by centrifugation and washed 5 times by suspension in pure dioxane, and then vacuum-dried. The weight of cyanuric chloride fixed to bentonite is of the order of 0.68%. The BCC complex is then added to an invertase (Serva)–distilled water solution and stirred for 1 hr at 4°. The bentonite–cyanuric chloride–invertase (BCCI) complex thus obtained is first washed 3 times in a sodium hydroxide solution (pH 10.5), to eliminate any enzyme adsorbed onto the bentonite, and a further 3 times in distilled water to remove the sodium hydroxide.

Comparison of Operational Stability of Free Invertase and Invertase Immobilized on Bentonite.[9] A flask containing an enzyme solution without substrate and two ultrafiltration reactors (Amicon, Model 52, UM10 membranes) is placed in a water bath thermostatted at 50°. One of the

[7] P. Monsan and D. Combes, *in* "The World Biotech Report 1984," Vol. 1, p. 379. Online, Pinner, England, 1984.
[8] P. Monsan and G. Durand, *FEBS Lett.* **16**, 39 (1971).
[9] P. Monsan and G. Durand, *C.R. Acad. Sci.* **273**, 33 (1971).

reactors contains a free enzyme solution whereas the other contains immobilized enzyme grafted onto bentonite (BCCI). At the beginning of the experiment, the three enzymatic activities are appreciably the same. The substrate (a 5 g/liter sucrose solution in acetate buffer) is fed into the ultrafiltration reactors using nitrogen at a pressure of 3 bars. Figure 1 shows the variations in enzymatic activity for free invertase without substrate, free invertase with sucrose, and immobilized invertase with substrate. The experiment lasted 46 days. In the reference solution, activity decreases very rapidly. In the reactor with free enzyme, activity decreases very rapidly at the beginning, then stabilizes. However, at the end of the experiment, it was noticed that residual activity is only due to the amount of enzyme adsorbed onto the ultrafiltration membrane. Finally, in the reactor containing the immobilized enzyme, activity remains constant throughout the experiment. Under continuous operation, immobilized invertase is thus greatly more stable than free invertase.

Inclusion of Microbial Cells

The stabilizing effect due to immobilization has been observed not only for isolated enzymes as described above, but also for the enzymatic

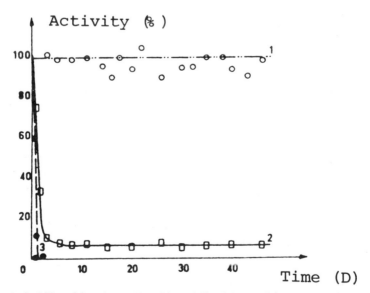

FIG. 1. Stability of free (curve 2) and immobilized (curve 1) invertase under continuous operation at 50°. Curve 3 corresponds to free invertase without substrate. Invertase was covalently grafted to bentonite. From Monsan and Durand.[9]

activity of whole cells. Chibata[10] studied the resistance to various denaturation factors of fumarase activity of whole cells of *Brevibacterium flavum,* both free and included in carrageenan gel, and of the enzyme extracted from this same strain. Figure 2 shows the resistance of fumarase activity to thermal denaturation: immobilized cells are a little more stable than intact cells and much more stable than extracted enzyme. Chibita also observed that immobilized cells are much more resistant than intact cells and extracted enzyme when subjected to 3 *M* ethanol treatment at 37° (Fig. 3) or 3 *M* urea treatment at 37° (Fig. 4). Carrageenan gel thus has a protective effect on fumarase activity of *Brevibacterium flavum* cells, equally with regard to thermal denaturation and to that of an organic solvent or protein-denaturing agent. This protective effect is related to the presence of carrageenan in the form of a gel, whereas liquid-state carrageenan has practically no protective effect whatsoever.[10]

Inclusion and/or Cross-Linking: Organelles

Biophotolysis of water is attracting considerable attention in view of its long-term potential for the production of energy-rich molecules from sun and water.[11] The application of such a system on a large scale is limited by problems of storage stability and photoinactivation of chloroplast membranes. Lettuce thylakoids (chloroplast membranes) have been immobilized by various methods on various carriers: cross-linked albumin polymer,[12] cross-linked gelatin polymer,[12] polyurethane matrix,[13] polyurethane–BSA matrix,[13] carrageenan gel,[14] alginate gel,[15] and photo-cross-linkable resin.[16] Storage and functional stability may be studied using comparative procedures of chloroplast membrane immobilization.

Storage Stability. Native and immobilized thylakoids are stored in the dark at 4° and periodically sampled, and their activity is assayed. Oxygen evolution is measured amperometrically using a Clark-type electrode.

[10] I. Chibata, *in* "Cellules Immobilisées" (J. M. Lebeault and G. Durand, eds.), p. 7. Soc. Fr. Microbiol., Paris, 1979.

[11] M. F. Cocquempot, B. Thomasset, J. N. Barbotin, G. Gellf, and D. Thomas, *Eur. J. Appl. Microbiol. Biotechnol.* **11**, 193 (1981).

[12] G. Brown, D. Thomas, G. Gellf, D. Domurado, A. M. Berjonneau, and C. Guillon, *Biotechnol. Bioeng.* **15**, 359 (1973).

[13] S. Fukushima, T. Nagai, K. Fujita, A. Tanaka, and S. Fukui, *Biotechnol. Bioeng.* **20**, 1465 (1978).

[14] A. Tanaka, S. Yasuhara, G. Gellf, M. Osumi, and S. Fukui, *Eur. J. Appl. Microbiol. Biotechnol.* **5**, 17 (1978).

[15] S. Ohlson, P. Larsson, and K. Mosbach, *Eur. J. Appl. Microbiol. Biotechnol.* **7**, 103 (1979).

[16] S. Fukui, A. Tanaka, T. Iida, and E. Hasegawa, *FEBS Lett.* **66**, 179 (1976).

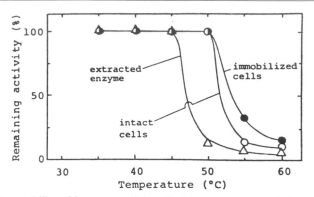

FIG. 2. Heat stability of fumarase activity. Heat treatment was carried out at pH 7.0 for 1 hr. From Chibata.[10]

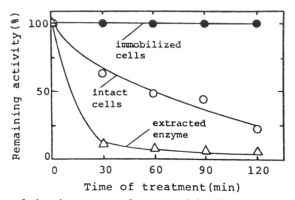

FIG. 3. Effect of ethanol treatment on fumarase activity. Treatment was carried out using 3 M ethanol at 37°. From Chibata.[10]

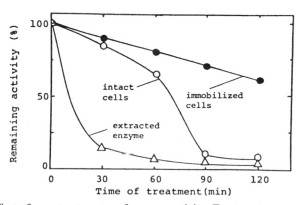

FIG. 4. Effect of urea treatment on fumarase activity. Treatment was carried out using 3 M urea at 37°. From Chibata.[10]

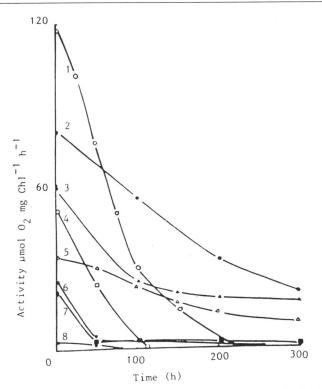

FIG. 5. Oxygen evolution measured as a function of time in storage in the dark at 4° for native thylakoids (curve 1) and thylakoids immobilized with cross-linked albumin polymer (curve 2), in a polyurethane–BSA matrix (curve 3), with cross-linked gelatin polymer (curve 4), in a polyurethane matrix (curve 5), in alginate gel (curve 6), in photo-cross-linkable resin (curve 7), and in carrageenan gel (curve 8). From Cocquempot et al.[11]

Ferricyanide (5 mM) is used as electron acceptor and ammonium chloride (5 mM) as the uncoupling reagent.[17] Reaction media are illuminated at a saturating intensity (30,000 lux) by a 100-W iode lamp equipped with focusing device and red filter. The temperature is maintained at 20°, and thylakoid activity (either free or immobilized) is expressed as μmol O_2/mg of chlorophyl/hr. Figure 5 shows the initial available activity and its evolution as a function of time. After 300 hr of storage, only polyurethane, polyurethane–BSA and cross-linked albumin polymers maintain residual activity.

Functional Stability. Oxygen production by native and immobilized thylakoids is continuously monitored at 20° under illumination (Fig. 6).

[17] M. F. Cocquempot, D. Thomas, M. L. Champigny, and A. Moyse, *Eur. J. Appl. Microbiol. Biotechnol.* **8,** 37 (1979).

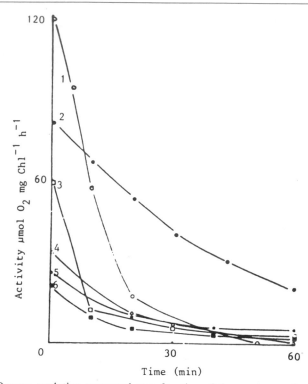

FIG. 6. Oxygen evolution measured as a function of time under continuous use with saturation illumination of native thylakoids (curve 1) and thylakoids immobilized with cross-linked albumin polymer (curve 2), with cross-linked gelatin polymer (curve 3), in a polyurethane matrix (curve 4), in alginate gel (curve 5), and in photo-cross-linkable resin (curve 6). From Cocquempot et al.[11]

After about 50 min the native thylakoids are completely inactivated while the immobilized thylakoids still maintain some residual activity. Continuous use under illumination thus accelerates the inactivation rate, compared with storage stability in the dark. Moreover, immobilization by the BSA–glutaraldehyde procedure would seem to protect thylakoids against photoinactivation to a considerable degree.

Chemically Modified Enzymes

The protective effect has also been observed in the case of covalent immobilization in a polyacrylamide gel. Martinek et al.[18] thus modified

[18] K. Martinek, A. M. Klibanov, V. S. Goldmacher, and I. V. Berezin, *Biochim. Biophys. Acta* **485**, 1 (1977).

enzyme molecules by acroylation using acryloyl chloride. They then performed inclusion of the derivatives thus obtained by copolymerization with acrylamide. Study of the influence of temperature on the initial reaction rate catalyzed by trypsin and chymotrypsin thus immobilized showed that denaturation phenomena by reversible modification of the conformation are therefore suppressed.[19] Moreover, these derivatives have a "life" over a thousand times longer than that of native enzymes. This immobilization method allows the optimum temperatures of trypsin and chymotrypsin to be increased by 25 and 30°, respectively.

Immobilization on Soluble Supports

Numerous proteins of eukaryotic organisms carry osidic residues covalently grafted onto certain amino acids. Such glycoproteins display a greatly increased resistance to denaturation factors, depending on the degree of glycosylation.[20] This has led to attempts at enzyme stabilization by the grafting of soluble polymers, dextran, CM–cellulose, DEAE–cellulose, PEG, polyamino acids, etc., resulting in derivatives more resistant to thermal and proteolytic denaturation[3,20] and displaying reduced antigenic properties. The grafting of enzymes onto soluble polymers forms the transition stage between stabilization by chemical modification and covalent immobilization on insoluble supports.

Immobilization on Insoluble Supports

The influence of immobilization on enzyme stability has also been studied using differential scanning calorimetry (DSC) techniques, which have above all been used to measure the denaturation resistance of free enzymes.[21] Mosbach's group used this technique to compare conformation modifications of free and immobilized enzymes and to determine in particular the effect of the number of enzyme–support covalent bonds per enzyme molecule.[22] The enzymes chosen for this study were ribonuclease A and α-chymotrypsin. They were immobilized by covalent grafting onto Sepharose CL-4B activated by varying concentrations of CNBr[22] in order to obtain varying reaction site densities on the activated support.

Investigation of the Regained Enzymatic Activity of Soluble and Immobilized Ribonuclease A after Heat Treatment. About 10 mg of aspi-

[19] V. V. Mozhaev, V. A. Siksnis, V. P. Torchilin, and K. Martinek, *Biotechnol. Bioeng.* **25,** 1937 (1983).

[20] T. B. Christensen, G. Vegarud, and A. J. Birkeland, *Process Biochem.* **11,** 25 (1976).

[21] Y. Fujita, Y. Iwasa, and Y. Noda, *Bull. Chem. Soc. Jpn.* **55,** 1896 (1982).

[22] A. C. Koch-Schmidt and K. Mosbach, *Biochemistry* **16,** 2105 (1977).

rated enzyme gel is suspended in 0.25 ml of the assay buffer (0.1 M phosphate buffer, pH 7.25) and then kept at 97° in a water bath for 10 min. The samples are then cooled on ice and washed with 0.1 M NaCl. The regained activity is assayed at 25°. Soluble ribonuclease A is treated in the same way using a 1% solution of enzyme.

Thermal Analysis of Soluble and Immobilized Enzymes Using DSC. Thermal analysis of the soluble and immobilized enzymes is performed using a DSC-2 differential scanning calorimeter (Perkin-Elmer) equipped with a cooling system. Aluminum pans, made for aqueous solutions, are used exclusively. The sample is filled with 8–12 mg of well-aspirated gel after which 5 μl of the buffer solution (0.1 M phosphate buffer, pH 7.25, for the ribonuclease A preparations; 0.1 M Tris buffer, 0.1 M in NaCl, pH 8.1, for α-chymotrypsin gels) is added to obtain a homogeneous gel without air bubbles between the beads. To study the soluble enzyme, 15 μl of a 1% enzyme solution is used. The pans are filled, pressure sealed, and weighed. All reference pans contain 15 μl of the corresponding buffer solution. The gel itself does not show any transition in the temperature interval studied. The thermograms are run from 290 to 370 K using a heating rate of 10°/min.

Differential scanning calorimetry was used to investigate the conformational state of immobilized ribonuclease A. With this method the endothermic unfolding process occurring in the protein molecule on heating can be traced. The thermograms thus obtained substantiate the results from enzyme activity assays and show that a loosely bound molecule retains its native properties. A strongly immobilized molecule with a relatively low specific activity behaves in a different manner. The broadening and the displacement of the profile illustrate that there exist immobilized protein molecules which show higher thermal stability than the native enzyme as a result of the introduction of covalent bonds between enzyme and matrix. It was also shown that the introduction of a few bonds does not adversely affect the reversibility of the refolding process. Several bonds caused less reversibility. Corresponding studies on α-chymotrypsin revealed that the unfolding process appeared irreversible for both soluble and immobilized enzyme. The thermograms and enzyme activity studies of the different ribonuclease A preparations revealed that immobilization of the enzyme by multiple points of attachment changed the reversibility of the refolding process.

Along the same lines, Iqbal and Saleemuddin[23] studied glycoenzyme immobilization by affinity adsorption on immobilized lectin preparations.

---

[23] J. Iqbal and M. Saleemuddin, *Biotechnol. Bioeng.* **25**, 3191 (1983).

They observed, in the case of invertase and glucose oxidase both immobilized by adsorption on Sepharose 4B to which concanavalin A had been grafted, that derivative stability is related to lectin density of supports.

Stabilization by Immobilization and/or Solute Addition

Effect of Substrate Addition

The production of invert sugar from sucrose may be achieved by acid hydrolysis or by using the enzyme invertase (β-D-fructofuranosidase, EC 3.2.1.26). The enzymatic process avoids the production of colored by-products which are obtained under acidic conditions. The covalent coupling of invertase from baker's yeast onto an agricultural by-product, corn grits, has been developed, and the influence of substrate concentration, sucrose, on immobilized invertase stability has been determined.

Preparation of Immobilized Invertase.[24] The osidic units of the corn grits (the lignocellulosic hard fraction of corn stover) are chemically modified by the following steps: (1) oxidation with sodium metaperiodate; (2) amination by condensation of ethylene diamine onto the aldehyde groups thus obtained; (3) reduction, using sodium cyanoborohydride, of the imine bonds into amine bonds; (4) activation of the amino groups using glutaraldehyde; and (5) immobilization of invertase onto the activated support. The standard conditions[25] for invertase (Grade VI, Sigma Chemical Co., St. Louis, MO) immobilization are given below. The reaction is carried out using 100 mg corn grits (Eurama) in a 25-ml screw-cap tube. The support particle size is 0.2 mm; its specific area is 0.6 m²/g, and its cellulose content is 30%. Rotative agitation is used for each step of the process. For washing, 20 ml distilled water is used, at 25°, for 24 hr. For oxidation, 20 ml 0.2 M sodium metaperiodate (Merck) solution in distilled water is used at 25°, for 24 hr, in the dark. For amination, 20 ml 3 M ethylenediamine (Prolabo) solution in methanol is used, at 25°, for 72 hr. For reduction, 20 ml 10 g/liter sodium cyanoborohydride (Merck) solution in 0.5 M phosphate buffer is used, at pH 6.5 and 25°, for 5 hr. For immobilization, 20 ml 2 g/liter invertase solution in 0.1 M acetate buffer is used, at pH 4.5 and 4°, for 30 hr.

Stability of Immobilized Invertase.[24] The half-life of immobilized invertase is determined using a packed-bed column (20 ml volume) continuously fed at a flow-rate of 40 ml/hr with sucrose solutions in 0.1 M acetate

---

[24] P. Monsan, D. Combes, and I. Alemzadeh, *Biotechnol. Bioeng.* **26**, 658 (1984).
[25] P. Monsan, *Brevet Fr.* 79-31382 (1979).

half-life (days)

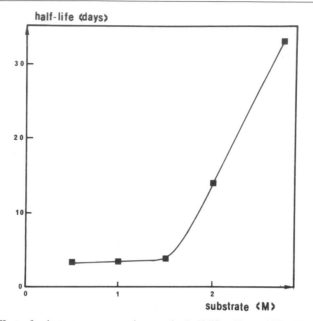

substrate (M)

FIG. 7. Effect of substrate concentration on the half-life of immobilized invertase. The temperature was 60° and pH 4.5. From Monsan *et al.*[24]

buffer, pH 4.5, at 60°. The reducing sugar content at the column outlet is determined by the dinitrosalicylic acid method.[26] Figure 7 shows the effect of sucrose concentration on the stability of immobilized invertase. At 60°, the half-life of immobilized invertase remains constant when the sucrose concentration is lower than 1.5 *M*, but there is a rapid linear increase with substrate concentration above this value: the half-life is thus multiplied by a factor 10 in the presence of a 2.75 *M* sucrose concentration.

It should be noted that in the case of acid phosphatase used continuously in an ultrafiltration membrane reactor, Greco *et al.*[27] observed that substrate (*p*-nitrophenyl phosphate) concentration had no marked effect on enzyme denaturation kinetics.

Effect of Cofactor Addition

The technique of differential scanning calorimetry, which has been applied to the study of thermal stability of free and immobilized RNase

---

[26] J. B. Sumner and S. F. Howell, *J. Biol. Chem.* **108,** 51 (1935).

[27] G. Greco, L. Gianfreda, D. Albanesi, and M. Cantarella, *J. Appl. Biochem.* **3,** 233 (1981).

and α-chymotrypsin (see above and Ref. 22), has also been used by Mosbach's group to determine the influence of various cofactors or cofactor fragments on the heat stability of soluble and immobilized dehydrogenases.[28]

The transition temperature (T_{tr}) of 82.5° obtained for soluble LADH was increased by 12.5° in the presence of a saturating concentration of NADH. In the presence of NAD$^+$, T_{tr} increased by 8.5°, whereas ADP–ribose and AMP caused an increase in T_{tr} of only 2 and 1°, respectively. The T_{tr} of 85.5° obtained for Sepharose-bound LADH was increased by about 12° after the addition of free NADH. Corresponding increases in heat stability were observed for LDH in solution in the presence of NADH, NAD$^+$, and AMP, leading to increases in T_{tr} from 72 to 79.5° and 74 and 73°, respectively.

Effect of Polymer Addition

It has long been known that the addition of certain compounds, e.g., sugars (sucrose, lactose), polyols (glycerol, sorbitol), salts (ammonium sulfate), and polymers, considerably increases the storage stability of mainly free enzymes.[3,7] In the case of immobilized enzymes, the stabilizing effect of polymers has been studied during continuous operation of enzymes in ultrafiltration membrane reactors.[29–32] Enzyme preparations, either in native form, or pre-cross-linked using glutaraldehyde in the presence of albumin, were dynamically immobilized by the formation of a polarization layer at the surface of an ultrafiltration membrane.

In a typical experiment[32] 0.4 mg acid phosphatase is diluted in phosphate buffer at 5° and fed to a membrane reactor equipped with Amicon PM10-type flat UF membranes (42 mm diameter, nominal molecular weight cutoff 10,000). Once the concentration profile of the protein stabilizes, 8 mg stabilizing polymer poly(vinyl alcohol), poly(vinylpyrrolidone) is injected through a multipart valve under nitrogen pressure. The cell is then connected, without modifying the pressure field, to a reservoir containing a 2 mM solution of p-nitrophenyl phosphate in 50 mM citrate buffer, pH 5.6, and the reaction started. The p-nitrophenol concentration in the permeate stream is measured by reading the outlet samples at 405 nm after alkalinization using 1 M NaOH. The stabilization effect of vari-

[28] A. C. Koch-Schmidt and K. Mosbach, *Biochemistry* **16**, 2101 (1977).

[29] G. Greco, Jr., D. Albanesi, M. Cantarella, L. Gianfreda, R. Palescandolo, and V. Scardi, *Eur. J. Appl. Microbiol. Biotechnol.* **8**, 249 (1979).

[30] L. Gianfreda and G. Greco, Jr., *Biotechnol. Lett.* **3**, 33 (1981).

[31] G. Greco, Jr., and L. Gianfreda, *Biotechnol. Bioeng.* **23**, 2199 (1981).

[32] F. Alfani, M. Cantarella, G. Cirielli, and V. Scardi, *Biotechnol. Lett.* **6**, 345 (1984).

TABLE I

STABILIZATION EFFECT OF POLYMER ADDITION ON
ACID PHOSPHATASE ACTIVITY[a]

Conditions	Half-life (hr)
Free enzyme	6.55
Enzyme cogelled with poly-HSA	28.75
Enzyme stabilized with PVA 125,000	96.20
Enzyme stabilized with PVP 44,000	45.89
Enzyme stabilized with PVP 700,000	52.10

[a] At 40°. From Alfani et al.[32]

ous polymers is given in Table I: it may be noticed that the stabilizing effect obtained by polymer addition is higher than that obtained by cross-linking acid phosphatase with albumin (HSA). The highest stabilizing effect is observed using poly(vinyl alcohol), which results in an increase by a factor of 15 of the half-life of the enzyme at 40°.

In earlier experiments, Gianfreda and Greco[30] used a similar experimental approach to determine the stabilizing effect of various polymers, dextran T 40 (MW 40,000), dextran T 500 (MW 500,000), polyacrylamide (Separan, MW above 1,000,000), and CM–cellulose, on β-galactose dehydrogenase, acid phosphatase, and β-glucosidase (Table II). Although most of the polymers used resulted in a significant decrease in initial enzyme activity (activity factor) which was not simply due to an increase in mass transfer resistance, an important stabilizing effect was obtained: a stabilizing factor (corresponding to the ratio of the deactivation constant of the dynamically immobilized enzyme to that of the stabilized enzyme)

TABLE II

EFFECT OF POLYMER ADDITION ON DYNAMICALLY IMMOBILIZED ENZYMES:
STABILITY AND ACTIVITY[a]

Enzyme	Polymer	Stabilization factor	Activity factor
β-Galactose dehydrogenase	Dextran T 40	4.21	0.22
β-Galactose dehydrogenase	Dextran T 500	3.38	0.15
Acid phosphatase	Polyacrylamide Separan MGL	12.5	0.52
β-Glucosidase	CM–Cellulose	21.8	0.91

[a] From Gianfreda and Greco.[30]

of up to 21.8 was obtained in the case of β-glucosidase using CM–cellulose as stabilizer. The data obtained show that the molecular weight of the stabilizing polymer would not seem to be a critical parameter. A similar conclusion was obtained after a detailed study of the stabilizing effect of different polyacrylamide polymers (Separan, Dow Chemical) with varying molecular weight and ionic characters[31]; no immediate correlation could be found between these two parameters and the extent of acid phosphatase stabilization.

However, study of the stabilizing effect of polyethylene glycol (PEG) on free invertase has allowed a direct correlation between the molecular weight of this polymer and its protective effect, defined as the ratio of enzyme half-life in the presence of PEG to that of the enzyme without PEG.[33] From our studies on the stabilizing effect of various additives (polyols, PEG, dextran) on free invertase,[33,34] we forwarded a stabilization mechanism using the effect of these molecules on the degree of organization of water molecules.

In fact, the respective interaction energy levels between enzyme, water, and additive molecules may be considered. Additives interacting more strongly with the enzyme than with water will tend to stabilize denatured states by the formation of additive–enzyme intermolecular bonds to the detriment of the intramolecular bonds initially present in the enzyme molecule. They will therefore have a denaturing effect. However, additives interacting more strongly with water molecules than with the enzyme will favor an increase in the degree of water molecule organization by the formation of clusters (as occurs in ice), and will thus limit the unfolding of the protein chain. Such compounds will have a stabilizing role. In fact, compounds as different as ions, sugars, polyols, polyethers, and polysaccharides, are known to increase by varying degrees the degree of water molecule organization.[35] This may be linked to the fact that the degree of organization of D_2O molecules is greater than that of H_2O molecules, resulting, in particular, in higher viscosity of D_2O. It is known that the substitution of D_2O for H_2O results in a higher resistance of homopeptides and proteins to denaturation by an increased contribution of water extrusion entropy change.[36] It may be observed that invertase is considerably more resistant to thermal denaturation in D_2O than in H_2O (Fig. 8):

[33] P. Monsan and D. Combes, *Enzyme Eng.* **7**, 48 (1984).

[34] D. Combes and P. Monsan, *Enzyme Eng.* **7**, 61 (1984).

[35] R. J. Dobbins, *in* "Industrial Gums" (R. L. Whistler, ed.), p. 19. Academic Press, New York, 1983.

[36] S. Lewin, "Displacement of Water and Its Control of Biochemical Reactions." Academic Press, London, 1974.

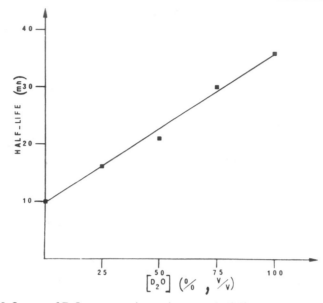

FIG. 8. Influence of D_2O concentration on invertase half-life. Denaturation temperature was 60°. Residual activity of invertase was measured at 40° using 0.4 M sucrose solution in 0.1 M acetate buffer, pH 4.5. From Combes and Monsan.[37]

the half-life of the enzyme in a D_2O solution is 4 times greater than that obtained in H_2O.[37]

Finally, additives with an interaction energy level near that of water and enzyme molecules will have no effect on enzyme stability. The fact that increased stability of enzymes after immobilization is a more or less general phenomenon, independent of the immobilization method adopted, links this effect with the previous discussion concerning the stabilization mechanisms using additives. It is, in fact, known that the degree of organization of water molecules in the vicinity of a solid–liquid interface is much greater than in the bulk of the solution.[35] It may thus be assumed that, as well as stabilization resulting from the formation of covalent or secondary interactions between enzyme and support, the fact that the enzyme molecules are localized at the solid–liquid interfaces (on the outside of and within the insoluble phase) has a not negligible positive effect on the stability of immobilized enzymes.

[37] D. Combes and P. Monsan, *Eur. Congr. Biotechnol., 3rd,* **I,** 233 (1984).

[54] Stabilization of Pyranose 2-Oxidase and Catalase by Chemical Modification

By ZE'EV SHAKED and SIDNEY WOLFE

The Cetus process for the conversion of D-glucose to crystalline D-fructose is composed of a two-enzyme coimmobilized system followed by a chemical step (Fig. 1). The process is based on the enzymatic transformation of D-glucose to D-glucosone, catalyzed by pyranose 2-oxidase (P-2-O). The by-product, hydrogen peroxide, is decomposed *in situ* in the reactor by a coimmobilized catalase. D-Glucosone is removed from the enzymatic reactor and reduced by dihydrogen to D-fructose over a Pt/C or Pd/C catalyst.[1]

The development of the described process to obtain over 96% pure D-fructose without the use of expensive cofactors established a unique challenge in the field of enzyme engineering. The formation of a reactive intermediate (D-glucosone) and a by-product which is a potent oxidant (hydrogen peroxide) required unique approaches in order to develop biological catalysts that have sufficiently economical half-lives. One of the strategies that was used to improve the stability of the enzymes under operating conditions was chemical modification prior to immobilization. The stabilization of proteins through chemical modification, and specifically by cross-linking methods, is not new and has been successfully employed in the past. Previous studies, however, have been mainly concerned with structure elucidation[2-5] and thermostabilization[6-8] of enzymes, and did not have to address inactivation çaused, for example, by highly chemically active intermediates (glucosone) or products (hydrogen peroxide) formed enzymatically.

Water-soluble imido and diimido esters that were introduced by Hunter and Ludwig[9] in 1962 have been since used by many investigators

---

[1] S. L. Neidleman, W. F. Amon, and J. Geigert, U.S. Patents 4,246,347 and 4,423,149.
[2] H. Zahm and J. Meienhofer, *Makromol. Chem.* **26,** 126 (1958).
[3] F. H. Carpenter and K. T. Harrington, *J. Biol. Chem.* **247,** 5580, (1972).
[4] F. Hucho and M. Yanda, *Biochem. Biophys. Res. Commun.* **57,** 1080, (1974).
[5] K. Bose and A. A. Bothmer-By, *Biochemistry* **22,** 1342 (1983).
[6] Y. R. Knowles and F. M. Richards, *J. Mol. Biol.* **37,** 231 (1968).
[7] R. Reiner, H. U. Siebeneick, I. Christensen, and H. Doring, *J. Mol. Catal.* **2,** 119 (1977).
[8] V. P. Torchilin, A. V. Maksimenko, V. N. Smirnov, L. V. Berezin, and A. M. Kilibanov, *Biochim. Biophys. Acta* **522,** 277 (1978).
[9] M. Y. Hunter and M. L. Ludwig, *J. Am. Chem. Soc.* **84,** 349 (1962).

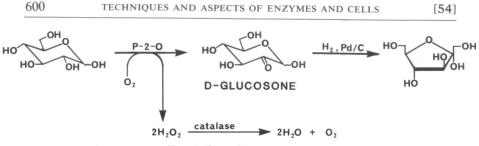

FIG. 1. Cetus fructose process.

(see, for example, Refs. 10–12). These reagents react specifically and under mild conditions with protein amino groups. The amidine bond can be cleaved by treatment with ammonium hydroxide to yield the original amino acid group under conditions that do not cause hydrolysis of peptide bonds. Moreover, since the pK_a of amidines is higher than that of ε-amino groups, an amidinated protein has the same net charge in the acid or neutral pH range as does the native protein. This has a beneficial effect on the stability of enzymes such as catalase and P-2-O that have pH optimums around 5–6. The introduction of the amidine bonds into catalase has an additional advantage, since it results in blocking the most accessible and active protein amino groups. These blocked amino groups cannot react with aldehydes to form Schiff bases. This is especially important in the Cetus process, since the enzymatic step catalyzed by P-2-O forms glucosone which is a very good electrophile that reacts readily with primary amino groups. Proton- and ^{13}C-NMR studies done in D_2O indicated that at room temperature only four tautomers are present (Fig. 2).

Experimental Procedure

Materials

Purified P-2-O (EC 1.1.3.10) from *Polyporus obtusus* (partially proteolyzed) and D-glucosone solutions over 94% pure were obtained in-house. Catalase (EC 1.11.1.6) from *Aspergillus niger* was obtained from Fermco Co. Dimethyl adipimidate, 2,4,6-trinitrobenzenesulfonic acid (TNBS), and bovine serum albumin (BSA) were obtained from Sigma. Ethyl acetimidate and dimethyl suberimidate were obtained from Aldrich. Amberlite DP-1 ion-exchange resin was obtained from Alpha Products.

[10] L. Wofsy and S. J. Singer, *Biochemistry* **2,** 104 (1963).
[11] Y. H. Reynolds, *Biochemistry* **7,** 3131 (1968).
[12] H. Peretz and D. Elson, *Eur. J. Biochem.* **63,** 77 (1976).

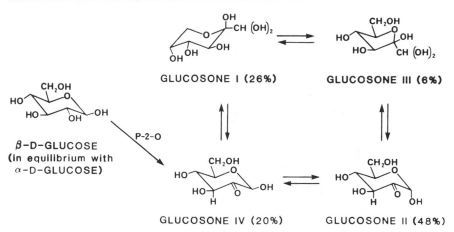

FIG. 2. Equilibrium forms of D-glucosone.

Assay Procedures

Catalase. Catalase is assayed by monitoring the absorbance of hydrogen peroxide at 215 nm. A sample (150 μl) is diluted with 1–3 ml of 50 mM phosphate buffer at pH 7. This diluted enzyme solution (50 μl) is added to 5 ml of 0.003% H_2O_2 in phosphate buffer (50 mM, pH 7). The blank sample contains 50 μl of diluted sample in 5 ml of the same phosphate buffer. The absorbance at 215 nm is measured every 12 sec for 3 min. The first-order rate constant is obtained by averaging $\ln(A_0/A_t)/t$, or by using a linear least-squares fit. The specific activity (U/mg) is obtained by dividing the rate constant by the milligrams of protein in the peroxide solution. A standard deviation of about 5% is obtained for a given sample. The catalase obtained from Fermco has an activity of about 500 U/mg protein.

P-2-O. P-2-O assays are done by following the formation of hydrogen peroxide using the *o*-dianisidine (ODAD) assay. Before the assay, a sample is diluted with phosphate buffer (50 mM, pH 6.0) to a concentration of 0.02–0.04 mg/ml of enzyme. (Caution: the half-life of the enzyme in dilute solutions at room temperature is about 1–2 hrs.) The dilute enzyme solution (0.1 ml) is added to a solution (0.9 ml) that contains 0.01% ODAD, 0.1 mg/ml horseradish peroxidase (HRP), and 4.2% glucose in air-saturated phosphate buffer (50 mM, pH 6.0). After 10 min at 25°, the reaction mixture is quenched by adding 1.0 ml of 2% sulfamic acid. The absorbance at 400 nm is determined against a blank made up with buffer instead of the enzyme solution. The amount of hydrogen peroxide produced is then determined by making up standard hydrogen peroxide solutions and mixing them with ODAD–glucose–HRP solution, followed by 2% sulfamic acid solution about 1 min later. Standards are measured at different concentrations, and the automatic concentration determination features of the Perkin-Elmer Lambda 5 spectrophotometer are used. The standard samples have a deviation of less than 2%, while samples generally have a standard deviation of less than 5%.

Cross-Linking of Catalase with Diimido Esters

A solution of 60 mg of catalase 3 mg/ml in citrate buffer (10 mM, pH 5) is run over a Sephadex G-25 column and then cooled down to 0–5°. The cross-linker (20% by weight of dimethyl suberimidate or dimethyl adipimidate dissolved in 2 ml of methanol) is added over a period of about 5 hr with a peristaltic pump (LKB). The pH is maintained at 9.4–9.7 with a pH controller (Horizon) with automatically adds 25 mM NaOH when the pH falls below 9.5. One hour after the addition, the reaction solution is again run over a Sephadex G-25 column (with 5–10 mM citrate, pH 5.0) to remove excess cross-linker and salts.

Cross-Linking of P-2-O with Diimido Esters

The crosslinking is carried out as described above for catalase. The reaction with dimethyl suberimidate is performed for about 12 hr.

Amidination of P-2-O

Amidination of P-2-O with ethyl acetimidate is done by two different methods. In the first method, ethyl acetimidate (23 mg, 0.2 mmol) is added slowly to an enzyme solution (20 ml, 6 mg/ml in 10 mM acetate). The pH

TABLE I
RESIDUAL ACTIVITIES OF CATALASE AND P-2-O UPON MODIFICATION

Enzyme	Reagent	Amount of reagent (% by weight)	Residual activity (%)
Catalase	Dimethyl adipimidate	20	98 (±2)
Catalase	Dimethyl suberimidate	20	98 (±2)
Catalase	Dimethyl suberimidate	30[a]	80 (±2)
P-2-O	Dimethyl adipimidate	2	98 (±2)
P-2-O	Dimethyl adipimidate	20	90 (±2)
P-2-O	Dimethyl suberimidate	20	90 (±2)
P-2-O	Ethyl acetimidate	20	93 (±2)
P-2-O	Ethyl acetimidate	200	80 (±2)

[a] The addition of the reagent is done in 30 min at pH 9.5 and 25°.

of the reaction mixture is maintained at 9.5 by adding 20 mM NaOH using a pH controller. After 5 hr, the mixture is run over a Sephadex G-25 column with dilute citrate buffer. The modified P-2-O retains about 93% ± 5% activity of the initial unmodified enzyme. In the second method, a solution of ethyl acetimidate (115 mg in 1 ml ethanol) is added in 0.1-ml portions every 30 min to a P-2-O solution (15 ml, 4 mg/ml, 10 mM acetate, pH 5.0). The pH is maintained at 10.0. After 2.5 hr from the final addition of ethyl acetimidate, the reaction is worked up as described before. This substantial modification of P-2-O still retains 80 ± 5% of the initial enzyme activity. Table I summarizes all the residual activities of catalase and P-2-O upon modification.

Determination of Amino Groups with 2,4,6-Trinitrobenzenesulfonic Acid (TNBS)

The TNBS determination of amino groups is done following the procedure described by Habeeb.[13] All catalase or P-2-O samples are run over a Sephadex G-25 column to remove any ammonium sulfate and small peptide fragments. The amount of catalase used is determined by weight. To 1 ml of protein solution (0.6–1 mg/ml), 1 ml of 4% NaHCO$_3$ (pH 8.5) and 1 ml of 0.1% TNBS are added. The solution is allowed to react at 40° for 2 hr; then 1 ml 10% SDS is added to solubilize the protein and prevent its precipitation on addition of HCl (0.6 ml, 1 N). The absorbance of the solution is read at 335 nm against a blank treated as above, but with 1 ml of water (or buffer) instead of the protein solution. A molar extinction coefficient of 1 × 10^4 M^{-1} cm^{-1} is used to calculate the residual amino groups. BSA is used as a standard to verify the results with catalase and P-2-O, and to compare our work with that of Habeeb. Our results for BSA are within 10% of those found by Habeeb, probably due to a different preparation that we use.

SDS–Polyacrylamide Electrophoresis of the Cross-Linked Catalase

A 6% polyacrylamide gel is prepared by a conventional procedure using a phosphate buffer (50 mM, pH 7.1). Samples are prepared by heating a solution of the protein, SDS, and 2-mercaptoethanol in a water bath (100°) for about 2 min. Then bromophenol blue and additional 2-mercaptoethanol are added. The protein concentration in a sample is about 1 mg/ml. The sample (10 μl) is placed on the gel and then run employing a potential of approximately 4 V/cm. The gel is stained with

---

[13] A. F. S. A. Habeeb, *Anal. Biochem.* **14**, 328 (1966).

Coomassie Brilliant Blue R, and BioRad high molecular weight SDS standards are used for molecular weight determination.

SDS–Polyacrylamide Electrophoresis of Cross-Linked P-2-O

A 10% polyacrylamide gel is prepared by conventional methods using an imidazole–phosphate buffer (0.1 M, pH 7.0). Sample and standards are prepared as described for catalase. Protein concentrations used are 0.6–1.2 mg/ml, and 10-μl samples are applied to this gel. Staining is done with Coomassie Brilliant Blue R.

Immobilization of Cross-Linked Catalase on DP-1 Methacrylate Supports

Catalase, cross-linked as described with dimethyl adipimidate, is adsorbed to an Amberlite DP-1 (methacrylic acid–divinylbenzene copolymer, Alfa Products) ion-exchange resin that has been previously equilibrated in 12 mM NaOAc at a pH between 4.8 and 5.2 by addition of concentrated HCl or NaOH as appropriate. Decanting, resuspension, and readjustment of pH of the exchange resin is continued until the pH stabilizes between 4.8 and 5.2 when fresh buffer is added.

Enzyme adsorption is carried out by one of two methods. In the first, a preweighed quantity of DP-1 resin is swirled or stirred with a 2–3 times larger volume of catalase in 10 mM NaOAc at pH 5 or 5–15 min and allowed to settle. After measuring the enzyme concentration in the supernatant by spectrophotometry, another aliquot of enzyme from a more concentrate stock solution is added and mixing resumed. The strategy is to keep adding enzyme until the supernatant adsorbance increases proportionally with enzyme added. A graph of supernatant enzyme adsorbance versus quantity of enzyme added per mass of exchanger gives a titration curve, initially horizontal because all or most protein is adsorbed, and finally rising linearly with a slope indicative of the protein extinction coefficient. This method depends on rapid equilibration of enzyme with exchanger.

As it was discovered that the DP-1 exchange resin equilibrated slowly, on the time scale of hours to days, a second immobilization strategy was used as follows. With swirling, the DP-1 is mixed with a quantity of the cross-linked catalase previously determined to be in excess of exchanger capacity. Supernatant spectra are taken at intervals of hours to days until the rate of adsorbance decline becomes negligible. Adsorption kinetics are followed at 25°, and the beakers (normally agitated at 100–200 rpm on a shaker table) are carefully sealed with Parafilm to minimize evaporation.

Supernatant samples are returned after spectral measurement to maintain a constant volume. Fines, generated through attrition during swirling, are clarified by spinning the sample for 5 min in an Eppendorf Model 5412 microcentrifuge before scanning if there is any sign of suspended matter. After it is clear that the DP-1 adsorbed significant amounts of catalase on the time scale of hours to days, nonkinetic adsorptions are performed without any agitation. Several milliliters of dimethyl adipimidate-cross-linked catalase solution is simply allowed to stand with a somewhat smaller volume of resin particles. Often adsorption is done with steri-filtered (0.2 μm pore size) cross-linked catalase and DP-1 previously auto-claved in a foil-covered small glass beaker, using sterile transfers to mini-mize bacterial contamination. When DP-1 exchanger resin has been loaded with immobilized cross-linked enzyme, it is washed repeatedly in 10 mM NaOAc to remove any bulk unadsorbed enzyme. Specific activity in M^{-1} sec^{-1} is calculated from specific activity in min^{-1} g dry weight support^{-1} liter by dividing by 60 x/min, dividing by g wet weight/g dry weight, dividing by mg enzyme adsorbed/g wet weight, and multiplying by 3.23 $\times$ 10^8 mg enzyme/mol enzyme.

Thermodenaturation of Native and Cross-Linked Catalase

Native catalase or cross-linked catalase (0.2 mg/ml in 10 mM NaCl, 10 mM phosphate buffer, pH 6.0) is placed in small polyethylene Eppendorf test tubes (300 μl) and put in an 81° water bath for a timed interval. Vials, when removed from the bath, are cooled quickly and then assayed after dilution. Adipimidate-cross-linked catalase is immobilized on DP-1 methacrylate resin beads as described above, and the thermostability studies of the immobilized cross-linked catalase are done at 75°.

Thermodenaturation of Native and Cross-Linked P-2-O

A P-2-O solution (1 ml, 5 mg/ml, 5 mM citrate, pH 4.9) is transferred to an Eppendorf test tube (1.5 ml capacity) and placed in a 65° water bath. At various time points, the residual activity of the enzyme is assayed after centrifugation.

Incubation of Cross-Linked and Native Enzymes with Glucosone

An enzyme solution (2.0 ml, 0.2–0.4 mg/ml, 45 mM citrate buffer, pH 4.5) and glucosone (5%) are placed in a small polypropylene tube and incubated in a 25° water bath. The control solution contains the same enzyme solution but no glucosone.

Results and Discussion

Effects of Glucosone on Native Catalase

In the presence of glucosone, buffers, temperature, enzyme concentration, and glucosone concentration all play important roles in the stability of native catalase. The incubation of the enzyme in glucosone at 25° strongly effects its stability. The half-life of catalase with glucosone (3.3% by weight) in citrate buffer is about 90–100 hr. With no glucosone present, the enzyme is stable for at least 150–170 hr (Fig. 3). Buffers have a substantial effect on the stability of catalase. In the presence of glucosone, the enzyme is twice as stable in citrate buffer than in acetate buffer. Perhaps somewhat surprisingly, higher glucosone concentrations somewhat improve the half-life of catalase (Fig. 4). A 4-fold increase in glucosone concentration improved the half-life of catalase by a factor of 3.

Glucosone, in addition to its α-ketoaldehyde moiety, contains four

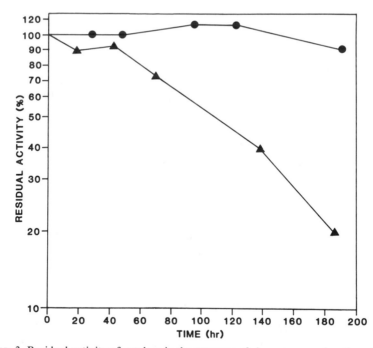

FIG. 3. Residual activity of catalase in the presence of glucosone as a function of time at 25°: (●) native catalase (0.34 mg/ml), 5 mM citrate, pH 5.0; (▲) native catalase (0.34 mg/ml), 5 mM citrate, pH 5.0, 3.3% glucosone.

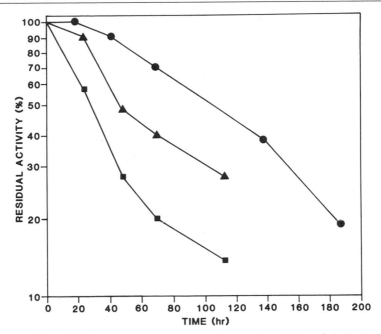

Fig. 4. Effect of glucosone concentration on native catalase (0.34 mg/ml, 5 mM citrate, pH 5.0): (●) 3.3%; (▲) 1.7%; (■) 0.8%.

hydroxyl groups. These hydroxyl groups might have similar effects to those of glycerol or other polyol compounds in stabilizing proteins. This stabilization mechanism, however, has not yet been firmly established. Studies on the effect of alcohols or polyols on the thermal transition of proteolytic enzymes (RNase, chymotrypsin) have indicated that the stabilizing effect of polyhydric alcohols is due to a decrease in the hydrogen-bond rupturing capacity as compared to monovalent alcohols. Solvents which decrease the dielectric constant due to weak solvation properties toward hydrophobic residues show better ligand binding to the native enzyme active center and have been shown to protect the enzyme. In addition, reduced surface energy of polyol solutions may also be a part of the stabilization process. Incubation of catalase in the presence of glucosone at higher temperatures accelerates the inactivation of the enzyme. For example, the half-life of the enzyme (45 hr) at 40° is about one-half the half-life at 25°. Higher temperatures may cause a further exposure of protein amino groups to glucosone. With no glucosone present the enzyme is practically stable at 40° for at least 180 hr.

Effect of Glucosone on Native P-2-O

In the presence of glucosone, the enzymatic activity of P-2-O decreases in a biphasic kinetic mode. During the first 48 hr, activity decreases with an apparent half-life of about 70 hr. After 48 hr, however, the activity has a half-life of about 250 hr (Fig. 5). Since the decrease in activity might be due to the presence of a contaminant in the pyranose dehydratase preparation, two different P-2-O preparations were used. The first P-2-O preparation had most of its pyranose dehydratase removed by a DEAE column (0.021 U/mg), while the second one was an unpurified enzyme preparation, also called fraction two (1.4 U/mg). Surprisingly, there were no significant differences seen in the inactivation of the enzyme in the presence of glucosone (Fig. 5). The biphasic decrease in activity suggests that there are at least two different inactivation mechanisms caused by glucosone. Alternatively, these biphasic kinetics may be due to two P-2-O populations that possess different reactivities. These results indicate that, analogous to catalase, P-2-O is inactivated in the presence of glucosone.

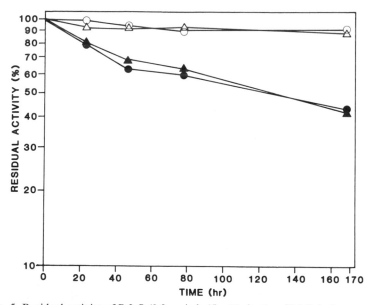

Fig. 5. Residual activity of P-2-O (0.2 mg/ml, 45 mM citrate, pH 4.5) in the presence of glucosone (5%) as a function of time at 25°: (○) DEAE-purified P-2-O (no glucosone); (△) unpurified P-2-O (no glucosone); (●) DEAE-purified P-2-O with glucosone; (▲) unpurified P-2-O with glucosone.

Characterization of Cross-Linked Catalase and P-2-O

In addition to monitoring the stability of catalase in the presence of glucosone, SDS electrophoresis and a spectrophotometric determination of amino groups were used to characterize the enzyme after cross-linking. SDS–polacrylamide electrophoresis with a 6% gel showed that inter-subunit cross-linking occurred only when 30% (by weight) of cross-linker was added to catalase at 25°. However, even in this case, a very small amount of intercross-linked catalase was obtained (Fig. 6). In the preparations made at low temperatures (0–5°) with the cross-linker (20–25%) added slowly, no evidence of intersubunit cross-linking could be observed. In the case of dimethyl adipimidate, the distance between amino groups must be less than 8.6 Å for the cross-linking to occur. This constraint apparently prevents the formation of intersubunit cross-linking with catalase. Other studies, however, do report intersubunit cross-linking with other enzymes using dimethyl suberimidate.[3]

The protein amino groups were determined spectrophotometrically by

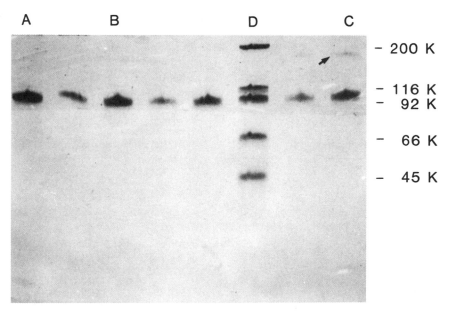

FIG. 6. SDS–PAGE of native and cross-linked catalase, 6% gel. (A) Native catalase; (B) catalase cross-linked with 20% by weight of dimethyl adipimidate at 0–5°, by slow addition; (C) catalase cross-linked with 30% by weight of dimethyl suberimidate at 25°; (D) BioRad high molecular weight standard.

employing 2,4,6-trinitrobenzenesulfonic acid (TNBS). Table II provides a summary of the TNBS determinations. Since this determination introduces a bulky trinitrobenzene group into the protein, only the most accessible amino groups of catalase are modified. Still, the large differences in the number of amino groups obtained for the various catalase preparations suggest that this method is useful at least on a comparative basis. All catalase modifications result in a decrease in the number of amino acids that are accessible to TNBS. More extensive modification of the amino groups does not result in a more stable catalase toward glucosone. In addition, the cross-linking with the enzyme of the diimido esters modifies only 20–30% of the amino acids available to TNBS in native catalase. All these results suggest that the differences between the amino acids are not just in accessibility, but more importantly, in reactivity. Once these amino acids (20–30% of total amino acids available to TNBS in the native enzyme) are blocked, either by chemical or physical methods, the enzyme is stable in glucosone solutions and can benefit from the stabilizing effects that glucosone has on proteins because of its hydroxyl groups. In the case of P-2-O, only 9 ± 2 amino groups per molecule (tetramer) of the extensively amidinated P-2-O react with TNBS, while with the native enzyme TNBS reacts with 170 ± 2 amino groups.

TABLE II

2,4,6-TRINITROBENZENESULFONIC ACID (TNBS) DETERMINATION OF AMINO GROUPS OF CATALASE AND P-2-O[a]

Preparation	Number of reactive amino groups per molecule	% of native value
Native catalase	44 (± 2)	—
Cross-linked catalase with 30% by weight of dimethyl suberimidate at 25°	16 (± 2)	36% (± 5)
Cross-linked catalase with 20% by weight of dimethyl suberimidate added slowly at 0–5°	36 (± 2)	82% (± 5)
Cross-linked catalase with 20% by weight of dimethyl adipimidate added slowly at 0–5°	30 (± 1)	68% (± 5)
Native P-2-O	170 (± 2)	—
Amidinated P-2-O with 200% by weight of ethyl acetimidate added slowly at 0–5°	9 (± 2)	6% (± 2)

[a] All the chemical modifications are done at pH 9.5 using a pH-stat.

Stabilization of Catalase toward Glucosone by Cross-Linking with Diimido Esters

The first cross-linking experiments were done with dimethyl subermidate. We found that, in addition to the amount of the cross-linker, the experimental procedure for cross-linking affects catalase stability.

When catalase was cross-linked with 2% (by weight) of dimethyl suberimidate added all at once at pH 7, the half-life of catalase in the presence of glucosone (4%) at 40° in 50 mM acetate improved only a factor of 2 (from 50 to about 100 hr). Higher amounts of the cross-linker (5–30% by weight), and performing the reaction at the higher pH values (9–10) did not significantly improve the stability of the enzyme in the presence of glucosone. The cross-linking amidine bonds are resistant to hydrolysis. The cross-linker, however, will react with water to form an ester, thus preventing it from cross-linking the protein. In pH 10 borate buffer (made with about 80% D$_2$O) at room temperature, the half-life of the dimethyl suberimidate is about 2.5 hr as measured by NMR.

The substantially modified experimental procedure described in the Experimental Procedures section gave a stable cross-linked catalase. A very small loss of the initial enzymatic activity (less than 10%) was caused by cross-linking the enzyme. The suberimidate- and the adipimidate-cross-linked catalases show an impressive stability in the presence of 3.3% glucosone. Moreover, their enzymatic activities over extensive periods (250–600 hr) show about 115–130% of their original activities (Fig. 7). This superactivity is probably due to the cross-linking. The molecular rigidity introduced into the enzyme via the covalent cross-links could result in superactivities toward hydrogen peroxide.

The incubation of the two cross-linked enzymes and native catalase in citrate buffer with no glucosone at 25° results in the same half-life (500 hr), and indicates that glucosone has a beneficial effect on the stability of the cross-linked catalase. The cross-linked catalase in the presence of glucosone at 25° is completely stable for at least 600 hr.

Effect of Glucosone on Cross-Linked P-2-O and on Amidinated P-2-O

P-2-O was cross-linked by diimido esters (dimethyl adipimidate and dimethyl suberimidate) analogously to the procedure described for catalase. The cross-linking of P-2-O does not result in an enzyme more stable toward glucosone (Fig. 8). Higher diimido ester concentrations and longer reaction times have not improved the stability of the enzyme in the presence of glucosone. If anything, the cross-linking of P-2-O destabilizes the enzyme by making all of the P-2-O behave like native P-2-O that loses its

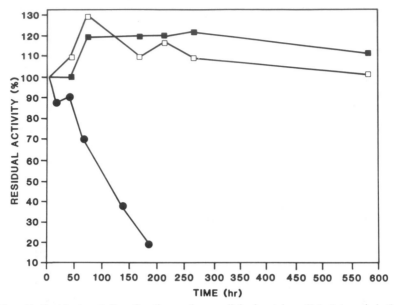

FIG. 7. Residual activity of native and cross-linked catalase (0.3–0.4 mg/ml, 5 m*M* citrate, pH 5.0) in the presence of glucosone (3.3%) as a function of time at 25°: (●) native catalase; (■) catalase cross-linked with dimethyl adipimidate; (□) catalase cross-linked with dimethyl suberimidate.

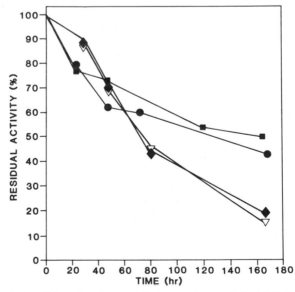

FIG. 8. Residual activity of native, amidinated, and cross-linked P-2-O (0.2 mg/ml, 45 m*M* citrate, pH 4.5) in the presence of glucosone (5%) as a function of time at 25°: (●) native P-2-O; (■) P-2-O amidinated with 200% by weight of ethyl acetimidate; (▽) heat treated P-2-O; (◆) P-2-O cross-linked with 2% by weight dimethyl adipimidate.

activity by the fast inactivation mechanism. Only few amino groups (~9) per tetramer react with TNBS, which may suggest that the reaction of glucosone with the amino groups of P-2-O plays only a minor role in the inactivation of the enzyme.

Thermostability of the Cross-Linked Catalase

The thermostability studies were done with the soluble diimido ester-cross-linked catalase at 81° (Fig. 9). In addition, the cross-linked enzyme was immobilized on Amberlite DP-1 support (methacrylic acid–divinyl-benzene copolymer). The thermostability studies of the supported cross-linked catalase were done at 75° (Fig. 10). Both thermostability tests suggest that cross-linking the enzyme with dimethyl adipimidate results in an appreciably improved thermostable catalase. Cross-linking the enzyme with dimethyl suberimidate, however, does not result in a better thermostable catalase. The introduction of the six-carbon chain of adipimidate probably results in a more rigid configuration of the protein than the introduction of the eight carbon chain by suberimidate.

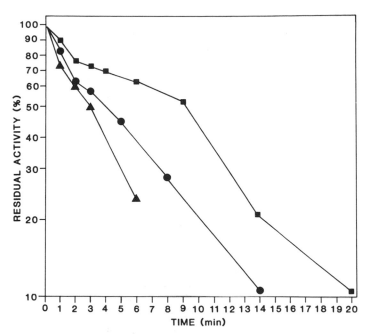

FIG. 9. Thermostability of native and cross-linked catalase (0.3 mg/ml, 10 mM phosphate, pH 6.0, 10 mM NaCl) at 81°: (●) native enzyme; (▲) catalase cross-linked with dimethyl suberimidate; (■) catalase cross-linked with dimethyl adipimidate.

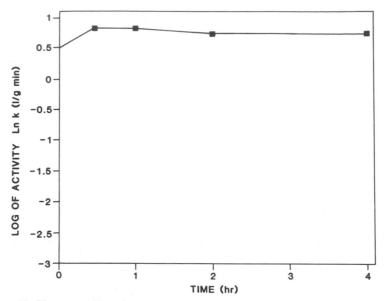

FIG. 10. Thermostability of dimethyl adipimidate-cross-linked catalase immobilized on Amberlite DP-1 at 75° (5 mM citrate, pH 5.0).

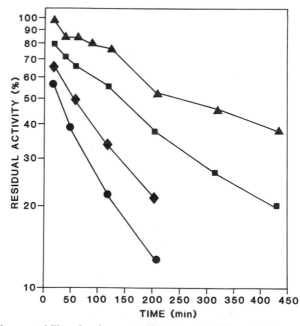

FIG. 11. Thermostability of native, cross-linked, and amidinated P-2-O (0.2 mg/ml, 5 mM citrate, pH 5.0) at 65°: (■) P-2-O amidinated with 20% by weight of ethyl acetimidate; (●) native P-2-O; (▲) P-2-O amidinated with 200% by weight of ethyl acetimidate; (◆) P-2-O cross-linked with 20% dimethyl adipimidate.

Thermodenaturation of Native, Amidinated, and Cross-Linked P-2-O

Native and cross-linked enzymes exhibit the same thermodenaturation kinetics (Fig. 11). The amidinated P-2-O preparations, however, show considerably greater thermostability. Apparently, amidination provides more opportunities for hydrogen bondings and hydrophobic interactions which may enhance the thermostability of the enzyme. The chemical modification of P-2-O with higher levels of amidination results in enzyme preparations that are 10 times more thermostable than native P-2-O. There are two effects on modification: (1) there is an elimination of the fast inactivation pattern of normal proteolyzed P-2-O; and (2) there is a 5-fold deceleration of the principal cause of thermal inactivation.

Acknowledgment

We thank our colleagues Mark Pemberton and Mike Kunitani for supplying us with purified P-2-O and glucosone, and S. Daniell for carrying out the cross-linked catalase immobilization.

[55] Stabilization of Enzymes by Intramolecular Cross-Linking Using Bifunctional Reagents

By Karel Martinek and V. P. Torchilin

The problem of enzyme stabilization has received considerable attention in recent years.[1-7] Enzyme immobilization has been used most frequently to solve the problem of enzyme stabilization. However, other methods have been suggested as well.[3] For example, enzyme stabilization has been achieved after (1) addition of low molecular weight compounds to enzymes free in solution, (2) chemical modification of enzymes by substitution with low molecular weight compounds, and (3) use of bifunctional reagents to produce enzymes containing artificial intramolecular cross-links. These methods are desirable in particular when the presence

[1] K. Martinek, A. M. Klibanov, and I. V. Berezin, *J. Solid-Phase Biochem.* **2**, 343 (1977).

[2] A. M. Klibanov, *Anal. Biochem.* **92**, 1 (1979).

[3] V. P. Torchilin and K. Martinek, *Enzyme Microb. Technol.* **1**, 74 (1979).

[4] R. D. Schmid, *Adv. Biochem. Eng.* **12**, 41 (1979).

[5] K. Martinek and V. V. Mozhaev, *Enzyme Eng.-Future Directions*, p. 3 (1980).

[6] A. M. Klibanov, *Biochem. Soc. Trans.* **11**, 19 (1983); *Science* **219**, 722 (1983).

[7] V. V. Mozhaev and K. Martinek, *Enzyme Microb. Technol.* **6**, 49 (1984).

of a support may decrease both the binding capacity and the reactivity of the enzyme. Also, in medical therapy applied enzyme must in many cases interact with receptors or other components of cellular membranes. In this instance, a support may change the key pathways dramatically.

In this chapter, stabilization of enzymes through intramolecular cross-linking will be discussed in detail. The principles of intramolecular cross-linking are shown schematically in Fig. 1. This approach is based on diminishing the polypeptide entropy which is the principal thermody-

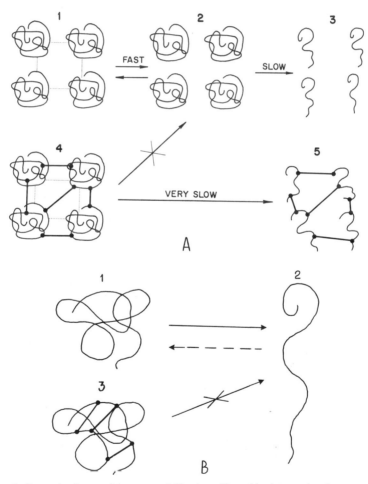

FIG. 1. General scheme of enzyme stabilization effected by intramolecular cross-linking. (A) **1**, Native oligomeric enzyme; **2**, reversibly dissociated subunits; **3**, irreversibly denatured subunits; **4**, cross-linked enzyme; **5**, irreversibly denatured cross-linked enzyme. (B) **1**, Native monomeric enzyme; **2**, denatured enzyme; **3**, cross-linked enzyme.

namic quantity stabilizing the denatured form.[8] In 1967 Hartman and Wold[9] introduced the use of bifunctional reagents in protein chemistry. Then, Husain and Lowe[10] used protein cross-linking with a bifunctional reagent as a means to study the tertiary structure of an enzyme molecule consisting of a single polypeptide chain. In addition, this technique has been applied to exploring the quaternary structure of oligomeric enzymes by Davies and Stark.[11] Since the publication of these pioneering works cross-linking of proteins has become a widely used technique.[12-21] Procedures have been developed for attachment of DNA and RNA molecules to proteins[22] with the aid of cross-linking methodology. Immunoanalysis and radioactive labeling have been used for identifying proteins in cross-linked protein complexes,[23] and, recently, the use of the cross-linking approach in fundamental studies in biochemistry has been reviewed.[24]

Cross-Linking Reagents

There are now many bifunctional compounds available for cross-linking of proteins.[12-21] Examples of such reagents are dialdehydes, diimido esters, diisocyanates, and bisdiazonium salts. Moreover, diamines such as $H_2N(CH_2)_nNH_2$ may be used for cross-linking of protein carboxyl groups, if the latter have been preactivated by treatment with carbodiimide.[25] Likewise, diacids such as $HOOC(CH_2)_nCOOH$ (after their preactivation with carbodiimide) could be used for cross-linking of protein

[8] P. J. Flory, *J. Am. Chem. Soc.* **78**, 5222 (1956).

[9] F. C. Hartman and F. Wold, *Biochemistry* **6**, 2439 (1967).

[10] S. S. Husain and G. Lowe, *Biochem. J.* **103**, 855 (1968).

[11] G. E. Davies and G. R. Stark, *Proc. Natl. Acad. Sci. U.S.A.* **66**, 651 (1970).

[12] H. Fasold, J. Klappenberger, and H. Remold, *Angew. Chem., Int. Ed. Engl.* **10**, 795 (1971).

[13] F. Wold, this series, Vol. 25, p. 623.

[14] O. R. Zaborsky, *Enzyme Eng.* **1**, 211 (1972).

[15] R. E. Peeney, G. Blankenborn, and H. B. F. Dixon, *Adv. Protein Chem.* **29**, 135 (1975).

[16] R. Uy and F. Wold, *in* "Biomedical Applications of Immobilized Enzymes and Proteins" (T. M. C. Chang, ed.), p. 15. Plenum, New York, 1976.

[17] K. Peters and F. M. Richards, *Annu. Rev. Biochem.* **46**, 523 (1977).

[18] R. B. Freedman, *Trends Biochem. Sci. (Pers. Ed.)* **4**, 193 (1979).

[19] M. Das and F. Fox, *Annu. Rev. Biophys. Bioeng.* **8**, 165 (1979).

[20] T. H. Ji, this series, Vol. 91, p. 580.

[21] K.-K. Han, C. Richard, and A. Delacourte, *Int. J. Biochem.* **16**, 129 (1984).

[22] K. C. Smith, *in* "Aging, Carcinogenesis and Radiation Biology," p. 67. Plenum, London, 1976.

[23] S. K. Sinha and K. Brew, *J. Biol. Chem.* **256**, 4193 (1981).

[24] K. Martinek and V. V. Mozhaev, *Adv. Enzymol.* **57**, 179 (1985).

[25] V. P. Torchilin, A. V. Maksimenko, A. M. Klibanov, I. V. Berezin, and K. Martinek, *Biochim. Biophys. Acta* **522**, 277 (1978).

amino groups.[26] Both diamines and diacids are commercially available and relatively inexpensive, factors that are of prime importance in biotechnology. In addition, application of heterobifunctional cross-linking reagents[13,20,21] offers the possibility of increasing the number of cross-links by reacting with different functional groups of the protein to be modified.

Photochemical activation provides another possibility in the use of cross-linking reagents[27] (for reviews, see Refs. 20, 21, and 28). Since cross-linking requires reaction with at least two functional groups, probably differing in chemical reactivity and/or spatial location, better control over the cross-linking reaction might be obtained in a stepwise cross-linking approach. This is possible if the reagent contains both a chemically reactive group and a light-activatable (photochemical) group (or two photochemical groups showing no overlap in their photoactivation spectra).[29]

Cleavable cross-linking reagents useful in some situations contain in the molecule a chemical bond that can be split readily under mild conditions (e.g., under mild oxidation or reduction conditions)[30]; for reviews, see Refs. 17–21. Also, water-insoluble (hydrophobic) cross-linking reagents have been used to modify membrane proteins.[17,18]

Reactions of Cross-Linking Reagents with Proteins

The reaction of a bifunctional reagent with an enzyme can in principle yield three different types of products: (1) a one-point modified enzyme, (2) an intramolecular cross-linked enzyme, and (3) an intermolecular cross-linked enzyme (see Fig. 2). The yields of one-point modification and intramolecular cross-linked products will depend on the length of the bifunctional reagent used and the distance between the functional groups on the protein to be modified. To increase the number of intramolecular cross-links in a protein molecule (and hence to decrease the degree of one-point modification) one can: (1) choose an optimal length of the cross-linking molecule[25,26]; (2) premodify the protein by substituting the protein surface with additional reactive groups[25,31]; (3) exploit the potentially re-

[26] V. P. Torchilin, V. S. Trubetskoy, and K. Martinek, *J. Mol. Catal.* **19,** 291 (1983).
[27] J. R. Knowles, *Acc. Chem. Res.* **5,** 155 (1972).
[28] P. Guire, this series, Vol. 44, p. 280.
[29] P. Guire, *in* "Enzyme Technology and Renewable Resources," p. 55. Univ. of Virginia, Charlottesville, Virginia, 1976.
[30] R. R. Taraut, A. Bollen, T. Sun, J. W. B. Hershey, J. Sundberg, and L. R. Pierce, *Biochemistry* **12,** 3266 (1973).
[31] V. P. Torchilin, A. V. Maksimenko, V. N. Smirnov, I. V. Berezin, and K. Martinek, *Biochim. Biophys. Acta* **568,** 1 (1979).

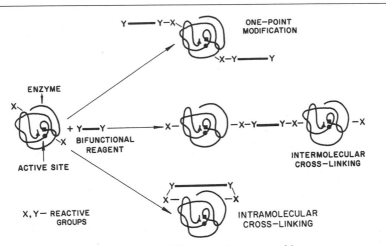

FIG. 2. Possible reactions of bifunctional reagents with enzymes.

versible character of chemical cross-linking by applying a mixture of bifunctional reagents of different chain lengths.[31] This means that in the course of the reaction the protein molecule itself will "select" intramolecular cross-linking in preference to one-point modification.

Furthermore, the probability of intermolecular cross-linking may be reduced by decreasing the enzyme concentration in the reaction medium. Alternatively, to suppress intermolecular cross-linking the protein could be attached to a solid support through a cleavable spacer arm, prior to cross-linking. After cross-linking, the spacer containing, for example, a disulfide linkage, is cleaved by reducing the S–S bond with thiol reagents.[32,33] In addition, a light-initiated heterobifunctional reagent can be used for cross-linking, resulting in no intermolecular side reactions.[27] In this case, first the bifunctional reagent reacts chemically at one end of the molecule, and then, after illumination of the premodified protein, it reacts photochemically at the other end of the cross-linking reagent (containing a diazo or an azide group). On illumination, a highly reactive carbene or nitrene is produced, reacting with the closest C–H linkage of the protein.

Thermostabilization of α-Chymotrypsin by Intramolecular Cross-Linking

Succinylation of α-Chymotrypsin. α-Chymotrypsin (EC 3.4.21.1) is succinylated according to the method of Goldstein.[34] α-Chymotrypsin

[32] G. P. Royer, S. Ikeda, and K. Aso, *FEBS Lett.* **80,** 89 (1977).
[33] S. Pillai and B. K. Bachhawat, *J. Mol. Biol.* **131,** 877 (1979).
[34] L. Goldstein, *Biochemistry* **11,** 4072 (1972).

(900 mg) is dissolved in 30 ml of 0.2 M phosphate buffer, pH 7.7. Succinic anhydride (300 mg) is then added in small portions while keeping the enzyme solution in the cold (4°) and maintaining the pH at 7.7. Under these conditions over 80% of available amino groups (14–15) of the enzyme are succinylated.[25] The reaction mixture is then passed through a column (2.6 × 60 cm) packed with Sephadex G-50 (Pharmacia). (The column is preequilibrated with 10 mM KCl.) The elution rate is 1.5 ml/min. The succinylated α-chymotrypsin preparation shows both catalytic activity and thermostability that are comparable to the same properties of the native enzyme.[25]

Carbodiimide Activation of Carboxyl Groups of α-Chymotrypsin. α-Chymotrypsin is treated with carbodiimide by a slightly modified version of the method described in Ref. 25. A solution (63 ml) containing α-chymotrypsin (10^{-6} M native or succinylated enzyme) is added to 7 ml of an aqueous solution containing 1-ethyl-3-(3-dimethylaminopropyl)carbodiimide (EDC) (10^{-2} M), and the mixture is left at a constant pH of 4.5 (using a pH-stat) for 1 hr at 20°. Under these conditions 15 out of 17 exposed carboxyl groups of α-chymotrypsin are modified. On treatment of α-chymotrypsin with carbodiimide, the relative catalytic activity of the enzyme drops 3-fold.

Reaction of Carbodiimide-Activated α-Chymotrypsin with Diamines. The solution (10 ml) containing α-chymotrypsin or succinylated α-chymotrypsin preactivated by carbodiimide treatment and 20 mM phosphate buffer (4 ml), pH 8.2, is added to a solution (1 ml) containing the amine reagent. The following amine concentrations are used: 10 mM hexamethylenediamine and dodecamethylenediamine; 0.1 M ethylenediamine, tetramethylenediamine, and pentamethylenediamine; 10 v/v% of hydrazine (or 1-amino-propan-3-ol). The reaction is carried out at pH 8.2 for 1 hr at 20°.

Thermoinactivation. To a solution (10 ml) containing cross-linked α-chymotrypsin (10^{-6} M), 20 mM phosphate buffer (5 ml), pH 7.0, is added, and the mixture is left at 50°. Aliquots (1 ml) are withdrawn at certain time intervals, and the enzyme activity is determined.

Activity Measurements of Native and Modified Enzyme. The catalytic activity of the native and modified enzyme is measured in a Radiometer TTT-1d pH-stat (Radiometer) by determining the initial rates of hydrolysis of 10 mM N-acetyl-L-tyrosine ethyl ester in 0.1 M KCl at pH 7.0, 20° (assay volume 10 ml).

Results. The rate of thermoinactivation of enzyme modified with diamines of different chain lengths showed a minimum in the inactivation curve (Fig. 3) when the cross-linking reagent contained 4 methylene

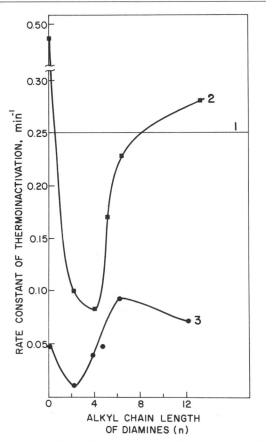

FIG. 3. Dependence of the first-order rate constant of thermoinactivation of cross-linked α-chymotrypsin on the chain length of the diamine reagents used for cross-linking: curve 2, cross-linked native α-chymotrypsin; curve 3, cross-linked succinylated α-chymotrypsin; curve 1, thermostability of native and succinylated α-chymotrypsin. From Torchilin et al.[25]

groups. It is worth adding that intermolecular cross-links were not formed under the experimental conditions, and that the monofunctional cross-linking analog, 1-aminopropan-3-ol, caused a certain destabilization of the enzyme. On the basis of the above, it is suggested that intramolecular cross-links were formed in α-chymotrypsin after treatment of the carbodiimide-activated enzyme with 1,4-tetramethylenediamine. Premodification of the enzyme with succinic anhydride resulted in additional reactive carboxyl groups on the protein surface. It was found that cross-linked succinylated preparations showed an increased thermostability compared

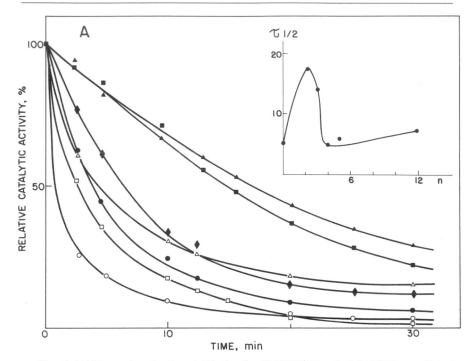

FIG. 4. (A) Thermoinactivation at 60° of native GAPDH (○) and of GAPDH cross-linked with oxalic acid (●), succinic acid (▲), glutaric acid (■), adipic acid (□), pimelic acid (△), and dodecandioic acid (◇). The inset shows the dependence of the half-life ($\tau_{1/2}$) of modified GAPDH on the number of methylene groups (n) of the diacid. (B) Densitometer traces for SDS–polyacrylamide gel electrophoresis of native GAPDH (1), GAPDH cross-linked with oxalic acid (2), GAPDH cross-linked with succinic acid (3), GAPDH cross-linked with glutaric acid (4), GAPDH cross-linked with adipic acid (5). Thirty micrograms of protein was applied to each gel. From Torchilin *et al.*[26]

with that of cross-linked unmodified enzyme (Fig. 3), indicating that a large quantity of cross-linkages had been formed (succinylation does not influence the thermostability of α-chymotrypsin, see above). It was also found that for cross-linked succinylated α-chymotrypsin, the maximal stabilizing effect is produced not by 1,4-tetramethylenediamine but by the shorter reagent 1,2-ethylenediamine (Fig. 3). This fact is an additional indication that the surface of succinylated α-chymotrypsin globule is more "populated" with carboxyl groups than that of the native enzyme. Thus, premodification of the enzyme makes possible regulation of the stabilization effect both with respect to the degree and the optimal length of the cross-link.

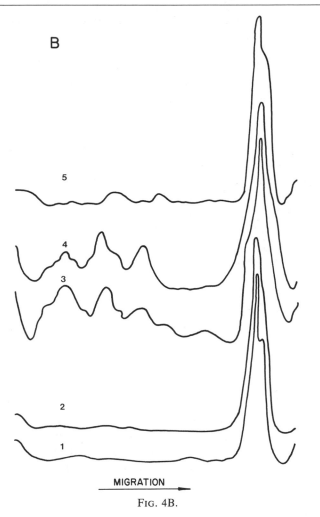

MIGRATION

Fig. 4B.

Thermostabilization of Glyceraldehyde-3-Phosphate Dehydrogenase by Intramolecular Cross-Linking

On heating or by action of a denaturant, oligomeric enzymes, such as glyceraldehyde-3-phosphate dehydrogenase (GAPDH) (EC 1.2.1.12) are reversibly dissociated into subunits, leading to inactivation of the enzyme.[35] The thermal stability of native GAPDH was studied and com-

[35] R. Rudolph, I. Heider, and R. Jaenicke, *Eur. J. Biochem.* **81,** 563 (1977).

pared with the thermostability of the enzyme modified with commercially available diacids such as $HOOC(CH_2)_nCOOH$ (using reagents with n varying from 0 to 10).

Experimental. On cross-linking two portions of solid carbodiimide (final concentration 2×10^{-3} M) are added at 45-min intervals to an aqueous solution containing different cross-linking reagents (5×10^{-4} M). However, dodecandioic acid (5×10^{-2} M) is activated in a solution containing dimethyl sulfoxide (DMSO) (1%, v/v). The reaction mixtures are allowed to incubate for 1.5 hr at pH 4.5, then the pH is increased to 8.2 and GAPDH is added to the reaction mixtures (final protein concentration 0.25 mg/ml). After reaction for 1.5 hr, the reaction is stopped by subjecting the mixtures to gel chromatography on Sephadex G-50 (packed in a minicolumn that is placed in a centrifuge or by dialyzing the reaction mixtures prior to preparative electrophoresis. All experiments are performed[26] at 20°. (The catalytic activity of the modified enzyme is found to be 20–40% of that of the native enzyme, depending on the bifunctional reagent used for cross-linking.)

In the thermoinactivation experiments, native or modified enzyme (2×10^{-6} M) in 50 mM phosphate buffer (pH 7.5) is incubated at 60°. Samples are withdrawn at appropriate time intervals, and the enzyme activity is measured spectrophotometrically at 60 or 25° according to the assay method described in Ref. 36.

Analytical sodium dodecyl sulfate (SDS)–polyacrylamide gel electrophoresis is performed as described by Laemmli.[37]

Results and Discussion. In Fig. 4A it can be seen that intramolecularly cross-linked GAPDH is more stable at 60° than native enzyme. Figure 4A also shows that the degree of stabilization is dependent on the chain length of the bifunctional reagent used. Figure 4B shows the results of SDS–gel electrophoresis of various cross-linked preparations. By comparing both figures it can be seen that the results of the SDS–polyacrylamide gel electrophoresis agree well with the results of the thermoinactivation experiments. Both native enzyme and the enzyme treated with the shortest bifunctional reagent, oxalic acid (this cross-linked preparation showed the same thermostability as native, untreated GAPDH), migrated in the gel as a single band corresponding to migration of the promoter of GAPDH (Fig. 4B). On the other hand, cross-linked enzyme preparations showing increased thermostability migrated as the dimer, trimer, tetramer, and/or higher oligomeric forms of the enzyme. Maximal thermostabilization was found when succinic acid (Fig. 4A) was used as a

[36] W. Ferdinand, *Biochem. J.* **92**, 578 (1964).
[37] U. K. Laemmli, *Nature (London)* **227**, 680 (1970).

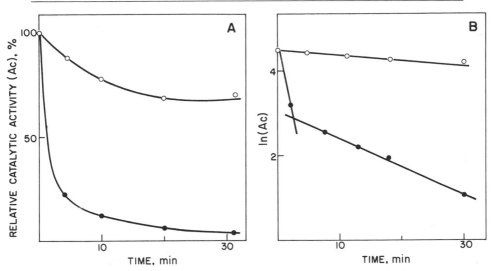

FIG. 5. (A) Thermoinactivation at 60° of GAPDH reconstituted from dimers cross-linked with succinic acid (O); native enzyme (●). (B) Semilogarithmic plot of the data in (A). Activity measurements were performed at 25°. From Torchilin et al.[26]

cross-linking reagent. In agreement with this observation is the finding that the same cross-linked enzyme yielded several SDS–polyacrylamide gel bands corresponding to different cross-linked forms of the enzyme (Fig. 4B). Thus, it is concluded that the chain length of succinic acid closely matches the distance between amino groups (located on different subunits of the enzyme) participating in the cross-linking reaction.

An interesting cross-linking experiment is thermoinactivation of GAPDH reconstituted from isolated cross-linked dimer molecules (Fig. 5). The dimers were prepared from succinic acid-treated GAPDH after preparative electrophoresis of the cross-linked enzyme in 8 M urea. From kinetic analysis of the thermoinactivation of native GAPDH, it can be concluded that the thermoinactivation is a two-step process in which the first step is reversible. It is interesting to note that the first step of the inactivation process was absent in thermoinactivation experiments of GAPDH obtained from cross-linked dimers. It should also be added that cross-linked GAPDH undergoes unfolding without prior dissociation of the enzyme into subunits and therefore cannot be reactivated.

Conclusion. The thermoinactivation of oligomeric enzymes is suggested[38] to be a two-step process in which the first step is protein dissocia-

[38] V. S. Trubetskoy and V. P. Torchilin, *Int. J. Biochem.* **17,** 661 (1985).

tion into subunits and the second step is unfolding of the subunits (Fig. 1). Thus, by cross-linking of protein structures, the first step of the inactivation process is prevented due to an increased barrier to enzyme dissociation.

[56] Long-Term Stability of Nongrowing Immobilized Cells of *Clostridium acetobutylicum* Controlled by the Intermittent Nutrient Dosing Technique

By Lena Häggström and Cecilia Förberg

Long-term stability in immobilized cell processes for continuous production of metabolites is an important factor in considering their practical applications. The problems encountered are different depending on the nature of the biological system, i.e., whether growing cells or nongrowing, but viable, cells are employed. The reactor design and the immobilization method also influence the stability of the process. This chapter focuses on nongrowing, but viable, cells of *Clostridium acetobutylicum* immobilized in alginate or adsorbed to beech wood shavings for the continuous production of acetone and butanol.

In any system where nongrowing cells are applied a loss of activity with time should be expected due to turnover of essential cell constituents. Addition of nutrients will restore the microbial activity, however, in order to maintain the cells in the nongrowing state and at the same time keep a constant productivity, the distribution of nutrients is critical. A technique for control of the activity in nongrowing immobilized cells has therefore been developed.[1]

Intermittent Nutrient Dosing Technique

The intermittent nutrient dosing technique is based on the pulsewise addition of nutrients to the reactor, which otherwise is continuously fed only with a nongrowth production medium. The nongrowth medium lacks a utilizable nitrogen source and growth factors. In order to maintain the organism in an active but nongrowing state the nutrient supply should be sufficient to enable the organism to restore essential cell constituents but not rich enough for reproduction to proceed. The addition of nutrients can

---

[1] C. Förberg, S.-O. Enfors, and L. Häggström, *Eur. J. Appl. Microbiol. Biotechnol.* **17**, 143 (1983).

be varied by varying the dosing interval, the dosage time, and the concentration of nutrients in the medium. The concentration of added nutrients in the reactor (complete mixing assumed) can be estimated as a function of time according to $f = e^{-t/t_r}$, where f is the fraction of material remaining after time t and t_r the retention time. The estimated concentrations will in practice be lower owing to the microbial consumption of nutrients.

The optimal nutrient dosing is likely to vary depending on the actual conditions in the reactor regarding, e.g., dilution rate, physiological state of the organism, and concentration of inhibitory metabolites (butanol and butyric acid) and has to be determined empirically. During continuous production of acetone and butanol (alginate-immobilized cells), at a dilution rate of 0.4 hr^{-1}, using the media below, and dosing in 15-min pulses, a dosing interval of 8 hr was found to be sufficient, while a 10-hr dosing interval resulted in decreased activity.

In starting up a continuous process the shift from a growth situation to a nongrowth situation is critical for the organism. It is recommended to make this transition gentle by slowly increasing the dosing intervals (2, 4, and 6 hr) each day, rather than starting directly with 8-hr dosing intervals.

Procedure for Alginate-Immobilized Cells

Organism. Stock cultures of *Clostridium acetobutylicum* ATCC 824 can be maintained as spores, stored at $-20°$. Preparation of spore cultures has been described elsewhere.[2]

Media

Nongrowth Medium. Add the following, in g/liter, to distilled water: KH$_2$PO$_4$, 0.4; Na$_2$HPO$_4 \cdot$ 2H$_2$O, 0.6; MgSO$_4 \cdot$ 7H$_2$O, 0.2; FeCl$_3 \cdot$ 6H$_2$O, 0.01; CaCl$_2$, 0.55 (for stabilization of the calcium alginate gel); butyric acid, 2 (can be excluded, but if done, pH control in the reactor is necessary, see below); glucose, 10; cysteine, 0.5. Add 0.5 ml/liter of a trace element solution containing (g/liter) the following: CaCl$_2 \cdot$ 2H$_2$O, 0.660; ZnSO$_4 \cdot$ 7H$_2$O, 0.180; CuSO$_4 \cdot$ 5H$_2$O, 0.160; MnSO$_4 \cdot$ 4H$_2$O, 0.160; CoCl$_2 \cdot$ 6H$_2$O, 0.180. Adjust the pH to 5.0 before sterilization. All medium components are autoclaved at 121°. Glucose and CaCl$_2$ are sterilized separately and mixed aseptically with the other medium components after cooling.

Nutrient Medium. The nutrient medium has the same basic composition but lacks extra CaCl$_2$ and contains further (g/liter) the following: NH$_4$Cl, 0.8; peptone (Difco), 10; yeast extract (Difco), 10.

[2] L. Häggström, *Adv. Biotechnol.* **2,** 79 (1981).

Growth Medium for Start-Up. As above, but butyric acid is always excluded and the glucose concentration is increased to 40 g/liter. The pH is adjusted to 6.7.

Immobilization. Spores, rather than vegetative cells,[3] are used because spores survive exposure to oxygen during the immobilization procedure. A 5% (w/v) sodium alginate (Sigma) solution (in 0.1 M phosphate buffer, pH 7) is autoclaved (121°, 20 min), cooled, and mixed with a frozen spore suspension to give a final concentration of 4% alginate and of 1.2×10^8 spores/ml. It is essential to keep the temperature of this mixture low (0–1°) during all steps since the spores otherwise germinate very readily. The spore–alginate suspension can be formed as beads, or it can be fixed as sheets onto a support of wire netting.[1] This is accomplished by dipping the supporting surfaces in the spore–alginate solution. A thin film of spore containing alginate is then retained on the wire netting. The film hardens on subsequent incubation (overnight) in cold 0.14 M $CaCl_2$. All solutions and materials are sterilized and sterile technique applied throughout.

Equipment. An example of experimental setup is shown in Fig. 1. A water-jacketed glass vessel (i.d. 5 cm; volume 185 ml) tempered to 37°, containing five concentrical cylinders (height 7 cm) of wire netting (onto which the alginate, about 35 g, is attached, see above) is used. Mixing is achieved by magnetic stirring under a supporting bottom. Other reactor configurations and immobilization methods may be used. Two media reservoirs (nongrowth and nutrient medium) are connected to the reactor by butyl rubber tubings. Anaerobic conditions are further established by a continuous flow of N_2 over the media surfaces. Two pumps and a timer for delivery of media to the reactor are also required. If the media contain butyric acid, the pH will stabilize at 4.3 in the reactor. If not, equipment for automatic pH control is required. pH is controlled by 0.5–1.0 M NaOH to a set point of 4.5.

Start-Up. The nets with the attached gel (or beads) are aseptically transferred from the cold $CaCl_2$ solution to the reactor, which immediately is filled with 95° growth medium (heat shocking of spores), and closed except for a gas outlet. After cooling, the reactor is incubated (28 hr, 35°), preferably in an anaerobic glove box for outgrowth of vegetative cells. The reactor is thereafter connected to the other equipment.

Continuous Operation. The continuous flow of nongrowth medium is started at a dilution rate of 0.4 hr^{-1}. The nutrient medium is dosed intermittently in 15-min pulses, at the same dilution rate as the nongrowth medium. During these pulses the flow of nongrowth medium may be

---

[3] L. Häggström and N. Molin, *Biotechnol. Lett.* **2**, 241 (1980).

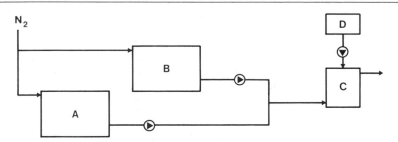

FIG. 1. Experimental setup. A, Nongrowth medium; B, nutrient medium; C, reactor with immobilized cells; D, pH control.

switched off or maintained. If maintained, the dilution rate is doubled during the pulses. If the dosing of nutrient medium follows a scheme of 2-hr dosing intervals for 2 days, 4-hr for 1 day, 6-hr for 1 day, and finally 8-hr dosing intervals, then the following results should be expected (Table I). During the first 2 days the organisms will grow since 2-hr dosing intervals are not growth limiting. After the gel is filled with cells, these are liberated into the surrounding medium, and the optical density of the outlet medium increases. At 4-hr dosing intervals growth is restricted, and

TABLE I
CONTINUOUS PRODUCTION OF BUTANOL WITH INTERMITTENT
NUTRIENT DOSING IN CALCIUM ALGINATE-IMMOBILIZED
C. acetobutylicum

Day	Dosing interval (hr)	OD_{580}[a]	Product concentration (g/liter)[b]	
			Butyric[c] acid	Butanol
1	2	0.06	2.44	0.73
2	2	0.23	2.71	0.75
3	4	0.24	2.60	1.17
4	6	0.12	2.28	1.43
6	8	0.06	2.12	1.63
9	8	0.04	1.87[d]	1.75
12	8	0.06	1.87[d]	1.75

[a] Free cells in the effluent. One optical density unit corresponds to 0.8 g/liter dry weight of cells.
[b] Production of acetic acid and acetone not shown.
[c] Media supplemented with 2.07 g/liter butyric acid.
[d] Uptake of butyric acid.

TABLE II
CONTINUOUS PRODUCTION[a] OF ETHANOL, ACETONE, AND BUTANOL
WITH INTERMITTENT NUTRIENT DOSING[b] IN C. acetobutylicum
ADSORBED TO BEECH WOOD SHAVINGS

	Product concentration (g/liter)				
Day[c]	Ethanol	Acetone	Butanol	Total solvents	Yield coefficient[d]
18	0.15	1.04	1.98	3.17	0.29
20	0.18	1.04	1.99	3.21	0.30
22	0.21	1.04	2.05	3.30	0.29
24	0.18	1.02	1.97	3.17	0.28
27	0.35	1.99	3.22	5.56	0.31
29	0.37	1.95	3.02	5.34	0.31
32	0.66	1.94	3.45	6.05	0.32
33	0.50	1.78	3.19	5.47	0.31

[a] The medium used was the same basic one as before, but no butyric acid or $CaCl_2$ was added. The glucose concentration was increased to 30 g/liter.
[b] The dosing interval was 7 hr.
[c] The dilution rate was initially 0.4 hr^{-1}, but from day 27 it was decreased to 0.2 hr^{-1}.
[d] Grams of total solvents formed per gram of glucose consumed.

at 6-hr dosing intervals the optical density starts to decrease. Finally, at 8-hr dosing intervals growth is almost retarded, and the steady-state level of cells in the effluent is in the range of 35 mg/liter.

Concomitant with the transition from growth to a nongrowth situation, a change in product formation pattern occurs. During growth the anaerobic metabolism yields mainly acetic and butyric acid, but in the nongrowth state the organism produces mainly the desired end products acetone and butanol. Constant activity has been maintained for 8 weeks of continuous production of acetone and butanol by means of the nutrient dosing technique.[1]

Adsorbed Cells

The nutrient dosing technique can also be used in controlling a process with adsorbed nongrowing cells. The same characteristics in terms of maintained activity, constant yield, long-time stability, and low cell leakage can be obtained (Table II).[4]

[4] C. Förberg and L. Häggström, Enzyme Microb. Technol. 7, 230 (1985).

The procedure is similar to that used with alginate-entrapped cells. The main differences are that in the start-up batch culture free spores (2.5 ml spore suspension/100 ml growth medium) are used and the organisms are adsorbed to the support material in the reactor during an initial period of the continuous phase. The reactor and the other equipment are the same as in Fig. 1, but the reactor is packed with sheets of a suitable support material (e.g., beech wood shavings). To achieve attachment of cells to the support special conditions are required. The adsorption procedure, which also is based on the effects of a restricted nutrient supply, is described elsewhere.[4] Once an adsorbed biofilm is obtained the system can be controlled by the nutrient dosing technique in the same way as for the alginate system.

Concluding Remarks

The intermittent nutrient dosing technique offers a means of controlling the activity of nongrowing immobilized cells during continuous production conditions at a steady-state level. Furthermore, the technique enhances growth of organisms at the support surfaces (due to accumulation of nutrients at the surfaces[5]) while suspended organisms are washed out if their growth rate is lower than the dilution rate. This effect can also be utilized for adsorption of organisms, and adsorbed cell systems can be controlled by the nutrient dosing technique with results comparable to or even better than that of an alginate system. By keeping the cells in a nongrowth state the production of biomass is reduced compared to a system with growing cells. Thereby part of the substrate, which otherwise would have been used for biomass production, is saved for product formation.

The optimal dosing rate was determined empirically in the procedures described here. This is a time-consuming way, and the results obtained may not be applicable if the conditions in the reactor are changed. If the microbial activity could be measured on-line, e.g., in terms of product concentration or substrate utilization, such a value could be used for distribution of nutrient pulses exactly when required by the organism. Further development of the intermittent nutrient dosing technique in that direction would facilitate its application to other microbial systems.

The products considered here (acetone and butanol) are end products of the anaerobic energy metabolism. Such products are usually considered as partially growth coupled, and can easily be obtained from non-

[5] K. C. Marshall, "Interfaces in Microbial Ecology." Harvard Univ. Press, Cambridge, Massachusetts, 1976.

growing cells. Another type of product that seems suitable for production with this technique are secondary metabolites, which usually are formed during the stationary phase of a batch culture. Many interesting microbial products are, however, primary metabolites in aerobic microbial processes. If the use of nongrowing immobilized cells can also be extended to these systems, this would be a further development of the utilization of microbes as biocatalysts for multistep enzymatic reactions. In fact, a recent investigation shows that the aromatic amino acid phenylalanine can be obtained from a nongrowing, plasmid-harboring *Escherichia coli* strain.[6]

[6] C. Förberg and L. Häggström, *Appl. Microbiol. Biotechnol.* **26,** 136 (1987).

[57] Production of Proinsulin by Entrapped Bacteria with Control of Cell Division by Inhibitors of DNA Synthesis

By L. Bülow, S. Birnbaum, and K. Mosbach

Introduction

With the development of recombinant DNA technology, novel microorganisms have become available which produce a variety of eukaryotic proteins such as hormones.[1,2] To date, most genetically engineered bacteria have been batch-fermented in a nonimmobilized form to produce the desired protein. However, owing to the inherent advantages of cell immobilization[3] it is on many occasions worthwhile investigating the potential of combining immobilized cell technology and gene technology.

The production of polypeptides by microorganisms requires that the transcriptional and translational machineries of the cell are functional. In the past, this has necessitated cell reproduction as well. Such reproduction often causes problems in fermentation processes based on immobilized cells as proliferating cells often clog the matrices in which they are embedded, thus impeding the flow of nutrients and eventually stopping

[1] L. Villa-Komaroff, A. Efstratiadis, S. Broome, P. Lomedico, R. Tizard, S. P. Naber, W. L. Chick, and W. Gilbert, *Proc. Natl. Acad. Sci. U.S.A.* **75,** 3727 (1978).
[2] K. Murray, *Philos. Trans. R. Soc. London B* **290,** 369 (1980).
[3] S. Birnbaum, P.-O. Larsson, and K. Mosbach, *in* "Solid Phase Biochemistry: Analytical and Synthetic Aspects" (W. H. Scouten, ed.), Chap. 15, pp. 679–762. Wiley, New York, 1983.

the process. Additionally, because of cell division, the cells become dislodged from the support and contaminate the product.

Therefore, we have investigated methods for inhibiting cell proliferation while still allowing polypeptide synthesis to continue. Addition of certain antibiotics which inhibit DNA replication to the growth medium, such as novobiocin and nalidixic acid, has proved useful. Protein synthesis is allowed to take place over a period of several days in spite of no or strongly reduced cell division.

The approach is one of several to limit cell release and yet utilize advantages of high cell concentrations. We think our contribution should act as a stimulus for further work in this area. The cells used in our experiments are strains of *Bacillus* sp. and *Escherichia coli* that are able to export their cloned gene products to the growth medium and the periplasmic space, respectively. From the periplasmic space the product diffuses to the growth medium. As a model system we have used bacteria carrying plasmids encoding rat or human proinsulin.

Materials and Reagents

Bacterial Strains

Bacillus subtilis 273 comprises the host strain SL 438 carrying plasmid pPCB6. This plasmid encodes rat preproinsulin.[4]

Escherichia coli EC703 comprises W3110 i^q carrying p*trc* 90K8 which specifies human proinsulin.

Both plasmids were kind gifts of Biogen S.A., Geneva, Switzerland.

Growth Media

L-broth: 10 g Bacto-tryptone, 5 g Bacto yeast extract, 5 g NaCl per liter

Supplemented M9: M9 salts (10×): Na_2HPO_4, 60 g; KH_2PO_4, 30 g; NaCl, 5 g; NH_4Cl, 10 g. These quantities, for 1 liter of a 10× solution, are dissolved in distilled water and autoclaved. To make 1 liter of medium, 100 ml of 10× M9 salts, 4 ml of 1 M $MgSO_4 \cdot 7H_2O$, 10 ml of 20% glucose, 100 ml of 20% casamino acids, and 10 ml of 10 mM $CaCl_2$ are added under sterile conditions to make a solution of 1 liter.

Materials and Reagents for Immobilization

Agarose: 4% (w/v) agarose (low gelling temperature, Sigma, Type VII) in phosphate-buffered saline (PBS).

[4] K. Mosbach, S. Birnbaum, K. Hardy, J. Davies, and L. Bülow, *Nature (London)* **302,** 543 (1983).

PBS: 8.0 g NaCl, 0.2 g KCl, 1.15 g $Na_2HPO \cdot 2H_2O$, 0.2 g KH_2PO_4 per liter.

Soybean oil: Soybean oil was obtained from Sigma.

Novobiocin and nalidixic acid: Stock solutions (1 mg/ml) of novobiocine and nalidixic acid (sodium salt, Sigma) are prepared in distilled water.

6-(p-Hydroxyphenylazo)uracil (ICI 3.854) was kindly provided by ICI-Pharma, Sweden. Stock solutions (1 mg/ml) are made up in 50 mM NaOH.

Methods

Immobilization Procedure

Bacillus subtilis 273 or *Escherichia coli* EC703 are grown in 25 ml L-broth or supplemented M9 (to which is added 12.5 μg/ml tetracycline or 40 μg/ml ampicillin to allow selection pressure) and collected during exponential growth at $OD_{550} = 1.0$. The cells are harvested by centrifugation (5,000 g) and washed twice with PBS. Twenty-five milliliters of the cell suspension is mixed with an equal volume of melted 4% (w/v) agarose, kept at 42°, and poured into a 1-liter beaker containing 500 ml soybean oil (prewarmed to 37°) under vigorous stirring. When the formed beads are 100–300 μm in diameter, the suspension is cooled to below the setting temperature of agarose (25°). After the beads have solidified, 300 ml of PBS is added, stirring is ceased, and the immobilized preparation is allowed to settle at room temperature. Subsequently, the oil layer is decanted and discarded. The mixture is then transferred to centrifuge tubes, after which PBS is added. The beads are spun down (2 min, 100 g), and the upper remaining oil phase and most of the aqueous phase are removed. Finally, the beads can be filtered and washed with PBS on a nylon net (250 mesh) to allow a narrow size distribution of the immobilized preparation.

Growth of Immobilized Bacteria

Batch Fermentation. The immobilized bacterial preparation 5 g (wet weight) is transferred to 20 ml of L-broth containing tetracycline or ampicillin in 100-ml Erlenmeyer flasks. The amounts can easily be scaled up. The flasks are incubated on a rotary shaker (200 rpm) at 37°. After incubation for 2 hr an inhibitor of DNA synthesis is added to the growth medium. At the same time 0.5 mM isopropyl-β-D-thiogalactoside (IPTG) is added to induce proinsulin synthesis. The amounts of antibiotics added

TABLE I

CONCENTRATIONS OF ANTIBIOTICS SUITABLE FOR
INHIBITING CELL DIVISION WHILE ALLOWING
PROTEIN SYNTHESIS

Bacteria	Concentration (μg/ml)		
	Nalidixic acid	Novo-biocin	Hydroxy-phenyl-azouracil
Bacillus sp.	20	5	10
Escherichia coli	5	25	—

must be chosen with care for each recipient strain. Table I serves as a guideline for appropriate additions. Proinsulin synthesis is then followed for about 24 hr.

Continuous Fermentation. After 24 hr of incubation the immobilized preparation can be transferred to fresh L-broth containing the same concentrations of antibiotics. The cycle can be repeated at least 3 times while maintaining satisfactory proinsulin synthesis. Alternatively, a continuous stirred tank reactor can be used. Five grams (wet weight) of the immobilized preparation is added to 20 ml of the medium in 100-ml Erlenmeyer flasks and shaken (150 rpm) at 37°. DNA synthesis inhibitors are added after incubation for 2 hr. Fresh medium is continuously added to the small tank reactor at a flow rate of 4 ml/hr. Proinsulin synthesis takes place over a period of several days.

Proinsulin Assay

Proinsulin is quantified by standard liquid radioimmunoassay (RIA).[5]

Comments

Inhibition of Cell Division

A number of potential solutions exist for controlling cellular growth. Techniques traditionally used in microbiology frequently focus on either autotroph mutants or nutrient limitation, usually of the carbon or nitrogen source (see this volume, Häggström and Förberg [56]). Since the primary aim is to stop cell division after reaching high cell concentrations more elegant techniques that specifically inhibit DNA synthesis without affect-

[5] L. G. Heding, *Diabetologia* **8**, 260 (1972).

ing normal cellular metabolism are highly attractive. Inhibitors of DNA synthesis, such as nalidixic acid or novobiocin, have proved to be useful in this respect. These antibiotics act by inhibiting supercoiling of DNA catalyzed by DNA gyrase and thus DNA replication.[6] 6-(p-Hydroxy-phenylazo)uracil on the other hand is a specific inhibitor of B. subtilis DNA polymerase.[7]

Alternatively, mutants that are blocked in DNA replication and cell division at elevated temperatures can be used because the inhibition is reversible on transferring the culture back to the permissive growth temperature. Temperature-sensitive mutants continue to synthesize protein under standard bacterial culture conditions for several hours, but the long-term effects on cellular metabolism are still unknown.[8]

Choice of Support Material

A number of support materials including alginate, polyacrylamide, and agarose have been tested for their ability to entrap the cells as well as to release proinsulin. Agarose (2%, w/v) is most effective in this latter respect as it allows rapid release of ^{125}I-labeled insulin; in addition it is nontoxic. However, in large-scale fermentations polyacrylamide or other related supports might be the first choice because of their higher mechanical stability.

Proinsulin Analysis

All processes based on microbial fermentations require that the product concentrations can be easily and quickly monitored. In many cases, as in insulin, these analysis are performed by conventional RIA or ELISA techniques which are quite time-consuming. A rapid thermometric enzyme-linked immunosorbent assay (TELISA) for insulin has been designed[9] (see also Birnbaum *et al.* [30]). The assay is completed in about 15 min. In all ELISA procedures there is a need for enzyme-labeled antigen or antibody, reagents that sometimes might be cumbersome to prepare. In analogy to the described fusion of two sequentially operating enzymes, construction of enzyme-labeled insulin by gene fusion might offer an attractive alternative.[10]

---

[6] M. Gellert, M. O'Dea, T. Itoh, and J.-I. Tomizawa, *Proc. Natl. Acad. Sci. U.S.A.* **73,** 4474 (1976).
[7] N. C. Brown, *J. Mol. Biol.* **59,** 1 (1971).
[8] G. E. Veomett and P. L. Kuempel, *Mol. Gen. Genet.* **123,** 17 (1973).
[9] B. Mattiasson, C. Borrebaeck, B. Sanfridson, and K. Mosbach, *Biochim. Biophys. Acta* **483,** 221 (1977).
[10] L. Bülow, P. Ljungcrantz, and K. Mosbach, *Biotechnology* **3,** 821 (1985).

[58] Permeabilized and Immobilized Cells

By HANSRUEDI FELIX

The study of macromolecular synthesis in intact cells is often hindered by permeability barriers due to the size and charge of the substrates. In recent years methods have been developed to make cells permeable to exogenous substrates.[1] Cells can be permeabilized without lysis of cells or destruction of the whole inner organization. Permeabilized cells are useful for the analysis of complicated metabolic processes such as DNA synthesis. After permeabilization with organic solvents cells usually are no longer viable. The effect of other permeabilization methods depends largely on the concentration of the agent and the time which is required for the procedure. It is possible to permeabilize cells reversibly, thus allowing the release of intracellularly stored products while preserving cell viability.[1,2] The following description of assays shows how to permeabilize different kinds of cells and how to test the permeabilizing effect. Immobilized cells that have been permeabilized may be useful for bioconversions.

Assay Methods

Principle. After healthy cells, either immobilized or free in solution, are treated with a permeabilizing agent, it is necessary to determine whether the treated cells are capable of carrying out biochemical reactions. This can be done by examining intracellular enzymes that are stable and need substrates not entering intact cells. Appropriate enzymes are hexokinase/glucose-6-phosphate dehydrogenase[3] or isocitrate dehydrogenase.[4] It is especially important to examine such enzyme systems before attempting to study enzymes that are less thoroughly characterized. The following methods describe how to permeabilize bacterial, fungal, and plant cells. Similar methods may be used to permeabilize mammalian cells, cells of invertebrates, and viruses.[1] Sometimes especially vulnerable cells can be permeabilized more efficiently by first immobilizing them. A hardening agent can be useful in such preparations for further stabilizing intracellular enzymes.

[1] H. R. Felix, *Anal. Biochem.* **120,** 211 (1982).
[2] P. Brodelius and K. Nilsson, *Eur. J. Appl. Microbiol. Biotechnol.* **17,** 275 (1983).
[3] H. R. Felix, J. Nüesch, and W. Wehrli, *Anal. Biochem.* **103,** 81 (1980).
[4] H. R. Felix, P. Brodelius, and K. Mosbach, *Anal. Biochem.* **116,** 462 (1981).

Permeabilization of Bacteria

Reagents

Starvation buffer: 67 mM KCl, 17 mM NaCl, 10 mM Tris–HCl, pH 7.4, 0.4 mM MgSO$_4$ · 7H$_2$O, 1 mM CaCl$_2$ · 2H$_2$O

Basic medium: 80 mM KCl, 40 mM Tris–HCl, pH 7.4, 7 mM magnesium acetate, 2 mM EGTA, 0.4 mM spermidine trihydrochloride, 0.5 M sucrose

Diethyl ether (stabilized by 7 ppm 2,6-di-*tert*-butyl-4-methylphenol)

Procedure. Bacterial cells, e.g., *Escherichia coli* and many other gram-negative bacteria, are grown to a density of 3 × 10^8 cells/ml. Fifty milliliters of this suspension is poured onto 20 ml starvation buffer, cooled with ice, and harvested by centrifugation (15 min, 8000 g, 4°). The pellet is resuspended in 1.5 ml basic medium and is manually shaken 1 min with 1.5 ml cold ether in a glass-stoppered tube. With careful handling, the ether and aqueous phases should separate immediately on standing in ice. The ether is removed and the cell suspension is layered into glass centrifuge tubes over 2-ml cushions of basic medium containing 0.8 M sucrose and centrifuged for 10 min at 8000 g, 4°. The cells are resuspended from the pellet in 0.43 ml basic medium to give 5 × 10^{10} cells/ml. This suspension is transferred to a series of tubes and frozen in dry ice. Even after being stored for months in a deep-freeze, the cells show no loss of activity. Each tube is thawed only once. If a basic medium containing 50% glycerol is used in the final suspension, the sample may be stored as a liquid at −18°, and several aliquots may be taken from the same tube. This ether permeabilization method[5] turned out to be the most effective for permeabilizing *E. coli* in order to measure RNA polymerase activity.[6] Other enzyme assays may require somewhat different suspension buffers.

Permeabilization of Fungal Cells

Reagents

Diethyl ether (stabilized by 7 ppm 2,6-di-*tert*-butyl-4-methylphenol)

Suspension buffer: 50 mM potassium phosphate buffer, pH 7.6, 1.5 mM MgCl$_2$ · 6H$_2$O; suitable for hexokinase/glucose-6-phosphate dehydrogenase and isocitrate dehydrogenase assay

Procedure. Cells (*Cephalosporium acremonium, Curvularia lunata, Saccharomyces cerevisiae,* and *Candida albicans* were tested) are grown to a certain density in either defined or production medium. A 20-ml

---

[5] H.-P. Vosberg and H. Hoffmann-Berling, *J. Mol. Biol.* **58,** 739 (1971).

[6] H. R. Felix, Ph.D. Dissertation. Univ. of Basel, Switzerland, 1980.

aliquot of the culture suspension is shaken gently by hand for 1 min with an equal volume of cold ether in a 50-ml glass-stoppered tube. The aqueous layer is centrifuged for 15 min at 12,000 g, 4°, and the pellet is resuspended in 6 ml buffer. The centrifugation is repeated. Finally the cells are resuspended in the same buffer either with or without glycerol. In the former case the suspension can be stored as a liquid at −18°.

Permeabilization of Plant Cells

Reagents

Diethyl ether (stabilized by 7 ppm 2,6-di-*tert*-butyl-4-methylphenol) or dimethyl sulfoxide (DMSO)
Suspension buffer: 50 mM potassium phosphate buffer pH 7.6, 1.5 mM MgCl$_2$ · 6H$_2$O

Procedure. Plant cells, e.g., *Catharantus roseus, Daucus carota, Abutilon theophrasti,* and *Datura innoxia,* are grown in suspension culture to a certain density. A 20-ml aliquot is treated with ether as described for fungal cells. After ether treatment, the cells are collected by filtration on a nylon net (50 μm) and washed with 100 ml of the suspension buffer. Treated cells on the nylon net are ready for use. For DMSO treatment, 2.2 ml DMSO is added to 20 ml of the plant cell culture suspension in an Erlenmeyer flask. The flask is shaken for 20 min at 110 rpm, 26°. Filtration and washing steps are the same as after ether treatment.

Immobilization of Permeabilized Cells

Reagents

5% agarose (60% agarose Type VII, 40% Type I, Sigma, liquid, 40°)
Hypol 3000 (polyurethane, W. R. Grace, Lexington, MA 02173)
Suspension buffer: 50 mM potassium phosphate buffer pH 7.6, 1.5 mM MgCl$_2$ · 6H$_2$O

Procedure 1.[7] One gram (wet weight) of plant cells, or 0.1 g of bacterial or fungal cells, either permeabilized or intact, is mixed with 5 g 5% agarose (equilibrated at 40°). This suspension is dispersed as quickly as possible in an oil phase (20 ml soy oil) by magnetic stirring. When droplets of appropriate size have been formed, the mixture is cooled on an ice bath with continuous stirring until the polymer has solidified. The mixture is then transferred to centrifuge tubes, and buffer is added. The beads are

---

[7] K. Nilsson, S. Birnbaum, S. Flygare, L. Linse, U. Schröder, U. Jeppsson, P.-O. Larsson, K. Mosbach, and P. Brodelius, *Eur. J. Appl. Microbiol. Biotechnol.* **17**, 319 (1983).

spun down (2 min, 100 g), the upper oil phase and most of the aqueous phase are removed with an aspirator, and the washing process is repeated if necessary. The same procedure may be used to immobilize intact cells. The intact cells may then be permeabilized in the following manner. Six grams of the bead preparation is suspended in 30 ml buffer and shaken at 110 rpm at room temperature either with 30 ml ether for 15 min or with 3.3 ml DMSO for 30 min. The beads are then shaken in suspension buffer alone for 20 min to remove residual DMSO or ether.

Procedure 2. Permeabilized or intact wet plant cells (10 g) are vigorously mixed with 3 g of Hypol 3000 and quickly poured into a glass column (1.6 × 9.95 cm). During polymerization the volume increases to 35 ml. Both ends of the foam columns must be cut off, as the surface of the foam is not porous. A reaction mixture can be pumped through the column, and the product solution composition may be monitored continuously. Immobilized intact cells can be permeabilized simply by adding 10% DMSO to the substrate solution. A gradual increase of the reaction rate (e.g., NADPH formation) is observed as permeabilization progresses.

Test of Permeabilization: Hexokinase/Glucose-6-Phosphate Dehydrogenase

Reagents

50 mM ATP
10 mM NADP$^+$
0.2 M glucose
15 mM MgCl$_2$ · 6H$_2$O
0.5 potassium phosphate buffer, pH 7.6

Procedure. An aliquot is taken from the preparation to be tested. The size of this aliquot depends on the type of cells and the method of preparation: ether-treated bacteria, 20 μl; ether- or DMSO-treated fungal cells, 80 μl; plant cells, 80 μl; immobilized cell beads, 0.2 g. This aliquot is mixed with 0.1 ml glucose, 0.1 ml ATP, 0.1 ml MgCl$_2$ · 6H$_2$O, 0.1 ml phosphate buffer, and the final volume is adjusted to 0.9 ml with water. The reaction is started by adding 0.1 ml NADP$^+$. At each time point a sample (100 μl) is withdrawn, diluted 1 : 10, filtered, and the absorbance at 340 nm is measured. If immobilized cells are used, a 15-ml assay mixture is prepared: Three grams immobilized cell beads is filled into a column, and 12 ml of a solution containing 1.5 ml glucose, 1.5 ml ATP, 1.5 ml MgCl$_2$ · 6H$_2$O, 1.5 ml phosphate buffer, 1.5 ml NADP$^+$, and 4.5 ml water is pumped through the column. The absorbance at 340 nm is followed continuously using a flow cuvette.

Test of Permeabilization: Isocitrate Dehydrogenase

Reagents

10 mM NADP$^+$
40 mM isocitric acid
15 mM MgCl$_2$ · 6H$_2$O
0.5 M potassium phosphate buffer, pH 7.6

Procedure. The procedure described above is used, but with the following substrates: 0.1 ml NADP$^+$, 0.1 ml isocitric acid, 0.1 ml MgCl$_2$ · 6H$_2$O, 0.1 ml phosphate buffer, permeabilized cells, and the appropriate amount of water (final volume 1 ml).

Hardening Procedure. Ten grams wet cells or 10 g cells embedded in agarose are suspended in 60 ml hexamethylenediamine solution for 10 min, then 5 ml of 12.5% glutardialdehyde is added, and the suspension is shaken at 110 rpm for 30 min. It is worth pointing out that the wet weight of the free cells dropped from 5 to 2 g during the hardening process. This loss was compensated for by the addition of water prior to immobilization.

Final Remarks

Simplification of the procedure is frequently possible depending on the enzyme system to be tested. Sometimes it is possible to omit a washing step. Enzymes in permeabilized, immobilized cells can be remarkably stable, especially after a hardening process.[8] The simplicity of these preparations makes them suitable for technical applications such as bioconversions.[2,8]

---

[8] H. R. Felix and K. Mosbach, *Biotechnol. Lett.* **4,** 181 (1982).

[59] Cells Coimmobilized with Enzymes

By Bärbel Hahn-Hägerdal

When two biocatalytic species are coimmobilized it is generally done in order to carry out in one step a reaction that otherwise would have been carried out in several steps. To put the problem in other words, with coimmobilization a substrate which otherwise is not available can be made available to the second catalytic species through the transformation by the first species. This situation resembles very much what in fermenta-

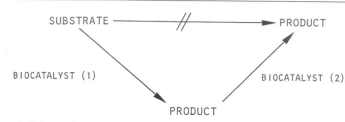

FIG. 1. Scheme for a one-step reaction employing coimmobilized biocatalysts.

tion technology is called a commensial mixed culture: the production of a growth factor (substrate) by one biocatalyst is essential and can be consumed by the second biocatalyst (Fig. 1).

This chapter deals with methods by which cells are coimmobilized with enzymes as the difference in size between these biocatalytic species necessitates special immobilization techniques. Essentially two in principle different methods have been reported for the coimmobilization of an enzyme and a cell: coentrapment in a gel and encapsulation of cells with enzymes (Fig. 2).

Coentrapment

If one wants to coentrap an enzyme and a microorganism one is faced with the problem that the two catalytic species differ considerably in size. In order to prevent leaking of the smaller species from the entrapment matrix it needs to be enlarged. This problem has been approached in essentially two different ways: the enzyme can be enlarged by binding it to a larger species or the enzyme can be enlarged through cross-linking (Fig. 2).

Enlargement through Binding to a Larger Species

The technique of enlargement through binding to a larger species was first reported by Häerdal and Mosbach[1] who covalently bound the enzyme β-glucosidase to the polymer alginate by the carbodiimide coupling procedure of Cuatrecasas and Parikh.[2] This enlarged enzyme preparation was then coentrapped with *Saccharomyces cerevisiae* in a calcium alginate gel and used for the direct conversion of cellobiose—a product from cellulose hydrolysis—to ethanol. A comparison of the coimmobilized preparation and separately immobilized and free biocatalysts has recently

---

[1] B. Hägerdal and K. Mosbach, *Food Process Eng.* **2**, 129 (1980).
[2] P. Cuatrecases and I. Parikh, *Biochemistry* **11**, 2291 (1972).

been reported.[3] The differently immobilized species were also studied under continuous operation over a 2-week period.

Typically the coimmobilized biocatalyst is prepared as follows[3]: 100 mg sodium alginate is carefully suspended in 2 ml distilled water, 32 mg N-hydroxysuccinimide, and 28 mg 1-ethyl-3-(3-dimethylaminopropyl)-carbodiimide–HCl (EDC) dissolved in 1 ml distilled water are added, and the alginate is activated for 15 min at room temperature. The β-glucosidase, dissolved in 1 ml water, is then added. Coupling is allowed to proceed overnight in the cold. The next day, a suspension of another 200 mg alginate in 6 ml distilled water is mixed with the alginate–β-glucosidase complex to give a final volume of 10 ml.

A sample of yeast cells (165 mg dry weight) is suspended in 5 ml 0.1 M acetate buffer, pH 4.9. The cell suspension is then mixed with the alginate sol containing both alginate and alginate with covalently bound β-glucosidase.

The suspension (~15 ml) is transferred to a disposable 5-ml syringe supplied with a needle having a diameter of 0.8 mm. The alginate sol is then slowly dropped from a height of 15 cm into a solution of 0.1 M CaCl$_2$ in 0.1 M acetate buffer, pH 4.9, whereby the beads are fixed. The beads are allowed to cure for at least 3 hr, after which they are transferred and stored in acetate buffer containing 10 mM CaCl$_2$. A suspension of about 15 ml alginate sol results in the formation of about 7.4 g (wet weight) beads. The average bead diameter is ~2 mm.

The optimal amount of enzyme to be bound in the alginate beads in this way was found to be 120 mg/g beads. Thirty percent of the added enzyme activity was immobilized, and 40% was recovered by washing the beads extensively with buffer. The alginate gel was found to impose severe diffusion limitations so that only 25% of the activity was found in the coimmobilized preparation comparing equal amount of free and immobilized biocatalysts. However, by obtaining the coimmobilized biocatalysts in beadlike particles it is possible to use them in a packed-bed-type reactor, which can be continuously operated. This was demonstrated with the above-described preparation over a 2-week period. At a dilution rate of 0.1 hr^{-1} a yield of 80% of ethanol was obtained from 50 gl^{-1} cellobiose.

The same technique for coimmobilizing an enzyme and a microorganism has also been used for the system β-galactosidase and Saccharomyces cerevisiae continuously converting concentrated acid whey (up to 15% lactose) to ethanol for more than 30 days.[4]

Instead of binding to alginate, enzymes can be enlarged prior to coen-

[3] B. Hahn-Hägerdal, Biotechnol. Bioeng. 26, 771 (1984).
[4] B. Hahn-Hägerdal, Biotechnol. Bioeng. 27 (1985).

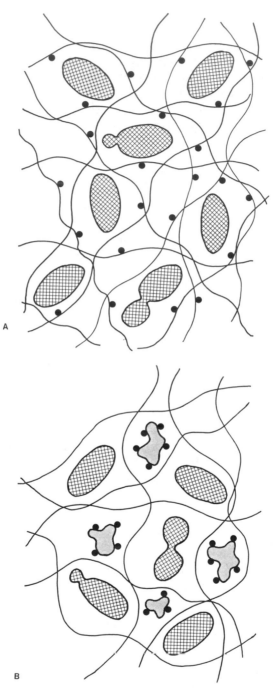

FIG. 2. Schematic representation of different ways to coimmobilize cells and enzymes. (A) Coentrapment, enzyme enlarged by binding to a polymer. Particle size ~2 mm. (B)

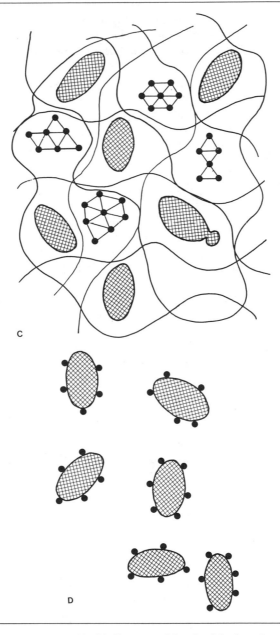

Coentrapment, enzyme enlarged by binding to particles. Particle size ~2 mm. (C) Coentrapment, enzyme enlarged by cross-linking. Particle size ~2 mm. (D) Encapsulation of cells with enzyme. Enzyme adsorbed and cross-linked around cells. Particle size equivalent to size of cell.

trapment with cells by coupling to dextran, as was shown for glucoamylase[5]; to concanavalin A–Sepharose via glucose or mannose residues, if the enzymes is a glycoprotein, as was demonstrated for β-D-glucosidase[6]; to CNBr-activated Sepharose 4B, as was shown for β-D-glucosidase[7]; to Spherosil beads with aromatic amino groups, activated by diazotation, as was shown for purified hydrogenase from *Desulfovibrio gigas*[8]; to controlled-pore glass or to amino-Spherosil as was shown for *Clostridium pasteurianum* hydrogenase and *Desulfovibrio desulfuricans* Norway hydrogenase, respectively (P. E. Gisby, K. K. Rao, and D. O. Hall, this series, Vol. 135 [39]).

Enlargement through Cross-Linking

Another way to enlarge the enzyme moiety of a coimmobilizate is to cross-link it in order to form larger aggregates. This was done with glucoamylase (glucan 1,4-α-glucosidase) using 1 and 2% glutaraldehyde at pH 6.0.[5] The polymeric products retained 55 and 25% of initial activity, respectively. On entrapment in calcium alginate gel only 1–4% of the initial activity was retained.

Glucoamylase was also polymerized with 0.04–0.18 M dimethyl suberimidate.[5] On entrapment in alginate gel only 0.5–1.5% of the original activity was displayed. Contrary to glutaraldehyde-polmyerized glucoamylase the dimethyl suberimidate-polymerized enzyme did not leak from the calcium alginate gel. The leakage of the glutaraldehyde-cross-linked preparation was, however, less than 2% of the initial entrapped enzyme activity.

In another study β-galactosidase was cross-linked with glutaraldehyde prior to coentrapment with *Zymomonas mobilis* in calcium alginate gel.[9] No information is available as to the amount of enzyme activity retained on cross-linking and coentrapment.

Encapsulation of Cells with Enzymes

To encapsulate cells with enzymes, one starts with an instant dried cell preparation which is rehydrated with an aqueous enzyme solution.[10]

---

[5] B. Svensson and M. Ottesen, *Carlsberg Res. Commun.* **46,** 13 (1981).

[6] J. M. Lee and J. Woodward, *Biotechnol. Bioeng.* **25,** 2441 (1983).

[7] M. Kierstan, A. McHale, and M. P. Coughlan, *Biotechnol. Bioeng.* **24,** 1461 (1982).

[8] M. F. Cocquempot, R. Aguirre, T. Lissolo, P. Monsan, E. C. Hatchikian, and D. Thomas, *Biotechnol. Lett.* **4,** 313 (1982).

[9] W. Hartmeir, E. D. Jankovic, U. Forster, and S. Tramm-Werner, *BioTech* **84,** p. 415 (1984).

[10] E. D. Jankovic, W. Hartmeier, and H. Dellweg (eds.), *Symp. Techn. Mikrobiol.* **5,** 377 (1982).

In a typical recipe 2 g dry weight baker's yeast cells, *Saccharomyces cerevisiae,* is rehydrated with 5 ml of an aqueous lactase solution (344 IU/ ml) for 15 min at 42°. This allows the enzyme to absorb on the yeast cell surface. The yeast cells are then suspended in 20 ml 2% tannin, after which the enzyme is cross-linked around the yeast cells by means of 0.01 ml 25% glutaraldehyde for 2 hr at 25°. This results in a preparation of 2.2 g dry weight holding 590 IU/g lactose activity (Fig. 2D).

The encapsulation method results in particles close to the size of separate cells, contrary to the coentrapment method which results in 2-mm spheres (Fig. 2A–C). The encapsulated preparations therefore show little diffusion limitation as compared to the coentrapped preparations. However, the encapsulated preparations cannot easily be used in continuous operations. This was demonstrated for *Aspergillus niger* mycelia encapsulated with glucoamylase for the deoxygenation of beer,[11] which had to be carried out in a frame reactor, where the coimmobilizate was packed between filter sheets.

The encapsulation method has been applied to a number of enzymes and microbial cells for use in the beverage industry.[12] However, for all these applications activities are only given as initial rates and little is known about the operational stability of the encapsulated preparations. At most they have been used in up to five repeated batch experiments, after which a substantial loss of activity occurs.

[11] W. Hartmeier and R. M. Lafferty (eds.) "Enzyme Technology," p. 207. Springer-Verlag, Berlin, 1983.
[12] W. Hartmeier, *Forum Mikrobiol.* **5,** 220, (1982).

[60] Affinity Immobilization

By Bo Mattiasson

Introduction

Conventional immobilization technology is focused on four main techniques: covalent coupling, entrapment, cross-linking and adsorption. These methods all have their advantages as well as disadvantages. Immobilization by covalent coupling or by cross-linking involves chemical modification of the protein to be immobilized. The entrapment technique, on the other hand, involves the formation of a three-dimensional lattice in which the enzyme molecules are captured. These three immobilized preparations have a limited lifetime, which is set by the lifetime of the catalyst.

There are no methods to regenerate the sorbents. In practical applications, this means that each preparation has a certain lifetime, and that storage time also has to be taken into account. In the research lab this is not normally a problem, even with labile enzymes, since a fresh preparation of immobilized enzymes can be prepared when needed. In other cases, however, this labile character of the preparations severely restricts the spectrum of enzymes that can be used. The fact that the support in many cases cannot be reused is also a severe limitation. In applications with technical grade enzymes, the cost of the support is often higher than that of the enzyme used. The fourth immobilization method, adsorption, in contrast to the three previously mentioned, offers the possibility for the desorption of inactive protein with a subsequent recharging of the support with fresh enzyme.

Provided that the experimental conditions are kept under strict control, ionic interaction may be exploited in the immobilization step.[1,2] Reports are also available on the use of hydrophobic interactions as a means of immobilizing an enzyme.[3,4] In these methods irrelevant proteins may be adsorbed along with the enzyme. Furthermore, displacement of the enzyme may take place during the subsequent operation due to competition for binding sites on the matrix.

Mentioned above are some of the limitations that are observed with the conventional methods of immobilization. Chemical modification and the fact that a large excess of enzyme is normally used to gain operational stability are two severe problems, especially when dealing with labile and/ or expensive enzymes or other biological structures. The degree of chemical modification of groups essential for the catalytic activity of the enzyme may in the covalent coupling procedure be reduced substantially by simply using another chemical coupling method.[5]

By using biospecific affinity interactions, all the advantages of the adsorption approach can be kept as well as be combined with biospecificity.[6] This gives conditions under which desorption can be better controlled and eliminates nonspecific displacement during operation. One drawback up to now, compared to conventional adsorption, has been the price for the sorbent. The affinity ligand used on the sorbent should be stable, or at least much more stable than the substance to be immobilized to it. Furthermore, in order to achieve a high operational stability of the sorbent, a large excess of ligand should be used whereas only a small

[1] T. Tosa, T. Mori, N. Fuse, and I. Chibata, *Agric. Biol. Chem.* **33**, 1047 (1969).
[2] R. A. Messing, this series, Vol. 44, p. 148.
[3] K. Dahlgren-Caldwell, R. Axén, and J. Porath, *Biotechnol. Bioeng.* **17**, 613 (1975).
[4] K. Dahlgren-Caldwell, R. Axén, and J. Porath, *Biotechnol. Bioeng.* **18**, 433 (1976).
[5] K. Mosbach (ed.), this series, Vol. 44.
[6] B. Mattiasson, *J. Appl. Biochem.* **3**, 183 (1981).

fraction of the binding positions are occupied by biocatalysts at each moment.

The operational stability that can be attained with a conventional immobilized preparation is limited due to the fact that the preparation, once formed, immediately starts to lose activity. It is very difficult to do anything that will prevent or retard this process.

The operational stability of a system based on affinity immobilization depends on the sorbent with its excess of ligands. The enzyme is stored under the most suitable conditions and applied to the sorbent just prior to use. The amount added is exactly what is needed at a certain time. No excess is required. By taking such an approach, a small amount of enzyme is sufficient at each time to create conditions that are usually obtained by the use of a large excess in the conventional method. One can also predict that affinity immobilized preparations cannot be used over extended periods of time as there is no built-in operational stability.

The sorbents used for affinity immobilization may be the same as those used in affinity chromatography. This is valid both for the ligands and, in some cases, for the matrix.[7] Affinity chromatography has to a large extent so far been focused on using soft hydrogels, whereas applications of immobilized enzymes often demand more pressure-stable gels.

The affinity sorbents used in chromatographic applications are selected because of low nonspecific binding—a prerequisite for high resolution in the separation process. This requirement is not so important in affinity immobilization, even if it is advantageous to have low or no nonspecific adsorption of other proteins.

The reactant pairs known from affinity chromatography may be applied to affinity immobilization. A basic principle when selecting the ligand–ligate pair, is that none of the molecular species should be present in the sample to be processed in a subsequent application.

Choice of the Reactant Pairs

One of the partners in the reactant pair is determined by the biomolecule to be immobilized. The other partner in the pair may be a low molecular weight ligand or a macromolecule such as lectin or an antibody. Some examples[8–13] are given in Table I together with indications of their respective binding constants.

[7] W. B. Jakoby and M. Wilchek (eds.), this series, Vol. 34.

[8] N. M. Green, *Biochem. J.* **89**, 585 (1963).

[9] C. W. Parker, "Radioimmunoassay of Biologically Active Compounds." Prentice-Hall, Englewood Cliffs, New Jersey, 1976.

[10] H. Schoemaker, M. Wall, and V. Zurawski, "Biotech 84," pp. 405–420. Online, Pinner, England, 1984.

TABLE I
ASSOCIATION CONSTANTS FOR SOME NATURALLY
OCCURRING REACTANT PAIRS

Reactant pair	K_{assoc} (liters/mol)	Reference
Avidin–biotin	10^{15}	8
Antibody–hapten	10^5–10^{11}	9
Antibody–antigen	10^5–10^{11}	10
Protein A–Fc region of IgG	10^6	11
Lectin–carbohydrate		
Simple sugars	10^3–10^4	12
Macromolecules and particulate structures	10^6–10^7	12
Triazine dyes–proteins	$\sim 10^4$	13

A limiting factor up to now in the use of this immobilization technique has been the cost of the ligand. The high prices for lectins, protein A, and antibodies, etc., have hampered their use. For bulk quantities, however, the prices are now decreasing markedly.

Examples

Use of Concanavalin A–Sepharose for Affinity Immobilization of Glycoprotein with Enzymatic Activity

Concanavalin A (Con A) is bound to CNBr-activated Sepharose (5–6 mg lectin/5 g wet gel) following conventional procedures.[14] Commercially available Con A-Sepharose (Pharmacia) is also used after appropriate washing.[15]

In the analytical applications, the Con A–Sepharose is packed into a small column and mounted in a continuous-flow analytical system. Enzyme is added as a pulse in the flow. After binding has taken place and any nonbound protein removed, the preparation is ready for use. The

[11] D. Lamet, D. Isenman, J. Sjödahl, J. Sjöquist, and I. Pecht, Biochem. Biophys. Res. Commun. 85, 608 (1978).
[12] A. L. Hubbard and Z. A. Cohn, in "Biochemical Analysis of Membranes" (A.-H. Muddy, ed.), pp. 427–501. Wiley, New York, 1976.
[13] S. Angal and I. D. G. Dean, Biochem. J. 167, 301 (1977).
[14] R. Axén, J. Porath, and S. Ernbach, Nature (London) 214, 1302 (1967).
[15] B. Mattiasson and C. Borrebaeck, FEBS Lett. 85, 119 (1978).

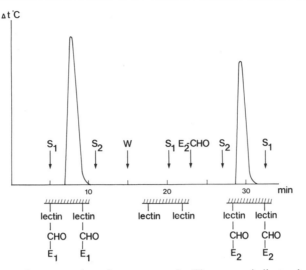

FIG. 1. Schematic presentation of an assay cycle. The arrows indicate changes in the perfusion medium, normally 0.1 M Tris–HCl, pH 7.0, with 1 M NaCl, 1 mM MgCl$_2$, 1 mM MnCl$_2$, and 1 mM CaCl$_2$, flow rate 0.75 ml/min. The cycle starts with glucoprotein (E$_1$-CHO) bound to the lectin-containing support material. At the arrows marked S$_1$ and S$_2$, substrate is introduced for enzymes E$_1$ and E$_2$, respectively. The heat signals obtained on substrate pulses are represented by peaks. At the arrow W a pulse of 0.2 M glycine–HCl, pH 2.2, is introduced in order to split the complex and to wash the system. A new enzyme E$_2$-CHO is then introduced, and substrate S$_2$ can be assayed. From Mattiasson and Borrebaeck[15] with permission.

examples discussed in this section deal with substrate analysis with either spectrophotometric[15] or thermal (enzyme thermistor) detection. For more information on the enzyme thermistor, see Chapters [16]–[19], this volume.

The general principle for a reaction cycle is shown in Fig. 1. It can be seen that it may be possible to reuse the same ligand column with a change in the enzyme loading.

Assay of L-Ascorbic Acid.[16] A 0.5-ml column filled with Con A–Sepharose, placed in an enzyme thermistor housing, is exposed to a 1-min pulse (flow rate of 0.85 ml/min) of L-ascorbate oxidase (4.5 U) (EC 1.10.3.3, Boehringer Mannheim, FRG) dissolved in a perfusing buffer of 0.1 M sodium acetate, pH 5.5. In a subsequent washing step using 1 mol/liter sodium chloride, unspecifically bound enzyme is removed.

The enzyme thermistor is then ready for use. In tests of the activity bound to the Con A–Sepharose column very small differences are ob-

[16] B. Mattiasson and B. Danielsson, *Carbohydr. Res.* **102,** 273 (1982).

served between separate experiments. The method used to quantify the bound enzyme is to expose the system to a short pulse of substrate of constant concentration and then read the response. Variations between different tests over several days is 1–2%. The sensitivity of the analyses using affinity immobilized L-ascorbate oxidase is satisfactory. A calibration curve is shown in Fig. 2.

Since only 4.5 U are used at a time, a new immobilization is done each day. Washing of the column is accomplished by a pulse of 0.1 mol/liter of pH 2 glycine–HCl. After reconditioning, a new pulse of enzyme can be introduced.

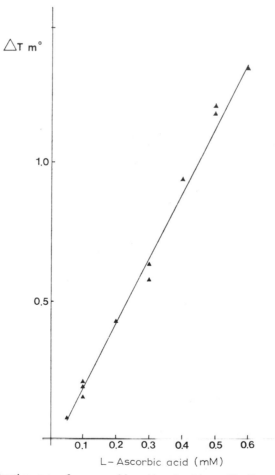

FIG. 2. Calibration curve for L-ascorbic acid, using biospecifically, reversibly immobilized L-ascorbate oxidase (4.5 U for the whole thermistor bed of 0.5 ml).

Other Examples of Con A Use. Con A–Sepharose is used for affinity immobilization of glucose oxidase,[15,16] invertase,[16] and peroxidase[15] as well, but these three enzymes are all well-known, stable enzymes available at reasonable prices. In those cases, therefore, conventional immobilization may be a better choice.

This gentle immobilization method may be used for immobilizing cells as well. Red blood cells are coimmobilized with glucose oxidase as an oxygen reservoir for oxidation processes,[15] and human lymphocytes are immobilized with retained ability to metabolize glucose. Using Con A–Sepharose, it was possible to immobilize *Trichosporon cutaneum* cells (a gift from Dr. H. Neujahr, Stockholm) and use them for quantification of phenols.[17]

Use of Antigen–Antibody Interactions for Affinity Immobilization

The antigen–antibody approach is very general and very selective. It was developed as a spin off from an immunochemical analytical system developed for continuous flow systems. In a competitive enzyme immunoassay developed for flow systems, a thermometric detection principle was applied. The procedure is called *t*hermometric *e*nzyme-*l*inked *i*mmuno*s*orbent *a*ssay (TELISA).[18] The general principle behind such an assay is illustrated in Fig 3. See also Chap. [30].

Using this procedure, it is possible to quantify macromolecular antigens as well as haptens.[18–20] By omitting the competitive step and instead letting the labeled enzyme bind to the immobilized antibody, an affinity immobilized enzyme preparation is obtained.[21] In the first example given, the enzyme is conjugated to human serum albumin (HSA), and the conjugate is then used for immobilization to a sorbent containing anti-HSA. The preparation and subsequent purification of conjugates between two macromolecular substances involves time-consuming and laborious steps. It is therefore easier to use low molecular weight substances for labeling the enzymes since it is then a matter of separating a small molecule from a substantially larger complex.

Affinity immobilization is used when setting up an analytical method

---

[17] B. Mattiasson, *in* "Immobilized Cells and Organelles" (B. Mattiasson, ed.), pp. 95–123. CRC Press, Boca Raton, Florida, 1983.

[18] B. Mattiasson, C. Borrebaeck, B. Sanfridsson, and K. Mosbach, *Biochim. Biophys. Acta* **483**, 221 (1977).

[19] B. Mattiasson, K. Svensson, C. Borrebaeck, S. Jonsson, and G. Kronvall, *Clin. Chem.* **24**, 1770 (1978).

[20] C. Borrebaeck, J. Börjesson, and B. Mattiasson, *Clin. Chem. Acta* **86**, 267 (1978).

[21] B. Mattiasson, *FEBS Lett.* **77**, 107 (1977).

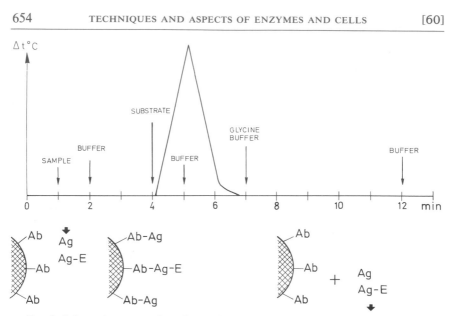

Fig. 3. Schematic presentation of a reaction cycle in the TELISA procedure. The arrows indicate changes in the perfusing medium (flow rate 0.8 ml/min). The cycle starts with potassium phosphate buffer, pH 7.0 (0.2 M). At this time the thermistor column contains only immobilized antibodies. At the arrow "sample" a mixture of antigen and catalase-labeled antigen is introduced. The system is then washed with potassium phosphate buffer for 2 min. Now the sites on the antibodies of the column are occupied by antigen as well as by catalase-labeled antigen. The amount of catalase bound is measured by registering the heat produced during a 1-min pulse of 1 mM H_2O_2. After the heat pulse is registered, the system is washed with glycine–HCl (0.2 M, pH 2.2) to split the complex. After 5 min of washing, phosphate buffer is introduced, and the system is ready for another assay.

for quantifying microbial cells. In the small column of a disposable plastic syringe is placed 1–2 ml of the affinity sorbent. Sample is introduced, either drawn into the syringe or pumped through when used as a column. After binding takes place, unspecifically retarded material is washed away prior to exposure of the cells to a developing solution. By supplying a substrate together with an indicator (pH or redox) it was possible to quantify the cells by monitoring their metabolism.

Affinity Binding of Saccharomyces cerevisiae to Sepharose-Bound Concanavalin A.[22] In a plastic syringe is filled 0.5 ml concanavalin A–Sepharose suspended in 2 ml buffer, pH 7.4 (0.004 mM KH$_4$PO$_4$, 0.054 mM Tris, 0.41 mM NaHCO$_3$, 0.95 mM CaCl$_2$, 0.80 mM MgSO$_4$, 5.36 mM KCl, 137 mM NaCl). The excess buffer is removed, and 2 ml of the sample containing the cells is drawn in. After binding for 1–3 min, the

[22] B. Mattiasson and P.-A. Johansson, *J. Immunol. Methods* **52**, 233 (1982).

excess solution is pressed out and fresh buffer added. After mixing, the buffer is removed and fresh buffer added. This procedure is repeated 4 times before substrate (50 mM glucose and neutral red, 0.2 mg/ml in the above buffer) is introduced. After careful mixing of substrate and gel the volume is decreased to 0.8 ml, and the incubation proceeds for 2 hr. The free solution is squeezed out of the syringe and the absorbance read. A correlation between cell number and absorbance is obtained.

Quantitation of Escherichia coli. The conditions are as in the example given above, except that concanavalin A is replaced by an antiserum against *Haemophilus influenzae* with cross-reactivity against *Escherichia coli.* After 20 min of incubation for binding, washings are carried out prior to 2 hr of incubation with the glucose–neutral red substrate solution. The readout of the used substrate gives a good correlation between the cell number and the observed change in absorbance.

Recently it has been demonstrated how gene technology can facilitate downstream processing by fusing the protein of interest with a protein or peptide of well-known properties that can be used as a specific handle in the purification process.[23,24] After isolation of the fused proteins a specific cleavage step is used to liberate the wanted protein. This procedure of forming conjugates by fusion seems very tempting when dealing with reversible immobilization, since the product isolated from the microorganism is directly ready for use.

Conclusions

The principle of affinity immobilization illustrated in this chapter is a suitable procedure when labile structures are going to be immobilized and when the enzyme is too expensive to be used in large excess in the immobilization step. Still another situation when reversible immobilization offers certain advantages is when strong inhibitors are to be quantified. In a conventional analytical system an excess of enzyme is used and inhibitory effects are easily compensated for by the resting enzyme molecules. This results in low sensitivity to inhibitors. If, however, a small amount of enzyme is used, then the inhibitory effect is much more easily quantified.

An additional benefit that may be obtained by the affinity immobilization procedure is stabilization of the bound molecule. Several reports in

[23] J. Germino, J. G. Gray, H. Charbonneau, T. Vanaman, and D. Bastia, *Proc. Natl. Acad. Sci. U.S.A.* **80,** 6848 (1983).
[24] M. Uhlén, B. Nilsson, B. Guss, M. Lindberg, S. Gatenbeck, and L. Philipson, *Gene* **23,** 369 (1983).

literature have demonstrated this.[25–28] The general availability of monoclonal antibodies in the future may make affinity immobilization an even more attractive method than it is today. Affinity immobilization certainly offers the possibility to expand the use of immobilization technology to labile and sensitive biological structures.

In a recent report[29] was described the immobilization of carboxypepsidase Y (serine carboxypeptidase) by biospecific interaction with immobilized concanavalin A. In a subsequent step the enzyme was covalently bound to the lectin by treatment with glutaraldehyde.

The crucial point in the procedure of reversible immobilization is the elution step. Dissociation of the complexes under denaturing conditions will rapidly ruin the operational stability of the affinity matrix. The same precautions needed in affinity chromatography have to be taken here. Use of antibodies with lower avidities is one possibility. It has been reported in the literature that certain clones of antibodies have very pH-sensitive antigen binding. A slight pH change may be enough to break the complex.[30]

Another very promising approach is the use of peptide-induced antibodies where elution may be achieved by the addition of a small peptide as the eluting agent.[31]

Acknowledgment

Support by the National Swedish Board for Technical Development is gratefully acknowledged.

[25] A. Ahman, S. Bishayee, and B. K. Bachhawat, *Biochem. Biophys. Res. Commun.* **53**, 730 (1973).

[26] E. Sulkowski and M. Laskowski, Sr., *Biochem. Biophys. Res. Commun.* **57**, 463 (1974).

[27] A. Surolia, S. Bishayee, A. Ahmad, K. A. Balasubramanian, D. Thambi-Dorai, S. K. Podder, and B. K. Bachhawat, *in* "Concanavalin A" (T. K. Chowdhyru and A. K. Weiss, eds.), pp. 95–115. Plenum, New York, 1975.

[28] E. Katchalski-Kazir, *in* "Affinity Chromatography and Biological Recognition" (I. M. Chaiken, M. Wilchek, and I. Parikh, eds.), pp. 7–26. Academic Press, New York, 1983.

[29] J. Turkova, M. Fusek, J. J. Maksimov, and Y. B. Alakholv, *Int. Symp. Bioaffinity Chromatogr. Relat. Techn. 6th,* Abstr. L20 (1985).

[30] R. Bartholomew, P. Neidler, and G. David, *Protides Biol. Fluids,* Abstr. 49 (1982).

[31] T. M. Shinnick, J. G. Sutcliffe, J. L. Gerin, R. H. Purcell, J. L. Bittle, H. Mexander, D. J. Rowlands, F. Brown, and R. A. Lerner, *in* "Affinity Chromatography and Biological Recognition" (I. M. Chaiken, M. Wilchek, and I. Parikh, eds.), pp. 343–353. Academic Press, New York, 1983.

[61] Bioconversions in Aqueous Two-Phase Systems: An Alternative to Conventional Immobilization

By BO MATTIASSON

Introduction

Immobilized systems based on solid supports have, beside many positive properties, also some severe limitations. Matrix-bound biocatalysts have often been shown to operate under severe diffusion restrictions, especially when the substrate is of high molecular weight. In the case of particulate substrates, the situation is even worse. Furthermore, within the pores of the matrix, product enrichment takes place causing concentrations that often are inhibitory.[1] These latter effects may be reduced in importance by, for example decreasing the bead size and also reducing the density of the catalyst in the beads. Still another limitation of the matrix-bound preparations is that when sequential enzyme reactions are to be catalyzed, the reaction runs smoothly only as long as all the enzymes are active. When one species denatures, then the whole preparation must be replaced.

As an alternative method to the conventional immobilization, a procedure for temporary immobilization was developed.[2,3] This method is based on the extractive bioconversion in aqueous polymer two-phase media and offers certain new properties in relation to those of the conventional methods. Aqueous two-phase systems have been known for a long time,[4] but only recently have they been exploited within the area of biotechnology.[5,6] When mixing two aqueous solutions of different polymers, an opaque solution is formed which spontaneously separates into a two-phase system. The phase systems are unique in the sense that both phases mainly consist of water (usually 85–95% each) and that the interfacial tension is extremely low. Down to 0.001 dyne/cm has been reported as compared with approximately 40 in oil/water systems.[7] The systems have

[1] R. Goldman, O. Kedem, I. H. Silman, S. R. Caplan, and E. Katchalski, *Biochemistry* **7**, 486 (1968).

[2] B. Hahn-Hägerdal, B. Mattiasson, and P.-Å. Albertsson, *Biotechnol. Lett.* **3**, 53 (1981).

[3] B. Mattiasson and B. Hahn-Hägerdal, *in* "Immobilized Cells and Organelles," (B. Mattiasson, ed.), Vol. 1, pp. 122–134. CRC Press, Boca Raton, Florida, 1983.

[4] M. N. Beijerinek, *Zentralbl. Bakteriol.* **2**, 627 (1896).

[5] B. Mattiasson, *Trends Biotechnol.* **1**, 16 (1983).

[6] B. Mattiasson, this series, Vol. 92, p. 498.

[7] P.-Å. Albertsson, "Partition of Cell Particles and Macromolecules." Almqvist & Wiksell, Uppsala, Sweden, 1971.

been proven to be biocompatible and, in some cases, also exert a stabilizing effect on the catalytic activity of an enzyme.[8]

A broad spectrum of polymers has been tested. In most cases, a difference in hydrophobicity has been the driving force. However, the hydrophilicities of all the polymer solutions are still quite close to that of water (Fig. 1). The partition behavior is given as a partition constant, K_{part}, which is the ratio between the activities in the top and the bottom phases, respectively [Eq. (1)]. Often the ratio between concentrations in the two phases is used.

$$K_{part} = C_{top}/C_{bottom} \tag{1}$$

Beside variations in hydrophobicity as a driving force for separation, contributions from variations in hydrophilicity in electrical charge, in conformation and so on have to be taken into account.[7] These contribu-

$$\ln K_{part} = \ln K_{el} + \ln K_{hydrophobic} + \ln K_{hydrophilic} + \ln K_{conformation} \tag{2}$$

tions are summarized in Eq. (2). In principle, any of the above factors can easily be used to create and maintain a phase system. However, if events take place that change the chemical composition in the phases, then the more subtle initial differences may not be enough to maintain the phase system.[9] Taking these factors into consideration, it is rather simple to create operable aqueous two-phase systems.

Basic Considerations

Most of the literature on separation in aqueous two-phase systems deals with methodological studies of the separation of various biochemical entities.[7] In those studies, it has been important to try to minimize the number of variables. Therefore, phase systems based on polyethylene glycol (PEG) and dextran of well characterized composition are the most abundant, even if the cost of polymer per liter of phase system, in some cases, has been considerable.[10] When dealing with small volumes and model studies, one can afford expensive phase systems, but when turning to processes in a larger scale a reduction in price is a prerequisite.

The choice of polymers is mainly governed by the desire to create favorable partition patterns in the system. Thus, in an ideal system, a situation such as that illustrated schematically in Fig. 2 is prevailing. It is difficult, if not impossible, to partition the low molecular weight sub-

[8] P. Monsan, *Eur. Congr. Biotechnol., 3rd,* (1984).
[9] B. G. Mattiasson and T. G. I. Ling, European Patent 0,011,837.
[10] R. Wennersten, F. Tjerneld, M. Larsson, and B. Mattiasson, *Proc. Int. Solvent Extract. Conf.,* p. 505 (1982).

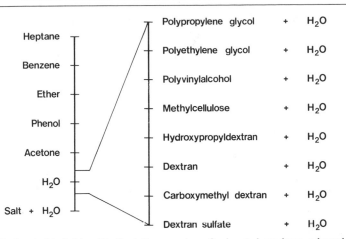

FIG. 1. Hydrophobic ladder. To the left, a number of solvents have been selected from a spectrum of solvents with increasing hydrophobicity. Aqueous solutions of the polymers to the right are mutually immiscible, but since they all consist mainly of water they fall within a narrow part of the solvent spectrum to the left. Reproduced with permission from Albertsson.[7]

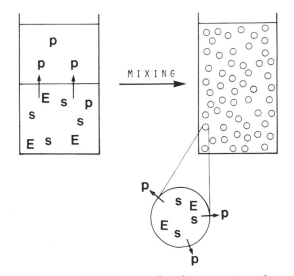

FIG. 2. Principle for extractive bioconversions in aqueous two-phase systems. E, Enzyme; S, substrate; P, product.

TABLE I
PARTITION CONSTANTS FOR VARIOUS BIOCHEMICAL ENTITIES

Species	K_{part}
Small molecules, substrates, products	0.5–2
Proteins	0.01–100
Particles (cells, etc.)	Extreme partitioning
Soluble separator molecules	0.001–1000
	(normally 0.01–100)
Separator particles	Extreme partitioning

strates and products as according to the idealized system. This means that a less efficient extraction process will be achieved. It is more important, however, to maintain the enzyme according to the partition rules as outlined in Fig. 2. Table I lists some biochemical species and typical partition constants.

Any deviation from ideal conditions means a loss of enzyme to the product phase. In batch processes this may not be critical, whereas if a continuous process is run then a substantial enzyme loss takes place. As a rule of thumb, partition constants of the enzyme of 100 : 1 or better between the "reaction phase" and the "extracting phase" should be used. When this is not attained, other methods have to be used. Among these, chemical modification of the enzyme to change its surface properties is a realistic possibility.[6,11] Other options are to recirculate the extracting phase over a membrane unit.[12] Such a method is discussed later in this chapter. Immobilized enzymes have a clear partition behavior, but the use of such preparations may only occasionally be advantageous.

Furthermore, when an extractive process is to be carried out and the partition behavior as such is not extreme, then the ratio between the volumes of the phases may be changed. Thus, in a system with a partition constant of 2, 66.7% of the product is extracted at equal phase volumes, while use of a 5 : 1 ratio of extracting to bioconversion phase, extracts 91% of the product formed.[12]

In cases of poor partition behavior, it may turn out that increasing the concentrations of the polymers may favor the partition. Such an action also increases the viscosity of the phase system and may simultaneously affect the water activity of the medium in such a way that the catalyzed reaction is influenced.[13,14]

[11] T. G. I. Ling and B. Mattiasson, *Talanta* **31,** 917 (1984).
[12] M. Larsson and B. Mattiasson, *Chem. Ind.* **June,** 428 (1984).
[13] B. Mattiasson, M. Suominen, E. Andersson, L. Häggström, P.-Å. Albertsson, and B. Hahn-Hägerdal, *Enzyme Eng.* **6,** 153 (1982).
[14] B. Mattiasson and B. Hahn-Hägerdal, *Eur. J. Appl. Microbiol. Biotechnol.* **16,** 52 (1982).

Examples

Degradation of a Macromolecular (Sometimes Particulate) Substrate

Production of Glucose from Starch. In the enzymatic conversion of starch to glucose, two enzymes are used: α-amylase (EC 3.2.1.1) and amylo-1,6-glucosidase (EC 3.2.1.33). In these studies corn starch (a gift from Stadex AB, Malmö, Sweden) was degraded using the thermostable α-amylase, Termamyl, and the amyloglucosidase, SAN 150. The enzymes were both gifts from Novo A/S, Bagsværd, Denmark.

It was shown earlier[15] that starch alone formed a two-phase system with PEG but that the properties of the system continuously changed as the degradation process continued. In order to stabilize the system crude dextran was used as the bottom phase constituent (3% w/w) in addition to the starch.[12]

This system was studied in batch experiments where starch, either gelatinized or native, was hydrolyzed into glucose. To the phase system consisting of 5% PEG 20M (Union Carbide) and 3% crude dextran (a gift from Sorigona AB, Sweden), both in 50 mM acetate buffer, pH 4.8, is added 100 g/liter starch and the enzymes, α-amylase and amyloglucosidase. The formation of monosaccharides was followed either by enzymatic analysis for glucose or by DNS assay[16] of reducing sugars.

The partition behavior of both the enzymes in the dextran/PEG phase system showed unfavorable partition patterns. However, on introducing starch to the system, α-amylase binds to the substrate and is then recovered from the bottom phase. Amyloglucosidase, on the other hand, showed a K_{part} of ~0.1. Variations in the buffer, polymer concentration, etc. did not change this to any substantial degree. As previously noted, it does not matter so much if the catalysts are partitioned to both phases in batch experiments.

In a process configuration where the top phase is continuously withdrawn and replaced with fresh top phase, the loss of enzyme with the top phase turns out to be too high to be economically acceptable.[10] Instead, the withdrawn top phase is passed over an ultrafiltration membrane unit where low molecular weight products are transported out with the eluate but polymer molecules and enzymes are kept in the retentat flow and thus could be recirculated to the reactor.[12] The process configuration is schematically shown in Fig. 3. Experimental results with such a unit clearly demonstrated that the enzymes could readily be recirculated to the reactor and a glucose-containing product stream isolated from the effluent. Productivity in such a system was double that in free solution. A time

[15] M. Larsson and B. Mattiasson, *Biotechnol. Bioeng.,* in press.
[16] G. L. Miller, R. Blum, W. E. Glennon, and A. L. Burton, *Anal. Biochem.* **2,** 127 (1960).

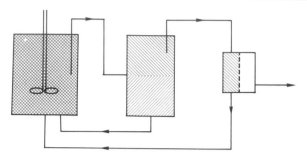

FIG. 3. Experimental setup when degrading starch in a reactor system consisting of a mixing chamber, a settling tank, and a membrane unit for product removal.

course as shown in Fig. 4 was obtained. The yield of glucose was 94% of the theoretical value.

Conversion of Low Molecular Weight Substrates

When dealing with reactions where the substrate and the product have similar molecular weights, one cannot predict a very different partition behavior between the substrate and the product. It is thus very important to optimize the system in order to develop as favorable a partition behavior as possible. This is especially true in cases where product inhibition

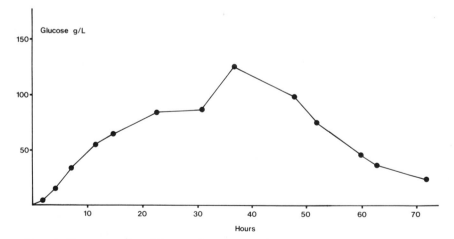

FIG. 4. Glucose produced from native starch as a function of time in 50 mM acetate buffer, pH 4.8 at 35° in a reactor volume of 21 liters. The reactor was fed with a 20% (w/w) starch slurry in the same buffer continuously for 38 hr. Enzymes used: amylase (Termamyl [Novo], total 2400 KNU) and glucoamylase (Spritamylase [Novo], total 6000 AGL); total amount of starch added, 5.0 kg; yield as glucose, >94%.

takes place. A classical example of a product-inhibited fermentation is described below:

Production of Acetone and Butanol from Glucose Using Clostridium acetobutylicum. Fermentative production of butanol and acetone by *Clostridium acetobutylicum* is severely inhibited by its products. At approximately 20 g/liter of solvents, the process was already completely inhibited. The costs in the downstream processing, together with a high substrate cost, have made this process uneconomical. Furthermore, the conversion was performed with a rather low productivity since the substrate level had to be kept low.

If a continuous extraction step is introduced, it would be possible to allow the cells to operate at high substrate concentrations but at low product concentrations. To achieve this extraction, biocompatible extraction systems have to be applied; therefore, the potential of aqueous two-phase systems was investigated.[13]

From Table II it can be seen that the selection of a proper phase system is important and that the extraction effect was by no means extreme; the K_{part} was at best 2.0. It should also, however, be borne in mind that by varying the ratio of the phase volumes, a rather efficient extraction can be achieved in spite of the low partition constants (Table III). It should be stressed in this context that the bacteria partitioned very nicely to the bottom phase.

Batch-wise Conversion of Glucose to Acetone and Butanol

Media Composition. Growth medium contains glucose, 40 g/liter; peptone, 10 g/liter; yeast extract, 10 g/liter; NH_4Cl, 0.8 g/liter; Na_2HPO_4 0.6 g/liter; KH_2PO_4 0.4 g/liter; $MgSO_4 \cdot 7H_2O$, 0.2 g/liter; and traces of Fe^{3+}, Ca^{2+}, Co^{2+}, Cu^{2+}, and Mn^{2+}. The phase system consists of 6% (w/w) Dextran T-40 (Pharmacia Fine Chemicals AB, Uppsala, Sweden) and 25% (w/w) PEG 8000 (Union Carbide, New York). This phase system gave a volume ratio between the phases of 6 : 1.

After inocculating the same number of *Clostridium* cells in this two-phase system and in a reference homogenous system, results as shown in Fig. 5 are obtained. During incubation a slight stirring is needed to keep the phases mixed. However, owing to the low surface tension between the phases a very gentle stirring is needed. It is seen from Fig. 5 that production of butyric acid preceded butanol formation. In the two-phase system, however, butanol formation started earlier and went to a higher value. If the reaction is allowed to proceed further, very little happens in a conventional batch experiment, whereas in the two-phase system the total concentration of butanol decreases substantially. Obviously, a meta-

TABLE II
EFFECTS OF VARIATIONS IN THE PHASE COMPOSITION ON THE PARTITION
BEHAVIOR OF ETHANOL, ACETONE, AND BUTANOL[a]

Dextran %(w/v)	PEG %(w/v)	$V_T : V_B$	$K = C_T/C_B$		
			Ethanol	Acetone	Butanol
3	15	8 : 1	1.2	1.3	1.6
3	17	9 : 1	1.3	1.5	1.9
3	20	12 : 1	1.2	1.5	1.9
4	15	7 : 1	1.2	1.3	1.6
6	10	3 : 1	1.5	1.0	1.3
6	15	4 : 1	1.3	1.5	1.8
6	25	6 : 1	1.9	1.9	2.0
	PVA				
2	8	8 : 1	1.0	1.2	1.2
2.25	8	4 : 1	1.1	1.1	1.3
2.5	8	3 : 1	1.1	1.3	1.3
	Pluronic				
4	14	5 : 1	1.4	1.4	1.5
6	14	4 : 1	1.3	1.4	1.4
	Ucon				
4	14	4 : 1	1.3	1.1	1.4
4	15	5 : 1	1.1	1.3	1.5
5	12	3 : 1	1.2	1.3	1.3

[a] PEG, Polyethylene glycol; PVA, poly(vinyl alcohol); Pluronic and Ucon, copolymers of PEG and poly(propylene glycol); V_T and V_B, volume of top and bottom phases; C_T and C_B, concentration of each product in top and bottom phases. From Mattiasson et al.[13] with permission.

TABLE III
INFLUENCE OF VOLUME RATIO BETWEEN TOP AND
BOTTOM PHASES ON EXTRACTIVE BIOCONVERSIONS AT
$K_{part} = 1.5$

Volume ratio	Product formed (calculated in relation to the amount in the bottom phase only)	% extracted of a fixed amount of substance
1 : 1	25	60
2 : 1	4	75
3 : 1	5.5	81.8
5 : 1	8.5	88.2
10 : 1	16	93.6
50 : 1	76	98.7

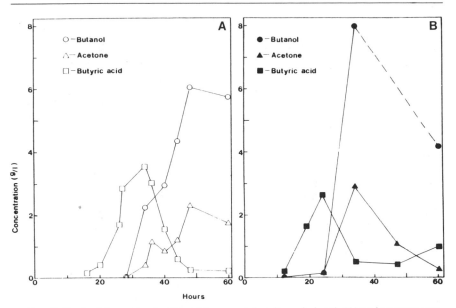

FIG. 5. Product formation by *Clostridium acetobutylicum* in batch (A) and aqueous two-phase (B) systems. Concentrations refer to the top phase.

bolic shift in the *Clostridium* has taken place. In more recent papers we have discussed this metabolic change as being due to changed water activity of the medium.[14] The conclusion drawn from these batch experiments was that butanol production should preferably be performed in a continuous process where the product is removed and fresh substrate added. Such a system is now being set up.[12]

Deacylation of Benzylpenicillin to 6-Aminopenicillanic Acid with Penicillin Acylase. The enzymatic deacylation of benzylpenicillin (BP) to 6-aminopenicillanic acid (6-APA) is a pH-dependent equilibrium which is shifted toward synthesis of BP when the pH decreases. This fact makes this reaction extremely suitable for study in a system with low diffusion-problems. The enzyme, penicillin acylase (EC 3.5.1.11, penicillin amidase), 10.4 U/mg protein, was a generous gift from Astra AB, Södertälje, Sweden. Selection of a two-phase system offering a suitable partition pattern for the enzyme turned out to be cumbersome. Table IV lists its partition behavior in some of the polymer/polymer aqueous two-phase systems studied, and Table V lists similar results from polymer/salt systems.[17]

[17] E. Andersson, B. Mattiasson, and B. Hahn-Hägerdal, *Enzyme Microb. Technol.* **6,** 301 (1984).

TABLE IV

PARTITION OF PENICILLIN ACYLASE IN AQUEOUS TWO-PHASE SYSTEMS
COMPOSED OF PEG AND DEXTRAN[a]

PEG type	% (w/w)	Dextran type	% (w/w)	Buffer	Molarity	K
8000	7.5	T 40	6.0	Tris	0.5	0.14
8000	7.5	T 40	6.0	Tris	0.3	0.13
8000	7.5	T 40	6.0	Sodium phosphate	0.2	0.10
8000	7.5	T 40	6.0	Sodium phosphate	0.5	0.04
8000	7.5	T 40	6.0	Potassium phosphate	0.5	0.35
8000	7.5	T 40	6.0	Sodium phosphate	0.05	0.15
20000	10.0	T 10	5.0	Sodium phosphate	0.12	0.03

[a] All data at pH 7.8 and 22°. K is based on activity measurements and is defined as the activity in the top phase divided by the activity in the bottom phase. From Andersson *et al.*[17] with permission.

As judged from Table V, a phase system of PEG 20,000 and potassium phosphate was the best. The $MgSO_4$-containing system gave a lower enzyme stability. In a typical experiment penicillin (243 mM final concentration) and penicillin acylase (0.3 mg/ml) are mixed. Starting pH is 8.1, and the running temperature is 37°. The results from one experiment as well as from a reference run in 0.5 M potassium phosphate buffer are shown in Fig. 6.

In spite of the high concentrations of buffering substances it turned out to be difficult to keep the pH constant, and titration had to be used. In order to influence the phase system as little as possible a strong base was used. When base was added to the top phase, only minor effects on the enzymes in the bottom phase were observed, whereas when titrating in

TABLE V

PARTITION OF PENICILLIN ACYLASE IN AQUEOUS TWO-PHASE SYSTEMS
COMPOSED OF PEG AND SALT[a]

PEG type	% (w/w)	Salt type	% (w/w)	Ratio (top/bottom)	K
8000	10.0	Potassium phosphate	10.0	0.69	0.02–0.03
20000	10.0	Potassium phosphate	10.0	0.64	0.02–0.03
20000*	8.9	Potassium phosphate	7.6	0.88	<0.01
3350	12.0	Magnesium sulfate	10.0	0.43	<0.01

[a] The experimental conditions are the same as in Table IV except for phase system marked with an asterisk, where the temperature was 37°. This phase system is formed at a temperature of ~25°. From Andersson *et al.*[17] with permission.

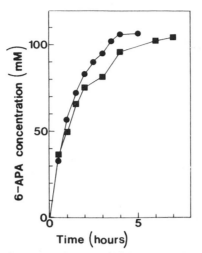

FIG. 6. Conversion of 243 mM benzylpenicillin to 6-aminopenicillanic acid (6-APA) (measured in samples from total phase system and buffer solution) with penicillin acylase at 37° and starting pH 8.1. Aqueous two-phase system: (●) 8.9% (w/w) PEG 20000/7.6% (w/w) potassium phosphate; (■) 0.5 M potassium phosphate buffer. The enzyme concentration was 0.3 mg protein/ml reaction solution (10.4 U/mg protein).

the bottom phase, severe protein denaturation took place. Repetitive incubations were carried out by changing the top phase and adding fresh substrate.

Summary

Aqueous two-phase systems offer certain advantages in bioconversion technology and might be a powerful tool in the future. This is especially probable in the case of cells, since in this case extreme partition behavior is easily obtained. When applied to pure enzymes it may be, as demonstrated, more time and labor consuming. The favorable characteristics of the aqueous two-phase systems that are exploited are the high water content in both phases in the extraction system; the low surface tension; the high biocompatability of the system; the potentially stabilizing effect on protein structures by the polymers used; the large surface area obtained in the emulsions created at a very low energy input; the spontaneous separation; and the fact that biomolecules partition in these systems.

Acknowledgment

Support by the National Swedish Board for Technical Development and the Biotechnology Research Foundation is gratefully acknowledged.

[62] Biochemical Energy Conversion by Immobilized Photosynthetic Bacteria

By Isao Karube and Shuichi Suzuki

Introduction

Hydrogen is attracting attention as one of the clean fuel resources. Various bacteria and algae produce hydrogen under anaerobic conditions.[1] However, because the hydrogen evolution system, especially, hydrogenase or nitrogenase, in bacteria is unstable, it is difficult to use whole cells for continuous hydrogen production.[2]

Recently, immobilization techniques for enzymes and bacteria have been developed for industrial application of these biocatalysts. Hydrogen-producing bacteria were immobilized in natural polymers and the immobilized whole cells continuously evolved hydrogen from diluted molasses.[3–6] Hydrogen exhibits excellent reactivity at electrodes. The hydrogen produced, 400–800 ml/min, when supplied to a fuel cell, gave about 10–12 W and a current of 10–12 A for 10 hr. However, hydrogen-producing bacteria also produce organic acids such as acetic acid, lactic acid, and propionic acid. The efficiency of the hydrogen production from molasses would be improved if hydrogen were to be produced from organic acids; in fact photosynthetic bacteria are known to do so, forming hydrogen from organic acids.

Rhodospirillum rubrum is a purple nonsulfur photosynthetic bacterium. Photosynthetic bacteria contain a single photosystem and do not evolve O_2. Photosynthetic cyclic electron transport is coupled to photophosphorylation. The purple nonsulfur Rhodosprillaceae utilize H_2, H_2S, and organic substrates. The ability of photosynthetic bacteria to produce large quantities of H_2 and CO_2 during photosynthetic growth on organic compounds has been known for several decades, and the pres-

[1] R. K. Thauer, K. Jungermann, and K. Dekker, *Bacteriol. Rev.* **41,** 100 (1977).

[2] H. Gest, M. D. Kamen, and H. M. Bregoff, *J. Biol. Chem.* **182,** 153 (1950).

[3] I. Karube, T. Matsunaga, S. Tsuru, and S. Suzuki, *Biochim. Biophys. Acta* **444,** 338 (1976).

[4] S. Suzuki, I. Karube, T. Matsunaga, S. Kuriyama, N. Suzuki, T. Shirogami, and T. Takamura, *Biochimie* **62,** 353 (1980).

[5] I. Karube, S. Suzuki, T. Matsunaga, and S. Kuriyama, *Ann. N.Y. Acad. Sci.* **369,** 91 (1981).

[6] S. Suzuki, I. Karube, H. Matsuoka, S. Ueyama, H. Kawakubo, S. Isoda, and T. Murahashi, *Ann. N.Y. Acad. Sci.* **413,** 133 (1983).

ence of hydrogenase has been demonstrated. However, H_2 production occurs only in the absence of NH_4^+ and is inhibited by N_2. Evolution of H_2 does not occur in the dark and is inhibited by uncouplers of phosphorylation. Therefore H_2 production is catalyzed by nitrogenase.[7,8] Nitrogenase catalyzes N_2 reduction and ATP-dependent H_2 evolution. *Rhodospirillum rubrum* utilizes malate, pyruvate, glutamate, succinate, acetate, fumarate, lactate, and oxalate.

Experimental Methods

Cultivation and Immobilization of Photosynthetic Bacteria

Rhodospirillum rubrum is used for hydrogen production and for the fuel cell system. *Rhodospirillum rubrum* IFO 3986 is cultivated in a medium containing 1% yeast extract, 0.05% $MgSO_4$, and 0.1% K_2HPO_4 (pH 7.0). The cultivation is performed for 3–4 days at 37° under fluorescent light (5000 lux). The bacteria are collected by centrifugation at 6000 g for 15 min. The wet cells are suspended in a Tris–HCl buffer solution (50 mM, pH 7.0) containing 2% sodium alginate. The bacterial suspension is dropped into 1% calcium chloride solution through a small capillary. The small beads (diameter 2–3 mm) formed are stored at 4°.

Preparation of Medium for R. rubrum

The wastewater from hydrogen-producing bacteria is used for the medium. This medium is prepared as follows. Diluted molasses (sugar content 0.87%) is transferred to the 200-liter bioreactor containing 3 kg of immobilized hydrogen-producing bacteria, *Citrobacter freundii* (*C. freundii* is absorbed in cellulose papers and coated with 1.5% agar). The wastewater is obtained from the outlet of the bioreactor and used as the medium for hydrogen production by photosynthetic bacteria. The wastewater contains 13 mM acetate, 6 mM propionate, and 30 mM butyrate.

Analytical Methods

Organic acids in the wastewater are determined by high-pressure liquid chromatography (Shimazu Seisakusho, Kyoto, Model LC-3A, column SCR-101H). Gas components are analyzed by gas chromatography (Shimazu Seisakusho, Kyoto, Model GC-3BT; column: molecular sieve 5 Å, 60–80 mesh).

[7] H. J. Shick, *Arch. Microbiol.* **75**, 102 (1971).
[8] W. J. Sweet and R. H. Burris, *J. Bacteriol.* **145**, 824 (1981).

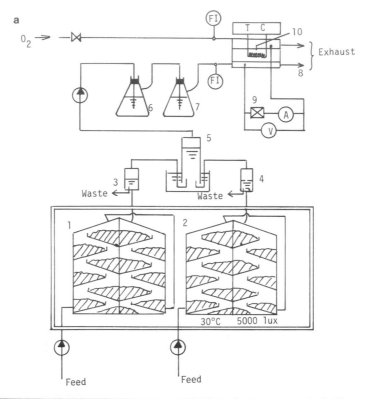

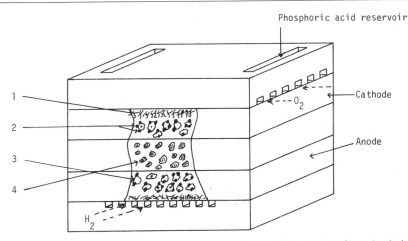

FIG. 2. Schematic diagram of the phosphoric acid fuel cell: 1, carbon fiber; 2, platinum black catalyst; 3, carbon pellet; 4, silicon carbide.

Bacterial Fuel Cell System

A photosynthetic bacterial fuel cell system is shown in Fig. 1a. Calcium alginate-immobilized photosynthetic bacteria (about 18 kg) are packed into the two bioreactors (20 liters each). The bioreactor is made of transparent acrylate. The immobilized cells are placed on the stage in the reactor (Fig. 1b) so that the hydrogen produced may easily flow to the hydrogen reservoirs. The reactor system is operated at pH 6.8–7.0, 30°, and with a light intensity of 5000 lux. The wastewater is continuously transferred to the reactors at a flow rate of 2.0 liter/hr. Biogas produced by this reactor is passed through 2 traps containing concentrated KOH solution (30%) and tap water, in order to remove compounds toxic for fuel cell, and then fed to the fuel cell system.

Figure 2 shows a schematic diagram of a phosphoric acid fuel cell. The fuel cell consisted of two carbon plate separators, a cathode, and an anode. The carbon plate separators have gas passages for hydrogen and oxygen. The anode and cathode consist of hydrophobic carbon fibers, platinum black catalysts, and carbon pellets. The matrix containing silicon carbide and phosphoric acid (100%) separates the anode and the cathode. Two fuel cells are connected in parallel. The fuel cell system is

FIG. 1. (a) Schematic diagram of a photosynthetic bacterial fuel cell system: 1 and 2, bioreactors for immobilized *R. rubrum*; 3–5, H$_2$ reservoirs; 6, KOH solution; 7, water; 8, phosphoric acid fuel cell; 9, variable electronic load; 10, heater; FI, flow indicator; TC, temperature control; A, ammeter; V, voltmeter. (b) Bioreactor for immobilized *R. rubrum*.

equipped with the electronic control system for a constant current. By adjusting an electronic load manually, a steady-state current can be maintained during the operation. The current and cell voltage are measured by an ammeter and a voltmeter and displayed on a recorder.

The effect of temperature on cell performance was examined. The current density and the cell voltage increased with rising temperature of the fuel cell. In the following experiments, the fuel cell was operated at 200°C.

Results and Discussion

Hydrogen was continuously evolved from organic acids by immobilized photosynthetic bacteria. The effect of cell content in calcium alginate beads on the H_2 evolution rate was examined. A 15 mM acetate solution was used as a substrate for hydrogen evolution. The highest hydrogen evolution rate was observed at a cell content of 1%. The optimum concentration of alginate was approximately 1%. However, this gel preparation was too soft for use. Thus, 2% gel and 1% cell content were employed in subsequent experiments.

The effect of initial pH on hydrogen evolution by R. $rubrum$ was examined. The optimum pH of the immobilized photosynthetic bacteria was 7.0, and further increase or decrease of the pH decreased the H_2 evolution rate. The H_2 production rate increased with increasing light intensity up to 5000 lux. Thus, initial conditions of pH 7.0, 30°, and light intensity 5000 lux were employed.

Hydrogen evolution from organic acids was performed in a batch system using immobilized photosynthetic bacteria. Concentrations of 15 mM acetate, propionate, and butyrate were employed. Hydrogen was evolved from these organic acids by immobilized R. $rubrum$. The H_2 evolution rate from butyrate was the lowest among these three organic acids.

Immobilized bacteria continuously evolved hydrogen over 90 hr from the wastewater containing organic acids (Fig. 3). Initially, the amount of hydrogen produced increased with increasing reaction time. Hydrogen produced by the reactors was transferred to an anode of a phosphoric acid fuel cell system at 19–31 ml/min. Oxygen was also transferred to a cathode at 0.1 liter/min.

Figure 4 shows the time course of the electric current, cell voltage, and electric power. A stable current from 0.5 to 0.6 A was obtained for 6 hr. A power of 0.16–0.18 W was observed during operation. The maximum power output of 0.4 W was obtained for 1 hr. During operation, the hydrogen productivity decreased gradually. This might be due to denaturation of the nitrogenase system.

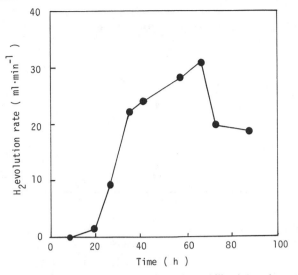

Fig. 3. Continuous production of hydrogen in an immobilized *R. rubrum* reactor. Reactor volume: 20 liters each; immobilized cells: 18 kg gel containing 180 g wet cells; wastewater feed rate: 2 liters/hr.

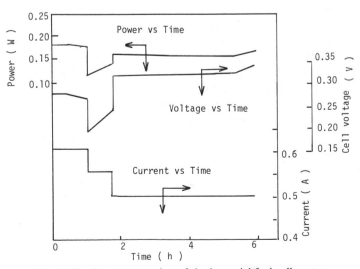

Fig. 4. Continuous operation of the bacterial fuel cell system.

TABLE I

CHANGES IN MAIN ORGANIC COMPOUND IN WASTEWATER DURING
THE OPERATION OF A PHOTOSYNTHETIC BACTERIAL FUEL
CELL SYSTEM

Compound	Inlet (mmol/liter)	Outlet (mmol/liter)	Difference (mmol/liter)
Acetate	14.0	6.4 (46%)	7.6 (54%)
Lactate	16.2	4.8 (30%)	11.4 (70%)
Propionate	11.0	1.5 (14%)	9.5 (86%)
Butyrate	0	0	0
COD	19000 ppm	13000 ppm (68%)	6000 ppm (32%)

Table I shows organic acid concentrations in the wastewater of the inlet and the outlet of the reactor. Organic acids such as acetic, lactic, and propionic acids were assimilated by the immobilized photosynthetic bacteria. About 86–54% of the organic acids were converted to hydrogen and carbon dioxide in the bioreactor. The chemical oxygen demand (COD) of the wastewater at the inlet of the reactors was 19,000 ppm and that at the outlet was 13,000 ppm. Only 32% of the COD was removed in the bioreactors. Therefore, the wastewater still contained organic compounds that could not be assimilated.

In this study, we have proposed the conversion of biomass by photosynthetic bacteria into hydrogen. The hydrogen produced was transferred to a fuel system and converted to electric energy. The utilization of organic acid as a source of hydrogen has also been shown. Considering the composition of wastewater of hydrogen-producing bacteria, such as from *C. freundii,* the combination of hydrogen-producing bacteria and photosynthetic bacteria may improve the efficiency of hydrogen production from biomass. As present, however, energy efficiency is low, but we are conducting a study on the genetic engineering of hydrogen-producing bacteria. Furthermore, optimization of the bioreactor system is also very important for continuous hydrogen production. Further study will make the realization of biochemical energy production by feasible these systems.

[63] Immobilization of a Cytochrome *P*-450 Enzyme from *Saccharomyces cerevisiae*

By David J. King, Mahmood R. Azari, and Alan Wiseman

Introduction

Cytochromes *P*-450 are a group of monooxygenase enzymes catalyzing the general reaction:

$$SH + O_2 + NADPH + H^+ \rightarrow SOH + H_2O + NADP^+$$

where SH is the substrate and SOH the monooxygenated product of the reaction. These heme-containing enzymes are best characterized from mammalian liver where they have been shown to be responsible for the monooxygenation of a very wide variety of substrates, including endogenous substrates, such as fatty acids and steroids, and xenobiotics such as drugs, carcinogens, and pesticides. Cytochrome *P*-450 enzymes are therefore of great interest not only for their biochemistry but also for their role in drug metabolism and toxicology, and as such have been intensively studied.[1-3]

Although cytochromes *P*-450 contain substrate and oxygen binding sites they are not capable of accepting electrons directly from NADPH and operate through one-electron reduction from NADPH–cytochrome *P*-450 reductase, in an electron transport chain as shown in Fig. 1. From many studies with purified cytochrome *P*-450 and NADPH–cytochrome *P*-450 reductase it has been shown that phospholipid is also required for activity of this monooxygenase system, although at present the role of phospholipid is unclear. In bacterial and mammalian mitochondrial cytochrome *P*-450 systems an extra protein component is involved, which shuttles electrons from the reductase to the cytochrome *P*-450. The requirement of these systems for oxygen and NADPH can be replaced by peroxides such as hydrogen peroxide or cumene hydroperoxide, though in these cases the stability of the enzyme is greatly reduced.[4,5]

[1] R. Sato and T. Omura (eds.), "Cytochrome P-450." Kodansha-Academic Press, Tokyo, 1978.
[2] J. B. Schenkman and D. Kupfer (eds.), "Hepatic Cytochrome *P*-450 Monooxygenase System." Pergamon, New York, 1982.
[3] E. Hietanen, M. Laitinen, and O. Hanninen (eds.), "Cytochrome *P*-450: Biochemistry, Biophysics and Environmental Implications." Elsevier, Amsterdam, 1982.
[4] G. D. Nordblom, R. E. White, and M. J. Coon, *Arch. Biochem. Biophys.* **175**, 524 (1976).
[5] E. G. Hrycay, J. A. Gustafsson, M. Ingelman-Sundberg, and L. Ernster, *FEBS Lett.* **56**, 161 (1975).

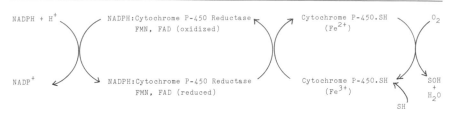

FIG. 1. Electron transport chain for substrate (S) monoxygenation by microsomal cytochrome *P*-450.

Cytochromes *P*-450 are widely distributed throughout nature, and we have used the enzyme from *Saccharomyces cerevisiae* as a model system. This yeast enzyme system has many properties in common with those of mammalian hepatic microsomal cytochrome *P*-450.[6] It consists of two protein components, cytochrome *P*-450 and NADPH–cytochrome *P*-450 reductase, and phospholipid and is located in the yeast endoplasmic reticulum.[7] The cytochrome *P*-450 enzyme from *S. cerevisiae* has a peak at 447–448 nm in purified form, and thus is a member of the narrower-specificity cytochrome *P*-448 class of cytochrome *P*-450 enzymes.[8] An endogenous substrate of yeast cytochromes *P*-450 is thought to be lanosterol, which undergoes 14α-demethylation as a key step in the formation of ergosterol, the major sterol component of yeast membranes.[9] A different yeast cytochrome *P*-450 enzyme may also be involved in ergosterol biosynthesis, catalyzing the Δ^{22}-desaturation of ergosta-5,7-dien-3β-ol.[10] Cytochrome *P*-450 from *S. cerevisiae* can also catalyze the monooxygenation of some xenobiotics including benzo[*a*]pyrene,[8,11] and aminopyrine, *p*-nitroanisole, and caffeine.[12] It has been demonstrated that this yeast cytochrome *P*-450 can metabolize many promutagens to their active mutagenic products.[13,14]

Several groups have attempted the immobilization of cytochrome *P*-450 enzymes, mainly to improve the stability of these unstable enzyme

[6] D. J. King, M. R. Azari, and A. Wiseman, *Xenobiotica* **14**, 187 (1984).

[7] K. Ishidate, K. Kawaguchi, K. Tagawa, and B. Hagihara, *J. Biochem.* **65**, 375 (1969).

[8] M. R. Azari and A. Wiseman, *Anal. Biochem.* **122**, 129 (1982).

[9] Y. Aoyama, Y. Yoshida, and R. Sato, *J. Biol. Chem.* **259**, 1661 (1984).

[10] S. Hata, T. Nishino, H. Katsuki, Y. Aoyama, and Y. Yoshida, *Biochem. Biophys. Res. Commun.* **116**, 162 (1983).

[11] A. Wiseman and L. F. J. Woods, *J. Chem. Technol. Biotechnol.* **29**, 320 (1979).

[12] M. Sauer, O. Kappeli, and A. Fiechter, *in* "Cytochrome *P*-450. Biochemistry, Biophysics and Environmental Implications" (E. Hietanen, M. Laitinen, and O. Hanninen, eds.), pp. 453–457. Elsevier, Amsterdam, 1982.

[13] D. F. Callen, C. R. Wolf, and R. M. Philpot, *Mutat. Res.* **77**, 55 (1980).

[14] D. Kelly and J. M. Parry, *Mutat. Res.* **108**, 147 (1983).

systems. Brunner and Losgen[15] have immobilized purified rabbit liver cytochrome *P*-450 together with NADPH–cytochrome *P*-450 reductase and UDPglucuronyltransferase to BrCN–Sepharose 4B. NADPH, UDPGA, and phospholipid were also placed in the enzyme beads, but hydroxylation activity was low. This work was an attempt to develop a system for extracorporeal drug detoxification, which was also the aim of Cohen *et al.*,[16] who immobilized microsomes from phenobarbital-induced rat liver by entrapment in hollow fibers. This immobilized system was capable of *p*-nitroanisole O-demethylation and hexobarbital metabolism.

Brunner *et al.*[17] have also immobilized rabbit cytochrome *P*-450 by copolymerization in acrylamide. Lehman *et al.*[18] have coimmobilized cytochrome *P*-450, NADPH–cytochrome *P*-450 reductase and glucuronyltransferase on BrCN–Sepharose 4B for applications in the synthesis of drug metabolites. Rabbit microsomal cytochrome *P*-450 has also been immobilized, by Schubert *et al.*,[19,20] to construct an enzyme electrode for the measurement of oxygen uptake by their system. More recently, Yawetz *et al.*[21] have immobilized rat liver microsomes, by entrapment in cross-linked prepolymerized polyacrylamide hydrazide, which were active toward *p*-nitroanisole O-demethylation for several hours. Our studies have concentrated on the immobilization of cytochrome *P*-450 from *S. cerevisiae* by several means.[22] The use of a microbial source for cytochrome *P*-450 has obvious advantages in any future applications, and therefore we have evaluated some immobilized yeast cytochrome *P*-450 systems.

Materials

The yeast strain *Saccharomyces cerevisiae* NCYC No. 240 was obtained from the National Collection of Yeast Cultures, Food Research Institute, Agricultural Research Council, Norwich, England. Mycological

[15] G. Brunner and H. Losgen, *Enzyme Eng.* **3**, 391 (1978).
[16] W. Cohen, W. H. Baricos, P. R. Kastl, and R. P. Chambers, *in* "Biomedical Applications of Immobilized Enzymes and Proteins" (T. M. S. Chang, ed.), pp. 319–328. Plenum, New York, 1977.
[17] G. Brunner, H. Losgen, G. B. Gawlik, and K. Belsner, *in* "Biochemistry, Biophysics and Regulation of Cytochrome *P*-450" (J. A. Gustafsson, ed.), pp. 573–576. Elsevier/North-Holland, Amsterdam, 1980.
[18] J. P. Lehman, L. Ferrin, C. Fenselau, and G. S. Yost, *Drug Metab. Dispos.* **9**, 15 (1981).
[19] F. Schubert, D. Kirstein, F. Scheller, and P. Mohr, *Anal. Lett.* **13**, 1167 (1980).
[20] F. Schubert, D. Kirstein, and F. Scheller, *Acta Biotechnol.* **2**, 187 (1982).
[21] A. Yawetz, A. S. Perry, A. Freeman, and E. Katchalski-Katzir, *Biochim. Biophys. Acta* **798**, 204 (1984).
[22] M. R. Azari and A. Wiseman, *Enzyme Microb. Technol.* **4**, 401 (1982).

peptone was obtained from London Analytical and Bacteriological Media Ltd., and yeast extract from Oxoid. Acrylamide, ammonium persulfate and N,N-methylenebisacrylamide were from BioRad laboratories, and Sepharose 4B was from Pharmacia Fine Chemicals. Cyanogen bromide and sodium cacodylate were from Aldrich Chemical Co. Dilaurylphosphatidylcholine, dimethylaminopropionitrile, dithiothreitol, agarose–concanavalin A, and benzo[a]pyrene were from Sigma Chemical Co. Standard 3-hydroxybenzo[a]pyrene was obtained from the National Institutes of Health, Bethesda, MD, and glutaraldehyde, sodium alginate, and all other chemicals were from BDH Chemicals Ltd.

Measurement of Enzyme Activity

The unique spectral properties of cytochrome P-450 enzymes in producing a peak at 450 nm in the reduced carbon monoxide difference spectrum allow for a direct spectral measurement of the enzyme without measuring enzyme activity. Thus there are two ways of quantifying cytochrome P-450, (1) by its reduced carbon monoxide difference spectrum and (2) by its enzymatic activity.

The spectral assay of cytochrome P-450 was achieved by the method of Omura and Sato[23] using a Varian Cary 219 spectrophotometer. The sample is divided between two cuvettes both, reduced by the addition of a small amount of sodium dithionite, and a base-line established between 400 and 500 nm. The sample cuvette is then gently bubbled through with carbon monoxide for 30 sec and the spectrum recorded. The height of the peak at 448 nm compared to 490 nm is then used to calculate the amount of enzyme using the extinction coefficient of 91 mM^{-1} cm^{-1}. In the case of purified cytochrome P-450, 8 min is allowed to elapse before recording the spectrum to allow for time dependency as described by Azari and Wiseman.[24]

Benzo[a]pyrene hydroxylase activity was determined using a modified version of the fluorimetric assay of Dehnen et al.[25] The method measures the formation of 3-hydroxybenzo[a]pyrene by its fluorimetric peak at 525 nm (467 nm excitation). Immobilized cytochrome P-450 (usually 1.5 nmol) is placed in a magnetically stirred flask at 37° and 0.1 M Tris–HCl buffer, pH 7.2, containing an NADPH regenerating system is added to give a final concentration of 1 mM NADP, 20 mM D-glucose 6-phosphate, 1 mM MgCl$_2$, and 8 IU of glucose-6-phosphate dehydrogenase. This sys-

[23] T. Omura and R. Sato, J. Biol. Chem. **239,** 2370 (1964).
[24] M. R. Azari and A. Wiseman, Anal. Biochem. **117,** 406 (1981).
[25] W. Dehnen, R. Tomingas, and J. Roos, Anal. Biochem. **53,** 373 (1973).

tem approximates a typical batch stirred-tank reactor. The reaction is initiated by the addition of benzo[*a*]pyrene from a stock solution of 2 mg/ml in dimethylformamide (this solvent gives much more reproducible results than acetone) to a final concentration in the range 0–160 μM, and the mixture is stirred at 37° for 15 min before stopping by the addition of an equal volume of ice-cold acetone and removing the immobilized enzyme by centrifugation or filtration. For free enzymes, incubations are carried out under the same conditions. Purified cytochrome *P*-450 and NADPH–cytochrome *P*-450 reductase are reconstituted to an active system with dilaurylphosphatidylcholine (1 nmol *P*-450 : 1 unit of reductase : 30 μg dilaurylphosphatidylcholine). After stopping the reaction with acetone, the mixture is centrifuged to remove protein precipitate, and a 0.6-ml aliquot of the supernatant is added to 1.4 ml of 10.7% (w/v) triethylamine in water. This is centrifuged again to remove any residual precipitate, a procedure that greatly reduces scatter in subsequent fluorescence measurements, and fluorescence is measured at 467 nm excitation, 525 nm emission. Fluorescence is calculated relative to a quinine sulfate solution (10 $\mu g/ml$ in 0.2 M H_2SO_4) which in turn is calibrated against a standard 3-hydroxybenzo[*a*]pyrene solution. Controls using enzyme preparations boiled for 15 min, compared with controls using complete reaction mixture stopped at zero time with acetone, are employed to show that nonenzymatic catalysis is not occurring (i.e., the two controls give the same fluorescence values).

Yeast Growth Conditions and Preparation of Enzyme

Saccharomyces cerevisiae NCYC No. 240 is grown in aerobic batch culture at 30° in a medium of 20% (w/v) glucose, 2% (w/v) mycological peptone, 1% (w/v) yeast extract, and 0.5% (w/v) sodium chloride. Yeast is harvested by centrifugation at late exponential phase, when the cytochrome *P*-450 level is maximal.[26] Cells are washed in 0.1 M potassium phosphate buffer, pH 7.0, and then used for permeabilization, or for the preparation of a microsomal fraction, and subsequent purification of cytochrome *P*-450.

Permeabilized whole cells are prepared by the method of Murakami *et al.*[27] Packed yeast cells (1 g wet weight) are resuspended in 4 ml of 0.1 M potassium phosphate buffer, pH 7.4, containing 0.4 M sorbitol. To this suspension 15 ml of toluene is added, and the mixture is incubated at 40° for 5 min with shaking. The mixture is cooled on ice, and the cells are

[26] A. Wiseman, T. K. Lim, and C. McCloud, *Biochem. Soc. Trans.* **3**, 276 (1975).
[27] K. Murakami, M. Nagura, and M. Yoshino, *Anal. Biochem.* **105**, 407 (1980).

collected by centrifugation at 10,000 g for 10 min. The cells are then washed twice in 20 mM sodium cacodylate buffer, pH 7.1, containing 0.4 M sorbitol and finally resuspended in this buffer to a concentration of 200 mg original yeast/ml.

For the preparation of microsomes, yeast is disrupted by milling with glass beads in a water-cooled Vibro-Mill disruptor. The broken yeast cells are removed by washing the glass beads with 0.1 M potassium phosphate buffer, pH 7.2, containing 20% (v/v) glycerol, 1 mM EDTA, and 1 mM dithiothreitol. The resulting suspension is centrifuged at 7,500 g for 15 min to remove unbroken cells, cell debris, and nuclear material, and the microsomal fraction is then obtained by centrifuging the supernatant at 160,900 g for 1 hr. The microsomal pellet is resuspended in the same buffer using a hand-operated Potter-type homogenizer. For purification of cytochrome P-450 and NADPH–cytochrome P-450 reductase, the microsomal fraction is prepared from approximately 2 kg of yeast (wet weight), and this is used for the solubilization of microsomal enzymes with sodium cholate, and purification as described in detail previously.[6,8]

Permeabilized Whole Cells and the Microsomal Fraction: Immobilization and Properties

Permeabilized yeast cells prepared as described above are entrapped by calcium alginate using a method modified from that of Kierstan and Bucke.[28] To 5 ml of suspended cells, 15 ml of 1% (w/v) sodium alginate is added, and the total suspension is slowly extruded into 50 mM CaCl$_2$ (containing 10% w/v glucose) with a sample pipet using a tip with an orifice of approximately 1 mm in diameter. The resultant fibers are removed from solution by filtration, washed with 10% (w/v) glucose solution, and excess liquid removed before use. The problem exists that, when immobilized in calcium alginate, the enzyme cannot be assayed spectrally, because of the rapid settling of the large enzyme/calcium alginate fibers, therefore the retention of the enzyme in the gel is estimated by difference, i.e., by the amount of cytochrome P-450 which is not removed by immobilization (this estimate does not take account of any loss of the 450-nm peak caused by denaturation). The microsomal fraction is also immobilized by entrapment in alginate gel using the above method (with the 5 ml of permeabilized cells replaced by 5 ml of microsomal fraction containing 1 nmol cytochrome P-450/ml).

On immobilization of permeabilized cells by this process, all of the cytochrome P-450 was retained in the immobilized fraction. No cells

---

[28] M. Kierstan and C. Bucke, *Biotechnol. Bioeng.* **19**, 387 (1977).

could be found which were not immobilized. However, during the permeabilization process itself, 50–60% of the cytochrome *P*-450 was lost so the cytochrome *P*-450 level in the cells was relatively low. Permeabilized cells were also immobilized in gelatin, but this was not useful as the gel depolymerized in our batch stirred-tank reactor at 37°. Permeabilized whole cells lost all of their activity when copolymerized with acrylamide and could not be immobilized on cellulose or BrCN-activated Sepharose 4B (for methods see below).

Cytochrome *P*-450 in the microsomal fraction is a membrane-bound "naturally immobilized" system, and as such is not easily bound to other supports. The only successful method found was again entrapment in alginate gel. This led to the incorporation of all of the spectrally detectable cytochrome *P*-450, with most of the activity in benzo[*a*]pyrene hydroxylation being retained (>75%). Table I shows the benzo[*a*]pyrene hydroxylase activity of permeabilized cells and of microsomes entrapped in alginate, and that of free microsomes. The apparent V_{max} (all V_{max} values are expressed per nanomole cytochrome *P*-450) of entrapped cells was very similar to that of free microsomal fraction, although the apparent K_m was higher. With the entrapped microsomal fraction the lowered apparent V_{max} and raised apparent K_m indicate that this system shows considerable diffusion limitation. During storage at 4°, the free microsomal fraction lost all of its activity within 2 weeks. However, the immobilized microsomes retained 84% of their activity after 2 weeks and 60% of their activity after

TABLE I
IMMOBILIZED YEAST MICROSOMAL FRACTION AND PERMEABILIZED CELLS

		Benzo[*a*]pyrene hydroxylase activity			
Enzyme form	% Retention to support of cytochrome *P*-450 (spectral assay)	Apparent V_{max} (pmol 3-hydroxybenzo[*a*]pyrene/min/nmol cytochrome *P*-450)	Apparent K_m (μM)	Activity (%) remaining after	
				2 weeks	4 weeks
Microsomal fraction "free"	—	11.1	111	0	0
Microsomal fraction entrapped in calcium alginate	100	8.3	200	84	60
Permeabilized cells entrapped in calcium alginate	100	11.1	222	35	20

4 weeks, i.e., they show considerable stabilization. The same stabilization was not observed for permeabilized cells, as only 20% of the activity remained after 4 weeks. Schubert *et al.*[20] have likewise found some stabilization of microsomal cytochrome *P*-450 activity on calcium alginate immobilization. They found that the cytochrome *P*-450-dependent aminopyrine demethylase activity of rabbit liver microsomes was stabilized compared to free microsomes, by immobilization in calcium alginate using a fixed-bed recirculation reactor.

Highly Purified Cytochrome P-450: Immobilization and Properties

Highly purified cytochrome *P*-450 (88–97% pure on a specific content basis[6]) was coimmobilized with NADPH–cytochrome *P*-450 reductase on several different supports.

Immobilization with Calcium Alginate. Immobilization with calcium alginate is carried out by the same method as for microsomal fraction and permeabilized cells, using a reconstituted system of purified cytochrome *P*-450 and the reductase.

Immobilization in Polyacrylamide, Copolymerization with Acrylamide. Immobilization in polyacrylamide and copolymerization with acrylamide are carried out by the method of Brunner *et al.*[17] Reconstituted purified enzymes (4.5 ml) are added to 2 ml of 0.2 *M* Tris–HCl buffer, pH 7.4, and nitrogen is passed through the solution for 20 min at 20°. The solution is cooled on ice, and 0.5 g acrylamide and 34 mg *N,N*-methylenebisacrylamide are added. The mixture is stirred under nitrogen until these compounds are dissolved, and copolymerization is initiated by the addition of 0.25 ml 5% (w/v) dimethylaminopropionitrile and 0.25 ml of 5% (w/v) ammonium persulfate. Stirring under nitrogen at 4° is continued until formation of the gel which is then granulated or meshed.

Immobilization on Cyanogen Bromide-Activated Sepharose 4B. Purified enzymes are immobilized to BrCN–Sepharose 4B by a method modified from that of Cuatrecasas *et al.*[29] Sepharose 4B is allowed to settle and excess buffer decanted. The Sepharose is then mixed with an equal volume of water, and cyanogen bromide (100 mg/ml of settled Sepharose 4B) is added in an equal volume of water. The pH is adjusted to 11 by titration with 4 *M* NaOH and maintained at 11 while the reaction proceeds. After completion, the Sepharose 4B beads are washed with approximately 20 volumes of ice-cold 0.1 *M* NaHCO$_3$. The beads are then suspended in cold 0.1 *M* potassium phosphate buffer, pH 7.5, containing 1 m*M*

[29] P. Cuatrecasas, M. Wilchek, and C. B. Anfinsen, *Proc. Natl. Acad. Sci. U.S.A.* **61,** 636 (1968).

dithiothreitol and 20% (v/v) glycerol, in a volume equal to that of the original Sepharose 4B. Then the purified reconstituted enzyme preparation is added at a ratio of 6 nmol cytochrome *P*-450/g of beads. After stirring overnight, the beads are washed with the above buffer and suspended again overnight in 0.1 *M* potassium phosphate buffer, pH 7.5, containing 1 m*M* dithiothreitol, 20% (v/v) glycerol, and 0.1 *M* glycine (to deactivate remaining binding groups on the beads). Beads are washed with 0.1 *M* potassium phosphate buffer, pH 7.0, containing 20% (v/v) glycerol before use.

Immobilization on Cellulose by Glutaraldehyde Cross-Linking. To 10 ml of the reconstituted system, 1.5 g of microcrystalline cellulose is added and mixed for 1 hr. Glutaraldehyde (0.3 ml of 0.25 *M*) is then added, and the suspension is mixed for a further hour. Immobilized enzymes are then collected by centrifugation at 12,000 *g* for 10 min and resuspended in 0.1 *M* potassium phosphate buffer, pH 7.0, containing 20% (v/v) glycerol.

Immobilization on Agarose–Concanavalin A by Glutaraldehyde Cross-Linking. The reconstituted enzyme system (3.3 ml of 1 nmol *P*-450/ml in 20 m*M* potassium phosphate buffer, pH 7.4, plus 0.15 *M* NaCl) is stirred at 4° for 30 min with 1 g of agarose–concanavalin A. The gel is washed with 300 ml 1 *M* KCl, and cross-linking is achieved by adding 0.01 ml of 0.1 *M* potassium phosphate buffer, pH 8, containing glutaraldehyde to a final concentration of 0.5% (w/v), followed by stirring for 30 min.

Results of the immobilization of purified cytochrome *P*-450 are shown in Table II. As with microsomal fractions and permeabilized cells, the purified cytochrome *P*-450 system was successfully immobilized by en-

TABLE II

IMMOBILIZATION OF HIGHLY PURIFIED CYTOCHROME *P*-450 TO ONE OF THE THREE SUPPORTS

Support	% Retention to support of cytochrome *P*-450 (spectral assay)	Benzo[*a*]pyrene hydroxylase activity			
		Apparent V_{max} (pmol 3-hydroxy-benzo[*a*]pyrene/min/nmol cytochrome *P*-450)	Apparent K_m (μM)	Activity (%) remaining after	
				2 weeks	4 weeks
None	—	16.7	33	ND[a]	7
Calcium alginate	90	12.3	50	45	15
BrCN–Sepharose 4B	100	11.1	66	70	58
Acrylamide	60	9.0	33	40	40

[a] ND, Not determined.

trapment in calcium alginate, with good retention of the cytochrome
P-450, estimated spectrally (90%). This system gave a relatively high
apparent V_{max} for benzo[a]pyrene hydroxylation of 12.3 pmol/min/nmol
cytochrome P-450 compared to the free system value of 16.7. The apparent K_m for benzo[a]pyrene was also good at 50 μM, being not much higher
than that of the free enzyme (33 μM). The stability of this immobilized
preparation was about the same as for permeabilized cells (see above), but
not as good as that of microsomal fraction on the same support.

Attempts to immobilize yeast cytochrome P-450 to cellulose by cross-
linking with glutaraldehyde were unsuccessful. Only up to 20% of the
enzyme was immobilized, and this had lost most of its activity toward
benzo[a]pyrene hydroxylation. In earlier work, Woods and Wiseman[30]
obtained similar results using a solubilized yeast cytochrome P-450 preparation. Here, a relatively low extent of binding to the cellulose was reported, and reduced apparent V_{max} and increased apparent K_m for benzo[a]pyrene hydroxylation was observed. During our experiments we
have found that glutaraldehyde has a destructive effect on cytochrome
P-450 enzyme activity.

Cytochrome P-450 enzymes as isolated have been reported to be glycoproteins, containing 1–2 molecules of sugar per enzyme molecule[31]
(although this may not be the case in native enzymes[32]). Our attempts to
immobilize yeast cytochrome P-450, through any carbohydrate moiety, to
agarose–concanavalin A were unsuccessful. Only 50% of the enzyme was
bound, and all of the benzo[a]pyrene hydroxylase activity was lost.

Purified cytochrome P-450 and reductase were successfully immobilized to CNBr-activated Sepharose 4B with 100% retention of spectrally
determined cytochrome P-450. The retention of benzo[a]pyrene hydroxylase activity was almost as good as when entrapped in alginate. This
support was much more effective at stabilizing the enzyme, such that only
42% was lost after 4 weeks of storage at 4°. This stability is about the same
as the high stability observed for microsomal cytochrome P-450 entrapped in alginate (Table I). (Note that Brunner et al.[17] immobilized
NADPH along with their cytochrome P-450 system to BrCN–Sepharose
4B, while we have added an NADPH-generating system for each assay.)

Purified cytochrome P-450 and reductase were also immobilized by
copolymerization with acrylamide monomer, resulting in the retention of
60% of the cytochrome P-450 in the gel. It was interesting that the appar-

[30] L. F. J. Woods and A. Wiseman, Biochim. Biophys. Acta 613, 52 (1980).
[31] D. Haugen and M. J. Coon, J. Biol. Chem. 251, 7929 (1976).
[32] R. N. Armstrong, C. Pinto-Coelho, D. E. Ryan, P. E. Thomas, and W. Levin, J. Biol. Chem. 258, 2106 (1983).

TABLE III
SOME IMMOBILIZED LIVER CYTOCHROME *P*-450 SYSTEMS

Form of enzyme	Source	Support	Reference
Microsomal	Rat liver	BrCN–Sepharose 4B	33
Microsomal	Rabbit liver	Gelatin	19
Microsomal	Rabbit liver	Calcium alginate	20
Microsomal	Rat liver	Polyacrylamide hydrazide	21
Microsomal	Rat liver	Hollow fibers	16, 34
Solubilized microsomal enzymes	Rabbit liver	BrCN–Sepharose 4B	18
Purified cytochrome *P*-450 and reductase and glucuronyltransferase	Rabbit liver	BrCN–Sepharose 4B	15
Purified cytochrome *P*-450 and reductase	Rabbit liver	BrCN–Sepharose 4B; copolymerization with acrylamide; encapsulation in liquid–lipid membrane	17

ent K_m for benzo[*a*]pyrene was the same as for the free enzyme, but a decrease in the apparent V_{max} to 9.0 pmol/min/nmol cytochrome *P*-450 was observed. This lower V_{max} might be due to diffusional limitation on 3-hydroxybenzo[*a*]pyrene in the swollen gel that forms during the assay.

Discussion

Previous studies on the immobilization of cytochrome *P*-450 enzymes have employed microsomal or purified cytochromes *P*-450 from rat or rabbit liver (as summarized in Table III[15–21,33,34]). Most of this work has been directed toward the development of a more stable liver microsomal enzyme system for use in an extracorporeal shunt system for drug detoxification.[15–18] In these cases stability was improved, though unfortunately not enough to facilitate this application. Yawetz *et al.*[21] have improved the stability of liver microsomal monoxygenase activity by entrapment in cross-linked prepolymerized polyacrylamide hydrazide. These workers found that their immobilized microsomes were capable of the continuous O-demethylation of *p*-nitroanisole for several hours at 37° provided that

[33] D. Baess, G. R. Janig, and K. Ruckpaul, *Acta Biol. Med. Ger.* **34,** 1745 (1975).
[34] P. R. Kastl, W. H. Baricos, R. P. Chambers, and W. Cohen, *Enzyme Eng.* **4,** 199 (1978).

the NADPH-generating system was added periodically. Nonimmobilized microsomes lost all of their activity within 90 min. We too have found that microsomes are more stable in immobilized form (in calcium alginate). In addition to entrapment of microsomes in calcium alginate, we have also found good stability in our purified preparation immobilized to BrCN–Sepharose 4B, as used by several other groups.[15,17,18,33]

There are other possible applications for a stable immobilized cytochrome P-450/448 system with good activity. One of these is the construction of an enzyme electrode, which could be arranged to provide a rapid assay for a substrate of the particulr cytochrome P-450 enzyme present. Although many substrates in foodstuffs and the environment are substrates for cytochromes P-450/P-448, and it is the requirement for specificity that is the problem in their assay.

An immobilized cytochrome P-450 system could be useful also in other applications such as the removal of carcinogens and other pollutants from air and water. Also, cytochromes P-450 could prove to be very useful as catalysts for the production of drug metabolites and in other stereospecific chemical synthetic reactions. However, any large-scale applications of cytochromes P-450 await not only improvements in enzyme stability but also the development of a cheap and efficient electron supply (other than NADPH, which is expensive even in the regenerating system). Possibilities include coimmobilization with glucose oxidase, peroxides, or NADPH itself, or electrochemical methods.

[64] Enzyme Production by Immobilized Cells

By Yu-Yen Linko, G.-X. Li, Li-Chan Zhong, Susan Linko, and P. Linko

The application of immobilized biocatalysts for bioconversions and for the production of useful low molecular weight compounds is well documented.[1] It is, however, less well known that extracellular enzymes may also be produced by microorganisms attached on a solid support.[2] Although a number of enzymes have been produced on an industrial scale by cultivating the microorganism on a semisolid medium, Kokubu et al.[3] first showed that polyacrylamide gel-entrapped Bacillus subtilis cells could

[1] P. Linko and Y.-Y. Linko, CRC Crit. Rev. Biotechnol. 1, 289 (1984).
[2] G.-X. Li, Y.-Y. Linko, and P. Linko, Biotechnol. Lett. 6, 645 (1984).
[3] T. Kokubu, I. Karube, and S. Suzuki, Eur. J. Appl. Microbiol. Biotechnol. 14, 7 (1982).

produce α-amylase in the medium. The use of immobilized biocatalysts allows either repeated-batch or continuous production of biologically active substances at high cell densities. Both process control and downstream operations are significantly simplified in comparison to conventional batch processing. However, the diffusion limitations in such systems have to be considered.

We have shown that not only bacteria but also fungi can produce extracellular enzymes in immobilized state.[4-9] In this chapter, we discuss the immobilization of *Aspergillus niger* for the simultaneous production of glucoamylase (glucan 1,4-α-glucosidase) and α-amylase, of *Penicillium funiculosum* for the production of cellulolytic enzymes, and of *Phanerochaete chrysosporium* for the production of lignin peroxidase (ligninase). In the latter case, in particular, the immobilization offers special advantages.

Culture of Microorganisms

Aspergillus niger for Glucoamylase and α-Amylase

Aspergillus niger A.S. 3.4303 is grown at 30° on Czapek agar slants for 6 days to obtain cultures with dense sporulation. The spores are collected and used for immobilization.

Penicillium funiculosum for Cellulolytic Enzymes

Penicillium funiculosum IMI 87160 ii employed in the production of cellulolytic enzymes is grown at 28° on potato dextrose agar (Difco) slants to obtain spores for immobilization.

Phanerochaete chrysosporium for Lignin Peroxidase

Phanerochaete chrysosporium ATCC 24755 employed in the production of lignin peroxidase is both maintained (at 4°) and sporulated (at

[4] P. Linko, Y.-Y. Linko, and G.-X. Li, *Kem.–Kemi* **12**, 203, (1985).

[5] G.-X. Li, Y.-Y. Linko, and P. Linko, *Biotechnol. Lett.* **6**, 645 (1984).

[6] Y.-Y. Linko, M. Leisola, N. Lindholm, J. Troller, P. Linko, and A. Fiechter, *J. Biotechnol.* **4**, 283 (1986).

[7] M. Leisola, J. Troller, A. Fiechter, and Y.-Y. Linko, *Proc. Symp. Biotechnol. Pulp Paper Industry*, 16–19 June 1986, Stockholm, pp. 46–48.

[8] S. Linko, L.-C. Zhong, M. Leisola, Y.-Y. Linko, A. Fiechter, and P. Linko, *Proc. Int. Sem. Lignin Enzym. Microbial Degradation*, 23–24 April 1987, Paris, in press.

[9] S. Linko, L.-C. Zhong, Y.-Y. Linko, M. Leisola, A. Fiechter, and P. Linko, *Proc. European Congress on Biotechnology, 4th*, 14–19 June 1987, Amsterdam, pp. 121–124.

FIG. 1. Multineedle large-scale apparatus for preparing biocatalyst gel beads.

37°) on 2% malt agar (E. Merck 5398) slants or flasks, depending on the enzyme production scale. Sporulation at 37° took about 7 days.

Immobilization

Aspergillus niger

Aspergillus niger spores are entrapped in calcium alginate gel beads under aseptic conditions essentially as described by Linko *et al.*[10] A simple apparatus for the production of biocatalyst gel beads has been developed[1] and scaled up for large-scale operations as shown in Fig. 1.

In a typical laboratory-scale experiment 100 g of sterilized (108°, 10

[10] Y.-Y. Linko, L. Weckström, and P. Linko, *Food Process Eng.* **2,** 81 (1980).

min) 3–10% (usually 6%) sodium alginate solution (BDH Chemicals) is mixed with 15 ml of spore suspension ($\sim 10^7$ spores/ml) and extruded under slight nitrogen pressure through a set of 27-gauge hollow (i.d. 0.6 mm) needles into 200 ml of 0.5 M calcium chloride solution. The biocatalyst beads formed are hardened for 20 min with mixing, and finally washed with 0.5% calcium chloride. The beads thus obtained have a relatively uniform diameter of about 2–3 mm.

Penicillium funiculosum

Penicillium funiculosum spores are immobilized in polyurethane gels by adding 4 ml of spore suspension ($\sim 10^7$ spores/ml of 0.9% saline) to 2 g of urethane prepolymer PU-3 (hydrophobic) or PU-6 (hydrophilic) (Toyo Rubber Industry Co., Japan) in a petri dish at 40°; the mixture is stirred rapidly and kept at 4° for 30 min for hardening. The spore-containing urethane polymers are subsequently cut to ~ 3 mm cubes, followed by thorough washing with 0.9% saline.

Phanerochaete chrysosporium

Phanerochaete chrysosporium spores (0.7 ml, $\sim 3 \times 10^6$ spores/ml) are inoculated for immobilization into 75 ml of growth medium in 250-ml Erlenmeyer flasks containing a suitable quantity (0.5–2.5 g) of washed nylon web carrier as ~ 7 mm cubes. Nylon web is used as a carrier because it was found superior to other investigated support materials, such as polyurethane and agar gel. The flasks are agitated on a rotary shaker at 150 rpm, 37°, for 2 days, when glucose is almost quantitatively consumed.

Enzyme Production

Glucoamylase and α-Amylase

The following media are used for the growth of the mycelium within the beads (medium A) and for the enzyme production (medium B):

Growth medium A (pH 6.0)		Enzyme production medium B (pH 6.0)	
Sucrose	4%	Soluble starch	5%
NH_4NO_3	1%	Corn steep liquor	2%
$MgSO_4 \cdot 7H_2O$	0.05%	Peptone	1.5%
KH_2PO_4	0.05%	Yeast extract	1%
$FeSO_4 \cdot 7H_2O$	0.001%	$MgSO_4 \cdot 7H_2O$	0.05%
		$CaCl_2 \cdot 2H_2O$	0.5%

The immobilized spores (10 g wet weight) or free spores are both first germinated and grown in the growth medium A (50 ml) at 28° with shaking (175 rev/min) for 5 days to obtain either biocatalyst beads with a dense layer of active growing mycelia on the surface of the bead[4] or free active mycelia. For enzyme production the activated biocatalyst beads for free mycelia (~14 g wet weight) are transferred into 50 ml of the enzyme production medium B, and incubated at 28° for 2–4 days per batch. For repeated batch experiments 10 mg/liter tetracycline is added to prevent microbial contamination.

Cellulolytic Enzymes

For the production of cellulolytic enzymes by *P. funiculosum* spores immobilized in polyurethane gel cubes the following media are employed:

Growth medium C (pH 5.5)

Glycerol	2%
KH_2PO_4	1.5%
$(NH_4)_2SO_4$	0.4%
$CaCl_2 \cdot 2H_2O$	0.03%
Urea	0.03%
$MgSO_4 \cdot 7H_2O$	0.06%
$FeSO_4 \cdot 7H_2O$	0.001%
$MnSO_4 \cdot H_2O$	0.00032%
$ZnSO_4 \cdot H_2O$	0.00028%
$CoCl_4 \cdot 6H_2O$	0.0004%

Enzyme production medium D
(pH 5.5)
As medium C without glycerol, plus

Bacto Casiton	0.75%
Avicel-SF	2.5%

The immobilized spores are germinated and grown to active mycelia in growth medium C, followed by enzyme production in medium D in 13-day batches as described for the *A. niger* spores above.

In all repeated batch fermentations with the immobilized biocatalyst systems the biocatalyst is thoroughly washed with 0.9% saline after the completion of each batch. The overgrown free mycelia released during the fermentation into the medium are removed during washing and, consequently, only immobilized mycelia are used for the subsequent batch.

Lignin Peroxidase

The following carbon-limited growth and enzyme production media are used in the production of lignin peroxidase by *Ph. chrysosporium:*

Growth medium E (pH 4.5)

Glucose $\cdot$ H_2O	0.2%
di-Ammoniumtartrate	0.066%
2,2-Dimethylsuccinic acid (purum, >99%)	0.142%

Enzyme production
medium F (pH 4.5)
Basic medium as
growth medium
E without glu-

MgSO$_4$ · 7H$_2$O	0.05%	cose; in some
FeSO$_4$ · H$_2$O	0.00184%	cases up to
Vitamins[a]	1 ml/liter	0.1% glucose
Trace elements[b]	1 ml/liter	was added
H$_3$PO$_4$ (2-N)	10.9 ml/liter	

[a] (In milligrams per liter) Biotin 4, folic acid 4, thiamine hydrochloride 100, riboflavine 10, pyridoxine hydrochloride 20, cyanocobalamin 0.2, nicotinic acid 10, Ca-D-pantothenate 10, p-aminobenzoic acid 10, thioctic acid 10.

[b] (In milligrams per liter) Na-nitrilotriacetate 1.5, MnSO$_4$ · H$_2$O 1.0, CoCl$_2$ · 6H$_2$O 1.0, ZnSO$_4$ · 7H$_2$O 3.0, CuSO$_4$ · 5H$_2$O 0.01, Alk-(SO$_4$)$_2$ · H$_2$O 0.1, H$_3$BO$_3$ 0.01, Na$_2$MoO$_4$ 0.01.

After the spores are germinated and the immobilized mycelium grown until glucose was exhausted, 25 ml of the growth medium is decanted off, and veratryl alcohol is added as an activator to a final concentration of 0.6–3.0 mM. The subsequent enzyme production is carried out in batch or repeated batch fermentations under 100% oxygen atmosphere and agitation (20–100 rpm).

Assays

α-Amylase Activity

α-Amylase activity is determined by the release of the blue dye from Remazol brilliant R (Farbwerke Hoechst) blue starch according to the method of Linko et al.[11] Two milliliters of 1% blue starch in 50 mM glycerophosphate buffer, pH 6.0, is added to 0.5 ml diluted enzyme solution and allowed to react at 30° for 15 min. Then 2.5 ml of ice-cold 96% ethanol is added, and the tubes are placed in an ice bath. The precipitated high molecular weight substrate is removed by filtration, and the absorbance of the filtrate is determined at 595 nm. One unit (U) of α-amylase activity is defined by absorbance (A) as 0.1 A/min.

Glucoamylase Activity

Glucoamylase activity is determined at 45°, pH 4.5, for 30 min using 2 ml of the enzyme solution and 2 ml of 60 mM maltose as substrate. The reaction is stopped by adding 1 ml of 0.18 M Tris solution, and the glucose produced is determined by the hexokinase/glucose-6-phosphate dehydrogenase method (Gluco-quant Kit, Boehringer Mannheim) at 340 nm. One

[11] Y.-Y. Linko, P. Saarinen, and M. Linko, *Biotechnol. Bioeng.* **17**, 153 (1975).

unit (U) of glucoamylase activity is defined as 1 μmol of glucose produced in 1 min.

Filter Paper Hydrolyzing Activity

The filter paper hydrolyzing activity is determined according to Ghose et al.[12] A 50-mg strip of Whatman 1 filter paper is added to 1 ml of sodium acetate buffer (50 mM, pH 4.8) at 50°. One milliliter of suitably diluted enzyme preparation is added, and the mixture is incubated for 1 hr at 50°. The reducing sugars formed are determined by the method of Nelson.[13] One activity unit (FPU) is defined as the quantity of enzyme liberating 1 μmol reducing sugars in 1 min.

CMC Hydrolyzing Activity

A modified method of Mandels and Weber[14] is used to determine the carboxymethylcellulose (CMC) hydrolyzing activity. One milliliter of 1% CMC (Fluka) in sodium acetate buffer (50 mM, pH 4.8) and 1 ml of appropriately diluted enzyme preparation are mixed and incubated at 50° for 10 min. The reaction is stopped in boiling water for 5 min, and the reducing sugars are determined as above from 0.1 ml of the hydrolysate. One unit of CMC activity is defined as the quantity of enzyme liberating 1 μmol reducing sugars in 1 min.

β-Glucosidase Activity

The determination of the β-glucosidase activity is carried out using p-nitrophenyl-β-glucopyranoside (Merck) as substrate, according to a slight modification of the method of Berghem and Pettersson.[15] One milliliter of 1 mM p-nitrophenyl-β-glucopyranoside in sodium acetate buffer (50 mM, pH 4.8) and 0.2 ml of the enzyme preparation are mixed and incubated at 50° for 10 min. One-half milliliter of 1 M sodium carbonate is added, and the mixture is diluted with 10 ml of deionized water. The p-nitrophenol liberated is determined at 400 nm. One unit of enzyme activity is defined as the quantity of enzyme liberating 1 μmol p-nitrophenol under the assay conditions in 1 min.

---

[12] T. K. Ghose, A. N. Pathak, and V. S. Bisaria, *Proc. Symp. Enzymat. Hydrol. Cellulose,* p. 111 (1975).
[13] N. Nelson, *J. Biol. Chem.* **153,** 375 (1944).
[14] M. Mandels and J. Weber, *Adv. Chem. Ser.* **95,** 391 (1969).
[15] L. E. R. Berghem and G. Pettersson, *Eur. J. Biochem.* **37,** 21 (1973).

Lignin Peroxidase Activity

Lignin peroxidase activity is determined essentially as described by Tien and Kirk[16] as follows: 50–200 μl of sample solution and, correspondingly, 550–400 μl of water, 300 μl of 333 mM Na-tartrate buffer of pH 3.0, 100 μl of 4.0 mM veratryl alcohol, and 6 μl of freshly prepared H_2O_2 (139 μl of 30% H_2O_2/25 ml of water) are added into a 1-ml quartz cuvette and mixed rapidly with a disposable plastic stirrer. The enzyme activity is obtained from the change in absorbance at 310 nm, and 1 unit (U) of lignin peroxidase activity is defined as 1 μmol of veratryl alcohol oxidized in 1 min. The activities are reported as U/liter.

General Observations

Glucoamylase and α-Amylase Production

We have previously shown that calcium alginate gel is an excellent support for the immobilization of many microorganisms for the production of a number of useful compounds,[1] and Samejima et al.[17] have recently shown the suitability of this carrier system for industrial-scale operations as well. However, if phosphate and/or alkali metal ions are present in the reaction medium in excess quantities the gel may soften and, in extreme cases, dissolve. A simple apparatus for large-scale biocatalyst bead production was developed. The method allows also the direct production of the biocatalyst into the bioreactor.

Inasmuch as the direct immobilization of mycelia was difficult to control, and the subsequent enzyme production was low because of the even distribution of the mycelia throughout the biocatalyst gel,[2] this approach was discarded. No problems were, however, encountered with the immobilization of microbial spores. The evenly distributed fungal spores germinated easily in the gel matrix in the growth medium A, and the mycelia grew as a dense thin layer near the surface of the beads. The subsequent enzyme production by using starch-containing medium B could be carried out with ease for at least 20 4-day batches.[4] The optimal pH and temperature for simultaneous glucoamylase and α-amylase production by the immobilized biocatalyst system were the same as those for the free mycelia, with somewhat higher stability at elevated temperatures for the immobilized spores.

---

[16] M. Tien and T. K. Kirk, *Proc. Natl. Acad. Sci. U.S.A.* **81,** 2280 (1984).

[17] H. Samejima, M. Nagashima, M. Azuma, S. Noguchi, and K. Inuzuka, *Ann. N.Y. Acad. Sci.* **434,** 394 (1984).

The optimal sodium alginate concentration for immobilization was 6%. It should be noted that the viscosity of such sodium alginate solution decreased during the sterilization at 108° for 10 min to the level of 4% alginate before sterilization. The addition of 0.5% calcium chloride to the enzyme production medium significantly increased the biocatalyst gel bead stability by preventing bead swelling and softening during repeated use. The drastic decrease in enzyme production activity noted occasionally after a few batches could be completely eliminated by 10 mg/liter tetracycline, as shown in Table I.

The mutant strain *A. niger* A.S. 3.4303 produced specific glucoamylase activity of about one-half and specific α-amylase activity about 7 times that of a typical commercial glucoamylase preparation. It is also of interest to note that the ratio of glucoamylase to α-amylase activity obtained decreased to about one-half on immobilization while total enzyme production increased by as much as 50% in a single 6-day batch, and the same biocatalyst could be used for extended periods while the activity of free mycelia decreased drastically already after the second batch.

Production of Celluolytic Enzymes

Although *Trichoderma reesei* is the most studied cellulolytic microorganism, it is relatively low in β-glucosidase activity. *Penicillium funiculosum,* on the other hand, is known to produce high β-D-glucosidase activities, in addition to endo- and exo-β-D-glucanases, consequently increasing the rate of cellulose hydrolysis by rapidly eliminating the inhibitory intermediate, cellobiose. In the present work, hydrophobic polyurethane PU-3 was found to be an excellent matrix for the entrapment of *P. funiculosum* for the production of cellulolytic enzymes.

It is of interest to compare the activities of the enzymes produced both by free mycelia and by the immobilized biocatalyst. Figure 2 shows clearly that the immobilized spores germinated and grown within the

TABLE I
Effect of Tetracycline on the Production of Amylolytic Enzymes by
Immobilized Germinated Spores after Contamination

Batch no.	Batch culture,[a] state of culture	Enzyme activity (U/ml)	
		Glucoamylase	α-Amylase
4	Original activity	58.8	17.4
5	After contamination	6.4	6.2
6	After adding tetracycline	65.6	17.0

[a] Batch cultures using the enzyme production medium B, 28°, 4 days.

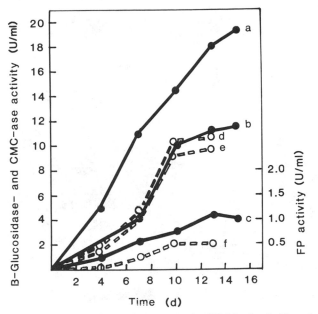

FIG. 2. The production of cellulolytic enzymes by PU-3 hydrophobic polyurethane gel cube-immobilized *Penicillium funiculosum* (solid line, immobilized, dotted line, free mycelia); a and d, β-glucosidase activity; b and e, carboxymethylcellulose hydrolyzing (CMCase) activity; c and f, filter paper hydrolyzing (FP) activity.

polyurethane gel produced significantly higher cellulolytic activities than the free cells. The preliminary experiments showed that the enzyme-producing activity could be maintained for at least three 13-day batches.

Production of Lignin Peroxidase

Figure 3 shows *Ph. chrysosporium* mycelium immobilized on the nylon web carrier. The immobilization appeared to stabilize the fungus against its well-known shear sensitivity in lignin peroxidase production, and thus, allowed efficient ligninase production in agitated cultures. This is a clear advantage, because it opens up the way to large-scale enzyme production in a bioreactor.[8] The maximum enzyme activity was usually obtained significantly faster than in static cultures with free mycelium, or with pellets, and the enzyme activities obtained were at least of the same order of magnitude up to about 600 U/liter. The immobilization also made repeated batch enzyme production possible, and as many as 20 subsequent 1- and 2-day batches could be fermented by using the same carrier.[8] Highest lignin peroxidase activities were obtained when the basic enzyme production medium F was supplemented with 1.0 g/liter glucose. If no

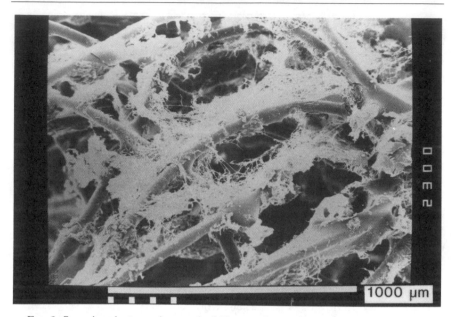

FIG. 3. Scanning electron micrograph of *Phanerochaete chrysosporium* mycelium immobilized on nylon web carrier.

glucose was added during the enzyme production phase, lignin peroxidase productivity decreased rapidly after two subsequent batches. It was, however, possible to reactivate enzyme production by intermittent incubation in the complete carbon-limited growth medium E for 24 hr at 150 rev/min. The relatively high-shaking speed during the reactivation period did not appear to affect adversely subsequent enzyme production.

By using a 2^3 factorial central composite experimental design with α-points and nine replicates at the center point, approximately 1.8 mM (for 0.5 g carrier) and 2.5 mM (for 1.7 g carrier), respectively, were obtained for the optimum veratryl alcohol concentration during the activation of the enzyme production.[9] The optimal shaking speed for 250-ml Erlenmeyer flasks was about 50 rev/min. The addition of Tween 80 surfactant increased lignin peroxidase production, with the optimum concentration of about 0.13%.

Conclusion

It has been demonstrated that the production of a number of extracellular enzymes by immobilized fungi is possible. Both gel entrapment and adherence to a polymer matrix can be employed.

[65] Application of Immobilized Captured Microorganisms in Water Purification: An Overview

By J. D. BRYERS and G. HAMER

Introduction

Immobilized microorganisms used in water purification are often classified into three general groups: (1) biofilm-entrapped cells (BEC), (2) biomass-retaining particles (BRP), and (3) artificially captured cell systems (ACC). (These three groups can be considered as forms of immobilized systems and the borderlines between them are diffuse.) Biofilm-entrapped cell and biomass-retaining particle systems rely on the natural ability (in pure or mixed culture) of the microorganisms to adhere either to each other or to an inert surface and produce extracellular polymeric material, thus forming either a biofilm or a biomass aggregate or floc. All three captured cell systems provide a means for retaining microorganisms within a reactor at concentrations well above those that would normally exist due to suspended growth alone. Therefore, captured cell systems allow biological reactors to operate at much higher loading rates than suspended culture reactors without the restriction of biomass "washout." Microorganisms in BEC and BRP systems are allowed to grow as they mediate desired biological reactions while artificially captured cells enzymatically mediate conversion processes with a loss in activity with time where growth is only encouraged periodically as a means to regenerate enzyme activity.

This chapter will focus more on the potential use of artificially captured cells, "immobilized whole cells," in wastewater treatment, with specific emphasis on methodology. However, the reader is reminded that BEC and BRP systems are currently the subject of intense research efforts and are more prevalent in pilot- and full-scale wastewater processes. Advantages BEC and BRP systems have over artificially captured cell systems at present are (1) costly entrapment procedures are not required, (2) greater hydrodynamic stability is afforded, and (3) no growth and contamination problems by undesired organisms occur.

BEC systems, more commonly referred to as "fixed-film" or "biofilm" reactors, are used extensively in experimental and full-scale wastewater treatment processes in the following configurations: (1) packed-bed or trickling filter reactors, (2) rotating biological contractors, and (3) fluidized-bed reactors as shown in Fig. 1. Table I provides an operating summary for each general type of BEC system.

METHODS IN ENZYMOLOGY, VOL. 137

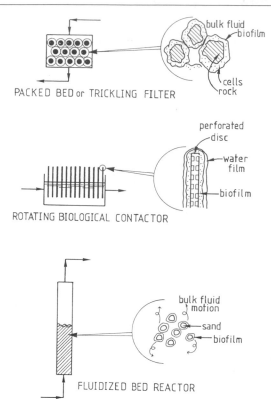

FIG. 1. General types of biofilm-entrapped cell systems.

One disadvantage of BEC systems is the lack of control over biofilm thickness once the system is operating which can lead to mass transfer limitations, and possibly excessive sloughing of the biomass. To remedy this lack of control, BRPs were developed by Atkinson and co-workers.[1,2] BRPs are small three-dimensional "webs" made of various materials forming a lattice in which cells and exopolymeric material are retained (Fig. 2). BRPs can be made of a variety of materials in a range of external and internal dimensions and geometry as summarized in Table II. Usually BRPs are fluidized within a reactor, so excess biomass is automatically removed from the particle by repeated collisions.

[1] B. Atkinson, G. M. Black, P. J. S. Lewis, and A. Pinches, *Biotechnol. Bioeng.* **21,** 193 (1979).
[2] B. Atkinson, G. M. Black, and A. Pinches, *in* "Biological Fluidized Bed Treatment of Water and Wastewater" (P. F. Cooper and B. Atkinson, eds.), Horwood, Chichester, England, 1981.

TABLE I

OPERATING CHARACTERISTICS OF BIOFILM ENTRAPPED CELL SYSTEMS

System	Support material	Surface area	Typical uses and operating details	System	
				Advantages	Disadvantages
Packed bed or trickling filter	Rock (crushed granite)	$S_v = 98$ m^2/m^3 reactor, porosity = 50%	Carbon oxidation	Cheap installation, capital, and maintenance costs	Aside from recycle, no operational control over performance
	Plastic media	$S_v = 98$ m^2/m^3 reactor, porosity = 98%	Nitrification	Ease of operation	No biofilm control
			Anaerobic CH$_4$ production	Not susceptible to toxins	Not good for low or intermittent flow
Rotating biological contractor	Reticulated plastic sheet in disk geometry	$S_v = 100$ cm^2/cm^3	Carbon oxidation	Cheap operation	Limited control over O$_2$ transfer rates
				Some control over biofilm thickness	Not good for very highly concentrated wastes
				Better suited than packed bed to low flow	
Fluidized-bed reactors	Sand (quartz)	$S_v \approx 2.9$ m^2/liter	Carbon oxidation	Good biofilm thickness control	Poor gas transfer to bed
	Charcoal	—	Nitrification	Well mixed operation	Solids reentrainment
	PVC	—	Denitrification	Small volume required	Loading limited by fluidization velocity
			Anaerobic CH$_4$ production		

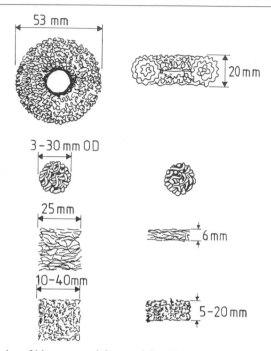

FIG. 2. Examples of biomass-retaining particles. Top to bottom, toroids, stainless steel spheres, mats, and cubes.[2]

TABLE II
BIOMASS-RETAINING PARTICLE PROPERTIES[a]

BRP type and material	Characteristic size	Material density (g/cm³)	Porosity	BRP particle density (g/cm³)	
				Clean	With biomass
Stainless steel spheres	6 mm o.d.	7.7	0.8	2.34	2.42
Polypropylene toroids	Overall o.d. 53 mm Torus o.d. 20 mm	0.9	0.9	0.99	1.08
Reticulated polyester foam	Cubes 10 × 10 × 2 mm	1.2	0.9	1.02	1.11
Matted reticulated polypropylene sheets	Cubes 25 × 25 × 6 mm	0.9	0.95	0.99	1.09

[a] From Atkinson et al.[2]

Biomass-retaining particles can be employed in either fixed, expanded, or fluidized-bed arrangements or within agitated tanks. Clearly, the advantage of BRPs, that of maintaining a controlled, preselected particle size of reactive biomass, must be preserved. Consequently, reactor operations must prevent excessive biomass accumulation outside the BRP to avoid blockage of the bed interstices. Biomass blockage is more likely to occur in fixed and expanded bed reactors. Periodic back-washing with high air/liquid velocities can act to remove excessive biomass. Such remedial measures also require the means to collect the excess sludge once it leaves the reactor. In fluidized-bed operation, biomass is continually removed due to particle–particle contact. Both means of operation accomplish the same goal, eliminating excess biomass, however, periodic back-washing of a fixed bed creates a reactor situation of variable hydrodynamic and biological character, while a fluidized bed operates at relatively constant biomass conditions. In this regard, fixed or expanded bed operation with periodic biomass control contradicts the main advantage of BRP.

Regarding the use of agitated tanks, BRPs can facilitate design of smaller, more reactive activated sludgelike systems which no longer require conventional clarification by gravity settling. Use of polyester sponge cubes allows for biomass separation via roller-squeezing action, returning the sponges to the vessel, and wastage of excess biomass. Such a system is currently patented as the Biomass Captor system by Simon-Hartley Ltd. Such BRP systems also negate the common problem of bulking and other biomass separation problems that limit the operation of conventional activated sludge systems.

Table III summarizes operating experiences for those few BRP systems currently employed for wastewater treatment. Description and applications of BRP systems can be found in papers by Atkinson and co-workers.[1,2]

Applications in Wastewater Treatment, Especially Analyses and Contaminant Removal Using Artificially Captured Cells

Artificially immobilized cells have been used in wastewater treatment for either analysis or specific contaminant removal. The following are examples of both applications with emphasis placed on methodology.

Karube and co-workers detail the construction of captured microbe (bacteria[3,4] and yeast[5]) probes for analysis of wastewater BOD (BOD is

[3] I. Karube, T. Matsunaga, and S. Suzuki, J. Solid-Phase Biochem. 2, 97 (1977).
[4] I. Karube, T. Matsunaga, S. Mitsuda, and S. Suzuki, Biotechnol. Bioeng. 19, 1535 (1977).
[5] M. Hikuma, H. Suzuki, T. Yasuda, I. Karube, and S. Suzuki, Eur. J. Appl. Microbiol. Biotechnol. 8, 289 (1979).

TABLE III

OPERATING EXPERIENCE WITH BRP SYSTEMS FOR WASTEWATER TREATMENT

BRP system	Purpose	Reactor			Details of reactor	BRP	Details
		Loading	Removal (%)				
Polyester foam cubes, polypropylene toroids[a]	Municipal wastewater 0.6 m³ pilot-scale plant study	3.0 kg BOD/m³ day Concentration (mg/liter): BOD = 70 COD = 200 NH₃ = 40 TSS = 80	— 80 NA 40 50		Volume 0.6 m³ Residence time 80 min Aeration in column itself Fluidized bed	8000 toroids/m³ (25 g/liter biomass) Polypropylene mat numbers = NA (2.9 g/liter biomass) Foam cubes numbers = NA (15–20 g/liter biomass)	Toroids: difficult to fluidize prior to biomass formation; poor fluidization with numbers greater than 8000/m³ Mats: better fluidization at higher number of mats; lower retained biomass; poorly fluidized until filled with biomass
Polyester foam cubes[b]	Mixed sanitary and industrial wastewater 5-liter laboratory scale study	Concentration (mg/liter): NH₄⁺ = 80 COD = 480 BOD = 120 SS = 2200 oil = 2% chromium = 0.2	NH₄⁺ 80–95 COD 60–90		Volume 5 liters Residence time 12 and 7 hr Fluidized tank	150 cubes (1.8 × 1.8 × 1.2 cm) No biomass concentration reported	Waste contained large amounts of oil and chromium which concentrated within cubes
Ceramic brick[c]	Two-phase anaerobic treatment of wastewater and CH₄ production	Concentration (mg/liter): COD = 800–2600 pH = 8.9	COD 60–90 Gas produced, 90% CH₄ and 5% CO₂		Operated at 20, 30, and 40° Residence times 2–5.5 hr Packed bed	Extruded ceramic brick (2 mm o.d. × 2–6 mm length; pore diameter 3 μm, porosity 60%) Reactor 1. 20 g brick Reactor 2. 50 g	Acidogenesis and methanogenesis separated; improved gas production; no plugging noted

[a] I. Walker and E. P. Austin, in "Biological Fluidized Bed Treatment of Water and Wastewater" (P. F. Cooper and B. Atkinson, eds.). Horwood, Chichester, England, 1981.

[b] S. T. Nesaratnam and F. H. Ghobrial, Proc. Int. Waste Treat. Waste Util. Symp., 3rd, in press.

[c] R. A. Messing, Biotechnol. Bioeng. 24, 1115 (1982).

roughly the oxygen required by a bacterial mixed culture to oxidize reduced organic matter in a water sample). Naturally culture techniques for
either the suspended yeast or bacteria are different, but the construction
of the microbial probes is quite similar. A schematic of the BOD probe is
shown in Fig. 3A.

Three milliliters of culture broth of the yeast *Trichosporon cutaneum*
(AJ 4816) is vacuum filtered onto porous acetylcellulose membrane (Type
HA, pore size 0.45 μm, diameter 47 mm, 150 μm thickness; Millipore) and

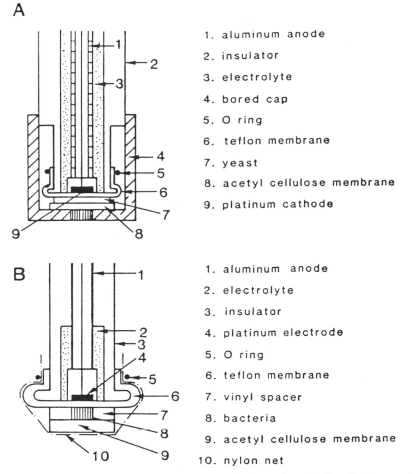

A

1. aluminum anode
2. insulator
3. electrolyte
4. bored cap
5. O ring
6. teflon membrane
7. yeast
8. acetyl cellulose membrane
9. platinum cathode

B

1. aluminum anode
2. electrolyte
3. insulator
4. platinum electrode
5. O ring
6. teflon membrane
7. vinyl spacer
8. bacteria
9. acetyl cellulose membrane
10. nylon net

FIG. 3. Schematic diagrams of captured cell microbe probes. (A) Specific diagram for
entrapped yeast–BOD probe. (B) Specific schematic for entrapped nitrifying bacteria–
ammonia probe.[3-6]

air dried. The porous microbial membrane is placed on the Teflon membrane of an oxygen probe so the yeast are sandwiched between the two membranes. On injection of a solution containing equal amounts of glucose and glutamic acid into the sensor system, current decreases with time and reaches a steady state within a response time of 18 min. A linear relationship was observed between current decrease and 5-day BOD below a maximum concentration of 60 mg/liter. Current output of the yeast probe was constant over a 17-day period and for 400 BOD tests.

For the bacterial probe,[3,4] a strip of nylon net (20 mesh, 700 μm thickness, 3.4 diameter) is attached to the platinum electrode with an adhesive. *Clostridium butyricum* (0.4 g of intact cells) is suspended in an acrylamide solution (0.36 g of 90% acrylamide and 10% N,N'-methyleneacrylamide in 10 ml saline buffer) in an ice bath. 0.5 ml of the solution is cast on the nylon net-covered electrode, and then polymerization is initiated with 0.03 ml of 10% dimethylaminopropionitrite and 0.1 mg of potassium persulfate and allowed to proceed anaerobically for 30 min at 37°. When placing the probe in a sample containing reduced organic matter, the immobilized microbes convert the organics into hydrogen and formate which develops, on diffusion to the cathode, a high current density. Response time is dependent on the concentration of organics in solution. A linear relationship between current density and the BOD of the solution was observed[3] for BOD values below 300 ppm. The lifetime of the electrode without loss of activity was approximately 30–40 days.

The reader is cautioned that the above BOD probes were calibrated to soluble BOD components of low molecular weights that are easily degraded. Apparently, these devices are not applicable to samples containing either high molecular weight components or particulate BOD.

Ammonia can be determined by a microbial sensor consisting of captured nitrifying bacteria (e.g., *Nitrosomonas europaea*) and a modified oxygen probe. Hikuma *et al.*[6] report entrapping the nitrifying bacteria between a porous acetylcellulose membrane and the Teflon membrane of a standard oxygen electrode (see Fig. 3b). Fifty milliliters of culture broth of *N. europaea* is filtered through a porous acetylcellulose membrane (Millipore Type HA, 0.45 μm, 25 mm diameter, 150 μm thickness) and surrounded with a vinyl ring spacer. The membrane is fixed to the Teflon membrane of a standard platinum cathode/aluminum anode oxygen probe and the entire assembly encased within a nylon net fastened to the probe by rubber rings. Calibration of current is by difference between the oxygen demand of the nitrifiers in a sample without versus one with ammonia. A linear response between this current difference and ammonia concen-

[6] M. Hikuma, T. Kubo, T. Yasuda, I. Karube, and S. Suzuki, *Anal. Chem.* **52,** 1020 (1980).

tration was reported by Hikuma *et al.* for concentrations between 0 and 1.5 mg/liter NH_3. The lifetime of the electrode was 14 days and 1400 assays.

Karube *et al.*[7] evaluated the production of methane gas from various wastewaters by methanogenic bacteria immobilized in three different artificial carriers: agar gel, polyacrylamide gel, and collagen membrane. Of the three carriers, agar gel provided the highest methane production activity (450 μmol CH_4/g dry cells/hr; 50% of freely suspended cell activity). Methanogenic bacteria were originally enriched from an industrial anaerobic digester. To entrap cells in agar gel, 0.02 g wet intact cells are added to 1 ml saline buffer containing 0.02 g agar in a Shrenk flask at 50° which is then quickly cooled to 37°. Polyacrylamide gel immobilization then proceeds as normal.

Collagen membrane entrapment was described by Karube *et al.*[8] A suspension of 0.01 g collagen fibrils and 0.02 g wet cells is treated with 1% glutaraldehyde solution (pH 7.0) for 1 min at 20°. Activity of cells in each capture material is determined in batch reactors where 5 ml wastewater (~1000 ppm BOD) are introduced to flasks containing equal weights of captured cell particles. No report of gas evolution and its effects on carrier integrity was given.

Nilsson *et al.*[9] report on the preparation of living *Pseudomonas denitrificans* cells immobilized in an alginate gel that were used to denitrify a wastewater. One gram of centrifuged *P. denitrificans,* containing 4×10^{11} living cells/g (wet weight), is mixed with 9 g of a 2% (w/v) sodium alginate solution. The mixture is pumped continuously through a hypodermic needle (0.1–1.00 mm i.d.) and gelled immediately to form either beads or fibers on contact with a solution of 0.1 M $CaCl_2$. In the presence of an exogenous carbon source the immobilized bacteria reduced nitrate to nitrite and gaseous products. Captured cells retained 75% of their initial nitrate reduction capacity after 21 days of storage at 4°. Operational stability of the catalysts were studied both in batch and continuous column reactors. In a subsequent study,[10] a column reactor reportedly produced 3 liters denitrified water/kg gel/hr for 2 months from a high nitrate (22 mg NO_3^-/liter)-containing drinking water. Nitrate reduction activity could be regenerated by periodic nutrient loading of the column. Growth and lysis of catalyst pellets and some release of cells into the fluid did occur.

[7] I. Karube, S. Kuriyama, T. Matsunaga, and S. Suzuki, *Biotechnol. Bioeng.* **22,** 847 (1980).

[8] I. Karube, S. Suzuki, S. Kinoshita, and J. Mizuguchi, *J. Ind. Eng. Chem. Product Res. Dev.* **10,** 160 (1971).

[9] I. Nilsson, S. Ohlson, L. Häggström, N. Molin, and K. Mosbach, *Eur. J. Appl. Microbiol. Biotechnol.* **10,** 261 (1980).

[10] I. Nilsson and S. Ohlson, *Eur. J. Appl. Microbiol. Biotechnol.* **14,** 86 (1982).

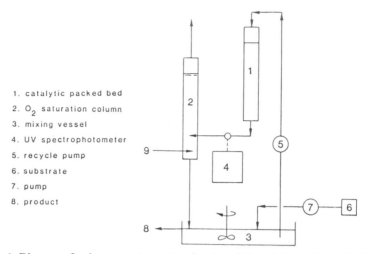

1. catalytic packed bed

2. O_2 saturation column

3. mixing vessel

4. UV spectrophotometer

5. recycle pump

6. substrate

7. pump

8. product

FIG. 4. Diagram of column reactor system for phenol degradation using a fixed bed of *Candida tropicalis* captured within an ionic polymer matrix.[12]

Nakajima *et al.*[11] report uranium recovery from sea and fresh water employing both *Streptomyces uiridochromogenes* and *Chlorella regularis* captured in polyacrylamide gels. Uranium adsorption by each culture was pH independent after capture. Adsorbed uranium could be desorbed and recovered using a solution of Na_2CO_3. Both cells had a high adsorptive capacity for uranium; *Chlorella* during 10 cycles adsorbed 160 mg uranium/g catalyst, and *Streptomyces* adsorbed 312 mg/g. *Chlorella* reportedly adsorbed 100% of the uranium in seawater samples (20.0 ppm concentration), with *Streptomyces* yielding 80% removal after four passes through the column.

Klein *et al.*[12] report phenol degradation by *Candida tropicalis* yeast captured within ionic polymer networks. The yeast is isolated and optimized for maximum phenol degradation rate in suspended culture. Resultant cells are spherical to ellipsoidal with a mean diameter of 3–5 μm. Polymer bead formation, using styrene-maleic acid copolymer, proceeds as follows: (1) preparation of copolymer solution in its sodium form (2) addition of cells, (3) dropping solution by a hypodermic needle in an Al^{3+} solution, and (4) separating and washing the particles.

[11] A. Nakajima, T. Horikoshi, and T. Sakaguchi, *Eur. J. Appl. Microbiol. Biotechnol.* **16,** 88 (1982).

[12] J. Klein, U. Häckel, and F. Wagner, *in* "Immobilized Microbial Cells," Am. Chem. Soc. Ser. 106. Am. Chem. Soc., Washington, D.C., 1979.

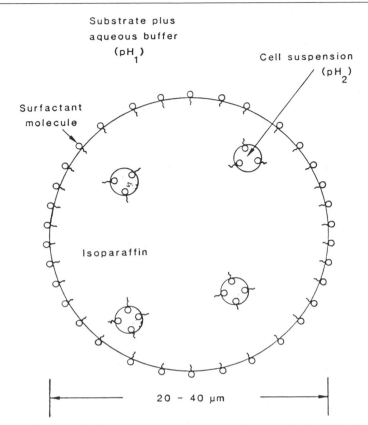

Substrate plus
aqueous buffer
(pH$_1$)

Cell suspension
(pH$_2$)

Surfactant
molecule

Isoparaffin

|← 20 - 40 μm →|

FIG. 5. Diagram of a liquid membrane entrapped cell system for denitrification.[13]

A fixed-bed recycle reactor system, employed in the kinetic study, consisted of a catalyst bed of ~140 g of beads, an oxygen saturation vessel, and phenol wastewater injection as shown in Fig. 4. The catalytic half-life for this particular column was estimated at 19 days. Reincubation of the catalysts within the original enrichment medium facilitates regrowth of cells and prolongs the life of the catalyst column.

Sequential reduction of nitrate and/or nitrite by whole *Micrococcus denitrificans* encapsulated in a liquid/surfactant membrane (see Fig. 5) was demonstrated by Mohan and Li.[13] In batch reactors, denitrification was studied as a function of substrate and cell concentration and pH in

[13] R. R. Mohan and N. N. Li, *Biotechnol. Bioeng.* **17,** 1137 (1975).

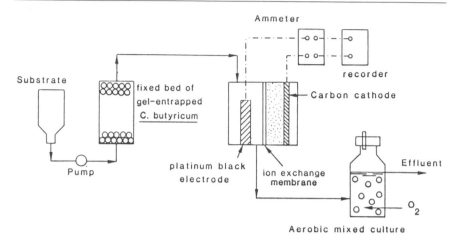

FIG. 6. Diagram of electrical generation system based on organic carbon waste oxidation by gel-entrapped *C. butyricum*.[15]

both the internal and external phases. The rate of denitrification was substrate diffusion limited. The captured cell system retained 80% of its original activity after 120 hr versus 0% for freely suspended cells after 16 hr.

Artificial capture of *C. butyricum* cells has been applied to a biochemical fuel cell which produced 1.2 mA continuously for 15 days with glucose as substrate.[14] Suzuki *et al.*[15] report the alternative use of an industrial alcoholic wastewater as carbon source for the anaerobic production of H_2 by *C. butyricum* entrapped within polyacrylamide gel. As shown in Fig. 6, H_2 produced by the packed column (300 g catalyst beads), at 0.75 mmol H_2/g beads/hr, passed to a fuel cell to generate a current of 8–9 mA for the 20 days at an inlet BOD of 660 ppm. The fuel cell consists of an anode chamber (10 × 3 cm) and a cathode chamber (10 × 10 × 0.5 cm) separated by an anion-exchange membrane. Anode material is platinum black (10 × 20 cm), and the cathode is a carbon electrode in 50 ml 0.1 *M* phosphate buffer. Effluent (BOD ~100 ppm) leaving the fuel cell passed directly to an aerobic biological reactor to further reduce its organic concentration. (See also the contribution by Karube and Suzuki [62] on biochemical fuel cells in this volume.)

[14] I. Karube, S. Suzuki, T. Matsunaga, and S. Kuriyama, *Ann. N.Y. Acad. Sci.* **369,** 321 (1981).
[15] S. Suzuki, I. Karube, and T. Matsunaga, *Biotechnol. Bioeng. Symp.* **8,** 501 (1978).

Summary

Table IV provides a comparison of the general operating details of the various immobilized cell systems; it is by no means comprehensive. BEC and BRP systems capitalize on mixed cultures of microorganisms, predominantly bacteria, to adhere to inert supports and form biofilms or biomass aggregates. The advantages of such systems are the following: (1) cheap, easily obtained support material (i.e., sand, quartz, plastic), (2) high reactor biomass concentrations, (3) proven capability to handle full-scale flow rates, and (4) freedom to operate with mixed cultures and nonsterile feed conditions. This last advantage allows BEC and BRP systems to operate in a growing biomass mode and to carry out a variety of biological reactions simultaneously (i.e., carbon oxidation and nitrification, carbon oxidation–methane production). Disadvantages of BEC systems are the lack of control over biofilm thickness which contributes to possible mass transfer limitations and which can lead to excessive biofilm sloughing. BRPs to a certain extent alleviate this lack of control over biomass growth. In both BRP and BEC systems, slow growing microorganisms (e.g., autotrophic nitrifiers, methane formers) do create the problem of lengthy (several months) reactor start-up times.

Artificially captured cell systems (ACCs) have the unique advantage that a biological process can be "stored" indefinitely then used, when desired, without wait. Thus, ACCs provide the advantage of relatively immediate use of the microorganism (once they are cultured and captured), thus eliminating system start-up problems. It may be a fruitful avenue of research to investigate the use of ACC systems to aid in the start-up of BEC and BRP that experience long initiation times. Disadvantages of ACCs are relatively short half-lives (around 20–40 days), the requirement of alternating nongrowth and growth periods of operation to maintain reactivity, and the physical breakage of encapsulating materials and the resultant contamination of downstream systems.

All ACC systems reviewed here were pure culture systems. One possible means of placing ACC systems on a more competitive basis with BECs and BRPs would be more research and development into defined mixed culture-multiple reaction ACC systems. For example, simultaneous nitrification–denitrification could be accomplished by first naturally attaching autotrophic denitrifiers to elemental sulfur, encapsulating the resultant particle, attaching a second culture of autotrophic nitrifiers, and finally encapsulating the entire aggregate a second time. Such research possibilities of bilayer and defined mixed culture microbial catalysts have been ignored.

TABLE IV

COMPARISON OF OPERATING DETAILS OF VARIOUS CAPTURED CELL SYSTEMS[a]

| | Biofilm entrapped cell systems | | | Biomass retaining particle systems fluidized-bed reactors | Artificially captured cell systems | |
Operating mode	Trickling filters, packed-bed reactors	Fluidized-bed reactors	Rotating disk reactors		Packed-bed column reactors	Batch systems
Biological process application	ACR, N, DN, AMP	ACR, N, DN, AMP	ACR, DN	ACR, N	DN, PD, ECG	AMP, URR
Biomass concentration	ACR = 0.2–3.0 kg/m² N = 0.05 kg/m² DN = 0.02 kg/m² AMP = 0.08 kg/m²	ACR = 15 g/liter N = NA DN = 0.02 kg/m² AMP = 0.02 kg/m²	ACR = NA N = NA	S.S.webs = 66.0 g/liter Sponge cubes = 38.5 g/liter Toroids = 35.5 g/liter	DNA = 45 g beads (9 g cells) PD = 0.8 g cells/g beads ECG = 40 g beads	AMP = 20 mg cells/g gel URR = 0.75 mg gel/ml media
Superficial fluid velocity (cm/sec)	0.01–0.33	0.8–1.0	300.0 (maximum peripheral speed)	0.08–0.17	0.01–0.07	—
Inlet substrate concentration (mg/liter)	ACR = 100–3000 as BOD N = 0–20 as NH₄ DN = 0–20 as NO₃ AMP = 10⁴ as CO₂	ACR = 0–1000 as BOD N = 0–200 DN = 10–1500 AMP = 10⁴–10⁵ COD	ACR = 50–3000 as BOD N = 0–20 as NH₄	ACR = 50–200 as BOD	ECG =3000 as BOD DN = 100 as NO₃⁻ PD = 250 as phenol	AMP = 1000 as BOD URR = 20 as uranium

[a] Abbreviations: ACR, aerobic carbon removal; N, nitrification; DN, denitrification; AMP, anaerobic methane production; PD, phenol degradation (via yeast); ECG, electrical current generation; URR, uranium recovery; NA, data not available.

In general, BEC and BRP systems are used more in all scales of water purification—from fundamental laboratory systems to full-scale municipal/industrial treatment facilities. Conversely, ACC systems find their most significant potential in the analysis of specific water/wastewater analysis using microbial probe devices. To extend the potential of ACC systems, serious research to answer questions of scale-up and particle longevity are sorely needed.

Author Index

Numbers in parentheses are footnote reference numbers and indicate that an author's work is referred to although the name is not cited in the text.

Subject Index

A

collagen membrane formation around, 78, 80, 82–83

continuous recording from two- and three-electrode implanted rats, 75

effect of glutaraldehyde treatment, 78

effect of scan rate on cyclic voltammogram, 76–77

fibrous cap formation on, 77–78

histology, 77

host response to, 71, 78–79

instrumentation, 71–73

intraperitoneal long-term stability testing, 74–75

in vivo stability, 78–81

lifetime of, 87

materials, 73–74

methods, 71–77

in mouse peritoneum, survival, 89

oxygen tension dependency, 87

principle, 71

in rat peritoneum, 70–71

reagents, 73

recordings obtainable from, 84–86

results with, 78–88

stabilization, and clinical applications, 87–88

submembrane leucodeposit, 81

industrial and commercial development of, 61

interfering compounds, 58

ionic strength dependence, 58

laboratory prototype, 45–47

flow block system in, 46–47

materials required, 20

measurement cell, 49

measurements, 49

mediated. *See* Mediator-modified electrode

oxygen dependence, 56–57

oxygen stabilized, 302–303

performance, factors determining, 99

pH dependence, 58

polarographic, 62

potentiometric, 15, 104. *See also* Biosensor, potentiometric

preparation, 20–27, 48–49

with chemically attached enzymes, 23–27

with chemically bound or soluble enzymes, 20–21

with physically entrapped enzyme, 20–23

selectivity, 58

self-contained, commercial sources, 29

soluble, 14

stability, 57–58

storage, 27

temperature dependence, 59–61

use, 27–28

using transient measurements, 103

voltammetric, 14–15

Enzyme immunoassay, 8, 288

Enzyme kinetics, pH-dependent, 249

Enzyme layers

gelatin-entrapped, preparation, 30

glutaraldehyde cross-linked, preparation, 30

Enzyme membrane, characterization, 31–35

Enzyme photometer, 202

for cephalosporin determinations, 210–216

Enzyme photometer–FIA

application to cephalosporin derivatives, 214–216

block switching diagram, 202

comparison with HPLC, method, 211–213

interface timing, 205

long-term stability of system, 212

Enzyme reactor electrode, 38–39

Enzyme reactors, 9, 288

from biological materials, 482–483

change in substrate affinity, 479–480

containing phenylalanine ammonia-lyase, preparation, 486–487

extracorporeal, 479

from hollow fiber artificial kidney, 483–488

preparation, 481

problems associated with, 479–481

for removal of amino acids from plasma, 479–491

stability of immobilized enzyme in, 480–481

sterilization, 481

Enzyme replacement therapy, 443

and endothelial barrier, 574

limitations, 572

Enzymes

coimmobilized, 172, 266, 267

coupling to hydrolyzed nylon, 290–291

feedback mode of operation, 248
structure, 257
theory of operation, 247–248
Iridium–MOS capacitor, 236, 246
selectivity, related to response for ammonia nitrogen, 242–243
Iridium–MOS structure, 233, 235
ammonia-sensitive, response characteristics, 235–236
ISFET. *See* Ion-sensitive field effect transistor
Islet cells
immunogenicity, 575
microencapsulated, 443, 451
insulin secretion from, 578
microencapsulation, 575–580
capsule material, 575
in vitro and *in vivo* assessment of
capsules, 578–580
materials, 576–577
method, 575–576
procedure, 577–578
variations, 580
syngeneic transplantation, 575
Isocitrate dehydrogenase, assay, in permeabilized cells, 637, 641
D-Isocitrate dehydrogenase, coimmobilized
with microsomes, in hybrid organelle
electrode, 159
Isoscope, 370, 372–373
Isotopes, stable, mass spectrometry, 429

K

Ketone, mass spectrometry, 429
Kidney, porcine
biocatalytic membrane electrode from,
140
in determination of glutamine in human CSF, 144–145
pH optimum, 143
membrane electrode from, 141–145
slices, integrity of, testing, 141–142
Kidney transplantation, rejection, 484
Kulbeka–Munk equation, 399

L

Laccase
electrode, amount incorporated into
active layer, 48

in electroreduction of oxygen, 432–435
immobilized, electrocatalysis, mechanism of, 436–437
source, 48
β-Lactamase penicillinase, electrode,
characteristics, 17
Lactase
electrode, amount of enzyme incorporated into active layer, 48
source, 48
Lactate
clinical analysis, using immobilized
lactate dehydrogenase, 266
determination, with enzyme thermistor,
191
enzyme electrode probe for, 69
immobilized enzyme nylon tube reactor
for analysis of, 293
performance, statistical parameters,
296
D-Lactate
assay, 6
enzyme electrode probe for, 48, 56
pH optimum, 58
L-Lactate
assay, 6
determination, with enzyme thermistor,
189–190
enzyme electrode probe for, 48
pH optimum, 58
results with, 54–56
ferrocene-modified electrode for, 101
Lactate dehydrogenase
binding to NAD chips, 393
dry reagent chemistry for, comparison
with reference method, 419
immobilized
for clinical analysis, 266
nylon tube reactor using, 294
performance characteristics, 293
statistical parameters of performance, 296
L-Lactate dehydrogenase
coimmobilized with microsomes, in
hybrid organelle electrode, 159
electrode, characteristics, 17
Lactate dehydrogenase–alanine aminotransferase, immobilized, nylon tube
reactor
performance characteristics, 293
statistical parameters of performance,
296